PAUL A. FOERSTER

Calculus

Concepts and Applications

Second Edition

Key Curriculum Press
Innovators in Mathematics Education

Project Editor: Christopher David
Consulting Editor: Josephine Noah
Editorial Assistants: Lori Dixon, Shannon Miller
Reviewer: Judith Broadwin
Accuracy Checkers: Jenn Berg, Dudley Brooks
Production Director: Diana Jean Ray
Production Editor: Kristin Ferraioli
Copyeditors: Tara Joffe, Luana Richards, Mary Roybal, Joan Saunders
Production Coordinator: Michael Hurtik
Text Designers: Adriane Bosworth, Jenny Somerville
Art Editors: Jason Luz, Laura Murray
Photo Researcher: Margee Robinson
Art and Design Coordinator: Kavitha Becker
Illustrator: Jason Luz
Technical Art: Matthew Perry
Cover Designer: Jenny Somerville
Cover Photo Credit: Alec Pytlowany/Masterfile
Composition and Prepress: The GTS Companies/York, PA
Printer: Von Hoffmann Printers

Executive Editor: Casey FitzSimons
Publisher: Steven Rasmussen

Key Curriculum Press
1150 65th Street
Emeryville, CA 94608
editorial@keypress.com
www.keypress.com

Printed in the United States of America
10 9 8 7 6 5 4 3 2 1   08 07 06 05 04
ISBN 1-55953-654-3

Photograph credits appear on the last two pages of the book.

Consultants to the First Edition

Donald J. Albers, Mathematical Association of America, Washington D.C.
Judith Broadwin, Jericho High School, Jericho, New York
Joan Ferrini-Mundy, University of New Hampshire, Durham, New Hampshire
Gregory D. Foley, Sam Houston State University, Huntsville, Texas
John Kenelly, Clemson University, Clemson, South Carolina
Dan Kennedy, Baylor School, Chattanooga, Tennessee
Deborah B. Preston, Keystone School, San Antonio, Texas

Field Testers of the First Edition

Betty Baker, Bogan High School, Chicago, Illinois
Glenn C. Ballard, William Henry Harrison High School, Evansville, Indiana
Bruce Cohen, Lick-Wilmerding High School, San Francisco, California
Christine J. Comins, Pueblo County High School, Pueblo, Colorado
Deborah Davies, University School of Nashville, Nashville, Tennessee
Linda E. de Sola, Plano Senior High School, Plano, Texas
Paul A. Foerster, Alamo Heights High School, San Antonio, Texas
Joan M. Gell, Palos Verdes Peninsula High School, Rolling Hills Estates, California
Valmore E. Guernon, Lincoln Junior/Senior High School, Lincoln, Rhode Island
David S. Heckman, Monmouth Academy, Monmouth, Maine
Don W. Hight, Pittsburg State University, Pittsburg, Kansas
Edgar Hood, Dawson High School, Dawson, Texas
Ann Joyce, Issaquah High School, Issaquah, Washington
John G. Kelly, Arroyo High School, San Lorenzo, California
Linda Klett, San Domenico School, San Anselmo, California
George Lai, George Washington High School, San Francisco, California
Katherine P. Layton, Beverly Hills High School, Beverly Hills, California
Debbie Lindow, Reynolds High School, Troutdale, Oregon
Robert Maass, International Studies Academy, San Francisco, California
Guy R. Mauldin, Science Hill High School, Johnson City, Tennessee
Windle McKenzie, Brookstone School, Columbus, Georgia
Bill Medigovich, Redwood High School, Larkspur, California
Sandy Minkler, Redlands High School, Redlands, California
Deborah B. Preston, Keystone School, San Antonio, Texas
Sanford Siegel, School of the Arts, San Francisco, California
Susan M. Smith, Ysleta Independent School District, El Paso, Texas
Gary D. Starr, Girard High School, Girard, Kansas
Tom Swartz, George Washington High School, San Francisco, California
Tim Trapp, Mountain View High School, Mesa, Arizona
Dixie Trollinger, Mainland High School, Daytona Beach, Florida
David Weinreich, Queen Anne School, Upper Marlboro, Maryland
John P. Wojtowicz, Saint Joseph's High School, South Bend, Indiana
Tim Yee, Malibu High School, Malibu, California

To people from the past, including James H. Marable of Oak Ridge National Laboratory, from whom I first understood the concepts of calculus; Edmund Eickenroht, my former student, whose desire it was to write his own calculus text; and my late wife, Jo Ann.

To my wife, Peggy, who shares my zest for life and accomplishment.

Author's Acknowledgments

This text was written during the period when graphing calculator technology was making radical changes in the teaching and learning of calculus. The fundamental differences embodied in the text have arisen from teaching my own students using this technology. In addition, the text has been thoroughly revised to incorporate comments and suggestions from the many consultants and field testers listed on the previous page.

Thanks in particular to the original field test people—Betty Baker, Chris Comins, Debbie Davies, Val Guernon, David Heckman, Don Hight, Kathy Layton, Guy Mauldin, Windle McKenzie, Debbie Preston, Gary Starr, and John Wojtowicz. These instructors were enterprising enough to venture into a new approach to teaching calculus and to put up with the difficulties of receiving materials at the last minute.

Special thanks to Bill Medigovich for editing the first edition, coordinating the field test program, and organizing the first two summer institutes for instructors. Special thanks also to Debbie Preston for drafting the major part of the *Instructor's Guide* and parts of the *Solutions Manual,* and for working with the summer institutes for instructors. By serving as both instructors and consultants, these two have given this text an added dimension of clarity and teachability.

Thanks also to my students for enduring all those handouts, and for finding things to be changed! Special thanks to my students Craig Browning, Meredith Fast, William Fisher, Brad Wier, and Matthew Willis for taking good class notes so that the text materials could include classroom-tested examples.

Thanks to the late Richard V. Andree and his wife, Josephine, for allowing their children, Phoebe Small and Calvin Butterball, to make occasional appearances in my texts.

Finally, thanks to Chris Sollars, Debbie Davies, and Debbie Preston for their ideas and encouragement as I worked on the second edition of *Calculus.*

Paul A. Foerster

About the Author

Paul Foerster enjoys teaching mathematics at Alamo Heights High School in San Antonio, Texas, which he has done since 1961. After earning a bachelor's degree in chemical engineering, he served four years in the U.S. Navy. Following his first five years at Alamo Heights, he earned a master's degree in mathematics. He has published five textbooks, based on problems he wrote for his own students to let them see more realistically how mathematics is applied in the real world. In 1983 he received the Presidential Award for Excellence in Mathematics Teaching, the first year of the award. He raised three children with the late Jo Ann Foerster, and he also has two grown stepchildren through his wife Peggy Foerster, as well as three grandchildren. Paul plans to continue teaching for the foreseeable future, relishing the excitement of the ever-changing content of the evolving mathematics curriculum.

Foreword

by John Kenelly, Clemson University

In the explosion of the information age and the resulting instructional reforms, we have all had to deal repeatedly with the question: "When machines do mathematics, what do mathematicians do?" Many feel that our historical role has *not* changed, but that the emphasis is now clearly on selection and interpretation rather than manipulation and methods. As teachers, we continue to sense the need for a major shift in the instructional means we employ to impart mathematical understanding to our students. At the same time, we recognize that behind any technology there must be human insight.

In a world of change, we must build on the past and take advantage of the future. Applications and carefully chosen examples still guide us through what works. Challenges and orderly investigations still develop mature thinking and insights. As much as the instructional environment might change, quality education remains our goal. What we need are authors and texts that bridge the transition. It is in this regard that Paul Foerster and his texts provide outstanding answers.

In *Calculus: Concepts and Applications*, Second Edition, Paul is again at his famous best. The material is presented in an easily understood fashion with ample technology-based examples and exercises. The applications are intimately connected with the topic and amplify the key elements in the section. The material is a wealth of both fresh items and ancient insights that have stood the test of time. For example, alongside Escalante's "cross hatch" method of repeated integration by parts, you'll find Heaviside's thumb trick for solving partial fractions! The students are repeatedly sent to their "graphers." Early on, when differentiation is introduced, Paul discusses local linearity, and later he utilizes the zoom features of calculators in the coverage of l'Hospital's rule—that's fresh. Later still, he presents the logistic curve and slope fields in differential equations. All of these are beautiful examples of how computing technology has changed the calculus course.

The changes and additions found in this second edition exhibit the timeliness of the text. Exponentials and logarithms have been given an even more prominent role that reflects their greater emphasis in today's calculus instruction. The narrative, problem sets, Explorations, and tests all support the position that the choice between technology and traditional methods is not exclusively "one or the other" but *correctly* both. Rich, substantive, in-depth questions bring to mind superb Advanced Placement free response questions, or it might be that many AP questions remind you of Foerster's style!

Throughout, you see how comprehensive Paul is in his study of the historical role of calculus and the currency of his understanding of the AP community and collegiate "calculus reform." Brilliant, timely, solid, and loaded with tons of novel applications—your typical Foerster!

John Kenelly has been involved with the Advanced Placement Calculus program for over 30 years. He was Chief Reader and later Chair of the AP Calculus Committee when Paul Foerster was grading the AP exams in the 1970s. He is a leader in development of the graphing calculator and in pioneering its use in college and school classrooms. He served as president of the IMO 2001 USA, the organization that acts as host when the International Mathematical Olympiad (IMO) comes to the United States.

Contents

CHAPTER 8

The Calculus of Plane and Solid Figures — 369

CHAPTER 9

Algebraic Calculus Techniques for the Elementary Functions — 431

CHAPTER 10

The Calculus of Motion—Averages, Extremes, and Vectors — 499

CHAPTER 11

The Calculus of Variable-Factor Products — 545

CHAPTER 12

The Calculus of Functions Defined by Power Series 587

A Note to the Student from the Author

In earlier courses you have learned about functions. Functions express the way one variable quantity, such as *distance you travel,* is related to another quantity, such as *time.* Calculus was invented over 300 years ago to deal with the rate at which a quantity varies, particularly if that rate does not stay constant.

In your calculus course you will learn the algebraic formulas for variable rates that will tie together the mathematics you have learned in earlier courses. Fortunately, computers and graphing calculators ("graphers") will give you graphical and numerical methods to understand the concepts even before you develop the formulas. In this way you will be able to work calculus problems from the real world starting on day one. Later, once you understand the concepts, the formulas will give you time-efficient ways to work these problems.

The time you save by using technology for solving problems and learning concepts can be used to develop your ability to write about mathematics. You will be asked to keep a written journal recording the concepts and techniques you have been learning, and verbalizing things you may not yet have mastered. Thus, you will learn calculus in four ways—algebraically, graphically, numerically, and verbally. In whichever of these areas your talents lie, you will have the opportunity to excel.

As in any mathematics course, you must learn calculus by doing it. Mathematics is not a "spectator sport." As you work on the Explorations that introduce you to new concepts and techniques, you will have a chance to participate in cooperative groups, learning from your classmates and improving your skills.

The Quick Review problems at the beginning of each problem set ask you to recall quickly things that you may have forgotten from earlier in the text or from previous courses. Other problems, marked by a shaded star, will prepare you for a topic in a later section. Prior to the Chapter Test at the end of each chapter, you will find review problems keyed to each section. Additionally, the Concept Problems give you a chance to apply your knowledge to new and challenging situations. So, keeping up with your homework will help to ensure your success.

At times you may feel you are becoming submerged in details. When that happens, just remember that calculus involves only four concepts:

- Limits
- Derivatives
- Integrals (one kind)
- Integrals (another kind)

Ask yourself, "Which of these concepts does my present work apply to?" That way, you will better see the big picture. Best wishes as you venture into the world of higher mathematics!

Paul A. Foerster
Alamo Heights High School
San Antonio, Texas

Limits, Derivatives, Integrals, and Integrals

Automakers have recently begun to produce electric cars, which utilize electricity instead of gasoline to run their engines. Engineers are constantly looking for ways to design an electric car that can match the performance of a conventional gasoline-powered car. Engineers can predict a car's performance characteristics even before the first prototype is built. From information about the acceleration, they can calculate the car's velocity as a function of time. From the velocity, they can predict the distance it will travel while it is accelerating. Calculus provides the mathematical tools to analyze quantities that change at variable rates.

Mathematical Overview

Calculus deals with calculating things that change at variable rates. The four concepts invented to do this are

- Limits
- Derivatives
- Integrals (one kind)
- Integrals (another kind)

In Chapter 1, you will study three of these concepts in four ways.

Graphically

The icon at the top of each even-numbered page of this chapter illustrates a limit, a derivative, and one type of integral.

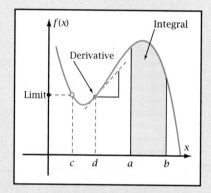

Numerically

x	$x - d$	Slope
2.1	0.1	1.071666...
2.01	0.01	1.007466...
2.001	0.001	1.000749...
⋮	⋮	⋮

Algebraically

Average rate of change $= \dfrac{f(x) - f(2)}{x - 2}$

Verbally

I have learned that a definite integral is used to measure the product of x and f(x). For instance, velocity multiplied by time gives the distance traveled by an object. The definite integral is used to find this distance if the velocity varies.

1-1 The Concept of Instantaneous Rate

If you push open a door that has an automatic closer, it opens fast at first, slows down, stops, starts closing, then slams shut. As the door moves, the number of degrees, d, it is from its closed position depends on how many seconds it has been since you pushed it. Figure 1-1a shows such a door from above.

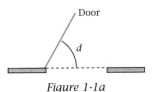

Figure 1-1a

The questions to be answered here are, "At any particular instant in time, is the door opening or closing?" and "How fast is it moving?" As you progress through this course, you will learn to write equations expressing the rate of change of one variable quantity in terms of another. For the time being, you will answer such questions graphically and numerically.

OBJECTIVE   Given the equation for a function relating two variables, estimate the instantaneous rate of change of the dependent variable with respect to the independent variable at a given point.

Suppose that a door is pushed open at time $t = 0$ s and slams shut again at time $t = 7$ s. While the door is in motion, assume that the number of degrees, d, from its closed position is modeled by this equation.

$$d = 200t \cdot 2^{-t} \quad \text{for} \quad 0 \le t \le 7$$

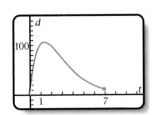

Figure 1-1b

How fast is the door moving at the instant when $t = 1$ s? Figure 1-1b shows this equation on a grapher (graphing calculator or computer). When t is 1, the graph is going *up* as t increases from left to right. So the angle is increasing and the door is opening. You can estimate the rate numerically by calculating values of d for values of t close to 1.

$$t = 1: \quad d = 200(1) \cdot 2^{-1} \quad = 100°$$
$$t = 1.1: \quad d = 200(1.1) \cdot 2^{-1.1} \quad = 102.633629...°$$

The door's angle increased by 2.633...° in 0.1 s, meaning that it moved at a rate of about (2.633...)/0.1, or 26.33... deg/s. However, this rate is an *average* rate, and the question was about an *instantaneous* rate. In an "instant" that is 0 s long, the door moves 0°. Thus, the rate would be 0/0, which is awkward because division by zero is undefined.

To get closer to the instantaneous rate at $t = 1$ s, find d at $t = 1.01$ s and at $t = 1.001$ s.

$$t = 1.01: \quad d = 200(1.01) \cdot 2^{-1.01} \quad = 100.30234..., \text{ a change of } 0.30234...°$$
$$t = 1.001: \quad d = 200(1.001) \cdot 2^{-1.001} \quad = 100.03064..., \text{ a change of } 0.03064...°$$

Here are the average rates for the time intervals 1 s to 1.01 s and 1 s to 1.001 s.

$$1 \text{ s to } 1.01 \text{ s:} \quad \text{average rate} = \frac{0.30234...}{0.01} = 30.234... \text{ deg/s}$$

$$1 \text{ s to } 1.001 \text{ s:} \quad \text{average rate} = \frac{0.03064...}{0.001} = 30.64... \text{ deg/s}$$

The important thing for you to notice is that as the time interval gets smaller and smaller, the number of degrees per second doesn't change much. Figure 1-1c shows why. As you zoom in on the point (1, 100), the graph appears to be straighter, so the change in d divided by the change in t becomes closer to the slope of a straight line.

If you list the average rates in a table, another interesting feature appears. The values stay the same for more and more decimal places.

t (s)	Average Rate
1 to 1.01	30.23420...
1 to 1.001	30.64000...
1 to 1.0001	30.68075...
1 to 1.00001	30.68482...
1 to 1.000001	30.68524...

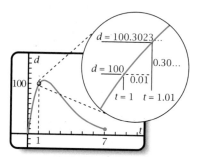

Figure 1-1c

There seems to be a *limiting* number that the values are approaching.

To estimate the instantaneous rate at $t = 3$ s, follow the same steps as for $t = 1$ s.

$t = 3$: $d = 200(3) \cdot 2^{-3}$ $= 75°$
$t = 3.1$: $d = 200(3.1) \cdot 2^{-3.1}$ $= 72.310056...°$
$t = 3.01$: $d = 200(3.01) \cdot 2^{-3.01}$ $= 74.730210...°$
$t = 3.001$: $d = 200(3.001) \cdot 2^{-3.001}$ $= 74.973014...°$

Here are the corresponding average rates.

3 s to 3.1 s: average rate $= \dfrac{72.310056... - 75}{3.1 - 3} = -26.899...$ deg/s

3 s to 3.01 s: average rate $= \dfrac{74.730210... - 75}{3.01 - 3} = -26.978...$ deg/s

3 s to 3.001 s: average rate $= \dfrac{74.973014... - 75}{3.001 - 3} = -26.985...$ deg/s

Again, the rates seem to be approaching some limiting number, this time, around -27. So the instantaneous rate at $t = 3$ s should be somewhere close to -27 deg/s. The negative sign tells you that the number of degrees, d, is *decreasing* as time goes on. Thus, the door is closing when $t = 3$ s. It is opening when $t = 1$ because the rate of change is positive.

For the door example shown above, the angle is said to be a **function** of time. Time is the **independent variable** and angle is the **dependent variable**. These names make sense, because the number of degrees the door is open *depends* on the number of seconds since it was pushed. The instantaneous rate of change of the dependent variable is said to be the **limit** of the average rates as the time interval gets closer to zero. This limiting value is called the **derivative** of the dependent variable with respect to the independent variable.

Problem Set 1-1

1. *Pendulum Problem:* A pendulum hangs from the ceiling (Figure 1-1d). As the pendulum swings, its distance, d, in centimeters from one wall of the room depends on the number of seconds, t, since it was set in motion. Assume that the equation for d as a function of t is

$$d = 80 + 30 \cos \frac{\pi}{3}t, \qquad t \geq 0$$

You want to find out how fast the pendulum is moving at a given instant, t, and whether it is approaching or going away from the wall.

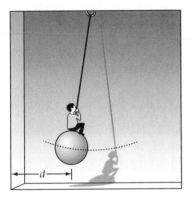

Figure 1-1d

a. Find d when $t = 5$. If you don't get 95 for the answer, make sure your calculator is in radian mode.

b. Estimate the instantaneous rate of change of d at $t = 5$ by finding the average rates from $t = 5$ to 5.1, $t = 5$ to 5.01, and $t = 5$ to 5.001.

c. Why can't the actual instantaneous rate of change of d with respect to t be calculated using the method in part b?

d. Estimate the instantaneous rate of change of d with respect to t at $t = 1.5$. At that time, is the pendulum approaching the wall or moving away from it? Explain.

e. How is the instantaneous rate of change related to the average rates? What name is given to the instantaneous rate?

f. What is the reason for the domain restriction $t \geq 0$? Can you think of any reason that there would be an *upper* bound to the domain?

2. *Board Price Problem:* If you check the prices of various lengths of lumber, you will find that a board twice as long as another of the same type does not necessarily cost twice as much. Let x be the length, in feet, of a $2'' \times 6''$ board (Figure 1-1e) and let y be the price, in cents, that you pay for the board. Assume that y is given by

$$y = 0.2x^3 - 4.8x^2 + 80x$$

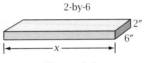

2-by-6

Figure 1-1e

a. Find the price of $2'' \times 6''$ boards that are 5 ft long, 10 ft long, and 20 ft long.

b. Find the average rate of change of the price in cents per foot for $x = 5$ to 5.1, $x = 5$ to 5.01, and $x = 5$ to 5.001.

c. The average number of cents per foot in part b is approaching an *integer* as the change in x gets smaller and smaller. What integer? What is the name given to this rate of change?

d. Estimate the instantaneous rate of change in price if x is 10 ft and if x is 20 ft. You should find that each of these rates is an integer.

e. One of the principles of marketing is that when you buy in larger quantities, you usually pay less per unit. Explain how the numbers in Problem 2 show that this principle does not apply to buying longer boards. Think of a reason why it does not apply.

1-2 Rate of Change by Equation, Graph, or Table

In Section 1-1, you explored functions for which an equation related two variable quantities. You found average rates of change of $f(x)$ over an interval of x-values, and used these to estimate the instantaneous rate of change at a particular value of x. The instantaneous rate is called the *derivative* of the function at that value of x. In this section you will estimate instantaneous rates for functions specified graphically or numerically, as well as algebraically (by equations).

OBJECTIVE   Given a function $y = f(x)$ specified by a graph, a table of values, or an equation, describe whether the y-value is increasing or decreasing as x increases through a particular value, and estimate the instantaneous rate of change of y at that value of x.

Background: Function Terminology and Types of Functions

The price you pay for a certain type of board depends on how long it is. In mathematics the symbol $f(x)$ (pronounced "f of x" or "f at x") is often used for the dependent variable. The letter f is the name of the function, and the number in parentheses is either a value of the independent variable or the variable itself. If $f(x) = 3x + 7$, then $f(5)$ is $3(5) + 7$, or 22.

The equation $f(x) = 3x + 7$ is the **particular equation** for a linear function. The **general equation** for a linear function is written $y = mx + b$, or $f(x) = mx + b$, where m and b represent the constants. The following box shows the names of some types of functions and their general equations.

DEFINITIONS: Types of Functions

Linear: $f(x) = mx + b$; m and b stand for constants, $m \neq 0$

Quadratic: $f(x) = ax^2 + bx + c$; a, b, and c stand for constants, $a \neq 0$

Polynomial: $f(x) = a_0 + a_1x + a_2x^2 + a_3x^3 + a_4x^4 + \cdots + a_nx^n$; $a_0, a_1, \ldots$ stand for constants, n is a positive integer, $a_n \neq 0$ (*n*th degree polynomial function)

Power: $f(x) = ax^n$; a and n stand for constants

Exponential: $f(x) = ab^x$; a and b stand for constants, $a \neq 0$, $b > 0$, $b \neq 1$

Rational Algebraic: $f(x) = (\text{polynomial})/(\text{polynomial})$

Absolute value: $f(x)$ contains $|(\text{variable expression})|$

Trigonometric or Circular: $f(x)$ contains $\cos x$, $\sin x$, $\tan x$, $\cot x$, $\sec x$, or $\csc x$

▶ **EXAMPLE 1**   Figure 1-2a shows the graph of a function. At $x = a$, $x = b$, and $x = c$, state whether y is increasing, decreasing, or neither as x increases. Then state whether the rate of change is fast or slow.

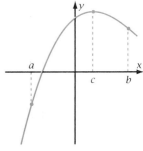

Figure 1-2a

Solution   At $x = a$, y is increasing quickly as you go from left to right.

At $x = b$, y is decreasing slowly because y is dropping as x goes from left to right, but it's not dropping very quickly.

At $x = c$, y is neither increasing nor decreasing, as shown by the fact that the graph has leveled off at $x = c$.

◀

▶ **EXAMPLE 2**   Figure 1-2b shows the graph of a function that could represent the height, $h(t)$, in feet, of a soccer ball above the ground as a function of the time, t, in seconds since it was kicked into the air.

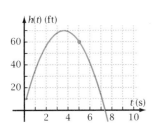

Figure 1-2b

a. Estimate the instantaneous rate of change of $h(t)$ at time $t = 5$.

b. Give the mathematical name of this instantaneous rate, and state why the rate is negative.

Solution   a. Draw a line tangent to the graph at $x = 5$ by laying a ruler against it, as shown in Figure 1-2c. You will be able to estimate the tangent line more accurately if you put the ruler on the *concave* side of the graph.

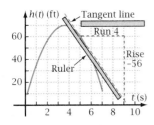

Figure 1-2c

The instantaneous rate is the slope of this tangent line. From the point where $t = 5$, run over a convenient distance in the t-direction, say 4 seconds. Then draw a vertical line to the tangent line. As shown in the figure, this rise is about 56 feet in the negative direction.

$$\text{Instantaneous rate} = \text{slope of tangent} \approx \frac{-56}{4} = -14 \text{ ft/s}$$

b. The mathematical name is *derivative*. The rate is negative because $h(t)$ is decreasing at $t = 5$.

◀

▶ **EXAMPLE 3**

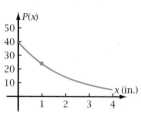

Figure 1-2d

Figure 1-2d shows a graph of $P(x) = 40(0.6^x)$, the probability that it rains a number of inches, x, at a particular place during a particular thunderstorm.

a. The probability that it rains 1 inch is $P(1) = 24\%$. By how much, and in which direction, does the probability change from $x = 1$ to $x = 1.1$? What is the average rate of change from 1 inch to 1.1 inches? Make sure to include units in your answer. Why is the rate negative?

b. Write an equation for $r(x)$, the average rate of change of $P(x)$ from 1 to x. Make a table of values of $r(x)$ for each 0.01 unit of x from 0.97 to 1.03. Explain why $r(x)$ is undefined at $x = 1$.

c. The instantaneous rate at $x = 1$ is the limit that the average rate approaches as x approaches 1. Estimate the instantaneous rate using information from part b. Name the concept of calculus that is given to this instantaneous rate.

Solution

a. To find the average rate, first you must find $P(1)$ and $P(1.1)$.

$$P(1) = 40(0.6^1) = 24$$
$$P(1.1) = 40(0.6^{1.1}) = 22.8048...$$
$$\text{Change} = 22.8048... - 24 = -1.1951...$$

Change is always final minus initial.

$$\text{Average rate} = \frac{-1.1951...}{0.1} = -11.1951 \text{ \%/in.}$$

The rate is negative because the probability decreases as the number of inches increases.

b. The average rate of change of $P(x)$ from 1 to x is equal to the change in $P(x)$ divided by the change in x.

$$r(x) = \frac{P(x) - 24}{x - 1} = \frac{40(0.6^x) - 24}{x - 1}$$

$$\frac{\text{change in } P(x)}{\text{change in } x}$$

Store $P(x)$ as y_1 and $r(x)$ as y_2 in your grapher. Make a table of values of x and $r(x)$.

x	$r(x)$
0.97	$-12.3542...$
0.98	$-12.3226...$
0.99	$-12.2911...$
1.00	Error
1.01	$-12.2285...$
1.02	$-12.1974...$
1.03	$-12.1663...$

Note that $r(1)$ is undefined because you would be dividing by zero. When $x = 1$, $x - 1 = 0$.

c. Average $r(0.99)$ and $r(1.01)$, the values in the table closest to $x = 1$.

$$\text{Instantaneous rate} \approx \frac{1}{2}[r(0.99) + r(1.01)] = -12.2598...$$

The percentage is decreasing at about 12.26% per inch. (The percentage decreases because it is less likely to rain greater quantities.) The name is *derivative*. ◀

▶ **EXAMPLE 4** A mass is bouncing up and down on a spring hanging from the ceiling (Figure 1-2e). Its distance, y, in feet, from the ceiling is measured by a calculator distance probe each 1/10 s, giving this table of values, in which t is time in seconds.

t (s)	y (ft)
0.2	3.99
0.3	5.84
0.4	7.37
0.5	8.00
0.6	7.48
0.7	6.01
0.8	4.16
0.9	2.63
1.0	2.00
1.1	2.52

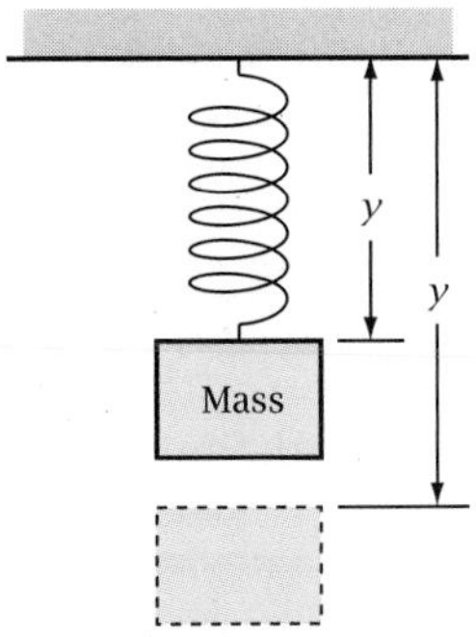

Figure 1-2e

a. How fast is y changing at each time?
   i. $t = 0.3$
   ii. $t = 0.6$
   iii. $t = 1.0$

b. At time $t = 0.3$, is the mass going up or down? Justify your answer.

Solution a. If data are given in numerical form, you cannot get better estimates of the rate by taking values of t closer and closer to 0.3. However, you can get a better estimate by using the closest t-values on both sides of the given value. A time-efficient way to do the computations is shown in the following table. If you like, do the computations mentally and write only the final answer.

t	y	Difference	Rate	Average Rate
0.2	3.99			
0.3	5.84	1.85	1.85/0.1 = 18.5	16.9
0.4	7.37	1.53	1.53/0.1 = 15.3	
0.5	8.00			
0.6	7.48	−0.52	−0.52/0.1 = −5.2	−9.95
0.7	6.01	−1.47	−1.47/0.1 = −14.7	
0.8	4.16			
0.9	2.63			
1.0	2.00	−0.63	−0.63/0.1 = −6.3	−0.55
1.1	2.52	0.52	0.52/0.1 = 5.2	

All you need to write on your paper are the results, as shown here.

   i. $t = 0.3$: increasing at about 16.9 ft/s

   ii. $t = 0.6$: decreasing at about 9.95 ft/s

   iii. $t = 1.0$: decreasing at about 0.55 ft/s   Write real-world answers with units.

  b. At $t = 0.3$, the rate is about 16.9 ft/s, a *positive* number. This fact implies that y is *increasing*. As y increases, the mass goes *downward*. ◀

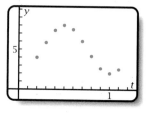

Figure 1-2f

Note that although a graph is not asked for in Example 4, plotting the data either on graph paper or by scatter plot on the grapher will help you understand what is happening. Figure 1-2f shows such a scatter plot.

The technique in Example 4 for estimating instantaneous rates by going forward and backward from the given value of x can also be applied to functions specified by an equation. The result is usually more accurate than the rate estimated by only going forward as you did in the last section.

As you learned in Section 1-1, the instantaneous rate of change of $f(x)$ at $x = c$ is the limit of the average rate of change over the interval from c to x as x approaches c. The value of the instantaneous rate is called the **derivative** of $f(x)$ with respect to x at $x = c$. The meaning of the word *derivative* is shown here. You will learn the precise definition when it is time to calculate derivatives exactly.

Meaning of Derivative

The derivative of function $f(x)$ at $x = c$ is the *instantaneous rate of change* of $f(x)$ with respect to x at $x = c$. It is found

- numerically, by taking the *limit* of the average rate over the interval from c to x as x approaches c
- graphically, by finding the slope of the line tangent to the graph at $x = c$

Note that "with respect to x" implies that you are finding how fast y changes *as x changes.*

Preview: Definition of Limit

In Section 1-1, you saw that the average rate of change of the y-value of a function got closer and closer to some fixed number as the change in the x-value got closer and closer to zero. That fixed number is called the **limit** of the average rate as the change in x approaches zero. The following is a verbal definition of limit. The full meaning will become clearer to you as the course progresses.

Verbal Definition of Limit

L is the limit of $f(x)$ as x approaches c
if and only if
L is the *one* number you can keep $f(x)$ arbitrarily close to
just by keeping x close enough to c, but not equal to c.

Problem Set 1-2

From now on, there will be ten short problems at the beginning of most problem sets. Some of the problems will help you review skills from previous sections or chapters. Others will test your general knowledge. Speed is the key here, not detailed work. You should be able to do all ten problems in less than five minutes.

Q1. Name the type of function: $f(x) = x^3$.

Q2. Find $f(2)$ for the function in Problem Q1.

Q3. Name the type of function: $g(x) = 3^x$.

Q4. Find $g(2)$ for the function in Problem Q3.

Q5. Sketch the graph: $h(x) = x^2$.

Q6. Find $h(5)$ for the function in Problem Q5.

Q7. Write the general equation for a quadratic function.

Q8. Write the particular equation for the function in Figure 1-2g.

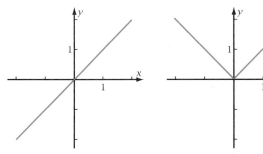

Figure 1-2g   Figure 1-2h

Q9. Write the particular equation for the function in Figure 1-2h.

Q10. What name is given to the instantaneous rate of change of a function?

Problems 1–10 show graphs of functions with values of x marked a, b, and so on. At each marked value, state whether the function is increasing, decreasing, or neither as x increases from left to right, and also whether the rate of increase or decrease is fast or slow.

1.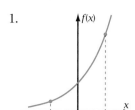

2.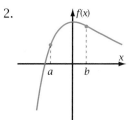

3.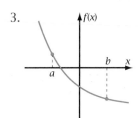

4.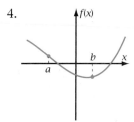

5.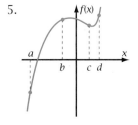

6.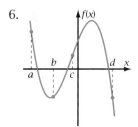

7.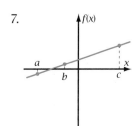

8.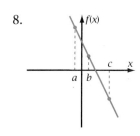

9.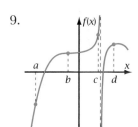

10.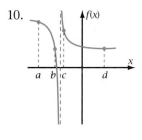

11. *Boiling Water Problem:* Figure 1-2i shows the temperature, $T(x)$, in degrees Celsius, of a kettle of water at time x, in seconds, since the burner was turned on.

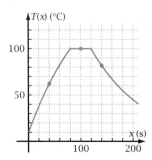

Figure 1-2i

a. On a copy of the figure, draw tangent lines at the points where $x = 40$, 100, and 140. Use the tangent lines to estimate the instantaneous rate of change of temperature at these times.

b. What do you suppose is happening to the water for $0 < x < 80$? For $80 < x < 120$? For $x > 120$?

12. *Roller Coaster Velocity Problem:* Figure 1-2j shows the velocity, $v(x)$, in ft/s, of a roller coaster car at time x, in seconds, after it starts down the first hill.

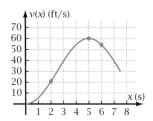

Figure 1-2j

a. On a copy of the figure, draw tangent lines at the points where $x = 2$, 5, and 6. Use the tangent lines to estimate the instantaneous rate of change of velocity at these times.

b. The instantaneous rates in part a are derivatives of $v(x)$ with respect to x. What units must you include in your answers? What physical quantity is this?

13. *Rock in the Air Problem:* A small rock is tied to an inflated balloon, then the rock and balloon are thrown into the air. While the rock and balloon are moving, the height of the rock is given by

$$h(x) = -x^2 + 8x + 2$$

where $h(x)$ is in feet above the ground at time x, in seconds, after the rock was thrown.

a. Plot the graph of function h. Sketch the result. Based on the graph, is $h(x)$ increasing or decreasing at $x = 3$? At $x = 7$?

b. How high is the rock at $x = 3$? At $x = 3.1$? What is the average rate of change of its height from 3 to 3.1 seconds?

c. Find the average rate of change from 3 to 3.01 seconds, and from 3 to 3.001 seconds. Based on the answers, what limit does the average rate seem to be approaching as the time interval gets shorter and shorter?

d. The limit of the average rates in part c is called the *instantaneous* rate at $x = 3$. It is also called the *derivative* of $h(x)$ at $x = 3$. Estimate the derivative of $h(x)$ at $x = 7$. Make sure to include units in your answer. Why is the derivative *negative* at $x = 7$?

14. *Fox Population Problem:* The population of foxes in a particular region varies periodically due to fluctuating food supplies. Assume that the number of foxes, $f(t)$, is given by

$$f(t) = 300 + 200 \sin t$$

where t is time in years after a certain date.

a. Store the equation for $f(t)$ as y_1 in your grapher, and plot the graph using a window with $[0, 10]$ for t. Sketch the graph. On the sketch, show a point where $f(t)$ is increasing, a point where it is decreasing, and a point where it is not changing much.

b. The change in $f(t)$ from 1 year to t is $(f(t) - f(1))$. So for the time interval $[1, t]$, $f(t)$ changes at the average rate $r(t)$ given by

$$r(t) = \frac{f(t) - f(1)}{t - 1}$$

Enter $r(t)$ as y_2 in your grapher. Then make a table of values of $r(t)$ for each 0.01 year from 0.97 through 1.03.

c. The instantaneous rate of change of $f(t)$ at $t = 1$ is the limit $f(t)$ approaches as t

approaches 1. Explain why your grapher gives an error message if you try to calculate $r(1)$. Find an estimate for the instantaneous rate by taking values of t closer and closer to 1. What special name is given to this instantaneous rate?

d. At approximately what instantaneous rate is the fox population changing at $t = 4$? Explain why the answer is negative.

15. *Bacteria Culture Problem:* Bacteria in a laboratory culture are multiplying in such a way that the surface area of the culture, $a(t)$, in mm^2, is given by

$$a(t) = 200(1.2^t)$$

where t is the number of hours since the culture was started.

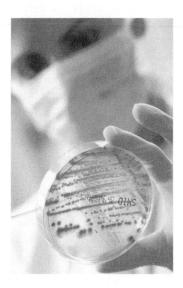

a. Find the average rate of increase of bacteria from $t = 2$ to $t = 2.1$.

b. Write an equation for $r(t)$, the average rate of change of $a(t)$, from 2 hours to t. Plot the graph of r using a friendly window that includes $t = 2$ as a grid point. What do you notice when you trace the graph of r to $t = 2$?

c. The instantaneous rate of change (the derivative) of $a(t)$ at $t = 2$ is $52.508608... \, mm^2/h$. How close to this value is $r(2.01)$? How close must t be kept to 2 on the positive side so that the average rate is within 0.01 unit of this derivative?

16. *Sphere Volume Problem:* Recall from geometry that the volume of a sphere is

$$V(x) = \frac{4}{3}\pi x^3$$

where $V(x)$ is volume in cubic centimeters and x is the radius in centimeters.

a. Find $V(6)$. Write the answer as a multiple of π.

b. Find the average rate of change of $V(x)$ from $x = 6$ to $x = 6.1$. Find the average rate from $x = 5.9$ to $x = 6$. Use the answers to find an estimate of the instantaneous rate at $x = 6$.

c. Write an equation for $r(x)$, the average rate of change of $V(x)$ from 6 to x. Plot the graph of r using a friendly window that has $x = 6$ as a grid point. What do you notice when you trace the graph to $x = 6$?

d. The derivative of $V(x)$ at $x = 6$ equals $4\pi 6^2$, the surface area of a sphere of radius 6 cm. How close is $r(6.1)$ to this derivative? How close to 6 on the positive side must the radius be kept for $r(x)$ to be within 0.01 unit of this derivative?

17. *Rolling Tire Problem:* A pebble is stuck in the tread of a car tire (Figure 1-2k). As the wheel turns, the distance, y, in inches, between the pebble and the road at various times, t, in seconds, is given by the table below.

t (s)	y (in.)
1.2	0.63
1.3	0.54
1.4	0.45
1.5	0.34
1.6	0.22
1.7	0.00
1.8	0.22
1.9	0.34
2.0	0.45

Figure 1-2k

a. About how fast is y changing at each time?

  i. $t = 1.4$

  ii. $t = 1.7$

  iii. $t = 1.9$

b. At what time does the stone strike the pavement? Justify your answer.

18. *Flat Tire Problem:* A tire is punctured by a nail. As the air leaks out, the distance, y, in inches, between the rim and the pavement (Figure 1-2l) depends on the time, t, in minutes, since the tire was punctured. Values of t and y are given in the table below.

t (min)	y (in.)
0	6.00
2	4.88
4	4.42
6	4.06
8	3.76
10	3.50
12	3.26
14	3.04
16	2.84

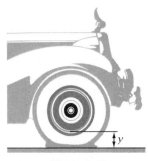

Figure 1-2l

a. About how fast is y changing at each time?

   i. $t = 2$ ii. $t = 8$ iii. $t = 14$

b. How do you interpret the *sign* of the rate at which y is changing?

For Problems 19–28,

   a. Give the type of function (linear, quadratic, and so on).

   b. State whether $f(x)$ is increasing or decreasing at $x = c$, and how you know this.

19. $f(x) = x^2 + 5x + 6, c = 3$

20. $f(x) = -x^2 + 8x + 5, c = 1$

21. $f(x) = 3^x, c = 2$

22. $f(x) = 2^x, c = -3$

23. $f(x) = \dfrac{1}{x - 5}, c = 4$

24. $f(x) = -\dfrac{1}{x}, c = -2$

25. $f(x) = -3x + 7, c = 5$

26. $f(x) = 0.2x - 5, c = 8$

27. $f(x) = \sin x, c = 2$ (Radian mode!)

28. $f(x) = \cos x, c = 1$ (Radian mode!)

29. *Derivative Meaning Problem:* What is the physical meaning of the derivative of a function? How can you estimate the derivative graphically? Numerically? How does the numerical computation of a derivative illustrate the meaning of limit?

30. *Limit Meaning Problem:* From memory, write the verbal meaning of limit. Compare it with the statement in the text. If you did not state all parts correctly, try writing it again until you get it completely correct. How do the results of Problems 13 and 14 of this problem set illustrate the meaning of limit?

1-3   One Type of Integral of a Function

The title of this chapter is *Limits, Derivatives, Integrals, and Integrals.* In Section 1-2, you estimated the derivative of a function, which is the instantaneous rate of change of y with respect to x. In this section you will learn about one type of integral, the definite integral.

Suppose you start driving your car. The velocity increases for a while, then levels off. Figure 1-3a shows the velocity, $v(t)$, increasing from zero, then approaching and leveling off at 60 ft/s.

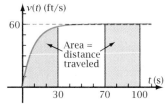

Figure 1-3a

In the 30 seconds between time $t = 70$ and $t = 100$, the velocity is a constant 60 ft/s. Because distance = rate × time, the distance you go in this time interval is

$$60 \text{ ft/s} \times 30 \text{ s} = 1800 \text{ ft}$$

Geometrically, 1800 is the area of the rectangle shown in Figure 1-3a. The width is 30 and the length is 60. Between 0 s and 30 s, where the velocity is changing, the area of the region under the graph also equals the distance traveled. Because the length varies, you cannot find the area simply by multiplying two numbers.

The process of evaluating a product in which one factor varies is called finding a **definite integral**. You can evaluate definite integrals by finding the corresponding area. In this section you will find the approximate area by counting squares on graph paper (by "brute force"!). Later, you will apply the concept of limit to calculate definite integrals *exactly*.

OBJECTIVE   Given the equation or the graph for a function, estimate on a graph the definite integral of the function between $x = a$ and $x = b$ by counting squares.

If you are given only the equation, you can plot it with your grapher's grid-on feature, estimating the number of squares in this way. However, it is more accurate to use a plot on graph paper to count squares. You can get plotting data by using your grapher's TRACE or TABLE feature.

▶ **EXAMPLE 1**   Estimate the definite integral of the exponential function $f(x) = 8(0.7)^x$ from $x = 1$ to $x = 7$.

Solution   You can get reasonable accuracy by plotting $f(x)$ at each integer value of x (Figure 1-3b).

x	$f(x)$
0	$f(0) = 8$
1	$f(1) = 5.6$
2	$f(2) \approx 3.9$
3	$f(3) \approx 2.7$
4	$f(4) \approx 1.9$
5	$f(5) \approx 1.3$
6	$f(6) \approx 0.9$
7	$f(7) \approx 0.7$

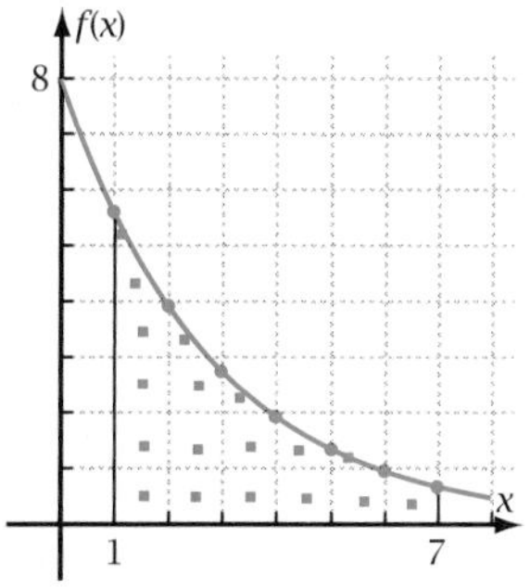

Figure 1-3b

The integral equals the area under the graph from $x = 1$ to $x = 7$. "Under" the graph means "between the graph and the x-axis." To find the area, first count the whole squares. Put a dot in each square as you count it to keep track, then estimate the area of each partial square to the nearest 0.1 unit. For instance, less than half a square is 0.1, 0.2, 0.3, or 0.4. You be the judge. You should get about 13.9 square units for the area, so the definite integral is approximately 13.9. Answers anywhere from 13.5 to 14.3 are reasonable. ◀

If the graph is already given, you need only count the squares. Be sure you know how much area each square represents! Example 2 shows you how to do this.

▶ **EXAMPLE 2**   Figure 1-3c shows the graph of the velocity function $v(t) = -100t^2 + 90t + 14$, where t is in seconds and $v(t)$ is in feet per second. Estimate the definite integral of $v(t)$ with respect to t from $t = 0.1$ to $t = 1$.

Solution   Notice that each space in the t-direction is 0.1 s and each space in the direction of $v(t)$ is 2 ft/s. Thus, each square represents (0.1)(2), or 0.2 ft. You should count about 119.2 squares for the area. So, the definite integral will be about

$$(119.2)(0.2) \approx 23.8 \text{ ft}$$

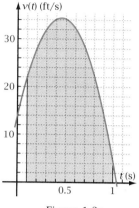

Figure 1-3c

◀

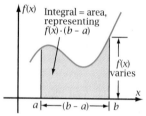

Figure 1-3d

The following box gives the meaning of definite integral. The precise definition is given in Chapter 5, where you will learn an algebraic technique for calculating *exact* values of definite integrals.

Meaning of Definite Integral

The definite integral of the function f from $x = a$ to $x = b$ gives a way to find the product of $(b - a)$ and $f(x)$, even if $f(x)$ is not a constant. See Figure 1-3d.

Problem Set 1-3

Quick Review   🕐 *5 min*

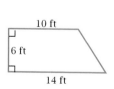

Figure 1-3e

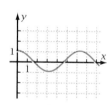

Figure 1-3f

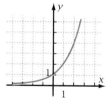

Figure 1-3g

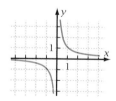

Figure 1-3h

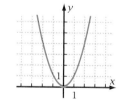

Figure 1-3i

Q1. Find the area of the trapezoid in Figure 1-3e.

Q2. Write the particular equation for the function graphed in Figure 1-3f.

Q3. Write the particular equation for the function graphed in Figure 1-3g.

Q4. Write the particular equation for the function graphed in Figure 1-3h.

Q5. Write the particular equation for the function graphed in Figure 1-3i.

Q6. Find $f(5)$ if $f(x) = x - 1$.

Q7. Sketch the graph of a linear function with positive y-intercept and negative slope.

Q8. Sketch the graph of a quadratic function opening downward.

Q9. Sketch the graph of a decreasing exponential function.

Q10. At what value(s) of x is $f(x) = (x - 4)/(x - 3)$ undefined?

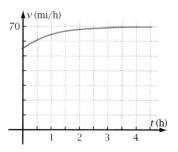

Figure 1-3k

For Problems 1–4, estimate the definite integral by counting squares on a graph.

1. $f(x) = -0.1x^2 + 7$
   a. $x = 0$ to $x = 5$
   b. $x = -1$ to $x = 6$

2. $f(x) = -0.2x^2 + 8$
   a. $x = 0$ to $x = 3$
   b. $x = -2$ to $x = 5$

3. $h(x) = \sin x$
   a. $x = 0$ to $x = \pi$
   b. $x = 0$ to $x = \pi/2$

4. $g(x) = 2^x + 5$
   a. $x = 1$ to $x = 2$
   b. $x = -1$ to $x = 1$

5. In Figure 1-3j, a car is slowing down from velocity $v = 60$ ft/s. Estimate the distance it travels from time $t = 5$ s to $t = 25$ s by finding the definite integral.

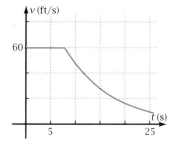

Figure 1-3j

6. In Figure 1-3k, a car slowly speeds up from $v = 55$ mi/h during a long trip. Estimate the distance it travels from time $t = 0$ h to $t = 4$ h by finding the definite integral.

For Problems 7 and 8, estimate the derivative of the function at the given value of x.

7. $f(x) = \tan x$, $x = 1$

8. $h(x) = -7x + 100$, $x = 5$

9. *Electric Car Problem:* You have been hired by an automobile manufacturer to analyze the predicted motion of a new electric car they are building. When accelerated hard from a standing start, the velocity of the car, $v(t)$, in ft/s, is expected to vary exponentially with time, t, in seconds, according to the equation

   $$v(t) = 50(1 - 0.9^t)$$

   a. Plot the graph of function v in the domain [0, 10]. What is the corresponding range of the function?

   b. Approximately how many seconds will it take the car to reach a velocity of 30 ft/s?

   c. Approximately how far will the car have traveled when it reaches 30 ft/s? Which of the four concepts of calculus is used to find this distance?

   d. At approximately what rate is the velocity changing when $t = 5$? Which of the four concepts of calculus is used to find this rate? What is the physical meaning of the rate of change of velocity?

10. *Slide Problem:* Phoebe sits atop the swimming pool slide (Figure 1-3l). At time $t = 0$ s she pushes off. Calvin finds that her velocity, $v(t)$, in ft/s, is given by

$$v(t) = 10 \sin 0.3t$$

Figure 1-3l

Phoebe splashes into the water at time $t = 4$ s.

a. Plot the graph of function v. Use radian mode.

b. How fast was Phoebe going when she hit the water? What, then, are the domain and range of the velocity function?

c. Find, approximately, the definite integral of the velocity function from $t = 0$ to $t = 4$. What are the units of the integral? What real-world quantity does this integral give you?

d. What, approximately, was the derivative of the velocity function when $t = 3$? What are the units of the derivative? What is the physical meaning of the derivative in this case?

11. *Negative Velocity Problem:* Velocity differs from speed in that it can be *negative.* If the velocity of a moving object is negative, then its distance from its starting point is *decreasing* as time increases. The graph in Figure 1-3m shows $v(t)$, in cm/s, as a function of t, in seconds, after its motion started. How far is the object from its starting point when $t = 9$?

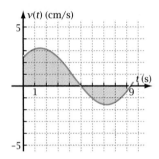

Figure 1-3m

12. Write the meaning of derivative.

13. Write the meaning of definite integral.

14. Write the verbal definition of limit.

1-4 **Definite Integrals by Trapezoids, from Equations and Data**

In Section 1-3, you learned that the definite integral of a function is the product of x- and y-values, where the y-values may be different for various values of x. Because the integral is represented by the area of a region under the graph, you were able to estimate it by counting squares. In this section you will learn a more efficient way of estimating definite integrals.

Figure 1-4a shows the graph of

$$f(x) = -100x^2 + 90x + 14$$

which is the function in Example 2 of Section 1-3 using $f(x)$ instead of $v(t)$. Instead of counting squares to find the area of the region under the graph, the region is divided into vertical strips. Line segments connect the points where the strip boundaries meet the graph. The result is a set of trapezoids whose areas add up to a number approximately equal to the area of the region.

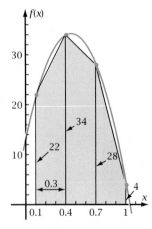

Figure 1-4a

Recall from geometry that the area of a trapezoid is the altitude times the average of the lengths of the parallel sides. Figure 1-4a shows that for three trapezoids, the parallel sides are $f(0.1) = 22, f(0.4) = 34, f(0.7) = 28$, and $f(1) = 4$. The "altitude" of each trapezoid is the change in x, which equals 0.3 in this case. Using T_3 to represent the sum of the areas of the three trapezoids,

$$T_3 = \tfrac{1}{2}(22 + 34)(0.3) + \tfrac{1}{2}(34 + 28)(0.3) + \tfrac{1}{2}(28 + 4)(0.3) = 22.5$$

which is approximately equal to the definite integral. The answer is slightly smaller than the 23.8 found by counting squares in Example 2 of Section 1-3 because the trapezoids are inscribed, leaving out small parts of the region. You can make the approximation more accurate by using more trapezoids. You can also do the procedure numerically instead of graphically.

OBJECTIVE Estimate the value of a definite integral by dividing the region under the graph into trapezoids and summing the areas.

To accomplish the objective in a time-efficient way, observe that each y-value in the sum appears *twice*, except for the first and the last values. Factoring out 0.3 leads to

$$T_3 = 0.3\left[\tfrac{1}{2}(22) + 34 + 28 + \tfrac{1}{2}(4)\right] = 22.5$$

Inside the brackets is the sum of the y-values at the boundaries of the vertical strips, using half of the first value and half of the last value. (There is one more boundary than there are strips.) The answer is multiplied by the width of each strip.

▶ **EXAMPLE 1** Use trapezoids to estimate the definite integral of $f(x) = -100x^2 + 90x + 14$ from $x = 0.1$ to $x = 1$. Use 9 increments (that is, 9 strips, with 9 trapezoids).

Solution From $x = 0.1$ to $x = 1$, there is 0.9 x-unit. So the width of each strip is $0.9/9 = 0.1$. An efficient way to compute this is to make a list of the y-values in your grapher, taking half of the first value and half of the last value. Then sum the list.

$L_1 = x$	$L_2 = f(x)$	
0.1	11	Half of $f(0.1)$
0.2	28	
0.3	32	
0.4	34	
0.5	34	
0.6	32	
0.7	28	
0.8	22	
0.9	14	
1.0	2	Half of $f(1)$
	237	

Integral $\approx T_9 = 0.1(237) = 23.7$ ◀

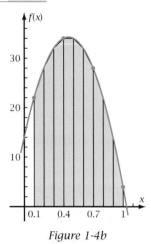

Figure 1-4b

Note that using more increments gives an answer closer to 23.8 than counting squares, which you used in Example 2 of Section 1-3. This is to be expected. As Figure 1-4b shows, using more trapezoids reduces the area of the region left out by the trapezoids.

You can generalize the preceding example to any number of increments, *n*. The result is called the *trapezoidal rule*.

PROPERTY: The Trapezoidal Rule

The definite integral of $f(x)$ from $x = a$ to $x = b$ is approximately equal to

$$T_n = \Delta x \left(\tfrac{1}{2}f(a) + f(x_1) + f(x_2) + f(x_3) + \cdots + f(x_{n-1}) + \tfrac{1}{2}f(b) \right)$$

where *n* is the number of increments (trapezoids), $\Delta x = (b - a)/n$ is the width of each increment, and the values of $x_1, x_2, x_3, \ldots$ are spaced Δx units apart.

Verbally: "Add the values of $f(x)$, taking half of the first value and half of the last value, then multiply by the width of each increment."

The exact value of the integral is the *limit* of the areas of the trapezoids as the number of increments becomes very large.

$$T_9 = 23.7$$
$$T_{20} = 23.819625$$
$$T_{100} = 23.8487850\ldots$$
$$T_{1000} = 23.84998785\ldots$$

The answers appear to be approaching 23.85 as *n* becomes very large. The exact value of the integral is the *limit* of the sum of the areas of the trapezoids as the number of trapezoids becomes infinitely large. In Chapter 5, you will learn how to calculate this limit algebraically.

PROPERTY: Exact Value of a Definite Integral

The exact value of a definite integral equals the limit of the trapezoidal rule sum T_n as *n* approaches infinity, provided the limit exists. The exact value can be estimated numerically by taking trapezoidal sums with more and more increments and seeing whether the sums approach a particular number.

The trapezoidal rule is advantageous if you must find the definite integral of a function specified by a table, rather than by equation. Example 2 shows you how to do this.

▶ **EXAMPLE 2**   On a ship at sea, it is easier to measure how fast you are going than it is to measure how far you have gone. Suppose you are the navigator aboard a supertanker. The velocity of the ship is measured every 15 min and recorded in the table. Estimate the distance the ship has gone between 7:30 p.m. and 9:15 p.m.

Time	mi/h	Time	mi/h
7:30	28	8:30	7
7:45	25	8:45	10
8:00	20	9:00	21
8:15	22	9:15	26

Solution   Figure 1-4c shows the given points. Because no information is known for times between the given ones, the simplest thing to assume is that the graph is a sequence of line segments. Because distance equals (miles/hour)(hours), the answer will equal a definite integral. You can find the integral from the area of the shaded region in Figure 1-4c, using the trapezoidal rule.

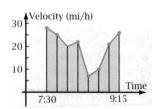

Figure 1-4c

$$T_7 = 0.25\left[\tfrac{1}{2}(28) + 25 + 20 + 22 + 7 + 10 + 21 + \tfrac{1}{2}(26)\right] = 33$$

Why 0.25? Why 7 increments?

∴ the distance is about 33 mi.   ◀

Problem Set 1-4

Q1. The value of y changes by 3 units when x changes by 0.1 unit. About how fast is y changing?

Q2. The value of y changes by -5 units when x changes by 0.01 unit. Approximately what does the derivative equal?

Q3. Sketch the graph of the absolute value function, $y = |x|$.

Q4. Find $f(3)$ if $f(x) = x^2$.

Q5. What is 50 divided by 1/2?

Q6. Evaluate: $\sin(\pi/2)$

Q7. How many days are there in a leap year?

Q8. The instantaneous rate of change of a function y is called the —?— of function y.

Q9. The product of x and y for a function is called the —?— of the function.

Q10. At what value(s) of x is $f(x) = (x - 4)/(x - 3)$ equal to zero?

1. *Spaceship Problem:* A spaceship is launched from Cape Canaveral. As the last stage of the rocket motor fires, the velocity is given by

$$v(t) = 1600 \times 1.1^t$$

where $v(t)$ is in feet per second and t is the number of seconds since the last stage started.

a. Plot the graph of $v(t)$ versus t, from $t = 0$ to $t = 30$. Sketch trapezoids with parallel sides at 5-s intervals, extending from the t-axis to the graph.

b. Find, approximately, the definite integral of $v(t)$ with respect to t from $t = 0$ to $t = 30$ by summing the areas of the trapezoids. Will the sum overestimate the integral or underestimate it? How can you tell?

c. Based on the units of the definite integral, explain why it represents the distance the spaceship traveled in the 30-s interval.

d. To go into orbit around Earth, the spaceship must be traveling at least 27,000 ft/s. Will it be traveling this fast when $t = 30$? How can you tell?

2. *Walking Problem:* Pace Walker enters an AIDS walkathon. She starts off at 4 mi/h, speeds up as she warms up, then slows down as she gets tired. She estimates that her speed is given by

$$v(t) = 4 + \sin 1.4t$$

where t is the number of hours since she started and $v(t)$ is in miles per hour.

a. Pace walks for 3 h. Plot the graph of $v(t)$ as a function of t for these three hours. Sketch the result on your paper. (Be sure your calculator is in radian mode!)

b. Explain why a definite integral is used to find the distance Pace has gone in 3 h.

c. Estimate the integral in part b, using six trapezoids. Show these trapezoids on your graph. About how far did Pace walk in the 3 h?

d. How fast was Pace walking at the end of the 3 h? When did her maximum speed occur? What was her maximum speed?

3. *Aircraft Carrier Landing Problem:* Assume that as a plane comes in for a landing on an aircraft carrier, its velocity, in ft/s, takes on the values shown in the table. Find, approximately, how far the plane travels as it comes to a stop.

t (s)	y (ft/s)
0.0	300
0.6	230
1.2	150
1.8	90
2.4	40
3.0	0

In 1993, Kara Hultgreen became one of the first female pilots authorized to fly navy planes in combat.

4. *Water over the Dam Problem:* The amount of water that has flowed over the spillway on a dam can be estimated from the flow rate and the length of time the water has been flowing. Suppose that the flow rate has been recorded every 3 h for a 24-h period, as shown in the table. Estimate the number of cubic feet of water that has flowed over the dam in this period.

Time	ft³/h	Time	ft³/h
12:00 a.m.	5,000	12:00 p.m.	11,000
3:00 a.m.	8,000	3:00 p.m.	7,000
6:00 a.m.	12,000	6:00 p.m.	4,000
9:00 a.m.	13,000	9:00 p.m.	6,000
		12:00 a.m.	9,000

5. *Program for Trapezoidal Rule Problem:* Download or write a program for your grapher to evaluate definite integrals using the

trapezoidal rule. Store the equation as y_1. For the input, use a and b, the initial and final values of x, and n, the number of increments. The output should be the value of n that you used, and the approximate value of the integral. Test your program by using it to find T_3 from Example 1. Then find T_{20} and see if you get 23.819625.

6. *Program for Trapezoidal Rule from Data Problem:* Download or modify the program from Problem 5 to evaluate an integral, approximately, for a function specified by a table of data. Store the data points in L_1 on your grapher. For the input, use n, the number of increments, and Δx, the spacing between x-values. It is not necessary to input the actual x-values. Be aware that there will be $n + 1$ data points for n increments. Test your program by finding T_7 for the function in Example 2.

7. For the definite integral of $f(x) = -0.1x^2 + 7$ from $x = 1$ to $x = 4$,

   a. Sketch the region corresponding to the integral.

   b. Approximate the integral by finding T_{10}, T_{20}, and T_{50} using the trapezoidal rule. Do these values overestimate the integral or underestimate it? How do you know?

   c. The exact value of the integral is 18.9. How close do T_{10}, T_{20}, and T_{50} come to this value? How many increments, n, do you need to use until T_n is first within 0.01 unit of the limit? Give evidence to suggest that T_n is within 0.01 unit of 18.9 for all values of n greater than this.

8. For the definite integral of $g(x) = 2^x$ from $x = 1$ to $x = 3$,

   a. Sketch the region corresponding to the integral.

   b. Approximate the integral by finding T_{10}, T_{20}, and T_{50} using the trapezoidal rule. Do these values overestimate the integral or underestimate it? How do you know?

   c. The exact value of the integral is 8.65617024.... How close do T_{10}, T_{20}, and T_{50} come to this value? How many increments, n, do you need to use until T_n is first within 0.01 unit of the limit? Give evidence to suggest that T_n is within 0.01 unit of the

exact value for all values of n greater than this.

9. *Elliptical Table Problem:* Figure 1-4d shows the top of a coffee table in the shape of an ellipse. The ellipse has the equation

$$\left(\frac{x}{110}\right)^2 + \left(\frac{y}{40}\right)^2 = 1$$

where x and y are in centimeters. Use the trapezoidal rule to estimate the area of the table. Will this estimate be too high or too low? Explain. What is the *exact* area of the ellipse?

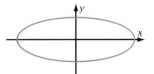

Figure 1-4d

10. *Football Problem:* The table shows the cross-sectional area, A, of a football at various distances, d, from one end. The distances are in inches and the areas are in square inches. Use the trapezoidal rule to find, approximately, the integral of area with respect to distance. What are the units of this integral? What, then, do you suppose the integral represents?

d (in.)	A (in.^{2})	d (in.)	A (in.^{2})
0	0.0	7	29.7
1	2.1	8	23.8
2	7.9	9	15.9
3	15.9	10	7.9
4	23.8	11	2.1
5	29.7	12	0.0
6	31.8		

11. *Exact Integral Conjecture Problem 1:* Now that you have a program to calculate definite integrals approximately, you can see what happens to the value of the integral as you use narrower trapezoids. Estimate the definite integral of $f(x) = x^2$ from $x = 1$ to $x = 4$, using 10, 100, and 1000 trapezoids. What number do the values seem to be approaching as the number of trapezoids gets larger and larger? Make a conjecture about the *exact* value of the definite integral as the width of each trapezoid

approaches zero. This number is the —?— of the areas of the trapezoids as their widths approach zero. What word goes in the blank?

12. *Exact Integral Conjecture Problem 2:* The *exact* definite integral of $g(x) = x^3$ from $x = 1$ to $x = 5$

is an *integer*. Make a conjecture about what this integer is equal to. Justify your answer.

13. *Trapezoidal Rule Error Problem:* How can you tell whether the trapezoidal rule overestimates or underestimates an integral? Draw a sketch to justify your answer.

1-5   Calculus Journal

You have been learning calculus by reading, listening, and discussing, and also by working problems. An important ability to develop for any subject you study is the ability to *write* about it. To gain practice in this technique, you will be asked to record what you've been learning in a **journal**. (*Journal* comes from the same source as the French word *jour*, meaning "day." *Journey* comes from the same source and means "a day's travel.")

OBJECTIVE   Start writing a journal in which you can record things you've learned about calculus and questions you still have about certain concepts.

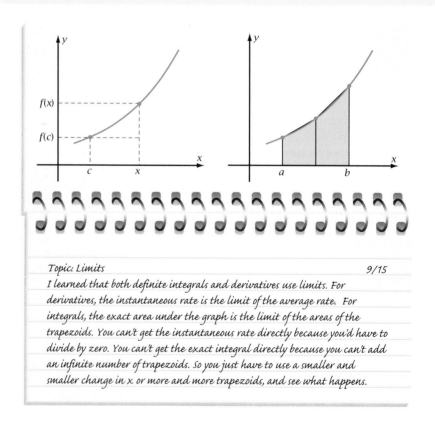

Topic: Limits   9/15

I learned that both definite integrals and derivatives use limits. For derivatives, the instantaneous rate is the limit of the average rate. For integrals, the exact area under the graph is the limit of the areas of the trapezoids. You can't get the instantaneous rate directly because you'd have to divide by zero. You can't get the exact integral directly because you can't add an infinite number of trapezoids. So you just have to use a smaller and smaller change in x or more and more trapezoids, and see what happens.

Use a bound notebook or spiral-bound notebook with large index cards for pages. You can write narrative and equations on the lined side of the card and draw graphs on the facing blank side. A typical entry might look like the index card on the previous page.

Your journal should *not* be a simple transcription of your class notes. Nor should you take notes directly in it. Save it for concise summaries of things you have learned, conjectures that have been made, and topics about which you are not yet certain.

Problem Set 1-5

1. Start a journal in which you will record your understandings about calculus. The first entry should include such things as
   - The four concepts of calculus
   - The distinctions among derivative, definite integral, and limit
   - The fact that you still don't know what the other kind of integral is
   - The techniques you know for calculating derivatives, definite integrals, and limits
   - Any questions that still aren't clear in your mind

1-6   Chapter Review and Test

In this chapter you have had a brief introduction to the major concepts of calculus.

Limits

Derivatives

Definite integrals

Another type of integrals

The derivative of a function is its instantaneous rate of change. A definite integral of a function involves a product of the dependent and independent variables, such as (rate)(time). A limit is a number that y can be kept close to, just by keeping x suitably restricted. The other type of integral is called an *indefinite integral,* also known as an *antiderivative.* You will see why two different concepts use the word "integral" when you learn the fundamental theorem of calculus in Chapter 5.

You have learned how to calculate approximate values of derivatives by dividing small changes in y by the corresponding change in x. Definite integrals can be found using areas under graphs and can thus be estimated by counting squares. Limits of functions can be calculated by finding the y-value of a removable discontinuity in the graph. Along the way you have refreshed your memory about the shapes of certain graphs.

The Review Problems are numbered according to the sections of this chapter. Answers are provided at the back of the book. The Concept Problems allow you to apply your knowledge to new situations. Answers are not provided, and in later chapters you may be required to do research to find answers to open-ended problems. The Chapter Test resembles a typical classroom test your instructor might give you. It has a calculator part and a no-calculator part, and the answers are not provided.

Review Problems

R1. *Bungee Problem:* Lee Per attaches himself to a strong bungee cord and jumps off a bridge. At time $t = 3$ s, the cord first becomes taut. From that time on, Lee's distance, d, in feet, from the river below the bridge is given by the equation

$$d = 90 - 80 \sin[1.2(t - 3)]$$

a. How far is Lee from the water when $t = 4$?

b. Find the average rate of change of d with respect to t for the interval $t = 3.9$ to $t = 4$, and for the interval $t = 4$ to $t = 4.1$. Approximately what is the instantaneous rate of change at $t = 4$? Is Lee going up or going down at time $t = 4$? Explain.

c. Estimate the instantaneous rate of change of d with respect to t when $t = 5$.

d. Is Lee going up or down when $t = 5$? How fast is he going?

e. Which concept of calculus is the instantaneous rate of change?

R2. a. What is the physical meaning of the derivative of a function? What is the graphical meaning?

b. For the function in Figure 1-6a, explain how $f(x)$ is changing (increasing or decreasing, quickly or slowly) when x equals -4, 1, 3, and 5.

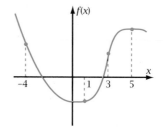

Figure 1-6a

c. If $f(x) = 5^x$, find the average rate of change of $f(x)$ from $x = 2$ to $x = 2.1$, from $x = 2$ to $x = 2.01$, and from $x = 2$ to $x = 2.001$. How close are these average rates to the instantaneous rate, $40.235947...$? Do the average rates seem to be approaching this instantaneous rate as the second value of x approaches 2? Which concept of calculus is the instantaneous rate? Which concept of calculus is used to find the instantaneous rate?

d. Mary Thon runs 200 m in 26 s! Her distance, d, in meters from the start at various times t, in seconds, is given in the table. Estimate her instantaneous velocity in m/s when $t = 2, t = 18$, and $t = 24$. For which time intervals did her velocity stay relatively constant? Why is the velocity at $t = 24$

reasonable in relation to the velocities at other times?

t (s)	d (m)	t (s)	d (m)
0	0	14	89
2	7	16	103
4	13	18	119
6	33	20	138
8	47	22	154
10	61	24	176
12	75	26	200

R3. Izzy Sinkin winds up his toy boat and lets it run on the pond. Its velocity is given by

$$v(t) = (2t)(0.8^t)$$

as shown in Figure 1-6b. Find, approximately, the distance the boat travels between $t = 2$ and $t = 10$. Which concept of calculus is used to find this distance?

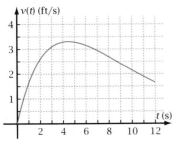

Figure 1-6b

R4. The graph in Figure 1-6c shows

$$f(x) = -0.5x^2 + 1.8x + 4$$

a. Plot the graph of f. Sketch your results. Does your graph agree with Figure 1-6c?

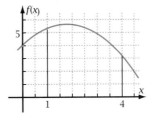

Figure 1-6c

b. Estimate the definite integral of $f(x)$ with respect to x from $x = 1$ to $x = 4$ by counting squares.

c. Estimate the integral in part b by drawing trapezoids each 0.5 unit of x and summing their areas. Does the trapezoidal sum overestimate the integral or underestimate it? How can you tell?

d. Use your trapezoidal rule program to estimate the integral using 50 increments and 100 increments. How close do T_{50} and T_{100} come to 15, the exact value of the integral? Do the trapezoidal sums seem to be getting closer to 15 as the number of increments increases? Which concept of calculus is used to determine this?

R5. In Section 1-5, you started a calculus journal. In what ways do you think keeping this journal will help you? How can you use the completed journal at the end of the course? What is your responsibility throughout the coming year to ensure that keeping the journal will be a worthwhile project?

Concept Problems

C1. *Exact Value of a Derivative Problem:* You have been calculating approximate values of derivatives by finding the change in y for a given change in x, then dividing. In this problem you will apply the concept of limit to the concept of derivative to find the *exact* value of a derivative. Let $y = f(x) = x^2 - 7x + 11$.

a. Find $f(3)$.

b. Suppose that x is slightly different from 3. Find an expression in terms of x for the amount by which y changes, $f(x) - f(3)$.

c. Divide the answer to part b by $x - 3$ to get an expression for the approximate rate of change of y. Simplify.

d. Find the limit of the fraction in part c as x approaches 3. The answer is the *exact* rate of change at $x = 3$.

C2. *Tangent to a Graph Problem:* If you worked Problem C1 correctly, you found that the instantaneous rate of change of $f(x)$ at $x = 3$ is exactly -1 y-unit per x-unit. Plot the graph of function f. On the same screen, plot a line through the point $(3, f(3))$ with slope -1. What do you notice about the line and the curve as you zoom in on the point $(3, f(3))$?

C3. *Formal Definition of Limit Problem:* In Chapter 2, you will learn that the formal definition of limit is

$L = \lim_{x \to c} f(x)$ if and only if

for any positive number epsilon, no matter how small

there is a positive number delta such that

if x is within delta units of c, but not equal to c,

then $f(x)$ is within epsilon units of L.

Notes: "$\lim_{x \to c} f(x)$" is read "the limit of $f(x)$ as x approaches c." Epsilon is the Greek lowercase letter ϵ. Delta is the Greek lowercase letter δ.

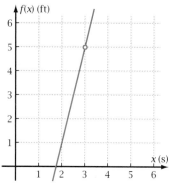

Figure 1-6d

Figure 1-6d shows the graph of the average velocity in ft/s for a moving object from 3 s to x s given by the function

$$f(x) = \frac{4x^2 - 19x + 21}{x - 3}$$

From the graph you can see that 5 is the limit of $f(x)$ as x approaches 3 (the instantaneous velocity at $x = 3$), but that $r(3)$ is undefined because of division by zero.

a. Show that the $(x - 3)$ in the denominator can be canceled out by first factoring the numerator, and that 5 is the value of the simplified expression when $x = 3$.

b. If $\epsilon = 0.8$ unit, on a copy of Figure 1-6d show the range of permissible values of $f(x)$ and the corresponding interval of x-values that will keep $f(x)$ within 0.8 unit of 5.

c. Calculate the value of δ to the right of 3 in part b by substituting $3 + \delta$ for x and 5.8 for $f(x)$, then solving for δ. Show that you get the same value of δ to the left of 3 by substituting $3 - \delta$ for x and 4.2 for $f(x)$.

d. Suppose you must keep $f(x)$ within ϵ units of 5, but you haven't been told the value of ϵ. Substitute $3 + \delta$ for x and $5 + \epsilon$ for $f(x)$. Solve for δ in terms of ϵ. Is it true that there is a positive value of δ for each positive value of ϵ, no matter how small, as required by the definition of limit?

e. In this problem, what are the values of L and c in the definition of limit? What is the reason for the clause "...but not equal to c" in the definition?

Chapter Test

T1. Write the four concepts of calculus.

T2. Write the verbal definition of limit.

T3. Write the physical meaning of derivative.

T4. Sketch the graph of a function that is increasing quickly at $(2, 3)$ and decreasing slowly at $(5, 6)$.

T5. Figure 1-6e shows the graph of the velocity, $v(t)$, in feet per second of a roller coaster as a function of time, t, in seconds since it started. Which concept of calculus is used to find the distance the roller coaster travels from $t = 0$ to $t = 35$? Estimate this distance graphically by counting squares.

T6. On a copy of Figure 1-6e, sketch trapezoids that you would use to find T_7, the distance in Problem T5 using the trapezoidal rule with 7 increments. Estimate T_7 by trapezoidal rule. Will T_7 overestimate or underestimate the actual distance? How do you know?

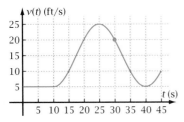

Figure 1-6e

T7. In Problem T5, which concept of calculus is used to find the rate of change of velocity at the instant when $t = 30$? Estimate this rate graphically. Give the units of this rate of change, and the physical name for this quantity.

T8. At what time is the roller coaster in Problem T5 first at the bottom of a hill? How do you explain the fact that the graph is horizontal between $t = 0$ and $t = 10$?

PART 2: Graphing calculators allowed (T9–T18)

On the no-calculator part of this test, you estimated graphically the distance a roller coaster traveled between 0 s and 35 s. The equation is

$$y = 5, \quad \text{if } 0 \le x \le 10$$
$$y = 15 + 10 \cos \tfrac{\pi}{15}(x - 25), \quad \text{if } 10 \le x \le 35$$

where x is in seconds and y is in ft/s. Use this information for Problems T9-T15.

T9. How far did the roller coaster go from $x = 0$ to $x = 10$?

T10. Use your trapezoidal rule program (radian mode) to estimate the integral of y with respect to x from $x = 10$ to $x = 35$ by finding the trapezoidal sums T_5, T_{50}, and T_{100}.

T11. The exact value of the integral in Problem T10 is 416.349667..., the *limit* of T_n as n approaches infinity. Give numerical evidence that T_n is getting closer to this limit as n increases.

T12. Find (without rounding) the average rate of change of y with respect to x from

$$x = 30 \text{ to } x = 31$$
$$x = 30 \text{ to } x = 30.1$$
$$x = 30 \text{ to } x = 30.01$$

T13. Explain why the average rates in Problem T12 are *negative*.

T14. The instantaneous rate of change of y at 30 s is −1.81379936..., the *limit* of the average rates from 30 to x as x approaches 30. Find the difference between the average rate and this limit for the three values in Problem T12. How does the result confirm that the average rate is approaching the instantaneous rate as x approaches 30?

T15. About how close would you have to keep x to 30 (on the positive side) so that the average rates are within 0.01 unit of the limit given in Problem T14?

T16. Name the concept of calculus that means instantaneous rate of change.

T17. You can estimate derivatives numerically from tables of data. Estimate $f'(4)$ (read "f-prime of four"), the derivative of $f(x)$ at $x = 4$.

x	$f(x)$
3.4	24
3.7	29
4.0	31
4.3	35
4.7	42

T18. What did you learn from this test that you did not know before?

Properties of Limits

Finding the average velocity of a moving vehicle requires dividing the distance it travels by the time it takes to go that distance. You can calculate the instantaneous velocity by taking the limit of the average velocity as the time interval approaches zero. You can also use limits to find exact values of definite integrals. In this chapter you'll study limits, the foundation for the other three concepts of calculus.

Mathematical Overview

Informally, the limit of a function f as x approaches c is the y-value that $f(x)$ stays close to when x is kept close enough to c but not equal to c. In Chapter 2, you will formalize the concept of limit by studying it in four ways.

Graphically

The icon at the top of each even-numbered page of this chapter shows that $f(x)$ is close to L when x is close enough to c but not equal to c.

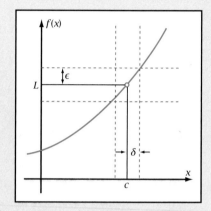

Numerically

x	$f(x)$
3.01	3.262015
3.001	3.251200...
3.0001	3.250120...
3.00001	3.250012...
⋮	⋮

Algebraically

$0 < |x - c| < \delta \Rightarrow |f(x) - L| < \epsilon$, $f(x)$ is within ϵ units of L whenever x is within δ units of c, the definition of limit.

Verbally

I have learned that a limit is a y-value that $f(x)$ can be kept arbitrarily close to just by keeping x close enough to c but not equal to c. Limits involving infinity are related to vertical and horizontal asymptotes. Limits are used to find exact values of derivatives.

2-1 Numerical Approach to the Definition of Limit

In Section 1-2 you learned that L is the limit of $f(x)$ as x approaches c if and only if you can keep $f(x)$ arbitrarily close to L by keeping x close enough to c but not equal to c. In this chapter you'll acquire a deeper understanding of the meaning and properties of limits.

OBJECTIVE   Find the limit of $f(x)$ as x approaches c if $f(c)$ is undefined.

Exploratory Problem Set 2-1

1. Figure 2-1a shows the function

$$f(x) = \frac{2x^2 - 5x + 2}{x - 2}$$

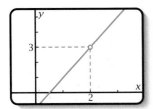

Figure 2-1a

a. Show that $f(2)$ takes the **indeterminate form** $0/0$. Explain why there is no value for $f(2)$.

b. The number $y = 3$ is the **limit** of $f(x)$ as x approaches 2. Make a table of values of $f(x)$ for each 0.001 unit of x from 1.997 to 2.003. Is it true that $f(x)$ stays close to 3 when x is kept close to 2 but not equal to 2?

c. How close to 2 would you have to keep x for $f(x)$ to stay within 0.0001 unit of 3? Within 0.00001 unit of 3? How could you keep $f(x)$ *arbitrarily* close to 3 just by keeping x close enough to 2 but not equal to 2?

d. The missing point at $x = 2$ is called a **removable discontinuity**. Why do you suppose this name is used?

2. Let $g(x) = (x - 3) \sin\left(\dfrac{1}{x - 3}\right) + 2$.

Plot the graph of g using a window that includes $y = 2$ with $x = 3$ as a grid point. Then zoom in on the point $(3, 2)$ by a factor of 10 in both the x- and y-directions. Sketch the resulting graph. Does $g(x)$ seem to be approaching a limit as x approaches 3? If so, what does the limit equal? If not, explain why not.

3. Let $h(x) = \sin\left(\dfrac{1}{x - 3}\right) + 2$.

Plot the graph of h using a window with a y-range of about 0 to 3 with $x = 3$ as a grid point. Then zoom in on the point $(3, 2)$ by a factor of 10 in the x-direction. Leave the y-scale the same. Sketch the resulting graph. Does $h(x)$ appear to approach a limit as x approaches 3? If so, what does the limit equal? If not, explain why not.

2-2 Graphical and Algebraic Approaches to the Definition of Limit

In Figure 2-1a, you saw a function for which $f(x)$ stays close to 3 when x is kept close to 2, even though $f(2)$ itself is undefined. The number 3 fits the verbal definition of limit you learned in Section 1-2: You can keep $f(x)$ arbitrarily close to 3 by keeping x close enough to 2 but not equal to 2. In this section you'll use two Greek letters: ϵ (lowercase epsilon) to specify how arbitrarily close $f(x)$ must be kept to the limit L, and δ (lowercase delta) to specify how close x must be kept to c in order to do this. The result leads to a formal definition of limit.

OBJECTIVE
Given a function f, state whether $f(x)$ has a limit L as x approaches c, and if so, explain how close you can keep x to c for $f(x)$ to stay within a given number ϵ units of L.

Here is the formal definition of limit. You should commit this definition to memory so that you can say it orally and write it correctly without having to look at the text. As you progress through the chapter, the various parts of the definition will become clearer to you.

DEFINITION: Limit

$L = \lim\limits_{x \to c} f(x)$ if and only if
for any number $\epsilon > 0$, no matter how small
there is a number $\delta > 0$ such that
if x is within δ units of c, but $x \neq c$,
then $f(x)$ is within ϵ units of L.

Notes:

- $\lim\limits_{x \to c} f(x)$ is pronounced "the limit of $f(x)$ as x approaches c."

- "For any number $\epsilon > 0$" can also be read as "for any positive number ϵ." The same is true for $\delta > 0$.

- The optional words "no matter how small" help you focus on keeping $f(x)$ *close* to L.

- The restriction "but $x \neq c$" is needed because the value of $f(c)$ may be undefined or different from the limit.

▶ **EXAMPLE 1**
State whether the function graphed in Figure 2-2a has a limit at the given x-value, and explain why or why not. If there is a limit, give its value.

a. $x = 1$ b. $x = 2$ c. $x = 3$

Solution

a. If x is kept close to 1 but not equal to 1, you can make $f(x)$ stay within ϵ units of 2, no matter how small ϵ is. This is true even though there is no value for $f(1)$.

$$\lim_{x \to 1} f(x) = 2$$

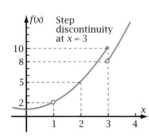

Figure 2-2a

b. If x is kept close to 2, you can make $f(x)$ stay within ϵ units of 5 no matter how small ϵ is. The fact that $f(2) = 5$ has no bearing on whether $f(x)$ has a limit as x approaches 2.

$$\lim_{x \to 2} f(x) = 5$$

c. If x is close to 3 on the left, then $f(x)$ is close to 10. If x is close to 3 on the right, then $f(x)$ is close to 8. Therefore, there is no *one* number you can keep $f(x)$ close to just by keeping x close to 3 but not equal to 3. The fact that $f(3)$ exists and is equal to 10 does *not* mean that 10 is the limit.

$$\lim_{x \to 3} f(x) \text{ does not exist.} \quad \blacktriangleleft$$

Note: The discontinuity at $x = 1$ in Figure 2-2a is called a **removable discontinuity**. If $f(1)$ were defined to be 2, there would no longer be a discontinuity. The discontinuity at $x = 3$ is called a **step discontinuity**. You cannot remove a step discontinuity simply by redefining the value of the function.

▶ **EXAMPLE 2**

Figure 2-2b shows the graph of function f for which $f(2)$ is undefined.

a. What number does $\lim_{x \to 2} f(x)$ equal? Write the definition of this limit using proper limit terminology.

b. If $\epsilon = 0.6$, estimate to one decimal place the largest possible value of δ you can use to keep $f(x)$ within ϵ units of the limit by keeping x within δ units of 2 (but not equal to 2).

c. What name is given to the missing point at $x = 2$?

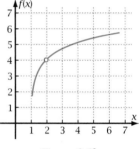

Figure 2-2b

Solution

a. $\lim_{x \to 2} f(x) = 4$

$4 = \lim_{x \to 2} f(x)$ if and only if for any number $\epsilon > 0$, no matter how small, there is a number $\delta > 0$ such that if x is within δ units of 2, but $x \neq 2$, then $f(x)$ is within ϵ units of 4.

Use separate lines for the various parts of the definition.

b. Because $\epsilon = 0.6$, you can draw lines 0.6 unit above and below $y = 4$, as in Figure 2-2c. Where these lines cross the graph, go down to the x-axis and estimate the corresponding x-values to get $x \approx 1.6$ and $x \approx 2.8$.

So, x can go as far as 0.4 unit to the left of $x = 2$ and 0.8 unit to the right. The smaller of these units, 0.4, is the value of δ.

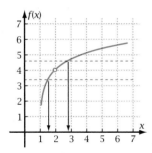

Figure 2-2c

c. The graphical feature is called a removable discontinuity. $\quad \blacktriangleleft$

▶ **EXAMPLE 3**   Figure 2-2d shows the graph of
$f(x) = (x - 3)^{1/3} + 2$. The limit of $f(x)$ as
x approaches 3 is $L = 2$, the same as the value
of $f(3)$.

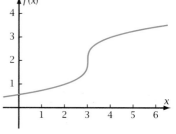

Figure 2-2d

   a. Find graphically the largest value of δ for
      which $f(x)$ is within $\epsilon = 0.8$ unit of 2
      whenever x is kept within δ units of 3.

   b. Find the value of δ in part a algebraically.

   c. Substitute $(3 + \delta)$ for x and $(2 + \epsilon)$ for $f(x)$.
      Solve algebraically for δ in terms of ϵ. Use the result to conclude that there
      is a *positive* value of δ for any positive value of ϵ, no matter how small ϵ is.

Solution   a. Figure 2-2e shows the graph of f with horizontal lines plotted 0.8 unit
             above $y = 2$ and 0.8 unit below. Using the intersect feature of your grapher,
             you will find

$$x = 2.488 \text{ for } y = 1.2 \quad \text{and} \quad x = 3.512 \text{ for } y = 2.8$$

             Each of these values is 0.512 unit away from $x = 3$, so the maximum value
             of δ is 0.512.

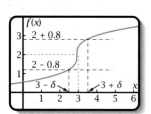

Figure 2-2e

   b. By symmetry, the value of δ is the same on either side of $x = 3$. Substitute
      2.8 for $f(x)$ and $(3 + \delta)$ for x.

$$[(3 + \delta) - 3]^{1/3} + 2 = 2.8$$
$$\delta^{1/3} = 0.8$$
$$\delta = 0.8^3 = 0.512$$

      which agrees with the value found graphically.

   c.
$$[(3 + \delta) - 3]^{1/3} + 2 = 2 + \epsilon$$
$$\delta^{1/3} = \epsilon$$
$$\delta = \epsilon^3$$

      ∴ there is a positive value of δ for any positive number ϵ, no matter
      how small ϵ is.   ◀

▶ **EXAMPLE 4**   Let $f(x) = 0.2(2^x)$.

   a. Plot the graph on your grapher.

   b. Find $\lim\limits_{x \to 3} f(x)$.

   c. Find graphically the maximum value of δ you can use for $\epsilon = 0.5$ at $x = 3$.

   d. Show algebraically that there is a positive value of δ for any $\epsilon > 0$, no
      matter how small.

Solution   a. Figure 2-2f shows the graph.

   b. By tracing to $x = 3$, you will find that $f(3) = 1.6$, which is the same as $\lim\limits_{x \to 3} f(x)$.

   c. Plot the lines $y = 1.1$ and $y = 2.1$, which are $\epsilon = 0.5$ unit above and below
      1.6, as shown in Figure 2-2f.

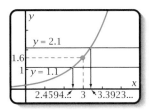

Figure 2-2f

Use the intersect feature of your grapher to find the *x*-values where these lines cross the graph of *f*.

$$x = 2.4594\ldots \qquad \text{and} \qquad x = 3.3923\ldots, \text{ respectively}$$

The candidates for δ are $\delta_1 = 3 - 2.4594\ldots = 0.5403\ldots$ and $\delta_2 = 3.3923\ldots - 3 = 0.3923\ldots$.

The largest possible value of δ is the *smaller* of these two, namely $\delta = 0.3923\ldots$.

d. Because the slope of the graph is positive and increasing as *x* increases, the more restrictive value of δ is the one on the positive side of $x = 3$. Substituting $(1.6 + \epsilon)$ for $f(x)$ and $(3 + \delta)$ for *x* gives

$$0.2(2^{3+\delta}) = 1.6 + \epsilon$$
$$(2^{3+\delta}) = 8 + 5\epsilon$$
$$\log(2^{3+\delta}) = \log(8 + 5\epsilon)$$
$$(3 + \delta)\log 2 = \log(8 + 5\epsilon)$$
$$\delta = \frac{\log(8 + 5\epsilon)}{\log 2} - 3$$

Because $(\log 8)/(\log 2) = 3$, because $8 + 5\epsilon$ is greater than 8 for any positive number ϵ, and because log is an increasing function, the expression $\log(8 + 5\epsilon)/(\log 2)$ is greater than the 3 that is subtracted from it. So δ will be positive for any positive number ϵ, no matter how small ϵ is. ◀

Problem Set 2-2

Quick Review

Q1. Sketch the graph of $y = 2^x$.

Q2. Sketch the graph of $y = \cos x$.

Q3. Sketch the graph of $y = -0.5x + 3$.

Q4. Sketch the graph of $y = -x^2$.

Q5. Sketch the graph of a function with a removable discontinuity at the point (2, 3).

Q6. Name a numerical method for estimating the value of a definite integral.

Q7. What graphical method can you use to estimate the value of a definite integral?

Q8. Write the graphical meaning of derivative.

Q9. Write the physical meaning of derivative.

Q10. If $\log_3 x = y$, then
A. $3^x = y$ B. $3^y = x$ C. $x^3 = y$
D. $y^3 = x$ E. $x^y = 3$

1. Write the definition of limit without looking at the text. Then check the definition in this section. If any part of your definition is wrong, write the entire definition over again. Keep doing this until you can write the definition from memory without looking at the text.

2. What is the reason for the restriction "... but $x \ne c \ldots$" in the definition of limit?

For Problems 3–12, state whether the function has a limit as *x* approaches *c*; if so, tell what the limit equals.

3.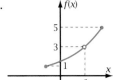

4.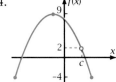

5.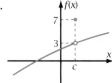

6.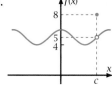

15. $x = 6, \epsilon = 0.7$ **16.** $x = 4, \epsilon = 0.8$

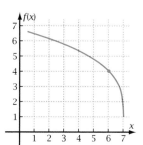 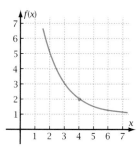

7.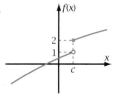

8.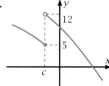

17. $x = 5, \epsilon = 0.3$ **18.** $x = 3, \epsilon = 0.4$

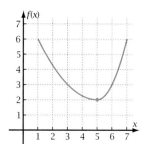 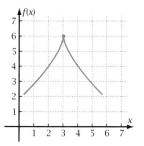

9.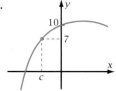

10.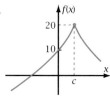

11.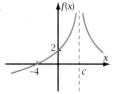

12.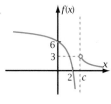

For Problems 19-24,

a. Plot the graph on your grapher. How does the graph relate to Problems 13-18?

b. Find the limit of the function as x approaches the given value.

c. Find the maximum value of δ that can be used for the given value of ϵ at the point.

d. Calculate algebraically a positive value of δ for any $\epsilon > 0$, no matter how small.

For Problems 13-18, photocopy or sketch the graph. For the point marked on the graph, use proper limit notation to write the limit of $f(x)$. For the given value of ϵ, estimate to one decimal place the largest possible value of δ that you can use to keep $f(x)$ within ϵ units of the marked point when x is within δ units of the value shown.

19. $f(x) = 5 - 2\sin(x - 3)$
$x = 3, \epsilon = 0.5$

20. $f(x) = (x - 2)^3 + 3$
$x = 2, \epsilon = 0.5$

21. $f(x) = 1 + 3\sqrt[3]{7 - x}$
$x = 6, \epsilon = 0.7$

13. $x = 3, \epsilon = 0.5$ **14.** $x = 2, \epsilon = 0.5$

22. $f(x) = 1 + 2^{4-x}$
$x = 4, \epsilon = 0.8$

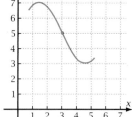 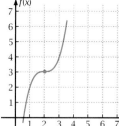

23. $f(x) = \begin{cases} 0.25(x - 5)^2 + 2, & \text{if } x < 5 \\ (x - 5)^2 + 2, & \text{if } x \geq 5 \end{cases}$
$x = 5, \epsilon = 0.3$

24. $f(x) = 6 - 2(x - 3)^{2/3}$
$x = 3, \epsilon = 0.4$

25. *Removable Discontinuity Problem 1:* Function

$$f(x) = \frac{(x^2 - 6x + 13)(x - 2)}{x - 2}$$

is undefined at $x = 2$. However, if $x \ne 2$, you can cancel the $(x - 2)$ factors, and the equation becomes

$$f(x) = x^2 - 6x + 13, \quad x \ne 2$$

So f is a quadratic function with a removable discontinuity at $x = 2$ (Figure 2-2g). The y-value at this missing point is the limit of $f(x)$ as x approaches 2.

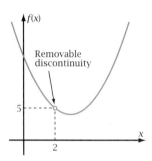

Figure 2-2g

a. Show that $f(2)$ has the indeterminate form $0/0$. What feature does the graph of f have at $x = 2$? Do an appropriate calculation to show that 5 is the limit of $f(x)$ as x approaches 2.

b. Find the interval of x-values close to 2, but not including 2, for which $f(x)$ is within 0.1 unit of 5. Keep at least six decimal places for the x-values at the ends of the interval. Based on your answer, what is the largest value of δ for which $f(x)$ is within $\epsilon = 0.1$ unit of 5 when x is kept within δ unit of 2?

c. Draw a sketch to show how the numbers L, c, ϵ, and δ in the definition of limit are related to the graph of f in this problem.

26. *Removable Discontinuity Problem 2:* Function

$$f(x) = \frac{4x^2 - 7x - 2}{x - 2}$$

is undefined at $x = 2$.

a. Plot the graph of f using a friendly window that includes $x = 2$ as a grid point. What do you notice about the shape of the graph? What feature do you notice at $x = 2$? What does the limit of $f(x)$ appear to be as x approaches 2?

b. Try to evaluate $f(2)$ by direct substitution. What form does your result take? What is

the name for an expression of the form taken by $f(2)$?

c. Algebraically find the limit of $f(x)$ as x approaches 2 by factoring the numerator, then canceling the $(x - 2)$ factors. How does the clause "... but $x \ne c$ ..." in the definition of limit allow you to do this canceling?

d. "If x is within —?— unit of 2, but not equal to 2, then $f(x)$ is within 0.001 unit of the limit." What is the largest number that can go in the blank? Show how you find this.

e. Write the values for L, c, ϵ, and δ in the definition of limit that appears in part d.

27. *Limits Applied to Derivatives Problem:* Suppose you start driving off from a traffic light. Your distance, $d(t)$, in feet, from where you started is given by

$$d(t) = 3t^2$$

where t is time, in seconds, since you started.

a. Figure 2-2h shows $d(t)$ versus t. Write the average speed, $m(t)$, as an algebraic fraction for the time interval from 4 seconds to t seconds.

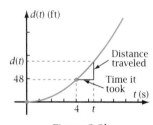

Figure 2-2h

b. Plot the graph of function m on your grapher. Use a friendly window that includes $t = 4$. What feature does this graph have at the point $t = 4$? Sketch the graph.

c. Your speed at the instant $t = 4$ is the limit of your average speed as t approaches 4. What does this limit appear to equal? What are the units of this limit?

d. How close to 4 would you have to keep t for $m(t)$ to be within 0.12 unit of the limit? (This is an easy problem if you simplify the algebraic fraction first.)

e. Explain why the results of this problem give the *exact* value for a derivative.

2-3 The Limit Theorems

Suppose that $f(x)$ is given by the algebraic fraction

$$f(x) = \frac{3x^2 - 48}{x - 4}$$

You may have seen this fraction in Problem 27 of Problem Set 2-2. There is no value for $f(4)$ because of division by zero. Substituting 4 for x gives

$$f(4) = \frac{3 \cdot 4^2 - 48}{4 - 4} = \frac{0}{0}$$

$f(4)$ is undefined because it has an indeterminate form.

Because the numerator is also zero, there may be a limit of $f(x)$ as x approaches 4. Limits such as this arise when you try to find exact values of derivatives. Simplifying the fraction before substituting 4 for x gives

$$f(x) = \frac{(3x + 12)(x - 4)}{x - 4}$$
$$= 3x + 12, \text{ provided } x \neq 4$$

Surprisingly, you can find the limit by substituting 4 for x in the *simplified* expression

$$\lim_{x \to 4} = 3(4) + 12 = 24$$

From Section 2-2, recall that $0/0$ is called an indeterminate form. Its limit can be different numbers depending on just what expressions go to zero in the numerator and denominator. Fortunately, several properties (called the **limit theorems**) allow you to find such limits by making substitutions, as shown above. In this section you will learn these properties so that you can find exact values of derivatives and integrals the way Isaac Newton and Gottfried Leibniz did more than 300 years ago.

OBJECTIVE

For the properties listed in the property box in this section, be able to *state* them, *use* them in a proof, and *explain* why they are true.

Limit of a Product or a Sum of Two Functions

Suppose that $g(x) = 2x + 1$ and $h(x) = 5 - x$. Let function f be defined by the product of g and h.

$$f(x) = g(x) \cdot h(x) = (2x + 1)(5 - x)$$

You are to find the limit of $f(x)$ as x approaches 3. Figure 2-3a shows the graphs of functions f, g, and h. Direct substitution gives

$$f(3) = (2 \cdot 3 + 1)(5 - 3) = (7)(2) = 14$$

The important idea concerning limits is that $f(x)$ stays close to 14 when x is kept close to 3. You can demonstrate this fact by making a table of values of x, $g(x)$, $h(x)$, and $f(x)$.

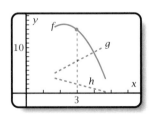

Figure 2-3a

x	$g(x)$	$h(x)$	$f(x) = g(x) \cdot h(x)$
2.95	6.9	2.05	14.145
2.96	6.92	2.04	14.1168
2.97	6.94	2.03	14.0882
2.98	6.96	2.02	14.0592
2.99	6.98	2.01	14.0298
3.01	7.02	1.99	13.9698
3.02	7.04	1.98	13.9392
3.03	7.06	1.97	13.9082
3.04	7.08	1.96	13.8768

When $g(x)$ and $h(x)$ are close to 7 and 2, respectively, $f(x)$ is close to 14.

You can keep the product as close to 14 as you like by keeping x close enough to 3, even if x is not allowed to equal 3. From this information you should be able to see that the limit of a product of two functions is the product of the two limits. A similar property applies to sums of two functions. By adding the values of $g(x)$ and $h(x)$ in the preceding table, you can see that the sum $g(x) + h(x)$ is close to 7 + 2, or 9, when x is close to, but not equal to, 3.

Limit of a Quotient of Two Functions

The limit of a quotient of two functions is equal to the quotient of the two limits, provided that the denominator does not approach zero. Suppose that function f is defined by

$$f(x) = \frac{g(x)}{h(x)} = \frac{2x + 1}{5 - x}$$

and you want to find the limit of $f(x)$ as x approaches 3. The values of $g(3)$ and $h(3)$ are 7 and 2, respectively. By graphing or by compiling a table of values, you can see that if x is close to 3, then $f(x)$ is close to $7/2 = 3.5$. You can keep $f(x)$ as close as you like to 3.5 by keeping x close enough to 3. (When x is equal to 3, $f(x)$ happens to equal 3.5, but that fact is of no concern when you are dealing with limits.)

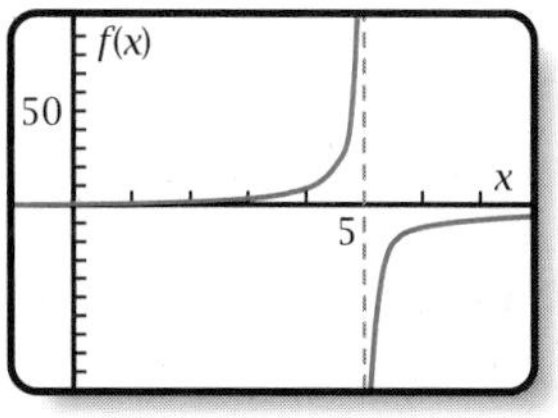

There is no limit of $f(x)$ as x approaches 5. The denominator goes to zero, but the numerator does not. Thus, the absolute value of the quotient becomes infinitely large, as shown in this table. Figure 2-3b shows that the graph of f has a vertical asymptote at $x = 5$.

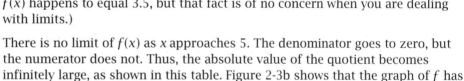

x	$g(x)$	$h(x)$	$f(x) = g(x)/h(x)$
4.96	10.92	0.04	273
4.97	10.94	0.03	364.6...
4.98	10.96	0.02	548
4.99	10.98	0.01	1098
5.00	11.00	0.00	None (infinite)
5.01	11.02	−0.01	−1102
5.02	11.04	−0.02	−552
5.03	11.06	−0.03	−368.6...
5.04	11.08	−0.04	−277

Figure 2-3b

Some important properties of limits are summarized in this box. You can prove the properties by the epsilon, delta techniques of Section 2-2, as you will see in later courses in mathematics.

SOME PROPERTIES OF LIMITS: *The Limit Theorems*

Limit of a Product of Two Functions: If $\lim_{x \to c} g(x) = L_1$ and $\lim_{x \to c} h(x) = L_2$,

$$\text{then } \lim_{x \to c}[g(x) \cdot h(x)] = \lim_{x \to c} g(x) \cdot \lim_{x \to c} h(x) = L_1 \cdot L_2.$$

Verbally: Limit distributes over multiplication, *or* the limit of a product equals the product of the limits.

Limit of a Sum of Two Functions: If $\lim_{x \to c} g(x) = L_1$ and $\lim_{x \to c} h(x) = L_2$,

$$\text{then } \lim_{x \to c}[g(x) + h(x)] = \lim_{x \to c} g(x) + \lim_{x \to c} h(x) = L_1 + L_2.$$

Verbally: Limit distributes over addition, *or* the limit of a sum equals the sum of the limits.

Limit of a Quotient of Two Functions: If $\lim_{x \to c} g(x) = L_1$ and $\lim_{x \to c} h(x) = L_2$, where $L_2 \neq 0$,

$$\text{then } \lim_{x \to c} \frac{g(x)}{h(x)} = \frac{\lim_{x \to c} g(x)}{\lim_{x \to c} h(x)} = \frac{L_1}{L_2}.$$

Verbally: Limit distributes over division, except for division by zero, *or* the limit of a quotient equals the quotient of the limits.

Limit of a Constant Times a Function: If $\lim_{x \to c} g(x) = L$,

$$\text{then } \lim_{x \to c}[k \cdot g(x)] = k \cdot \lim_{x \to c} g(x) = kL.$$

Verbally: The limit of a constant times a function equals the constant times the limit.

Limit of the Identity Function (Limit of x): $\lim_{x \to c} x = c$

Verbally: The limit of x as x approaches c is simply c.

Limit of a Constant Function: If $f(x) = k$, where k is a constant,

$$\text{then } \lim_{x \to c} f(x) = k.$$

Verbally: The limit of a constant is that constant.

▶ **EXAMPLE 1**   Use the limit properties to prove that $\lim_{x \to 3} \dfrac{x^3 - x^2 - 2x - 12}{x - 3} = 19$.

Solution The expression approaches the indeterminate form 0/0 as x approaches 3.

$$\frac{x^3 - x^2 - 2x - 12}{x - 3} \to \frac{0}{0} \quad \text{as } x \to 3$$

Use synthetic substitution to factor the numerator, then cancel the $(x - 3)$ factors.

$$
\begin{array}{r|rrrr}
3 & 1 & -1 & -2 & -12 \\
  &   & 3 & 6 & 12 \\
\hline
  & 1 & 2 & 4 & 0
\end{array}
$$

$$\lim_{x \to 3} \frac{x^3 - x^2 - 2x - 12}{x - 3}$$

$$= \lim_{x \to 3} \frac{(x - 3)(x^2 + 2x + 4)}{x - 3} \qquad \text{Use the results of synthetic substitution.}$$

$$= \lim_{x \to 3} (x^2 + 2x + 4), \quad x \neq 3 \qquad \begin{array}{l}\text{Canceling is allowed because the definition}\\\text{of limit says, ``\ldots but not equal to 3.''}\end{array}$$

$$= \lim_{x \to 3} x^2 + \lim_{x \to 3} (2x) + \lim_{x \to 3} 4 \qquad \text{Limit of a sum (applied to three terms).}$$

$$= 3^2 + 2 \cdot 3 + 4 \qquad \begin{array}{l}\text{Limit of a product } (x \cdot x)\text{, limit of } x\text{, limit of a}\\\text{constant times a function, and limit of a constant.}\end{array}$$

$$= 19, \text{Q.E.D.} \qquad \begin{array}{l}\text{Q.E.D. stands for the Latin } \textit{quod erat demonstrandum,}\\\text{``which was to be demonstrated.''}\end{array} \blacktriangleleft$$

This proof reveals a simple way to find a limit of the indeterminate form 0/0. If you can remove the expression that makes the denominator equal zero, you can substitute the value $x = c$ into the remaining expression. The result is the limit.

Problem Set 2-3

Quick Review

Q1. Find the limit of $13x/x$ as x approaches zero.

Q2. Sketch the graph of a function if 3 is the limit as x approaches 2 but $f(2)$ is undefined.

Q3. Sketch the graph of a function that is decreasing slowly when $x = -4$.

Q4. Sketch the graph of a quadratic function.

Q5. Sketch the graph of $y = x^3$.

Q6. Factor: $x^2 - 100$

Q7. Thirty is what percentage of 40?

Q8. What is meant by definite integral?

Q9. Divide quickly, using synthetic substitution:

$$\frac{x^3 - 8x + 22x - 21}{x - 3}$$

Q10. When simplified, the expression $(12x^{30})/(3x^{10})$ becomes

A. $9x^3$ B. $9x^{20}$ C. $4x^3$ D. $4x^{20}$

E. None of these

1. *Limit of a Function Plus a Function Problem:* Let $g(x) = x^2$ and $h(x) = 12/x$. Plot the two graphs on your grapher, along with the graph of $f(x) = g(x) + h(x)$. Sketch the result, showing that the limit of $f(x)$ as x approaches 2 is equal to the sum of the limits of $g(x)$ and $h(x)$ as x approaches 2. Make a table of values that shows $f(x)$ is close to the limit when x is close, but not equal, to 2.

2. *Limit of a Constant Times a Function Problem:* Plot $g(x) = x^2$ and $f(x) = 0.2x^2$ on your grapher. Sketch the result. Find the limit of $f(x)$ as

x approaches 3 and the limit of $g(x)$ as x approaches 3. Show that the limit of $f(x)$ is 0.2 multiplied by the limit of $g(x)$. Make a table that shows $f(x)$ is close to the limit when x is close, but not equal, to 3.

3. *Limit of a Constant Problem:* Let $f(x) = 7$. Sketch the graph of f. (Don't waste time using your grapher!) On the graph, show that the limit of $f(x)$ as x approaches 3 is 7. Does it bother you that $f(x) = 7$, even if x is not equal to 3?

4. *Limit of x Problem:* Let $f(x) = x$. Sketch the graph of f. (Don't waste time using your grapher!) Then explain why the limit of $f(x)$ as x approaches 6 must be equal to 6.

5. *Limit of a Product Problem:* Let $f(x) = (2x^2)(3 \sin \frac{\pi}{6} x)$. Plot the graphs of $y_1 = 2x^2$ and $y_2 = 3 \sin \frac{\pi}{6} x$, and $y_3 = y_1 \cdot y_2$ on your grapher. Use a window with $x \in [0, 2]$. Sketch the results, showing the limits of each of the three functions as x approaches 1. Demonstrate that the limit of $f(x)$ as x approaches 1 is the product of the other two limits by making a table of values that shows that y_3 is close to the product of the other two limits when x is close to 1 but not equal to 1.

6. *Limit of a Quotient Problem:* Let

$$y_3 = r(x) = \frac{2^x}{\sin \frac{\pi}{3.6} x} = \frac{y_1}{y_2}$$

Write the values of 2^3 and $\sin\left(\frac{3\pi}{3.6}\right)$, then divide them to find $r(3)$. Make a table of values of $r(x)$ starting at $x = 2.9997$ and stepping by 0.0001 unit. Use the results to show that y_3 stays close to the quotient of the limits when x is close to 3 but not equal to 3. Explain why the limit of a quotient property *cannot* be applied to $\lim\limits_{x \to 3.6} f(x)$.

For Problems 7 and 8, find the limit as x approaches the given value. Prove that your answer is correct by naming the limit theorems used at each step.

7. $f(x) = x^2 - 9x + 5$, $\quad x \to 3$

8. $f(x) = x^2 + 3x - 6$, $\quad x \to -1$

For Problems 9–14, plot the graph using a friendly window with the given value of x as a grid point. Sketch the result. Show that the function takes the indeterminate form 0/0 as x approaches the given

value. Then simplify the given fraction, find the limit, and prove that your answer is correct by naming the limit theorems used at each step.

9. $r(x) = \dfrac{x^2 - 4x - 12}{x + 2}$, $\quad x \to -2$

10. $f(x) = \dfrac{x^2 + 3x - 40}{x - 5}$, $\quad x \to 5$

11. $f(x) = \dfrac{x^3 - 3x^2 - 4x - 30}{x - 5}$, $\quad x \to 5$

12. $f(x) = \dfrac{x^3 + x^2 - 5x - 21}{x - 3}$, $\quad x \to 3$

13. $f(x) = \dfrac{x^3 - 4x^2 - 2x + 3}{x + 1}$, $\quad x \to -1$

14. $f(x) = \dfrac{x^4 - 11x^3 + 21x^2 - x - 10}{x - 2}$, $\quad x \to 2$

15. *Check the Answer by Table Problem:* For Problem 11, make a table of values of $f(x)$ for each 0.001 unit of x starting at $x = 4.990$. Use the table to find the largest interval of x-values around $x = 5$ for which you can say that $f(x)$ is within 0.1 unit of the limit whenever x is within the interval, but not equal to 5.

16. *Check the Answer by Graph Problem:* For the graph you plotted in Problem 13, use TRACE on both sides of $x = -1$ to show that $f(x)$ is close to the limit when x is close to -1.

For Problems 17 and 18, show that even though the function takes on the indeterminate form 0/0, you cannot find the limit by the techniques of Example 1.

17. $f(x) = \dfrac{x^2 - 5x + 6}{x^2 - 6x + 9}$, $\quad x \to 3$

18. $f(x) = \dfrac{x^3 - 8}{x^2 - 4x + 4}$, $\quad x \to 2$

19. *Pizza Delivery Problem:* Ida Livermore starts off on her route. She records her truck's speed, $v(t)$, in mi/h, at various times, t, in seconds, since she started.

a. Show Ida that these data fit the equation $v(t) = 5t^{1/2}$.

t (s)	$v(t)$ (mi/h)
0	0
1	5
4	10
9	15
16	20

b. The truck's acceleration, $a(t)$, is the instantaneous rate of change of $v(t)$. Estimate $a(9)$ by using $v(9)$ and $v(9.001)$. Make a conjecture about the exact value of $a(9)$. What are the units of $a(t)$?

c. Note that $a(9)$ is exactly equal to the limit of $[v(t) - v(9)]/(t - 9)$ as t approaches 9. Factor the denominator as a difference of two "squares." Then find the limit as t approaches 9 by applying the limit properties. Does this limit agree with your conjecture in part b?

d. Approximately how far did Ida's truck travel from $t = 1$ to $t = 9$?

20. *Exact Derivative Problem:* Let $f(x) = x^3$.

a. Find, approximately, the derivative of f at $x = 2$ by dividing the change in $f(x)$ from $x = 2$ to $x = 2.1$ by the corresponding change in x.

b. In part a, you evaluated the fraction $[f(x) - f(2)]/(x - 2)$ to get an approximate value of the derivative. The exact value is the limit of this fraction as x approaches 2. Find this limit by first simplifying the fraction. Prove that your answer is correct by citing limit properties.

c. Plot the graph of f. Construct a line through the point $(2, 8)$, whose slope is the value of the derivative in part b. What relationship does the line seem to have to the graph?

21. Find, approximately, the derivative of $f(x) = 0.7^x$ when $x = 5$.

22. Find, approximately, the definite integral of $f(x) = 1.4^x$ from $x = 1$ to $x = 5$.

23. *Mathematical Induction Problem—The Limit of a Power:* Recall that $x^2 = x \cdot x$, so you can use the limit of a product property to prove that

$$\lim_{x \to c} x^2 = c^2$$

Prove by mathematical induction that

$$\lim_{x \to c} x^n = c^n$$

for any positive integer value of n. The recursive definition of x^n, which is $x^n = x \cdot x^{n-1}$, should be helpful in doing the induction part of the proof.

24. *Journal Problem:* Update your calculus journal. You should consider

- The one most important thing you have learned since your last journal entry
- What you now understand more fully about the definition of limit
- How the shortened definition of limit corresponds to the definition you learned in Chapter 1
- Why the limit properties for sums, products, and quotients are so obviously true
- What may still bother you about the definition of limit

2-4 Continuity and Discontinuity

A function such as

$$g(x) = \frac{x^2 - 9}{x - 3}$$

has a discontinuity at $x = 3$ because the denominator is zero there. It seems reasonable to say that the function is "continuous" everywhere else because the graph seems to have no other "gaps" or "jumps" (Figure 2-4a). In this section you will use limits to define the property of continuity precisely.

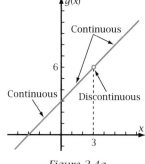

Figure 2-4a

OBJECTIVE   Define continuity. Learn the definition by using it several ways.

Figures 2-4b through 2-4g show graphs of six functions, some of which are continuous at $x = c$ and some of which are not.

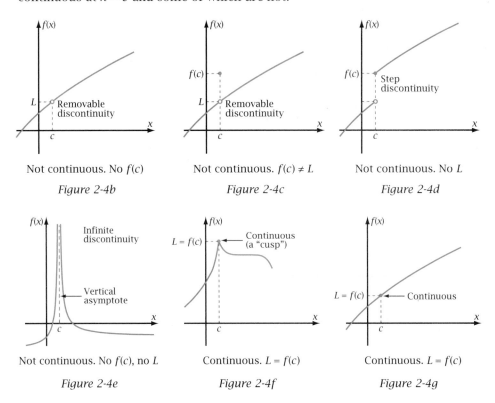

Not continuous. No $f(c)$

Figure 2-4b

Not continuous. $f(c) \neq L$

Figure 2-4c

Not continuous. No L

Figure 2-4d

Not continuous. No $f(c)$, no L

Figure 2-4e

Continuous. $L = f(c)$

Figure 2-4f

Continuous. $L = f(c)$

Figure 2-4g

The first two functions have a limit as x approaches c. In Figure 2-4b, f is discontinuous at c because there is no value for $f(c)$. In Figure 2-4c, f is discontinuous at c because $f(c) \neq L$. Both are **removable discontinuities**. You can define or redefine the value of $f(c)$ to make f continuous there.

In Figure 2-4d, f has a **step discontinuity** at $x = c$. Although there is a value for $f(c), f(x)$ approaches different values from the left of c and the right of c. So, there is no limit of $f(x)$ as x approaches c. You cannot remove a step discontinuity simply by redefining $f(c)$.

In Figure 2-4e, function f has an **infinite discontinuity** at $x = c$. The graph approaches a **vertical asymptote** there. As x gets closer to c, the value of $f(x)$ becomes large without bound. Again, the discontinuity is not removable just by redefining $f(c)$. In Section 2-5, you will study such **infinite limits**.

Figures 2-4f and 2-4g show graphs of functions that are continuous at $x = c$. The value of $f(c)$ equals the limit of $f(x)$ as x approaches c. The branches of the graph are "connected" by $f(c)$.

These examples lead to a formal definition of continuity.

Note that the graph can have a **cusp** (an abrupt change in direction) at $x = c$ and still be continuous there (Figure 2-4f). The word *bicuspid* in relation to a tooth comes from the same root word.

Figures 2-4h, 2-4i, and 2-4j illustrate why a function must satisfy all three parts of the continuity definition. In Figure 2-4h, the graph has a limit as x approaches c, but it has no function value. In Figure 2-4i, the graph has a function value, $f(c)$, but no limit as x approaches c. In Figure 2-4j, the graph has both a function value and a limit, but they are not equal.

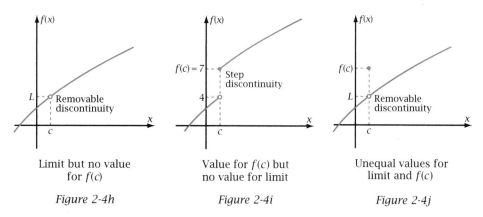

Limit but no value for $f(c)$	Value for $f(c)$ but no value for limit	Unequal values for limit and $f(c)$
Figure 2-4h	*Figure 2-4i*	*Figure 2-4j*

One-Sided Limits and Piecewise Functions

The graph in Figure 2-4i is an example of a function that has different **one-sided limits** as x approaches c. As x approaches c from the left side, $f(x)$ stays close to 4. As x approaches c from the right side, $f(x)$ stays close to 7. This box shows the symbols used for such one-sided limits, and a property relating one-sided limits to the limit from both sides.

PROPERTY: *One-Sided Limits*

$\lim\limits_{x \to c^-} f(x)$ $x \to c$ from the left (through values of x on the negative side of c)

$\lim\limits_{x \to c^+} f(x)$ $x \to c$ from the right (through values of x on the positive side of c)

$L = \lim\limits_{x \to c} f(x)$ if and only if $L = \lim\limits_{x \to c^-} f(x)$ and $L = \lim\limits_{x \to c^+} f(x)$

A step discontinuity can result if $f(x)$ is defined by a different rule for x in the piece of the domain to the right of c than it is for the piece to the left. Figure 2-4k shows such a **piecewise function**.

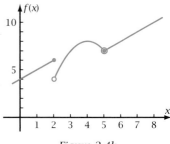

$$f(x) = \begin{cases} x + 4, & \text{if } x \le 2 \\ -x^2 + 8x - 8, & \text{if } 2 < x < 5 \\ x + 2, & \text{if } x \ge 5 \end{cases}$$

Figure 2-4k

Each part of the function is called a **branch**. You can plot the three branches on your grapher by entering the three equations, then dividing by the appropriate **Boolean variable** (named for British mathematician George Boole, 1815–1864). A Boolean variable, such as $(x \le 2)$, equals 1 if the condition inside the parentheses is true and 0 if the condition is false. Example 1 shows you how to plot a piecewise function on your grapher and how to decide whether the function has a limit at the transition points where the rule changes.

▶ **EXAMPLE 1**   For the piecewise function f shown in Figure 2-4k,

   a. Plot the function f on your grapher.

   b. Does $f(x)$ have a limit as x approaches 2? Explain. Is f continuous at $x = 2$?

   c. Does $f(x)$ have a limit as x approaches 5? Explain. Is f continuous at $x = 5$?

Solution   Notice where the open and closed points are in Figure 2-4k.

   a. Use a friendly window that includes $x = 2$ and $x = 5$ as grid points. Enter these three functions in the Y= menu:

$$y_1 = x + 4/(x \le 2)$$ Divide any term of the equation by the Boolean variable.

$$y_2 = -x^2 + 8x - 8/(2 < x \text{ and } x < 5)$$

$$y_3 = x + 2/(x \ge 5)$$

   Dividing by the Boolean variable $(x \le 2)$ in y_1 divides by 1 when $x \le 2$, and divides by 0 when x is *not* less than or equal to 2. So the grapher plots the left branch by plotting y_1 in the appropriate part of the domain, and plots nothing for y_1 elsewhere.

b. $\lim\limits_{x \to 2^-} f(x) = 6$ and $\lim\limits_{x \to 2^+} f(x) = 4$ The left and right limits are unequal.

$\therefore \lim\limits_{x \to 2} f(x)$ does not exist.   There is a step discontinuity.

The function f is discontinuous at $x = 2$.

c. $\lim\limits_{x \to 5^-} f(x) = 7$ and $\lim\limits_{x \to 5^+} f(x) = 7$ The left and right limits are equal.

$\lim\limits_{x \to 5} f(x) = 7 = f(5)$ The open circle at the right end of the middle branch is filled with the closed dot on the left end of the right branch.

The function f is continuous at $x = 5$ because the limit as x approaches 5 is equal to the function value at 5.   ◀

▶ **EXAMPLE 2**   Let the function $h(x) = \begin{cases} kx^2, & \text{if } x < 2 \\ |x - 3| + 4, & \text{if } x \geq 2 \end{cases}$

a. Find the value of k that makes the function continuous at $x = 2$.

b. Plot and sketch the graph.

Solution   a. $\lim\limits_{x \to 2^-} h(x) = k \cdot 2^2 = 4k$

$\lim\limits_{x \to 2^+} h(x) = |2 - 3| + 4 = 5$

For h to be continuous at $x = 2$, the two limits must be equal.
$4k = 5 \quad \Rightarrow \quad k = 1.25$

b. Enter:

$$y_1 = 1.25x^2 / (x < 2)$$
$$y_2 = \text{abs}(x - 3) + 4 / (x \geq 2)$$

The graph is shown in Figure 2-4l. The missing point at the end of the left branch is filled by the point at the end of the right branch, showing graphically that h is continuous at $x = 2$.   ◀

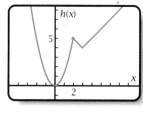

Figure 2-4l

Problem Set 2-4

Quick Review

Q1. What is meant by the derivative of a function?

Q2. What is meant by the definite integral of a function?

Q3. If $f(x) = 200x + 17$, what is the maximum value of δ that ensures $f(x)$ is within 0.1 unit of $f(3)$ when x is within δ units of 3?

Q4. Draw a pair of alternate interior angles.

Q5. What type of function has a graph like that in Figure 2-4m?

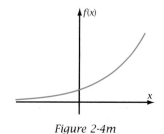

Figure 2-4m

Q6. Sketch the graph of $y = \cos x$.

Q7. Factor: $x^2 + 5x - 6$

Q8. Evaluate: $53^{2001}/53^{2000}$

Q9. Evaluate: $5!$

Q10. Quick! Divide 50 by $\frac{1}{2}$ and add 3.

For Problems 1–10, state whether the graph illustrates a function that

    a. Has left and right limits at the marked value of x.

    b. Has a limit at the marked value of x.

    c. Is continuous at the marked value of x. If it is not continuous there, explain why.

1.
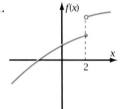

2.
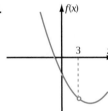

3.
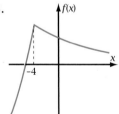

4.
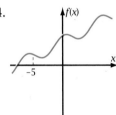

5.
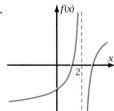

6.
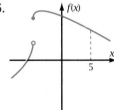

7.
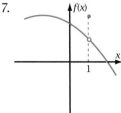

8.
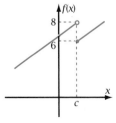

9.
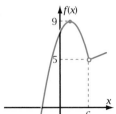

10.
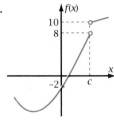

For Problems 11–20, sketch the graph of a function that has the indicated features.

11. Is continuous at $x = 3$ but has a cusp there.

12. Is continuous at $x = 4$ and is "smooth" there.

13. Has no value for $f(5)$ but has a limit as x approaches 5.

14. Has a value for $f(-2)$ but has no limit as x approaches -2.

15. Has a vertical asymptote at $x = 6$.

16. Has a value for $f(2)$ and a limit as x approaches 2, but is not continuous at $x = 2$.

17. Has a step discontinuity at $x = -2$, and $f(-2) = 10$.

18. The limit of $f(x)$ as x approaches 5 is -2, and the value for $f(5)$ is also -2.

19. The limit of $f(x)$ as x approaches 1 is 4, but $f(1) = 6$.

20. $f(3) = 5$, but $f(x)$ has no limit as x approaches 3 and no vertical asymptote there.

For Problems 21–24, state where, if anywhere, the function is discontinuous.

21. $f(x) = \dfrac{x - 4}{x + 3}$

22. $f(x) = \dfrac{x + 5}{x - 11}$

23. $g(x) = \tan x$

24. $g(x) = \cos x$

For Problems 25–30, the function is discontinuous at $x = 2$. State which part of the definition of continuity is not met at $x = 2$. Plot the graph on your grapher. (*Note:* The symbol int(n) indicates the greatest integer less than or equal to n. Graph in dot mode.) Sketch the graph.

25. $f(x) = x + \text{int}(\cos \pi x)$

Chapter 2: Properties of Limits

26. $g(x) = x + \text{int}(\sin \pi x)$

27. $s(x) = 3 + \sqrt{x - 2}$

28. $p(x) = \text{int}(x^2 - 6x + 9)$

29. $h(x) = \dfrac{\sin(x - 2)}{x - 2}$

30. $f(x) = \begin{cases} x + (2 - x)^{-1}, & \text{if } x \ne 2 \\ 3, & \text{if } x = 2 \end{cases}$

For the piecewise functions graphed in Problems 31 and 32, make a table showing these quantities for each value of c, or stating that the quantity does not exist.

- $f(c)$ • $\lim\limits_{x \to c^-} f(x)$ • $\lim\limits_{x \to c^+} f(x)$ • $\lim\limits_{x \to c} f(x)$
- Continuity or kind of discontinuity

31. $c = \{1, 2, 4, 5\}$

$$f(x) = \begin{cases} x + 1, & \text{if } x < 1 \\ 4, & \text{if } x = 1 \\ (x - 2)^2 + 1, & \text{if } 1 < x \le 4 \\ \dfrac{1}{x - 5} + 3, & \text{if } x > 4 \text{ and } x \ne 5 \end{cases}$$

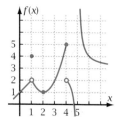

32. $c = \{1, 2, 3, 5\}$

$$f(x) = \begin{cases} x^2 + 1 & \text{if } x < 1 \\ \dfrac{x^2 - 4}{x - 2}, & \text{if } 1 \le x \le 3 \text{ and } x \ne 2 \\ 1, & \text{if } x = 2 \\ 5, & \text{if } 3 < x \le 5 \\ \dfrac{1}{\sin(x - 4)}, & \text{if } x > 5 \end{cases}$$

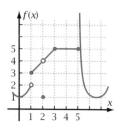

For the piecewise functions in Problems 33–36,

a. Plot the graph using Boolean variables to restrict the branches. Use a friendly window including as a grid point any transition point where the rule changes. Sketch the graph.

b. Find the left and right limits at the transition point, and state whether the function is continuous at the transition point.

33. $d(x) = \begin{cases} 7 - x^2, & \text{if } x \le 2 \\ 5 - x, & \text{if } x > 2 \end{cases}$

34. $h(x) = \begin{cases} 4 - x^2, & \text{if } x < 1 \\ x + 1, & \text{if } x \ge 1 \end{cases}$

35. $m(x) = \begin{cases} 3^x, & \text{if } x < 2 \\ 9 - x, & \text{if } x \ge 2 \end{cases}$

36. $q(x) = \begin{cases} 2^{-x}, & \text{if } x \le -1 \\ x + 3, & \text{if } x > -1 \end{cases}$

For the piecewise functions in Problems 37–40, use one-sided limits in an appropriate manner to find the value of the constant k that makes the function continuous at the transition point where the defining rule changes. Plot the graph using Boolean variables. Sketch the result.

37. $g(x) = \begin{cases} 9 - x^2, & \text{if } x < 2 \\ kx, & \text{if } x \ge 2 \end{cases}$

38. $f(x) = \begin{cases} 0.4x + 1, & \text{if } x < 1 \\ kx + 2, & \text{if } x \ge 1 \end{cases}$

39. $u(x) = \begin{cases} kx^2, & \text{if } x \le 3 \\ kx - 3, & \text{if } x > 3 \end{cases}$

40. $v(x) = \begin{cases} kx + 5, & \text{if } x < -1 \\ kx^2, & \text{if } x \ge -1 \end{cases}$

41. *Two Constants Problem:* Let a and b stand for constants and let

$$f(x) = \begin{cases} b - x, & \text{if } x \le 1 \\ a(x - 2)^2, & \text{if } x > 1 \end{cases}$$

a. Find an equation relating a and b if f is to be continuous at $x = 1$.

b. Find b if $a = -1$. Show by graphing that f is continuous at $x = 1$ for these values of a and b.

c. Pick another value of a and find b. Show that f is continuous for these values of a and b.

42. *Two Values of Constant Problem:* For function f, use one-sided limits in an appropriate way to find the *two* values of k that make f continuous at $x = 2$.

$$f(x) = \begin{cases} k^2 - x^2, & \text{if } x < 2 \\ 1.5kx, & \text{if } x \geq 2 \end{cases}$$

43. *River Crossing Problem:* Calvin stands at the beginning of a bridge that is perpendicular to the banks of a 120-ft-wide river (Figure 2-4n). He can walk across the bridge at 5 ft/s, or take a scenic trip in a rowboat at 3 ft/s, making an angle θ, in degrees, with the riverbank. The time he takes to get to the other side of the river is a piecewise function of θ. Write an equation for this function. Plot the graph in a suitable domain and sketch the result.

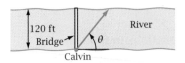

Figure 2-4n

44. *Surprise Function Problem!* Let

$$f(x) = x + 3 + \frac{10^{-20}}{x - 1}$$

a. Plot the graph on your grapher.

b. What appears to be the limit of $f(x)$ as x approaches 1?

c. Show that $f(x)$ is very close to the number in part b when $x = 1.0000001$.

d. Function f is not continuous at $x = 1$ because there is no value for $f(1)$. What type of discontinuity occurs at $x = 1$? (Be careful!)

45. *Continuity of Polynomial Functions:* The general polynomial function of degree n has an

equation of the form

$$P(x) = a_0 + a_1 x + a_2 x^2 + a_3 x^3 + \cdots + a_n x^n$$

Based on the closure axioms for real numbers and the properties of limits you have learned, explain why any polynomial function is continuous for all real values of x.

46. *The Signum Function:* Figure 2-4o shows the graph of the signum function, $f(x) = \text{sgn } x$. The value of the function is 1 when x is positive, -1 when x is negative, and 0 when x is zero. This function is useful in computing for testing a value of x to see what sign it has (hence the name *signum*). Here is the formal definition:

$$\text{sgn } x = \begin{cases} 1, & \text{if } x > 0 \\ 0, & \text{if } x = 0 \\ -1, & \text{if } x < 0 \end{cases}$$

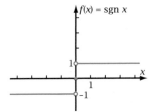

Figure 2-4o

In this problem you will explore various compositions of the signum function.

a. Does $r(x) = |\text{sgn } x|$ have a limit as x approaches 0? Does it have a function value at $x = 0$? Is it continuous at $x = 0$?

b. Sketch the graph of $g(x) = 3 \text{ sgn}(x - 2)$.

c. Sketch the graph of $h(x) = x^2 - \text{sgn } x$.

d. Show that the function $a(x) = |x|/x$ is equal to sgn x for all x except zero.

e. Sketch the graph of $f(x) = \cos x + \text{sgn } x$.

2-5 Limits Involving Infinity

Suppose the demand for doctors in a particular community has increased. The number of people who choose to pursue that career will increase to meet the demand. After a while there may be too many doctors, causing the number of people who want to enter the medical profession to decrease. Eventually, the

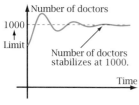

Figure 2-5a

number of doctors stabilizes, say at 1000 (Figure 2-5a). This steady-state value is the limit of the number of doctors as time approaches infinity.

You can visualize another type of limit involving infinity by imagining you are pointing a flashlight straight at a wall (Figure 2-5b). If you begin to turn with the flashlight in your hand, the length of the light beam, L, increases. When the angle, x, is $\pi/2$ radians ($90°$), the beam is parallel to the wall, so its length becomes infinite. The graph of the function has a *vertical asymptote* at $x = \pi/2$, so it has an *infinite discontinuity* there.

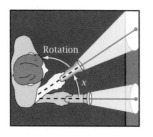
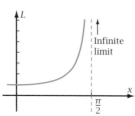

Figure 2-5b

In this section you'll learn some terminology to use if the value of $f(x)$ becomes infinitely large as x approaches c, or if x itself becomes infinitely large.

OBJECTIVE Find limits of functions where either x becomes infinite or the limit is infinite.

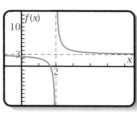

Figure 2-5c

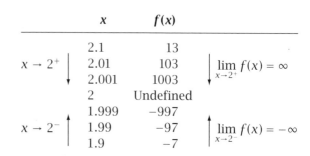

	x	$f(x)$	
$x \to 2^+$	2.1	13	
	2.01	103	$\lim\limits_{x \to 2^+} f(x) = \infty$
	2.001	1003	
	2	Undefined	
	1.999	−997	
$x \to 2^-$	1.99	−97	$\lim\limits_{x \to 2^-} f(x) = -\infty$
	1.9	−7	

Figure 2-5c shows the graph of

$$f(x) = \frac{1}{x-2} + 3$$

As x approaches 2, the denominator approaches zero. The reciprocal of a number close to zero is very large. The table shows that as x approaches 2 from the positive side, $f(x)$ becomes infinitely large in the positive direction. As x approaches 2 from the negative side, $f(x)$ becomes infinitely large in the negative direction. The symbol ∞ represents infinity. You can express the behavior of $f(x)$ for values of x close to 2 by writing one-sided limits this way:

$$\lim_{x \to 3^+} f(x) = \infty \qquad \text{and} \qquad \lim_{x \to 3^-} f(x) = -\infty$$

Note that the ∞ symbol does not stand for a number. It represents the fact that the value of $f(x)$ increases without bound as x gets closer to 2. Be careful to use ∞ only in conjunction with limits. Because the "=" sign connects two *numbers,* it is *not* correct to write statements like "$f(2) = \infty$."

Function f in Figure 2-5c gets closer and closer to 3 as x gets larger and larger. This happens because the reciprocal of a large number is close to zero, leaving a bit more than 3 for $f(x)$. A similar thing happens as x becomes very large in the negative direction.

x	$f(x)$
-1000	$2.99900199\ldots$
-100	$2.990196\ldots$
-10	$2.9166\ldots$
$\vdots$	$\vdots$
10	3.125
100	$3.0102\ldots$
1000	$3.001002\ldots$

$x \to -\infty$ $\qquad \lim\limits_{x \to -\infty} f(x) = 3$

$x \to \infty$ $\qquad \lim\limits_{x \to \infty} f(x) = 3$

You can express this behavior by writing

$$\lim_{x \to \infty} f(x) = 3 \qquad \text{and} \qquad \lim_{x \to -\infty} f(x) = 3$$

Again, the ∞ symbol is used only in conjunction with limits.

DEFINITIONS: Limits Involving Infinity

$\lim\limits_{x \to c} f(x) = \infty$ if and only if $f(x)$ can be kept arbitrarily far away from zero in the positive direction just by keeping x close enough to c but not equal to c. $\left(\lim\limits_{x \to c} f(x) = -\infty \text{ is similarly defined.}\right)$

$\lim\limits_{x \to c} f(x)$ is infinite if and only if $f(x)$ can be kept arbitrarily far away from zero just by keeping x close enough to c but not equal to c.

$\lim\limits_{x \to \infty} f(x) = L$ if and only if $f(x)$ can be made to stay arbitrarily close to L just by making x large enough in the positive direction. $\left(\lim\limits_{x \to -\infty} f(x) = L \text{ is similarly defined.}\right)$

$\lim\limits_{x \to \infty} f(x) = \infty$ if and only if $f(x)$ can be kept arbitrarily far away from zero in the positive direction just by making x large enough. $\left(\lim\limits_{x \to \infty} f(x) = -\infty, \lim\limits_{x \to -\infty} f(x) = \infty, \text{ and } \lim\limits_{x \to -\infty} f(x) = -\infty \text{ are similarly defined.}\right)$

The lines $y = 3$ and $x = 2$ in Figure 2-5c are horizontal and vertical asymptotes, respectively. The definition of limits involving infinity can be used to give a more precise definition of the concept of asymptote.

DEFINITIONS: Horizontal and Vertical Asymptotes

If $\lim\limits_{x \to \infty} f(x) = L$, then the line $y = L$ is a horizontal asymptote. The same applies if x approaches $-\infty$.

If $\lim\limits_{x \to c} |f(x)| = \infty$, then the line $x = c$ is a vertical asymptote.

Note that the graph of a function never crosses a vertical asymptote because functions have only *one* value of *y* for any one value of *x*. This fact agrees with the origin of the name, the Greek *asymptotos*, meaning "not due to coincide." It is customary to use "asymptote" for a horizontal limit line like the one in Figure 2-5a, even though the graph does cross the line.

▶ **EXAMPLE 1**   For $f(x) = \dfrac{1}{x-2} + 3$ in Figure 2-5c, $\lim_{x \to 2^+} f(x) = \infty$ and $\lim_{x \to \infty} f(x) = 3$.

a. What value of $x > 2$ makes $f(x) = 500$? Choose several values of *x* closer to 2 than this number and show numerically that $f(x) > 500$ for each of them. What does it mean to say that the limit of $f(x)$ is infinity as *x* approaches 2 from the positive side? What feature does the graph of *f* have at $x = 2$?

b. What value of $x > 2$ makes $f(x) = 3.004$? Choose several values of *x* greater than this number and show numerically that $f(x)$ is within 0.004 unit of 3 for each of these numbers. What does it mean to say that the limit of $f(x)$ is 3 as *x* approaches infinity? How is the line $y = 3$ related to the graph?

Solution

a.
$$500 = \frac{1}{x-2} + 3 \qquad \text{Substitute 500 for } f(x).$$

$$\frac{1}{x-2} = 497$$

$$x - 2 = \frac{1}{497} = 0.002012\ldots$$

$$x = 2.002012\ldots \qquad \text{This value of } x \text{ makes } f(x) = 500.$$

Pick $x = 2.002$, $x = 2.001$, and $x = 2.0005$, for example.

x	*f(x)*
2.002	503
2.001	1003
2.0005	2003

In each case, $f(x) > 500$.

The equation $\lim_{x \to 2^+} f(x) = \infty$ means that you can make $f(x)$ arbitrarily far away from zero by picking values of *x* close enough to 2 on the positive side (but not equal to 2).

There is a *vertical asymptote* at $x = 2$.

b.
$$3.004 = \frac{1}{x-2} + 3 \qquad \text{Substitute 3.004 for } f(x).$$

$$\frac{1}{x-2} = 0.004$$

$$x = 252 \qquad \text{This value of } x \text{ makes } f(x) = 3.004.$$

Pick $x = 260$, $x = 500$, and $x = 1000$.

x	*f(x)*
260	3.00387
500	3.002008…
1000	3.001002…

Particle accelerators, like this one located at the Center for European Nuclear Research in Meyrin, Switzerland, make subatomic particles move close to the speed of light. According to the Theory of Relativity, the particle would need an infinite amount of energy to reach the speed of light.

In each case, $f(x)$ is within 0.004 unit of 3.

Thus, $\lim_{x \to \infty} f(x) = 3$ means that you can keep $f(x)$ arbitrarily close to 3 by making x large enough.

The line $y = 3$ is a *horizontal asymptote*.

◀

▶ **EXAMPLE 2**   For the piecewise function f in Figure 2-5d, what do the following limits appear to be?

- $\lim_{x \to \infty} f(x)$
- $\lim_{x \to 3^-} f(x)$
- $\lim_{x \to 3^+} f(x)$
- $\lim_{x \to 1^+} f(x)$
- $\lim_{x \to 1^-} f(x)$
- $\lim_{x \to 0} f(x)$
- $\lim_{x \to -\infty} f(x)$

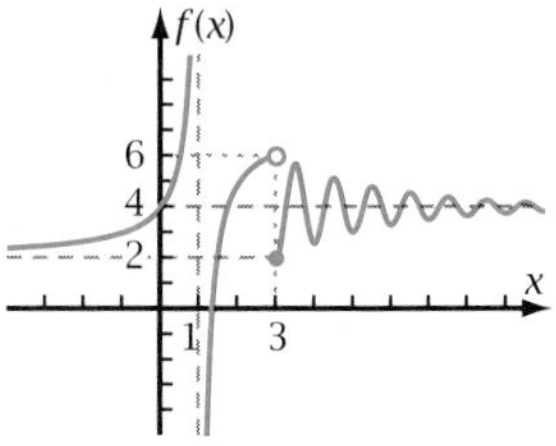

Figure 2-5d

Solution

- $\lim_{x \to \infty} f(x) = 4$ $f(x)$ stays close to 4 when x is very large. (There is a horizontal asymptote at $y = 4$.)

- $\lim_{x \to 3^-} f(x) = 6$ $f(x)$ stays close to 6 when x is close to 3 on the negative side.

- $\lim_{x \to 3^+} f(x) = 2$ $f(x)$ stays close to 2 when x is close to 3 on the positive side.

- $\lim_{x \to 1^+} f(x) = -\infty$ $f(x)$ becomes infinitely large in the negative direction.

- $\lim_{x \to 1^-} f(x) = \infty$ $f(x)$ becomes infinitely large in the positive direction.

- $\lim_{x \to 0} f(x) = 4$ f is continuous at $x = 0$.

- $\lim_{x \to -\infty} f(x) = 2$ $f(x)$ stays close to 2 when x is very large in the negative direction. (There is a horizontal asymptote at $y = 2$.)

◀

Notes on Undefined, Infinite, and Indeterminate

- An expression, A, is *undefined* at any value $x = c$ that causes division by zero. (An expression can also be undefined for other reasons, such as $\sin^{-1} 3$ or $\log(-4)$.)

- If the undefined expression, A, takes the form (nonzero)/(zero), A is said to be an *infinite form*, or simply to be *infinite*. Its absolute value gets larger than any real number as the denominator gets closer to zero. The graph of $f(x) = A$ has an *infinite discontinuity* (a *vertical asymptote*) at $x = c$.

- If the undefined expression, A, takes the form 0/0, A is said to be an *indeterminate form*. You can't determine the limit of A as x approaches c just by looking at 0/0.

- If an indeterminate form has a finite limit, L, as x approaches c, then the function $f(x) = A$ has a *removable discontinuity* at the point (c, L).

Problem Set 2-5

Quick Review

Refer to Figure 2-5e for Problems Q1–Q10.

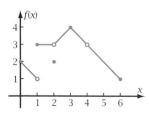

Figure 2-5e

Q1. $\lim\limits_{x\to 1} f(x) = $ —?—

Q2. $\lim\limits_{x\to 2} f(x) = $ —?—

Q3. $\lim\limits_{x\to 3} f(x) = $ —?—

Q4. $\lim\limits_{x\to 4} f(x) = $ —?—

Q5. $\lim\limits_{x\to 5} f(x) = $ —?—

Q6. Is f continuous at $x = 1$?

Q7. Is f continuous at $x = 2$?

Q8. Is f continuous at $x = 3$?

Q9. Is f continuous at $x = 4$?

Q10. Is f continuous at $x = 5$?

1. For piecewise function f in Figure 2-5f, what do these limits appear to be?

- $\lim\limits_{x\to -\infty} f(x)$
- $\lim\limits_{x\to -3^-} f(x)$
- $\lim\limits_{x\to -3^+} f(x)$
- $\lim\limits_{x\to 1} f(x)$
- $\lim\limits_{x\to 2} f(x)$
- $\lim\limits_{x\to 3^-} f(x)$
- $\lim\limits_{x\to 3^+} f(x)$
- $\lim\limits_{x\to \infty} f(x)$

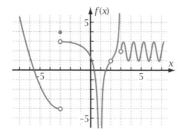

Figure 2-5f

2. For piecewise function g in Figure 2-5g, what do these limits appear to be?

- $\lim\limits_{x\to -\infty} g(x)$
- $\lim\limits_{x\to -2^-} g(x)$
- $\lim\limits_{x\to -2^+} g(x)$
- $\lim\limits_{x\to 1^-} g(x)$
- $\lim\limits_{x\to 2} g(x)$
- $\lim\limits_{x\to 3^-} g(x)$
- $\lim\limits_{x\to \infty} g(x)$

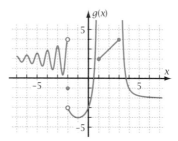

Figure 2-5g

For Problems 3–6, sketch the graph of a function that has the given features.

3. $\lim\limits_{x\to 2^-} f(x) = \infty$ and $\lim\limits_{x\to 2^+} f(x) = \infty$

4. $\lim\limits_{x\to 2^-} f(x) = \infty$ and $\lim\limits_{x\to 2^+} f(x) = -\infty$

5. $\lim\limits_{x\to \infty} f(x) = -5$ and $\lim\limits_{x\to -\infty} f(x) = 7$

6. $\lim\limits_{x\to \infty} f(x) = \infty$ and $\lim\limits_{x\to -\infty} f(x) = \infty$

7. Let $f(x) = 2 + \dfrac{1}{x - 3}$.

a. Sketch the graph of f.

b. Find $\lim_{x\to 3^+} f(x)$, $\lim_{x\to 3^-} f(x)$, $\lim_{x\to 3} f(x)$, $\lim_{x\to \infty} f(x)$, and $\lim_{x\to -\infty} f(x)$.

c. Find a value of x on the positive side of 3 for which $f(x) = 100$. Choose several values of x closer to 3 than this, and show numerically that $f(x) > 100$ for all of these values. What does it mean to say that the limit of $f(x)$ is infinity as x approaches 3 from the positive side? How is the line $x = 3$ related to the graph of f?

d. What value of $x > 3$ makes $f(x) = 2.001$? Choose several values of x greater than this number and show numerically that $f(x)$ is within 0.001 unit of 2 for each of these numbers. What does it mean to say that the limit of $f(x)$ is 2 as x approaches infinity? How is the line $y = 2$ related to the graph?

8. Let $g(x) = \sec x$.

   a. Sketch the graph of g.

   b. Find $\lim_{x \to \pi/2^-} g(x)$ and $\lim_{x \to \pi/2^+} g(x)$. Explain why, even though $\lim_{x \to \pi/2} g(x)$ is infinite, it is *not* correct to say that $\lim_{x \to \pi/2} g(x) = \infty$. What feature does the graph of g have at $x = \pi/2$?

   c. Find a value of x close to $\pi/2$ on the positive side for which $g(x) = -1000$. Choose several values of x closer to $\pi/2$ than this, and show numerically that $g(x) < -1000$ for all of these values. What does it mean to say that the limit of $g(x)$ is negative infinity as x approaches $\pi/2$ from the positive side? How is the line $x = \pi/2$ related to the graph of g?

9. Let $r(x) = 2 + \dfrac{\sin x}{x}$.

   a. Plot the graph of r. Use a friendly window with an x-range of about -20 to 20 for which $x = 0$ is a grid point. Sketch the result.

   b. Find the limit, L, of $r(x)$ as x approaches infinity.

   c. Show that $r(28)$ is within 0.01 unit of 2, but that there are values of $x > 28$ for which $r(x)$ is *more than* 0.01 unit away from 2. Use a suitable window to show this graphically, and sketch the result. Find a value $x = D$ large enough so that $r(x)$ is within 0.01 unit of 2 for *all* $x > D$.

   d. In part b, if you draw a horizontal line at $y = L$, will it be an asymptote? Explain.

   e. Make a conjecture about the limit of $r(x)$ as x approaches zero. Give evidence to support your conjecture.

10. Let $h(x) = \left(1 + \dfrac{1}{x}\right)^x$.

    a. Plot the graph of h. Use a friendly window with an x-range of 0 to about 100. You will have to explore to find a suitable y-range. Sketch the result.

    b. As x becomes large, $1/x$ approaches zero, so $h(x)$ takes on the form 1^∞. You realize that 1 to any power is 1, but the base is always

greater than 1, and a number greater than 1 raised to a large positive power becomes infinite. Which phenomenon "wins" as x approaches infinity: 1, infinity, or some "compromise" number in between?

11. Figure 2-5h shows the graph of

$$y = \log x$$

Does the graph level off and approach a finite limit as x approaches infinity, or is the limit infinite? Justify your answer. The definition of logarithm is helpful here ($y = \log x$ if and only if $10^y = x$).

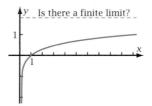

Figure 2-5h

12. Wanda Wye wonders why the form $1/0$ is infinite and why the form $1/\infty$ is zero. Explain to her what happens to the size of fractions such as $1/0.1$, $1/0.0001$, and so on, as the denominator gets close to zero. Explain what happens as the denominator becomes very large.

13. *Limits Applied to Integrals Problem:* Rhoda starts riding down the driveway on her tricycle. Being quite precocious, she figures her velocity, v, in ft/s is

$$v = \sqrt{t}$$

where t is time, in seconds, since she started. Figure 2-5i shows v as a function of t.

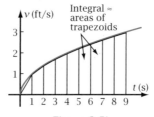

Figure 2-5i

a. Explain why the definite integral from $t = 0$ to $t = 9$ represents the distance Rhoda rode in the first 9 seconds.

b. Use the trapezoidal rule to estimate the integral in part a. Try 9, 45, 90, and 450 trapezoids. Record all the decimal places your program gives you.

c. What number (an integer in this case) do you think is the exact value of the integral? Explain why this number is a limit. Why are the approximate answers by trapezoids all smaller than this number?

d. Figure out how many trapezoids are needed so that the approximation of the integral is within 0.01 unit of the limit. Explain how you go about getting the answer.

14. *Work Problem:* The work done as you drag a box across the floor is equal to the product of the force you exert on the box and the distance the box moves. Suppose that the force varies with distance, and is given by

$$F(x) = 10 - 3\sqrt{x}$$

where $F(x)$ is the force, in pounds, and x is the distance, in feet, the box is from its starting point. Figure 2-5j shows the graph of F.

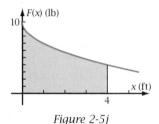

Figure 2-5j

a. Explain why a definite integral is used to calculate the amount of work done.

b. Use the trapezoidal rule with $n = 10$ and $n = 100$ increments to estimate the value of the integral from $x = 0$ to $x = 4$. What are the units of work in this problem?

c. The exact amount of work is the limit of the trapezoidal sums as n approaches infinity. In this case the answer is an integer. What do you suppose the integer is?

d. What is the minimum number, D, such that the trapezoidal sums are closer than 0.01 unit to the limit in part c whenever $n > D$?

15. *Searchlight Problem:* A searchlight shines on a wall as shown in Figure 2-5k. The perpendicular distance from the light to the wall is 100 ft. Write an equation for the length, L, of the beam of light as a function of the angle, x, in radians, between the perpendicular and the beam. How close to $\pi/2$ must the angle be for the length of the beam to be at least 1000 ft, assuming that the wall is long enough?

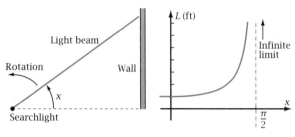

Figure 2-5k

16. *Zero Times Infinity Problem:* You have learned that $0/0$ is called an indeterminate form. You can't determine what it equals just by looking at it. Similarly, $0 \cdot \infty$ is an indeterminate form. In this problem you will see three possibilities for the limit of a function whose form goes to $0 \cdot \infty$. Let f, g, and h be functions defined as follows.

$$f(x) = 5x(x - 2) \cdot \frac{1}{x - 2}$$

$$g(x) = 5x(x - 2) \cdot \frac{1}{(x - 2)^2}$$

$$h(x) = 5x(x - 2)^2 \cdot \frac{1}{x - 2}$$

a. Show that each of the three functions takes the form $0 \cdot \infty$ as x approaches 2.

b. Find the limit of each function as x approaches 2.

c. Describe three things that the indeterminate form $0 \cdot \infty$ could approach.

2-6 The Intermediate Value Theorem and Its Consequences

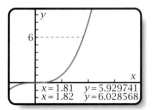

Figure 2-6a

Suppose you try to find a solution of the equation $x^3 = 6$ by tracing the graph of $y = x^3$ (Figure 2-6a). The cursor never quite hits a value of x that makes y equal exactly 6. That's because graphers plot **discrete points** that only approximately represent the **continuous** graph. However, because $y = x^3$ is continuous, there really is a value of x (an irrational number) that, when cubed, gives exactly 6.

The property of continuous functions that guarantees there is an exact value is called the **intermediate value theorem**. Informally, it says that if you pick a value of y between any two values of $f(x)$, there is an x-value in the domain that gives exactly that y-value for $f(x)$. Because $1.81^3 = 5.929741$ and $1.82^3 = 6.028568$, and because 6 is between 5.929741 and 6.028568, there must be a number x between 1.81 and 1.82 for which $f(x) = 6$ exactly. The function $y = x^3$ must be continuous for this property to apply.

PROPERTY: *The Intermediate Value Theorem*

If the function f is continuous for all x in the closed interval $[a, b]$, and y is a number between $f(a)$ and $f(b)$, then there is a number $x = c$ in (a, b) for which $f(c) = y$.

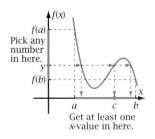

Figure 2-6b

Figure 2-6b illustrates the theorem. Pick y between $f(a)$ and $f(b)$. If f is continuous, you can go over to the graph, then go down to the x-axis and find c, a corresponding value of x. The value of $f(c)$ will thus equal y exactly. The proof of this theorem relies on the **completeness axiom**. This axiom, which comes in several forms, says that there is a real number corresponding to every point on the number line, and vice versa. Thus, the set of real numbers is "complete." It has no "holes," as does the set of rational numbers.

A formal proof of the intermediate value theorem usually appears in later courses on analysis of real numbers. The gist of the proof is that for any y-value you pick in the interval, there will be a point on the graph because the graph is continuous. Going vertically to the x-axis gives a point on the number line. This point corresponds to a real number $x = c$, because the set of real numbers is complete. Reversing the steps shows that $f(c)$ really does equal y.

OBJECTIVE　Given an equation for a continuous function f and a value of y between $f(a)$ and $f(b)$, find a value of $x = c$ between a and b for which $f(c) = y$.

In addition, you will investigate a corollary of the intermediate value theorem, called the **image theorem**, which relies on the **extreme value theorem** for its proof.

▶ **EXAMPLE 1**

a. If $f(x) = x^3 - 4x^2 + 2x + 7$, use the intermediate value theorem to conclude that a value of $x = c$ occurs between 1 and 3 for which $f(c)$ is exactly equal to 5.

b. Find an approximation for this value of c numerically.

Solution

a. Because polynomial functions are continuous for all values of x (see Problem 45 in Section 2-4), f is continuous on $[1, 3]$. Because $f(1) = 6$ and $f(3) = 4$, the intermediate value theorem applies, and a value of $x = c$ occurs in $(1, 3)$ for which $f(c) = 5$.

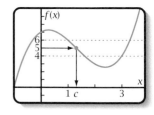

b. By solver or by plotting the line $y = 5$ and using the intersect feature, you can find that $c \approx 1.3111078...$, which is between 1 and 3 (as Figure 2-6c shows).

Figure 2-6c

◀

Problem Set 2-6

Quick Review

Q1. Evaluate $f(2)$ if $f(x) = 3x^4 + 5$.

Q2. Find $\lim_{x \to 2} f(x)$ if $f(x) = 3x^4 + 5$.

Q3. Evaluate $h(3)$ if $h(x) = 5(x - 3)/(x - 3)$.

Q4. Find $\lim_{x \to 3} h(x)$ if $h(x) = 5(x - 3)/(x - 3)$.

Q5. Evaluate $s(0)$ if $s(x) = |x|/x$.

Q6. Find $\lim_{x \to 0} s(x)$ if $s(x) = |x|/x$.

Q7. Evaluate: $\sin(\pi/2)$

Q8. Fill in the blank with the correct operation: $\log(xy) = \log x$ —?— $\log y$.

Q9. The expression $0/0$ is called a(n) —?— form.

Q10. Which of these definitely is *not* true if $\lim_{x \to c} f(x)$ takes the form $0/0$ as x approaches c?

    A. $f(c)$ is undefined.

    B. f is discontinuous at $x = c$.

    C. $f(x) \to 1$ as $x \to c$.

    D. $f(x)$ may approach 0 as $x \to c$.

    E. $f(x)$ may approach 3 as $x \to c$.

For Problems 1 and 2, explain why the intermediate value theorem applies on the given closed interval. Then find an approximation for the value of c in the corresponding open interval for which $f(c)$ is exactly equal to the given y-value. Illustrate by graph.

1. $f(x) = (x - 3)^4 + 2$, $[1, 4]$, $y = 8$

2. $f(x) = 0.001x^5 - 8$, $[0, 6]$, $y = -1$

3. *Converse of the Intermediate Value Theorem?* The intermediate value theorem is not an "if and only if" theorem. The conclusion can be true even if the hypotheses are not met.

a. The graph on the left in Figure 2-6d (on the next page) shows

$$f(x) = 2 + x + \frac{|x - 2|}{x - 2}$$

Explain why the conclusion of the intermediate value theorem could be true or false for the interval $[1, 5]$, depending on the value of y you pick in the interval $[2, 8]$.

b. The graph on the right in Figure 2-6d shows

$$g(x) = 2 + x - \frac{|x - 2|}{x - 2}$$

Explain why the intermediate value theorem is always true for the interval [1, 5], no matter what value of y you pick between $g(1)$ and $g(5)$, even though the function is discontinuous at $x = 2$.

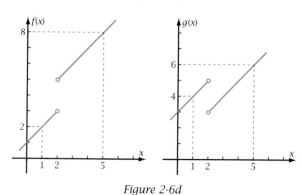

Figure 2-6d

4. Figure 2-6e shows the graph of

$$f(x) = \begin{cases} 2^x, & \text{if } x \text{ is rational} \\ 8, & \text{if } x \text{ is irrational} \end{cases}$$

a. Find $f(2)$, $f(3)$, $f(0.5)$, and $f(\sqrt{5})$.

b. Is f continuous at $x = 3$? Explain.

c. Where else is f continuous? Surprising?

d. Because $f(0) = 1$ and $f(2) = 4$, is the intermediate value theorem true for all values of y between 1 and 4? Explain.

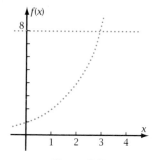

Figure 2-6e

5. Use the intermediate value theorem to prove that there is a real number equal to $\sqrt{3}$. That is, prove that there is a number c such that $c^2 = 3$.

6. Use the intermediate value theorem to prove that if f is continuous, and if $f(a)$ is positive and $f(b)$ is negative, then there is at least one zero of $f(x)$ between $x = a$ and $x = b$. (Recall that a zero of a function is a value of x that makes $f(x) = 0$.)

7. The intermediate value theorem is an example of an **existence theorem**. Why do you suppose this term is used? What does an existence theorem *not* tell you how to do?

8. *Sweetheart Problem:* You wish to visit your sweetheart, but you don't want to go all the way over to their house if your sweetheart isn't home. What sort of "existence proof" could you do beforehand to decide whether it is worthwhile to make the trip? What sort of information will your proof *not* give you about making the trip? Why do you suppose mathematicians are so interested in doing existence proofs before they spend a lot of time searching for solutions?

9. *Foot Race Problem:* Jesse and Kay run the 1000-m race. One minute after the race begins, Jesse is running 20 km/h and Kay is running 15 km/h. Three minutes after the race begins, Jesse has slowed to 17 km/h and Kay has speeded up to 19 km/h.

Assume that each runner's speed is a continuous function of time. Prove that there is a time between 1 min and 3 min after the race begins at which each one is running exactly the same speed. Is it possible to tell what that speed is? Is it possible to tell when that speed occurred? Explain.

10. *Postage Stamp Problem:* U.S. postage rates in 2003 for first-class letters were 37¢ for the first ounce and 23¢ per ounce thereafter. Figure 2-6f, an example of a **step function**, shows the cost of mailing a first-class letter versus its weight in ounces. Does the function meet the hypotheses of the intermediate value theorem? Is there a weight of letter that you can mail for exactly $2.00? Justify your answers.

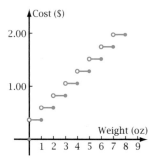

Figure 2-6f

11. *Cosine Function Problem:* Figure 2-6g is a graph of $f(x) = \cos x$. Recall that $\cos 0 = 1$ and $\cos \pi = -1$. What assumption must you make about the cosine function to be able to use the intermediate value theorem? Find as accurate an approximation as possible for a value of $x = c$ between 0 and π for which $\cos x = 0.6$. Explain how you found this approximation.

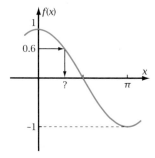

Figure 2-6g

12. *Exponential Function Problem:* Figure 2-6h is a graph of $f(x) = 2^x$. What assumption must you make about exponential functions in order to use the intermediate value theorem? Why does $f(0) = 1$? Find as accurate an approximation as possible for a value of $x = c$ between 0 and 2 for $2^c = 3$. Explain how you found this approximation.

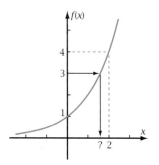

Figure 2-6h

13. *The Extreme Value Theorem:* The **extreme value theorem** expresses the property that if f is continuous on the closed interval $[a, b]$, then there are numbers c_1 and c_2 in $[a, b]$ for which $f(c_1)$ and $f(c_2)$ are the maximum and minimum values of $f(x)$ for that interval. Think about what this means, and express this theorem with a graph. Then draw a graph that shows why the conclusion might not be true for a function that has a discontinuity somewhere in $[a, b]$.

14. *The Image Theorem:* The **image theorem,** a corollary of the intermediate value theorem, expresses the property that if f is continuous on the interval $[a, b]$, then the image (the set of y-values) of f on $[a, b]$ is all real numbers between the minimum of $f(x)$ and the maximum of $f(x)$ on $[a, b]$, inclusive. Use the extreme value theorem as a *lemma* (a preliminary result) to prove the image theorem.

Chapter Review and Test

In this chapter you have gained further insight into the meaning of limit. You broadened the idea of limit to include one-sided limits and infinite limits, and you applied these ideas to define the concept of a continuous function. By now you should understand that a limit is a number that $f(x)$ stays close to when x is kept close to c but not equal to c. You should be able to find a limit of a function given by graph or by equation and also to show that $f(x)$ really does stay within ϵ units of the limit when x is within δ units of c ($x \neq c$). You should be able to use the limit properties to prove that the number you've found is the limit. If you think of "lim" as an operation that acts on functions, many of these properties can be thought of as distributive properties. This box summarizes the properties.

SUMMARY: Limit Property

Limit distributes over addition, subtraction, multiplication, and division (denominator $\neq 0$) with any finite number of terms or factors.

Finally, you learned another major theorem of calculus, the intermediate value theorem, which expresses a property of continuous functions.

Review Problems

R0. You have learned that calculus involves four concepts. Your goal for this course is to be able to do four things with each of these concepts.

	Define it.	Understand it.	Do it.	Apply it.
Limit				
Derivative				
Integral				
Integral				

In your journal, make a table like the one shown above. Check each concept you have worked on as you studied this chapter. Make journal entries for such things as

- The one most important thing you learned in studying Chapter 2

- A statement explaining what you now understand a limit to be

- How limits apply to derivatives and definite integrals

- Your understanding of continuity and the intermediate value theorem

- Anything you need to ask about in class before your test on Chapter 2

R1. Let $f(x) = \dfrac{4x^2 - 17x + 15}{x - 3}$.

a. What numerical form does $f(3)$ take? What name is given to this numerical form?

b. Plot the graph of f using a friendly window that includes $x = 3$ as a grid point. Sketch the graph of f taking into account the fact that $f(3)$ is undefined because of division by zero. What graphical feature appears at $x = 3$?

c. The number 7 is the limit of $f(x)$ as x approaches 3. How close to 3 would you have to keep x in order for $f(x)$ to be within 0.01 unit of 7? Within 0.0001 unit of 7? How could you keep $f(x)$ *arbitrarily* close to 7 just by keeping x close to 3 but not equal to 3?

R2. a. State the epsilon-delta definition of limit.

b. For $x = 1, 2, 3, 4$, and 5, state whether or not the function f in Figure 2-7a has a limit, and if so, what the limit appears to be.

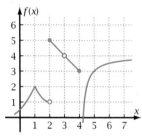

Figure 2-7a   Figure 2-7b

c. For the function in Figure 2-7b, write the limit of $f(x)$ as x approaches 2. From the graph, estimate the largest possible value of δ that can be used to keep $f(x)$ within 0.4 unit of the limit when x is within δ units of 2.

d. In part c, $f(x) = 2 + \sqrt{x - 1}$. Calculate the maximum value of δ that you estimated in part c.

e. For the function in part d, show that δ is positive for any $\epsilon > 0$ where $f(x)$ is within ϵ units of the limit when x is within δ units of 2 (but $x \neq 2$).

R3. a. State these limit properties.
 • Limit of a sum of two functions
 • Limit of a constant times a function
 • Limit of a quotient of two functions

b. For $g(x) = \dfrac{x^3 - 13x^2 + 32x - 6}{x - 3}$,

 • Plot the graph using a friendly window that includes $x = 3$ as a grid point. Sketch the graph.

 • Explain why you cannot use the limit of a quotient property to find $\lim\limits_{x \to 3} g(x)$.

• Simplify the fraction. Why can you cancel $(x - 3)$ without worrying about dividing by zero?

• Find $\lim\limits_{x \to 3} g(x)$, naming the limit theorems used at each step.

c. Figure 2-7c shows the graphs

$$f(x) = 2^x \text{ and } g(x) = \frac{x^2 - 8x + 15}{3 - x}$$

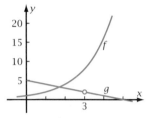

Figure 2-7c

• Find $\lim\limits_{x \to 3} f(x)$ and $\lim\limits_{x \to 3} g(x)$.
• Let $p(x) = f(x) \cdot g(x)$. Show that

$$\lim_{x \to 3} p(x) = \lim_{x \to 3} f(x) \cdot \lim_{x \to 3} g(x)$$

by showing numerically that $p(x)$ stays close to 16 when x is kept close to 3 but not equal to 3.

• Let $r(x) = f(x)/g(x)$. Plot the graph of r. Sketch the graph, showing that

$$\lim_{x \to 3} r(x) = \frac{\lim\limits_{x \to 3} f(x)}{\lim\limits_{x \to 3} g(x)}$$

d. *Chuck's Rock Problem:* Chuck throws a rock high into the air. Its distance, $d(t)$, in meters, above the ground is given by $d(t) = 35t - 5t^2$, where t is the time, in seconds, since he threw it. Find the average velocity of the rock from $t = 5$ to $t = 5.1$. Write an equation for the average velocity from 5 seconds to t seconds. By taking the limit of the expression in this equation, find the instantaneous velocity of the rock at $t = 5$. Was the rock going up or down at $t = 5$? How can you tell? What mathematical quantity is this instantaneous velocity?

R4. a. Write the definitions of continuity at a point and continuity on a closed interval.

b. For the piecewise function in Figure 2-7d, make a table showing these quantities for $c = 1, 2, 3, 4$, and 5, or stating that the quantity does not exist.

- $f(c)$
- $\lim\limits_{x \to c^-} f(x)$
- $\lim\limits_{x \to c^+} f(x)$
- $\lim\limits_{x \to c} f(x)$
- Continuity or kind of discontinuity

$$f(x) = \begin{cases} 2 + \dfrac{2}{|x-1|}, & \text{if } x < 3 \text{ and } x \neq 2 \\ 1, & \text{if } x = 2 \\ 3 + 2\cos\frac{\pi}{2}(x-3), & \text{if } 3 \leq x < 5 \\ 4 - 3(x-6)^2, & \text{if } x \geq 5 \end{cases}$$

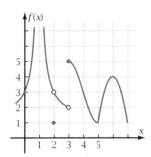

Figure 2-7d

c. Sketch graphs of the functions described.
- Has a removable discontinuity at $x = 1$
- Has a step discontinuity at $x = 2$
- Has a vertical asymptote at $x = 3$
- Has a cusp at $x = 4$
- Is continuous at $x = 5$
- Has a limit as $x \to 6$ and a value of $f(6)$, but is discontinuous at $x = 6$
- Has a left limit of -2 and a right limit of 5 as $x \to 1$

d. For the piecewise function

$$f(x) = \begin{cases} x^2, & \text{if } x \leq 2 \\ x^2 - 6x + k, & \text{if } x > 2 \end{cases}$$

- Sketch the graph of f if $k = 10$.
- Show that f is discontinuous at $x = 2$ if $k = 10$.
- Find the value of k that makes f continuous at $x = 2$.

R5. a. Use the appropriate limit definitions to write the meanings of

$$\lim_{x \to 4} f(x) = \infty \quad \text{and} \quad \lim_{x \to \infty} f(x) = 5$$

b. For piecewise function f in Figure 2-7e, what do these limits appear to be?

- $\lim\limits_{x \to -\infty} f(x)$
- $\lim\limits_{x \to 2^-} f(x)$
- $\lim\limits_{x \to \infty} f(x)$
- $\lim\limits_{x \to -2} f(x)$
- $\lim\limits_{x \to 2^+} f(x)$

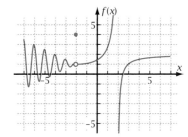

Figure 2-7e

c. Let $f(x) = 6 - 2^{-x}$. Find a value of $x = c$ for which $f(x)$ is exactly 0.001 unit below the limit of $f(x)$ as x approaches infinity. Choose several values of $x > c$ and show that $f(x)$ is within 0.001 unit of the limit for all of these values.

d. Let $g(x) = x^{-2}$. Find $\lim_{x \to 0} g(x)$. Find a positive value of $x = c$ for which $g(x) = 10^6$. Choose several values of x closer to 0 than c and show that $g(x) > 10^6$ for all of these values.

e. The distance you travel at a variable velocity, $v(t)$, is the definite integral of $v(t)$ with respect to time, t. Suppose that your car's velocity is given by $v(t) = 40 + 6\sqrt{t}$, where t is in seconds and $v(t)$ is in ft/s. Use the trapezoidal rule with varying numbers of increments, n, to estimate the distance traveled from $t = 0$ to $t = 9$. What limit do these sums seem to approach as n approaches infinity? Find a number D for which the trapezoidal sum is within 0.01 unit of this limit when $n > D$.

R6. a. State the intermediate value theorem. What axiom forms the basis for the proof of the intermediate value theorem? State the extreme value theorem. What word describes how the extreme value theorem relates to the intermediate value theorem?

b. For $f(x) = -x^3 + 5x^2 - 10x + 20$, find $f(3)$ and $f(4)$. Based on these two numbers, how can you tell immediately that a zero of $f(x)$ occurs between $x = 3$ and $x = 4$? What property of polynomial functions allows you to make this conclusion? Find as accurate a value of this zero as possible.

c. Plot the function

$$f(x) = \frac{x^2 + 11x + 28}{x + 4}$$

Use a friendly window that includes $x = -4$. Show that $f(-6) = 1$ and $f(-2) = 5$. Based on the intermediate value theorem, if you pick a number y between 1 and 5, will you always get a value of $x = c$ between -6 and -2 for which $f(c) = y$? If so, explain why. If not, give a counterexample.

Concept Problems

C1. *Squeeze Theorem Introduction Problem:* Suppose that $g(x)$ and $h(x)$ both approach 7 as x approaches 4, but that $g(x) < h(x)$ for all other values of x. Suppose another function, f, has a graph that is bounded above by the graph of h and bounded below by the graph of g. That is, $g(x) \le f(x) \le h(x)$ for all values of x. Sketch possible graphs of the three functions on the same set of axes. Make a conjecture about the limit of $f(x)$ as $x \to 4$.

C2. *Derivatives and Continuity Problem:* Figure 2-7f shows the graph of

$$f(x) = \begin{cases} x^2 + 3, & \text{if } x < 1 \\ x^2 - 6x + 9, & \text{if } x \ge 1 \end{cases}$$

Find the value of $f(1)$. Is f continuous at $x = 1$? Find the limit of $[f(x) - f(1)]/(x - 1)$ as $x \to 1^-$ and as $x \to 1^+$. Based on your work, explain how a function can be continuous at a point but not have a derivative there.

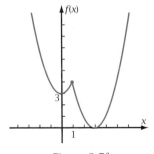

Figure 2-7f

C3. *Equation from Graph Problem:* Figure 2-7g is the graph of a discontinuous function. Write a single equation whose graph could be that shown in the figure.

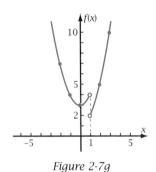

Figure 2-7g

C4. *Absolute Value Definition of Limit:* Later in your mathematical career, you may encounter the definition of limit written in the form shown in the box.

> **Algebraic (Absolute Value) Definition: Limit**
> $L = \lim\limits_{x \to c} f(x)$ if and only if, for any $\epsilon > 0$, there is a $\delta > 0$ such that
> if $0 < |x - c| < \delta$, then $|f(x) - L| < \epsilon$.

Explain how this algebraic definition of limit is equivalent to the "within" definition you have learned.

Chapter Test

PART 1: No calculators allowed (T1–T9)

T1. State the definitions of continuity at a point and on a closed interval.

T2. a. For function f in Figure 2-7h, find

- $\lim\limits_{x \to 2^-} f(x)$
- $\lim\limits_{x \to 2^+} f(x)$
- $\lim\limits_{x \to 2} f(x)$
- $\lim\limits_{x \to 6^-} f(x)$
- $\lim\limits_{x \to 6^+} f(x)$
- $\lim\limits_{x \to 6} f(x)$

b. Is f continuous on the closed interval $[2, 6]$? Explain.

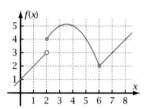

Figure 2-7h

T3. State the property for the limit of a quotient.

For the functions graphed in Problems T4–T7, state the following for $x = c$.

a. Left and right limits if they exist

b. The limit if it exists

c. Whether the function is continuous

T4.

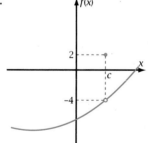

T5.

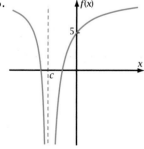

T6.

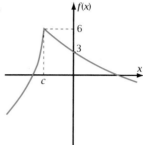

T7.

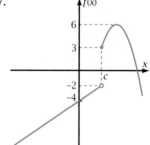

T8. Sketch the graph of a function for which $\lim\limits_{x \to \infty} f(x) = 2$.

T9. Sketch graphs that show you understand the difference in the behaviors of the following functions at $x = 0$.

a. $f(x) = \dfrac{1}{x^2}$

b. $g(x) = \dfrac{x}{x}$

c. $h(x) = \dfrac{|x|}{x}$

d. $s(x) = \sin \dfrac{1}{x}$

PART 2: Graphing calculators allowed (T10–T19)

T10. Figure 2-7i shows the graph of function f in a neighborhood of $x = 3$.

$$f(x) = \frac{(x^2 - 5x + 8)(x - 3)}{x - 3}$$

a. What form does $f(3)$ take? What name is given to an expression of this form?

b. Use the limit properties to prove algebraically that $\lim\limits_{x \to 3} f(x) = 2$.

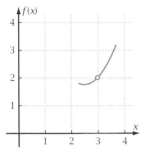

Figure 2-7i

For Problems T11–T13, let $f(x) = \begin{cases} kx^2, & \text{if } x \le 2 \\ x + k, & \text{if } x > 2 \end{cases}$.

T11. Show that f is discontinuous at $x = 2$ if $k = 1$. Sketch the graph.

T12. Find the value of k that makes f continuous at $x = 2$.

T13. On your grapher, plot the graph using the value of k from Problem T11. Use Boolean variables to restrict the domains of the two branches. Sketch the graph.

T14. *Temperature Versus Depth Problem:* During the day the soil at Earth's surface warms up. Heat from the surface penetrates to greater depths. But before the temperature lower down reaches the surface temperature, night comes and Earth's surface cools. Figure 2-7j shows what the temperature, T, in degrees Celsius, might look like as a function of depth, x, in feet.

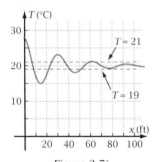

Figure 2-7j

a. From the graph, what does the limit of T seem to be as x approaches infinity? How deep do you think you would have to be so that the temperature varies no more than 1 degree from this limit? What feature does the graph have when T equals this limit?

b. The equation of the function T in Figure 2-7i is $T(x) = 20 + 8(0.97^x) \cos 0.5x$. Use this equation to calculate a positive number $x = c$ such that T does not vary more than 0.1 unit from the limit whenever $x > c$.

c. Just for fun, see if you can figure out the approximate time of day for which the graph in Figure 2-7j applies.

T15. *Glacier Problem:* To determine how far a glacier has traveled in a given time interval, naturalists drive a metal stake into the surface of the glacier. From a point not on the glacier, they measure the distance, $d(t)$, in centimeters, from its original position that the stake has moved in time t, in days. Every ten days they record this distance, getting the values shown in the table.

t (days)	$d(t)$ (cm)
0	0
10	6
20	14
30	24
40	36
50	50

a. Show that the equation $d(t) = 0.01t^2 + 0.5t$ fits all the data points in the table. Use the most time-efficient way you can think of to do this problem.

b. Use the equation to find the average rate the glacier is moving during the interval $t = 20$ to $t = 20.1$.

c. Write an equation for the average rate from 20 days to t days. Perform the appropriate algebra, then find the limit of the average rate as t approaches 20. What is the instantaneous rate the glacier is moving at $t = 20$? What mathematical name is given to this rate?

d. Based on the table, does the glacier seem to be speeding up or slowing down as time goes on? How do you reach this conclusion?

T16. *Calvin and Phoebe's Acceleration Problem:* Calvin and Phoebe are running side by side along the jogging trail. At time $t = 0$, each one starts to speed up. Their speeds are given by the following, where $p(t)$ and $c(t)$ are in ft/s and t is in seconds.

$c(t) = 16 - 6(2^{-t})$ For Calvin.

$p(t) = 10 + \sqrt{t}$ For Phoebe.

Show that each is going the same speed when $t = 0$. What are the limits of their speeds as t approaches infinity? Surprising?!

T17. Let $f(x) = \begin{cases} kx^2, & \text{if } x \le 2 \\ 10 - kx, & \text{if } x > 2 \end{cases}$.

What value of k makes f continuous at $x = 2$? What feature will the graph of f have at this point?

T18. Let $h(x) = x^3$. Show that the number 7 is between $h(1)$ and $h(2)$. Since h is continuous on the interval $[1, 2]$, what theorem allows you to conclude that there is a real number $\sqrt[3]{7}$ between 1 and 2?

T19. What did you learn as a result of taking this test that you did not know before?

Derivatives, Antiderivatives, and Indefinite Integrals

During the free-fall part of a sky diver's descent, her downward acceleration is influenced by gravity and air resistance. The velocity increases at first, then approaches a limit called the terminal velocity. At any instant in time the acceleration is the derivative of the velocity. The distance she has fallen is the antiderivative, or indefinite integral, of the velocity. Velocity, acceleration, and displacement all vary, and depend on time.

Mathematical Overview

In Chapter 3 you will apply the concept of *limit* to find formulas you can use to calculate exact values of derivatives. Then you will work backward to find the function equation if the derivative equation is given. This is the antiderivative, which will give you clues as to why there are *two* concepts of calculus that both have the name "integral." You will work with these concepts four ways.

Graphically   The icon at the top of each even-numbered page of this chapter shows the run, Δx, and the rise, Δy, from one point on a graph to another. The limit of $\Delta y/\Delta x$ is the instantaneous rate, or derivative, of $f(x)$.

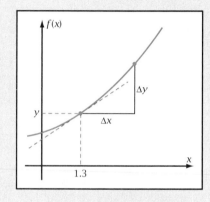

Numerically

Δx	$\Delta y/\Delta x$
0.1	0.705
0.01	0.6915
0.001	0.69015
0.0001	0.690015
$\vdots$	$\vdots$

Algebraically $f'(1.3) = \displaystyle\lim_{x \to 1.3} \frac{f(x) - f(1.3)}{x - 1.3}$, the definition of derivative

Verbally   *I had learned about limits, derivatives, and definite integrals already. But it wasn't until I learned how to find an algebraic equation for a derivative that I understood the meaning of indefinite integral, which is a name for antiderivative.*

3-1 Graphical Interpretation of Derivative

You have learned physical and graphical meanings of derivative—instantaneous rate of change and slope of the tangent line, respectively. In this section you will see a way to calculate the value of a derivative algebraically and verify the answer graphically.

OBJECTIVE Given the equation of a function, calculate the derivative at a given point by taking the limit of the quantity (change in y)/(change in x).

Exploratory Problem Set 3-1

Spaceship Problem: A spaceship approaches a far-off planet. At time x, in minutes after its retrorockets fire, its distance, $f(x)$, in kilometers from the surface of the planet, is given by

$$f(x) = x^2 - 8x + 18$$

1. Figure 3-1a shows the graph of f. Confirm by grapher that this graph is correct.

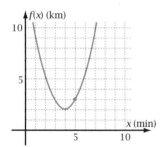

Figure 3-1a

2. Find the average rate of change of $f(x)$ with respect to x from $x = 5$ to 5.1. What are the units of this rate of change?

3. This function gives the average rate of change of function f from 5 to x.

$$r(x) = \frac{f(x) - f(5)}{x - 5}$$

Plot the graphs of $y_1 = f(x)$ and $y_2 = r(x)$ on the same screen. Use a friendly window for x

that includes $x = 5$ as a grid point. Use grid-off format. Sketch the result.

4. What feature do you notice in the graph of r at $x = 5$? Show that $r(5)$ takes the indeterminate form 0/0.

5. Simplify the expression for $r(x)$ by factoring and canceling. Then use the limit properties to find the *derivative* of f at $x = 5$ by taking the *limit* of $r(x)$ as x approaches 5. What physical quantity does the derivative represent? What are its units?

6. Find an equation of the linear function containing the point $(5, f(5))$ and having slope equal to the derivative you calculated in Problem 5. Plot this linear function as y_3 on the same screen as y_1, with y_2 inactive. Sketch the result on the graph you drew in Problem 3. What relationship does the line y_3 have to the graph of function f?

7. Zoom in on the point where y_1 and y_3 appear to intersect. Describe the relationship between the line and the graph of f as you zoom in.

8. In your journal, record what you learned as a result of this problem set. Mention anything that you are still unsure about.

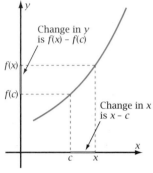

3-2 Difference Quotients and One Definition of Derivative

You have been finding derivatives by taking a change in x, dividing it into the corresponding change in y, and taking the limit of the resulting fraction as the change in x approaches zero. Figure 3-2a illustrates what you have been doing. The change in the y-value is equal to $f(x) - f(c)$. The change in the x-value is $x - c$. Thus, the derivative is approximately equal to

$$\frac{f(x) - f(c)}{x - c}$$

This fraction is called a **difference quotient**.

The derivative is the function you get by taking the limit of the difference quotient as the denominator approaches zero. If the function's name is f, then the symbol f' (pronounced "f prime") is often used for the derivative function. The symbol shows that there is a relationship between the original function and the function "derived" from it (hence the name *derivative*), which indicates its rate of change. One way to write the definition of derivative is shown in the box. An alternative way is given in Section 3-4.

Figure 3-2a

DEFINITION: *Derivative at a Point (derivative at x = c form)*

$$f'(c) = \lim_{x \to c} \frac{f(x) - f(c)}{x - c}$$

Meaning: The instantaneous rate of change of $f(x)$ with respect to x at $x = c$.

OBJECTIVE   Given the equation of a function and a value of x, use the definition of derivative to calculate algebraically the value of the derivative at that point, and confirm your answer numerically and graphically.

Example 1 shows how you can use the definition of derivative to accomplish this objective.

▶ **EXAMPLE 1**   If $f(x) = x^2 - 3x - 4$, find $f'(5)$, the value of the derivative if $x = 5$. Check by graphing the difference quotient and finding the limit.

Solution

$$f(5) = 5^2 - 3(5) - 4 = 6$$

$$\therefore f'(5) = \lim_{x \to 5} \frac{f(x) - f(5)}{x - 5} = \lim_{x \to 5} \frac{(x^2 - 3x - 4) - 6}{x - 5} = \lim_{x \to 5} \frac{x^2 - 3x - 10}{x - 5}$$

$$= \lim_{x \to 5} \frac{(x - 5)(x + 2)}{x - 5} = \lim_{x \to 5}(x + 2) \qquad \text{Why can you cancel without dividing by zero?}$$

$$= 5 + 2 = 7 \qquad \text{Use the limit of a sum and the limit of } x \text{ properties from Chapter 2.}$$

As a check, plot the difference quotient (Figure 3-2b):

$$y = \frac{f(x) - f(5)}{x - 5} = \frac{x^2 - 3x - 10}{x - 5}$$

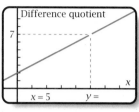

x	quotient
4.97	6.97
4.98	6.98
4.99	6.99
5.00	undefined
5.01	7.01
5.02	7.02
5.03	7.03

Figure 3-2b

Use a friendly window with $x = 5$ as a grid point. A removable discontinuity occurs at $x = 5$ and the function takes the indeterminate form $0/0$.

Use TRACE or TABLE to give the values shown for x on either side of 5. You should be able to see from the pattern that the limit of the quotient is 7 as x approaches 5.

Another way to check the answer in Example 1 is to graph the original function, then draw a line with slope 7 (the derivative) through the point on the graph at $x = 5$ (that is, $(5, 6)$). To plot the line on your grapher, find its equation.

$$y = 7x + b \Rightarrow 6 = 7(5) + b \Rightarrow -29 = b$$
$$\therefore y = 7x - 29$$

Figure 3-2c shows that the line will be **tangent** to the graph of f. The **tangent line** to a curve at a given point is the line with the same slope as the curve at that point.

You've already discovered this property from Problem Set 3-1. The fact that the line is tangent to the graph is a graphical interpretation of the meaning of derivative.

One of the factors engineers take into account when evaluating the safety of a roller coaster design is the steepness of the slope.

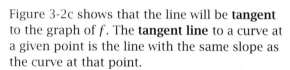

Figure 3-2c

GRAPHICAL INTERPRETATION OF DERIVATIVE: *Slope of a Tangent Line*

The derivative of a function at a point equals the slope of the tangent line to the graph of the function at that point. Both equal the instantaneous rate of change.

Problem Set 3-2

Q1. Quick! What does derivative mean?

Q2. Simplify: $(x^2 - 81)/(x - 9)$

Q3. Find: $\lim_{x \to 9}(x^2 - 81)/(x - 9)$

Q4. Sketch the graph of $y = 2^x$.

Q5. Expand: $(3x - 7)^2$

Q6. Fill in the blank with the appropriate operation: $\log(x/y) = \log x -?- \log y$.

Q7. Sketch a graph with a step discontinuity at $x = 3$.

Q8. Sketch a graph with a cusp at the point $(5, 2)$.

Q9. Who is credited with inventing calculus?

Q10. $\lim_{x \to c} kf(x) = k \lim_{x \to c} f(x)$ is a statement of which limit property?

    A. Limit of a sum

    B. Limit of a product

    C. Limit of a constant

    D. Limit of a constant times a function

    E. Limit of the identity function

1. Write the definition of derivative.

2. What are the physical and the graphical meanings of the derivative of a function?

For Problems 3 and 4,

    a. Use the definition of derivative to calculate the value of $f'(c)$ exactly.

    b. Plot the difference quotient in a window that includes c and sketch the result.

    c. Plot the graph of the function in a window that includes c.

    d. Plot a line through $(c, f(c))$ with slope $f'(c)$. Sketch the graph and the line.

3. $f(x) = 0.6x^2$, $c = 3$

4. $f(x) = -0.2x^2$, $c = 6$

For Problems 5–12, use the definition of derivative to calculate $f'(c)$ exactly.

5. $f(x) = x^2 + 5x + 1$, $c = -2$

6. $f(x) = x^2 + 6x - 2$, $c = -4$

7. $f(x) = x^3 - 4x^2 + x + 8$, $c = 1$

8. $f(x) = x^3 - x^2 - 4x + 6$, $c = -1$

9. $f(x) = -0.7x + 2$, $c = 3$

10. $f(x) = 1.3x - 3$, $c = 4$

11. $f(x) = 5$, $c = -1$

12. $f(x) = -2$, $c = 3$

13. From the results of Problems 9 and 10, what can you conclude about the derivative of a linear function? How does this conclusion relate to derivatives and tangent lines?

14. From the results of Problems 11 and 12, what can you conclude about the derivative of a constant function? How does this conclusion relate to derivatives and tangent lines?

15. *Local Linearity Problem:* Figure 3-2d shows the graph of $f(x) = x^2$, and a line of slope $f'(1)$ passing through the point on the graph of f where $x = 1$.

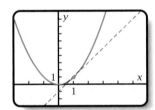

Figure 3-2d

    a. Reproduce this graph on your grapher. Use a friendly window that includes the point $(1, 1)$. Describe how you plotted the tangent line.

    b. Zoom in on the point $(1, 1)$. What do you notice about the line and the curve?

    c. Zoom in several more times. How do the line and the curve seem to be related now?

    d. The graph of f possesses a property called **local linearity** at $x = 1$. Why do you suppose these words are used to describe this property?

    e. Explain why you could say that the value of the derivative at a point equals the "slope of the graph" at that point if the graph has local linearity.

Chapter 3: Derivatives, Antiderivatives, and Indefinite Integrals

16. *Local Nonlinearity Problem:* Figure 3-2e shows the graph of

$$f(x) = x^2 + 0.1(x - 1)^{2/3}$$

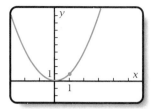

Figure 3-2e

a. Show that the point (1, 1) is on the graph of f. Does the graph seem to possess local linearity at that point? (See Problem 15.)

b. Zoom in on the point (1, 1) several times. Sketch what you see.

c. Explain why the graph of f does not have local linearity at $x = 1$.

d. Explain why f does not have a value for the derivative at $x = 1$.

17. Let f be the piecewise function

$$f(x) = \begin{cases} \dfrac{x^2 - x - 6}{x - 3}, & \text{if } x \neq 3 \\ 7, & \text{if } x = 3 \end{cases}$$

a. Plot the graph on your grapher, using a friendly window with $x = 3$ as a grid point. Sketch the result, showing clearly what happens at $x = 3$.

b. Write the difference quotient for $f'(3)$ and plot it on your grapher. (See if you can find a time-efficient way to enter the equation!) Sketch the result.

c. Make a short table of values of the difference quotient for values of x close to 3 on both sides of 3. Based on your work, explain why the function has no derivative at $x = 3$.

18. Let $s(x) = 2 + |\sin(x - 1)|$.

a. Plot the graph of s. Sketch the result.

b. Plot the difference quotient for $s'(1)$. Sketch the graph.

c. Explain why s does not have a value for the derivative at $x = 1$.

A line through the points where each tire touches the half-pipe is a secant line.

19. *Tangent Lines as Limits of Secant Lines:* Figure 3-2f shows the graph of

$$f(x) = 0.25x^2 - 2.5x + 7.25$$

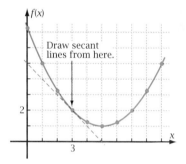

Figure 3-2f

a. Show that the tangent line on the diagram at $x = 3$ has slope equal to $f'(3)$.

b. Recall that a secant line intersects a curve at a minimum of two points. On a copy of Figure 3-2f, draw secant lines starting at the point (3, 2) and going through the points on the graph where $x = 9, 8, 7, 6, 5,$ and 4. Describe what happens to the secant lines as the x-distance between those points and the point (3, 2) decreases.

c. Does the same thing happen with the secant lines from the point (3, 2) to the points on the graph where $x = 0, 1,$ and 2?

d. Figure 3-2g shows the graph of

$$g(x) = 4 - 6|\cos \tfrac{\pi}{6} x|$$

On a copy of Figure 3-2g, draw secant lines from the cusp at the point $(3, 4)$ through the points on the graph where $x = 8, 7, 6, 5, 4$ and where $x = 0, 1, 2$.

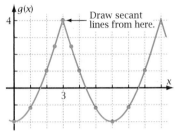

Draw secant lines from here.

Figure 3-2g

e. The slopes of the lines in part d approach a different limit as x approaches 3 from the left than they do when x approaches 3 from the right. Explain why g has no derivative at $x = 3$.

f. Make a conjecture about what numbers the two limits in part e equal.

20. Based on your work in Problem 19, explain why the following property is true.

> **Property: Tangent Line as a Limit of Secant Lines**
>
> The tangent to a graph at $(c, f(c))$ is the limit of the secant lines from $(c, f(c))$ to $(x, f(x))$ as x approaches c. The slope of the tangent line equals $f'(c)$.

3-3 Derivative Functions, Numerically and Graphically

You have been calculating the derivative of a given function at one particular point, $x = c$. Now it is time to turn your attention to finding the derivative for *all* values of x. That is, you seek a new function whose values are the derivatives of the given function. In this section you will use the numerical derivative feature of your grapher to do this. In the next sections you will find formulas that will help you easily write the derivative function for various types of given functions.

OBJECTIVE   Given the equation for a function, graph the function and its (numerical) derivative function on the same set of axes, and make conjectures about the relationship between the derivative function and the original function.

Graphers calculate numerical derivatives by using difference quotients, just as you have been doing. Often they use a **symmetric difference quotient**. As illustrated in the third figure in Figure 3-3a, points on the graph $\pm h$ units (meaning "horizontal" distance) from x are found. The corresponding difference in y-values, Δy (read "delta y," meaning difference between y-values), is divided by the difference in x-values, $2h$, to get an estimate of the rate of change of the function. You have used a **forward difference quotient** or a **backward difference quotient**, as shown in the first and second figures of Figure 3-3a. A symmetric difference quotient usually gives a more accurate answer.

Chapter 3: Derivatives, Antiderivatives, and Indefinite Integrals

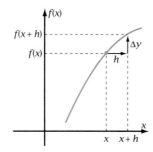 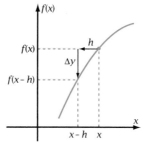 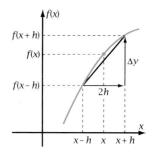

| Forward difference quotient $\Delta y/h$ | Backward difference quotient $\Delta y/h$ | Symmetric difference quotient $\Delta y/(2h)$ |

Figure 3-3a

To see how to calculate forward, backward, and symmetric difference quotients, suppose you want to estimate the derivative of $f(x) = x^3$ at $x = 2$ (abbreviated $f'(2)$ and read "f-prime of 2"). The x-increment, h, is often called the **tolerance**. If a tolerance of 0.01 is specified, for example, the calculations are as shown.

Forward: $\dfrac{f(2.01) - f(2)}{2.01 - 2} = \dfrac{2.01^3 - 2^3}{0.01} = \dfrac{8.120601 - 8}{0.01} = 12.0601$

Backward: $\dfrac{f(1.99) - f(2)}{1.99 - 2} = \dfrac{1.99^3 - 2^3}{-0.01} = \dfrac{7.880599 - 8}{-0.01} = 11.9401$

Symmetric: $\dfrac{f(2.01) - f(1.99)}{2.01 - 1.99} = \dfrac{2.01^3 - 1.99^3}{0.02} = \dfrac{8.120601 - 7.880599}{0.02} = 12.0001$

As you might guess from the answers, the exact value of the derivative is 12. The symmetric difference quotient gives a more accurate answer in this case. The numerical derivative instruction on a typical grapher is shown here.

$$\text{nDeriv}(x^3, x, 2)$$

It says to take the numerical derivative of x^3 with respect to x, evaluated where $x = 2$. You will need to check your grapher's manual to find out exactly how to write the instruction and specify the desired tolerance.

Because there is a value of the derivative for each value of x, there is a *function* whose independent variable is x and whose dependent variable is the value of the derivative. This function is called (obviously!) the **derivative function** and is usually abbreviated $f'(x)$. The grapher plots the (numerical) derivative function by using an instruction such as

$$y_2 = \text{nDeriv}(y_1, x, x)$$

This instruction "tells" your grapher to find the numerical derivative of y_1 with respect to x (the first x in the parentheses) and evaluate it at whatever the value of x happens to be (the second x in the parentheses). Example 1 shows what you can learn.

▶ **EXAMPLE 1**   Given $f(x) = x^3 - 5x^2 - 8x + 70$,

a. Plot the graphs of f and f' (the numerical derivative of f) on the same screen.

b. Function f is a *cubic* function. What type of function does f' appear to be?

c. Trace to find values of $f(3)$ and $f'(3)$. Describe how $f'(3)$ relates to the graph of f at $x = 3$.

d. For what values of x does $f'(x) = 0$? What feature does the graph of f have at these x-values?

e. Plot $g(x) = f(x) + 10$. On the same screen, plot g'. How is the graph of g related to the graph of f? How is the graph of g' related to the graph of f'?

Solution

a. Figure 3-3b shows the graphs of f and f' (dashed) on the same screen. Type in instructions such as those shown (depending on your grapher).

$$y_1 = x^3 - 5x^2 - 8x + 70$$
$$y_2 = \mathrm{nDeriv}(y_1, x, x)$$

b. The graph of f' looks like a parabola. Conjecture: f' is quadratic.

c. $f(3) = 28$ and the value of $f'(3) = -11$. The negative value of the derivative says that $f(x)$ is decreasing when $x = 3$, which the graph shows.

d. $f'(x) = 0$ for $x = 4$ and for $x \approx -0.7$. The graph of f appears to have a high or low point when $f'(x) = 0$.

e. Figure 3-3c shows the graphs of g and g', entered as $y_3 = y_1 + 10$ and $y_4 = \mathrm{nDeriv}(y_3, x, x)$, respectively, along with the graphs of f and f'. The graphs of f and g are congruent to each other, separated vertically by ten units space. The graphs of f' and g' are identical. This is to be expected because f and g are changing at the same rate.

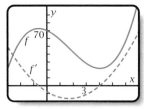

Figure 3-3b

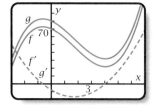

Figure 3-3c

▶ **EXAMPLE 2**

Figure 3-3d shows the graph of a function f, defined for x in the open interval $(-4, 6)$. On a copy of the figure, sketch the graph of the derivative function, f'. Describe your process.

Solution

The value of $f'(x)$ equals the *slope* of the line *tangent* to the graph. The tangent will be horizontal at $x = -3$, $x = 1$, and $x = 5$. Draw dots on the x-axis at these three places, as shown in Figure 3-3e, to indicate that the derivative, $f'(x)$, is zero there.

Between $x = -3$ and $x = 1$ the tangent will have positive slope, increasing to a bit more than 1

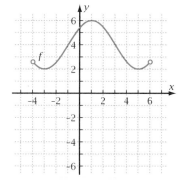

Figure 3-3d

at $x = -1$, then decreasing to zero at $x = 1$. Draw the part of the graph of f' that is in this interval (the graph of f' is shown dashed in Figure 3-3e).

Between $x = 1$ and $x = 5$ the tangent has negative slope, decreasing to a bit less than -1 at $x = 3$, then increasing to 0 at $x = 5$. The tangent also has negative slope between the endpoint at $x = -4$ and $x = -3$, and positive slope between $x = 5$ and the endpoint at $x = 6$. Draw the rest of the f' graph as shown in Figure 3-3e.

Note that because function f is not defined outside the interval $(-4, 6)$, you should not extend the derivative graph beyond this interval.

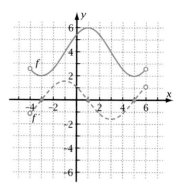

Figure 3-3e

▶ **EXAMPLE 3** On your grapher, explore the cubic function $f(x) = -x^3 + 3x^2 + 9x + 20$ and its numerical derivative, f'. Make some conclusions about how certain features on the graph of f', such as high and low points and x-intercepts, are related to features on the original function's graph. Write your conclusions in your journal.

Solution In this problem you have considerable freedom to explore, to conjecture, to discuss, and to write. Problems such as this are best done with your study group so that you may share ideas. Here is a typical (correct!) response to such a problem as it might appear in your journal.

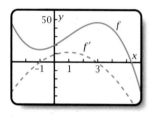

Figure 3-3f

I plotted the graphs of f and f' and got something like this (Figure 3-3f). The high and the low points of the cubic graph come where the derivative graph crosses the x-axis. I suppose this is reasonable. Where the derivative graph crosses the x-axis, the derivative is zero, meaning the rate of change of the function graph is zero. This would be true at a high or low point. One of the group members saw that wherever the derivative graph is positive, the function graph is going up. Another one saw that the high point on the derivative graph is at x = 1, where the function graph has its steepest up slope. I still don't understand why the derivative graph is going up at x = -2, for instance, but the function graph is decreasing.

Problem Set 3-3

Quick Review

Q1. Find: $\lim_{x \to 0} 3x/x$

Q2. Sketch a graph with a removable discontinuity at the point $(-2, 5)$.

Q3. Sketch the graph of $y = \tan x$.

Q4. What percent of 70 is 14?

Q5. Multiply: $(3x + 4)(x - 2)$

Q6. Expand: $(5x - 7)^2$

Q7. Write as the log of a single argument: $\log 36 - \log 12 + \log 2$.

Q8. Draw a tangent line to a circle.

Q9. Draw a secant line to a circle.

Q10. Explain what is missing in this formula:

$$f'(c) = \frac{f(x) - f(c)}{x - c}$$

1. *Cubic Function Problem I:* Let

$$f(x) = -x^3 + 12x + 25$$

a. Plot the graphs of f and f' (the numerical derivative of f) on the same screen.

b. For what values of x is $f'(x)$ positive? What is the graph of f doing for these values of x?

c. For what values of x is $f(x)$ decreasing? What is true about $f'(x)$ for these values of x?

d. What does the graph of f do at values of x where the f' graph crosses the x-axis?

e. Sketch the graphs of f and f', showing clearly the relationships you have stated in parts a–d.

f. Make a conjecture about what type of function f' is.

2. *Cubic Function Problem II:* Explore the cubic function $g(x) = x^3 - 2x^2 + 2x - 15$ and its numerical derivative, g'. Does the graph have a high point and a low point as is typical for cubic functions? How does the graph of the derivative function reveal this behavior?

3. *Quartic Function Problem I:* Let

$$h(x) = x^4 - 2x^3 - 9x^2 + 20x + 80$$

a. Plot the graphs of h and h' on the same screen.

b. What type of function has a graph that is the same shape as the graph of h'? Make a conjecture about what type of function the derivative of a seventh-degree function would be.

c. What are the zeros of $h'(x)$? That is, what values of x make $h'(x) = 0$?

d. What features does the graph of h have if $h'(x) = 0$? Based on the meaning of derivative, explain why this observation is reasonable.

e. Sketch the graphs of h and h', consistent with your answers to parts a–d.

4. *Quartic Function Problem II:* Explore the function $q(x) = -x^4 + 8x^3 - 24x^2 + 32x - 25$ and its numerical derivative, q'. Does the function have the shape you expect of a fourth-degree function (that is, three high or low points)? How does the derivative graph compare with the derivative graph for the quartic function in Problem 3?

5. *Sinusoid Problem I:* Let $f(x) = 4 + \sin x$.

a. Plot the graphs of f and f' on the same screen. (Be sure your calculator is set in radian mode.)

b. The graph of f is a *sinusoid*. Recall that the amplitude of a sinusoid is the distance from the middle to a high point, and the period is the distance along the x-axis from one high point to the next. What are the amplitude and the period of the graph of f?

c. The graph of f' is also a sinusoid. What are its amplitude and period?

d. Let $g(x) = 3 + \sin x$. Plot the graphs of g and g' on the same screen, along with the graphs of f and f'. How are the graphs of f and g related to each other? How can you explain the relationship between the graphs of f' and g'?

6. *Sinusoid Problem II:* Let $f(x) = 4 + \sin x$, as in Problem 5. Make a conjecture about which function on your grapher the numerical derivative will turn out to be. Give both graphical and numerical evidence to support your conjecture.

For Problems 7–10, sketch the derivative graph on a copy of the given graph.

7.

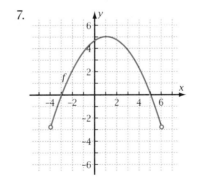

Chapter 3: Derivatives, Antiderivatives, and Indefinite Integrals

8.

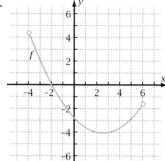

9.

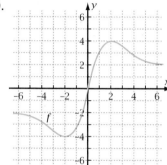

10.

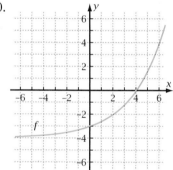

11. *Exponential Function Problem:* Figure 3-3g shows the graphs of the exponential function $f(x) = 2^x$ and its numerical derivative, f'. They are similar in shape. For instance, each has

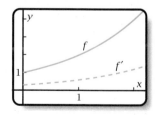

Figure 3-3g

y-values that are four times as large at $x = 2$ as they are at $x = 0$. By experimenting with bases other than 2, see if you can find an exponential function that is identical to its derivative.

12. *How Your Grapher Works Problem:* The text in this section shows that by symmetric difference quotient, if $f(x) = x^3$, then $f'(2) \approx 12.0001$. Find $f'(2)$ by using the numerical derivative feature on your grapher. Set the tolerance for x at 0.01, as you did earlier in the book. Based on the result, state whether your grapher seems to find numerical derivatives by symmetric difference quotient or by some other method.

13. *Tolerance Problem (Epsilon and Delta):* Figure 3-3h shows a square floor tile that is to be manufactured with a certain tolerance for error. Suppose the dimensions are 12 in. × 12 in.

Figure 3-3h

a. If the tolerances on the dimensions are ± 0.01 in., what range of areas can the tile have? Within how many square inches of the ideal area (144 in.^{2}) is the tile?

b. In order to manufacture tiles that are within 0.02 in.^{2} of the ideal area, to what tolerance must the dimensions of the tiles be kept?

c. How does your work in this problem relate to the epsilon-delta definition of limit?

14. *Symmetric Difference Quotient Problem:* In the box are formulas for forward, backward, and symmetric difference quotients. Show algebraically that the symmetric difference quotient is always the average of the forward and backward difference quotients.

Formulas: Difference Quotients

Forward difference quotient =
$$\frac{f(x + h) - f(x)}{h}$$
Backward difference quotient =
$$\frac{f(x) - f(x - h)}{h}$$
Symmetric difference quotient =
$$\frac{f(x + h) - f(x - h)}{2h}$$

The tolerance, h, is positive in each case.

---

15. *Difference Quotient Accuracy Problem:* Figure 3-3i shows the graph of
$$f(x) = x^3 - x + 1$$

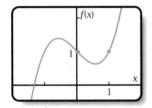

Figure 3-3i

a. By the method of Section 3-2, find the exact value of $f'(1)$.

b. Using a tolerance of $h = 0.1$, find the forward, backward, and symmetric difference quotients for $f'(1)$. Explain why the symmetric difference quotient is so much closer to the actual derivative than either the forward or backward difference quotient.

c. Find the exact value of $f'(0)$.

d. Using a tolerance of $h = 0.1$, find the forward, backward, and symmetric difference quotients for $f'(0)$. How do you explain the fact that all three are equal, and thus that the symmetric difference quotient is not more accurate than the other two?

16. *Numerical Derivative Error Problem:* Figure 3-3j shows the graph of
$$f(x) = (x - 2)^{2/3} + x - 1$$

Use various tolerances, h, to explore the forward, backward, and symmetric difference quotients at $x = 2$. Do all three values seem to be approaching the same number as h approaches zero? What do you conclude about $f'(2)$? Does the numerical derivative function on your grapher give a value for $f'(2)$?

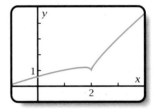

Figure 3-3j

17. *Journal Problem:* Update your journal with things you've learned. Include such things as

- The one most important thing you have learned since your last journal entry
- What you now better understand about the derivative of a function
- How the grapher calculates numerical derivatives
- How the grapher could give a wrong answer for a numerical derivative
- Any conjectures you have made concerning derivatives of various types of functions
- Anything you're still unsure of concerning derivatives and want to ask about during your next class

3-4 Derivative of the Power Function and Another Definition of Derivative

In Section 3-2, you learned a definition of derivative. Another form of this definition lets you find an equation that gives values of the derivative without having to resort to the definition each time. In this section you will find the derivative of a **power function**, such as

$$f(x) = x^5$$

or a **linear combination** of power functions, such as

$$g(x) = 7x^{4/5} - 11x^{-2} + 13$$

In each term the exponent is a **constant**. (You will study algebraic derivatives of exponential functions with variable exponents in Section 3-9.) If each exponent is a *nonnegative integer* constant, the function is called a **polynomial function**.

OBJECTIVE Given a power function, $f(x) = x^n$, where n stands for a constant, or given a linear combination of power functions, find an equation expressing $f'(x)$ in terms of x.

The derivative of a function at a point $x = c$ is

$$f'(c) = \lim_{x \to c} \frac{f(x) - f(c)}{x - c}$$

This form of the definition is easy to remember because it ties in with the slope formula for a line you will recall from algebra, namely,

$$\text{slope} = \frac{\text{rise}}{\text{run}}$$

It also calls attention to the fact that you are finding the derivative at the one fixed point where $x = c$ (Figure 3-4a).

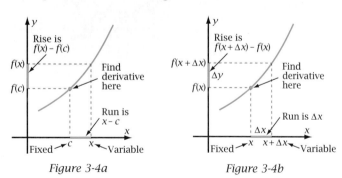

Figure 3-4a　　　　　*Figure 3-4b*

Another form of the definition leads more directly to an equation for a derivative. As shown in Figure 3-4b, c is replaced by x and x is replaced by $x + \Delta x$. The change in x thus takes the simpler form Δx (pronounced "delta x," meaning "difference between x-values"). The rise, Δy, can be written $\Delta y = f(x + \Delta x) - f(x)$. As x approaches c in Figure 3-4a, Δx approaches zero in Figure 3-4b. The definition of derivative reduces to that shown in the box.

DEFINITION: *Derivative as a Function (Δx or h form)*

$$f'(x) = \lim_{\Delta x \to 0} \frac{\Delta y}{\Delta x} = \lim_{\Delta x \to 0} \frac{f(x + \Delta x) - f(x)}{\Delta x} = \lim_{h \to 0} \frac{f(x + h) - f(x)}{h}$$

Note that the definition of derivative uses Δ, the uppercase form of the Greek letter delta, and that the definition of limit uses δ, the lowercase form of delta. Sometimes the single letter h (for *horizontal*) is used in place of Δx to make the algebra easier to write.

▶ **EXAMPLE 1**   Given $f(x) = x^5$, use the definition of derivative to find an equation for $f'(x)$.

Solution
$$f'(x) = \lim_{h \to 0} \frac{(x + h)^5 - x^5}{h} \qquad \text{By the definition of derivative.}$$

$$= \lim_{h \to 0} \frac{(x^5 + 5x^4 h + 10x^3 h^2 + 10x^2 h^3 + 5xh^4 + h^5) - x^5}{h} \qquad \text{Expand.}$$

$$= \lim_{h \to 0} \frac{5x^4 h + 10x^3 h^2 + 10x^2 h^3 + 5xh^4 + h^5}{h} \qquad \text{Combine like terms.}$$

$$= \lim_{h \to 0}(5x^4 + 10x^3 h + 10x^2 h^2 + 5xh^3 + h^4) \qquad \begin{array}{l}\text{Distribute } 1/h \text{ to each term.} \\ \text{You can cancel because } h \neq 0, \\ \text{by the definition of limit.}\end{array}$$

$$= 5x^4 + 0 + 0 + 0 + 0 = 5x^4 \qquad \text{Use the limit properties.} \quad ◀$$

You should see both the source of the answer and a pattern.

If $f(x) = x^5$, then $f'(x) = 5x^4$.

The answer, $5x^4$, is what remains of the second term in the binomial expansion of $(x + h)^5$.

Source: The answer comes from the *second* term.

$$x^5 + \widetilde{5x^4}h + 10x^3 h^2 + 10x^2 h^3 + 5xh^4 + h^5$$

Patterns: One *less* than the original exponent.
The *original* exponent.

Generalizing these patterns gives a property for finding the derivative of any power function.

PROPERTY: *Derivative of the Power Function*

If $f(x) = x^n$, then $f'(x) = nx^{n-1}$. Restriction: The exponent n is a constant.

Example 1 illustrated the binomial theorem makes this property true for nonnegative integer exponents. In Problem 28 of Problem Set 3-4, you will demonstrate that this property works for powers with constant exponents but not for variable exponents. In Chapter 4, you will prove that it works for negative and fractional exponents. In Section 3-9, you will work with variable exponents.

The process of finding an equation for the derivative of a function is called **differentiation**. The word reflects the fact that $\Delta y/\Delta x$ is a *difference* quotient. The corresponding verb is **differentiate**. It may be easier to remember this property by what it allows you to do.

PROCEDURE: Differentiating the Power Function

To differentiate the power function, $f(x) = x^n$, multiply by the original exponent, n, then reduce the exponent by 1 to get the new exponent.

To find a formula for differentiating a linear combination of powers, you must be able to differentiate a sum of two functions, and a constant times a function. An algebraic proof for a sum property is shown here. You will prove a property about the derivative of a constant times a function in Problem 35 of Problem Set 3-4.

▶ **THEOREM** If $f(x) = g(x) + h(x)$, where g and h are differentiable functions, then $f'(x) = g'(x) + h'(x)$.

Algebraic Proof By the definition of derivative,

$$f'(x) = \lim_{\Delta x \to 0} \frac{f(x + \Delta x) - f(x)}{\Delta x} \qquad \text{Definition of derivative.}$$

$$= \lim_{\Delta x \to 0} \frac{[g(x + \Delta x) + h(x + \Delta x)] - [g(x) + h(x)]}{\Delta x} \qquad \text{Substitute for } f(x).$$

$$= \lim_{\Delta x \to 0} \frac{[g(x + \Delta x) - g(x)] + [h(x + \Delta x) - h(x)]}{\Delta x} \qquad \begin{array}{l}\text{Associate the } g\text{'s} \\ \text{and the } h\text{'s.}\end{array}$$

$$= \lim_{\Delta x \to 0} \left[\frac{g(x + \Delta x) - g(x)}{\Delta x} + \frac{h(x + \Delta x) - h(x)}{\Delta x} \right] \qquad \text{Fraction addition.}$$

$$= \lim_{\Delta x \to 0} \frac{g(x + \Delta x) - g(x)}{\Delta x} + \lim_{\Delta x \to 0} \frac{h(x + \Delta x) - h(x)}{\Delta x} \qquad \begin{array}{l}\text{Limit of the sum of} \\ \text{two functions.}\end{array}$$

$$= g'(x) + h'(x) \qquad \begin{array}{l}\text{Definition of derivative} \\ \text{(backward).}\end{array} \quad ◀$$

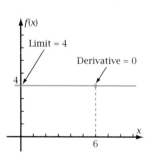

Figure 3-4c

The derivative of a constant function is zero. Figure 3-4c shows the function

$$f(x) = 4$$

As you can see, the function value is always 4. It doesn't change, so its rate of change is zero. Don't confuse the derivative of a constant function with the limit of a constant function. The limit of $f(x)$ as x approaches 6, for instance, is still 4, not zero.

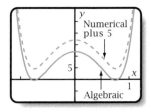

> **PROPERTIES:** *Differentiation*

Derivative of a Sum of Two Functions: If $f(x) = g(x) + h(x)$, where g and h are differentiable functions of x, then $f'(x) = g'(x) + h'(x)$.

Verbally: The derivative of a sum equals the sum of the derivatives. Differentiation distributes over addition.

Derivative of a Constant Times a Function: If $f(x) = kg(x)$, where g is a differentiable function of x, then $f'(x) = kg'(x)$, provided k is a constant.

Verbally: The derivative of a constant times a function equals the constant times the derivative of the function.

Derivative of a Constant Function: If $f(x) = C$, where C stands for a constant, then $f'(x) = 0$ for all values of x.

Verbally: Constants don't change, so their rate of change is zero.

Combining these properties with the derivative of a power function allows you to differentiate any linear combination of power functions in one step!

▶ **EXAMPLE 2**   If $f(x) = 5x^7 - 11x^3 + 12x - 47$, find $f'(x)$.

Solution $f'(x) = 5(7x^6) - 11(3x^2) + 12(1) - 0 = 35x^6 - 33x^2 + 12$ ◀

The answer to Example 2 is called the **algebraic derivative**, which distinguishes it from the numerical derivative found by difference quotients. The algebraic derivative gives the exact values. The values by numerical derivative are only approximate.

▶ **EXAMPLE 3**   Check the answer to Example 2 by graphing both the algebraic derivative and the numerical derivative on the same screen.

Solution   Enter: $y_1 = 5x^7 - 11x^3 + 12x - 47$
$$y_2 = 35x^6 - 33x^2 + 12$$
$$y_3 = \text{numerical derivative of } y_1$$

Plot only y_2 and y_3. One graph appears on the screen. By using TRACE, you can find small differences between y_2 and y_3. Another option is to separate the graphs by adding a constant to y_3, as shown below.

$$y_3 = \text{nDeriv}(y_1, x, x) + 5$$

Figure 3-4d shows the two graphs, using a window with $[-1.2, 1.2]$ for x and $[-10, 30]$ for y. The check is performed by observing that the two graphs have the same shape. ◀

Figure 3-4d

Other symbols are widely used for the derivative, and you should become familiar with them. Each has advantages and disadvantages, depending on where it is used.

TERMINOLOGY: *dy/dx and y'*

If $y = f(x)$, then instead of writing $f'(x)$ you can write any of the following.

y', read "y prime" (a short form of $f'(x)$)

$\dfrac{dy}{dx}$, read "dy, dx" (a single symbol, not a fraction)

$\dfrac{d}{dx}(y)$, read "d, dx, of y" (an operation done on y)

The symbol dy/dx comes from the difference quotient $\Delta y / \Delta x$. It means that the limit is to be taken as both Δy and Δx go to zero. German mathematician Gottfried Leibniz (1646–1716) started using this terminology in about 1675. Later in this book you will learn that dy and dx are called *differentials.* For the time being, regard dy/dx as a single symbol that cannot be taken apart—avoid saying "dy over dx."

You can write the expression $\dfrac{dy}{dx}$ as $\dfrac{d}{dx}(y)$ where $\dfrac{d}{dx}$ is the **operator** that tells you to take the derivative with respect to x, of y. Thus, $\dfrac{d}{dx}$ is useful if you want to show the expression for the function rather than just y. For example,

$$\frac{d}{dx}(x^5) = 5x^4$$

▶ **EXAMPLE 4**   If $y = 7x^{-4/5}$, write an equation for the derivative function, dy/dx. Assume the power rule works for negative and rational exponents.

Solution   Use the power rule.

$$\frac{dy}{dx} = 7\left(-\frac{4}{5}x^{-9/5}\right) = -5.6x^{-9/5}$$ Subtract 1 from $-4/5$ to get $-9/5$.   ◀

▶ **EXAMPLE 5**   If $y = 7^5$, find y'.

Solution $y' = 0$ Because 7^5 is a constant and the derivative of a constant is zero.   ◀

It is possible to sketch the graph of a derivative function just by looking at the graph of the function. For example, at a smooth high or low point in the function graph, the derivative will be zero. If the graph is going up, the derivative will be positive, and so forth.

Problem Set 3-4

Q1. Expand: $(3x - 4)^2$

Q2. Expand: $(a + b)^3$

Q3. Write the definition of derivative (either form).

Q4. Write the formula for a symmetric difference quotient.

Q5. Find: $\lim_{x \to -3}(x - 5)/(x + 3)$

Q6. Write as the log of a power: $3 \log 7$.

Q7. Write the exact value of $\tan(\pi/3)$.

Q8. Name the theorem whose conclusion is $r^2 = s^2 + t^2$.

Q9. Evaluate: $100^{1/2}$

Q10. For function $f(x)$, shown in Figure 3-4e, the derivative, $f'(x)$, has a minimum at

    A. $x = 1$ B. $x = 2$ C. $x = 3$

    D. $x = 4$ E. $x = 5$

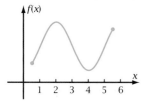

Figure 3-4e

For Problems 1–18, find the equation of the derivative function.

1. $f(x) = 5x^4$

2. $y = 11x^8$

3. $v = 0.007t^{-83}$

4. $v(x) = \dfrac{x^{-9}}{18}$

5. $M(x) = 1215$

6. $f(x) = 4.77^{23}$

7. $y = 0.3x^2 - 8x + 4$

8. $r = 0.2x^2 + 6x - 1$

9. $\dfrac{d}{dx}(13 - x)$

10. $f(x) = 4.5x^2 - x$

11. $y = x^{2.3} + 5x^{-2} - 100x + 4$

12. $\dfrac{d}{dx}(x^{2/5} - 4x^2 - 3x^{-1} + 14)$

13. $v = (3x - 4)^2$ (Expand first.)

14. $u = (5x - 7)^2$ (Expand first.)

15. $f(x) = (2x + 5)^3$ (Be clever!)

16. $f(x) = (4x - 1)^3$ (Be clever!)

17. $P(x) = \dfrac{x^2}{2} - x + 4$

18. $Q(x) = \dfrac{x^3}{3} + \dfrac{x^2}{2} - x + 1$

For Problems 19–22, use the definition of derivative in the $f(x + h)$ form to find the equation of the derivative function. In each case, show that your answer is consistent with the answer you would get if you used the formula for the derivative of the power function.

19. $f(x) = 7x^4$

20. $g(x) = 5x^3$

21. $v(t) = 10t^2 - 5t + 7$

22. $s(t) = t^4 - 6t^2 + 3.7$

23. *Misconception Problem:* Mae needs to find $f'(3)$, where $f(x) = x^4$. She substitutes 3 for x, gets $f(3) = 81$, differentiates 81, and gets zero for the answer. Explain why she also gets zero for her grade.

24. *Higher Math Problem:* Chuck stands atop a cliff (Figure 3-4f). He throws his math book into the air with an initial upward velocity of 20 m/s. As a result, its height, $h(x)$, in meters above where he threw it after time x, in seconds, is

$$h(x) = -5x^2 + 20x$$

  a. The upward velocity of the book is the derivative of height with respect to time. Write an equation for the upward velocity.

  b. How fast was the book going at time $x = 3$? Was it going up or going down? Explain.

c. At time $x = 3$, was the book above or below where Chuck threw it? How far?

d. At what time was the book at its highest point? How do you know?

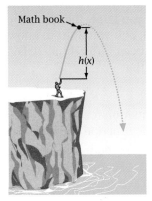

Figure 3-4f

For Problems 25 and 26, copy or sketch the graph of the function. On the same set of axes, sketch a reasonable graph of the derivative function.

25.

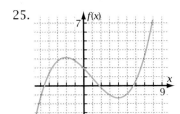

26.

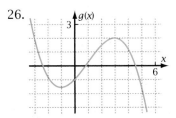

27. *Numerical Versus Exact Derivative Problem:* Figure 3-4g shows the graph of

$$f(x) = 0.4x^3 - 7x + 4$$

along with two graphs of $f'(x)$. One derivative graph is found numerically as in Section 3-3. The other derivative graph is found algebraically using this section's technique.

a. Plot these graphs on your grapher.

b. Explain which graph is f and which is f'.

c. Why do there appear to be only two graphs on the screen, not three?

d. Trace to $x = 3$. Write the y-value you get for each of the three graphs. How closely does the numerical derivative fit the algebraic derivative at $x = 3$?

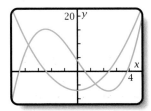

Figure 3-4g

28. *Power Formula for Various Types of Exponents:* The derivation of the power formula,

$$f(x) = x^n \text{ implies } f'(x) = nx^{n-1}$$

was based on Example 1 of this section, where a pattern was found. In Example 1, the binomial formula was used to expand $(x + h)^5$ as a sum. Unfortunately, such sums have an infinite number of terms unless the exponent is a nonnegative integer. Your objective in this problem is to show by example that the power formula for derivatives works if the exponent is a negative or fractional constant but clearly does not work if the exponent is a variable. To accomplish this objective, predict the derivative for each function given below, assuming that the formula works for the type of exponent shown. Then show that the answer is correct or incorrect by plotting three graphs on the same screen.

y_1 = the given function

y_2 = your answer from using the derivative of a power formula

y_3 = the numerical derivative of y_1

a. $g(x) = x^{-1}$

b. $h(x) = x^{1/2}$

c. $e(x) = 2^x$

For Problems 29–32, explain whether $f(x)$ is increasing or decreasing at $x = c$, and at what rate.

29. $f(x) = x^{1/2} + 2x - 13, \quad c = 4$

30. $f(x) = x^{-2} - 3x + 11, \quad c = 1$

31. $f(x) = x^{1.5} - 6x + 30, \quad c = 9$

32. $f(x) = -3\sqrt{x} + x + 1, \quad c = 2$

For Problems 33 and 34, plot the graphs of f and f' on the same screen. Show that each place where the f' graph crosses the x-axis corresponds to a high or low point on the f graph.

33. $f(x) = \dfrac{x^3}{3} - x^2 - 3x + 5$

34. $f(x) = \dfrac{x^3}{3} - 2x^2 + 3x + 9$

35. *Formula Proof Problem I:* Use the definition of derivative to prove that if g is a differentiable function of x, and $f(x) = k \cdot g(x)$, then $f'(x) = k \cdot g'(x)$. Explain why this property can be stated verbally as, "Dilating a function vertically by a factor of k dilates the derivative function vertically by a factor of k."

36. *Formula Proof Problem II:* Given $f(x) = x^5$, derive the formula for $f'(c)$ directly from the $x = c$ form of the definition of derivative,

$$f'(c) = \lim_{x \to c} \frac{f(x) - f(c)}{x - c}$$

It will be helpful to remember how you factor differences of like, odd powers.

$$a^5 - b^5 = (a - b)(a^4 + a^3b + a^2b^2 + ab^3 + b^4)$$

37. *Derivative of a Power Formula:* Show that the formula for the derivative of a power, namely, $f'(x) = nx^{n-1}$, comes from the second term in the binomial series expansion of $(x + \Delta x)^n$.

38. *Derivative of a Sum of n Functions Problem:* Prove by mathematical induction that the derivative of a sum of n functions is equal to the sum of the derivatives.

39. *Introduction to Antiderivatives and Indefinite Integrals:* Suppose you know the answer to a derivative problem, and you want to know just what function has been differentiated. For example, suppose that

$$f'(x) = 3x^2 - 10x + 5$$

a. Figure out an equation for a function f for which $f'(x)$ is the function shown above.

b. If $f(x)$ is the answer you found in part a, explain why $g(x) = f(x) + 13$ is also an answer to part a.

c. The functions f and g in parts a and b are called **antiderivatives** of f'. Why do you suppose they are called antiderivatives?

d. If $g(x) = f(x) + C$, where C is any constant, explain why $g'(x) = f'(x)$. The antiderivative function g is also called the **indefinite integral** of $f'(x)$, the fourth of the four concepts of calculus. In Chapter 5, you will learn the reason for the word *integral*. Why do you suppose the word *indefinite* is used?

40. *Indefinite Integral Problem:* Given the derivative function $f'(x)$, write the indefinite integral function as explained in Problem 39. Find the value of the constant C for the given value of $f(x)$.

a. $f'(x) = 5x^4, \quad f(2) = 23$

b. $f'(x) = 0.12x^2, \quad f(1) = 500$

c. $f'(x) = x^3, \quad f(5) = 2$

3-5 Displacement, Velocity, and Acceleration

You have found the velocity or acceleration of a moving object several times in this course. Now that you know an algebraic method to find derivatives, you can find an *equation* for the velocity or acceleration if you know an equation for the position of the object.

OBJECTIVE   Given an equation for the displacement of a moving object, find an equation for its velocity and an equation for its acceleration, and use the equations to analyze the motion.

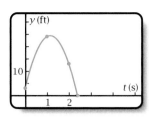

Figure 3-5a

Suppose a football is punted into the air. As it rises and falls, its **displacement** (directed distance) from the ground is a function of the number of seconds since it was punted. Suppose that by experiment it is found that

$$y = -16t^2 + 37t + 3$$

where y is the football's displacement in feet and t is the number of seconds since it was punted. Figure 3-5a shows the graph of this function.

The **velocity** of the ball gives its speed and the direction in which it's going. Because velocity is an instantaneous rate of change (change at any instant), it is a derivative. For the above equation,

$$\text{velocity} = \frac{dy}{dt} = -32t + 37 \qquad \text{Differentiate } y = -16t^2 + 37t + 3.$$

To find the velocity at any particular time, you simply substitute for t. Here are some examples.

$$t = 1: \quad \frac{dy}{dt} = -32(1) + 37 = 5 \text{ ft/s}$$

$$t = 2: \quad \frac{dy}{dt} = -32(2) + 37 = -27 \text{ ft/s}$$

The dy/dt symbol for the derivative helps you remember the units of velocity. Because y is in feet and t is in seconds, dy/dt is in feet/second.

Speed is the absolute value of velocity. Speed tells how fast an object is going, without regard to its direction. When $t = 1$, the ball is traveling 5 ft/s. When $t = 2$, it is traveling faster, at 27 ft/s. The negative velocity when $t = 2$ indicates that the displacement is decreasing, which means that the ball is coming back down. Although the ball is traveling faster when $t = 2$, its velocity is less because -27 is less than 5.

Note that the velocity changes from $t = 1$ to $t = 2$. The instantaneous rate of change in velocity is called **acceleration**. It is the derivative of the velocity. Using v for velocity, $v = -32t + 37$. Thus,

$$\text{acceleration} = \frac{dv}{dt} = -32 \qquad \text{Differentiate } v = -32t + 37.$$

The dv/dt symbol for the derivative gives you the units of acceleration. Because v is in feet per second and t is in seconds, dv/dt is in (feet/second)/second. This quantity is read, "feet per second, per second," with a pause at the comma, and written "ft/s^2." You can also say "feet per second squared."

Negative acceleration means that the velocity is decreasing. Be sure you know how to interpret this idea. In this instance, v decreases from 5 ft/s when $t = 1$ to -27 ft/s when $t = 2$, even though the ball is traveling faster at $t = 2$. Note that the acceleration is *constant*, -32 (ft/s)/s, for an object acted on only by gravity.

Also, at $t = 2$, the speed is increasing even though the velocity is decreasing. The ball is going faster in the negative direction. To tell quickly whether an object is speeding up or slowing down, compare the signs of the velocity and acceleration, as shown in the box.

TECHNIQUE: Speeding Up or Slowing Down

- If velocity and acceleration have the *same* sign, the object is *speeding up.*

- If velocity and acceleration have *opposite* signs, the object is *slowing down.*

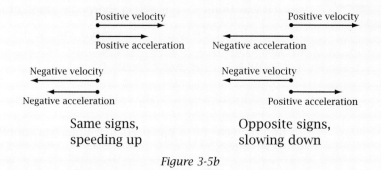

Same signs,
speeding up

Opposite signs,
slowing down

Figure 3-5b

The acceleration of a moving object is the derivative of a derivative. The term **second derivative** is used for the derivative of a derivative. The symbol for the second derivative of y with respect to t is

$$\frac{d^2y}{dt^2}$$

which is read "d squared y, dt squared." The symbol comes from performing d/dt on dy/dt, then using "algebra" on the symbols.

$$\frac{d}{dt}\left(\frac{dy}{dt}\right) \quad \text{becomes} \quad \frac{d^2y}{dt^2}$$

If $x = f(t)$, then the symbol $f''(t)$, read "f double prime of t," is used for the second derivative. Sometimes y'' is used if it is clear what the independent variable is.

▶ **EXAMPLE 1**   An object moves in the x-direction in such a way that its displacement from the y-axis is

$$x = 3t^3 - 30t^2 + 64t + 57, \qquad \text{for } t \geq 0$$

where x is in miles and t is in hours. With your grapher in parametric mode, plot the given equation as x_1 and some convenient constant such as 1 for y_1. Use path style and a t-range from 0 to 10 with an increment of 0.1. Use a window with an x-range of [0, 200] and a y-range of [0, 2]. Write a paragraph describing the motion. Include the approximate times and places at which the object reverses direction, and the time intervals during which the object is traveling to the right and to the left. A sketch may help.

Solution   *The object starts out at x = 57 mi when t = 0 h. It goes right, slowing down until t is about 1.3 h, at which time it stops where x is about 96 mi, as shown in Figure 3-5c. Then it turns around and starts off going to the left, speeding up for a while, then slowing down and stopping when t is about 5.3 h, at which time x is about 0.1 mi. After that it starts off to the right again and continues speeding up. The final graph looks like a straight line (Figure 3-5d).*

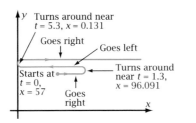

Figure 3-5c

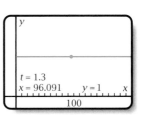

Figure 3-5d

In Example 2, you will use y versus t graphs to analyze the same motion.

▶ **EXAMPLE 2**   The object in Example 1 moves in the x-direction with displacement from the y-axis given by

$$x = 3t^3 - 30t^2 + 64t + 57, \qquad \text{for } t \geq 0$$

where x is in miles and t is in hours.

a. Find equations for its velocity and acceleration.

b. Find the velocity and acceleration at $t = 2$, $t = 4$, and $t = 6$. At each time, state
   • Whether x is increasing or decreasing, and at what rate
   • Whether the object is speeding up or slowing down, and how you decided

c. At what times in the interval $[0, 8]$ is x at a maximum? Is x ever negative if t is in this interval?

Solution   a. Velocity is the first derivative of position. Acceleration is the second derivative of position,

$$v = \frac{dx}{dt} = 9t^2 - 60t + 64 \qquad \text{Velocity equation.}$$

$$a = \frac{dv}{dt} = \frac{d^2x}{dt^2} = 18t - 60 \qquad \text{Acceleration equation.}$$

b. Enter x as y_1, v as y_2, and a as y_3. Make a table.

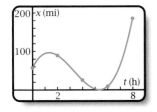

Figure 3-5e

t	$v(t)$	$a(t)$	x is...	Speeding/Slowing?
2	−20	−24	Decreasing at 20 mi/h	Speeding up; a and v are same sign.
4	−32	12	Decreasing at 32 mi/h	Slowing down; a and v are opposite.
6	28	48	Increasing at 28 mi/h	Speeding up; a and v are same sign.

c. Figure 3-5e shows $y_1 = x(t)$ and Figure 3-5f shows $y_2 = v(t)$. Note that x and v are plotted vertically because they are dependent variables. The maximum or minimum value of x occurs when the object stops going one way and starts going the other, or at an endpoint of the given interval. To find out when the object stops, set $v = 0$ and solve for t.

$$9t^2 - 60t + 64 = 0 \qquad \text{Set the velocity equal to zero.}$$

$$t = 5.3333... \quad \text{or} \quad t = 1.3333... \qquad \begin{array}{l}\text{Solve numerically or by} \\ \text{quadratic formula.}\end{array}$$

A **relative maximum** occurs when $t \approx 1.33$.   Velocity goes from positive to negative.

The **absolute maximum** occurs when $t = 8$.   An endpoint maximum.

The minimum occurs when $t = 5.3333\ldots$, where v changes from negative to positive. Because $x(5.3333\ldots) = 0.1111\ldots$, which is greater than zero, the value of x is never negative for t in the interval $[0, 8]$. ◀

▶ **EXAMPLE 3**   The acceleration of a moving object is given by $a(t) = 12t - 5$, where $a(t)$ is in centimeters per minute and t is in minutes. If the velocity at time $t = 1$ is 8 cm/min, find an equation for $v(t)$, the velocity as a function of t. Use the equation to find the velocity at the instant $t = 3$.

Solution   Because $a(t) = v'(t)$, v is an *antiderivative* of a. You can write $v'(t) = 12t - 5$. The $12t$ comes from differentiating $6t^2$ and the -5 comes from differentiating $-5t$. So $v(t)$ could equal $6t^2 - 5t$. But $v(t)$ could also equal $6t^2 - 5t + 17$ or $6t^2 - 5t - 31$, or in general

$v(t) = 6t^2 - 5t + C$	C stands for a constant.
$8 = 6(1)^2 - 5(1) + C$	Substitute 1 for t and 8 for $v(1)$.
$C = 7$	Solve for C.
$v(t) = 6t^2 - 5t + 7$	Substitute $C = 7$ into the equation for $v(t)$.
$v(3) = 6(9) - 5(3) + 7 = 46$ cm/min	Substitute 3 for t. ◀

A hot-air balloon accelerating upward

Note that in Example 3 you found a function whose derivative was given. Such a function is called an **antiderivative**. An antiderivative is also called an **indefinite integral**. The word *indefinite* is used because an antiderivative always has an unspecified constant, C, added. This constant is called the **constant of integration**. The reason *integral* is used will become clear when you learn the fundamental theorem of calculus in Chapter 5. The velocity of a moving object is the antiderivative of the acceleration, and the displacement of the object is the antiderivative of the velocity.

DEFINITION: Antiderivative, or Indefinite Integral

Function g is an antiderivative (or indefinite integral) of function f if and only if $g'(x) = f(x)$.

PROPERTIES: Velocity, Speed, and Acceleration

If x is the displacement of a moving object from a fixed plane (such as the ground), and t is time, then

Velocity: $v = x' = \dfrac{dx}{dt}$

Acceleration: $a = v' = \dfrac{dv}{dt} = x'' = \dfrac{d^2x}{dt^2}$

Speed: $|v|$

Problem Set 3-5

Quick Review

Q1. Quick! For what values of t does $9t^2 - 30t + 64$ equal zero? (Be clever!)

Q2. Find an equation for dy/dx: $y = 5x^2$.

Q3. Find an equation for y': $y = 17x^{-3}$.

Q4. Find an equation for $f'(x)$: $f(x) = x^{1.7}$.

Q5. Find: $(d/dx)(3x + 5)$

Q6. Find $f(3)$: $f(x) = 5x^2$

Q7. Find $f'(3)$: $f(x) = 5x^2$

Q8. Find: $\lim_{x \to 3} 5x^2$

Q9. In the definition of limit, which goes with $f(x)$, ϵ or δ?

Q10. The concept of calculus used to find the product of x and y if y varies with x is
    A. Limit
    B. Derivative
    C. Definite integral
    D. Indefinite integral
    E. Continuity

For Problems 1 and 2, find equations for the velocity, v, and the acceleration, a, of a moving object if y is its displacement.

1. $y = 5t^4 - 3t^{2.4} + 7t$

2. $y = 0.3t^{-4} - 5t$

For Problems 3 and 4, an object is moving in the x-direction with displacement, $x(t)$, in feet from the y-axis, where t is in seconds. With your grapher in parametric mode and path style, plot the path of the object. Write a paragraph describing the motion. Include the approximate times and places at which the object reverses direction, and the time intervals during which the object is traveling to the right and to the left. A sketch may help.

3. $x = -t^3 + 13t^2 - 35t + 27, \quad t \geq 0$

4. $x = t^4 - 11t^3 + 38t^2 - 48t + 50, \quad t \geq 0$

5. An object moves as in Problem 3, with displacement, x, given by

$$x = -t^3 + 13t^2 - 35t + 27$$

where x is in feet and t is in seconds.

a. Find equations for the velocity and the acceleration.

b. Find the velocity and acceleration at $t = 1$, $t = 6$, and $t = 8$. At each time, state
    • Whether x is increasing or decreasing, and at what rate
    • Whether the object is speeding up or slowing down, and how you decided

c. At what times in the interval $[0, 9]$ is x a relative maximum? Is x ever negative if t is in this interval?

6. A particle (a small object) moves as in Problem 4, with displacement, x, given by

$$x = t^4 - 11t^3 + 38t^2 - 48t + 50$$

where x is in meters and t is in minutes.

a. Write equations for the velocity, v, and acceleration, a.

b. Is the object speeding up or slowing down at $t = 1$? At $t = 3$? At $t = 5$? How can you tell?

c. Find all values of t for which $v = 0$.

d. Plot the graphs of x and v on the same screen, then sketch the results. Explain what is true about the displacement whenever $v = 0$.

e. Plot the graph of a on the same screen you used in part d. At each time the acceleration is zero, what is true about v? About x?

7. *Car Problem:* Calvin's car runs out of gas as it is going up a hill. The car rolls to a stop, then starts rolling backward. As it rolls, its displacement, d, in feet from the bottom of the hill, at t, in seconds since Calvin's car ran out of gas, is given by

$$d(t) = 99 + 30t - t^2$$

a. Plot graphs of d and d' on the same screen. Use a window large enough to include the point where the graph of d crosses the positive t-axis. Sketch the result.

b. For what range of times is the velocity positive? How do you interpret this answer in terms of Calvin's motion on the hill?

c. At what time did Calvin's car stop rolling up and start rolling back? How far was it from the bottom of the hill at this time?

d. If Calvin doesn't put on the brakes, when will he be back at the bottom of the hill?

e. How far was Calvin from the bottom of the hill when his car ran out of gas?

8. *Sky Diver's Acceleration Problem:* Phoebe jumps from an airplane. While she free-falls, her downward velocity, v, in feet per second, as a function of t, in seconds since the jump, is

$$v(t) = 251(1 - 0.88^t)$$

a. Plot the velocity, v, and acceleration, a, on the same screen. Use an x-window (actually a t-window) of 0 s to 30 s, and use your grapher's numerical derivative function. Sketch the results.

b. What is Phoebe's acceleration when she first jumps? Why do you suppose the acceleration decreases as she moves faster and faster?

c. What does the limit of $v(t)$ seem to be as t approaches infinity? (This limit is called the *terminal velocity.*)

d. How many seconds does it take Phoebe to reach 90% of her terminal velocity? Explain how you found your answer.

e. When Phoebe reaches 90% of her terminal velocity, is her acceleration equal to 10% of its initial value? Justify your answer.

9. *Velocity from Displacement Problem:* If you place-kick a football, its displacement above the ground, d, in meters, is given by

$$d(t) = 18t - 4.9t^2$$

where t is time in seconds since it was punted (Figure 3-5g). (The coefficient 18 is the initial upward velocity in meters per second.)

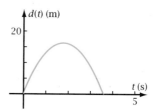

Figure 3-5g

a. Find the algebraic derivative, $d'(t)$. Use your answer to find $d'(1)$ and $d'(3)$. What name from physics is given to d'?

b. At these times, is the football going up or down? How fast? How does the graph in Figure 3-5g confirm these conclusions?

c. Use the d' equation to find the velocity at time $t = 4$. Explain why the answer has a meaning in the mathematical world but not in the real world.

10. *Clock Pendulum Acceleration from Velocity Problem:* Figure 3-5h shows the graph of the velocity function, $y = v(t)$, in feet per second, of the pendulum on a grandfather clock as a function of time t, in seconds.

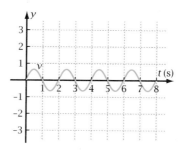

Figure 3-5h

a. On a copy of Figure 3-5h, sketch the acceleration graph.

b. The pendulum is moving fastest when it is at the bottom of its swing. Its velocity is zero at the instant it is at either end of its swing. What is the acceleration at the bottom of the swing? What is the acceleration at either end of its swing?

11. *Acceleration Antiderivative from Graph Problem:* Figure 3-5i shows the graph of the acceleration function for a moving object, $y = a(t)$, in (cm/s)/s. Because acceleration is the derivative of velocity, velocity is the *antiderivative* (or *indefinite integral*) of acceleration. On a copy of the figure, sketch the velocity graph, $y = v(t)$, given that $v(1) = 2$ cm/s. In particular, show the behavior of the velocity function at points where the acceleration is zero.

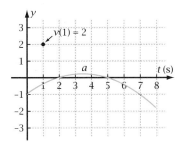

Figure 3-5i

12. *Displacement from Velocity Indefinite Integral Problem:* Recall that the velocity of a moving object is the derivative of the displacement. So the displacement is the *antiderivative,* or *indefinite integral,* of the velocity. Suppose a sports car accelerates in such a way that its velocity as a function of time is given by

$$v(t) = 15t^{0.6}$$

where $v(t)$ is in feet per second and t is in seconds. Find an equation for $x(t)$, the displacement of the car from a fixed point. Assume that the car is 50 ft from the fixed point at time $t = 0$. Use the equation for $x(t)$ to find the position of the car at $t = 10$ s. How far does it travel between $t = 0$ and $t = 10$ s?

13. *Average Versus Instantaneous Velocity Problem:* Suppose that $f(t)$ is the displacement of a moving object from a fixed plane. Explain why the difference quotient,

$$\frac{f(b) - f(a)}{b - a}$$

is the average velocity from $t = a$ to $t = b$. How is the instantaneous velocity at $t = a$ related to this average velocity? The definition of derivative should help you answer this question.

14. Figure 3-5j shows the displacement, y, in centimeters from a tabletop, for a mass bouncing up and down on a spring. Time t is in seconds.

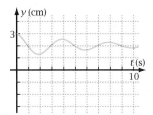

Figure 3-5j

a. On a copy of Figure 3-5j, sketch the velocity graph.

b. For what values of t is y a relative maximum? For what values of t is y a relative minimum?

c. At what times is the velocity a relative maximum? Make some observations about the displacement graph at these times.

d. Just for fun, see if you can duplicate the graph in Figure 3-5j on your grapher. If you can, then check your answer to part a by plotting the numerical derivative.

For Problems 15–18, find the second derivative, $\dfrac{d^2y}{dx^2}$.

15. $y = 5x^3$

16. $y = 7x^4$

17. $y = 9x^2 + x^5$

18. $y = 10x^2 - 15x + 42$

19. *Compound Interest Second Derivative Problem:* If you invest $1000 in an IRA (individual retirement account) that pays 10% APR (annual percentage rate), then the amount, $m(t)$, in dollars, you have in the account after time t, in years, is given by the exponential function

$$m(t) = 1000(1.1^t)$$

Figure 3-5k shows the graph of this function. Use the numerical derivative feature of your grapher to find the values of the derivatives $m'(5)$ and $m'(10)$. Explain what these numbers represent, and why $m'(10) > m'(5)$. Use the numerical derivative feature again to find the second derivatives, $m''(5)$ and $m''(10)$. You can do this by storing the numerical derivative function as y_2, then finding the numerical derivative of y_2. What are the units of the second derivative? What real-world quantity does the second derivative represent?

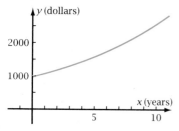

Figure 3-5k

20. *Radioactive Decay Second Derivative Problem:* The isotope nitrogen 17 generated when a nuclear reactor is running decays with a half-life of 7 s. The percentage $p(t)$ of nitrogen 17 remaining after time t, in seconds, after the reactor is shut down is given by the exponential function

$$p(t) = 100(0.5^{t/7})$$

Figure 3-5l shows the graph of function p. Use the numerical derivative feature of your grapher to show that $p'(14)$ is half of $p'(7)$.

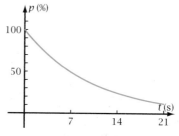

Figure 3-5l

What is the significance of the fact that these derivatives are negative? Find the second derivatives, $p''(7)$ and $p''(14)$, numerically. What rates do these derivatives represent? What are their units?

For Problems 21 and 22, sketch the graph of the derivative function on a copy of the figure. Make a conjecture about an equation for the derivative function.

21.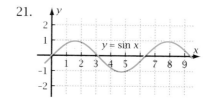

22.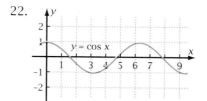

3-6 Introduction to Sine, Cosine, and Composite Functions

You have learned how to write a formula for the derivative of a linear combination of power functions. Now you will do this for the *transcendental* functions sine and cosine.

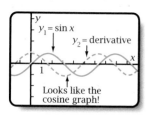

Figure 3-6a

Figure 3-6a shows the graph of $y = \sin x$ and its numerical derivative as they might appear on your grapher. Note that the derivative graph is congruent to the sine graph, just shifted sideways. Each graph is a *sinusoid*. Recall from trigonometry that the cosine function has a graph that looks just like the derivative graph. A conjecture is as follows.

$$\text{If } f(x) = \sin x, \text{ then } f'(x) = \cos x.$$

Figure 3-6b shows similar relationships for the graphs of $g(x) = \cos x$ and its instantaneous rate of change. This time, the pattern seems to be the opposite of $\sin x$, so the conjecture is as follows.

$$\text{If } g(x) = \cos x, \text{ then } g'(x) = -\sin x.$$

These two conjectures turn out to be true, as you will prove in Section 3-8.

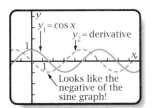

Figure 3-6b

PROPERTIES: Derivatives of Sine and Cosine Functions

$$\frac{d}{dx}(\sin x) = \cos x \qquad \frac{d}{dx}(\cos x) = -\sin x$$

In this section you will try to discover how to find the derivative of a **composite function**, such as

$$f(x) = \sin(x^5) \qquad \text{or} \qquad g(x) = (\cos x)^4$$

where an operation is performed on the answer to another function. The function performed first is called the **inside function** (the fifth-power function in f and the cosine function in g, above). The function performed second, on the answer to the inside function, is called the **outside function** (the sine in f and the fourth power in g, above).

OBJECTIVE   Work with your study group to form conjectures on how to differentiate a composite function.

Problem Set 3-6

1. Plot the graph of $y = \sin x$ on your grapher. Use a window as shown in Figure 3-6a. Sketch the result.

2. You have conjectured that the derivative of $f(x) = \sin x$ is $f'(x) = \cos x$. On the same screen as in Problem 1, plot $y_2 = \cos x$ and $y_3 = $ (numerical derivative of $\sin x$). Use thick style for y_3. Does the result confirm your conjecture or refute it?

3. Figure 3-6c shows $g(x) = \sin 3x$. Make a conjecture about what an equation for $g'(x)$ might be. Enter $g(x)$ as y_1 on your grapher. Then plot $y_2 = $ your conjectured $g'(x)$

and $y_3 = $ the numerical derivative of g. Does the result verify or refute your conjectured $g'(x)$? Sketch the correct graph of g'. If your conjecture was wrong, write the correct equation for $g'(x)$.

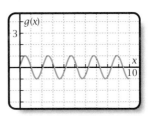

Figure 3-6c

4. Figure 3-6d shows $h(x) = \sin x^2$. Make a conjecture about what an equation for $h'(x)$ might be. Then verify (or refute!) your conjecture by appropriately graphing on your grapher. Sketch the correct h' graph. If your conjecture was wrong, see if you can find a correct equation for $h'(x)$.

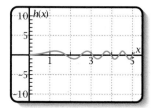

Figure 3-6d

5. You have assumed that the derivative of a power formula works for any constant exponent. Use this fact and the patterns you have observed in Problems 1–4 to make a conjecture about the derivative function for $t(x) = \sin x^{0.7}$. Verify your conjecture by plotting graphs as you did in Problems 3 and 4.

6. Suppose that $f(x) = \sin[g(x)]$ for some differentiable function g. What special name is given to function f in this case? What special name is given to function g? What special name is given to the sine function in this case? Write a statement explaining how you could find an equation for $f'(x)$.

7. For each function below, state which is the inside function and which is the outside one.

a. $f(x) = \sin 3x$

b. $h(x) = \sin^3 x$

c. $g(x) = \sin x^3$

d. $r(x) = 2^{\cos x}$

e. $q(x) = \dfrac{1}{\tan x}$

f. $L(x) = \log(\sec x)$

8. Write a paragraph in your journal explaining what you learned as a result of doing this problem set.

3-7 Derivatives of Composite Functions—The Chain Rule

A function with an equation such as

$$f(x) = \sin(x^2)$$

is called a **composite function**. It is composed of two other functions. The function x^2 in parentheses is called the **inside function**. The sine function outside the parentheses is called the **outside function**. It operates on the answer to the first function. Your purpose in this section is to learn how to find the derivative of a composite function.

OBJECTIVE Given the equation for a composite function, write the equation for its derivative function.

Suppose you set out to differentiate

$$y = \sin(x^2)$$

Because the derivative of sine is cosine, your first thought might be to say $y' = \cos(x^2)$. However, the numerical derivative shown in Figure 3-7a reveals that the amplitude increases. The high points of the derivative graph follow the line $y = 2x$. This happens because the cycles of $y = \sin(x^2)$ occur more frequently as x increases. As you saw in Problem Set 3-6, the derivative is really

$$y' = \cos(x^2) \cdot 2x$$

The anticipated result, $\cos(x^2)$, is multiplied by the derivative of the inside function, $2x$. You can use the **chain rule** to differentiate composite functions.

Verbally, the chain rule says to differentiate the outside function, while the inside function stays the same. Then multiply the result by the derivative of the inside function.

To see why the chain rule works, realize that there are really three functions involved in $f(x) = \sin(x^2)$: the inside function, x^2; the outside function, sine; and the composite function, $\sin(x^2)$. Each one has its own derivative. For simplicity, let $u = x^2$ and let $y = \sin u$. Graphs of the three functions are shown in Figure 3-7b.

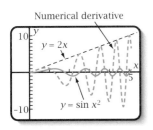

Numerical derivative

Figure 3-7a

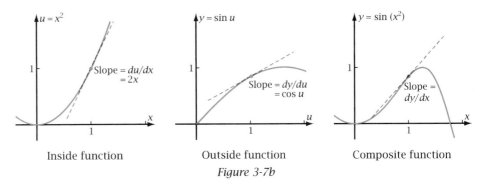

Inside function   Outside function   Composite function

Figure 3-7b

Observe that y depends on u and that u depends on x. The derivative of u with respect to x, du/dx, is $2x$. It describes the slope of the xu-graph (left). Similarly, the derivative of y with respect to u, $\cos u$, describes the slope of the uy-graph (center). (The phrase *with respect to u* means that u is the independent variable.) You seek the slope of the xy-graph, dy/dx (right). It is the limit of $\Delta y / \Delta x$.

Although derivatives are not fractions, you can use the definition of derivative to write them that way.

$$\frac{dy}{dx} = \lim_{\Delta x \to 0} \frac{\Delta y}{\Delta x} \qquad \frac{dy}{du} = \lim_{\Delta u \to 0} \frac{\Delta y}{\Delta u} \qquad \frac{du}{dx} = \lim_{\Delta x \to 0} \frac{\Delta u}{\Delta x}$$

To find the fraction for dy/dx in terms of the fractions for dy/du and du/dx, use some algebra. Multiplying by a clever form of 1, namely $\Delta u/\Delta u$, gives the following.

$$\frac{dy}{dx} = \lim_{\Delta x \to 0} \left(\frac{\Delta y}{\Delta x} \cdot \frac{\Delta u}{\Delta u} \right)$$

$$\frac{dy}{dx} = \lim_{\Delta x \to 0} \left(\frac{\Delta y}{\Delta u} \cdot \frac{\Delta u}{\Delta x} \right)$$

$$\frac{dy}{dx} = \lim_{\Delta x \to 0} \frac{\Delta y}{\Delta u} \cdot \lim_{\Delta x \to 0} \frac{\Delta u}{\Delta x} \qquad \text{Limit of a product of two functions.}$$

If u is a *continuous* function of x, then $\Delta x \to 0$ implies $\Delta u \to 0$. So you can replace $\Delta x \to 0$ with $\Delta u \to 0$.

$$\frac{dy}{dx} = \lim_{\Delta u \to 0} \frac{\Delta y}{\Delta u} \cdot \lim_{\Delta x \to 0} \frac{\Delta u}{\Delta x}$$

The two limits now fit the definition of derivative. Therefore,

$$\frac{dy}{dx} = \frac{dy}{du} \cdot \frac{du}{dx}$$

The name *chain rule* is used because when a composite function has several inside functions, you get a whole "chain" of derivatives multiplied together. Note that Δu is a change in the dependent variable, so it might go to zero somewhere before Δx gets to zero. There is a way to get past this difficulty by using the extended mean value theorem, which you will learn later in your mathematical career.

PROPERTY: The Chain Rule

***dy/dx* form:** If y is a differentiable function of u and u is a differentiable function of x, then the derivative of y with respect to x is given by

$$\frac{dy}{dx} = \frac{dy}{du} \cdot \frac{du}{dx}$$

***f(x)* form:** If $f(x) = g(h(x))$, then $f'(x) = g'(h(x)) \cdot h'(x)$.

Outside function, inside function form: To differentiate a composite function, differentiate the outside function with respect to the inside function (the inside function does not change). Then multiply by the derivative of the inside function with respect to x.

It is easiest to remember the chain rule if you think of it as a procedure, in its outside function, inside function form. This form indicates something to do. It is also easy to remember the dy/dx form: When you "cancel" the du's, you are left with dy/dx.

Use the chain rule to differentiate the functions in Examples 1 and 2.

▶ **EXAMPLE 1** If $f(x) = (x^3 + 7)^2$, find $f'(x)$.

Solution $f(x) = (x^3 + 7)^2$ Write the given function's equation.

$f'(x) = \underline{2(x^3 + 7)^1} \cdot \underline{3x^2}$ Apply the chain rule.

Derivative of the inside function

Derivative of the outside function

$= 6x^2(x^3 + 7)$ Simplify. ◀

▶ **EXAMPLE 2** If $y = \cos^4 x$, find dy/dx.

Solution $y = \cos^4 x = (\cos x)^4$ Write the fourth power as an outside function.

$\dfrac{dy}{dx} = 4(\cos x)^3 \cdot (-\sin x)$ Differentiate the outside and inside functions.

$= -4\cos^3 x \sin x$ Simplify. ◀

Problem Set 3-7

Quick Review

Q1. Write one condition for f to be continuous at $x = c$.

Q2. Write another condition for f to be continuous at $x = c$.

Q3. Write the third condition for f to be continuous at $x = c$.

Q4. Is the signum function, $f(x) = \operatorname{sgn} x$, continuous at $x = 0$?

Q5. Find dy/dx if $y = 20x^{4/5}$.

Q6. Find $f(x)$ if $f'(x) = 30x^{-4}$.

Q7. Function f in Problem Q6 is called a(n) —?—.

Q8. Sketch the graph of $y = \sin x$.

Q9. Sketch the graph of $y = \cos x$.

Q10. The first positive value of x for which $\sin x = \dfrac{\sqrt{3}}{2}$ is

A. π B. $\dfrac{\pi}{2}$ C. $\dfrac{\pi}{3}$

D. $\dfrac{\pi}{4}$ E. $\dfrac{\pi}{6}$

1. State the chain rule in each form.
   a. Using dy/dx terminology
   b. Using $f'(x)$ terminology
   c. Verbally, using the words *inside function* and *outside function*

2. Given $f(x) = (x^2 - 1)^3$:
   a. Differentiate using the chain rule.
   b. Expand the power, then differentiate term by term.
   c. Show that the answers to parts a and b are equivalent.

For Problems 3–22, find an equation for the derivative function. You may check your answer by comparing its graph with the numerical derivative graph.

3. $f(x) = \cos 3x$

4. $f(x) = \sin 5x$

5. $g(x) = \cos(x^3)$

6. $h(x) = \sin(x^5)$

7. $y = (\cos x)^3$

8. $f(x) = (\sin x)^5$

9. $y = \sin^6 x$

10. $f(x) = \cos^7 x$

11. $y = -6 \sin 3x$

12. $f(x) = 4 \cos(-5x)$

13. $\dfrac{d}{dx}(\cos^4 7x)$

14. $\dfrac{d}{dx}(\sin^9 13x)$

15. $f(x) = 24 \sin^{5/3} 4x$

16. $f(x) = -100 \sin^{6/5}(-9x)$

17. $f(x) = (5x + 3)^7$

18. $f(x) = (x^2 + 8)^9$

19. $y = (4x^3 - 7)^{-6}$

20. $y = (x^2 + 3x - 7)^{-5}$

21. $y = [\cos(x^2 + 3)]^{100}$

22. $y = [\cos(5x + 3)^4]^5$

23. Find $\dfrac{d^2y}{dx^2}$ if $y = 4 \cos 5x$.

24. Find $\dfrac{d^2y}{dx^2}$ if $y = 7 \sin(2x + 5)$.

25. If $f'(x) = \cos 5x$, find an antiderivative $f(x)$.

26. If $f'(x) = 10 \sin 2x$, find an antiderivative $f(x)$.

27. *Graphical Verification Problem:* For $f(x) = 5 \cos 0.2x$, plot the graph of function f. Where $x = 3$, plot a line on the graph with slope equal to $f'(3)$. Is the line really tangent to the graph?

28. *Beanstalk Problem:* Jack's beanstalk grows in spurts. Its height, y, in feet above the ground, at time t, in hours since he planted it, is given by

$$y = 7 \sin \pi t + 12t^{1.2}$$

Write an equation for dy/dt. Plot the graph of y and the velocity graph on the same screen. Use a window with $[0, 10]$ for t. Do there appear to be times when the beanstalk is shrinking? Justify your answer.

29. *Balloon Volume Problem:* A spherical balloon is being inflated with air (see Figure 3-7c). The volume of the sphere depends on the radius, and the radius depends on time. Thus, the volume is a composite function of time.

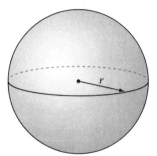

Figure 3-7c

a. The volume of a sphere is $V = (4\pi/3)r^3$, where r is the radius in cm. Find an equation for dV/dr. What are the units of dV/dr?

b. At time $t = 0$, the radius is 10 cm. If r increases at 6 cm/min, write r as a function of t.

c. Find an equation for dr/dt. Surprising? What are the units of dr/dt?

d. By appropriate use of the chain rule, find dV/dt when $t = 5$ min. Based on the units of dV/dr and dr/dt, explain why the units of dV/dt are cm^3/min.

e. Find dV/dt directly by substituting r from part b into the equation for V. Show that you get the same answer you did in part d for dV/dt when $t = 5$.

30. Δu *and* Δx *Problem:* The derivation of the chain rule states that if u is a continuous function of x, then $\Delta x \to 0$ implies $\Delta u \to 0$.

a. Sketch a graph showing that this may not be true if u has a step discontinuity.

b. Sketch a graph showing why this is true if u is continuous.

3-8 Proof and Application of Sine and Cosine Derivatives

In Section 3-6 you discovered that the derivative of the sine function is the cosine function and that the derivative of the cosine function is the opposite of the sine function. In this section you will prove algebraically and graphically that these properties are true. You will also apply the derivatives of these sinusoidal functions to some problems from the real world.

OBJECTIVE Be able to derive algebraically the formulas for the derivatives of $\sin x$ and $\cos x$. Find rates of change of sinusoidal functions in real-world problems.

Background: The Limit of $(\sin x)/x$

The function

$$y = \frac{\sin x}{x}$$

takes on the indeterminate form $0/0$ as x approaches zero. If you use a friendly window that includes $x = 0$ to graph the function, you will see that the fraction gets very close to 1 as x gets close to zero (Figure 3-8a). By using geometry, it is possible to prove that 1 is the limit of y as x approaches zero.

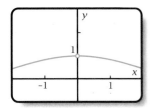

Figure 3-8a

Figure 3-8b shows a unit circle with an arc of length x cut by an angle of x radians. The line segment tangent to the arc has length $\tan x$ (hence the term *tangent*). The half-chord perpendicular to the u-axis has length $\sin x$. The distance along the u-axis from the origin to this chord is $\cos x$.

Consider the areas of three regions shown in Figure 3-8b.

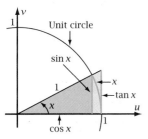

Figure 3-8b

Large triangle: Area $= \frac{1}{2}(1)(\tan x)$ Area $= \frac{1}{2}$(base)(altitude).

Sector inscribed in the large triangle: Area $= \frac{1}{2}(x)$ A fraction of a unit circle, $(x/2\pi) \cdot (\pi \cdot 1^2)$.

Small triangle inscribed in the sector: Area $= \frac{1}{2}(\sin x)(\cos x)$

From the relationship between the areas of these figures, you can write the following three-member inequality, then use the appropriate algebra.

$$\tfrac{1}{2}(\sin x)(\cos x) < \tfrac{1}{2}(x) < \tfrac{1}{2}(1)(\tan x)$$

$$\cos x < \frac{x}{\sin x} < \frac{\tan x}{\sin x}$$ Divide all three members of the inequality by $\sin x$, then multiply by 2.

$$\frac{1}{\cos x} > \frac{\sin x}{x} > \frac{\sin x}{\tan x}$$ The order reverses because all three members of the inequality are the same sign.

$$\sec x > \frac{\sin x}{x} > \cos x$$ Because $1/(\cos x) = \sec x$, and $(\sin x)/(\tan x) = \cos x$.

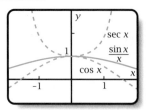

Figure 3-8c

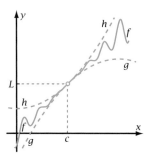

Figure 3-8d

Figure 3-8c shows how the graphs of the three members of this inequality would appear on your grapher. As you can see, $\sec x$ and $\cos x$ both approach 1 as x approaches zero. So $(\sin x)/x$ is also "squeezed" to 1. Be careful not to read too much into this answer. Note that $(\sin 0)/0$ is undefined, but has 1 for a limit as the argument, x, approaches zero. The indeterminate form $0/0$ can approach numbers other than 1.

PROPERTY: *Limit of (sin x)/x*

$$\lim_{x \to 0} \frac{\sin x}{x} = 1$$

The function $(\sin x)/x$ approaches 1 as x approaches zero because it stays between two other functions, each of whose limits equals 1 as x approaches zero. This fact is an example of the **squeeze theorem**. The theorem follows directly from the definition of limit. Figure 3-8d shows functions f, g, and h. If $g(x)$ and $h(x)$ have limit L as x approaches c, and if $f(x)$ is always between $g(x)$ and $h(x)$, then L is also the limit of $f(x)$ as x approaches c.

PROPERTY: *The Squeeze Theorem*

If

1. $g(x) \le h(x)$ for all x in a neighborhood of c, where $x \ne c$

2. $\lim_{x \to c} g(x) = \lim_{x \to c} h(x) = L$

3. f is a function for which $g(x) \le f(x) \le h(x)$ for all x in a neighborhood of c

then

$$\lim_{x \to c} f(x) = L$$

Note: A **neighborhood** of a number c is an open interval containing c. The interval is open, so c cannot be at an endpoint. Thus, there are numbers in the interval on both sides of c.

Derivative of the Sine Function

Let $f(x) = \sin x$. You suspect that $f'(x) = \cos x$. By the definition of derivative,

$$f'(x) = \lim_{h \to 0} \frac{\sin(x + h) - \sin x}{h}$$

Algebra with fractions is usually easier if the numerator and the denominator have one term each, perhaps consisting of a product of several factors. The following property from trigonometry can be used to transform the numerator above from a sum to a product.

108

$$\sin A - \sin B = 2\cos\tfrac{1}{2}(A + B)\sin\tfrac{1}{2}(A - B)$$ A sum and a product property from trigonometry.

$$\therefore f'(x) = \lim_{h \to 0} \frac{2\cos\tfrac{1}{2}[(x + h) + x]\sin\tfrac{1}{2}[(x + h) - x]}{h}$$ Use $(x + h)$ for A and x for B.

$$= \lim_{h \to 0} \frac{2\cos\left(x + \tfrac{h}{2}\right)\sin\tfrac{h}{2}}{h}$$

$$= \lim_{h \to 0}\cos\left(x + \tfrac{h}{2}\right) \cdot \lim_{h \to 0}\frac{2\sin\tfrac{h}{2}}{h}$$ Limit of a product property.

$$= (\cos x) \cdot \lim_{h \to 0}\frac{2\sin\tfrac{h}{2}}{h}$$ Take limit of first factor, assuming cosine is continuous.

$$= (\cos x) \cdot \lim_{h \to 0}\frac{\sin\tfrac{h}{2}}{\tfrac{h}{2}}$$ Express the second factor as $\dfrac{\sin(\text{argument})}{\text{argument}}$, whose limit is 1.

$$= (\cos x) \cdot (1) = \cos x$$

This confirms algebraically the property that you discovered graphically in Section 3-6.

Derivative of the Cosine Function

Using the cofunction properties from trigonometry, this new problem becomes an old problem.

$$y = \cos x = \sin\left(\frac{\pi}{2} - x\right) \Rightarrow y' = \cos\left(\frac{\pi}{2} - x\right) \cdot (-1) = -\cos\left(\frac{\pi}{2} - x\right) = -\sin x$$

So the property you discovered graphically in Section 3-6 is true.

Sinusoidal Equations from Real-World Information

The graph of a sine or cosine function is called a **sinusoid**. Such functions occur in the real world, particularly in connection with periodic motion. To fit a sinusoid to a particular real-world situation, you must recall how to transform a function. Figure 3-8e shows the *parent* sinusoid $y = \cos x$ (dashed) and the transformed sinusoid

$$y = 5 + 3\cos\frac{2}{3}(x - 1)$$

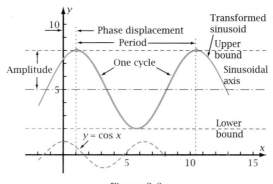

Figure 3-8e

The additive constant 5 is the **vertical translation** and gives the location of the **sinusoidal axis**. The multiplicative constant 3 is the **vertical dilation** factor, also called the **amplitude** for sinusoids. It states the distance that the high and low points are above and below the sinusoidal axis. The fraction $\frac{2}{3}$ is the *reciprocal* of the **horizontal dilation** factor. Because the period of $y = \cos x$ is 2π radians (one revolution), the period of the transformed sinusoid is

$$\text{Period} = \frac{3}{2} \cdot 2\pi = 3\pi \qquad \text{Multiply the normal period by the dilation factor.}$$

The constant 1 subtracted from x is the **horizontal translation**, also called the **phase displacement** for sinusoids. It is the value of x that makes the argument of the cosine equal to zero, and shows the horizontal location of a high point. These properties are summarized here.

PROPERTIES: *Graph of a Sinusoid*

If $y = C + A \cos B(x - D)$, then   General equation of a sinusoidal function.

- The sinusoidal axis is the line $y = C$. C is the vertical translation.

- The amplitude equals $|A|$. A is the vertical dilation factor.

- The period equals $2\pi \cdot \dfrac{1}{|B|}$. $1/B$ is the horizontal dilation factor.

- The phase displacement equals D. D is the horizontal translation.

▶ **EXAMPLE 1**

A mass is bouncing up and down on a spring hanging from the ceiling (Figure 3-8f). Its distance, y, in feet from the ceiling, varies sinusoidally with time, t, in seconds, making a complete cycle every 1.6 s. At $t = 0.4$, y reaches its greatest value, 8 ft. The smallest y gets is 2 ft.

     a. Write an equation for y in terms of t.

     b. Write an equation for the derivative, y'.

     c. How fast is the mass moving for these values of t?

         i. $t = 1$ ii. $t = 1.5$ iii. $t = 2.7$

     d. At $t = 2.7$ s, is the mass moving up or down? Justify your answer.

     e. What is the fastest the mass moves? Where is the mass when it is moving this fast?

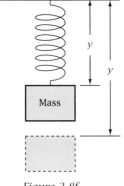

Figure 3-8f

Solution

First sketch the graph. Show the high point at $t = 0.4$. The next high point will be at $t = 0.4$ plus the period, or 2.0. Halfway between the two high points, at

---

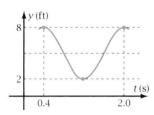

Figure 3-8g

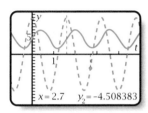

Figure 3-8h

$t = 1.2$, is a low point. From these points you can sketch a reasonable graph, as shown in Figure 3-8g.

a. $C = \frac{1}{2}(2 + 8) = 5$ Sinusoidal axis is halfway between upper and lower bounds.

$A = 8 - 5 = 3$ Amplitude is from sinusoidal axis to upper bound.

$B = \dfrac{2\pi}{1.6} = 1.25\pi$ Period = 1.6, so $B = 2\pi/1.6$.

$D = 0.4$ A high point occurs at $x = 0.4$.

∴ the equation is $y = 5 + 3\cos 1.25\pi(t - 0.4)$.

b. $y' = -3\sin 1.25\pi(t - 0.4) \cdot 1.25\pi$ Derivative of cos is $-\sin$. Use the chain rule!

$y' = -3.75\pi \sin 1.25\pi(t - 0.4)$

c. Plot the graphs of y and y' on the same screen (Figure 3-8h), then trace the graph of y' to find the values.

  i. $t = 1$: $y' = -8.3304... \approx -8.3$ ft/s

  ii. $t = 1.5$: $y' = 10.884... \approx 10.9$ ft/s

  iii. $t = 2.7$: $y' = -4.508... \approx -4.5$ ft/s

d. At $t = 2.7$, the mass is going up, because y' is negative, meaning y (the distance between the mass and the ceiling) is getting smaller.

e. The fastest the mass moves is 3.75π, or about 11.8 ft/s, which equals the amplitude of y'. Tracing shows a high point of y' at $t = 1.6$ (dashed line in Figure 3-8h). At this time the mass is halfway between its high and low points. ◀

Problem Set 3-8

Quick Review

Q1. Differentiate: $f(x) = x^9$

Q2. Find dy/dx: $y = 3\cos x$

Q3. Find y': $y = (5x^6 + 11)^{2.4}$

Q4. Find $s'(x)$: $s(x) = 2^{-13}$ (Be careful!)

Q5. Find: $\displaystyle\lim_{x \to 7} \frac{(x + 5)(x - 7)}{x - 7}$

Q6. Find: $\lim_{x \to 0} 2^x$

Q7. Is $f(x) = 1/(x - 3)$ continuous at $x = 4$?

Q8. If $f'(x) = 2x \cdot \sin x^2$, find $f(x)$.

Q9. Sketch the graph of the derivative of the function shown in Figure 3-8i.

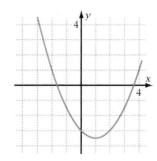

Figure 3-8i

Q10. If $y'' = 30x^2$, then y could equal

    A. $60x$ B. 60 C. $10x^3$ D. $15x^3$

    E. $2.5x^4$

1. *Ferris Wheel Problem:* When you ride a Ferris wheel, your distance, $y(t)$, in feet from the ground, varies sinusoidally with time t, in seconds since the wheel started rotating. Suppose that the Ferris wheel has a diameter of 40 ft and that its axle is 25 ft above the ground (see Figure 3-8j). Three seconds after it starts, your seat is at its high point. The wheel makes 3 rev/min.

Figure 3-8j

a. Sketch the graph of function y. Figure out the particular equation for $y(t)$.

b. Write an equation for $y'(t)$.

c. When $t = 15$, is $y(t)$ increasing or decreasing? How fast?

d. What is the fastest $y(t)$ changes? Where is the seat when $y(t)$ is changing the fastest?

2. *Pendulum Problem:* A pendulum hung from the ceiling makes a complete back-and-forth swing each 6 s (see Figure 3-8k). As the pendulum swings, its distance, d, in cm, from one wall of the room, depends on the time, t, in seconds, since it was set in motion. At $t = 1.3$, d is at its maximum of 110 cm from the wall. The

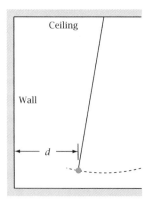

Figure 3-8k

lower bound of d is 50 cm. Assume that d is a sinusoidal function of t.

a. Write an equation expressing d as a function of t.

b. Write an equation for the derivative function.

c. How fast is the pendulum moving when $t = 5$? When $t = 11$? How do you explain the relationship between these two answers?

d. When $t = 20$, is the pendulum moving toward the wall or away from it? Explain.

e. What is the fastest the pendulum swings? Where is the pendulum when it is swinging its fastest?

f. What is the first positive value of t at which the pendulum is swinging 0 cm/s? Where is the pendulum at this time?

3. *Playground Problem:* Opportunity Park, a playground in Midland, Texas, has a sinusoidal ramp for children to walk on (see photo). A brick curb starts 0.75 ft above the ground and slopes up to 3.25 ft above the ground, 44 ft away. In every 8 ft, the concrete walkway varies from 0.75 ft below this curb to 0.25 ft below it, as shown in Figure 3-8l. Let x be the number of feet from the beginning of the ramp.

a. Write an equation for $f(x)$, the height from the ground to the top of the brick curb.

b. Write an equation for $g(x)$, the height from the ground to the top of the concrete walkway. Take into account the slope of the ramp.

c. Write an equation for $g'(x)$. What is the slope of the ramp at $x = 9$? What is the slope at $x = 15$? What are the units of the slope? At these places, would a child walking up

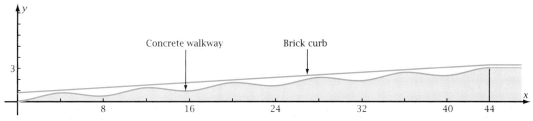

Figure 3-8l

the ramp be going upward, downward, or on the level? Explain.

  d. As a child walks up the ramp, what are the steepest up and steepest down slopes?

4. *Daylight Problem:* The length of daylight (sunrise to sunset) is approximately a sinusoidal function of the day of the year. The longest day is around June 21, the 172nd day of the year. In San Antonio, Texas, the longest day has 14 h 3 min of daylight. The shortest day has 10 h 15 min of daylight.

  a. Write an equation expressing the length of daylight, in minutes per day, for San Antonio as a function of day of the year. How much daylight is there on August 7?

  b. Write an equation for the derivative of the function in part a. At what rate is the length of daylight changing on August 7? What are the units of this rate of change?

  c. What is the greatest rate of change of daylight length? On what two days of the year is the rate equal to this number?

5. *Pendulum Experiment:* Swing a weight on a string from the ceiling of your classroom. Make appropriate measurements to find an equation for the distance from the wall as a function of time since you let the pendulum go. Based on your equation, calculate the fastest the pendulum moves as it swings.

6. *Daylight Research Project:* From some source such as the Naval Observatory's Web site, *http://aa.usno.navy.mil/data/*, obtain a chart of sunrise and sunset times for various days of the year in your locality. Use this chart to derive a sinusoidal equation for the number of minutes of daylight in a day as a function of the day of the year. See Problem 4 for ideas.

Draw a graph of predicted length of daylight versus day. Use your grapher to get the plotting data. On the same axes, plot the actual length of daylight derived from your chart. Discuss how well the sinusoidal model fits the real data. What is the greatest change in length of daylight from the table? How does this number compare with the derivative of the sinusoidal function at that time of year?

7. *Squeeze Theorem, Numerically:* Define functions f, g, and h as shown.

$$f(x) = -2x^2 + 8x - 2$$
$$g(x) = 2x^2 + 2$$
$$h(x) = 4x$$

  a. Plot the three functions on your grapher. What is the limit of each function as x approaches 1?

  b. Show that the squeeze theorem applies to the three functions. Which function is the upper bound, which is the lower bound, and which is in between?

  c. Make a table of values of $f(x)$, $g(x)$, and $h(x)$ for each 0.01 unit of x from 0.95 to 1.05. Use a time-efficient method, such as the TABLE or TRACE feature of your grapher.

  d. From the table, determine an interval of values of x around 1 such that if x is in this interval, then both the upper- and lower-bound functions will be within 0.1 unit of the limit. Show that the third function is also within 0.1 unit of the limit for values of x in this interval. How does this work illustrate the squeeze theorem?

8. *Limit of* $(\sin x)/x$ *Problem:* Without looking at the text, see if you can prove that 1 is the limit of $(\sin x)/x$ as x approaches zero. If you get stuck, look at the text only enough to get going.

9. *Limit of* $(\sin x)/x$, *Numerically:* The table below shows values of $(\sin x)/x$ as x gets closer to zero.

x	$(\sin x)/x$
0.5	0.9588510...
0.4	0.9735458...
0.3	0.9850673...
0.2	0.9933466...
0.1	0.9983341...

a. Use the TABLE or TRACE feature to verify that these numbers are correct. Don't forget to use radian mode!

b. Make a table of values for each 0.01 unit of x from 0.05 to 0.01. Do the values seem to be getting closer to 1?

c. Use smaller increments for x to find the value of x at which your grapher rounds the answer to exactly 1.

d. The National Bureau of Standards lists the value of $\sin 0.001$ to 23 places as

$$\sin 0.001 = 0.00099\ 99998\ 33333\ 34166\ 667$$

(The spaces here are for ease of reading.) How close to 1 is $(\sin 0.001)/0.001$? How does this answer compare with what your calculator gives you?

e. Just for fun, see if you can explain why $\sin 0.001$ has so many repeated digits.

10. *Derivative of the Sine Function:* Without looking at the text, see if you can prove directly from the definition of derivative that $\cos x$ is the derivative of $\sin x$. If you get stuck, look at the text only enough to get started again.

11. *Derivative of the Cosine Function:* By a clever application of the chain rule and the appropriate trigonometric properties, prove that $-\sin x$ is the derivative of $\cos x$.

12. *Squeeze Theorem Problem:*

a. State the squeeze theorem.

b. Prove directly from the definition of limit that the squeeze theorem is true.

c. Sketch a graph that shows you understand what the squeeze theorem means.

13. *Group Discussion Problem:* Figure 3-8m shows the graph of

$$y = 2 + (x - 1) \sin\left(\frac{1}{x - 1}\right)$$

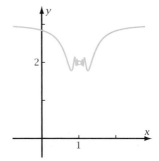

Figure 3-8m

The function has a discontinuity at $x = 1$ because of division by zero.

a. What does the limit of y seem to be as x approaches 1?

b. Plot the graph using a friendly window that has $x = 1$ as a grid point and has bounds approximately as shown in Figure 3-8m. Then zoom in on the point $(1, 2)$ several times. Sketch the result.

c. Find two linear functions that are upper and lower bounds for the graph, and use the number in part a for their limits as x approaches 1. Plot these on your grapher, then sketch them.

d. Prove that your answer in part a is correct. The squeeze theorem will help!

e. Why do you suppose the two linear functions are said to form an *envelope* for the y-graph?

f. If x is far away from 1, y seems to approach the constant 3. Based on what you have learned in this section, see if you can explain why this is true.

14. *Journal Problem:* Update your journal with things you have learned. You should include such things as

- The one most important thing you have learned since your last journal entry

- What you now better understand about formulas for derivatives of functions

- Properties you have learned and what they're named

- Any properties or techniques about which you are still unsure

3-9 Exponential and Logarithmic Functions

Power functions, such as $y = x^2$, where x is raised to a power, model such things as the area of a circle as a function of its radius. You have learned the power rule for derivatives that works with these functions. Exponential functions, on the other hand, such as $y = 2^x$, have the variable in the exponent. They model such things as population growth or radioactive decay where the rate of growth or decay of the quantity at a given instant is proportional to the amount of that quantity present. Figures 3-9a and 3-9b show the fundamental difference in the behavior of power and exponential functions.

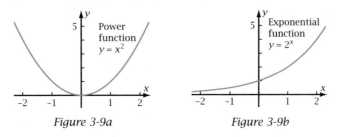

Figure 3-9a   Figure 3-9b

Unfortunately, the power rule for derivatives does not work for exponential functions, as you may have noticed in earlier problems. In this section you will learn an algebraic technique for differentiating exponential functions and their inverses, logarithmic functions.

OBJECTIVE   Given an exponential or logarithmic function, find its derivative function algebraically.

Exponential Functions

Figure 3-9c shows $f(x) = 2^x$ and its numerical derivative. The derivative graph suggests that $f'(x)$ is a *vertical dilation* of $f(x)$, that is, a constant times $f(x)$. Dividing y_2 by y_1, you find that this constant is approximately 0.6931, as shown in the table. To see why this is true, apply the definition of derivative.

$$f'(x) = \lim_{h \to 0} \frac{2^{x+h} - 2^x}{h}$$

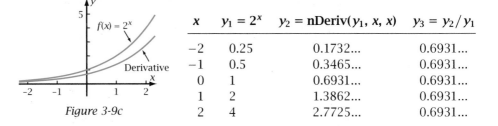

Figure 3-9c

x	$y_1 = 2^x$	$y_2 = \mathbf{nDeriv}(y_1, x, x)$	$y_3 = y_2/y_1$
-2	0.25	0.1732...	0.6931...
-1	0.5	0.3465...	0.6931...
0	1	0.6931...	0.6931...
1	2	1.3862...	0.6931...
2	4	2.7725...	0.6931...

You can write the 2^{x+h} in the numerator as $2^x \cdot 2^h$ so you can factor out 2^x.

$$f'(x) = \lim_{h \to 0} \frac{2^x \cdot 2^h - 2^x}{h}$$ Product of powers with equal bases, backward.

$$f'(x) = \lim_{h \to 0} \frac{2^x(2^h - 1)}{h}$$

Because 2^x does not depend on h, it is a *constant* with respect to h. Applying the limit of a constant property gives

$$f'(x) = 2^x \cdot \lim_{h \to 0} \frac{2^h - 1}{h}$$ Limit of a constant.

The fraction takes the indeterminate form $0/0$ as h approaches zero. By making a table of values, you can see numerically that the limit seems to be the constant 0.6931... shown earlier.

h	$(2^h - 1)/h$	
-0.0003	0.693075...	
-0.0002	0.693099...	
-0.0001	0.693123...	
0	undefined	Approaches 0.693147180... as a limit.
0.0001	0.693171...	
0.0002	0.693195...	
0.0003	0.693219...	

By using different values for the bases, b, you get different limits for the fraction $(b^h - 1)/h$. Figure 3-9d shows what happens for $b = 2$, 3, and 4. If you use the number $e = 2.71828...$ that appears on your grapher, then the limit of $(e^h - 1)/h$ equals exactly 1, as shown in Figure 3-9e.

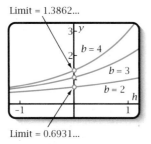

Figure 3-9d

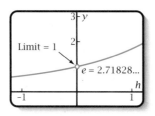

Figure 3-9e

So the derivative of $f(x) = e^x$ is $f'(x) = e^x \cdot 1 = e^x$. The function is its own derivative! This fact explains why the exponential function with base e is important enough to have its own key on your grapher. The name *natural exponential function* is used for $f(x) = e^x$ because the number e arises "naturally" in various places, as does the number $\pi = 3.14159...$.

The natural exponential function is

$$f(x) = e^x$$

where e is an irrational number equal to 2.71828182845....

If $f(x) = e^x$, then

$$f'(x) = e^x$$

Note: The symbol $\exp(x)$ is often used for e^x in higher-level mathematics and engineering courses to clarify that the x is the argument of a function, similar to $\cos(x)$.

▶ **EXAMPLE 1**   Find $f'(x)$ if $f(x) = e^{\cos x}$. Confirm the answer by graphing the algebraic and numerical derivatives on the same screen.

Solution

$$f'(x) = e^{\cos x} \cdot (-\sin x) \qquad \text{Use the chain rule on the inside function, } \cos x.$$
$$f'(x) = -e^{\cos x} \sin x \qquad \text{Simplify.}$$

The graph in Figure 3-9f shows that the algebraic derivative (solid) and numerical derivative (dotted) coincide.

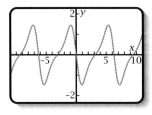

Figure 3-9f   ◀

Logarithmic Functions

From previous courses you should recall that a logarithm is an exponent. For instance,

If $5^x = 70$,   then the exponent $x = \log_5 70$.

The expression $\log_5 70$ is read "log to the base 5 of 70." If the exponential expression has e as its base, the corresponding base e logarithms are called *natural logarithms*. For instance,

If $e^x = 17$,   then $x = \log_e 17$, or $x = \ln 17$.

The l stands for "logarithm" and the n stands for "natural." Figure 3-9g shows the graph of $f(x) = \ln x$. For comparison, the figure also shows the graph of $g(x) = e^x$. You can see the

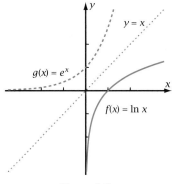

Figure 3-9g

inverse relationship between the two functions—the graphs are reflections of each other across the line $y = x$.

The definition and some properties of logarithms are shown in the box.

DEFINITION AND PROPERTIES: Logarithm

Definition:

$a = \log_b c$ if and only if $b^a = c, b > 0$, and $b \neq 1$.

Verbally: b is the base,

a is the exponent,

and c is the "answer" to b^a.

Properties:

- The **log of a power** equals the exponent times the log of the base.

$$\log_b(c^d) = d \cdot \log_b c$$

- The **log of a product** equals the sum of the logs of the factors.

$$\log_b(cd) = \log_b c + \log_b d$$

- The **log of a quotient** equals log of numerator minus log of denominator.

$$\log_b \frac{c}{d} = \log_b c - \log_b d$$

To find the derivative of $y = \ln x$, start by turning this new problem into an old problem. Because $\ln x$ means $\log_e x$, you can use the definition of logarithm to write

$y = \ln x = \log_e x$	$\ln x$ means base-e log of x.
$e^y = x$	By definition, e is the base, y is the exponent, x is the answer.
$e^y \cdot y' = 1$	Take the derivative of both sides. Use the chain rule on the inside function y.
$y' = \dfrac{1}{e^y} = \dfrac{1}{x}$	Divide by e^y, then substitute x for it.

The natural logarithm has the remarkable property that its derivative is the **reciprocal function**, $1/x$. Taking the derivative of both sides in the third line is justified because if two functions are identical for all values of x in a given domain, then their derivatives are also equal for all such x. The domain of $\ln x$ is the positive real numbers, $x > 0$, as you can see from Figure 3-9g.

Both the left and right sides of the equation $e^y = x$ are differentiated with respect to x. The technique of differentiating both sides of such an equation with respect to x is called **implicit differentiation**. You will study this technique more extensively in Section 4-8.

Chapter 3: Derivatives, Antiderivatives, and Indefinite Integrals

> **DEFINITION AND PROPERTY: Natural Logarithm and Its Derivative**
>
> $\ln x = \log_e x$, where $e = 2.71828182845...$.
>
> If $f(x) = \ln x$, then $f'(x) = \dfrac{1}{x}$.
>
> *Verbally:* The derivative of the natural logarithm function is the reciprocal function.

▶ **EXAMPLE 2** Find $f'(x)$ if $f(x) = \ln x^3$. Confirm the answer by comparing values with the numerical derivative.

Solution
$$f'(x) = \frac{1}{x^3} \cdot 3x^2 \qquad \text{Apply the chain rule to the inside function, } x^3.$$

$$= \frac{3}{x}$$

Numerical confirmation:

x	$3/x$	Numerical Derivative
1	3	3.0000...
2	3/2	1.5000...
3	1	1.0000...
4	3/4	0.7500...

◀

Inverse Relationship between $\ln x$ and e^x

Because $\ln x = \log_e x$, the definition of $\ln x$ lets you conclude that $y = \ln x$ if and only if $e^y = x$. This is the equation for the exponential function $y = e^x$, with the variables interchanged. Interchanging the variables in a function produces the *inverse* of that function. So the functions $f(x) = \ln x$ and $y = e^x$ are inverses of each other. That is,

If $f(x) = \ln x$, then $f^{-1}(x) = e^x$.

If $f(x) = e^x$, then $f^{-1}(x) = \ln x$.

The inverse of a function "undoes" what the function does. For instance, squaring the square root of a number gives back the original number,

$$\left(\sqrt{3}\right)^2 = 3$$

The same property applies to $\ln x$ and e^x.

$$\ln e^x = x \qquad \text{and} \qquad e^{\ln x} = x$$

Derivatives of Exponential Functions with Other Bases

Suppose you are asked to find y' if $y = 2^x$. This new problem can be transformed to an old problem by taking the natural log of both sides first.

$$\ln y = \ln 2^x = x \cdot \ln 2$$

The log of a power equals the exponent times the log of the base.

$$\frac{1}{y} \cdot y' = 1 \cdot \ln 2$$

Use the chain rule on the left; $\ln 2$ on the right is a constant.

$$y' = y \ln 2 = 2^x \ln 2$$

Solve for y', then substitute for y.

By calculator you will find that $\ln 2 = 0.6931\ldots$, which is the constant you saw at the beginning of this section. In general, if $f(x) = b^x$, then $f'(x) = b^x \ln b$. If you replace b with e, you get $f'(x) = e^x \ln e = e^x$, which proves algebraically that the derivative of e^x is e^x.

PROPERTY: *Derivative of Base-b Exponential Functions*

If $f(x) = b^x$, then $f'(x) = b^x \ln b$.

▶ **EXAMPLE 3**   *APR Problem:* The annual percentage rate, APR, is the average rate at which money in an account would increase if interest were compounded only once a year. If \$10,000 is placed in a savings account that earns interest at a rate of 7% APR compounded continuously, then $n(x)$, the dollar amount in the account after a number of years, x, is given by

$$n(x) = 10000(1.07^x)$$

Show that at time $x = 1$ year, the amount of money has actually increased by 7%. Find the instantaneous rate at which the money is increasing at $x = 0$ and at $x = 1$. How do these rates compare to the 7% APR?

Solution $n(0) = 10000$ and $n(1) = 10700$

∴ the amount has increased by \$700, which is 7% of 10,000.

$$n'(x) = 10000(1.07^x) \cdot \ln 1.07$$
$$n'(0) = 676.58\ldots \text{dollars per year, or } 6.7658\ldots\% \text{ of } 10,000$$
$$n'(1) = 723.94\ldots \text{dollars per year, or } 7.2394\ldots\% \text{ of } 10,000$$

The instantaneous rate is less than 7% at the beginning of the year and more than 7% at the end of the year, compared to an *average* rate of 7%.   ◀

Problem Set 3-9

Q1. Find: $\displaystyle\lim_{x \to 0} \frac{\sin x}{x}$

Q2. Find: $\dfrac{d}{dx}(\cos x)$

Q3. Find: $\dfrac{d^2}{dx^2}(\cos x)$

Q4. If $y' = \cos x$, what function could y equal?

Q5. Simplify: x^{12}/x^4

Q6. Simplify: $(x^{12})^4$

Q7. Write $\log 32$ in terms of $\log 2$.

Q8. Sketch the graph of $y = |x|$.

Q9. What is the outside function in $y = \sin^3 x$?

Q10. "If $f(x)$ is between $g(x)$ and $h(x)$ for all x close to c" is a hypothesis of

    A. The mean value theorem

    B. The intermediate value theorem

    C. The squeeze theorem

    D. The derivative of a product property

    E. The limit of a product property

---

1. *Compound Interest Problem:* Juan deposits $1000 in a savings account that earns interest at a rate of 6% per year, compounded continuously. He reads in a handbook that the amount of money, $M(x)$, in the account after a number of years, x, is given by the function

$$M(x) = 1000e^{0.06x}$$

    a. Find the derivative function, $M'(x)$. At what instantaneous rate is the amount of money increasing at $x = 1$, 10, and 20 years?

    b. Find the amount of money in Juan's account at the end of 0, 1, 2, and 3 years. Does the number of dollars in the account increase by the same amount each year?

    c. Find the APR for 0 to 1 year, for 1 to 2 years, and for 2 to 3 years. Are the rates equal? How does the APR compare to the instantaneous rate of 6%? Why do you suppose savings institutions prefer to advertise APR rather than instantaneous interest rate?

2. *Radioactive Tracer Problem:* Doctors use radioactive fluorine, in a sugar called 18-fluorodeoxyglucose (18-F), to trace the metabolism of glucose in the heart. Suppose that a patient has 10 mCi (microcuries) of 18-F injected at time $t = 0$ hours. Due to radioactive decay, the amount, $f(t)$, in microcuries, remaining in the patient's body after time t, in hours, is given by $f(t) = 10e^{-0.34t}$.

    a. Find the derivative function, $f'(t)$. Use it to find the instantaneous rate of change of 18-F at $t = 0$, 2, 4, and 6. By what factor does the

rate change between 0 and 2 hours? Between 4 and 6 hours?

    b. Use the original equation to find the number of microcuries of 18-F remaining at $t = 0$, 2, 4, and 6. By what factor do the amounts change between 0 and 2 hours? Between 4 and 6 hours? How do these factors compare with those in part a?

    c. Plot graphs of $f(t)$ and $f'(t)$ on the same screen. Use a window with [0, 10] for t. Sketch the result. Explain why the values of $f'(t)$ are negative.

3. *Altimeter Problem:* The altimeter on an airplane measures altitude indirectly by measuring the pressure of the air through which the plane is flying. The altitude is a logarithmic function of the pressure. Suppose that the altitude, $A(p)$, is given by the function

$$A(p) = 63 - 23.5 \ln p$$

where $A(p)$ is in thousands of feet above sea level and p is pressure in psi (pounds per square inch).

    a. Find the algebraic derivative, $A'(p)$. Plot functions A and A' on the same screen. Use a window with [0, 20] for p and [−50, 100] for A and A'. Sketch the graphs.

    b. Based on the graphs, what does an increase in pressure indicate about the altitude? At what rate is the altitude changing if $p = 10$ psi? How do you explain the sign of $A'(10)$?

    c. How do you interpret the fact that $A'(5)$ has a greater absolute value than $A'(10)$?

    d. What is the pressure of air at sea level, where $A(p) = 0$? How do you interpret the fact that $A(p)$ is negative for values of p greater than this?

4. *Time-for-Money Interest Problem:* If $3000 is invested in a savings account that earns 5% per year interest, compounded continuously, then the dollar amount, x, in the account after a number of years, y, is given by the exponential function

$$x = 3000e^{0.05y}$$

a. By taking the natural log of both sides of this equation, show that the number of years it takes to reach a particular dollar amount, x, in the account is given by the logarithmic function

$$y = 20\ln x - 20\ln 3000$$

You will need to use the properties for log of a product and log of a power.

b. Using the equation in part a, find the numbers of years it will take the amount to reach $3000, $4000, $5000, and $6000, and the numbers of years required to get from $3000 to $4000, from $4000 to $5000, and from $5000 to $6000. Explain why these time intervals decrease as the amount of money increases.

c. Find the derivative function, y'. What are its units? Evaluate y' when the amount is $3000, $4000, $5000, and $6000. How are these values consistent with the conclusion you reached in part b?

For Problems 5–28, find the indicated derivative.

5. $f(x) = 5e^{3x}$

6. $f(x) = 7e^{-6x}$

7. $g(x) = -4e^{\cos x}$

8. $h(x) = 8e^{-\sin x}$

9. $y = 2\sin(e^{4x})$

10. $y = 6\cos(e^{-0.5x})$

11. $f(x) = 10\ln 7x$

12. $g(x) = 9\ln 4x$

13. $T = 18\ln x^3$

14. $P = 1000\ln x^{0.7}$

15. $y = 3\ln(\cos 5x)$

16. $y = 11\ln(\sin 0.2x)$

17. $u = 6\ln(\sin x^{0.5})$

18. $v = 0.09\ln(\cos x^8)$

19. $r(x) = \ln e^x$ (Surprising?)

20. $c(x) = e^{\ln x}$ (Evaluate $c'(x)$ for $x = 2, 3,$ and 4.)

21. $f(x) = 3^x$

22. $g(x) = 0.007^x$

23. $y = 1.6^{\cos x}$

24. $y = \sin(5^x)$

25. Find $\dfrac{d^2y}{dx^2}$ if $y = \ln x^5$.

26. Find $\dfrac{d^2y}{dx^2}$ if $y = e^{7x}$.

27. Find y'' if $y = e^{-0.7x}$.

28. Find y'' if $y = \ln 8x$.

29. If $f'(x) = 12e^{2x}$, find the antiderivative function $f(x)$.

30. If $y' = 5^x \ln 5$, find the antiderivative function y.

3-10 Chapter Review and Test

In this chapter you have learned some algebraic techniques for differentiating functions. You can differentiate powers, sums, linear combinations, sine and cosine, exponential and logarithmic functions, and composite functions. You have learned about velocity and acceleration and that acceleration is the second derivative of position. You have also been exposed to the fourth concept of calculus, the indefinite integral, which is an antiderivative.

The advantage of algebraic techniques over numerical and graphical ones is that the exact values of a derivative can be found fairly quickly. The disadvantage of algebraic techniques is that the way you calculate a derivative has nothing to do with what derivative means. You must always bear in mind that a derivative is an instantaneous rate of change.

Review Problems

R0. Update your journal with things you've learned since your last entry. You should include such things as

- The one most important thing you have learned in studying Chapter 3
- Which boxes you have been working on in the "define, understand, do, apply" table (See page 64.)
- Key words, such as *chain rule, difference quotient, squeeze theorem, second derivative, antiderivative, indefinite integral, sinusoid*
- Ways in which you have used graphs, tables, algebra, and writing to understand concepts
- Any ideas about calculus that you're still unclear about

R1. Let $f(x) = x^3$.

a. Estimate the derivative of f at $x = 2$ numerically by taking average rates of change from 2 to x as x gets closer and closer to 2.

b. Write an equation for $r(x)$, the average rate of change of $f(x)$ from 2 to x. What form does $r(2)$ take? What seems to be the limit of $r(x)$ as x approaches 2?

c. Find the exact value of the derivative algebraically by simplifying the expression for $r(x)$, then using the simplified form to find the limit of $r(x)$ as x approaches 2. It might help for you to recall how to factor a difference of two cubes, $a^3 - b^3 = (a - b)(a^2 + ab + b^2)$.

d. Show that the numerical derivative in part a and the limit in part b agree with the exact value of the derivative you calculated in part c.

R2. a. Write the definition of derivative at a point $x = c$.

b. If $f(x) = 0.4x^2 - x + 5$, find $f'(3)$ directly from the definition of derivative.

c. Plot the difference quotient in a neighborhood of 3. Sketch the result.

d. On the same screen, plot the graph of function f in part b and a line through the point $(3, f(3))$ with slope $f'(3)$. Sketch the result.

e. How is the line in part d related to the graph?

f. Does function f seem to have the property of local linearity at $x = 3$? How do you know?

R3. a. Plot $y_1 = x^4 - 4x^3 - 7x^2 + 34x - 24$. Use a window that goes from at least $x = -4$ to $x = 5$ and has a fairly large range of y-values. Sketch the result.

b. Plot $y_2 = $ the numerical derivative of y_1. Sketch the result.

c. What feature does the y_1 graph have where the derivative graph equals zero?

d. *Leaky Tire Problem:* The air pressure in a leaky tire decreases according to the equation $p(t) = 35(0.9)^t$, where $p(t)$ is pressure, in pounds per square inch, and t is time, in hours, since it was last inflated. Plot $p(t)$ and its numerical derivative on your grapher. Then sketch the result. At approximately what rate is the pressure changing when $t = 3$? When $t = 6$? When the tire was just filled? What are the units of pressure change? How do you explain the sign of the pressure change? Is the rate of change of pressure getting closer to zero as time goes on?

R4. a. Write the definition of derivative (the h or Δx form).

b. What single word means "find the derivative"?

c. Write the property for the derivative of a power function.

d. Prove the property of the derivative of a constant times a function.

e. Prove the property of the derivative of a sum of two functions.

f. How do you pronounce $\frac{dy}{dx}$? How do you pronounce $\frac{d}{dx}(y)$? What do these symbols mean?

g. Find an equation for the derivative function.

   i. $f(x) = 7x^{9/5}$

   ii. $g(x) = 7x^{-4} - \dfrac{x^2}{6} - x + 7$

   iii. $h(x) = 7^3$

h. Compare the exact value of $f'(32)$ in part g with the numerical derivative at that point.

i. On a copy of Figure 3-10a, sketch the graph of the derivative function for the function shown.

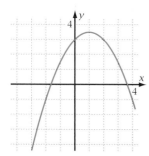

Figure 3-10a

R5. a. If x is the displacement of a moving object from a fixed plane as a function of time, t, write the appropriate calculus symbol for the velocity, v. Write two forms of the symbol for the acceleration, a, one in terms of v and the other in terms of x.

b. What is the meaning of the symbol $\dfrac{d^2y}{dx^2}$? Find $\dfrac{d^2y}{dx^2}$ if $y = 10x^4$.

c. If $f'(x) = 12x^3$, write an equation for $f(x)$. Give a name for $f(x)$ that expresses its relationship to $f'(x)$. Which of the four concepts of calculus is this?

d. Figure 3-10b shows the graph of a derivative function $y = f'(x)$. What is the slope of the antiderivative graph, $y = f(x)$, at $x = 1$? At $x = 5$? At $x = -1$? If $f(1) = 3$, sketch the graph of $y = f(x)$ consistent with these slopes on a copy of Figure 3-10b.

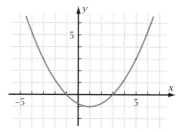

Figure 3-10b

e. *Spaceship Problem:* A spaceship is approaching Mars. It fires its retrorockets, which causes it to slow down, stop, rise up again, then come back down. Its displacement, y, in kilometers, from the surface is found to be

$$y = -0.01t^3 + 0.9t^2 - 25t + 250$$

where t is time, in seconds, since the retrorocket was fired.

   i. Write equations for the velocity and acceleration of the spaceship.

   ii. Find the acceleration at time $t = 15$. At that time, is the spaceship speeding up or slowing down? How do you tell?

   iii. Find by direct calculation the values of t at which the spaceship is stopped.

   iv. When does the spaceship touch the surface of Mars? What is its velocity at that time? Describe what you think will happen to the spaceship at that time.

R6. a. Plot and sketch $y_1 = \cos x$ and $y_2 =$ the numerical derivative of y_1.

b. Explain how the graphs show that $\cos' x = -\sin x$.

c. Verify numerically that $\cos' 1 = -\sin 1$.

d. What name is given to a function such as $f(x) = \cos(x^2)$ for which a function such as cosine is performed on an inside function, x^2? What does $f'(x)$ equal?

R7. a. State the chain rule in each form.

   i. dy/dx

   ii. $f(x)$

   iii. Using the outside and inside functions

b. Prove the chain rule. What assumption must you make about the quantity Δu so that the proof is valid?

c. Differentiate $f(x) = (x^2 - 4)^3$ in the two ways described and show that the two answers are equivalent.

    i. Use the chain rule.

    ii. Expand the binomial power first.

d. Differentiate.

    i. $f(x) = \cos(x^3)$ ii. $g(x) = \sin 5x$

    iii. $h(x) = \cos^6 x$ iv. $k(x) = \sin 3$

e. If $f'(x) = 12 \cos 3x$, find $f''(x)$ and $f(x)$. What special name is given to $f''(x)$? What special name is given to $f(x)$ in relation to $f'(x)$?

f. *Shark Problem:* The weight of a great white shark, W, in pounds, is given approximately by $W = 0.6x^3$, where x is the length of the shark in feet. Suppose that a baby shark is growing at the rate of $dx/dt = 0.4$ ft/day. How heavy is the shark when it is 2 ft long? 10 ft long? At what rate is it gaining weight (pounds per day) when it is these lengths? Explain how the chain rule allows you to calculate these answers.

R8. a. What number does $\lim_{x \to 0}[(\sin x)/x]$ equal? Give numerical evidence that the expression really does approach this limit as x approaches zero.

   b. State the squeeze theorem. Between what two quantities can $(\sin x)/x$ be "squeezed" to prove that the limit is the number you wrote in part a?

   c. Prove directly from the definition of derivative that $\sin' x = \cos x$.

   d. Given that $\sin' x = \cos x$, use appropriate trigonometric properties to prove that $\cos' x = -\sin x$.

e. *Clock Problem:* A clock with a sweep second hand is fastened to the wall. As the sweep hand turns, the distance from its tip to the floor varies sinusoidally with time. The center of the clock is 180 cm from the floor, and the sweep hand is 20 cm long from the center to its tip. Write an equation for the distance from tip to floor as a function of the number of seconds since the hand was at the 12. Write an equation for the instantaneous rate of change of this distance with respect to time. How fast is the distance changing when the sweep hand points at 2? At 3? At 7? How do you explain the signs you get for these rates?

R9. *Biological Half-Life Problem:* If you take a dose of medication, the percentage of medicine remaining in your body decreases exponentially with time. Suppose that the equation for a particular dose is

$$p(x) = 100e^{-0.1x}$$

where $p(x)$ is the percentage a number of hours, x, after you took the medication.

a. Find the instantaneous rate of change of the percentage at $x = 0$, 10, and 20 hours. Explain why these rates are negative. Find the biological half-life of the medication, the time it takes for the percentage to drop to half its original value. Find the percentage at a time twice this biological half-life.

b. Find the indicated derivatives:

    i. $f'(x)$ if $f(x) = 5e^{2x}$

    ii. $\dfrac{dy}{dx}$ if $y = 7^x$

    iii. $\dfrac{d}{dx}[\ln(\cos x)]$

    iv. $\dfrac{d^2y}{dx^2}$ if $y = \ln x^8$ (Be clever!)

c. If $f'(x) = 12e^{3x}$, find the antiderivative function, $y = f(x)$.

d. On the same screen, plot $y_1 = e^x$, $y_2 = \ln x$, and $y_3 = x$. Use the same scales on both axes. Sketch the results, and explain the relationship among the three graphs.

Concept Problems

C1. *Introduction to the Derivative of a Product:* Let $f(x) = x^7$ and $g(x) = x^9$. Let

$$h(x) = f(x) \cdot g(x)$$

   a. Write an equation for $h(x)$ in the simplest possible form.

   b. Find $h'(x)$.

   c. Find $f'(x)$ and $g'(x)$. Does $h'(x)$ equal $f'(x) \cdot g'(x)$?

   d. Show that for these particular functions, $h'(x) = f'(x) \cdot g(x) + f(x) \cdot g'(x)$.

C2. *Graph of an Interesting Function:* Let

$$f(x) = \frac{x - \sin 2x}{\sin x}$$

   a. What form does $f(0)$ take? What name is given to a form like this? Explain why f is discontinuous at $x = 0$.

   b. The discontinuity in f at $x = 0$ is removable. What number should $f(0)$ be defined to equal so that the function is continuous at $x = 0$?

   c. Make a conjecture about whether or not the function as defined in part b is differentiable at $x = 0$. If you think it is differentiable, make a conjecture about what $f'(0)$ equals. If you think it is not differentiable, explain why not.

   d. Use the definition of derivative (derivative at $x = c$ form) to establish whether or not your conjecture in part c is true.

Chapter Test

PART 1: No calculators allowed (T1–T14)

T1. Write the definition of derivative at a point. Write the definition of derivative as a function.

T2. Prove directly from the definition of derivative that if $f(x) = 3x^4$, then $f'(x) = 12x^3$.

T3. Sketch the graph of a function for which $f'(5) = 2$. Sketch a line tangent to the graph at that point. What would happen to the graph and the tangent line if you were to zoom in on the point where $x = 5$? What is the name of the property that expresses this relationship between the graph and the tangent line?

T4. Amos must evaluate $f'(5)$ where $f(x) = 7x$. He substitutes 5 for x, getting $f(5) = 35$. Then he differentiates the 35 and gets $f'(5) = 0$ (which equals the score his instructor gives him for the problem!). What mistake did Amos make? What is the correct answer?

For Problems T5–T11, find an equation for the derivative function.

T5. $f(x) = (7x + 3)^{15}$

T6. $g(x) = \cos(x^5)$

T7. $\dfrac{d}{dx}[\ln(\sin x)]$

T8. $u = 3^{6x}$

T9. $u = \cos(\sin^5 7x)$

T10. $y = 60x^{2/3} - x + 2^5$

T11. $\dfrac{d^2y}{dx^2}$ if $y = e^{9x}$

T12. Estimate the value of the derivative, y', at $x = 1$ for the graph shown in Figure 3-10c.

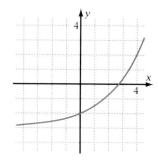

Figure 3-10c

T13. If $y = 3 + 5x^{-1.6}$, where y is the displacement of a moving object, find equations for the velocity function and the acceleration function. What special name is given to the acceleration function that expresses its relationship to the original displacement function?

T14. Find an equation for the antiderivative of $f'(x) = 72x^{5/4}$.

PART 2: Graphing calculators allowed (T15–T23)

T15. Find an equation for $f(x)$ if $f'(x) = 5 \sin x$ and $f(0) = 13$.

T16. Is the function $f(x) = \cos 3x$ increasing or decreasing when $x = 5$? At what rate?

T17. Plot the graph of $f(x) = (\sin x)/x$. Then sketch the result. In particular, show what happens to $f(x)$ as x approaches zero. In the proof of the limit of this function as x approaches zero, the function is bounded below by $y = \cos x$ and bounded above by $y = \sec x$. Name and state (without proof) the theorem used in this proof.

T18. Show numerically that the fraction $\frac{5^h - 1}{h}$ approaches $\ln 5$ as a limit as h approaches zero. Use the result and the definition of derivative to prove that $\frac{d}{dx}(5^x) = 5^x \ln 5$.

T19. *Sky Diver Problem Revisited:* In Problem 8 of Section 3-5, Phoebe's velocity t seconds after she started a sky dive was $v(t) = 251(1 - 0.88^t)$. Find an equation for her acceleration, $a(t)$. Use the equation to find $a(10)$. Show that the numerical derivative gives the same answer.

T20. How can you tell from the velocity and acceleration of a moving object that the object is slowing down at a particular value of t?

T21. The velocity of a moving object, $v(t)$, in centimeters per second, where time, t, is in seconds, is given by

$$v(t) = t^{1.5} + 3$$

a. Find an equation for the acceleration, $a(t)$.

b. Find an equation for the displacement, $d(t)$, if the object is 20 cm from the reference point when $t = 1$.

c. Find $d(9) - d(1)$. What does the answer represent?

T22. *Carbon Dioxide Problem:* The concentration of carbon dioxide in the atmosphere fluctuates slightly. The concentration is at a minimum in the summer when tree leaves absorb the carbon dioxide and at a maximum in the winter when trees are bare. The fluctuation is only about ± 2 ppm (parts per million) with an average concentration of 300 ppm. Assuming that the concentration varies sinusoidally with time, t, in days since the first of the year, the concentration, $c(t)$, is

$$c(t) = 300 + 2 \cos \frac{2\pi}{365} t$$

whose graph is shown in Figure 3-10d.

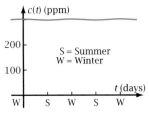

Figure 3-10d

a. Write an equation for $c'(t)$.

b. The greatest rate of increase occurs at $t = 273$, which is at the autumnal equinox. What is this rate in parts per million per day?

c. Earth's atmosphere contains a total of about 6×10^{15} tons of air. Show that the amount of carbon dioxide in the atmosphere is increasing by about 2390 tons/s at $t = 273$. Surprising, isn't it?

T23. What did you learn as a result of taking this test that you did not know before?

Products, Quotients, and Parametric Functions

Unlike the pendulum of a grandfather clock, this Foucault pendulum is free to swing in any direction. The path of the pendulum is as constant as possible with respect to space, and thus seems to move with respect to Earth as Earth rotates underneath it. Parametric functions are used as mathematical models that show the paths of objects whose x- and y-coordinates both vary with time.

Mathematical Overview

You have already learned a way to differentiate power functions. In Chapter 4, you will learn algebraic methods for differentiating products and quotients. You will also apply these methods to parametric functions in which both x and y depend on a third variable such as time. You will gain this knowledge in four ways.

Graphically
The icon at the top of each even-numbered page of this chapter shows the path followed by a pendulum that is allowed to swing in both the x- and y-directions.

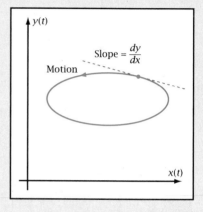

Numerically

t	x	y
0	7	4
1	5.620...	5.501...
2	1.030...	4.176...
3	2.039...	3.053...
⋮	⋮	⋮

Algebraically
$$\frac{dy}{dx} = \frac{dy/dt}{dx/dt},$$ the parametric chain rule

Verbally
I had my first major surprise in calculus! The derivative of a product of two functions is not the product of the two derivatives. The same thing goes for quotients. I also proved that the power rule works when the exponent is negative and not an integer.

4-1   Combinations of Two Functions

You have learned to differentiate algebraically a function that is a linear combination of two other functions, such as $f(x) = 3\cos x + 2\sin x$. In this section you will explore derivatives of a **product** and a **quotient** of two other functions, and of a **parametric function**. For example,

$$p(x) = (3\cos x)(2\sin x), \quad q(x) = \frac{3\cos x}{2\sin x}, \quad \text{and} \quad \begin{array}{l} x = 3\cos t \\ y = 2\sin t \end{array}$$

OBJECTIVE   On your own or with your study group, explore the derivatives of functions formed by multiplying, dividing, and composing two functions.

Exploratory Problem Set 4-1

Let $f(x) = 3\cos x$ and $g(x) = 2\sin x$, as shown in Figures 4-1a and 4-1b.

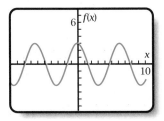

Figure 4-1a

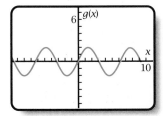

Figure 4-1b

1. Find equations for $f'(x)$ and $g'(x)$.

2. *Derivative of a Product of Two Functions:* Plot $p(x) = f(x) \cdot g(x) = (3\cos x)(2\sin x)$. Use a friendly window that is about the scale shown in Figures 4-1a and 4-1b. Sketch the result. Find $p'(2)$ numerically. How can you tell whether $p(x)$ is increasing or decreasing at $x = 2$? Does $p'(2)$ equal $f'(2) \cdot g'(2)$?

3. *Derivative of a Quotient of Two Functions:* Plot $q(x) = f(x)/g(x) = (3\cos x)/(2\sin x)$. Sketch the result. What familiar function is q? Find $q'(2)$ numerically. Is $q(x)$ increasing or decreasing at $x = 2$? Does $q'(2)$ equal $f'(2)/g'(2)$?

4. *Parametric Function:* Let $x = 3\cos t$ and $y = 2\sin t$. If you draw an xy-graph for various values of t, the result is called a **parametric function**. The elliptical path traced on the floor by a pendulum that is free to swing in any direction is an example of such a function. Put your grapher in parametric mode. Enter $x = 3\cos t$ and $y = 2\sin t$. Use a window with x- and y-ranges like those shown in Figure 4-1a. Use a t-range of 0 to 2π, with a t-step of 0.1. Plot and sketch the graph. Does the graph really seem to be an ellipse?

5. *Derivative of a Parametric Function:* Trace the graph in Problem 4 to the point where $t = 2$. Find Δx and Δy as t goes from 1.9 to 2.1 and use the results to find a symmetric difference quotient that approximates dy/dx at $t = 2$. Give evidence to show that dy/dx at this point could equal $(dy/dt)/(dx/dt)$ given that the derivatives are taken at $t = 2$.

6. *Conjectures?:* Just for fun, see if you can calculate the correct values of $p'(2)$ and $q'(2)$ from Problems 2 and 3, using the values $f'(2)$, $g'(2)$, $f(2)$, and $g(2)$.

4-2   Derivative of a Product of Two Functions

The derivative of a sum of two functions is equal to the sum of the two derivatives. As you learned in Exploratory Problem Set 4-1, there is no such distributive property for the derivative of a product of two functions. Here is an example.

Derivative of the product $\longleftarrow$ $f(x) = (x^5)(x^8)$ $\longrightarrow$ Product of the derivatives

$$f(x) = x^{13}$$

$$(5x^4)(8x^7)$$

$$f'(x) = 13x^{12} \qquad \longleftarrow \quad \text{Not the same.} \quad \longrightarrow \qquad = 40x^{11}$$

In this section you will learn how to differentiate a product of two functions without having to multiply first.

OBJECTIVE   Given a function that is a product of two other functions, find, in one step, an equation for the derivative function.

What follows involves some fairly complicated algebra. Therefore, it helps to streamline the symbols somewhat. Instead of writing $f(x) = g(x)\,h(x)$, write

$$y = uv$$

where u and v stand for differentiable functions of x. The idea is to find dy/dx in terms of du/dx and dv/dx.

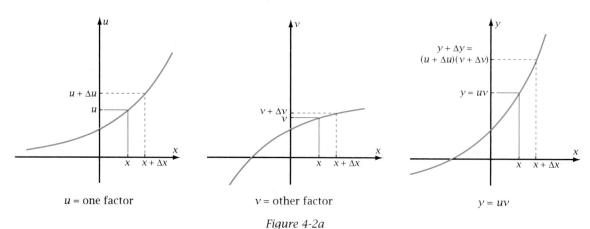

| u = one factor | v = other factor | $y = uv$ |

Figure 4-2a

The graphs of two functions, u and v, and their product, $y = uv$, are shown in Figure 4-2a. By the definition of derivative,

$$\frac{dy}{dx} = \lim_{\Delta x \to 0} \frac{(y + \Delta y) - y}{\Delta x}$$

Because $f(x) = g(x) \cdot h(x)$, $f(x + \Delta x)$ is equal to $g(x + \Delta x) \cdot h(x + \Delta x)$. So you can replace $(y + \Delta y)$ in the dy/dx equation with $(u + \Delta u)(v + \Delta v)$, as indicated on the right side in Figure 4-2a, and you can replace y with uv. Therefore,

$$\frac{dy}{dx} = \lim_{\Delta x \to 0} \frac{(u + \Delta u)(v + \Delta v) - uv}{\Delta x}$$

$$= \lim_{\Delta x \to 0} \frac{uv + \Delta uv + u\Delta v + \Delta u\Delta v - uv}{\Delta x} \qquad \text{Δuv means $(\Delta u) \cdot v$.}$$

$$= \lim_{\Delta x \to 0} \frac{\Delta uv + u\Delta v + \Delta u\Delta v}{\Delta x}$$

$$= \lim_{\Delta x \to 0} \left(\frac{\Delta uv}{\Delta x} + \frac{u\Delta v}{\Delta x} + \frac{\Delta u\Delta v}{\Delta x} \right)$$

$$= \lim_{\Delta x \to 0} \left(\frac{\Delta u}{\Delta x}v + u\frac{\Delta v}{\Delta x} + \Delta u\frac{\Delta v}{\Delta x} \right)$$

Using the properties for limit of a sum and limit of a product, and also the fact that the limits of $\Delta u/\Delta x$ and $\Delta v/\Delta x$ are du/dx and dv/dx, respectively, gives

$$\frac{dy}{dx} = \frac{du}{dx}v + u\frac{dv}{dx} + 0\frac{dv}{dx} \qquad \text{Because u is a continuous function, $\Delta u \to 0$ as $\Delta x \to 0$.}$$

$$= \frac{du}{dx}v + u\frac{dv}{dx} \qquad \text{Derivative of a product property.}$$

$$y' = u'v + uv' \qquad \text{Short form, where u' and v' are derivatives with respect to x.}$$

This property is best remembered as a rule, as shown in the box.

PROPERTY: Derivative of a Product of Two Functions—The Product Rule

If $y = uv$, where u and v are differentiable functions of x, then $y' = u'v + uv'$.

Verbally: Derivative of first times second, plus first times derivative of second.

With this rule you can accomplish this section's objective of differentiating a product in one step.

▶ **EXAMPLE 1**   If $y = x^4 \cos 6x$, find dy/dx.

Solution   Use the product rule where $u = x^4$ and $v = \cos 6x$.

$$\frac{dy}{dx} = 4x^3 \cos 6x + x^4(-\sin 6x) \cdot 6$$
$$= 4x^3 \cos 6x - 6x^4 \sin 6x \qquad ◀$$

As you write the derivative, say to yourself, "Derivative of first times second, plus first times derivative of second." Don't forget the chain rule!

▶ **EXAMPLE 2**   If $y = (3x - 8)^7(4x + 9)^5$, find y' and simplify.

Solution   Use both the product rule and the chain rule.

$$y' = 7(3x - 8)^6(3)(4x + 9)^5 + (3x - 8)^7(5)(4x + 9)^4(4)$$

The two terms have common binomial factors. Simplifying includes factoring out these common factors.

$$y' = (3x - 8)^6(4x + 9)^4[21(4x + 9) + 20(3x - 8)]$$
$$= (3x - 8)^6(4x + 9)^4(144x + 29)$$ ◀

The factored form in Example 2 is considered simpler because it is easier to find values of x that make the derivative equal zero.

▶ **EXAMPLE 3**   If $y = e^{\sin x}$, find $\dfrac{d^2y}{dx^2}$.

Solution   First use the chain rule on the inside function.

$$\frac{dy}{dx} = e^{\sin x}\cos x$$

$$\frac{d^2y}{dx^2} = e^{\sin x}\cos x \cdot \cos x + e^{\sin x}(-\sin x) \qquad \text{Use the product and chain rules.}$$

$$= e^{\sin x}(\cos^2 x - \sin x) \qquad\qquad \text{Simplify.}$$ ◀

Problem Set 4-2

Q1. Differentiate: $y = x^{3/4}$

Q2. Find y': $y = \ln x$

Q3. Find dy/dx: $y = (5x - 7)^{-6}$

Q4. Find: $\dfrac{d}{dx}(\sin 2x)$

Q5. Differentiate: $v = \cos^3 t$

Q6. Differentiate: $L = m^2 + 5m + 11$

Q7. If $dy/dx = \cos x^3 \cdot 3x^2$, find y.

Q8. In Figure 4-2b, if $x = -2$, $y' \approx$ —?—.

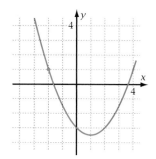

Figure 4-2b

Q9. If $u = v^2/6$, where u is in feet and v is in seconds, how fast is u changing when $v = 12$ s?

Q10. If $y' = e^{3x}$, then $y'' =$

    A. e^{3x} B. $3e^{3x}$ C. $9e^{3x}$

    D. $3e^{2x}$ E. $6e^x$

For Problems 1–20, differentiate and simplify. You may check your answer by comparing its graph with the graph of the numerical derivative.

1. $f(x) = x^3 \cos x$

2. $f(x) = x^4 \sin x$

3. Find $g'(x)$: $g(x) = x^{1.5}e^{2x}$

4. Find $h'(x)$: $h(x) = x^{-6.3} \ln 4x$

5. Find $\dfrac{dy}{dx}$: $y = x^7(2x + 5)^{10}$

6. Find $\dfrac{dy}{dx}$: $y = x^8(3x + 7)^9$

7. Find z': $z = \ln x \sin 3x$

8. Find v': $v = e^{5x} \cos 2x$

9. $y = (6x + 11)^4 (5x - 9)^7$

10. $y = (7x - 3)^9(6x - 1)^5$

11. $P = (x^2 - 1)^{10}(x^2 + 1)^{15}$

12. $P(x) = (x^3 + 6)^4(x^3 + 4)^6$

13. $a(t) = 4 \sin 3t \cos 5t$

14. $v = 7 \cos 2t \sin 6t$

15. $y = \cos(3 \sin x)$

16. $y = \sin(5 \cos x)$

17. Find $\dfrac{d^2y}{dx^2}$: $y = \cos e^{6x}$

18. Find $\dfrac{d^2y}{dx^2}$: $y = \ln(\sin x)$ (Be clever!)

19. $z = x^3(5x - 2)^4 \sin 6x$ (Be clever!)

20. $u = 3x^5(x^2 - 4) \cos 10x$ (Be clever!)

21. *Product of Three Functions Problem:* Prove that if $y = uvw$, where $u, v,$ and w are differentiable functions of x, then $y' = u'vw + uv'w + uvw'$.

22. *Product of n Functions Conjecture Problem:* Make a conjecture about what an equation for y' would be if $y = u_1 u_2 u_3 \ldots u_n$, where $u_1, \ldots, u_n$ are differentiable functions of x.

For Problems 23–26, differentiate and simplify.

23. $z = x^5 \cos^6 x \sin 7x$

24. $y = 4x^6 \sin^3 x \cos 5x$

25. $y = x^4 (\ln x)^5 \sin x \cos 2x$

26. $u = x^5 e^{2x} \cos 2x \sin 3x$

27. *Bouncing Spring Problem:* A weight is suspended on a spring above a table top. As the weight bounces up and down, friction decreases the amplitude of the motion. Figure 4-2c shows the displacement, $y(t)$, in centimeters, as a function of time, t, in seconds.

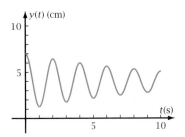

Figure 4-2c

a. If $y(t) = 4 + 3e^{-0.1t} \cos \pi t$, find an equation for the velocity, $v(t)$.

b. There appears to be a high point at $t = 2$. Show that this is *not* a high point by showing that $v(2) \neq 0$. Set $v(t) = 0$ and solve numerically to find the t-value of the high point close to $t = 2$.

28. *Tacoma Narrows Bridge Problem:* If a structure is shaken at its natural resonant frequency, it vibrates with an increasing amplitude.

a. Suppose that an object vibrates in such a way that its displacement, $y(t)$, in feet, from its rest position is given by $y(t) = t \sin t$ where t is time in seconds. Find an equation for $v(t)$. Plot $v(t)$ on your grapher. If it does not agree with Figure 4-2d, go back and correct your work.

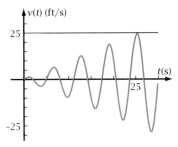

Figure 4-2d

b. The velocity seems to reach a high point of 25 ft/s close to $t = 25$. Zoom in on point $(25, 25)$ and state whether the graph of v just falls short of 25, just exceeds 25, or reaches 25 at this high point. Illustrate your conclusion with a sketch.

c. Look up the Tacoma Narrows Bridge on the Internet or other source. How do your findings relate to this problem? Give the Web site or other source you found.

29. *Odd and Even Functions Derivative Problem:* A function is called an **odd function** if it has the property $f(-x) = -f(x)$. Similarly, f is called an **even function** if $f(-x) = f(x)$. For instance, sine is odd because $\sin(-x) = -\sin x$, and cosine is even because $\cos(-x) = \cos x$. Use the chain rule appropriately to prove that the derivative of an odd function is an even function and that the derivative of an even function is an odd function.

30. *Double Argument Properties Problem:* Let $f(x) = 2\sin x \cos x$ and let $g(x) = \sin 2x$. Find $f'(x)$ and $g'(x)$. Use the appropriate trigonometric properties to show that $f'(x)$ and $g'(x)$ are equivalent. Then show that $f(x)$ and $g(x)$ are also equivalent. Do the same for the functions $f(x) = \cos^2 x - \sin^2 x$ and $g(x) = \cos 2x$.

31. *Derivative of a Power Induction Problem:* Prove by mathematical induction that for any positive integer n, if $f(x) = x^n$, then $f'(x) = nx^{n-1}$.

32. *Derivative Two Ways Problem:* You can differentiate the function $y = (x + 3)^8(x - 4)^8$ by either of two methods: First, consider the function as a product of two composite functions; second, multiply and differentiate the result, $y = (x^2 - x - 12)^8$. Show that both methods give the same result for the derivative.

33. *Confirmation of the Product Rule:* Let $f(x) = x^3 \cdot \sin x$ (Figure 4-2e).

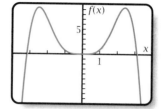

Figure 4-2e

a. Sketch this graph, then draw what you think the derivative graph, f', would look like. Show, especially, places where $f'(x)$ would equal zero.

b. Write an equation for $f'(x)$. Plot f and f' on your grapher. How does the graph you predicted compare with the actual graph?

c. Plot the numerical derivative on your grapher. How does this graph confirm that your equation for $f'(x)$ is correct?

34. *Repeated Roots Problem:* In this problem you will sketch the graph of the function $f(x) = (5x - 7)^4(2x + 3)^5$.

a. On your grapher, plot the graph of f. Use an x-window from -2 to 2 and a y-window that sufficiently fits the graph. Sketch the result.

b. Find $f'(x)$. Simplify. For example, factor out any common factors.

c. Find all three values of x for which $f'(x) = 0$. The factored form of $f'(x)$ from part b is convenient for this purpose.

d. Find $f(x)$ for each of the values of x in part c. Show these three points on your graph.

e. The graph of f should have a horizontal tangent line at each of the points in part d. State whether the following statement is true or false: The graph has a high or low point where $f'(x) = 0$.

136

35. *Pole Dance Problem:* In a variation of a Filipino pole dance, two pairs of bamboo poles are moved back and forth at floor level. The dancer steps into and out of the region between the poles (Figure 4-2f), trying to avoid being pinched between them as they come together. The area, $A = LW$, varies with time, t, where distances are in feet and time is in seconds.

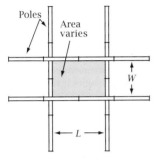

Figure 4-2f

Pole dancing in the Mokantarak village of Flores, Indonesia

a. Write dA/dt in terms of $L, W, dL/dt,$ and dW/dt.

b. Suppose that $W = 2 + 2\cos t$ and that $L = 3 + 2\sin 2t$. At what rate is the area changing when $t = 4$? When $t = 5$? At these times is the area increasing or decreasing?

4-3   Derivative of a Quotient of Two Functions

In this section you will learn how to find the derivative of a quotient of two functions. For example,

$$f(x) = \frac{\sin 5x}{8x - 3}$$

OBJECTIVE   Given a function whose equation is a quotient of two other functions, find an equation for the derivative function in one step and simplify the answer.

Suppose that $f(x) = g(x)/h(x)$. Using $y, u,$ and v for the three function values lets you write the equation in a simpler form.

$$y = \frac{u}{v}$$

By the definition of derivative,

$$\frac{dy}{dx} = \lim_{\Delta x \to 0} \frac{\Delta y}{\Delta x} = \lim_{\Delta x \to 0} \left(\frac{\dfrac{u + \Delta u}{v + \Delta v} - \dfrac{u}{v}}{\Delta x} \right)$$

Multiplying the numerator and the denominator by $(v + \Delta v)(v)$ eliminates the complex fraction.

$$\frac{dy}{dx} = \lim_{\Delta x \to 0} \left[\frac{(u + \Delta u)v - u(v + \Delta v)}{\Delta x(v + \Delta v)(v)} \right]$$

Further algebra allows you to take the limit.

$$\frac{dy}{dx} = \lim_{\Delta x \to 0} \left[\frac{uv + \Delta u v - uv - u\Delta v}{\Delta x(v + \Delta v)(v)} \right]$$

$$= \lim_{\Delta x \to 0} \left[\frac{1}{(v + \Delta v)(v)} \cdot \frac{\Delta u v - u\Delta v}{\Delta x} \right]$$
Simplify, then associate Δx with the numerator.

$$= \lim_{\Delta x \to 0} \left[\frac{1}{(v + \Delta v)(v)} \cdot \left(\frac{\Delta u}{\Delta x} \cdot v - u \cdot \frac{\Delta v}{\Delta x} \right) \right]$$
Distribute Δx, then associate it with Δu and Δv.

$$= \frac{1}{v^2} \cdot \left(\frac{du}{dx} \cdot v - u \cdot \frac{dv}{dx} \right)$$
Limits of products, quotient, and sum; definition of derivative; $\Delta v \to 0$ as $\Delta x \to 0$.

$$y' = \frac{1}{v^2}(u'v - uv') = \frac{u'v - uv'}{v^2}$$
Where u' and v' are derivatives with respect to x.

PROPERTY: Derivative of a Quotient of Two Functions—The Quotient Rule

If $y = \dfrac{u}{v}$, where u and v are differentiable functions, and $v \neq 0$, then

$$y' = \frac{u'v - uv'}{v^2}$$

Verbally: Derivative of top times bottom, minus top times derivative of bottom, all divided by bottom squared.

Note that the numerator has the same pattern as the product rule, namely $u'v + uv'$, except that a subtraction sign $(-)$ appears instead of an addition sign $(+)$.

▶ **EXAMPLE 1**   Differentiate: $f(x) = \dfrac{\sin 5x}{8x - 3}$

Solution   Use the quotient rule, where $u = \sin 5x$ and $v = 8x - 3$

$$f'(x) = \frac{\cos 5x(5)(8x - 3) - \sin 5x(8)}{(8x - 3)^2}$$

$$f'(x) = \frac{5 \cos 5x(8x - 3) - 8 \sin 5x}{(8x - 3)^2}$$ ◀

As you differentiate, say to yourself, "Derivative of top times bottom, minus top times derivative of bottom, all divided by bottom squared." You must also, of course, apply the chain rule when necessary.

▶ **EXAMPLE 2**   Differentiate: $y = \dfrac{(5x - 2)^7}{(4x + 9)^3}$

Solution

$$y' = \frac{7(5x-2)^6(5) \cdot (4x+9)^3 - (5x-2)^7 \cdot 3(4x+9)^2(4)}{(4x+9)^6}$$

$$= \frac{(5x-2)^6(4x+9)^2[35(4x+9) - 12(5x-2)]}{(4x+9)^6} \qquad \text{Factor the numerator.}$$

$$= \frac{(5x-2)^6(80x+339)}{(4x+9)^4} \qquad \text{Cancel common factors.} \qquad \blacktriangleleft$$

▶ **EXAMPLE 3**   Differentiate: $\dfrac{d}{dx}\left(\dfrac{5}{7x^3}\right)$

Solution   Although you can use the quotient rule here, it is simpler to transform the expression to a power and use the power rule.

$$\frac{d}{dx}\left(\frac{5}{7x^3}\right) = \frac{d}{dx}\left(\frac{5}{7}x^{-3}\right) = -\frac{15}{7}x^{-4} \qquad \blacktriangleleft$$

Problem Set 4-3

Q1. Find: $\dfrac{d}{dx}(x^{1066})$

Q2. Antidifferentiate: $f'(x) = 60x^4$

Q3. Find y': $y = x^3 \sin x$

Q4. Find $\dfrac{dy}{dx}$: $y = \cos(x^7)$

Q5. Differentiate: $f(x) = (3^5)(2^8)$

Q6. Find $a'(t)$: $a(t) = 6e^{9t}$

Q7. Write the definition of derivative at a point.

Q8. Write the physical meaning of derivative.

Q9. Factor: $(x-3)^5 + (x-3)^4(2x)$

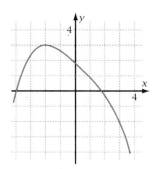

Figure 4-3a

Q10. Sketch the graph of the derivative of the function shown in Figure 4-3a.

For Problems 1–26, differentiate and simplify. You may check your answer by comparing its graph with the graph of the numerical derivative.

1. $f(x) = \dfrac{x^3}{\sin x}$

2. $f(x) = \dfrac{x^4}{\cos x}$

3. $g(x) = \dfrac{\cos^3 x}{\ln x}$

4. $h(x) = \dfrac{\sin^5 x}{e^{3x}}$

5. $y = \dfrac{\sin 10x}{\cos 20x}$

6. $y = \dfrac{\cos 12x}{\sin 18x}$

7. Find y' if $y = \dfrac{3x-7}{6x+5}$.

8. Find y' if $y = \dfrac{10x+9}{5x-3}$.

9. Find $\dfrac{dz}{dx}$ if $z = \dfrac{(8x+1)^6}{(5x-2)^9}$.

10. Find $\dfrac{dA}{dx}$ if $A = \dfrac{(4x - 1)^7}{(7x + 2)^4}$.

11. Find Q' if $Q = \dfrac{e^{x^3}}{\sin x}$.

12. Find r' if $r = \dfrac{\ln x^4}{\cos x}$.

13. Find: $\dfrac{d}{dx}(60x^{-4/3})$

14. Find: $\dfrac{d}{dx}(24x^{-7/3})$

15. $r(x) = \dfrac{12}{x^3}$ (Be clever!)

16. $t(x) = \dfrac{51}{x^{17}}$ (Be clever!)

17. $v(x) = \dfrac{14}{\cos 0.5x}$

18. $a(x) = \dfrac{20}{\sin^2 x}$

19. $r(x) = \dfrac{1}{x}$

20. $s(x) = \dfrac{1}{x^2}$

21. Find $W'(x)$ if $W(x) = \dfrac{10}{(x^3 - 1)^{-5}}$.

22. Find $T'(x)$ if $T(x) = \dfrac{1}{\cos x \sin x}$.

23. $T(x) = \dfrac{\sin x}{\cos x}$

24. $C(x) = \dfrac{\cos x}{\sin x}$

25. Find C' if $C = \dfrac{1}{\sin x}$.

26. Find S' if $S = \dfrac{1}{\cos x}$.

27. *Black Hole Problem:* Ann's spaceship gets trapped in the gravitational field of a black hole! Her velocity, $v(t)$, in miles per hour, is given by

$$v(t) = \dfrac{1000}{3 - t}$$

where t is time in hours.

The black hole Sagittarius A, shown here, is located at the center of the Milky Way galaxy.

a. How fast is she going when $t = 1$? When $t = 2$? When $t = 3$?

b. Write an equation for her acceleration, $a(t)$. Use this equation to find her acceleration when $t = 1$, $t = 2$, and $t = 3$. What are the units of acceleration in this problem?

c. Using the same screen, plot graphs of her velocity and acceleration as functions of time. Sketch the results.

d. Ann will be in danger if the acceleration exceeds 500 (mi/h)/h. For what range of times is Ann's acceleration below the danger point?

28. *Catch-Up Rate Problem:* Willie Ketchup is out for his morning walk. He sees Betty Wont walking ahead of him and decides to catch up to her. He quickly figures that she is walking at a rate of 5 ft/s. He lets x, in ft/s, stand for his walking rate.

a. Explain why Willie catches up at the rate of $(x - 5)$ ft/s.

b. Betty is 300 ft ahead of Willie when he first begins to catch up. Recall that distance = rate × time to write an equation for $t(x)$, the number of seconds it will take him to catch up to her. Use the equation to find out how long it will take Willie to catch up if he walks at rates of 6, 8, 10, 5, 4, and 5.1 ft/s. What is a reasonable domain for x?

c. What is the instantaneous rate of change of Willie's catch-up time if he is going 6 ft/s? What are the units of this rate?

d. Explain why $t(x)$ has no value for the derivative at $x = 5$.

29. *Confirmation of Quotient Rule Problem:* Find $f'(x)$ for the function

$$f(x) = \frac{3x + 7}{2x + 5}$$

Use the answer to evaluate $f'(4)$. Use $x = 4.1, 4.01,$ and 4.001 to show that the difference quotient $[f(x) - f(4)]/(x - 4)$ gets closer and closer to $f'(4)$.

30. *Derivative Graph and Table Problem:*

a. For the function f in Figure 4-3b, sketch what you think the graph of the derivative, f', looks like.

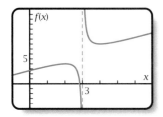

Figure 4-3b

b. The equation for $f(x)$ is

$$f(x) = \frac{x^2 - 8}{x - 3}$$

Find the algebraic derivative, $f'(x)$. On the same screen, plot the graphs of $y_1 = f(x)$, $y_2 = f'(x)$, and $y_3 =$ the numerical derivative. Use a friendly window that includes $x = 3$ as a grid point. Does your algebraic derivative agree with the numerical derivative? Does your sketch in part a agree with the graph of the actual derivative?

c. Make a table of values of $f(x)$ and $f'(x)$ for each 0.01 unit of x from 2.95 to 3.05. Based on your table and graphs, describe the way $f(x)$ changes as x approaches 3.

d. Use the maximum and minimum features on your grapher to find the relative maximum and minimum points in Figure 4-3b. Show that $f'(x) = 0$ at these points.

e. What are the domain and the range of f? Of f'?

31. *Proof of the Power Rule for Negative Exponents:* The proof you used in Section 3-4 for the power rule for derivatives assumes that the exponent is a positive integer. You have seen by example that the rule also works for some powers with negative exponents. Suppose that $y = x^{-5}$. To prove that $y' = -5x^{-6}$, use the quotient rule of this section and write y as

$$y = \frac{1}{x^5}$$

Prove that, in general, if $y = x^n$, where n is a negative constant, then $y' = nx^{n-1}$. To do this, it helps to write y as

$$y = \frac{1}{x^p}$$

where p is a positive number equal to the opposite of n.

32. Figures 4-3c and 4-3d show the graphs of $y = \sec x$ and $y = \tan x$. Using a copy of each graph, sketch what you think the graph of the derivative looks like. Check your answers by grapher, using the numerical derivative feature.

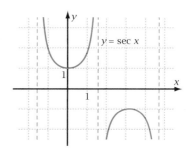

Figure 4-3c

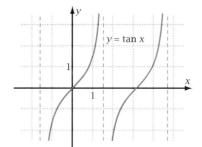

Figure 4-3d

33. *Journal Problem:* Update your journal with what you've learned. Include such things as

- The one most important thing you have learned since your last journal entry
- The primary difference between differentiating a sum of two functions and a product or a quotient of two functions

- The meaning of a parametric function
- What you better understand about the meaning of derivative
- Any technique or idea about derivatives that is still unclear to you

4-4 Derivatives of the Other Trigonometric Functions

Recall from trigonometry that you can write the tangent, cotangent, secant, and cosecant functions in terms of sine and cosine.

$$\tan x = \frac{\sin x}{\cos x} \qquad \cot x = \frac{\cos x}{\sin x} \qquad \sec x = \frac{1}{\cos x} \qquad \csc x = \frac{1}{\sin x}$$

Each of these is a quotient. Now that you can differentiate quotients, you can differentiate the other four trigonometric functions.

OBJECTIVE   Given a function whose equation contains any of the six trigonometric functions, find the equation for the derivative function in one step.

Derivative of Tangent and Cotangent Functions

$$y = \tan x = \frac{\sin x}{\cos x}$$

$$\therefore y' = \frac{(\cos x)(\cos x) - (\sin x)(-\sin x)}{\cos^2 x} = \frac{\cos^2 x + \sin^2 x}{\cos^2 x}$$

$$= \frac{1}{\cos^2 x} = \sec^2 x \qquad \text{Derivative of tangent.}$$

In Problem 29 of Problem Set 4-4, you will show that if $y = \cot x$, then $y' = -\csc^2 x$.

Derivative of Secant and Cosecant Functions

$$y = \sec x = \frac{1}{\cos x}$$

$$\therefore y' = \frac{(0)(\cos x) - (1)(-\sin x)}{\cos^2 x} = \frac{\sin x}{\cos^2 x} = \frac{1}{\cos x} \cdot \frac{\sin x}{\cos x}$$

$$= \sec x \tan x \qquad \text{Derivative of secant.}$$

In Problem 30 of Problem Set 4-4, you will show that if $y = \csc x$, then $y' = -\csc x \cot x$.

PROPERTIES: Derivatives of the Six Trigonometric Functions

$$\sin' x = \cos x \qquad\qquad \cos' x = -\sin x$$
$$\tan' x = \sec^2 x \qquad\qquad \cot' x = -\csc^2 x$$
$$\sec' x = \sec x \tan x \qquad \csc' x = -\csc x \cot x$$

Note: x must be in radians because this was assumed for the sine derivative.

Memory Aids:

- The derivatives of the "co-" functions have a negative sign $(-)$.

- To find the derivatives of the "co-" functions in the right column, replace each function in the left column with its cofunction (for example, sec with csc).

▶ **EXAMPLE 1** Differentiate: $y = 3 \tan^5 7x$

Solution
$$y' = 3(5 \tan^4 7x)(\sec^2 7x)(7)$$
$$= 105 \tan^4 7x \sec^2 7x$$ ◀

Note that this example involves two applications of the chain rule. The outermost function is the fifth power function. This fact is easier to see if you write the original function as

$$y = 3(\tan 7x)^5$$

The next function in is tan. Its derivative is $\sec^2$. The innermost function is $7x$; its derivative is 7. You should begin to see the "chain" of derivatives that gives the chain rule its name.

$$y' = 3(5 \tan^4 7x) \cdot (\sec^2 7x) \cdot (7)$$
$$\qquad\quad \uparrow \qquad\qquad \uparrow \qquad \uparrow$$

Here are the three "links" in the chain.

Problem Set 4-4 gives you practice in differentiating all six trigonometric functions.

Problem Set 4-4

Quick Review 5 min

Q1. $(\sin x)/(\tan x) = $ —?—

Q2. $1/(\sec x) = $ —?—

Q3. $\cos^2 3 + \sin^2 3 = $ —?—

Q4. Differentiate: $f(x) = e^x \sin x$

Q5. Differentiate: $g(x) = x/(\cos x)$

Q6. Differentiate: $h(x) = (3x)^{-5/7}$

Q7. Find dy/dx: $y = (\cos x)^{-3}$

Q8. Find: $\lim_{x \to 2}(x^2 - 7x + 10)/(x - 2)$

Q9. Sketch the graph of the derivative for the function shown in Figure 4-4a.

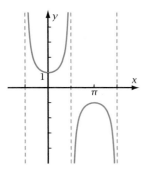

Figure 4-4a

Q10. The equation $\lim_{\Delta p \to 0} \dfrac{\Delta j}{\Delta p} = \dfrac{dj}{dp}$ is an expression of

    A. Derivative of a quotient property

    B. Limit of a quotient property

    C. Definition of derivative

    D. Definition of limit

    E. None of these

For Problems 1–28, differentiate and perform obvious simplification. You may check each answer by comparing its graph with the graph of the numerical derivative.

1. $f(x) = \tan 5x$

2. $f(x) = \sec 3x$

3. $y = \sec x^7$

4. $z = \tan x^9$

5. Find $g'(x)$: $g(x) = \cot e^{11x}$

6. Find $h'(x)$: $h(x) = \csc e^{10x}$

7. $r(x) = \ln(\csc x)$

8. $p(x) = \ln(\cot x)$

9. Find $\dfrac{d}{dx}(y)$: $y = \tan^5 4x$

10. Find $\dfrac{d}{dx}(y)$: $y = \tan^7 9x$

11. $\dfrac{d}{dx}(\sec x \tan x)$

12. $\dfrac{d}{dx}(\csc x \cot x)$

13. $y = \sec x \csc x$

14. $y = \tan x \cot x$

15. $y = \dfrac{\tan x}{\sin x}$

16. $y = \dfrac{\cot x}{\cos x}$

17. $y = \dfrac{5 \ln 7x}{\cot 14x}$

18. $y = \dfrac{4 \csc 10x}{e^{40x}}$

19. $w = \tan(\sin 3x)$

20. $t = \sec(\cos 4x)$

21. $S(x) = \sec^2 x - \tan^2 x$

22. $m(x) = \cot^2 x - \csc^2 x$

23. $A(x) = \sin x^2$

24. $f(x) = \cos x^3$

25. $F(x) = \sin^2 x$

26. $g(x) = \cos^3 x$

27. Find $\dfrac{d^2 y}{dx^2}$: $y = \tan x$

28. Find y'': $y = \sec x$

29. *Derivative of Cotangent Problem:* Derive the formula for y' if $y = \cot x$. You may write $\cot x$ either as $(\cos x)/(\sin x)$ or as $1/(\tan x)$.

30. *Derivative of Cosecant Problem:* Derive the formula for y' if $y = \csc x$ by first transforming $\csc x$ into a function you already know how to differentiate.

31. *Confirmation of Tangent Derivative Formula:* Figure 4-4b shows the graph of $f(x) = \tan x$ as it might appear on your grapher.

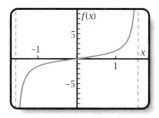

Figure 4-4b

a. Sketch the graph. Without using your grapher, sketch what you think the graph of the derivative, f', would look like.

b. Find an equation for $f'(x)$. Plot the graphs of f and f' on the same screen. Did the f' you predicted look like the actual one? If not, write down where your thinking went astray and what you learned from this problem.

c. Calculate the symmetric difference quotient for $f'(1)$, using $\Delta x = 0.01$. How close is the answer to the actual value of $f'(1)$?

32. *Confirmation of Secant Derivative Formula:* Figure 4-4c shows an accurate graph of $f(x) = \sec x$ in the closed interval $[-\pi/2, \pi/2]$.

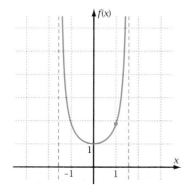

Figure 4-4c

a. Write the formula for $f'(x)$. Use it to find the value of $f'(1)$.

b. Plot $y_1 = f(x)$. On the same screen, plot the line y_2 through the point $(1, f(1))$ with slope equal to $f'(1)$. Sketch the result. How do the two graphs confirm that the derivative formula for secant gives the correct answer at $x = 1$?

c. Deactivate y_2 and on the same screen as y_1 plot the graph of $y_3 = f'(x)$ from part a. Sketch the result. What is happening to the graph of f at values of x where $f'(x)$ is negative?

33. *Light on the Monument Problem:* Suppose you stand 10 ft away from the base of the Washington Monument and shine a flashlight at it (Figure 4-4d). Let x be the angle, in radians, the light beam makes with the horizontal line. Let y be the vertical distance, in feet, from the flashlight to the spot of light on the wall.

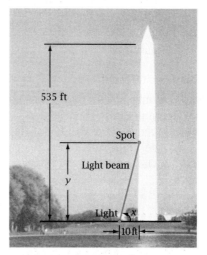

Figure 4-4d

a. Show that $y = 10 \tan x$.

b. As you rotate the flashlight upward, at what rate is y increasing with respect to x when $x = 1$? What are the units of this rate? What is this rate in feet per *degree*?

c. At what rate does y change when your light points at the top of the vertical monument wall, where $y = 535$?

34. *Point of Light Problem:* As the beacon light at an airport rotates, the narrow light beam from it forms a moving point of light on objects in its path. The north-south wall of a building is

500 ft east of the beacon (Figure 4-4e). Let x be the angle, in radians, the light beam makes with the east–west line. Let y be the distance, in feet, north of this line where the light beam is shining when the angle is x.

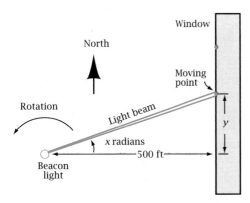

Figure 4-4e

a. Show that $y = 500 \tan x$.

b. Let t be the number of seconds since the beacon was pointed due east. Find dy/dt. (Because x depends on t, x is an inside function. By the chain rule, the answer will contain dx/dt.)

c. Suppose that the beacon is rotating counterclockwise at 0.3 rad/s. What does dx/dt equal? How fast is the point of light moving along the wall when it passes the window located at $y = 300$?

35. *Antiderivative Problem:* Write an equation for the antiderivative of each function. Remember "$+C$"!

   a. $y' = \cos x$

   b. $y' = \sin 2x$

   c. $y' = \sec^2 3x$

   d. $y' = \csc^2 4x$

   e. $y' = 5 \sec x \tan x$

36. *Journal Problem:* Update your journal with what you've learned. Include such things as

   • The one most important thing you have learned since your last journal entry

   • The extension you have made in the power rule

   • How to differentiate all six trigonometric functions, using what you have recently learned

   • Any questions you need to ask in class

4-5   Derivatives of Inverse Trigonometric Functions

You have learned how to differentiate the six trigonometric functions. In this section you will explore the inverses of these functions.

OBJECTIVE   Differentiate each of the six inverse trigonometric functions.

Background Item 1: Inverse of a Function

The graph on the left in Figure 4-5a shows how the population of a certain city might grow as a function of time. If you are interested in finding the time when the population reaches a certain value, it may be more convenient to reverse the variables and write time as a function of population. The relation you get by interchanging the two variables is called the **inverse** of the original function. The graph of the inverse is the graph on the right in Figure 4-5a.

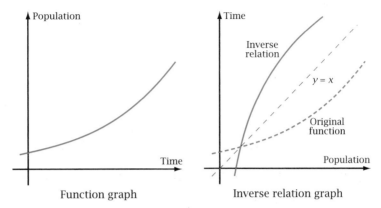

Figure 4-5a

For a linear function such as $y = 2x + 6$, interchanging the variables gives $x = 2y + 6$, which is equivalent to $y = 0.5x - 3$. So the inverse relation for a linear function is another function. However, the inverse relation of a trigonometric function is not a function. As shown in Figure 4-5b, the relation $x = \tan y$ has multiple values of y for the same value of x.

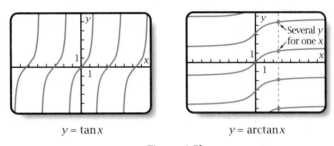

Figure 4-5b

In this text, $y = \arctan x$ is used to denote the inverse *relation* for $y = \tan x$. The notation "arctan x" means "an arc (or angle in radians) whose tangent is x." You can use parametric mode to plot $y = \arctan x$ on your grapher. To get the graph on the right in Figure 4-5b, use a suitable t-range and enter

$$x = \tan t \qquad \text{and} \qquad y = t$$

If an inverse relation turns out to be a function, as for linear functions, then the function is said to be **invertible,** and the inverse relation is called the **inverse function.** In this case, you can use $f(x)$ terminology for the function and its inverse. For example,

If $f(x) = 2x + 6$, then $f^{-1}(x) = 0.5x - 3$.

The notation $f^{-1}(x)$ is read "f-inverse of x." The -1 exponent used with a function name means *function* inverse, unlike the -1 exponent used with a numeral, which means "reciprocal," or *multiplicative* inverse. For functions, the -1 exponent expresses the fact that f^{-1} "undoes" what was done by the function. That is, $f^{-1}(f(x)) = x$ and $f(f^{-1}(x)) = x$. For instance,

If $f(x) = \sqrt{x}$, then $f^{-1}(x) = x^2$, so that $f^{-1}(f(x)) = (\sqrt{x})^2 = x$.

Notice in Figure 4-5a that if the same scales are used for both axes, then the graphs of f and f^{-1} are mirror images with respect to the 45-degree line $y = x$.

DEFINITIONS, SYMBOL, AND PROPERTIES: *Inverse of a Function*

Definition: If $y = f(x)$, then the **inverse** of function f has the equation $x = f(y)$.

Definition: If the inverse relation is a function, then f is said to be **invertible.**

Notation: If f is invertible, then you can write the **inverse function** as $y = f^{-1}(x)$.

Property: If f is invertible, then $f(f^{-1}(x)) = x$, and $f^{-1}(f(x)) = x$.

Property: The graphs of f and f^{-1} are mirror images across the line $y = x$.

Background Item 2: Inverse Trigonometric Functions

You can make an inverse trigonometric relation such as $y = \arctan x$ a function by restricting the range. Each cycle of $y = \arctan x$ shown in Figure 4-5b is called a branch. The branch that contains the origin is called the **principal branch**. If y is restricted to the open interval $(-\pi/2, \pi/2)$, then the inverse relation is a function and you can use $f(x)$ terminology.

$$y = \tan^{-1} x \qquad \text{if and only if} \qquad \tan y = x \text{ and } y \in (-\pi/2, \pi/2)$$

If you evaluate the inverse tangent function on your grapher, the answer given is the value on this principal branch. Figure 4-5c shows the principal branch of $y = \tan x$ and the function $y = \tan^{-1} x$, and the fact that the graphs are mirror images with respect to the line $y = x$.

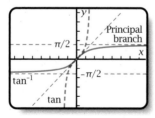

Figure 4-5c

The other five inverse trigonometric functions are defined the same way. For each one, a principal branch is a function. The principal branch is near the origin, continuous if possible, and positive if there is a choice between two branches. The graphs are shown in Figure 4-5d.

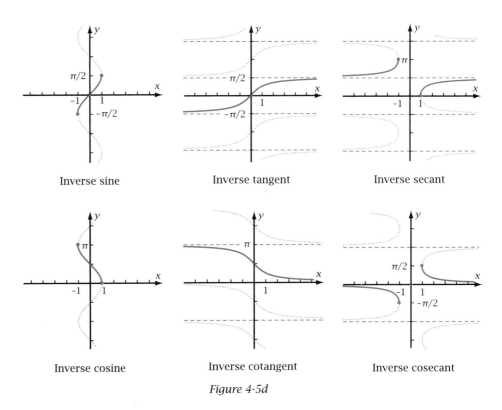

Inverse sine

Inverse tangent

Inverse secant

Inverse cosine

Inverse cotangent

Inverse cosecant

Figure 4-5d

The definitions and ranges of the inverse trigonometric functions are summarized in this box.

DEFINITIONS: *Inverse Trigonometric Functions (Principal Branches)*

$y = \sin^{-1} x$ if and only if $\sin y = x$ and $y \in \left[-\frac{\pi}{2}, \frac{\pi}{2} \right]$

$y = \cos^{-1} x$ if and only if $\cos y = x$ and $y \in [0, \pi]$

$y = \tan^{-1} x$ if and only if $\tan y = x$ and $y \in \left(-\frac{\pi}{2}, \frac{\pi}{2} \right)$

$y = \cot^{-1} x$ if and only if $\cot y = x$ and $y \in (0, \pi)$

$y = \sec^{-1} x$ if and only if $\sec y = x$ and $y \in [0, \pi]$, but $x \neq \frac{\pi}{2}$

$y = \csc^{-1} x$ if and only if $\csc y = x$ and $y \in \left[-\frac{\pi}{2}, \frac{\pi}{2} \right]$, but $x \neq 0$

Algebraic Derivative of the Inverse Tangent Function

Example 1 shows how to differentiate the inverse tangent function. The definition of inverse function lets you turn this new problem into the old problem of differentiating the tangent.

▶ **EXAMPLE 1**   Differentiate: $y = \tan^{-1} x$

Solution   Use the definition of $\tan^{-1}$ to write the equation in terms of tangent.

$$y = \tan^{-1} x \Rightarrow \tan y = x$$
$$\sec^2 y \cdot y' = 1$$

The derivative of tan is $\sec^2$. Because y depends on x, it is an *inside* function. The y' is the derivative of this inside function (from the chain rule).

$$y' = \frac{1}{\sec^2 y}$$

Use algebra to solve for y'.

$$= \cos^2 y$$

To find y' in terms of x, consider that y is an angle whose tangent and cosine are being found. Draw a right triangle with angle y in standard position (Figure 4-5e). By trigonometry,

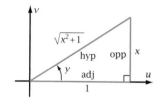

Figure 4-5e

$$\tan y = \frac{\text{opposite leg}}{\text{adjacent leg}}$$

Because $\tan y = x$, which equals $x/1$, put x on the opposite leg and 1 on the adjacent leg. The hypotenuse is thus $\sqrt{x^2 + 1}$. Cosine equals (adjacent)/(hypotenuse), so you can write

$$y' = \frac{1}{\left(\sqrt{x^2 + 1}\right)^2} = \frac{1}{x^2 + 1}$$ ◀

In Example 1, both the left and right sides of the equation $\tan y = x$ are functions of x. As you learned in Section 3-9, the technique of differentiating both sides of such an equation with respect to x is called **implicit differentiation**. You will study this technique more extensively in Section 4-8.

Derivative of the Inverse Secant Function

The derivative of the inverse secant is tricky. Example 2 shows what happens.

▶ **EXAMPLE 2**   Differentiate: $y = \sec^{-1} x$

Solution   Transform the new problem into an old problem.

$$y = \sec^{-1} x \Rightarrow \sec y = x$$
$$\sec y \tan y \cdot y' = 1$$

Remember the chain rule! That's where y' comes from.

$$y' = \frac{1}{\sec y \tan y}$$

Use algebra to solve for y'.

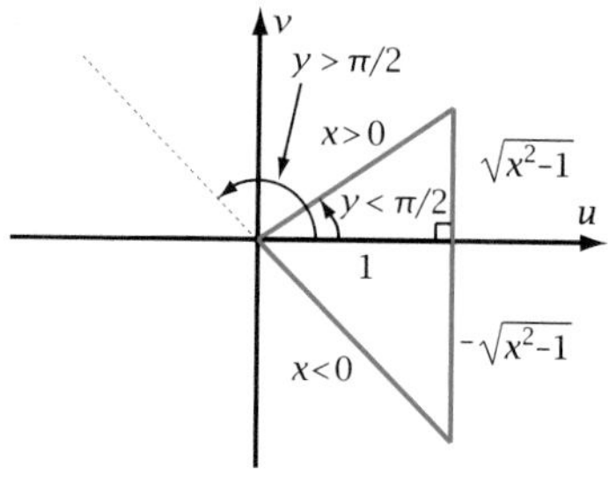

Consider y to be an angle in standard position as shown in Figure 4-5f. Secant equals (hypotenuse)/(adjacent) and $\sec y = x = x/1$, so write x on the hypotenuse and 1 on the adjacent leg. By the Pythagorean theorem, the third side is $\pm\sqrt{x^2 - 1}$. The range of $\sec^{-1}$ is $y \in [0, \pi]$ (excluding $y = \pi/2$), so y can terminate in Quadrant I or II. Where $\sec y$ is negative, y terminates in Quadrant II. In this case, the hypotenuse, x, is negative, so you must draw it in the negative direction, opposite the terminal side of y, as you do with negative radii in polar coordinates. (You'll learn about polar coordinates in

Figure 4-5f

Chapter 8.) As shown in Figure 4-5f, you pick the negative square root for the vertical coordinate. So the denominator of the derivative, sec y tan y, is

$$x\sqrt{x^2 - 1} \quad \text{if } x > 0 \qquad \text{or} \qquad -x\sqrt{x^2 - 1} \quad \text{if } x < 0$$

In both cases, this quantity is positive. To avoid two different representations, use the notation $|x|$ and write the derivative in this way.

$$y' = \frac{1}{|x|\sqrt{x^2 - 1}} \qquad \blacktriangleleft$$

The derivatives of the six inverse trigonometric functions are shown here. In Problem Set 4-5 you will derive the four properties not yet discussed.

PROPERTIES: *Derivatives of the Six Inverse Trigonometric Functions*

$$\frac{d}{dx}(\sin^{-1} x) = \frac{1}{\sqrt{1 - x^2}} \qquad \frac{d}{dx}(\cos^{-1} x) = -\frac{1}{\sqrt{1 - x^2}}$$

$$\frac{d}{dx}(\tan^{-1} x) = \frac{1}{1 + x^2} \qquad \frac{d}{dx}(\cot^{-1} x) = -\frac{1}{1 + x^2}$$

$$\frac{d}{dx}(\sec^{-1} x) = \frac{1}{|x|\sqrt{x^2 - 1}} \qquad \frac{d}{dx}(\csc^{-1} x) = -\frac{1}{|x|\sqrt{x^2 - 1}}$$

Note: Your grapher must be in radian mode.

Memory Aid: The derivative of each "co-" inverse function is the *opposite* of the derivative of the corresponding inverse function because each co-inverse function is decreasing at $x = 0$ or 1 (Figure 4-5d).

▶ **EXAMPLE 3**   Differentiate: $y = \cos^{-1} e^{3x}$

Solution $y = \cos^{-1} e^{3x}$

$$y' = -\frac{1}{\sqrt{1 - (e^{3x})^2}} \cdot 3e^{3x} = -\frac{3e^{3x}}{\sqrt{1 - e^{6x}}} \qquad \text{Use the chain rule on the inside function.} \qquad \blacktriangleleft$$

Problem Set 4-5

Quick Review

Q1. $\sin' x = $ —?—

Q2. $\cos' x = $ —?—

Q3. $\tan' x = $ —?—

Q4. $\cot' x = $ —?—

Q5. $\sec' x = $ —?—

Q6. $\csc' x = $ —?—

Refer to Figure 4-5g for Problems Q7–Q10.

Q7. $f'(1) = $ —?—

Q8. $f'(3) = $ —?—

Q9. $f'(4) = $ —?—

Q10. $f'(6) = $ —?—

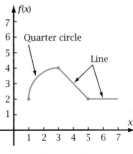

Figure 4-5g

For Problems 1–4, duplicate the graphs in Figure 4-5d on your grapher. For Problems 3 and 4, recall that $\csc y = 1/\sin y$ and $\cot y = 1/\tan y$.

1. $y = \cos^{-1} x$

2. $y = \sin^{-1} x$

3. $y = \csc^{-1} x$

4. $y = \cot^{-1} x$

5. Explain why the principal branch of the inverse cotangent function goes from 0 to π rather than from $-\pi/2$ to $\pi/2$.

6. Explain why the principal branch of the inverse secant function cannot be continuous.

7. Evaluate: $\sin(\sin^{-1} 0.3)$

8. Evaluate: $\cos^{-1}(\cos 0.8)$

For Problems 9–12, derive the formula.

9. $\dfrac{d}{dx}(\sin^{-1} x) = \dfrac{1}{\sqrt{1 - x^2}}$

10. $\dfrac{d}{dx}(\cos^{-1} x) = -\dfrac{1}{\sqrt{1 - x^2}}$

11. $\dfrac{d}{dx}(\csc^{-1} x) = -\dfrac{1}{|x|\sqrt{x^2 - 1}}$

12. $\dfrac{d}{dx}(\cot^{-1} x) = -\dfrac{1}{1 + x^2}$

For Problems 13–24, find the derivative algebraically.

13. $y = \sin^{-1} 4x$

14. $y = \cos^{-1} 10x$

15. $y = \cot^{-1} e^{0.5x}$

16. $y = \tan^{-1}(\ln x)$

17. $y = \sec^{-1} \frac{x}{3}$

18. $y = \csc^{-1} \frac{x}{10}$

19. $y = \cos^{-1} 5x^2$

20. $f(x) = \tan^{-1} x^3$

21. $g(x) = (\sin^{-1} x)^2$

22. $u = (\sec^{-1} x)^2$

23. $v = x\sin^{-1} x + (1 - x^2)^{1/2}$ (Surprise!)

24. $I(x) = \cot^{-1}(\cot x)$ (Surprise!)

25. *Radar Problem:* An officer in a patrol car sitting 100 ft from the highway observes a truck approaching (Figure 4-5h).

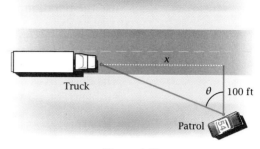

Figure 4-5h

a. At a particular instant, t, in seconds, the truck is at a distance x, in feet, down the highway. The officer's line of sight to the truck makes an angle θ, in radians, to a perpendicular to the highway. Explain why $\theta = \tan^{-1}(x/100)$.

b. Find $d\theta/dx$. Use the chain rule to write an equation for $d\theta/dt$.

c. When the truck is at $x = 500$ ft, the officer notes that angle θ is changing at a rate $d\theta/dt = -0.04$ rad/s. How fast is the truck going? How many miles per hour is this?

26. *Exit Sign Problem:* The base of a 20-ft-tall exit sign is 30 ft above the driver's eye level (Figure 4-5i). When cars are far away, the sign is hard to read because of the distance. When they are very close, the sign is hard to read because drivers have to look up at a steep angle. The sign is easiest to read when the distance x is such that the angle θ at the driver's eye is as large as possible.

Figure 4-5i

152

a. Write θ as the difference of two inverse cotangents.

b. Write an equation for $d\theta/dx$.

c. The sign is easiest to read at the value of x where θ stops increasing and starts decreasing. This happens when $d\theta/dx = 0$. Find this value of x.

d. To confirm that your answer in part c is correct, plot θ as a function of x and thus show that the graph really does have a high point at that value.

27. *Numerical Answer Check Problem:* For $f(x) = \cos^{-1} x$, make a table of values that show $f'(x)$ both numerically and by the formula. Start at $x = -0.8$ and go to $x = 0.8$, with $\Delta x = 0.2$. Show that the formula and the numerical derivative give the same answers for each value of x.

28. *Graphical Analysis Problem:* Figure 4-5j shows the graph of $y = \sec^{-1} x$.

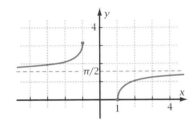

Figure 4-5j

a. Calculate the derivative at $x = 2$. Based on the graph, why is the answer reasonable?

b. What does y equal when x is 2? What does $(d/dy)(\sec y)$ equal for this value of y?

c. In what way is the derivative of the inverse secant function related to the derivative of the secant function?

29. *General Derivative of the Inverse of a Function:* In this problem you will derive a general formula for the derivative of the inverse of a function.

a. Let $y = \sin^{-1} x$. Show that $\dfrac{dy}{dx} = \dfrac{1}{\cos y}$.

b. By directly substituting $\sin^{-1} x$ for y in part a, you get $\dfrac{dy}{dx} = \dfrac{1}{\cos(\sin^{-1} x)}$. Show that this equation and the derivative of $\sin^{-1} x$ given in this section give the same value when $x = 0.6$.

c. Show that this property is true for the derivative of the inverse of a function:

> **Property: Derivative of the Inverse of a Function**
>
> If $y = f^{-1}(x)$, then $\dfrac{d}{dx}(f^{-1}(x)) = \dfrac{1}{f'(f^{-1}(x))}$.

d. Suppose that $f(x) = x^3 + x$. Let h be the inverse function of f. Find x if $f(x) = 10$. Use the result and the property given above to calculate $h'(10)$.

30. Quick! Which of the inverse trigonometric derivatives are preceded by a negative $(-)$ sign?

4-6 Differentiability and Continuity

It's time to pause in your study of derivatives and take care of some unfinished business. If a function f has a value for $f'(c)$, then f is said to be **differentiable** at $x = c$. If f is differentiable at every value of x in an interval, then f is said to be differentiable on that interval.

DEFINITIONS: Differentiability

Differentiability at a point: Function f is **differentiable at** $x = c$ if and only if $f'(c)$ exists. (That is, $f'(c)$ is a real number.)

Differentiability on an interval: Function f is **differentiable on an interval** if and only if it is differentiable for every x-value in the interval.

Differentiability: Function f is **differentiable** if and only if it is differentiable at every value of x in its domain.

Note: If a function is defined only on a *closed* interval $[a, b]$, then it can be differentiable only on the *open* interval (a, b) because taking the limit at a point requires being able to approach the point from both sides.

In Section 2-4, you learned that a function f is continuous at $x = c$ if $\lim_{x \to c} f(x) = f(c)$. A function can be continuous at $x = c$ without being differentiable at that point. But a function that is differentiable at $x = c$ is always continuous at that point. Figure 4-6a illustrates the two cases.

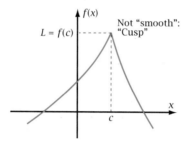

Continuous, but not differentiable   Differentiable, so continuous

Figure 4-6a

OBJECTIVE   Prove that a differentiable function is continuous, and use this property to prove that certain functions are continuous.

To prove that a function f is continuous at $x = c$, you must show that $\lim_{x \to c} f(x) = f(c)$. The definition of derivative at a point contains all of these ingredients.

$$f'(c) = \lim_{x \to c} \frac{f(x) - f(c)}{x - c}$$

The trick is to perform some mathematically correct operations that lead from the hypothesis to the conclusion. In this case it is easier to start somewhere "in the middle" and pick up the hypothesis along the way. Here goes!

▶ **PROPERTY**   Prove that if f is differentiable at $x = c$, then f is continuous at $x = c$.

Proof You must prove that $\lim_{x \to c} f(x) = f(c)$.

$$\lim_{x \to c} [f(x) - f(c)]$$
Start with something that contains limit, $f(x)$, and $f(c)$.

$$= \lim_{x \to c} \left[\frac{f(x) - f(c)}{x - c} \cdot (x - c) \right]$$
Multiply by $\frac{x - c}{x - c}$.

$$= \lim_{x \to c} \frac{f(x) - f(c)}{x - c} \cdot \lim_{x \to c} (x - c)$$
Limit of a product.

$$= f'(c) \cdot 0$$
Definitions of derivative and limit of a linear function.

$$= 0$$
Because f is differentiable at $x = c$, $f'(c)$ is a real number; (number) $\cdot 0 = 0$.

$$\therefore \lim_{x \to c} [f(x) - f(c)] = 0$$
Transitive property.

$$\lim_{x \to c} [f(x) - f(c)] = [\lim_{x \to c} f(x)] - [\lim_{x \to c} f(c)]$$

$$[\lim_{x \to c} f(x)] - [\lim_{x \to c} f(c)] = [\lim_{x \to c} f(x)] - f(c)$$

$$\therefore [\lim_{x \to c} f(x)] - f(c) = 0, \text{ also.}$$
Transitive property again.

$$\therefore \lim_{x \to c} f(x) = f(c)$$

$$\therefore f \text{ is continuous at } x = c, \text{ Q.E.D.}$$
Definition of continuity. ◀

The secret to this proof is to multiply by 1 in the form of $(x - c)/(x - c)$. This transformation causes the difference quotient $[f(x) - f(c)]/(x - c)$ to appear inside the limit sign. The rest of the proof involves algebra and limit properties and the definitions of derivative and continuity.

PROPERTY: Differentiability Implies Continuity

If function f is differentiable at $x = c$, then f is continuous at $x = c$.

Contrapositive of the Property: If function f is not continuous at $x = c$, then f is not differentiable at $x = c$.

(The converse and the inverse of this property are false.)

This property and its contrapositive provide a simple way to prove that a function is continuous or not differentiable, respectively.

▶ **EXAMPLE 1** Prove that $f(x) = x^2 - 7x + 13$ is continuous at $x = 4$.

Solution Find the derivative and substitute 4 for x.

$$f'(x) = 2x - 7$$
$$f'(4) = 2(4) - 7 = 1, \text{ which is a real number.}$$
$$\therefore f \text{ is differentiable at } x = 4.$$
$$\therefore f \text{ is continuous at } x = 4, \text{ Q.E.D.}$$
Differentiability implies continuity. ◀

Note that you could prove that f is continuous by applying the limit theorems. The technique in Example 1 is faster if you can find the derivative easily.

▶ **EXAMPLE 2**   Is the function $g(x) = \dfrac{(x-2)(x+3)}{x-2}$ differentiable at $x = 2$? Justify your answer.

Solution   The function g has a (removable) discontinuity at $x = 2$.

∴ g is not differentiable at $x = 2$.   Contrapositive of differentiability
implies continuity.   ◀

The most significant thing for you to understand here is the distinction between the concepts of differentiability and continuity. To help you acquire this understanding, it helps to look at graphs of functions and state which, if either, of the two properties applies.

▶ **EXAMPLE 3**   State whether the functions in Figure 4-6b are differentiable or continuous at $x = c$.

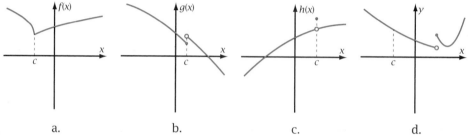

Figure 4-6b

Solution   Remember the definition of differentiability at a point and that differentiability implies continuity.

a. The function is continuous but not differentiable at $x = c$. At the cusp, the rate of change approaches a different number as $x \to c$ from the left side than it does as $x \to c$ from the right.

b. The function is neither continuous nor differentiable at $x = c$. There is no limit for $g(x)$ as $x \to c$.

c. The function is neither continuous nor differentiable at $x = c$. Although the graph appears "smooth" as x goes through c, the difference quotient $[h(x) - h(c)]/(x - c)$ approaches $+\infty$ (positive infinity) as x approaches c from the left side, and approaches $-\infty$ as x approaches c from the right.

d. The function is continuous and differentiable at $x = c$. The discontinuity elsewhere has no effect on the behavior of the function at $x = c$.   ◀

▶ **EXAMPLE 4**   Use one-sided limits to find the values of the constants a and b that make piecewise function f differentiable at $x = 2$. Check your answer by graphing.

$$f(x) = \begin{cases} ax^3, & \text{if } x \le 2 \\ b(x-3)^2 + 10, & \text{if } x > 2 \end{cases}$$

Solution   For f to be differentiable at $x = 2$, the derivatives of the two pieces must approach the same limit, and the function must be continuous there.

$$f'(x) = \begin{cases} 3ax^2, & \text{if } x < 2 \\ 2b(x - 3), & \text{if } x > 2 \end{cases}$$

Note $x < 2$ because there is no derivative at an endpoint.

Equate the left and right limits of the derivative.

$$\lim_{x \to 2^-} f'(x) = 12a \qquad \text{and} \qquad \lim_{x \to 2^+} f'(x) = 2b(2 - 3) = -2b$$

$$\therefore 12a = -2b \Rightarrow b = -6a$$

Equate the left and right limits of the function values.

$$\lim_{x \to 2^-} f(x) = 8a \qquad \text{and} \qquad \lim_{x \to 2^+} f(x) = b(2 - 3)^2 + 10 = b + 10$$

$$\therefore 8a = b + 10$$

$$8a = -6a + 10 \qquad\qquad \text{By substitution.}$$

$$a = \frac{5}{7}$$

$$b = -6 \cdot \frac{5}{7} = -\frac{30}{7}$$

$$\therefore f(x) = \begin{cases} \frac{5}{7}x^3, & \text{if } x \le 2 \\ -\frac{30}{7}(x - 3)^2 + 10, & \text{if } x > 2 \end{cases}$$

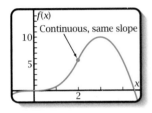

Figure 4-6c

Figure 4-6c shows the graph. Use Boolean variables to restrict the domains of the two pieces. To make them connect, use $x \le 2$ and $x \ge 2$. As you can see, the two pieces have the same slope at $x = 2$, and they are continuous. This means that f is differentiable at $x = 2$. ◀

Problem Set 4-6

Quick Review

Q1. Write the definition of continuity.

Q2. Write the definition of derivative.

Q3. Find y: $y' = 12x^{-3}$

Q4. Find: $\cos' x$

Q5. Find dy/dx: $y = \tan x$

Q6. Find: $\dfrac{d}{dx} \sec^{-1} x$

Q7. If $f(x) = x^4$, find $f''(2)$.

Q8. Find dy/dx: $y = (x^3 + 1)^5$

Q9. Estimate the definite integral from -2 to 2 of the function in Figure 4-6d.

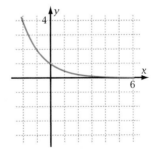

Figure 4-6d

Q10. For the function in Figure 4-6d,

A. $f'(-1) > 2$

B. $1 < f'(-1) < 2$

C. $0 < f'(-1) < 1$

D. $-1 < f'(-1) < 0$

E. $-2 < f'(-1) < -1$

For Problems 1–12, state whether the function is continuous, differentiable, both, or neither at $x = c$.

1.
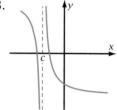

2.
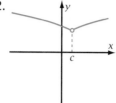

3.
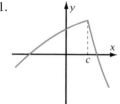

4.
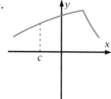

5.
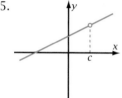

6.
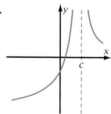

7.
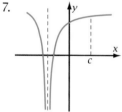

8.
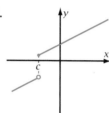

9.
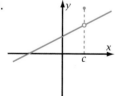

10.
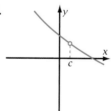

11.
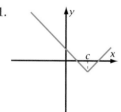

12.
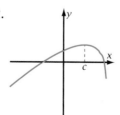

For Problems 13–20,

    a. Sketch the graph of a function that has the indicated features.

    b. Write the equation for a function that has these features.

13. Is differentiable and continuous at the point $(3, 5)$

14. Is differentiable and continuous at the point $(-2, 4)$

15. Has a finite limit as x approaches 6, but is not continuous at that point because $f(6)$ is undefined

16. Has a finite limit as x approaches 1, has a value for $f(1)$, but still is not continuous at that point

17. Has a value for $f(-5)$ but has no limit as x approaches -5

18. Has a cusp at the point $(-1, 3)$

19. Is continuous at the point $(4, 7)$ but is not differentiable at that point

20. Is differentiable at the point $(3, 8)$ but is not continuous at that point

For Problems 21–24, sketch the graph. State whether the function is differentiable, continuous, neither, or both at the indicated value of $x = c$.

21. $f(x) = |x - 3|$, $c = 3$

22. $f(x) = 4 + |x|$, $c = 2$

23. $f(x) = \sin x$, $c = 1$

24. $f(c) = \tan x$, $c = \pi/2$

For Problems 25–30, use one-sided limits to find the values of the constants a and b that make the piecewise function differentiable at the point where the rule for the function changes. Check your answer by plotting the graph. Sketch the results.

25. $f(x) = \begin{cases} x^3, & \text{if } x < 1 \\ a(x - 2)^2 + b, & \text{if } x \geq 1 \end{cases}$

26. $f(x) = \begin{cases} -(x - 3)^2 + 7, & \text{if } x \geq 2 \\ ax^3 + b, & \text{if } x < 2 \end{cases}$

27. $f(x) = \begin{cases} ax^2 + 10, & \text{if } x < 2 \\ x^2 - 6x + b, & \text{if } x \geq 2 \end{cases}$

28. $f(x) = \begin{cases} a/x, & \text{if } x \le 1 \\ 12 - bx^2, & \text{if } x > 1 \end{cases}$

29. $f(x) = \begin{cases} e^{ax}, & \text{if } x \le 1 \\ b + \ln x, & \text{if } x > 1 \end{cases}$

30. $f(x) = \begin{cases} a \sin x, & \text{if } x < \frac{2\pi}{3} \\ e^{bx}, & \text{if } x \ge \frac{2\pi}{3} \end{cases}$

31. *Railroad Curve Problem:* Curves on a railroad track are in the shape of cubic parabolas. Such "parabola tracks" have the property that the curvature starts as zero at a particular point and increases gradually, easing the locomotive into the curve slowly so that it is less likely to derail. Figure 4-6e shows curved and straight sections of track defined by the piecewise function

$$y = \begin{cases} ax^3 + bx^2 + cx + d, & \text{if } 0 \le x \le 0.5 \\ x + k, & \text{if } x > 0.5 \end{cases}$$

where x and y are coordinates in miles. The dashed portions show where the cubic and linear functions would go if they extended into other parts of the domain.

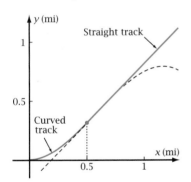

Figure 4-6e

a. The curved left branch of the graph contains the origin and has $y' = 0$ at that point. At the transition point where $x = 0.5$, $y' = 1$ so that the curve goes the same direction as the straight section. At this point, $y'' = 0$ so that the curvature is zero. Find the coefficients $a, b, c,$ and d in the cubic branch of the function.

b. Find the value of k in the linear branch of the function that makes the piecewise function continuous.

32. *Bicycle Frame Design Problem:* Figure 4-6f shows a side view of a bicycle frame's front fork, holding the front wheel. To make the bike track properly, the fork curves forward at the bottom where the wheel bolts on. Assume that the fork is bent in the shape of a cubic parabola, $y = ax^3 + bx$. What should the constants a and b be so that the curve joins smoothly to the straight part of the fork at the point (10 cm, 20 cm) with slope equal to 5?

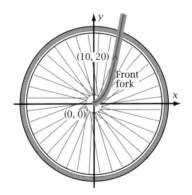

Figure 4-6f

33. Let $f(x) = \begin{cases} x^2 - \dfrac{|x - 2|}{x - 2}, & \text{if } x \ne 2 \\ 4, & \text{if } x = 2 \end{cases}$

Find an equation for $f'(x)$. Use the definition of derivative to show that function f is not differentiable at $x = 2$, even though the left and right limits of $f'(x)$ are equal as x approaches 2.

34. *Baseball Line Drive Problem:* Milt Famey pitches his famous fastball. At time $t = 0.5$ s after Milt releases the ball, Joe Jamoke hits a line drive to center field. The distance, $d(t)$, in

feet, of the ball from home plate is given by the two-rule function.

$$d(t) = \begin{cases} 60.5 \left(\dfrac{0.5 - t}{0.5 + t} \right), & \text{if } t \le 0.5 \\ 150 \left(2 - \dfrac{1}{t} \right), & \text{if } t \ge 0.5 \end{cases}$$

Figure 4-6g shows the graph of function d.

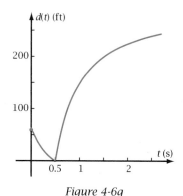

Figure 4-6g

a. Find an equation for $d'(t)$. (Be careful about the inequality signs at $t = 0.5$).

b. Prove quickly that d is continuous at $t = 1$.

c. Find the limit of $d'(t)$ as $t \to 0.5^-$ and as $t \to 0.5^+$. State the real-world meanings of these two numbers.

d. Explain why d is continuous, but not differentiable, at $t = 0.5$.

e. What is the significance of the number 60.5 in the first rule for $d(t)$?

35. *Continuity Proof Problem:* Use the fact that differentiability implies continuity to prove that the following types of functions are continuous.

   a. Linear function, $y = mx + b$

   b. Quadratic function, $y = ax^2 + bx + c$

   c. Reciprocal function, $y = 1/x$, provided $x \ne 0$

   d. Identity function, $y = x$

   e. Constant function, $y = k$

36. *Differentiability Implies Continuity Proof:* Prove that if f is differentiable at $x = c$, then it is continuous at that point. Try to write the proof without looking at the text. Consult the text only if you get stuck.

4-7   Derivatives of a Parametric Function

Figure 4-7a shows how a pendulum hung from the ceiling of a room might move if it were to swing in both the x- and y-directions. It is possible to calculate its velocity in both the x- and y-directions, and along its curved path. These rates help you determine facts about the path of a moving object. In this section you will use **parametric functions** to make these determinations.

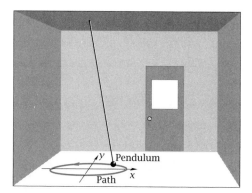

Figure 4-7a

OBJECTIVE   Given equations for x and y in terms of t, find dx/dt, dy/dt, and dy/dx.

The pendulum in Figure 4-7a swings back and forth sinusoidally in both the x- and y-directions. Using the methods you learned in Section 3-8, you can find equations for these sinusoids. Suppose that the equations for a particular pendulum are

$$x = 50 \cos 1.2t$$
$$y = 30 \sin 1.2t$$

where x and y are in centimeters and t is time, in seconds. The variable t on which x and y both depend is called a **parameter**. The two equations, one for x and one for y, are called **parametric equations**.

You can use the parametric mode on your grapher to plot the xy-graph of the pendulum's path. The result is an ellipse, as shown in Figure 4-7b. The ellipse goes from -50 to 50 cm in the x-direction and from -30 to 30 cm in the y-direction. The numbers 50 and 30 in the parametric equations are x- and y-dilations, respectively, equal to the amplitudes of the two sinusoids. (The x- and y-translations are zero, which indicates that the pendulum hangs over the origin when it is at rest.)

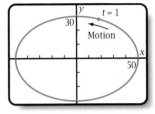

Figure 4-7b

You can find the rates of change of x and y with respect to t by differentiating.

$$\frac{dx}{dt} = -60 \sin 1.2t \qquad \text{and} \qquad \frac{dy}{dt} = 36 \cos 1.2t$$

Evaluating these derivatives at a certain value of t, say $t = 1$, shows that the pendulum is moving at about -55.9 cm/s in the x-direction and at about 13.0 cm/s in the y-direction. If you divide dy/dt by dx/dt, the result is the slope of the ellipse, dy/dx, at the given point.

$$\frac{dy}{dx} = \frac{13.044...}{-55.922...} = -0.233...$$

Note that although dx/dt and dy/dt have units such as cm/s, dy/dx is dimensionless because the units cancel.

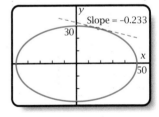

Figure 4-7c

A line with this slope at the point where $t = 1$ is tangent to the graph (Figure 4-7c). The property illustrated by this example is called the **parametric chain rule**.

PROPERTY: *The Parametric Chain Rule*

If x and y are differentiable functions of t, then the slope of the xy-graph is

$$\frac{dy}{dx} = \frac{dy/dt}{dx/dt}$$

▶ **EXAMPLE 1**   Given: $x = 3 \cos 2\pi t$
$y = 5 \sin \pi t$

a. Plot the xy-graph. Use path style and a t-range that generates at least two complete cycles of x and y. A t-step of 0.05 is reasonable. Sketch the result and describe the behavior of the xy-graph as t increases.

b. Find an equation for dy/dx in terms of t. Use the equation to evaluate dy/dx when $t = 0.15$. Show how your answer corresponds to the graph.

c. Show that dy/dx is indeterminate when $t = 0.5$. Find, approximately, the limit of dy/dx as t approaches 0.5. How does your answer relate to the graph?

d. Make a conjecture about what geometric figure the graph represents. Then confirm your conjecture by eliminating the parameter t and analyzing the resulting Cartesian equation.

e. How do the range and domain of the parametric function relate to the range and domain of the Cartesian equation in part d?

Solution   a. Figure 4-7d shows the graph as it might appear on your grapher. The period for x is $2\pi/2\pi = 1$; for y it is $2\pi/\pi = 2$. Thus, a minimal range is $0 \le t \le 4$.

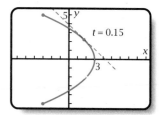

Figure 4-7d

If you watch the graph being generated, you will see that the points start at (3, 0), go upward to the left, stop, retrace the path through the point (3, 0), then go downward to the left, eventually coming back to the point (3, 0).

b. $\dfrac{dx}{dt} = -6\pi \sin 2\pi t$ and $\dfrac{dy}{dt} = 5\pi \cos \pi t$

$\therefore \dfrac{dy}{dx} = \dfrac{5\pi \cos \pi t}{-6\pi \sin 2\pi t} = \dfrac{5 \cos \pi t}{-6 \sin 2\pi t}$

$t = 0.15 \Rightarrow \dfrac{dy}{dx} = \dfrac{5 \cos 0.15\pi}{-6 \sin 0.3\pi} = \dfrac{4.455...}{-4.854...} = -0.917...$

As shown in Figure 4-7d, a tangent line to the graph at the point where $t = 0.15$ has slope of about -1, which corresponds to the exact value of $-0.917...$.

c. $t = 0.5 \Rightarrow \dfrac{dy}{dx} = \dfrac{5 \cos 0.5\pi}{-6 \sin \pi} = \dfrac{0}{0}$, which is indeterminate. A graph of dy/dx versus t (Figure 4-7e) shows a removable discontinuity at $t = 0.5$.

To find the limit of dy/dx more precisely, either zoom in on the discontinuity or use the TABLE feature. The limit appears to be $-0.4166666...$, which equals $-5/12$.

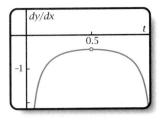

t	dy/dx
0.498	−0.4166748...
0.499	−0.4166687...
0.500	undefined
0.501	−0.4166687...
0.502	−0.4166748...

Figure 4-7e

d. The graph appears to be a parabola. Eliminating the parameter t involves solving one equation for t in terms of x (or y) and substituting the result into the other equation. Sometimes certain shortcuts will let you do this more easily, as shown here.

$x = 3 \cos 2\pi t$ and $y = 5 \sin \pi t$ The given parametric equations.

$x = 3(1 - 2 \sin^2 \pi t)$ The double argument property gets $\cos 2\pi t$ in terms of $\sin \pi t$, which appears in the original parametric equation for y.

But $\sin \pi t = y/5$.   From the original parametric equations.

$\therefore x = 3[1 - 2(y/5)^2]$ Substituting $y/5$ for $\sin \pi t$ eliminates the parameter t.

$x = -\dfrac{6}{25}y^2 + 3$ By algebra.

As conjectured, this is the equation of a parabola opening in the negative x-direction.

e. The Cartesian equation has domain $x \le 3$, which is unbounded in the negative x-direction. The parametric graph stops at $x = -3$.   ◀

Second Derivative of a Parametric Function

Unfortunately, the parametric chain rule does not extend to second derivatives. Example 2 shows you how to find a second derivative for a parametric function.

▶ **EXAMPLE 2**   Find $\dfrac{d^2y}{dx^2}$ for the parametric function:

$x = t^3$

$y = e^{2t}$

Solution   Use the parametric chain rule to differentiate.

$$\frac{dy}{dx} = \frac{2e^{2t}}{3t^2}$$

$$\frac{d^2y}{dx^2} = \frac{d}{dx}\left(\frac{2e^{2t}}{3t^2}\right)$$

$$= \frac{4e^{2t}(dt/dx) \cdot 3t^2 - 2e^{2t} \cdot 6t(dt/dx)}{(3t^2)^2}$$ Derivative of a quotient. Chain rule on inside function t.

The dt/dx in the numerator can be factored out. Because dt/dx is the reciprocal of dx/dt, you can divide by dx/dt, which equals $3t^2$, making the denominator $(3t^2)^3$.

$$\therefore \frac{d^2y}{dx^2} = \frac{12t^2e^{2t} - 12te^{2t}}{(3t^2)^2} \cdot \frac{1}{3t^2}$$

$$= \frac{12t^2e^{2t} - 12te^{2t}}{(3t^2)^3}$$

$$= \frac{12te^{2t}(t - 1)}{27t^6}$$

You can generalize the second derivative from the results of Example 2, as shown in this box.

◀

PROPERTY: Second Derivative of a Parametric Function

If $x = u$ and $y = v$, where u and v are twice-differentiable functions of t, then

$$\frac{d^2y}{dx^2} = \frac{u'v'' - u''v'}{(u')^3}$$

where the derivatives of u and v are with respect to t.

Problem Set 4-7

Quick Review

Q1. Differentiate: $y = 0.2x^{1215}$

Q2. Find $\dfrac{dy}{dx}$: $y = \dfrac{x - 3}{x - 1}$

Q3. Find $f'(x)$: $f(x) = x \ln x$

Q4. Find y': $y = \sin e^{5x}$

Q5. Find $\dfrac{d^2}{dx^2}(y)$: $y = \dfrac{x^8}{x^5}$

Q6. Differentiate: $y = \dfrac{\sin 8}{\cos 3}$

Q7. Find θ': $\theta = \cos^{-1} x$

Q8. If $v'(5) = -3$, what can you conclude about $v(t)$ at $t = 5$?

Q9. For Figure 4-7f, sketch the graph of y'.

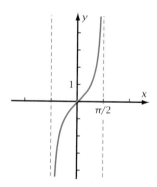

Figure 4-7f

Q10. If $u'(7) = 4$, which of these can you conclude about $u(x)$ at $x = 8$?

    A. u is continuous at $x = 8$.

    B. u is differentiable at $x = 8$.

    C. u has a limit as x approaches 8.

    D. All of A, B, and C

    E. None of A, B, and C

For Problems 1 and 2, find $\dfrac{dy}{dx}$ and $\dfrac{d^2y}{dx^2}$.

1. $x = t^4$

   $y = \sin 3t$

2. $x = 6 \ln t$

   $y = t^3$

3. *Parabola Problem:* A parametric function has the equations

   $x = 2 + t$

   $y = 3 - t^2$

   a. Make a table of values of x and y for each integer value of t from -3 through 3.

   b. Plot the graph of this function on graph paper, using the points found in part a.

   c. Find dy/dx when $t = 1$. Show that the line through the point (x, y) from part a, with slope dy/dx, is tangent to the graph at that point.

   d. Eliminate the parameter t and show that the resulting Cartesian equation is that of a parabola.

   e. Find dy/dx by direct differentiation of the equation in part d. Show that the value of dy/dx calculated this way is equal to the value you found in part c using the parametric chain rule.

4. *Semicubical Parabola Problem:* A parametric function has the equations

   $x = t^2$

   $y = t^3$

   a. Make a table of values of x and y for each integer value of t from -3 through 3.

   b. Plot the graph of this function on graph paper, using the points found in part a.

   c. Find dy/dx when $t = 1$. Show that the line through the point (x, y) from part a, with slope dy/dx, is tangent to the graph at that point.

   d. Eliminate the parameter t. Find y in terms of x. From the result, state why this graph is called a **semicubical parabola.**

e. Find dy/dx by direct differentiation of the equation in part d. Show that the value of dy/dx calculated in this way is equal to the value you found in part c by using the parametric chain rule.

5. *Ellipse Problem:* The ellipse in Figure 4-7g has the parametric equations

   $x = 3 \cos t$

   $y = 5 \sin t$

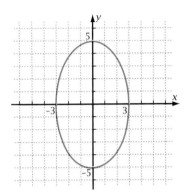

Figure 4-7g

   a. Confirm by grapher that these equations give the graph in Figure 4-7g.

   b. Find an equation for dy/dx.

   c. Evaluate the point (x, y) when $t = \pi/4$, and find dy/dx when $t = \pi/4$. On a copy of Figure 4-7g, draw a line at this point (x, y) that has slope dy/dx. Is the line tangent to the graph?

   d. Determine whether this statement is true or false: When $t = \pi/4$, the point (x, y) is on a line through the origin that makes a 45-degree angle with the x- and y-axes.

   e. Use your equation for dy/dx from part b to find all the points where the tangent line is vertical or horizontal. Show these points on your graph.

   f. Eliminate the parameter t and thus confirm that your graph actually is an ellipse. This elimination can be done by cleverly applying the Pythagorean property for sine and cosine.

6. *Astroid Problem:* The star-shaped curve in Figure 4-7h is called an **astroid**. Its parametric equations are

$$x = 8\cos^3 t$$
$$y = 8\sin^3 t$$

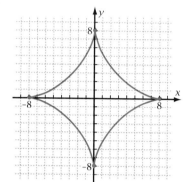

Figure 4-7h

a. Confirm by grapher that these equations give the graph in Figure 4-7h.

b. Find an equation for dy/dx.

c. Evaluate the point (x, y) when $t = 1$, and find dy/dx when $t = 1$. On a copy of Figure 4-7h, draw a line at this point (x, y) that has slope dy/dx. Is the line tangent to the graph?

d. At each cusp, dy/dx has the indeterminate form $0/0$. Explain the difference in behavior at the cusp at the point $(8, 0)$ and at the cusp at the point $(0, 8)$.

e. Eliminate the parameter t. To do this transformation, solve the two equations for the squares of $\cos t$ and $\sin t$ in terms of x and y, then use the Pythagorean property for sine and cosine.

7. *Circle Problem:* A parametric function has the equations

$$x = 6 + 5\cos t$$
$$y = 3 + 5\sin t$$

a. Plot the graph of this function. Sketch the result.

b. Find an equation for dy/dx in terms of t.

c. Find a value of t that makes dy/dx equal zero. Find a value of t that makes dy/dx infinite. Show a point on the graph for which dy/dx is infinite. What is true about dx/dt

and about dy/dt at a point where dy/dx is infinite?

d. Eliminate the parameter t. To do this, express the squares of cosine and sine in terms of x and y, then apply the Pythagorean property for sine and cosine.

e. From the equation in part d, you should be able to tell that the graph is a circle. How can you determine the center and the radius of the circle just by looking at the original equations?

8. *Line Segment Problem:* Plot the graph of the parametric function

$$x = \cos^2 t$$
$$y = \sin^2 t$$

Show that dy/dx is constant. How does this fact correspond to what you observe about the graph? Confirm your observation by eliminating the parameter to get an xy-equation. Describe the difference in domain and range between the parametric function and the xy-equation.

9. *Deltoid Problem:* The graph shown in Figure 4-7i is called a **deltoid.** The parametric function of this deltoid is

$$x = 2\cos t + \cos 2t$$
$$y = 2\sin t - \sin 2t$$

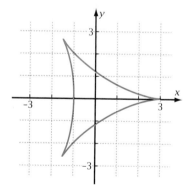

Figure 4-7i

a. Confirm by grapher that these equations give the deltoid in Figure 4-7i.

b. Find an equation for dy/dx in terms of t.

c. Show that at two of the cusps, the tangent line is neither horizontal nor vertical, yet the derivative dy/dx fails to exist. Find the limit of dy/dx as t approaches the value at the cusp in Quadrant II.

10. *Witch of Agnesi Problem:* The **Witch of Agnesi,** named for Italian mathematician Maria Gaetana Agnesi (1718–1799), has the equations

$$x = 2a \tan t$$
$$y = 2a \cos^2 t$$

where a is a constant.

a. Figure 4-7j shows a curve for which $a = 3$. Confirm by grapher that these equations, for $a = 3$, give the graph in Figure 4-7j.

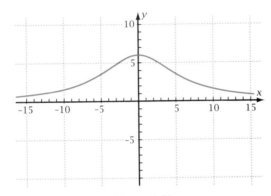

Figure 4-7j

b. Find dy/dx in terms of t.

c. Eliminate the parameter to get an equation for y in terms of x.

d. Differentiate the equation in part c to get an equation for dy/dx in terms of x.

e. Show that both equations for dy/dx give the same answer at $t = \pi/4$, and that a line through the point where $t = \pi/4$ with this value of dy/dx as its slope is tangent to the curve.

11. *Involute Problem:* A string is wrapped around a circle with radius 1 in. As the string is unwound, its end traces a path called the **involute of a circle** (Figure 4-7k). The parametric equations of this involute are

$$x = \cos t + t \sin t$$
$$y = \sin t - t \cos t$$

where t is the number of radians from the positive x-axis to the radius drawn to the point of tangency of the string.

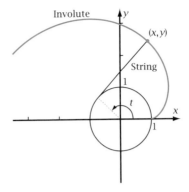

Figure 4-7k

a. Use your grapher to confirm that these parametric equations give the graph shown in Figure 4-7k.

b. Find dy/dx in terms of t. Simplify as much as possible.

c. Show that the value you get for dy/dx at $t = \pi$ is consistent with the graph.

12. *Clock Problem:* A clock sits on a shelf close to a wall (Figure 4-7l). As the second hand turns, its distance from the wall, x, and from the shelf, y, both in centimeters, depend on the number of seconds, t, since the second hand was pointing straight up.

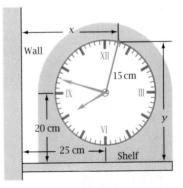

Figure 4-7l

a. Write parametric equations for x and y in terms of t.

b. At what rates are x and y changing when $t = 5$ s?

c. What is the slope of the circular path traced by the second hand when $t = 5$ s?

d. Confirm that the path really is a circle by finding an xy-equation.

13. *Pendulum Experiment:* Suspend a small mass from the ceiling on the end of a nylon cord. Place metersticks on the floor, crossing them at the point below which the mass hangs at rest, as shown in Figure 4-7m. Determine the period of the pendulum by measuring the time for 10 swings. Then start the pendulum in an elliptical path by pulling it 30 cm in the *x*-direction and pushing it sideways just hard enough for it to cross the *y*-axis at 20 cm. Write parametric equations for the path this pendulum traces on the floor. Predict where the pendulum will be at time *t* = 5 s and place a coin on the floor at that point. (Lay the coin on top of a ruler tilted at an angle corresponding to the slope of the path at that time.) Then set the pendulum in motion again. How close do your predicted point and slope come to those you observe by experiment?

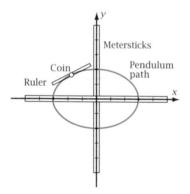

Figure 4-7m

The Foucault pendulum at the Griffith Observatory in Los Angeles, California

14. *Spring Problem:* Figure 4-7n shows a "spring" drawn by computer graphics. Find equations for a parametric function that generates this graph. How did you verify that your equations are correct? Use your equations to find values of *x* and *y* at which the graph has interesting features, such as horizontal or vertical tangents and places where the graph seems to cross itself.

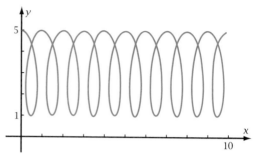

Figure 4-7n

15. *Lissajous Curves:* You can make a pendulum swing with different periods in the *x*- and *y*-directions. The parametric equations of the path followed by the pendulum can have the form

$$x = \cos nt$$
$$y = \sin t$$

where *n* is a constant. The resulting paths are called **Lissajous curves**, or sometimes **Bowditch curves**. In this problem you will investigate some of these curves.

a. Figure 4-7o shows the Lissajous curve with the parametric equations

$$x = \cos 3t$$
$$y = \sin t$$

Use your grapher to confirm that these equations generate this graph.

168

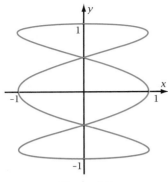

Figure 4-7o

b. Plot the Lissajous curve with the equations

$$x = \cos 4t$$
$$y = \sin t$$

Sketch the resulting curve. In what way do the curves differ for $n = 3$ (an odd number) and for $n = 4$ (an even number)?

c. Sketch what you think these curves would look like. Then plot the graphs on your grapher. Do they confirm or refute your sketches?

  i. $x = \cos 5t$

   $y = \sin t$

  ii. $x = \cos 6t$

   $y = \sin t$

d. What two familiar curves are special cases of Lissajous curves when $n = 1$ and $n = 2$?

e. Look up Jules Lissajous and Nathaniel Bowditch on the Internet or other source. When and where did they live? Give the sources you used.

4-8   Graphs and Derivatives of Implicit Relations

If y equals some function of x, such as $y = x^2 + \sin x$, then there is said to be an **explicit relation** between x and y. The word *explicit* comes from the same root as the word *explain*. If x and y appear in an equation such as

$$x^2 + y^2 = 25$$

then there is an **implicit relation** between x and y because it is only "implied" that y is a function of x. In this section you will see how to differentiate an implicit relation without first solving for y in terms of x. As a result, you will be able to prove that the power rule for derivatives works when the exponent is a rational number, not simply an integer. In Section 3-9, you used implicit differentiation to find the derivative of the natural logarithmic function. In Section 4-5, you used it to find derivatives of the inverse trigonometric functions.

> **OBJECTIVE**   Given the equation for an implicit relation, find the derivative of y with respect to x, and show by graph that the answer is reasonable.

► **EXAMPLE 1**   Consider the implicit relation $x^2 + y^2 = 25$ plotted in Figure 4-8a.

     a. Explain why the graph is a circle.

     b. Differentiate implicitly to find dy/dx.

     c. Calculate the two values of y when $x = 3$.

     d. Draw a line with slope dy/dx through the point found in part c with the lower value of y. How is this line related to the graph?

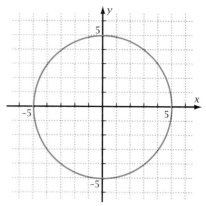

Figure 4-8a

Solution   a. The graph is a circle by the Pythagorean theorem. Because $x^2 + y^2 = 25 = 5^2$, all points on the graph are five units from the origin, implying that the graph is a circle.

     b. For simplicity, use y' for dy/dx.

$$2x + 2yy' = 0 \qquad \text{The } y' \text{ comes from the chain rule.}$$
$$2yy' = -2x$$
$$y' = \frac{-x}{y}$$

     c. $x = 3: 9 + y^2 = 25$
       $y^2 = 16 \Rightarrow y = \pm 4$

     d. At the point $(3, -4)$, slope $= y' = \dfrac{-3}{-4} = 0.75$. The line goes through the point $(3, -4)$ and is tangent to the graph (Figure 4-8b).

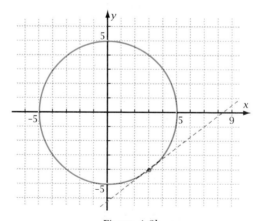

Figure 4-8b ◄

► **EXAMPLE 2**   Find dy/dx for the implicit relation $y^4 + x^3y^5 - 2x^7 = 13$.

Solution   For simplicity, use y' for dy/dx. Be on the lookout for places where you'll need to use the product rule. Of course, you must also obey the chain rule wherever y appears as an inside function.

$$4y^3y' + 3x^2y^5 + x^3 \cdot 5y^4y' - 14x^6 = 0$$

Note that y' shows up as a result of the chain rule and is only to the first power, so you can use relatively easy algebra to isolate y'.

$$4y^3y' + x^3 \cdot 5y^4y' = -3x^2y^5 + 14x^6 \qquad \text{Make sure all } y' \text{ terms are on the}$$
$$\text{same side of the equation.}$$

$$(4y^3 + 5x^3y^4)y' = -3x^2y^5 + 14x^6$$

$$y' = \frac{-3x^2y^5 + 14x^6}{4y^3 + 5x^3y^4} \qquad \blacktriangleleft$$

The solution to Example 2 expresses y' in terms of both x and y, which is acceptable because x and y are together in the original relation. Given a point (x, y) on the graph, the expression above could be used to find y'. In practice, it is usually harder to find a point on the graph than it is to do the calculus! So you will work with problems like that in Example 2 mainly for practice in differentiating.

TECHNIQUE: Implicit Differentiation

To find dy/dx for a relation whose equation is written implicitly:

1. Differentiate both sides of the equation with respect to x. Obey the chain rule by multiplying by dy/dx each time you differentiate an expression containing y.

2. Isolate dy/dx by getting all of the dy/dx terms onto one side of the equation, and all other terms onto the other side. Then factor, if necessary, and divide both sides by the coefficient of dy/dx.

▶ **EXAMPLE 3**   If $y = x^{7/3}$, prove that the power rule, which you proved for integer exponents, gives the correct answer for y'.

Solution

$$y^3 = x^7 \qquad \text{Cube both sides of the given equation.}$$

$$3y^2y' = 7x^6 \qquad \text{Use the power rule to differentiate}$$
$$\text{implicitly with respect to } x.$$

$$y' = \frac{7x^6}{3y^2}$$

$$y' = \frac{7}{3}\frac{x^6}{(x^{7/3})^2} = \frac{7}{3}\frac{x^6}{x^{14/3}} = \frac{7}{3}x^{4/3}$$

This is the answer you would get by directly applying the power rule using fractional exponents, Q.E.D. $\blacktriangleleft$

In the following problem set you will prove, in general, the property shown in Example 3. You will also use implicit differentiation to verify the results you obtained by differentiating parametric functions in the preceding section.

Problem Set 4-8

Q1. Differentiate: $y = x^{2001}$

Q2. Differentiate: $y = 2001^x$

Q3. $\lim_{x \to 5}(x^2 - 5x)/(x - 5) = $ —?—

Q4. Differentiate: $f(u) = \cot u$

Q5. A definite integral is a —?— of x and y.

Q6. $y = \tan^{-1} 3x \Rightarrow y' = $ —?—

Q7. If $dy/dx = 3x^2$, what could y equal?

Q8. A derivative is an —?—.

Q9. Sketch the derivative of the function graphed in Figure 4-8c.

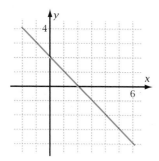

Figure 4-8c

Q10. If the position $x(t)$, in feet, of a moving object is given by $x(t) = t \sin t$, what is the acceleration of the object when $t = 3$ s?

For Problems 1–20, differentiate implicitly to find y' in terms of x and y.

1. $x^3 + 7y^4 = 13$

2. $3x^5 - y^4 = 22$

3. $x \ln y = 10^4$

4. $ye^x = 21^3$

5. $x + xy + y = \sin 2x$

6. $\cos xy = x - 2y$

7. $x^{0.5} - y^{0.5} = 13$

8. $x^{1.2} + y^{1.2} = 64$

9. $e^{xy} = \tan y$

10. $\ln xy = \tan^{-1} x$

11. $(x^3 y^4)^5 = x - y$

12. $(xy)^6 = x + y$

13. $\cos^2 x + \sin^2 y = 1$

14. $\sec^2 y - \tan^2 x = 1$

15. $\tan xy = xy$

16. $\cos xy = xy$

17. $\sin y = x$

18. $\cos y = x$

19. $\csc y = x$

20. $\cot y = x$

21. Derive the formula for y' if $y = \cos^{-1} x$ by writing the given function as $\cos y = x$, then differentiating implicitly with respect to x. Transform the answer so that y' is expressed as an algebraic function of x.

22. Derive the formula for y' if $y = \ln x$ by writing the given equation as $e^y = x$, then differentiating implicitly with respect to x. Transform the answer so that y' is expressed as an algebraic function of x.

23. If $y = x^{11/5}$, prove that the power rule for powers with integer exponents gives the correct answer for y'.

24. *Derivative of a Rational Power:* Suppose that $y = x^n$, where $n = a/b$ for integers a and b. Write the equation $y = x^{a/b}$ in the form $y^b = x^a$. Then use the power rule for integer exponents to prove that the power rule also works for rational constant exponents.

25. *Circle Problem:* Consider the circle
$x^2 + y^2 = 100$ (Figure 4-8d).

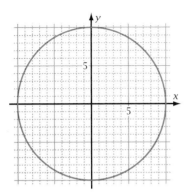

Figure 4-8d

a. Show that the point $(-6, 8)$ is on the graph.

b. Evaluate dy/dx at the point $(-6, 8)$. Explain why your answer is reasonable.

c. Show that the parametric equations

$$x = 10 \cos t$$
$$y = 10 \sin t$$

give the same value for dy/dx at the point $(x, y) = (-6, 8)$.

26. *Hyperbola Problem:* Consider the hyperbola
$x^2 - y^2 = 36$ (Figure 4-8e).

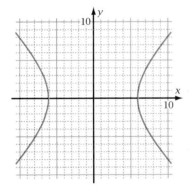

Figure 4-8e

a. Show that the point $(10, -8)$ is on the graph.

b. Evaluate dy/dx at the point $(10, -8)$. Explain why your answer is reasonable.

c. Show that the parametric equations

$$x = 6 \sec t$$
$$y = 6 \tan t$$

give the same value for dy/dx at $x = 10$.

27. *Cubic Circle Problem:* Figure 4-8f shows the **cubic circle**

$$x^3 + y^3 = 64$$

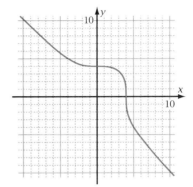

Figure 4-8f

a. Find dy/dx at the points where $x = 0$, $x = 2$, and $x = 4$. Show that your answers are consistent with the graph.

b. Find dy/dx at the point where $y = x$.

c. Find the limit of dy/dx as x approaches infinity.

d. Why do you suppose this graph is called a cubic circle?

28. *Ovals of Cassini Project:* Figure 4-8g shows the **ovals of Cassini**,

$$[(x - 6)^2 + y^2][(x + 6)^2 + y^2] = 1200$$

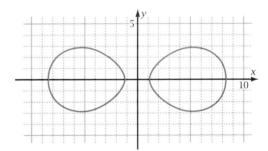

Figure 4-8g

In 1680, Italian astronomer Giovanni Domenico Cassini (1625–1712) used these figures for the relative motions of Earth and the Sun.

a. Find the two values of dy/dx when $x = 8$. Show that your answers are reasonable.

b. Find the four x-intercepts. What does dy/dx seem to be at these points? Confirm your conclusion for the largest intercept.

c. Starting with the original equation on the previous page, use the quadratic formula to find an equation for y^2 explicitly in terms of x. Use this equation to duplicate Figure 4-8g.

d. Replace the number 1200 in the original equation with the number 1400, then plot the graph. In what way or ways does this graph differ from the one shown in Figure 4-8g?

e. Show that for any point on the graph, the product of its distances from the points $(6, 0)$ and $(-6, 0)$ is constant.

An interpretation from 1661 of Earth revolving around the Sun

4-9 Related Rates

With parametric functions you analyzed rates of two variable quantities that depended on a third variable such as time. With implicit differentiation you found derivatives without first having to solve explicitly for one variable in terms of the other. In this section you will use these techniques together to find rates of two or more variable quantities whose values are related.

OBJECTIVE Given a situation in which several quantities vary, predict the rate at which one of them is changing when you know other related rates.

▶ **EXAMPLE 1** An airplane is flying 600 mi/h on a horizontal path that will take it directly over an observer. The airplane is 7 mi high (Figure 4-9a).

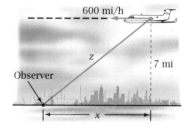

Figure 4-9a

a. Write an equation for the rate of change of the line-of-sight distance, z, in miles, between the observer and the airplane in terms of the horizontal displacement, x, in miles, from observer to airplane.

b. Plot dz/dt as a function of x. Sketch the result.

c. How fast is z changing when x is 10 mi? When x is -5 mi (that is, 5 mi beyond the observer)? Interpret the answers.

d. Interpret the graph in part b when $x = 0$ and when $|x|$ is very large.

Solution

a. You know the rate of change of x with respect to time t, and you want the rate of change of z. So the key to this problem is establishing a relationship between z and x. Here's how.

$$\text{Know: } \frac{dx}{dt} = -600 \qquad\qquad \text{The airplane is getting closer to the observer.}$$

$$\text{Want: } \frac{dz}{dt}$$

$$z^2 = x^2 + 7^2 \qquad\qquad \text{Find a relationship between } x \text{ and } z \text{ (Pythagorean theorem).}$$

$$2z \frac{dz}{dt} = 2x \frac{dx}{dt} \qquad\qquad \text{Differentiate implicitly with respect to } t\text{. Remember the chain rule!}$$

$$\frac{dz}{dt} = \frac{x}{z} \frac{dx}{dt} = -600 \frac{x}{z} = \frac{-600x}{\sqrt{x^2 + 49}} \qquad \text{By algebra and the Pythagorean theorem.}$$

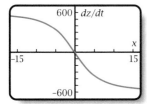

Figure 4-9b

b. Figure 4-9b shows the graph of dz/dt.

c. When $x = 10$, $dz/dt = -491.539...$, meaning that the distance is decreasing at about 492 mi/h. When $x = -5$, $dz/dt = 348.742...$, which means that the distance is increasing at about 349 mi/h.

d. At $x = 0$, the airplane is directly overhead. Although it is still moving at 600 mi/h, the distance between the airplane and the observer is not changing at that instant. So, $dz/dt = 0$, as shown by the graph. For large values of $|x|$ the rate approaches ± 600 mi/hr, the speed of the airplane. This happens because z and x are very nearly equal when the plane is far away. ◀

▶ **EXAMPLE 2**   Suppose you are drinking root beer from a conical paper cup. The cup has diameter 8 cm and depth 10 cm. As you suck on the straw, root beer leaves the cup at the rate of 7 cm³/s. At what rate is the level of the liquid in the cup changing

a. When the liquid is 6 cm deep?

b. At the instant when the last drop leaves the cup?

Solution

a. The secret to getting started is drawing an appropriate diagram, then identifying the known rate and the wanted rate. Figure 4-9c shows a cross section through the cup. Since the level of the liquid varies, you should label the depth with a variable, even though you are looking for the rate when the depth is 6 cm. Let y = number of cm deep. Let x = number of cm radius of liquid surface.

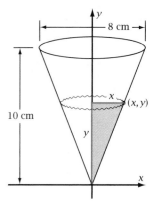

Figure 4-9c

$$\text{Know: } \frac{dV}{dt} = -7 \text{ cm}^3/\text{s} \qquad \text{Want: } \frac{dy}{dt}$$

$$V = \tfrac{1}{3}\pi x^2 y \qquad \text{Volume of a cone.}$$

By the properties of similar triangles, you can express x in terms of y, and thus reduce the problem to the two variables whose rates you know or want.

$$\frac{x}{4} = \frac{y}{10} \Rightarrow x = 0.4y$$

$$\therefore V = \frac{1}{3}\pi(0.4y)^2(y) = \frac{4}{75}\pi y^3$$

$$\frac{dV}{dt} = \frac{4}{25}\pi y^2 \frac{dy}{dt}$$

$$\frac{dy}{dt} = \frac{25}{4\pi y^2} \cdot \frac{dV}{dt} \qquad \text{Find a general formula before you substitute particular values.}$$

$$\frac{dy}{dt} = \frac{25}{4\pi \cdot 6^2}(-7) = \frac{-175}{144\pi} = -0.3868... \approx -0.39 \text{ cm/s}$$

Note that the rate is negative, consistent with the fact that the volume is decreasing as time increases.

b. At the instant the last drop leaves the cup, $y = 0$. Substituting $y = 0$ into the equation for dy/dt you found in part a leads to division by zero. So the liquid level is changing infinitely fast! ◀

The techniques for working related-rates problems like those in Examples 1 and 2 are summarized in this box.

TECHNIQUE: Related-Rates Problems

- Write the given rate(s) and the rate(s) you are asked to find as derivatives.
- Write an equation that relates the variables that appear in the given and desired rates.
- Differentiate implicitly with respect to time, then solve for the desired rate.
- Answer the question asked in the problem statement.

Problem Set 4-9

Quick Review

Q1. $\dfrac{d}{dx}(xy^2) = $ —?—

Q2. Name the technique you used in Problem Q1.

Q3. Name a property of derivatives you used in Problem Q1.

Q4. Name another property of derivatives you used in Problem Q1.

Q5. If velocity and acceleration are both negative, is the object speeding up or slowing down?

Q6. If dy/dx is negative, then y is getting —?—.

Q7. If $y = x \cos x$, then $y' = $ —?—.

Q8. If $y = \ln x$, then $dy/dt = $ —?—.

Q9. If $y = xe^{-x}$, then $y'' = $ —?—.

Q10. If $f(x) = \sin x$, then $f'(\pi) = $

     A. 1   B. 0.5   C. 0   D. −0.5   E. −1

1. *Bacteria Spreading Problem:* Bacteria are growing in a circular colony one bacterium thick. The bacteria are growing at a constant rate, thus making the area of the colony increase at a constant rate of 12 mm^2/h. Find an equation expressing the rate of change of the radius as a function of the radius, r, in millimeters, of the colony. Plot dr/dt as a function of r. How fast is r changing when it equals 3 mm? Describe the way dr/dt changes with the radius of the circle.

2. *Balloon Problem:* Phil blows up a spherical balloon. He recalls that the volume is $(4/3)\pi r^3$. Find dV/dt as a function of r and dr/dt. To make the radius increase at 2 cm/s, how fast must Phil blow air into the balloon when $r = 3$? When $r = 6$? Plot the graph of dV/dt as a function of r under these conditions. Sketch the result and interpret the graph.

3. *Ellipse Problem:* Recall that the area of an ellipse is $A = \pi ab$, where a and b are the lengths of the semiaxes (Figure 4-9d). Suppose that an ellipse is changing size but always keeps the same proportions, $a = 2b$. At what rate is the length of the major axis changing when $b = 12$ cm and the area is decreasing at 144 cm^2/s?

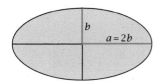

Figure 4-9d

4. *Kinetic Energy Problem:* The kinetic energy of a moving object equals half the product of its mass and the square of its velocity.

$$K = \tfrac{1}{2}mV^2$$

As a spaceship is fired into orbit, all three of these quantities vary. Suppose that the kinetic energy of a particular spaceship is increasing at a constant rate of 100,000 units per second and that mass is decreasing at 20 kg/s because its rockets are consuming fuel. At what rate is the spaceship's velocity changing when its mass is 5000 kg and it is traveling at 10 km/s?

5. *Base Runner Problem:* Milt Famey hits a line drive to center field. As he rounds second base, he heads directly for third, running at 20 ft/s (Figure 4-9e). Write an equation expressing the rate of change of his distance from home plate as a function of his displacement from third base. Plot the graph in a suitable domain. How fast is this distance changing when he is halfway to third? When he is at third? Is the latter answer reasonable? Explain.

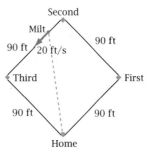

Figure 4-9e

6. *Tugboat Problem:* A tugboat moves a ship up to the dock by pushing its stern at a rate of 3 m/s (Figure 4-9f). The ship is 200 m long. Its bow remains in contact with the dock and its stern remains in contact with the pier. At what rate is the bow moving along the dock when the stern is 120 m from the dock? Plot the graph of this rate as a function of the distance between the stern and the dock.

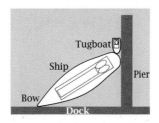

Figure 4-9f

7. *Luke and Leia's Trash Compactor Problem:* Luke and Leia are trapped inside a trash compactor on the Death Star (Figure 4-9g). The side walls are moving apart at 0.1 m/s, but the end walls are moving together at 0.3 m/s. The volume of liquid inside the compactor is 20 m^3, a constant.

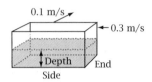

Figure 4-9g

a. Write an equation expressing the rate of change of depth of the liquid in terms of the width and length of the region inside the compactor.

b. When the side walls are 5 m long and the end walls are 2 m long, is the depth of liquid increasing or decreasing? At what rate?

8. *Darth Vader's Problem:* Darth Vader's spaceship is approaching the origin along the positive y-axis at 50 km/s. Meanwhile, Han Solo's spaceship is moving away from the origin along the positive x-axis at 80 km/s. When Darth is at $y = 1200$ km and Han is at $x = 500$ km, is the distance between them increasing or decreasing? At what rate?

9. *Barn Ladder Problem:* A 20-ft-long ladder in a barn is constructed so that it can be pushed up against the wall when it is not in use. The top of the ladder slides in a track on the wall, and the bottom is free to roll across the floor on wheels (Figure 4-9h). To make the ladder easier to move, a counterweight is attached to the top of the ladder by a rope over a pulley. As the ladder goes away from the wall, the counterweight goes up and vice versa.

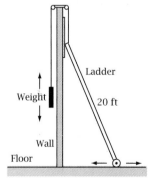

Figure 4-9h

a. Write an equation expressing the velocity of the counterweight as a function of the distance the bottom of the ladder is from

the wall and the velocity at which the bottom of the ladder moves away from the wall.

b. Find the velocity of the counterweight when the bottom is 4 ft from the wall and is being pushed toward the wall at 3 ft/s..

c. If the ladder drops all the way to the floor with its bottom moving at 2 ft/s and its top still touching the wall, how fast is the counterweight moving when the top just hits the floor? Surprising?

10. *Rectangle Problem:* A rectangle of length L and width W has a constant area of 1200 in.^{2}. The length changes at a rate of dL/dt inches per minute.

a. Find dW/dt in terms of W and dL/dt.

b. At a particular instant, the length is increasing at 6 in./min and the width is decreasing at 2 in./min. Find the dimensions of the rectangle at this instant.

c. At the instant in part b, is the length of the diagonal of the rectangle increasing or decreasing? At what rate?

11. *Conical Water Tank Problem:* The water tank shown in Figure 4-9i has diameter 6 m and depth 5 m.

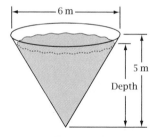

Figure 4-9i

a. If the water is 3 m deep, and is rising at 5 m/h, at what rate is the volume changing?

b. If the water is pumped out at 2 m^3/h, at what rate is the depth changing

i. When the water is 4 m deep?

ii. When the last drop is pumped out?

c. If water flows out under the action of gravity, the rate of change in the volume is directly proportional to the square root of the water's depth. Suppose that its volume V is decreasing at 0.5 m^3/h when its depth is 4 m.

i. Find dV/dt in terms of the depth of the water.

ii. Find dV/dt when the water is 0.64 m deep.

iii. Find the rate of change of depth when the water is 0.64 m deep.

12. *Cone of Light Problem:* A spotlight shines on the wall, forming a cone of light in the air (Figure 4-9j). The light is being moved closer to the wall, making the cone's altitude decrease by 6 ft/min. At the same time, the light is being refocused, making the radius increase by 7 ft/min. At the instant when the altitude is 3 ft and the radius is 8 ft, is the volume of the cone increasing or decreasing? How fast?

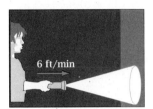

6 ft/min

Figure 4-9j

13. *Planetary Motion Problem:* On August 27, 2003, Mars was at its closest position to Earth. It then receded at an increasing rate (Figure 4-9k). In this problem you will analyze the rate at which the distance between the two planets changes. Assume that the orbits of the two planets are both circular and are both in the same plane. The radius of Earth's orbit is 93 million mi and the radius of Mars' orbit is 141 million mi. Assume also that the speed each planet moves along its orbit is constant. (This would be true if the orbits were exactly circular.) Mars orbits

the Sun once each 687 Earth-days. The Earth, of course, orbits once each 365 Earth-days. Answer the following questions.

a. What are the angular velocities of Earth and Mars about the Sun in radians per day? What is their relative angular velocity? That is, what is $d\theta/dt$?

b. What is the period of the planets' relative motion? On what day and date were the two planets next at their closest position?

c. Write an equation expressing the distance, D, between the planets as a function of θ.

d. At what rate is D changing today? Convert your answer to miles per hour.

e. Will D be changing its fastest when the planets are 90 degrees apart? If so, prove it. If not, find the angle θ at which D is changing fastest. Convert the answer to degrees.

f. Plot the graph of D versus time for at least two periods of the planets' relative motion. Is the graph a sinusoid?

14. *Sketchpad*™ *Project:* Use dynamic geometry software such as The Geometer's Sketchpad® to construct a segment AB that connects the origin of a coordinate system to a point on the graph of $y = e^{0.4x}$, as shown in Figure 4-9l. Display the length of $\overline{AB}$. Then drag point B on the graph. Describe what happens to the length of $\overline{AB}$ as B moves from negative values of x to positive values of x. If you drag B in such a way that x increases at 2 units/s, at what rate is the length of $\overline{AB}$ changing at the instant $x = -5$? At $x = 2$? At what value of x does the length of AB stop decreasing and start increasing? Explain how you arrived at your answer.

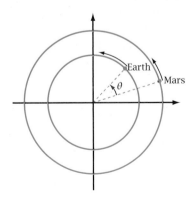

Figure 4-9k

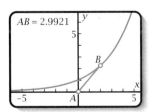

Figure 4-9l

4-10 Chapter Review and Test

In this chapter you have learned more techniques for differentiating functions algebraically. You can now differentiate products, quotients, parametric functions, and implicit relations.

These techniques enabled you to differentiate all six trigonometric functions and their inverses, prove that the power rule for derivatives works for non-integer exponents, and analyze complicated geometrical figures and related rates.

Review Problems

R0. Update your journal with what you've learned since your last entry. Include such things as

- The one most important thing you have learned by studying Chapter 4
- Which boxes you've been working on in your "define, understand, do, apply" table
- Key terms, such as *parametric function, implicit differentiation*, and *differentiability*
- Surprising properties, such as the product rule and the parametric chain rule
- Ways in which you've used graphs, tables, algebra, and writing to understand concepts
- Any ideas about calculus that you're still unclear about

R1. Let $x = g(t) = t^3$ and $y = h(t) = \cos t$.

a. If $f(t) = g(t) \cdot h(t)$, show by counterexample that $f'(t)$ does not equal $g'(t) \cdot h'(t)$.

b. If $f(t) = g(t)/h(t)$, show by counterexample that $f'(t)$ does not equal $g'(t)/h'(t)$.

c. Show by example that dy/dx equals $(dy/dt)/(dx/dt)$.

R2. a. State the product rule.

b. Prove the product rule, using the definition of derivative.

c. Differentiate and simplify.

i. $f(x) = x^7 \ln 3x$

ii. $g(x) = \sin x \cos 2x$

iii. $h(x) = (3x - 7)^5(5x + 2)^3$

iv. $s(x) = x^8 e^{-x}$

d. Differentiate $f(x) = (3x + 8)(4x + 7)$ in two ways.

i. As a product of two functions

ii. By multiplying the binomials, then differentiating

Show that your answers are equivalent.

R3. a. State the quotient rule.

b. Prove the quotient rule, using the definition of derivative.

c. Differentiate and simplify.

i. $f(x) = \dfrac{\sin 10x}{x^5}$

ii. $g(x) = \dfrac{(2x + 3)^9}{(9x - 5)^4}$

iii. $h(x) = (100x^3 - 1)^{-5}$

d. Differentiate $y = 1/x^{10}$ as a quotient and as a power with a negative exponent. Show that both answers are equivalent.

e. Find $t'(x)$ if $t(x) = (\sin x)/(\cos x)$. Use the result to find $t'(1)$.

f. Using $t(x)$ given in part e, plot the difference quotient $[t(x) - t(1)]/(x - 1)$, using a window centered at $x = 1$, with $\Delta x = 0.001$. (Enter $t(1)$ as $(\sin 1)/(\cos 1)$ to avoid rounding errors.) Sketch the result. Show that the difference quotient approaches $t'(1)$ by tracing the graph and making a table of values for several x-values on either side of 1.

R4. a. Differentiate.

    i. $y = \tan 7x$

    ii. $y = \cot x^4$

    iii. $y = \sec e^x$

    iv. $y = \csc x$

  b. Derive the property $\cot' x = -\csc^2 x$.

  c. Plot the graph of $y = \tan x$. Make a connection between the slope of the graph and the fact that the derivative is equal to the square of a function.

  d. Suppose that $f(t) = 7 \sec t$. How fast is $f(t)$ changing when $t = 1$? When $t = 1.5$? When $t = 1.57$? How do you explain the dramatic increase in the rate of change of $f(t)$ even though t doesn't change very much?

R5. a. Differentiate.

    i. $y = \tan^{-1} 3x$

    ii. $\dfrac{d}{dx}(\sec^{-1} x)$

    iii. $c(x) = (\cos^{-1} x)^2$

  b. Plot the graph of $f(x) = \sin^{-1} x$. Use a friendly window that includes $x = -1$ and $x = 1$ as grid points, and the same scales on both axes. Sketch the graph. Then use the algebraic derivative to explain how $f'(0)$ and $f'(1)$ agree with the graph, and why $f'(2)$ is undefined.

R6. a. State the relationship between differentiability and continuity.

  b. Sketch a graph for each function described.

    i. Function f is neither differentiable nor continuous at $x = c$.

    ii. Function f is continuous but not differentiable at $x = c$.

    iii. Function f is differentiable but not continuous at $x = c$.

    iv. Function f is differentiable and continuous at $x = c$.

  c. Let $f(x) = \begin{cases} x^2 + 1, & \text{if } x < 1 \\ -x^2 + 4x - 1, & \text{if } x \geq 1 \end{cases}$

    i. Sketch the graph.

    ii. Show that f is continuous at $x = 1$.

    iii. Is f differentiable at $x = 1$? Justify your answer.

  d. Let $g(x) = \begin{cases} \sin^{-1} x, & \text{if } 0 \leq x \leq 1 \\ x^2 + ax + b, & \text{if } x < 0 \end{cases}$

  Use one-sided limits to find the values of the constants a and b that make g differentiable at $x = 0$. Confirm your answer by graphing.

R7. a. For this parametric function, find $\dfrac{dy}{dx}$ and $\dfrac{d^2 y}{dx^2}$.

$$x = e^{2t}$$
$$y = t^3$$

  b. Figure 4-10a shows the spiral with the parametric equations

$$x = (t/\pi)\cos t$$
$$y = (t/\pi)\sin t$$

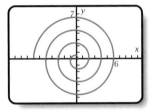

Figure 4-10a

  Write dy/dx in terms of t. The graph appears to pass vertically through the point $(6, 0)$. Does the graph contain this point? If not, why not? If so, is it really vertical at that point? Justify your answer.

  c. *Ferris Wheel Problem:* Figure 4-10b shows a Ferris wheel of diameter 40 ft. Its axle is 25 ft above the ground. (The same Ferris wheel appeared in Problem 1 of Problem Set 3-8.)

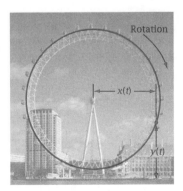

Figure 4-10b

The seat shown is a distance $y(t)$ from the ground and $x(t)$ out from the axle, both in feet. The seat first reaches a high point when $t = 3$ s. The wheel makes 3 rev/min. Given that x and y vary sinusoidally with time, t, as the Ferris wheel rotates clockwise, write parametric equations for x and y in terms of t. When $t = 0$, is the seat moving up or down? How fast? How can you determine these facts? When $t = 0$, is the seat moving to the left or to the right? How fast? How can you determine this? What is the first positive value of t for which dy/dx is infinite?

R8. a. Given $y = x^{8/5}$, transform the equation so that it has only integer exponents. Then differentiate implicitly with respect to x, using the power rule, which you know works for powers with positive integer exponents. Show that the answer you get for y' this way is equivalent to the answer you get by using the power rule directly on $y = x^{8/5}$.

b. Find dy/dx: $y^3 \sin xy = x^{4.5}$.

c. *Cissoid of Diocles Problem:* The cissoid of Diocles in Figure 4-10c has the equation $4y^2 - xy^2 = x^3$. (The word *cissoid* is Greek for "ivylike.") In *Curves and Their Properties* (National Council of Teachers of Mathematics, 1974), Robert C. Yates reports that the Greek mathematician Diocles (ca. 250–100 B.C.E.) used cissoids for finding cube roots.

i. Find dy/dx when $x = 2$. Show that your answers are reasonable.

ii. Find dy/dx at the point $(0, 0)$. Interpret your answer.

iii. Find the vertical asymptote.

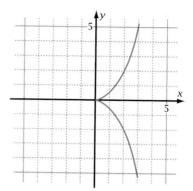

Figure 4-10c

R9. *Rover's Tablecloth Problem:* Rover grabs the tablecloth and starts backing away from the table at 20 cm/s. A glass near the other end of the tablecloth (Figure 4-10d) moves toward the edge and finally falls off. The table is 80 cm high, Rover's mouth is 10 cm above the floor, and 200 cm of tablecloth separate Rover's mouth from the glass. At the instant the glass reached the table's edge, was it going faster or slower than Rover's 20 cm/s? By how much?

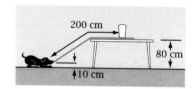

Figure 4-10d

Concept Problems

C1. *Historical Problem: Newton's Method:* Suppose you are trying to find a zero (an x-intercept) of function f, and you can't solve the equation $f(x) = 0$ by using algebra. English mathematician Isaac Newton (1642–1727) is credited with finding a way to solve such equations approximately by using derivatives. This process is called **Newton's method**. Figure 4-10e shows the graph of function f for which a zero is to be found. Pick a convenient value $x = x_0$ close to the zero. Calculate $y_0 = f(x_0)$. The tangent line to the graph of f at that point will have slope $m = f'(x_0)$. If you extend this tangent line to the x-axis, it should cross at a place x_1 that is closer to the desired zero than your original choice, x_0. By repeating this process, you can find the zero to as many decimal places as you like!

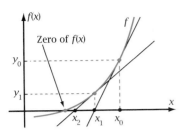

Figure 4-10e

a. Let m be the slope of the tangent line at the point (x_0, y_0). Write an equation for the tangent line.

b. The point $(x_1, 0)$ is on the tangent line. Substitute these coordinates for x and y in the equation from part a. Use the result to show that $x_1 = x_0 - (y_0/m)$.

c. If a tangent line is drawn at (x_1, y_1), it will cross the x-axis at x_2. Explain why

$$x_2 = x_1 - \frac{f(x_1)}{f'(x_1)}$$

d. Write a program that uses the equation from part c to calculate values of x iteratively. The equation for $f(x)$ can be stored in the Y= menu. You can use the numerical derivative feature to calculate the value of $f'(x)$. The input should be the value of x_0. The program should allow you to press ENTER, then read the next value of x. Test the program on $f(x) = x^2 - 9x + 14$. Start with $x_0 = 1$, then again with $x_0 = 6$.

e. Use the program to find the three zeros of $g(x) = x^3 - 9x^2 + 5x + 10$. Compare your answers with those you get using your grapher's built-in solver feature.

f. The equation of the graph in Figure 4-10e is $f(x) = \sec x - 1.1$. Use Newton's method to estimate the zero shown in the figure. If you start with $x_0 = 1$, how many iterations, n, does it take to make x_{n+1} indistinguishable from x_n on your calculator?

C2. *Speeding Piston Project:* A car engine's crankshaft is turning 3000 revolutions per minute (rpm). As the crankshaft turns in the xy-plane, the piston moves up and down inside the cylinder (Figure 4-10f). The radius of the crankshaft is 6 cm. The connecting rod is 20 cm long and fastens to a point 8 cm below the top of the piston. Let y be the distance from the top of the piston to the center of the crankshaft. Let θ be an angle in standard position measured from the positive x-axis and increasing as the crankshaft rotates counterclockwise.

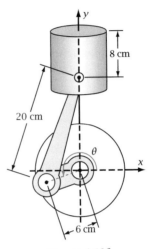

Figure 4-10f

a. Write an equation expressing y in terms of θ. The law of cosines is helpful here.

b. Find an equation for the rate of change of y with respect to t in terms of θ and $d\theta/dt$.

c. Find an equation for the acceleration of the piston. Remember that $d\theta/dt$ is a constant because the engine is turning at a constant 3000 rpm.

d. Between what two values of θ is the piston turning down with an acceleration greater than that of gravity (980 cm/s^2)?

Chapter Test

PART 1: No calculators allowed (T1–T5)

T1. Write the power rule for derivatives. That is, write y' if $y = uv$, where u and v are differentiable functions of x.

T2. In deriving the quotient rule for derivatives, where $y = u/v$, you encountered

$$y' = \lim_{\Delta x \to 0} \left[\frac{\dfrac{u + \Delta u}{v + \Delta v} - \dfrac{u}{v}}{\Delta x} \cdot \frac{(v + \Delta v)v}{(v + \Delta v)v} \right]$$

Show how the quotient rule follows from this complex fraction by multiplying, then taking the limit.

T3. Use the quotient rule and the fact that $\cot x = (\cos x)/(\sin x)$ to prove that $\cot' x = -\csc^2 x$.

T4. Use the definition of the inverse trigonometric functions and appropriate trigonometry and calculus to show that

$$y = \sin^{-1} x \Rightarrow y' = \frac{1}{\sqrt{1 - x^2}}$$

T5. For this function, find $\dfrac{dy}{dx}$ and $\dfrac{d^2 y}{dx^2}$.

$$x = t^2$$
$$y = t^4$$

PART 2: Graphing calculators allowed (T6–T24)

T6. Is the function $c(x) = \cot 3x$ increasing or decreasing when $x = 5$? At what rate?

T7. If $f(x) = \sec x$, find $f'(2)$. Show that the difference quotient $\dfrac{\sec x - \sec 2}{x - 2}$ approaches $f'(2)$ by making a table of values for each 0.001 unit of x from $x = 1.997$ through $x = 2.003$.

T8. Sketch the graph of a function that has all of these features.

• The value $f(-2) = 7$.
• The function $f(x)$ increases slowly at $x = -2$.
• The value $f(1)$ is positive, but $f(x)$ decreases rapidly at $x = 1$.
• The function $f(x)$ is continuous at $x = 2$ but is not differentiable at that point.

T9. Use the most time-efficient method to prove that the general linear function, $f(x) = mx + b$, is continuous for every value of $x = c$.

For Problems T10–T15, find the derivative.

T10. $f(x) = \sec 5x$

T11. $y = \tan^{7/3} x$

T12. $f(x) = (2x - 5)^6 (5x - 1)^2$

T13. $f(x) = \dfrac{e^{3x}}{\ln x}$

T14. $x = \sec 2t$
$y = \tan 2t^3$

T15. $y = 4 \sin^{-1}(5x^3)$

T16. *Rotated Ellipse Problem:* Figure 4-10g shows the graph of

$$9x^2 - 20xy + 25y^2 - 16x + 10y - 50 = 0$$

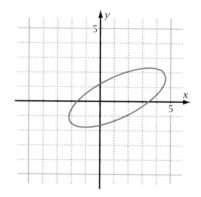

Figure 4-10g

The figure is an ellipse that is rotated with respect to the x- and y-axes. Evaluate dy/dx at both points where $x = -2$. Show that each answer is reasonable.

T17. Let $f(x) = \begin{cases} x^3 + 1, & \text{if } x \le 1 \\ a(x - 2)^2 + b, & \text{if } x > 1 \end{cases}$

Use one-sided limits to find the values of the constants a and b that make function f

differentiable at $x = 1$. Show that the limits of $f(x)$ as x approaches 1 from the right and from the left can equal each other but that f is still not differentiable at $x = 1$.

T18. Let $y = x^{7/3}$. Raise both sides to the third power, getting $y^3 = x^7$, and differentiate both sides implicitly with respect to x using the power rule with positive integer exponents. Show that the answer you get for y' this way is equivalent to the one you get using the power rule directly on $y = x^{7/3}$.

Airplane Problem: For Problems T19–T23, an airplane is flying level at 420 mi/h, 5 miles above the ground (Figure 4-10h). Its path will take it directly over an observation station on the ground (see figure). Let x be the horizontal distance, in miles, between the plane and the station, and let θ be the angle, in radians, to the line of sight.

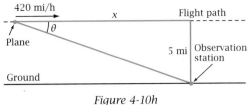

Figure 4-10h

T19. Show that $\theta = \cot^{-1} \dfrac{x}{5}$.

T20. Find: $\dfrac{d\theta}{dx}$

T21. If t is time in hours, what does dx/dt equal?

T22. Find $d\theta/dt$ as a function of x. Plot the graph on your grapher and sketch the result.

T23. Where is the plane when θ is changing fastest?

T24. What did you learn as a result of taking this test that you did not know before?

Definite and Indefinite Integrals

Cable cars travel the steep streets of San Francisco. For every 100 ft the car goes horizontally, it might rise as much as 20 ft vertically. The slope of the street is a derivative, dy/dx, a single quantity. The differentials dy and dx can be defined separately in such a way that their ratio is the slope. These differentials play a crucial role in the definition, computation, and application of definite integrals.

Mathematical Overview

In your study of calculus so far, you have learned that a definite integral is a product of *x* and *y*, where *y* varies with *x*. In Chapter 5, you will learn the formal definition of definite integral, that an indefinite integral is an antiderivative, and that the two are related by the fundamental theorem of calculus. You will gain this knowledge in four ways.

Graphically   The icon at the top of each even-numbered page of this chapter shows how you can analyze a definite integral by slicing a region into vertical strips.

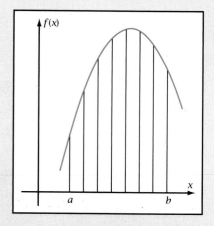

Numerically

x	f(x)	Integral
1.0	5	0.0
1.5	8	2.5
2.0	10	7.0
⋮	⋮	⋮

Algebraically $\int_{1}^{4} \cos x \, dx = \sin 4 - \sin 1$, the fundamental theorem

Verbally   *Finally I found out why definite and indefinite integrals share a name. Indefinite integrals can be used to calculate definite integrals exactly. I also learned a way to analyze many different problems involving the product of x and y by slicing the region under the graph into vertical strips.*

5-1 A Definite Integral Problem

So far you have learned three of the four concepts of calculus. In this chapter you will learn about the fourth concept—indefinite integral—and why its name is so similar to that of definite integral. In this section you will refresh your memory about definite integrals.

OBJECTIVE   Work the problems in this section, on your own or with your study group, as an assignment after your last test on Chapter 4.

Exploratory Problem Set 5-1

Oil Well Problem: An oil well that is 1000 ft deep is to be extended to a depth of 4000 ft. The drilling contractor estimates that the cost in dollars per foot, $f(x)$, for doing the drilling is

$$f(x) = 20 + 0.000004x^2$$

where x is the number of feet below the surface at which the drill is operating. Figure 5-1a shows an accurate graph of function f.

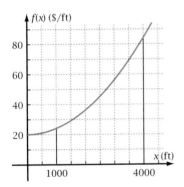

Figure 5-1a

1. How much does it cost per foot to drill at 1000 ft? At 4000 ft? Why do you suppose the cost per foot increases with increasing depth?

2. The actual cost of extending the well is given by the definite integral of $f(x)$ with respect to x from $x = 1000$ to $x = 4000$. Estimate this integral by finding T_6, the trapezoidal rule sum found by dividing the interval [1000, 4000] into six strips of equal width. Will this estimate be higher than the actual value or lower? How can you tell?

3. Estimate the cost again, this time estimating the integral by using rectangles whose altitudes are measured at the midpoint of each strip (that is, $f(1250)$, $f(1750)$, $f(2250)$, and so on). The sum of the areas of these rectangles is called a **Riemann sum**, R_6. Is R_6 close to T_6?

4. The actual value of the integral is the *limit* of T_n, the trapezoidal rule sum with n increments, as n becomes infinite. Find T_{100} and T_{500}. Based on the answers, make a conjecture about the exact value of the integral.

5. Let $g(x)$ be an antiderivative of $f(x)$. Find an equation for $g(x)$. Use the equation to evaluate the quantity $g(4000) - g(1000)$. What do you notice about the answer? Surprising? Write another name for antiderivative.

6. Find these antiderivatives.
   a. $f(x)$ if $f'(x) = 7x^6$
   b. y if $y' = \sin x$
   c. u if $u' = e^{2x}$
   d. v if $v' = (4x + 5)^7$

5-2   Linear Approximations and Differentials

In Section 5-1, you saw a connection between definite integrals and indefinite integrals. So far you have not learned mathematical symbols for these quantities. In this section you will define the **differentials** dx and dy that appear in the symbol for derivative and that will also appear in the symbol for integral. The ratio $dy \div dx$ is the slope of the **tangent line**, the linear function that best fits the graph of a function at the point of tangency (Figure 5-2a).

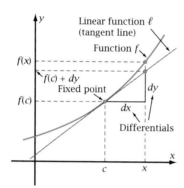

Figure 5-2a

OBJECTIVE   Given the equation of a function f and a fixed point on its graph, find an equation of the linear function that best fits the given function. Use this equation to find approximate values of $f(x)$ and values of the differentials dx and dy.

If function f is differentiable at the point $x = c$, then it is also **locally linear** there. As Figure 5-2a shows, the tangent line, $\ell(x)$, is almost indistinguishable from the curved graph when x is close to c. You can see this clearly by zooming in on the point $(c, f(c))$. The tangent line and the curved graph seem to merge.

Example 1 shows you how to find the particular equation of the tangent line at a given point and illustrates that it is a good approximation of the function for values of x close to that point.

▶ **EXAMPLE 1**   For $f(x) = 15 - x^3$, find the particular equation for $\ell(x)$, the linear function whose graph is tangent to the graph of f at the fixed point $(2, 7)$. Show graphically that f is locally linear at $x = 2$. Show numerically that $\ell(x)$ is a close approximation to $f(x)$ for values of x close to 2 by calculating the error in the approximation $\ell(x) \approx f(x)$.

Solution
$$f(2) = 15 - 2^3 = 7, \text{ so the point } (2, 7) \text{ is on the graph of } f.$$
$$f'(x) = -3x^2 \Rightarrow f'(2) = -12, \text{ so the line } \ell \text{ has slope } m = -12.$$

$$\ell(x) - 7 = -12(x - 2) \qquad \text{Point-slope form of the linear equation.}$$
$$\ell(x) = 7 - 12(x - 2) \qquad \text{Solve explicitly for } \ell(x).$$

Figure 5-2b shows f and ℓ before and after zooming in on the point $(2, 7)$. The graph of f becomes indistinguishable from the tangent line, illustrating local linearity.

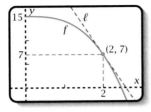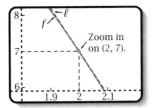

Figure 5-2b

This table shows that the linear function fits the graph of f perfectly at $x = 2$ and has small but increasing errors for values of x farther from 2.

x	$f(x)$	$\ell(x)$	Error $= f(x) - \ell(x)$
1.9	8.141	8.2	-0.059
1.95	7.585125	7.6	-0.014875
1.99	7.119401	7.12	-0.000599
2	7	7	0
2.01	6.879399	6.880	-0.000601
2.05	6.384875	6.4	-0.015125
2.1	5.739	5.8	-0.061

Notice the parts of the linear equation $\ell(x) = 7 - 12(x - 2)$ from Example 1 that show up in Figure 5-2b.

- 2 is the value of $x = c$ at the fixed point.

- 7 is $f(2)$, the value of $f(c)$ at the fixed point.

- -12 is $f'(2)$, the value of the derivative at the fixed point.

- $(x - 2)$ is the differential dx, the horizontal displacement of x from the fixed point.

- The quantity $-12(x - 2)$ is the differential dy, the vertical displacement from the fixed point to the point on the tangent line (the linear function graph).

These properties are summarized in this box.

PROPERTIES: *Local Linearity and Linearization of a Function*

If f is differentiable at $x = c$, then the linear function, $\ell(x)$, containing $(c, f(c))$ and having slope $f'(c)$ (the tangent line) is a close approximation to the graph of f for values of x close to c.

Linearizing a function f means approximating the function for values of x close to c using the linear function

$$\ell(x) = f(c) + f'(c)(x - c) \quad \text{or, equivalently,}$$
$$\ell(x) = f(c) + f'(c)\,dx \quad \text{or} \quad \ell(x) = f(c) + dy$$

Note that the differentials dx and dy are related to the quantities Δx and Δy that you have used in connection with derivatives. Figure 5-2c shows that dx is a change in x, the same as Δx. The differential dy is the corresponding change in y along the tangent line, whereas Δy is the actual change in y along the curved graph of f. The figure also shows the error between the actual value of $f(x)$ and the linear approximation of $f(x)$.

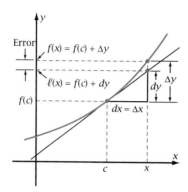

Figure 5-2c

Example 2 shows you how to use dy to approximate Δy.

▶ **EXAMPLE 2** If $f(x) = \sin x$, write an equation for the linear function $\ell(x)$ that best fits at $x = 1$. Use the linear function to approximate $f(1.02)$. What is the error in the approximation? What are the values of $dx, dy, \Delta x$, and Δy? Sketch a graph that shows these quantities and the error.

Solution

$f(1) = \sin 1 = 0.84147...$

$f'(x) = \cos x \Rightarrow f'(1) = \cos 1 = 0.54030...$

$\therefore \ell(x) = (\sin 1) + (\cos 1)(x - 1)$ $\qquad$ $\ell(x) = f(c) + f'(c)(x - c)$

$\qquad y = 0.84147... + 0.54030... \, (x - 1)$

If $x = 1.02$,

then $\ell(x) = 0.84147... + 0.54030... \, (1.02 - 1) = 0.852277030...$

$f(1.02) = \sin 1.02 = 0.852108021...$

$\therefore$ error $= \sin 1.02 - 0.852277030... = -0.000169008...$ $\qquad$ Small error. 1.02 is close to 1.

$\qquad dx = 1.02 - 1 = 0.02$

$\qquad dy = (\cos 1)(0.02) = 0.01080604...$

$\qquad \Delta x = dx = 0.02$

$\qquad \Delta y = \sin 1.02 - \sin 1 = 0.01063703... \approx dy$

Figure 5-2d illustrates the differentials dx and dy and shows their relationship to Δx and Δy. The error is equal to $\Delta y - dy$.

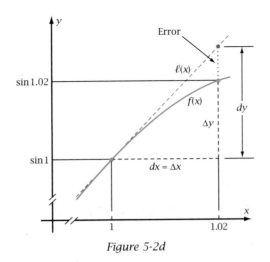

Figure 5-2d ◀

To find a differential algebraically in one step, multiply the derivative by dx. Examples 3 and 4 show you how to do this and how to interpret the answer in terms of the chain rule.

▶ **EXAMPLE 3**   If $y = e^{3x} \cos x$, find dy.

Solution $dy = (3e^{3x} \cos x - e^{3x} \sin x)\, dx$ Multiply the derivative by dx.

$= e^{3x}(3 \cos x - \sin x)\, dx$ ◀

Note that finding a differential of a given function is comparable to using the chain rule. In Example 3, you can think of x as the *inside function*. So, to find the differential of a function, you multiply the derivative by the *differential of the inside function*.

▶ **EXAMPLE 4**   If $y = \ln(\tan 5x)$, find dy.

Solution $dy = \dfrac{1}{\tan 5x}(5 \sec^2 5x\, dx)$ The quantity $(5 \sec^2 5x\, dx)$ is the differential of the inside function.   ◀

The definitions and some properties of dx and dy are given in this box.

DEFINITION AND PROPERTIES: *Differentials*

Algebraically: The differentials dx and dy are defined as

$$dx = \Delta x$$
$$dy = f'(x)\, dx$$

Verbally: Multiply the derivative by the differential of the inside function.

Notes: $dy \div dx$ is equal to $f'(x)$

The differential dy is usually *not* equal to Δy.

$$\Delta y = f(x + \Delta x) - f(x)$$

Example 5 shows you how to find the antiderivative if the differential is given.

▶ **EXAMPLE 5**   Given $dy = (3x + 7)^5 \, dx$, find an equation for the antiderivative, y.

Solution $$\frac{dy}{dx} = (3x + 7)^5$$ Divide both sides of the given equation by dx.

Thought process:
It looks as if someone differentiated $(3x + 7)^6$.
But the differential of $(3x + 7)^6$ is
$6(3x + 7)^5 \cdot 3 \cdot dx$, or $18(3x + 7)^5 \, dx$.
So the function must be only $1/18$ as big
as $(3x + 7)^6$, and the answer is

$$y = \frac{1}{18}(3x + 7)^6 + C$$ Why is $+ C$ needed?   ◀

Problem Set 5-2

Quick Review   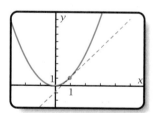 5 min

Q1. Sketch a graph that illustrates the meaning of definite integral.

Q2. Write the physical meaning of derivative.

Q3. Differentiate: $f(x) = 2^{-x}$

Q4. Find the antiderivative: $y' = \cos x$

Q5. If $y = \tan t$ where y is in meters and t is in seconds, how fast is y changing when $t = \pi/3$ s?

Q6. Find $\lim_{\Delta x \to 0} \frac{f(x+\Delta x)-f(x)}{\Delta x}$ if $f(x) = \sec x$.

Q7. Find $\lim_{x \to 0} \sec x$.

Q8. What is the limit of a constant?

Q9. What is the derivative of a constant?

Q10. If $\lim_{x \to c} g(x) = g(c)$, then g is —?— at $x = c$.

A. Differentiable

B. Continuous

C. Undefined

D. Decreasing

E. Increasing

1. For $f(x) = 0.2x^4$, find an equation of the linear function that best fits f at $x = 3$. What is the error in this linear approximation of $f(x)$ if $x = 3.1$? If $x = 3.001$? If $x = 2.999$?

2. For $g(x) = \sec x$, find an equation of the linear function that best fits g at $x = \pi/3$. What is the error in this linear approximation of $g(x)$ if $dx = 0.04$? If $dx = -0.04$? If $dx = 0.001$?

3. *Local Linearity Problem I:* Figure 5-2e shows the graph of $f(x) = x^2$ and the line tangent to the graph at $x = 1$.

Figure 5-2e

a. Find the equation of this tangent line, and plot it and the graph of f on the same screen. Then zoom in on the point of tangency. How does your graph illustrate that function f is locally linear at the point of tangency?

194

b. Make a table of values showing $f(x)$, the value of y on the tangent line, and the error $f(x) - y$ for each 0.01 unit of x from 0.97 to 1.03. How does the table illustrate that function f is locally linear at the point of tangency?

4. *Local Linearity Problem II:* Figure 5-2f shows the graph of

$$f(x) = x^2 - 0.1(x - 1)^{1/3}$$

Explore the graph for x close to 1. Does the function have local linearity at $x = 1$? Is f differentiable at $x = 1$? If f is differentiable at $x = c$, is it locally linear at that point? Is the converse of this statement true or false? Explain.

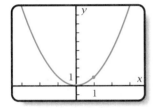

Figure 5-2f

5. *Steepness of a Hill Problem:* On roads in hilly areas, you sometimes see signs like this.

> Steep hill
> 20% grade

The grade of a hill is the slope (rise/run) written as a percentage, or, equivalently, as the number of feet the hill rises per hundred feet horizontally. Figure 5-2g shows the latter meaning of grade.

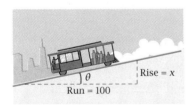

Figure 5-2g

a. Let x be the grade of a hill. Explain why the angle, θ degrees, that a hill makes with the horizontal is given by

$$\theta = \frac{180}{\pi} \tan^{-1} \frac{x}{100}$$

b. Find an equation for $d\theta$ in terms of x and dx. Then find $d\theta$ in terms of dx for grades of $x = 0\%$, 10%, and 20%.

c. You can estimate θ at $x = 20\%$ by simply multiplying $d\theta$ at $x = 0$ by 20. How much error is there in the value of θ found by using this method rather than by using the exact formula that involves the inverse tangent function?

d. A rule of thumb you can use to estimate the number of degrees a hill makes with the horizontal is to divide the grade by 2. Where in your work for part c did you divide by approximately 2? When you use this method to determine the number of degrees for grades of 20% and 100%, how much error is there in the number?

6. *Sphere Expansion Differential Problem:* The volume of a sphere is given by $V = \frac{4}{3}\pi r^3$ where r is the radius. Find dV in terms of dr for a spherical ball bearing with radius 6 mm. If the bearing is warmed so that its radius expands to 6.03 mm, find dV and use it to find a linear approximation of the new volume. Find the actual volume of the warmed bearing by substituting 6.03 for r. What does ΔV equal? What error is introduced by using dV instead of ΔV to estimate the volume?

7. *Compound Interest Differential Problem:* Lisa Cruz invests $6000 in an account that pays 5% annual interest, compounded continuously.

a. Lisa wants to estimate how much money she earns each day, so she finds 5% of 6000 and divides by 365. About how much will she earn the first day?

b. The actual amount of money, m, in dollars, in the account after the time, t, in days, is given by

$$m = 6000e^{(.05/365)t}$$

Find dm in terms of t and dt. Does dm give you about the same answer as in part a for the first day, starting at $t = 0$ and using $dt = 1$? How much interest would the differential dm predict Lisa would earn in the first 30 days? The first 60 days?

c. Use the equation in part b to find Δm for the first 1 day, 30 days, and 60 days. How does the linear approximation $\Delta m \approx t\,dm$ compare to the actual value, $\Delta m = (m - 6000)$, as t increases?

8. *Sunrise Time Differential Problem:* Based on a table of sunrise times from the U.S. Naval Observatory (*http://aa.usno.navy.mil/*), the time of sunrise in Chicago for the first few days of March 2003 was changing at a rate of $dS/dt \approx -1.636$ minutes per day. For March 1, the table lists sunrise at 6:26 a.m. However, no equation is given explaining how the table was computed.

a. Write the differential dS as a function of dt. Use your result to estimate the times of sunrise 10 days and 20 days later, on March 11 and March 21. How do your answers compare to the tabulated times of 6:10 a.m. and 5:53 a.m., respectively?

b. Explain why dS would *not* give a good approximation for the time of sunrise on September 1, 2003.

For Problems 9–26, find an equation for the differential dy.

9. $y = 7x^3$

10. $y = -4x^{11}$

11. $y = (x^4 + 1)^7$

12. $y = (5 - 8x)^4$

13. $y = 3x^2 + 5x - 9$

14. $y = x^2 + x + 9$

15. $y = e^{-1.7x}$

16. $y = 15 \ln x^{1/3}$

17. $y = \sin 3x$

18. $y = \cos 4x$

19. $y = \tan^3 x$

20. $y = \sec^3 x$

21. $y = 4x \cos x$

22. $y = 3x \sin x$

23. $y = \dfrac{x^2}{2} - \dfrac{x}{4} + 2$

24. $y = \dfrac{x^3}{3} - \dfrac{x}{5} + 6$

25. $y = \cos(\ln x)$

26. $y = \sin(e^{0.1x})$

For Problems 27–40, find an equation for the antiderivative y.

27. $dy = 20x^3\,dx$

28. $dy = 36x^4\,dx$

29. $dy = \sin 4x\,dx$

30. $dy = \cos 0.2x\,dx$

31. $dy = (0.5x - 1)^6\,dx$

32. $dy = (4x + 3)^{-6}\,dx$

33. $dy = \sec^2 x\,dx$

34. $dy = \csc x \cot x\,dx$

35. $dy = 5\,dx$

36. $dy = -7\,dx$

37. $dy = (6x^2 + 10x - 4)\,dx$

38. $dy = (10x^2 - 3x + 7)\,dx$

39. $dy = \sin^5 x \cos x\,dx$ (Be clever!)

40. $dy = \sec^7 x \tan x\,dx$ (Be very clever!)

For Problems 41 and 42, do the following.

a. Find dy in terms of dx.

b. Find dy for the given values of x and dx.

c. Find Δy for the given values of x and dx.

d. Show that dy is close to Δy.

41. $y = (3x + 4)^2(2x - 5)^3$, $x = 1$, $dx = -0.04$

42. $y = \sin 5x$, $x = \pi/3$, $dx = 0.06$

5-3 Formal Definition of Antiderivative and Indefinite Integral

You have learned that an antiderivative is a function whose derivative is given. As you will learn in Section 5-6, the antiderivative of a function provides an algebraic way to calculate exact definite integrals. For this reason an antiderivative is also called an indefinite integral. The word *indefinite* is used because there is always a "+ C" whose value is not determined until an initial condition is specified.

RELATIONSHIP: Antiderivative and Indefinite Integral

Indefinite integral is another name for antiderivative.

In this section you will learn the symbol that is used for an indefinite integral or an antiderivative.

OBJECTIVE Become familiar with the symbol used for an indefinite integral or an antiderivative by using this symbol to evaluate indefinite integrals.

In Section 5-2, you worked problems such as "If $dy = x^5\, dx$, find y." You can consider **indefinite integration**, which you now know is the same thing as antidifferentiation, to be the operation performed on a differential to get the expression for the original function. The **integral sign** used for this operation is a stretched-out S, as shown here.

$$\int$$

As you will see in Section 5-4, the S shape comes from the word *sum*. To indicate that you want to find the indefinite integral of $x^5\, dx$, write

$$\int x^5\, dx$$

The whole expression, $\int x^5\, dx$, is the **integral**. The function x^5 "inside" the integral sign is called the **integrand**. These words are similar, for example, to the words *radical*, used for the expression $\sqrt{7}$, and *radicand*, used for the number 7 inside. Note that although dx must appear in the integral, only the function x^5 is called the integrand.

Writing the answer to an indefinite integral is called **evaluating** it or **integrating**. Having seen the techniques of Section 5-2, you should recognize that

$$\int x^5\, dx = \tfrac{1}{6}x^6 + C$$

From this discussion, you can understand the formal definition.

> ### DEFINITION: Indefinite Integral
>
> $$g(x) = \int f(x)\,dx \text{ if and only if } g'(x) = f(x)$$
>
> That is, an indefinite integral of $f(x)\,dx$ is a function whose derivative is $f(x)$. The function $f(x)$ inside the integral sign is called the integrand.
>
> *Verbally:* The expression $\int f(x)\,dx$ is read "The integral of $f(x)$ with respect to x."
>
> *Notes:* An indefinite integral is the same as an **antiderivative**. The symbol $\int$ is an **operator**, like cos or the minus sign, that acts on $f(x)\,dx$.
>
> An equivalent equation for $g'(x) = f(x)$ is $\dfrac{d}{dx}\int f(x)\,dx = f(x)$.

Integral of a Constant Times a Function and of a Sum of Several Functions

To develop a systematic way of integrating, knowing some properties of indefinite integrals, like those shown below, is helpful.

$$\int 5\cos x\,dx \qquad \text{and} \qquad \int(x^5 + \sec^2 x - x)\,dx$$

$$= 5\int \cos x\,dx \qquad\qquad\quad = \int x^5\,dx + \int \sec^2 x\,dx - \int x\,dx$$

$$= 5\sin x + C \qquad\qquad\quad = \tfrac{1}{6}x^6 + \tan x - \tfrac{1}{2}x^2 + C$$

In the first case, 5, a constant, can be multiplied by the answer to $\int \cos x\,dx$. In the second case, the integral of each term can be evaluated separately and the answers added together. The "only if" part of the definition of indefinite integral means that all you have to do to prove these facts is differentiate the answers. Following is a proof of the integral of a constant times a function property. You will prove the integral of a sum property in Problem 35 of Problem Set 5-3.

▶ **PROPERTY**   If k stands for a constant, then $\int k f(x)\,dx = k\int f(x)\,dx$.

Proof

Let $g(x) = k\int f(x)\,dx$.

Then $g'(x) = k \cdot \dfrac{d}{dx}\left(\int f(x)\,dx\right)$ Derivative of a constant times a function.

$= k f(x)$ Definition of indefinite integral (the "only if" part).

$\therefore g(x) = \int k f(x)\,dx$ Definition of indefinite integral (the "if" part).

$\therefore \int k f(x)\,dx = k\int f(x)\,dx$, Q.E.D.   Transitive property of equality. ◀

> **Two Properties of Indefinite Integrals**
>
> **Integral of a Constant Times a Function:** If f is a function that can be integrated and k is a constant, then
>
> $$\int k f(x)\, dx = k \int f(x)\, dx$$
>
> *Verbally:* You can pull a constant multiplier out through the integral sign.
>
> **Integral of a Sum of Two Functions:** If f and g are functions that can be integrated, then
>
> $$\int [f(x) + g(x)]\, dx = \int f(x)\, dx + \int g(x)\, dx$$
>
> *Verbally:* Integration distributes over addition.

Be sure you don't read too much into these properties! You can't pull a variable through the integral sign. For instance,

$$\int x \cos x\, dx \quad \text{does not equal} \quad x \int \cos x\, dx$$

The integral of the equation on the right is $x \sin x + C$. Its differential is $(\sin x + x \cos x)\, dx$, not $x \cos x\, dx$. The integral $\int x \cos x\, dx$ is the integral of a product of two functions. Recall that the derivative of a product does not equal the product of the derivatives. In Section 9-2, you will learn integration by parts, a technique for integrating the product of two functions.

The "*dx*" in an Indefinite Integral

There is a relationship between the differential, dx, at the end of the integral and the argument of the function in the integrand. For example,

$$\int \cos x\, dx = \sin x + C$$

It does not matter what letter you use for the variable.

$$\int \cos r\, dr = \sin r + C$$

$$\int \cos t\, dt = \sin t + C$$

$$\int \cos u\, du = \sin u + C$$

The phrase "with respect to" identifies the variable in the differential that follows the integrand function. Note that in each case dx, dr, dt, or du is the *differential of the argument* (or inside function) of the cosine. This observation provides you with a way to integrate some composite functions.

▶ **EXAMPLE 1**   Evaluate: $\int 5 \cos(5x + 3)\, dx$

   Solution
$$\int \cos(5x + 3)(5\, dx) \qquad \text{Commute the 5 on the left, and associate it with } dx.$$
$$= \sin(5x + 3) + C \qquad \text{The differential of the inside function, } 5x + 3, \text{ is } 5\, dx. \quad ◀$$

From Example 1, you can see that in order to integrate the cosine of a function of x, everything inside the integral that is *not* a part of the cosine (in this case,

$5\,dx$) must be the differential of the *inside* function (the argument of the cosine, in this case $5x + 3$). If it is not, you must transform the integral so that it is. In Example 2, you'll see how to do this.

▶ **EXAMPLE 2**   Evaluate: $\int (7x + 4)^9\,dx$

Solution   This is the integral of the ninth power function. If dx were the differential of the inside function, $7x + 4$, then the integral would have the form

$$\int u^n\,du$$

Transform the integral to make dx the differential of $(7x + 4)$. Your work will then look like this.

$$\int (7x + 4)^9\,dx$$

$$= \int (7x + 4)^9 \left(\frac{1}{7} \cdot 7\,dx \right) \qquad \text{Multiply by 1, using the form } (1/7)(7).$$

$$= \int \frac{1}{7}(7x + 4)^9 (7\,dx) \qquad \begin{array}{l}\text{Associate } 7\,dx, \text{ the differential of the inside function.} \\ \text{Commute } 1/7.\end{array}$$

$$= \frac{1}{7} \int (7x + 4)^9 (7\,dx) \qquad \begin{array}{l}\text{Integral of a constant times a function. (Pull } 1/7 \text{ out} \\ \text{through the integral sign.)}\end{array}$$

$$= \frac{1}{7} \cdot \frac{1}{10}(7x + 4)^{10} + C \qquad \text{Integrate the power function, } \int u^n\,du.$$

$$= \frac{1}{70}(7x + 4)^{10} + C \qquad\qquad\qquad\qquad\qquad ◀$$

Once you understand the process you can leave out some of the steps, thus shortening your work. Example 2 makes use of the *integral of a power* property, which reverses the steps in differentiating a power.

PROPERTY: Integral of the Power Function

For any constant $n \neq -1$ and any differentiable function u,

$$\int u^n\,du = \frac{1}{n + 1}u^{n+1} + C$$

Verbally: To integrate a power, increase the exponent by 1, then divide by the new exponent (and add C).

Notes:

- In order for the property to apply, the differential du must be the differential of the inside function (the base of the power). If it is not, you must transform the integral so that it is.

- The reason for $n \neq -1$ is to avoid division by zero.

▶ **EXAMPLE 3**   Evaluate: $\int e^{-2x}\,dx$

Solution   In order to integrate the natural exponential function, everything that is not part of the exponential must be the differential of the inside function (in this case, the exponent). The differential of the inside function is $-2\,dx$. Because -2 is a constant, you can multiply inside the integral by -2 and divide outside the integral by -2.

$$\int e^{-2x}\,dx = \frac{1}{-2}\int e^{-2x}(-2\,dx) \qquad \text{Make the differential of the inside function.}$$

$$= -0.5e^{-2x} + C$$

◀

The result of Example 3 is summarized in this box.

PROPERTY: *Integral of the Natural Exponential Function*

If u is a differentiable function, then

$$\int e^u\,du = e^u + C$$

Verbally: The natural exponential function is its own integral (plus C).

Note: Anything inside the integral sign that is *not* part of the exponential must be the differential of the *inside* function (the exponent). If it is not, you must transform the integral so that it is.

▶ **EXAMPLE 4**   Evaluate: $\int 5^x\,dx$

Solution   Recall that if $y = 5^x$, then $dy = 5^x \ln 5\,dx$. Because $\ln 5$ is a constant, you can transform this new problem into a familiar problem by multiplying by $\ln 5$ inside the integral and dividing by $\ln 5$ outside the integral.

$$\int 5^x\,dx$$

$$= \frac{1}{\ln 5}\int 5^x \ln 5\,dx \qquad \text{Get the familiar } 5^x \ln 5\,dx \text{ inside the integral sign.}$$

$$= \frac{1}{\ln 5}\cdot 5^x + C \qquad \text{Write the antiderivative.}$$

$$= \frac{5^x}{\ln 5} + C$$

◀

From Example 4, you can see a relationship between differentiating an exponential function and integrating an exponential function.

> **PROPERTY: Derivative and Integral of Base-b Exponential Functions**
>
> Given $y = b^x$ where b stands for a positive constant, $b \neq 1$,
>
> To differentiate, multiply b^x by $\ln b$.
>
> To integrate, divide b^x by $\ln b$ (and add C).

Note: The reason for $b \neq 1$ is to avoid division by zero.

Problem Set 5-3

Quick Review

Q1. Find the antiderivative: $3x^2\, dx$

Q2. Find the indefinite integral: $x^5\, dx$

Q3. Find the derivative: $y = x^3$

Q4. Find the derivative, y': $y = 3^x$

Q5. Find the differential, dy: $y = 3^x$

Q6. Find the second derivative, y'': $y = (1/6)x^6$

Q7. Integrate: $\cos x\, dx$

Q8. Differentiate: $y = \cos x$

Q9. $\lim_{j \to 0}(\sin j)/(j) = -?-$

Q10. If the graph of a function seems to merge with its tangent line as you zoom in, the situation illustrates —?—.

A. Continuity

B. Differentiability

C. Definition of derivative

D. Definition of differential

E. Local linearity

For Problems 1–32, evaluate the indefinite integral.

1. $\int x^{10}\, dx$

2. $\int x^{20}\, dx$

3. $\int 4x^{-6}\, dx$

4. $\int 9x^{-7}\, dx$

5. $\int \cos x\, dx$

6. $\int \sin x\, dx$

7. $\int 4 \cos 7x\, dx$

8. $\int 20 \sin 9x\, dx$

9. $\int 5e^{0.3x}\, dx$

10. $\int 2e^{-0.01x}\, dx$

11. $\int 4^m\, dm$

12. $\int 8.4^r\, dr$

13. $\int (4v + 9)^2\, dv$

14. $\int (3p + 17)^5\, dp$

15. $\int (8 - 5x)^3\, dx$

16. $\int (20 - x)^4\, dx$

17. $\int (\sin x)^6 \cos x\, dx$

18. $\int (\cos x)^8 \sin x\, dx$

19. $\int \cos^4 \theta \sin \theta\, d\theta$

20. $\int \sin^5 \theta \cos \theta\, d\theta$

21. $\int (x^2 + 3x - 5)\, dx$

22. $\int (x^2 - 4x + 1)\, dx$

23. $\int (x^2 + 5)^3\, dx$
   (Beware!)

24. $\int (x^3 - 6)^2\, dx$
   (Beware!)

25. $\int e^{\sec x} \sec x \tan x\, dx$

26. $\int e^{\tan x} \sec^2 x\, dx$

27. $\int \sec^2 x\, dx$

28. $\int \csc^2 x\, dx$

29. $\int \tan^7 x \sec^2 x\, dx$

30. $\int \cot^8 x \csc^2 x\, dx$

31. $\int \csc^9 x \cot x\, dx$

32. $\int \sec^7 x \tan x\, dx$

33. *Distance from Velocity Problem:* As you drive along the highway, you step hard on the accelerator to pass a truck (Figure 5-3a).

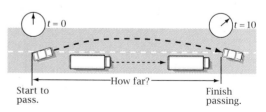

Figure 5-3a

Assume that your velocity $v(t)$, in feet per second, is given by

$$v(t) = 40 + 5\sqrt{t}$$

where t is the number of seconds since you started accelerating. Find an equation for $D(t)$, your displacement from the starting point, that is, from $D(0) = 0$. How far do you go in the 10 s it takes to pass the truck?

34. *Definite Integral Surprise!* Figure 5-3b shows the region that represents the definite integral of $f(x) = 0.3x^2 + 1$ from $x = 1$ to $x = 4$.

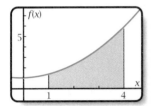

Figure 5-3b

    a. Evaluate the integral by using the trapezoidal rule with $n = 100$ increments.

    b. Let $g(x) = \int f(x)\,dx$. Integrate to find an equation for $g(x)$.

    c. Evaluate the quantity $g(4) - g(1)$. What is interesting about your answer?

35. *Integral of a Sum Property:* Prove that if f and g are functions that can be integrated, then $\int [f(x) + g(x)]\,dx = \int f(x)\,dx + \int g(x)\,dx$.

36. *Integral Table Problem:* Calvin finds the formula $\int x \cos x\,dx = x \sin x + \cos x + C$ in a table of integrals. Phoebe says, "That's right!" How can Phoebe be sure the formula is right?

37. *Introduction to Riemann Sums:* Suppose the velocity of a moving object is given by

$$v(t) = t^2 + 10$$

where $v(t)$ is in feet per minute and t is in minutes. Figure 5-3c shows the region that represents the integral of $v(t)$ from $t = 1$ to $t = 4$. Thus the area of the region equals the distance traveled by the object. In this problem you will find the integral, approximately, by

dividing the region into rectangles instead of into trapezoids. The width of each rectangle will still be Δt, and each rectangle's length will be $v(t)$ found at the t-value in the middle of the strip. The sum of the areas of these rectangles is called a *Riemann sum.*

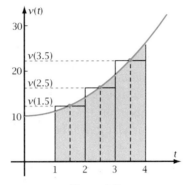

Figure 5-3c

    a. Use a Riemann sum with $n = 3$ strips, as shown in Figure 5-3c, to find an approximation for the definite integral of $v(t) = t^2 + 10$ from $t = 1$ to $t = 4$.

    b. Use a Riemann sum with $n = 6$ strips, as in Figure 5-3d, to find another approximation for the definite integral. The altitudes of the rectangles will be values of $v(t)$ for t at the midpoints of the intervals, namely, $t = 1.25$, $t = 1.75$, $t = 2.25$, $t = 2.75$, $t = 3.25$, and $t = 3.75$. The widths of the strips are, of course, $\Delta t = 0.5$.

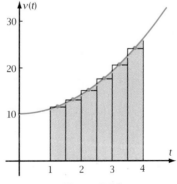

Figure 5-3d

c. Explain why the Riemann sum you used in part b should be a better approximation of the integral than that in part a.

d. Make a conjecture about the exact value of the definite integral. Check your conjecture by evaluating the integral using the trapezoidal rule with $n = 100$ increments.

e. How far did the object travel between $t = 1$ min and $t = 4$ min? What was its average velocity for that time interval?

38. *Journal Problem:* Update your journal with what you've learned. Include such things as

   • The one most important thing you've learned since your last journal entry

   • The difference, as you understand it so far, between definite integral and indefinite integral

   • The difference between a differential and a derivative

   • Any insight you may have gained about why two different concepts are both named *integral*, and how you gained that insight

   • Algebraic techniques you have learned for finding equations for indefinite integrals

   • Questions you plan to ask during the next class period

5-4   Riemann Sums and the Definition of Definite Integral

Recall that a definite integral is used for the product of $f(x)$ and x, such as (rate)(time), taking into account that $f(x)$ varies. The integral is equal to the area of the region under the graph of f, as shown in Figure 5-4a.

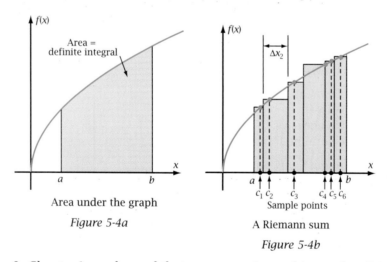

Area under the graph

Figure 5-4a

A Riemann sum

Figure 5-4b

In Chapter 1, you learned that you can estimate this area by slicing the region into strips whose areas you can approximate using trapezoids. As you saw in Problem Set 5-1, you can also estimate the areas by *Riemann sums* using rectangles instead of trapezoids. Figure 5-4b illustrates a Riemann sum for the integral in Figure 5-4a. In this section you will use Riemann sums as the basis for defining a definite integral.

OBJECTIVE Use Riemann sums and limits to define and estimate values of definite integrals.

Figure 5-4b shows the region under the graph of function f from $x = a$ to $x = b$ divided into n strips of variable width $\Delta x_1, \Delta x_2, \Delta x_3, \ldots$ ($n = 6$, in this case). These strips are said to **partition** the interval $[a, b]$ into n **subintervals**, or **increments**. Values of $x = c_1, c_2, c_3, \ldots, c_n$, called **sample points**, are picked, one in each subinterval. At each of the sample points, the corresponding function values $f(c_1), f(c_2), f(c_3), \ldots$ are the altitudes of the rectangles. The area of any one rectangle is

$$A_{\text{rect}} = f(c_k) \Delta x_k$$

where $k = 1, 2, 3, \ldots, n$. The integral is approximately equal to the sum of the areas of the rectangles.

$$\text{Area} \approx \sum_{k=1}^{n} f(c_k) \Delta x_k$$

This sum is called a Riemann sum (read "ree-mahn") after German mathematician G. F. Bernhard Riemann (1826–1866).

DEFINITIONS: Riemann Sum and Its Parts

A partition of the interval $[a, b]$ is an ordered set of $n + 1$ discrete points

$$a = x_0 < x_1 < x_2 < x_3 < \cdots < x_n = b$$

The partition divides the interval $[a, b]$ into n subintervals of width

$$\Delta x_1, \Delta x_2, \Delta x_3, \ldots, \Delta x_n$$

Sample points for a partition are points $x = c_1, c_2, c_3, \ldots, c_n$, with one point in each subinterval.

A Riemann sum R_n for a function f on the interval $[a, b]$ using a partition with n subintervals and a given set of sample points is a sum of the form

$$R_n = \sum_{k=1}^{n} f(c_k) \Delta x_k$$

Note: A Riemann sum is an approximation of the definite integral of $f(x)$ with respect to x on the interval $[a, b]$ using rectangles to estimate areas.

A weaver uses thin strips of straw to create a straw mat. You can calculate the area of the mat by adding up the areas of each strip in one direction.

Figure 5-4c illustrates a **left Riemann sum**, a **midpoint Riemann sum**, and a **right Riemann sum**. As the names suggest, the sample points are taken at the left endpoint, the midpoint, and the right endpoint of each subinterval, respectively. In this text the partition points are usually chosen to give increments of equal width, Δx, as in this figure. In this case "Δx approaches zero" is equivalent to "n approaches infinity."

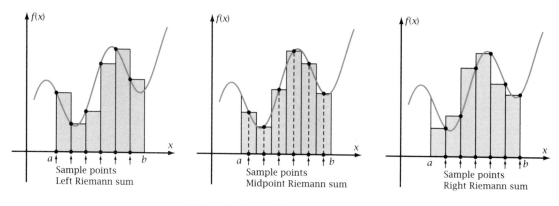

Left, midpoint, and right Riemann sums

Figure 5-4c

A definite integral is defined in terms of Riemann sums. The symbol used is the same as for an indefinite integral, but with the numbers a and b as subscript and superscript, respectively. This symbol is read "the (definite) integral of $f(x)$ with respect to x from a to b."

$$\int_a^b f(x)\,dx \qquad \text{Definite integral notation.}$$

The definition of definite integral uses a **lower Riemann sum**, L_n, and an **upper Riemann sum**, U_n. As shown in Figure 5-4d, these sums use sample points at values of x where $f(x)$ is the lowest or the highest in each subinterval, not necessarily at the end or middle of a subinterval.

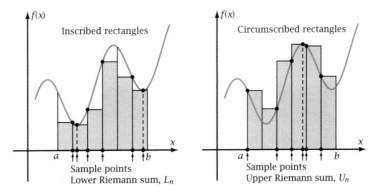

Lower and upper Riemann sums

Figure 5-4d

A lower sum is a lower bound for the value of the integral, and an upper sum is an upper bound. Therefore

$$L_n \le \int_a^b f(x)\,dx \le U_n$$

If the lower and upper sums approach the same limit as the largest value of Δx approaches zero, then the value of the integral is squeezed (recall the *squeeze theorem*) to this common limit. When this happens, function f is **integrable** on $[a, b]$, and the common limit is defined as the **definite integral**.

DEFINITIONS: Definite Integral and Integrability

If the lower sums, L_n, and the upper sums, U_n, for a function f on the interval $[a, b]$ approach the same limit as Δx approaches zero (or as n approaches infinity in the case of equal-width subintervals), then f is integrable on $[a, b]$. This common limit is defined to be the definite integral of $f(x)$ with respect to x from $x = a$ to $x = b$. The numbers a and b are called **lower** and **upper limits of integration**, respectively. Algebraically,

$$\int_a^b f(x)\,dx = \lim_{\Delta x \to 0} L_n = \lim_{\Delta x \to 0} U_n$$

provided the two limits exist and are equal.

The advantage of using Riemann sums to define definite integrals is that, if a function is integrable, the limit of *any* Riemann sum, R_n, equals the value of the integral. From the definition of upper and lower sums

$$L_n \le R_n \le U_n$$

Now take the limit of the three members of the inequality.

$$\lim_{\Delta x \to 0} L_n \le \lim_{\Delta x \to 0} R_n \le \lim_{\Delta x \to 0} U_n$$

For an integrable function, the left and right members of the inequality approach the definite integral. By the squeeze theorem, so does the middle member.

PROPERTY: Limit of a Riemann Sum

If f is integrable on $[a, b]$ and if R_n is any Riemann sum for f on $[a, b]$, then

$$\int_a^b f(x)\,dx = \lim_{\Delta x \to 0} R_n$$

▶ **EXAMPLE 1**   For the integral $\int_1^4 \frac{1}{x}\,dx$, complete steps a and b.

a. Find the lower sum L_6, the upper sum U_6, and the midpoint sum M_6. Show that M_6 is between L_6 and U_6.

b. Find the trapezoidal rule approximation, T_6. Show that T_6 is an overestimate for the integral and that M_6 is an underestimate.

Solution   a. Figure 5-4e shows the graph of the integrand, $f(x) = 1/x$, and rectangles for each of three sums. With equal increments, Δx will equal $(4 - 1)/6$, or 0.5, so each rectangle will have area $f(c)(0.5)$ for the sample points $x = c$. Because $f(x)$ is decreasing on the interval $[1, 4]$, the low point in each subinterval is at the right side of the rectangle and the high point is at the left side. Therefore L_6 is equal to the right Riemann sum, and U_6 is equal to the left Riemann sum. The common factor $\Delta x = 0.5$ can be factored out of each term.

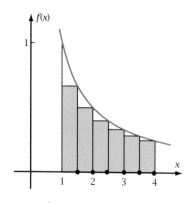
Lower Riemann sum

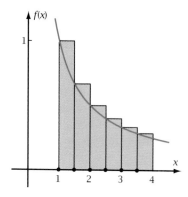
Upper Riemann sum

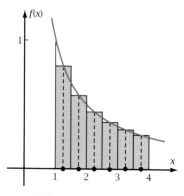
Midpoint Riemann sum

Figure 5-4e

$$L_6 = 0.5[f(1.5) + f(2) + f(2.5) + f(3) + f(3.5) + f(4)]$$
$$= 0.5(1/1.5 + 1/2 + 1/2.5 + 1/3 + 1/3.5 + 1/4) = 1.2178571\ldots$$
$$U_6 = 0.5[f(1) + f(1.5) + f(2) + f(2.5) + f(3) + f(3.5)]$$
$$= 0.5(1/1 + 1/1.5 + 1/2 + 1/2.5 + 1/3 + 1/3.5) = 1.5928571\ldots$$
$$M_6 = 0.5[f(1.25) + f(1.75) + f(2.25) + f(2.75) + f(3.25) + f(3.75)]$$
$$= 0.5(1/1.25 + 1/1.75 + 1/2.25 + 1/2.75 + 1/3.25 + 1/3.75)$$
$$= 1.3769341\ldots$$

Observe that $1.2178571\ldots < 1.3769341\ldots < 1.5928571\ldots$, so M_6 is between L_6 and U_6.

b. Using your trapezoidal rule program, you get $T_6 = 1.405357\ldots$.

As shown on the left in Figure 5-4f, each trapezoid will be circumscribed about the corresponding strip and therefore will overestimate the area. This happens because the graph is *concave up*, a property you will explore in more detail in Chapter 8. As shown on the right in Figure 5-4f, the midpoint sum rectangle leaves out more area of the strip on one side of the midpoint than it includes on the other side and therefore underestimates the area, Q.E.D.

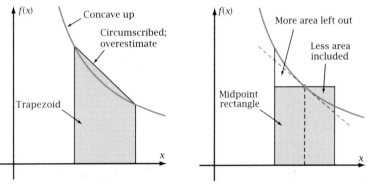

Figure 5-4f

In Example 1, part a, you saw that for a strictly decreasing function the lower Riemann sum is equal to the right Riemann sum, and the upper Riemann sum is equal to the left Riemann sum. For a strictly increasing function the opposite is true: The lower Riemann sum is equal to the left Riemann sum, and the upper Riemann sum is equal to the right Riemann sum. For a function that is not strictly increasing or decreasing, finding the lower and upper Riemann sums can be much more difficult.

Problem Set 5-4

Quick Review

Q1. Differentiate: $y = x \sin x$

Q2. Integrate: $\int \sec^2 x \, dx$

Q3. Differentiate: $f(x) = \tan x$

Q4. Integrate: $\int x^3 \, dx$

Q5. Differentiate: $z = \cos 7x$

Q6. Integrate: $\int \sin u \, du$

Q7. Find $\lim_{x \to 5}(x^2 - 2x - 15)/(x - 5)$.

Q8. Sketch a function graph with a cusp at the point $(4, 7)$.

Q9. Write the converse of this statement: If $a = 2$ and $b = 3$, then $a + b = 5$.

Q10. The statement in Problem Q9 is true. Is the converse true?

For Problems 1–6, calculate approximately the given definite integral by using a Riemann sum with n increments. Pick the sample points at the midpoints of the subintervals.

1. $\int_1^4 x^2 \, dx$, $n = 6$

2. $\int_2^6 x^3 \, dx$, $n = 8$

3. $\int_{-1}^{3} 3^x \, dx, \ n = 8$

4. $\int_{-1}^{2} 2^x \, dx, \ n = 6$

5. $\int_{1}^{2} \sin x \, dx, \ n = 5$

6. $\int_{0}^{1} \cos x \, dx, \ n = 5$

For Problems 7 and 8, calculate the Riemann sums U_4, L_4, and M_4, and the sum found by applying the trapezoidal rule, T_4. Show that M_4 and T_4 are between the upper and lower sums.

7. $\int_{0.4}^{1.2} \tan x \, dx$

8. $\int_{1}^{3} (10/x) \, dx$

For Problems 9 and 10, tell whether a midpoint Riemann sum and a trapezoidal rule sum will overestimate the integral or underestimate it. Illustrate each answer with a graph.

9. $\int_{1}^{5} \ln x \, dx$

10. $\int_{0}^{2} e^x \, dx$

11. *Sample Point Problem:* Figure 5-4g shows the graph of the sinusoid

$$h(x) = 3 + 2 \sin x$$

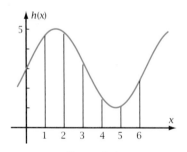

Figure 5-4g

a. At what values of x should the sample points be taken to get an upper sum for the integral of h on $[0, 6]$ with $n = 6$?

b. Where should the sample points be taken to get a lower Riemann sum for the integral?

c. Calculate these upper and lower sums.

12. *Program for Riemann Sums Problem:* Write a program to compute Riemann sums for a given

function. If you write the program on your grapher, you can store the integrand as y_1, just as you did for the trapezoidal rule program. You should be able to input a and b, the lower and upper bounds of integration; n, the number of increments to use; and p, the percentage of the way through each subinterval to take the sample point. For instance, for a midpoint sum, p would equal 50. Test the program by using it for Problem 7. If your program gives you the correct answers for U_4, L_4, and M_4, you may assume that it is working properly.

13. *Limit of Riemann Sums Problem:* In Problem 1, you evaluated

$$\int_{1}^{4} x^2 \, dx$$

by using midpoint sums with $n = 6$. In this problem you will explore what happens to approximate values of this integral as n gets larger.

a. Use your programs to show that $L_{10} = 18.795, U_{10} = 23.295, M_{10} = 20.9775$, and $T_{10} = 21.045$.

b. Calculate L_{100} and L_{500}. What limit does L_n seem to be approaching as n increases?

c. Calculate U_{100} and U_{500}. Does U_n seem to be approaching the same limit as L_n? What words describe a function $f(x)$ on the interval $[1, 4]$ if L_n and U_n have the same limit as n approaches infinity (and thus as Δx approaches zero)?

d. Explain why the trapezoidal sums are always slightly greater than your conjectured value for the exact integral and why the midpoint sums are always less than your conjectured value.

14. *Exact Integral of the Square Function by Brute Force Project:* In this problem you will find, exactly, the integral $\int_{0}^{3} x^2 \, dx$ by calculating algebraically the limit of the upper sums.

a. Find, approximately, the value of the integral by calculating the upper and lower Riemann sums with 100 increments, U_{100} and L_{100}. Make a conjecture about the exact value.

b. If you partition the interval $[0, 3]$ into n subintervals, each one will be $\Delta x = 3/n$ units wide. The partition points will be at x equals

$$0, \ 1 \cdot \tfrac{3}{n}, \ 2 \cdot \tfrac{3}{n}, \ 3 \cdot \tfrac{3}{n}, \ \dots, \ n \cdot \tfrac{3}{n}$$

Which of these partition points would you pick to find the upper sum?

c. The y-values will be the values of $f(x)$ at these sample points, namely, the squares of these numbers. For instance, at the end of the fifth increment, $x = 5(3/n)$ and $f(x) = [5(3/n)]^2$ (Figure 5-4h). Write a formula for U_n, an upper sum with n increments.

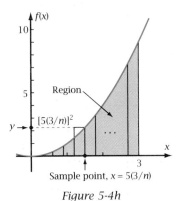

Figure 5-4h

d. You can rearrange the formula you wrote in part c by factoring out common factors, leaving only the squares of the counting numbers inside parentheses. From algebra this sum is

$$1^2 + 2^2 + 3^2 + \cdots + n^2 = \tfrac{n}{6}(n + 1)(2n + 1)$$

Use this information to write a closed formula (no ellipsis: ...) for U_n. Confirm that this formula gives the right answer for U_{100}.

e. Use the formula to predict U_{1000}. Does U_n seem to be approaching the limit you conjectured in part a?

f. Find algebraically the limit of U_n from part e as n approaches infinity. This limit is the definite integral of x^2 from $x = 0$ to $x = 3$ and equals the exact area and integral.

15. *Exact Integral of the Cube Function Project:* Find the exact value of $\int_0^2 x^3 \, dx$. Use as a guide the technique you used in Problem 14. Recall that the sum of the cubes of the counting numbers is given by

$$\sum_{k=1}^{n} k^3 = 1^3 + 2^3 + 3^3 + 4^3 + \cdots + n^3 = \left[\tfrac{n}{2}(n + 1) \right]^2$$

5-5 The Mean Value Theorem and Rolle's Theorem

Suppose that you go 500 ft in 10 s as you slow down on a freeway exit. Your average velocity for the 10-s time interval is 50 ft/s. It seems reasonable to conclude that sometime in that interval your instantaneous velocity is also equal to 50 ft/s (Figure 5-5a). In this section you will learn the **mean value theorem**, which states conditions under which this conclusion is true. You will also learn **Rolle's theorem** and use it as a lemma to prove the mean value theorem. In Section 5-8, you will see how the mean value theorem leads to an algebraic method for finding exact definite integrals.

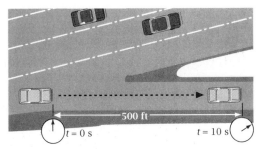

Figure 5-5a

OBJECTIVE Learn the mean value theorem and Rolle's theorem, and learn how to find the point in an interval at which the instantaneous rate of change of the function equals the average rate of change.

Let $f(x)$ be your displacement as you exit the freeway. As shown in Figure 5-5b, your average velocity from time $x = a$ to $x = b$ is displacement divided by time.

$$\text{Average velocity} = \frac{f(b) - f(a)}{b - a}$$

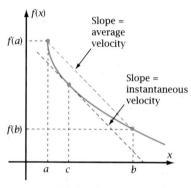

Figure 5-5b

The average velocity is the slope of the secant line connecting two points on the graph. As the figure shows, there is a time $x = c$ between the points a and b at which the tangent line parallels the secant line. At this point the instantaneous velocity, $f'(c)$, equals the average velocity.

The mean value theorem gives two **sufficient conditions** for there to be an instantaneous rate of change equal to the average rate from $x = a$ to $x = b$. First, the function is differentiable for all values of x between a and b. Second, the function is continuous at the endpoints $x = a$ and $x = b$, even if it is not differentiable there. You can express these two hypotheses by saying that f is differentiable on the open interval (a, b) and continuous on the *closed* interval $[a, b]$. Recall that continuity on a closed interval requires only that the function have one-sided limits $f(a)$ and $f(b)$ as x approaches a and b from *within* the interval. Recall also that differentiability implies continuity, so differentiability on the interval (a, b) automatically gives continuity everywhere except, perhaps, at the endpoints. The function graphed in Figure 5-5b is differentiable on the interval (a, b) but not at $x = a$, because the tangent line is vertical there.

PROPERTY: The Mean Value Theorem

If   1. f is differentiable for all values of x in the open interval (a, b), and

  2. f is continuous for all values of x in the closed interval $[a, b]$,

then there is at least one number $x = c$ in (a, b) such that

$$f'(c) = \frac{f(b) - f(a)}{b - a}$$

Note: The value $f'(c)$ is the *mean value* from which the theorem gets its name, because $f'(c)$ is the mean of the derivatives on the interval $[a, b]$.

The two conditions in the "if" part are the **hypotheses** of the mean value theorem. The "then" part is the **conclusion**. The left and center graphs of

Figure 5-5c show why the conclusion might not be true if a function does not meet the hypotheses. For the function on the left, there is a point between a and b at which the function is not differentiable. You cannot draw a unique tangent line at the cusp, and there is no other place at which the tangent line parallels the secant line. In the center graph, the function is differentiable for all values of x between a and b, but the function is not continuous at $x = a$. At no place is there a tangent line parallel to the secant line. The third function is continuous but not differentiable at $x = a$. There is a tangent line parallel to the secant line.

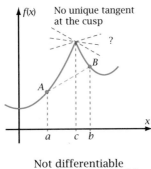

Not differentiable
at an x between a and b

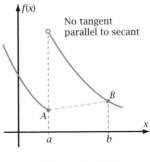

Not continuous
at one of the endpoints

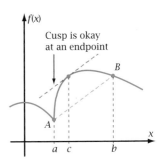

Not differentiable at one
endpoint, but continuous there

Figure 5-5c

Rolle's Theorem

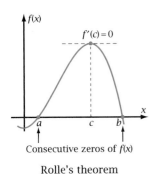

Consecutive zeros of $f(x)$

Rolle's theorem

Figure 5-5d

The proof of the mean value theorem uses as a lemma another theorem from the analysis of real numbers, Rolle's theorem. Named after 17th-century French mathematician Michel Rolle (read "roll"), the theorem states that between two consecutive zeros of a function there is a place where the derivative is zero (Figure 5-5d). Sufficient conditions for this conclusion to be true are the same as the hypotheses of the mean value theorem, with the additional hypothesis $f(a) = f(b) = 0$.

PROPERTY: Rolle's Theorem

If 1. f is differentiable for all values of x in the open interval (a, b) and

2. f is continuous for all values of x in the closed interval $[a, b]$ and

3. $f(a) = f(b) = 0$,

then there is at least one number $x = c$ in (a, b) such that $f'(c) = 0$.

Graphical Proof of Rolle's Theorem

Suppose that $f(x)$ is positive somewhere between $x = a$ and $x = b$. As Figure 5-5e on the following page shows, $f(x)$ must reach a maximum for some value of x ($x = c$) between a and b. The fact that there is such a number relies on the *completeness axiom* (Section 2-6) and on the fact that f is continuous (because it is differentiable).

If x is less than c, then Δx is negative. If x is greater than c, then Δx is positive. Because $f(c)$ is the maximum value of $f(x)$, Δy will always be negative (or possibly zero). Thus the difference quotient

$$\frac{\Delta y}{\Delta x} = \frac{f(x) - f(c)}{x - c}$$

is positive (or zero) when x is less than c and negative (or zero) when x is greater than c. The limit as x approaches c of this difference quotient is, of course, $f'(c)$. Thus

$$f'(c) \geq 0 \quad \text{and} \quad f'(c) \leq 0$$

Therefore, $f'(c) = 0$.

A similar case can be made if $f(x)$ is negative for some x between a and b or if $f(x)$ is always zero.

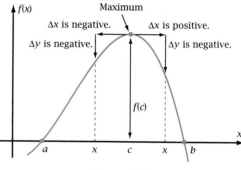

Figure 5-5e

So, if the three hypotheses are true, then there is always a number $x = c$ between a and b such that $f'(c) = 0$, Q.E.D.

Algebraic Proof of the Mean Value Theorem

You can prove the mean value theorem using Rolle's theorem as a lemma. The left graph in Figure 5-5f shows function f, the secant line through points A and B on the graph of f, and the tangent line you hope to prove is parallel to the secant line.

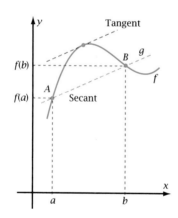

Secant line through A and B

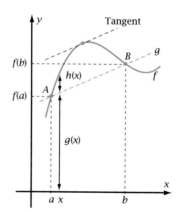

Define functions g and h.

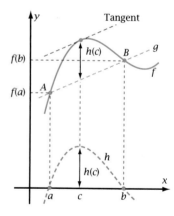

c is at maximum of $h(x)$.

Figure 5-5f

Let g be the linear function whose graph is the secant line through points A and B.

Let h be the function defined by

$$h(x) = f(x) - g(x)$$

As shown in the center graph of Figure 5-5f, the values of $h(x)$ are the vertical distances between the graph of f and the graph of g. Because the graphs of f and g coincide at both $x = a$ and $x = b$, it follows that $h(a) = h(b) = 0$. The graph of h is shown in the right graph of Figure 5-5f.

Function h is continuous and differentiable at the same places as function f. This is true because g is a linear function, which is continuous and differentiable everywhere, and because a difference such as $h(x)$ between continuous, differentiable functions is also continuous and differentiable.

Therefore h satisfies the three hypotheses of Rolle's theorem. Function h is differentiable on (a, b), is continuous at $x = a$ and $x = b$, and has $h(a) = h(b) = 0$. By the conclusion of Rolle's theorem, there is a number $x = c$ in (a, b) such that $h'(c) = 0$. All that remains to be done is the algebra.

$$h'(c) = 0$$

However, $h'(x) = f'(x) - g'(x)$, which implies that
$h'(c) = f'(c) - g'(c) = 0$.

Therefore $f'(c) = g'(c)$.

However, g is a linear function. Thus $g'(x)$ is everywhere equal to the slope of the graph. This slope is $[f(b) - f(a)]/(b - a)$. Hence

$$f'(c) = \frac{f(b) - f(a)}{b - a}, \quad \text{Q.E.D.}$$

Be careful! Do not read more into the mean value theorem and Rolle's theorem than these theorems say. The hypotheses of these theorems are *sufficient* conditions to imply the conclusions. They are not *necessary* conditions. Figure 5-5g shows two functions that satisfy the conclusions of the mean value theorem even though the hypotheses are not true.

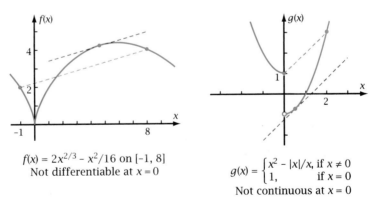

$f(x) = 2x^{2/3} - x^2/16$ on $[-1, 8]$
Not differentiable at $x = 0$

$$g(x) = \begin{cases} x^2 - |x|/x, & \text{if } x \neq 0 \\ 1, & \text{if } x = 0 \end{cases}$$
Not continuous at $x = 0$

Figure 5-5g

▶ **EXAMPLE 1**

Given $f(x) = x^{1/3}$, plot the graph. Explain why f satisfies the hypotheses of the mean value theorem on the interval $[0, 8]$. Find a value of $x = c$ in the open interval $(0, 8)$ at which the conclusion of the theorem is true, and show on your graph that the tangent really is parallel to the secant.

Solution

Figure 5-5h suggests that f is differentiable everywhere except, perhaps, at $x = 0$.

$$f'(x) = \frac{1}{3}x^{-2/3}$$

so $f'(0)$ would be $0^{-2/3} = 1/0^{2/3} = 1/0$, which is infinite. But f is differentiable on the *open* interval $(0, 8)$. Function f is continuous at $x = 0$ and $x = 8$ because the limits of $f(x)$ as $x \to 0$ and as $x \to 8$ are 0 and 2, the values of $f(0)$ and $f(8)$, respectively, and f is continuous for all other x in $[0, 8]$ because differentiability implies continuity.

Thus f satisfies the hypotheses of the mean value theorem on $[0, 8]$. The slope of the secant line (Figure 5-5h) is

$$m_{\sec} = \frac{8^{1/3} - 0^{1/3}}{8 - 0} = \frac{1}{4}$$

Setting $f'(c) = 1/4$ gives

$$\frac{1}{3}c^{-2/3} = \frac{1}{4} \Rightarrow c^{-2/3} = \frac{3}{4}$$

$$\therefore c = \pm\left(\frac{3}{4}\right)^{-3/2} = \pm 1.5396...$$

Only the positive value of c is in $(0, 8)$. Thus $c = 1.5396...$.

To plot the tangent line, find the particular equation of the line with slope $1/4$ through the point $(1.5396..., f(1.5396...))$.

$$f(1.5396...) = (1.5396...)^{1/3} = 1.1547...$$

$$\therefore y - 1.1547... = 0.25(x - 1.5396...) \Rightarrow y = 0.25x + 0.7698...$$

Enter this equation into your grapher as y_2. Figure 5-5i shows that the line is tangent to the graph and also parallel to the secant line. ◀

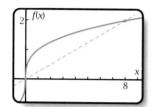

Figure 5-5h

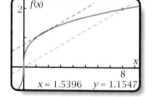

Figure 5-5i

▶ **EXAMPLE 2**

Given $f(x) = x \sin x$, find the first interval of nonnegative x-values on which the hypotheses of Rolle's theorem are true. Then find the point $x = c$ in the corresponding open interval at which the conclusion of Rolle's theorem is true. Illustrate with a graph.

Solution

As Figure 5-5j shows, $f(x) = 0$ at $x = 0$ and at $x = \pi$ because $\sin x$ is zero at those points. To establish differentiability,

$$f'(x) = \sin x + x \cos x$$

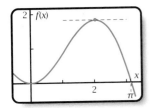

Figure 5-5j

which exists for all x. Thus f is differentiable on $(0, \pi)$. The function is continuous at 0 and π because it is differentiable there. So the hypotheses are met for the interval $[0, \pi]$.

$$f'(c) = 0 \iff \sin c + c \cos c = 0$$

$$\therefore c = 2.02875...$$
Use your grapher's solver feature.

Note that $f'(0)$ is also zero, but $c \neq 0$ because 0 is not in the open interval $(0, \pi)$.

◀

Problem Set 5-5

Quick Review

5 min

Q1. Integrate: $\int (x + 2)\, dx$

Q2. Integrate: $\int 10^t\, dt$

Q3. Integrate: $\int \csc^2 x\, dx$

Q4. Differentiate: $g(x) = \csc x$

Q5. Differentiate: $p(x) = (\ln x)^5$

Q6. Sketch a graph that shows $\int_1^2 x^2\, dx$.

Q7. Sketch a graph with a removable discontinuity at the point $(3, 5)$.

Q8. Sketch a graph of a function that is continuous at the point $(2, 1)$ but not differentiable at that point.

Q9. Find $\lim_{x \to 0} 1/x$.

Q10. If $f(x) = 2x + 6$, then $f^{-1}(x) = $ —?—.

A. $\dfrac{1}{2x + 6}$ B. $\dfrac{1}{2x - 6}$ C. $-2x - 6$

D. $0.5x - 3$ E. $0.5x + 3$

1. State the mean value theorem. If you have to refer to its definition in this book to find out what it says, then practice stating it until you can do so without looking.

2. State Rolle's theorem. If you have to refer to its definition in this book to find out what it says, then practice stating it until you can do so without looking.

For Problems 3–6, plot the graph. Find a point $x = c$ in the given interval at which the mean value theorem conclusion is true. Plot the secant line and the tangent line, showing that they really are parallel. Sketch the resulting graphs.

3. $g(x) = \frac{6}{x}$, $[1, 4]$

4. $f(x) = x^4$, $[-1, 2]$

5. $c(x) = 2 + \cos x$, $\left[0, \frac{\pi}{2}\right]$

6. $h(x) = 5 - \sqrt{x}$, $[1, 9]$

For Problems 7–10, plot the graph. Find an interval on which the hypotheses of Rolle's theorem are met. Then find a point $x = c$ in that interval at which the conclusion of Rolle's theorem is true. Plot a horizontal line through the point $(c, f(c))$ and show that the line really is tangent to the graph. Sketch the result.

7. $f(x) = x \cos x$ $\left(\text{Use } \left[0, \frac{\pi}{2}\right].\right)$

8. $f(x) = x^2 \sin x$

9. $f(x) = (6x - x^2)^{1/2}$

10. $f(x) = x^{4/3} - 4x^{1/3}$

11. *Compound Interest Problem:* Suppose you invest $1000 in a retirement account. The account pays interest continuously at a rate that makes the annual percentage rate (APR) equal 9%. Thus the number of dollars, $d(t)$, in your account at time t years is given by

$$d(t) = 1000(1.09^t)$$

a. When you retire 50 years from now, how much money will be in the account? Surprising?

b. At what average rate does your money increase?

c. Differentiate algebraically to calculate the instantaneous rate at which your money is increasing now, at $t = 0$, and when you retire, at $t = 50$. Is the average of these two rates equal to the average rate you found in part b?

d. Solve algebraically to find the time at which the instantaneous rate equals the average rate. Is this time halfway between now and the time you retire?

12. *Softball Line Drive Problem:* Suppose the displacement, $d(t)$, in feet, of a softball from home plate was given to be

$$d(t) = \begin{cases} 43\left(\dfrac{1-t}{1+t}\right), & \text{if } t \le 1 \\ 200\left(1 - \dfrac{1}{t}\right), & \text{if } t \ge 1 \end{cases}$$

where t is the number of seconds since the ball was pitched (Figure 5-5k).

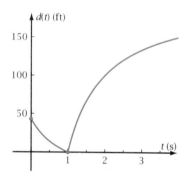

Figure 5-5k

Discuss how the mean value theorem applies to function d. For instance, do the hypotheses apply on the interval $[0, 1]$? $[0, 2]$? $[1, 2]$? Is the conclusion true anywhere in $(0, 1)$? $(0, 2)$? $(1, 2)$? How do your answers illustrate the fact that the hypotheses of the mean value theorem are sufficient conditions, not necessary ones?

13. Sketch a graph that clearly shows that you understand both the hypotheses and the conclusion of Rolle's theorem.

For Problems 14–16, sketch a graph that shows why the conclusion of Rolle's theorem might not be true if f meets all the hypotheses of the theorem on the interval $[a, b]$ except for the exception noted.

14. The function f is discontinuous at $x = b$.

15. The function f is continuous, but not differentiable, at $x = d$ in (a, b).

16. The value of $f(a)$ is not equal to zero.

17. Sketch a graph that shows that a function may satisfy the conclusion of Rolle's theorem even though the function does not meet all the hypotheses.

18. Look up Michel Rolle on the Internet or some other source. Where and when did he live? Give your source of information.

For Problems 19–28, plot (if necessary) and sketch the graph. State which hypotheses of Rolle's theorem are not met on the given interval. Then state whether the conclusion of Rolle's theorem is true on the corresponding open interval.

19. $f(x) = x^2 - 4x$ on $[0, 1]$

20. $f(x) = x^2 - 6x + 5$ on $[1, 2]$

21. $f(x) = x^2 - 4x$ on $[0, 2]$

22. $f(x) = x^2 - 6x + 5$ on $[1, 4]$

23. $f(x) = x^2 - 4x$ on $[0, 3]$

24. $f(x) = |x - 2| - 1$ on $[1, 3]$

25. $f(x) = \dfrac{1}{x}$ on $[0, 5]$

26. $f(x) = x - [x]$ on $[1, 2]$
($[x]$ is the greatest integer less than or equal to x.)

27. $f(x) = 1 - (x - 3)^{2/3}$ on $[2, 4]$

28. $f(x) = \dfrac{x^3 - 6x^2 + 11x - 6}{x - 2}$ on $[1, 3]$

29. Given $g(x) = \dfrac{x^3 - 7x^2 + 13x - 6}{x - 2}$, explain why the hypotheses of the mean value theorem are

not met on any interval containing $x = 2$. Is the conclusion true if the theorem is applied on the interval $[1, 3]$? On $[1, 5]$? Justify your answers. A graph may help.

30. Given $h(x) = x^{2/3}$, explain why the hypotheses of the mean value theorem are met on $[0, 8]$ but are not met on $[-1, 8]$. Is the conclusion of the mean value theorem true for any $x = c$ in $(-1, 8)$? Justify your answer. A graph may help.

31. Suppose $f(x) = |x - 3| + 2x$.

    a. Use the definition of absolute value to write an equation for $f(x)$ as a piecewise function, one part for $x \geq 3$ and one part for $x < 3$. Sketch the graph.

    b. Is f continuous at $x = 3$? Is f differentiable at $x = 3$? Justify your answers.

    c. Which hypothesis of the mean value theorem is not met on the interval $[1, 6]$? Show that there is no point $x = c$ in $(1, 6)$ at which the conclusion of the mean value theorem is true.

    d. Is f integrable on $[0, 5]$? If not, why? If so, evaluate $\int_0^5 f(x)\, dx$ graphically.

32. *New Jersey Turnpike Problem:* When you enter the New Jersey Turnpike, you receive a card that indicates your entrance point and the time at which you entered. When you exit, therefore, it can be determined how far you went and how long it took, and thus what your average speed was.

    a. Let $f(t)$ be the number of miles you traveled in t hours. What assumptions must you make about f so that it satisfies the hypotheses of the mean value theorem on the interval from $t = a$, when you entered the turnpike, to $t = b$, when you exited?

    b. Suppose that your average speed is 60 mi/h. If f meets the hypotheses of the mean value theorem, prove that your speed was exactly 60 mi/h at some time between $t = a$ and $t = b$.

33. *Rolle's Theorem Proof Illustrated by Graph and by Table:* The proof of Rolle's theorem shows that at a high point, $f(c)$, for the open interval (a, b), the difference quotient

$$\frac{f(x) - f(c)}{x - c}$$

is always positive (or zero) when $x < c$ and always negative when $x > c$. In this problem you will show graphically and numerically that this fact is true for a fairly complicated function.

a. Figure 5-5l shows the graph of

$$f(x) = 25 - (x - 5)^2 + 4 \cos\left[2\pi(x - 5)\right]$$

Plot the graph as y_1. Does your graph agree with the figure?

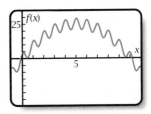

Figure 5-5l

b. Find $f'(x)$. How is the value of $f'(5)$ consistent with the fact that the high point of the graph is at $x = 5$?

c. Let y_2 be the difference quotient

$$y_2 = \frac{y_1 - f(5)}{x - 5}$$

Plot y_2 on the same screen as y_2 using a different style. Sketch the result. Then make a table of values of the difference quotient for each 0.5 unit of x from $x = 3$ to $x = 7$. What do you notice from the table and graph about values of y_2 for $x < 5$ and for $x > 5$?

d. Read the proof of Rolle's theorem, which appears in this section. Explain how the work you've done in this problem relates to the proof. Tell which hypothesis of Rolle's theorem has not been mentioned so far in this problem. Is this hypothesis true for function f? Can the conclusion of Rolle's theorem be true for a function if the hypotheses aren't? Explain.

34. *Mean Value Theorem Proof Illustrated by Graph and by Table:* In the proof of the mean value theorem, a linear function, g, and a difference function, h, were created. Rolle's theorem was then applied to function h. In this problem you will derive equations for $g(x)$ and $h(x)$ and illustrate the proof by graph and by a table of values.

a. Figure 5-5m shows the graph of

$$f(x) = 1 + x + \cos \pi x$$

and a chord drawn between the endpoints of the graph for the interval $[2, 4.5]$. Find an equation, $g(x)$, for the chord. Plot $y_1 = f(x)$ and $y_2 = g(x)$ on the same screen. Use a Boolean variable to restrict y_2 to the interval $[2, 4.5]$. Do your graphs agree with the figure?

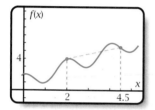

Figure 5-5m

b. Find a point $x = c$ in the interval $(2, 4.5)$ at which the conclusion of the mean value theorem is true. (There are three such points.)

c. Let y_3 be the function h mentioned in the proof of the mean value theorem. Plot y_3 on the same screen as y_1 and y_2. Then make a table of values of y_3 for each 0.05 unit of x from $x = c - 0.25$ to $x = c + 0.25$. Show that $h(c)$ is an upper (or lower) bound for the values of $h(x)$ in the table.

d. Show that function f meets the hypotheses of the mean value theorem on the interval $[2, 4.5]$. Explain how functions g and h in this problem illustrate the proof of the mean value theorem.

35. *Corollary of the Mean Value Theorem:* A **corollary** of a theorem is another theorem that follows easily from the first. Explain why the corollary of the mean value theorem stated in this box is true.

Property: Corollary of the Mean Value Theorem

If f is differentiable on the closed interval $[a, b]$, then there is at least one number $x = c$ in (a, b) such that

$$f'(c) = \frac{f(b) - f(a)}{b - a}$$

---

36. *Converse of a Theorem:* It is easy to show that if two differentiable functions differ by a constant, then their derivatives are equal for all values of x in the domain. For instance, if

$$f(x) = \sin x \quad \text{and} \quad g(x) = 2 + \sin x$$

then $f'(x) = g'(x)$ for all x (Figure 5-5n).

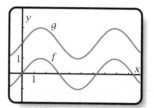

Figure 5-5n

The **converse** of a theorem is the statement that results from interchanging the hypothesis and the conclusion of the theorem, as in this box.

Property: Converse of the Equal-Derivative Theorem

Theorem: If $f(x) = g(x) + D$ for all x in the domain, where D is a constant, then $f'(x) = g'(x)$ for all x in the domain.

Converse of the Theorem: If $f'(x) = g'(x)$ for all x in the domain, then $f(x) = g(x) + D$ for all x in the domain, where D is a constant.

As you probably realize, the converse of a theorem is not necessarily true. For instance, the converse of the mean value theorem is false. However, the converse above *is* true and can be proved by contradiction with the help of the mean value theorem.

a. Suppose there are two values of x such that $f(a) = g(a) + D_1$ and $f(b) = g(b) + D_2$, where

$D_1 \neq D_2$. Let $h(x) = f(x) - g(x)$. Explain why the mean value theorem applies to h on the interval $[a, b]$.

b. The mean value theorem lets you conclude that there is a number $x = c$ in (a, b) such that

$$h'(c) = \frac{h(b) - h(a)}{b - a}$$

Show that $h'(c)$ also equals $(D_2 - D_1)/(b - a)$.

c. Show that $h'(x) = 0$ for all x in (a, b) and thus $h'(c) = 0$. Use the result to show that D_1 and D_2 are equal, which proves the converse by contradicting the assumption you made in part a.

37. *Antiderivative of Zero:* Prove as a corollary of the property in Problem 36 that if $f'(x) = 0$ for all values of x, then $f(x)$ is a constant function.

38. Let $f(x) = (\cos x + \sin x)^2$. Let $g(x) = \sin 2x$. On the same screen, plot graphs of f and g. Sketch the result. Make a table of values of the two functions for convenient values of x, say $0, 1, 2, \ldots$. What seems to be true about values of $f(x)$ and $g(x)$ at the same values of x? Prove algebraically that $f'(x) = g'(x)$ for all values of x.

39. *Maximum and Minimum Values of Continuous Functions:* When you study the analysis of real numbers you will show that if a function f is continuous on a closed interval $[a, b]$, then $f(x)$ has a maximum value and a minimum value at values of x in that interval. Explain why a function that meets the hypotheses of Rolle's theorem automatically meets these conditions.

40. *Intermediate Value Theorem Versus Mean Value Theorem:* The words *intermediate* and *mean* both connote the concept of betweenness. Both the intermediate value theorem and the mean value theorem assert the existence of a number $x = c$ that is between two numbers a and b. However, the hypotheses and conclusions of the theorems are quite different. Write one or two paragraphs describing how the two theorems differ and how they are alike. Graphs will help.

41. *Journal Problem:* Update your journal with what you've learned. Include such things as

   • The one most important thing you have learned since your last journal entry

   • The fact that you can now fill in one more square in your chart of calculus concepts

   • The two new theorems you have just learned and how they are related to each other

   • The difference between the mean value theorem and the intermediate value theorem

   • The difference between definite integral and indefinite integral

   • The different kinds of Riemann sums

   • Any technique or idea you plan to ask about at the next class period

---

5-6   The Fundamental Theorem of Calculus

In Section 5-4, you learned the definition of definite integral as a limit of Riemann sums. In this section you will use this definition and the mean value theorem to derive an algebraic method for calculating a definite integral. The resulting theorem is called the *fundamental theorem of calculus.*

OBJECTIVE   Derive the fundamental theorem of calculus, and use it to evaluate definite integrals algebraically.

Background: Some Very Special Riemann Sums

Suppose you want to evaluate the integral

$$I = \int_1^4 \sqrt{x}\, dx$$

Figure 5-6a shows two Riemann sums, R_3 and R_6, for this integral with sample points close to, but not exactly at, the midpoints of the subintervals. Here are the sample points used.

For R_3:

$c_1 = 1.4858425557$

$c_2 = 2.4916102607$

$c_3 = 3.4940272163$

For R_6:

$c_1 = 1.2458051304$

$c_2 = 1.7470136081$

$c_3 = 2.2476804000$

$c_4 = 2.7481034438$

$c_5 = 3.2483958519$

$c_6 = 3.7486100806$

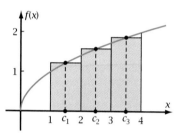 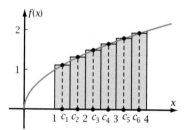

Figure 5-6a

Using these sample points and factoring out $\Delta x = 1$, you get

$$R_3 = \left(\sqrt{1.4858425557} + \sqrt{2.4916102607} + \sqrt{3.4940272163}\right)(1)$$
$$= 4.666666667$$

Surprisingly, R_6 turns out to be the same number. (In this case, $\Delta x = 0.5$.)

$$R_6 = \left(\sqrt{1.2458051304} + \sqrt{1.7470136081} + \cdots + \sqrt{3.7486100806}\right)(0.5)$$
$$= 4.666666667$$

Both answers are equal and look suspiciously like $4\frac{2}{3}$.

To see how the sample points were chosen, let

$$g(x) = \int \sqrt{x}\, dx = \frac{2}{3}x^{1.5}$$

(letting $C = 0$). Because $g'(x) = \sqrt{x}$, g is differentiable on the interval $[1, 1.5]$ (the first interval in R_6), and thus the mean value theorem applies. Therefore there is a number $x = c$ in the interval $(1, 1.5)$ for which

$$g'(c) = \sqrt{c} = \frac{g(1.5) - g(1)}{0.5} = 1.1161564094...$$

Squaring gives

$$c = (1.1161564094)^2 = 1.2458051304$$

which is the value of c_1 used in R_6. This observation leads to a conjecture.

CONJECTURE: Constant Riemann Sums

Let g be the indefinite integral $g(x) = \int f(x)\, dx$.

Let R_n be the Riemann sum for the definite integral $\int_a^b f(x)\, dx$, using as sample points a value of $x = c$ in each subinterval at which the conclusion of the mean value theorem is true for g on that subinterval.

Conjecture: The value of R_n is a constant, independent of the value of n.

Proof of the Conjecture

Figure 5-6b shows a definite integral, $\int_a^b f(x)\, dx$, that is to be evaluated. The top graph in Figure 5-6c shows the indefinite integral $g(x) = \int f(x)\, dx$. Point c_2 is the point in the second subinterval at which the conclusion of the mean value theorem is true for function g on that subinterval. The bottom graph in Figure 5-6c shows the corresponding Riemann sum for $f(x)$ using c_2 as the sample point in the second subinterval.

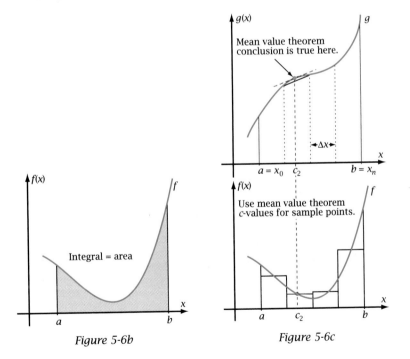

Figure 5-6b

Figure 5-6c

Applying the mean value theorem to g on each of the subintervals gives

$$g'(c_1) = \frac{g(x_1) - g(a)}{\Delta x}$$

$$g'(c_2) = \frac{g(x_2) - g(x_1)}{\Delta x}$$

$$g'(c_3) = \frac{g(x_3) - g(x_2)}{\Delta x}$$

$$\vdots$$

$$g'(c_n) = \frac{g(b) - g(x_{n-1})}{\Delta x}$$

Now use the points $c_1, c_2, c_3, \ldots, c_n$ as sample points for a Riemann sum of the original definite integral, as in the bottom graph of Figure 5-6c. That is,

$$R_n = f(c_1)\,\Delta x + f(c_2)\,\Delta x + f(c_3)\,\Delta x + \cdots + f(c_n)\,\Delta x$$

However, $f(x) = g'(x)$ for any value of x. Therefore

$$R_n = g'(c_1)\,\Delta x + g'(c_2)\,\Delta x + g'(c_3)\,\Delta x + \cdots + g'(c_n)\,\Delta x$$

Replacing $g'(c_1)$ with $\dfrac{g(x_1) - g(a)}{\Delta x}$, and so forth, canceling the Δx's, and arranging in column form gives

$$
\begin{aligned}
R_n = {}& g(x_1) - g(a) \\
& + g(x_2) - g(x_1) \\
& + g(x_3) - g(x_2) \\
& \quad\vdots \\
& \underline{+ g(b) - g(x_{n-1})} \\
R_n = {}& g(b) - g(a)
\end{aligned}
$$

All the middle terms "telescope," leaving only $-g(a)$ from the first term and $g(b)$ from the last. The result is *independent* of the number of increments. In fact, the quantity $g(b) - g(a)$ is the exact value of the definite integral. This is what the fundamental theorem of calculus says. You can calculate the definite integral by evaluating the antiderivative at the upper limit of integration and then subtracting from it the value of the antiderivative at the lower limit of integration. The formal statement of this theorem and its proof are given next.

PROPERTY: The Fundamental Theorem of Calculus

If f is an integrable function and if $g(x) = \int f(x)\,dx$,
then $\int_a^b f(x)\,dx = g(b) - g(a)$.

Proof of the Fundamental Theorem of Calculus

Partition the interval $[a, b]$ into n subintervals of equal width Δx. Let L_n and U_n be lower and upper sums for $\int_a^b f(x)\,dx$. Let R_n be the Riemann sum equal to

$g(b) - g(a)$, as derived previously. Because any Riemann sum is between the upper and lower sums, you can write

$$L_n \leq R_n \leq U_n$$

$$\lim_{n \to \infty} L_n \leq \lim_{n \to \infty} R_n \leq \lim_{n \to \infty} U_n \qquad \text{Take the limit of all three members of the inequality.}$$

$$\lim_{n \to \infty} L_n \leq g(b) - g(a) \leq \lim_{n \to \infty} U_n \qquad \text{R_n is a constant, $g(b) - g(a)$.}$$

$$\int_a^b f(x)\,dx \leq g(b) - g(a) \leq \int_a^b f(x)\,dx \qquad \text{Definition of definite integral.}$$

$$\therefore \int_a^b f(x)\,dx = g(b) - g(a), \text{Q.E.D.}$$

Example 1 shows you a way to use the fundamental theorem to evaluate the integral presented at the beginning of this section, showing that the area of the region under the graph of the function $y = \sqrt{x}$ from $x = 1$ to $x = 4$ really is $4\frac{2}{3}$, which you suspected.

▶ **EXAMPLE 1**   Use the fundamental theorem of calculus to evaluate $\int_1^4 \sqrt{x}\,dx$.

Solution $\int_1^4 \sqrt{x}\,dx = \int_1^4 x^{1/2}\,dx$

Let $g(x) = \int x^{1/2}\,dx = \frac{2}{3}x^{3/2} + C$.

$g(4) - g(1) = \frac{2}{3}(4^{3/2}) + C - \frac{2}{3}(1^{3/2}) - C = \frac{16}{3} - \frac{2}{3} = \frac{14}{3} = 4\frac{2}{3}$ ◀

Problem Set 5-6

Quick Review 5 min

Q1. $r(x) = \int m(x)\,dx$ if and only if —?—.

Q2. Write the definition of derivative.

Q3. How fast is $f(x) = x^2$ changing when $x = 3$?

Q4. Find dy: $y = \sec x$

Q5. Find y': $y = (x^2 + 3)^4$

Q6. Find d^2z/du^2: $z = \sin 5u$

Q7. Find $f'(x)$: $f(x) = 7^3$

Q8. Using equal increments, find the upper Riemann sum, R_4, for $\int_1^3 x\,dx$.

Q9. Sketch a graph that shows the conclusion of the mean value theorem.

Q10. $\int 5^x\,dx = $ —?—

A. $x5^{x-1} + C$

B. $5^x + C$

C. $5^x \ln x + C$

D. $5^x \ln 5 + C$

E. $\dfrac{5^x}{\ln 5} + C$

1. For the integral $I = \int_4^9 10x^{-1.5}\,dx$, do the following.

a. Find the exact value of I using the fundamental theorem of calculus. What happens to "$+ C$" from the indefinite integral?

b. Sketch a graph that shows an upper sum with $n = 5$ increments for this integral.

c. Find U_5, L_5, and the average of these two sums. Does the average overestimate or underestimate the integral? Explain.

d. Find midpoint Riemann sums M_{10}, M_{100}, and M_{1000} for I. Do the Riemann sums seem to be converging to the exact value of I?

2. Let $I = \int_0^{1.5} \sin x \, dx$. Find the exact value of I by using the fundamental theorem of calculus. Show that midpoint Riemann sums approach this value as the number of increments approaches infinity. Sketch an appropriate graph to show why midpoint Riemann sums overestimate the value of the definite integral.

3. State the fundamental theorem of calculus.

4. Prove that if f is an integrable function, then for any partition of the interval $[a, b]$ into n subintervals of equal width there is a Riemann sum for $\int_a^b f(x)\, dx$ whose value is independent of the number n.

5. Prove the fundamental theorem of calculus by using the result of Problem 4 as a lemma.

6. You have proved that it is possible to pick a Riemann sum whose value is independent of the number of increments. Suppose the interval $[a, b]$ is taken as a whole. That is, suppose there is just one "subinterval." Where should the sample point for this interval be picked so that the corresponding Riemann "sum" is exactly equal to the definite integral from a to b? How does your answer relate to the fundamental theorem of calculus?

7. *Freeway Exit Problem:* In the design of freeway exit ramps it is important to allow enough distance for cars to slow down before they enter the frontage road. Suppose a car's velocity is given by $v(t) = 100 - 20(t + 1)^{1/2}$, where $v(t)$ is in feet per second and t is the number of seconds since it started slowing down. Write a definite integral equal to the number of feet the car goes from $t = 0$ to $t = 8$. Evaluate the integral exactly by using the fundamental theorem of calculus.

8. *The Fundamental Theorem Another Way:* Let h be the square root function $h(x) = x^{1/2}$. Let P

be the region under the graph of h from $x = 4$ to $x = 9$ (Figure 5-6d).

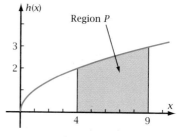

Figure 5-6d

a. Evaluate the Riemann sum R_{10} for P, picking sample points at the midpoints of the subintervals. Don't round your answer.

b. Let u be a value of x in the interval $[4, 9]$. Let $A(u)$ be the area of the portion of the region from $x = 4$ to $x = u$ (Figure 5-6e). Let Δu be a small change in u. The area of the strip from $x = u$ to $x = u + \Delta u$ equals $A(u + \Delta u) - A(u)$. Explain why this area is between $h(u)\Delta u$ and $h(u + \Delta u)\Delta u$. Write your result as a three-member inequality.

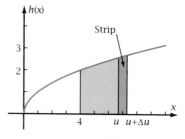

Figure 5-6e

c. Use the inequality you found in part b to prove that $dA/du = h(u)$. This equation is called a **differential equation**.

d. Multiply both sides of the differential equation given in part c by du. Then take the indefinite integral of both sides. Find the constant of integration by observing that $A(4)$ must equal zero.

e. Find the area of region P by evaluating $A(9)$. Explain why your answer to part a is consistent with your answer to this problem.

9. *Riemann Sum Sketching Problem:* Sketch appropriate rectangles to show the following.

a. An upper sum with four increments, where the sample points are taken at the left endpoint of each subinterval

b. A lower sum with three increments, where the sample points are taken at the left endpoint of each subinterval

c. A Riemann sum with five increments, where the sample points are taken at the midpoint of each subinterval

d. A Riemann sum with four increments, where no sample point is taken at the middle or at either end of a subinterval

e. A Riemann sum with three increments, where two different sample points are at the same x-value

f. A subinterval for an upper sum in which the sample point must be somewhere between the two endpoints

10. *Journal Problem:* Update your journal with what you've learned. Include such things as

• A statement of the fundamental theorem of calculus

• How sample points can be chosen so that a Riemann sum is independent of the number of increments, and how this fact leads to the fundamental theorem

• Evidence to show that Riemann sums really do get close to the value of a definite integral found by the fundamental theorem as n approaches infinity

• What you now better understand about the meaning of a Riemann sum

• Anything about the fundamental theorem of calculus that you're still unclear about

---

5-7   Definite Integral Properties and Practice

The table shows the four concepts of calculus—limit, derivative, and two kinds of integral—and the four things you should be able to do with each concept. You now know precise definitions of the four concepts and have a reasonably good understanding of their meanings.

	Define it.	Understand it.	Do it.	Apply it.
Limit				
Derivative				
Definite integral				
Indefinite integral				

In this section you will work on the "Do it" box for the definite integral. You will be using the fundamental theorem of calculus, so you will also be working on the "Do it" box for indefinite integrals.

OBJECTIVE   Evaluate quickly a definite integral, in an acceptable format, using the fundamental theorem of calculus.

To evaluate a definite integral such as

$$\int_1^4 x^2 \, dx$$

you could start by writing an indefinite integral, $g(x) = \int x^2 \, dx$, integrating, then finding $g(4) - g(1)$. Here is a compact format that is customarily used.

$$\int_1^4 x^2 \, dx$$

$$= \frac{1}{3}x^3 \Big|_1^4 \qquad\qquad \text{Read "}(1/3)x^3 \text{ evaluated from } x = 1 \text{ to } x = 4."$$

$$= \frac{1}{3} \cdot 4^3 - \frac{1}{3} \cdot 1^3$$

$$= 21$$

In the first step you find the indefinite integral. The vertical bar at the right reminds you that the upper and lower limits of integration are to be substituted into the expression to its left.

As mentioned in Section 5-4, the values 1 and 4 are called *limits of integration.* This terminology is unfortunate because these values have nothing to do with the concept of limit as you have defined it. The term *lower and upper bounds of integration* would be more suitable. However, the word *limit* is firmly entrenched in mathematical literature, so you should get used to the ambiguity and interpret the word in its proper context.

Following are some properties of definite integrals that are useful both for evaluating definite integrals and for understanding what they mean.

Integral with a Negative Integrand

Suppose you must evaluate $\int_1^4 (x^2 - 5x + 2) \, dx$. The result is

$$\left(\frac{1}{3}x^3 - \frac{5}{2}x^2 + 2x \right) \Big|_1^4$$

$$= \left(\frac{64}{3} - 40 + 8 \right) - \left(\frac{1}{3} - \frac{5}{2} + 2 \right)$$

$$= -10.5$$

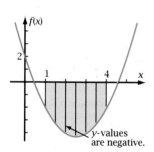

Figure 5-7a

The answer should surprise you. How could an area be negative? Figure 5-7a reveals the reason: The region lies below the x-axis. The Riemann sum has the form

$$\sum f(x) \, \Delta x$$

Each value of $f(x)$ is negative, and each Δx is positive. Thus each term in the sum is negative, and as a result the integral is negative.

Integral from a Higher Number to a Lower Number

Suppose an integral has a lower limit of integration that is greater than the upper limit, such as

$$\int_{\pi/3}^0 \cos x \, dx$$

Key Curriculum Press
Innovators in Mathematics Education

1150 65th Street • Emeryville • California • 94608-1109
1•888•881•8885

Whse: 01 Key Main Warehouse

B
I YELM COMM SCH DIST 2
L PEGGY SEIDEL
L 401 COATES ST NW
T YELM WA 98597
O

S
H YELM COMM SCH DIST 2
I PEGGY SEIDEL
P 401 COATES ST NW
T YELM WA 98597
O

DATE	CUSTOMER NO.	CUSTOMER P/O NO.		TERMS	F.O.B.	SHIP DATE	SHIP VIA
07/01/04	02421850			30 DAYS		07/02/04	UPS Commercial

QUANTITY ORDERED	QUANTITY BACK-ORDERED	QUANTITY SHIPPED	PRODUCT NUMBER	DESCRIPTION
1	1	1	53654	Calculus: Concepts and Applicatio $1-55953-654-3$ Order # 782392
1	1	1	53657	Calculus Solutions Manual 2nd Ed. $1-55953-657-8$ Order # 782392
1	1	1	88051	Precalculus/Calculus Brochure Order # 782392

SPECIAL INSTRUCTIONS:

PLEASE NOTE: This picklist is for shipping purposes only. An itemeized invoice with prices and payment information will be mailed separately.

FED. ID NO. 94-2668279

SEE REVERSE SIDE FOR ADDITIONAL INFORMATION

Our Sincere Thanks

Your order is important to us. To show our appreciation, we're committed to providing you with superior products and excellent customer service.

We're so confident that we offer the best in mathematics materials that we give you up to a year to make sure our products meet your needs.

College and University Bookstore Customers!

See special insert for return policy statement.

NEED TO RETURN MERCHANDISE?

If you decide within 60 days that you're not satisfied with your order, you may return any or all of it, unstamped and in saleable condition, for a full refund. If you find that you want to return an order after 60 days, we will gladly accept it within one year and will charge a 10% restocking fee to process the return.

1. Please call 1-888-881-8885 to request a Return Authorization (RA) number. In order to issue an RA number, please have your reference number or invoice number available when you call.
2. Make sure you put your RA number on the outside of your return shipment box, on the mailing label.
3. To ensure traceable delivery, please ship via UPS or insured U.S. mail.
4. Include a copy of your invoice or packing list in the package.
5. Address your shipment to the following address:

Attn: Returns Department—RA # _____
Key Curriculum Press
1150 65th Street
Emeryville, CA 94608-1109

For questions regarding orders, please call 1-888-881-8885

QUESTIONS ABOUT YOUR SHIPMENT OR INVOICE?

Our friendly customer service representatives are available to help you from 8:00 AM to 5:00 PM Pacific Time, Monday through Friday. Please call with your questions about an outstanding shipment you've already received, or an invoice.

If you have questions regarding an outstanding shipment, please have your billing address ready. For questions or inquiries regarding shipments already received, please have your invoice available when you contact us. Key Curriculum Press must be notified of any shipment discrepancies within 30 days upon receipt of goods.

Evaluating the integral gives

$$\sin x \Big|_{\pi/3}^{0}$$

$$= \sin 0 - \sin \frac{\pi}{3}$$

$$= 0 - \frac{\sqrt{3}}{2}$$

$$= -0.86602\ldots$$

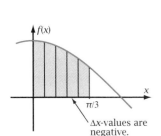

Figure 5-7b

The integral is negative, yet the integrand is positive, as shown in Figure 5-7b. The reason is the same as that for the previous property but is more subtle. In this case each Δx is negative. Because $\Delta x = (b - a)/n$, Δx will be negative whenever b is less than a. Thus each term in the sum

$$\sum f(x)\Delta x$$

will be negative, making the integral itself negative. Combining these observations leads you to conclude that if both $f(x)$ and Δx are negative, the integral is positive. For instance,

$$\int_{3}^{1} -x\, dx = -\tfrac{1}{2}x^2 \Big|_{3}^{1} = -\tfrac{1}{2} + \tfrac{9}{2} = 4$$

Sum of Integrals with the Same Integrand

Suppose you integrate a function, such as $x^2 + 1$, from $x = 1$ to $x = 4$ and then integrate the same function from $x = 4$ to $x = 5$. Figure 5-7c shows the two regions whose areas equal the two integrals. The sum of the two areas equals the area of the region from $x = 1$ all the way to $x = 5$. This fact suggests that the two integrals should sum to the integral of $x^2 + 1$ from $x = 1$ to $x = 5$. This turns out to be true, as you can see here.

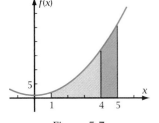

Figure 5-7c

$$\int_{1}^{4} (x^2 + 1)\, dx = \left(\tfrac{1}{3}x^3 + x\right)\Big|_{1}^{4} = \tfrac{64}{3} + 4 - \tfrac{1}{3} - 1 = 24$$

$$\int_{4}^{5} (x^2 + 1)\, dx = \left(\tfrac{1}{3}x^3 + x\right)\Big|_{4}^{5} = \tfrac{125}{3} + 5 - \tfrac{64}{3} - 4 = 21\tfrac{1}{3}$$

$$\int_{1}^{5} (x^2 + 1)\, dx = \left(\tfrac{1}{3}x^3 + x\right)\Big|_{1}^{5} = \tfrac{125}{3} + 5 - \tfrac{1}{3} - 1 = 45\tfrac{1}{3}, \text{ which equals } 24 + 21\tfrac{1}{3}$$

In general, $\int_{a}^{b} f(x)\, dx = \int_{a}^{c} f(x)\, dx + \int_{c}^{b} f(x)\, dx$.

Integrals Between Symmetric Limits

The integral $\int_{-a}^{a} f(x)\, dx$ is called an **integral between symmetric limits**. If f happens to be either an odd function (such as $\sin x$ or x^5) or an even function (such as $\cos x$ or x^6), then the integral has properties that make it easier to evaluate.

The graph on the left in Figure 5-7d at the top of page 230 shows an **odd function**, where $f(-x) = -f(x)$. The area of the region from $x = -a$ to $x = 0$ equals the area of the region from $x = 0$ to $x = a$, but the signs of the integrals are opposite. Thus the integral equals zero! For instance,

$$\int_{-2}^{2} x^3\, dx = \tfrac{1}{4}x^4 \Big|_{-2}^{2} = \tfrac{1}{4}(2)^4 - \tfrac{1}{4}(-2)^4 = 0$$

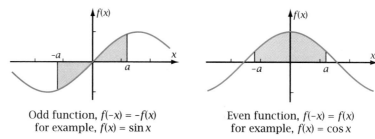

Odd function, $f(-x) = -f(x)$
for example, $f(x) = \sin x$

Even function, $f(-x) = f(x)$
for example, $f(x) = \cos x$

Figure 5-7d

The graph on the right in Figure 5-7d shows an **even function**, where $f(-x) = f(x)$. The areas of the regions from $x = -a$ to 0 and from $x = 0$ to a are again the same. This time the integrals have the same sign. Thus you can integrate from zero to a and then double the result. For instance,

$$\int_{-3}^{3} x^4 \, dx = \tfrac{1}{5}x^5 \Big|_{-3}^{3} = \tfrac{1}{5}(3)^5 - \tfrac{1}{5}(-3)^5 = 48.6 - (-48.6) = 97.2, \text{ and}$$

$$2\int_{0}^{3} x^4 \, dx = 2 \cdot \tfrac{1}{5}x^5 \Big|_{0}^{3} = 2 \cdot \tfrac{1}{5}(3)^5 - 2 \cdot \tfrac{1}{5}(0)^5 = 2(48.6) = 97.2$$

Integral of a Sum, and Integral of a Constant Times a Function

In Section 5-3, you learned that the indefinite integral of a sum of two functions equals the sum of the integrals and that the integral of a constant times a function equals the constant times the integral of the function. By the fundamental theorem of calculus, these properties also apply to definite integrals

$$\int_{a}^{b} [f(x) + g(x)] \, dx = \int_{a}^{b} f(x) \, dx + \int_{a}^{b} g(x) \, dx, \text{ and}$$

$$\int_{a}^{b} k\, f(x) \, dx = k\int_{a}^{b} f(x) \, dx$$

Figures 5-7e and 5-7f show graphically what these two properties say. In Figure 5-7e, the regions representing the two integrals are shaded differently. The second region sits on top of the first without a change in area. Thus the integral of the sum of the two functions is represented by the sum of the two areas. In Figure 5-7f, the region representing the integral of f is dilated by a factor of k in the vertical direction. The region representing the integral of kf thus has k times the area of the region representing the integral of f.

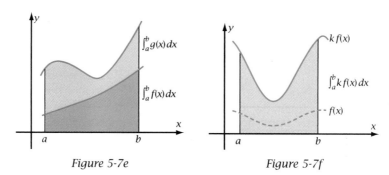

Figure 5-7e   Figure 5-7f

Upper Bounds for Integrals

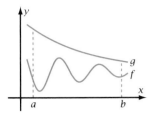

Figure 5-7g

Suppose that the graph of one function is always below the graph of another. Figure 5-7g shows functions for which

$$f(x) < g(x) \text{ for all } x \text{ in } [a, b]$$

The area of the region corresponding to the integral of f is smaller than that corresponding to the integral of g. Thus

$$\int_a^b f(x)\, dx < \int_a^b g(x)\, dx$$

The properties described above are summarized here.

Properties of Definite Integrals

1. **Positive and Negative Integrands:** The integral $\int_a^b f(x)\, dx$ is positive if $f(x)$ is positive for all values of x in $[a, b]$ and negative if $f(x)$ is negative for all values of x in $[a, b]$, provided $a < b$.

2. **Reversal of Limits of Integration:** $\int_b^a f(x)\, dx = -\int_a^b f(x)\, dx$

3. **Sum of Integrals with the Same Integrand:**
   $\int_a^b f(x)\, dx = \int_a^c f(x)\, dx + \int_c^b f(x)\, dx$

4. **Integrals Between Symmetric Limits:** If f is an odd function, then $\int_{-a}^a f(x)\, dx = 0$.
   If f is an even function, then $\int_{-a}^a f(x)\, dx = 2\int_0^a f(x)\, dx$.

5. **Integral of a Sum and of a Constant Times a Function:**
   $\int_a^b [f(x) + g(x)]\, dx = \int_a^b f(x)\, dx + \int_a^b g(x)\, dx$
   $\int_a^b k f(x)\, dx = k\int_a^b f(x)\, dx$

6. **Upper Bounds for Integrals:** If $f(x) < g(x)$ for all x in $[a, b]$, then $\int_a^b f(x)\, dx < \int_a^b g(x)\, dx$.

Problem Set 5-7

Quick Review

Q1. Evaluate: $\int x^5\, dx$

Q2. Evaluate: $\int (3x + 7)^5\, dx$

Q3. Evaluate: $\int x^{-4}\, dx$

Q4. Evaluate: $\int \sin^5 x \cos x\, dx$

Q5. Evaluate: $\int \cos 5x\, dx$

Q6. Evaluate: $\int (\cos^2 x + \sin^2 x)\, dx$

Q7. Evaluate: $\int \sec^2 x\, dx$

Q8. Find y'': $y = \ln x$

Q9. $\int_a^b f(x)\, dx$ is a(n) —?— integral.

Q10. $\int f(x)\, dx$ is a(n) —?— integral.

For Problems 1–26, evaluate the integral exactly by using the fundamental theorem of calculus. You may check your answers by Riemann sums or by the trapezoidal rule.

1. $\int_1^4 x^2 \, dx$

2. $\int_2^5 x^3 \, dx$

3. $\int_{-2}^3 (1 + 3x)^2 \, dx$

4. $\int_{-1}^4 (5x - 2)^2 \, dx$

5. $\int_1^8 60x^{2/3} \, dx$

6. $\int_1^4 24x^{3/2} \, dx$

7. $\int_2^8 5 \, dx$

8. $\int_{20}^{50} dx$

9. $\int_{-2}^0 (x^2 + 3x + 7) \, dx$

10. $\int_{-3}^0 (x^2 + 4x + 10) \, dx$

11. $\int_{-1}^1 \sqrt{4x + 5} \, dx$

12. $\int_{-3}^3 \sqrt{2x + 10} \, dx$

13. $\int_0^\pi 4 \sin x \, dx$

14. $\int_{-\pi/2}^{\pi/2} 6 \cos x \, dx$

15. $\int_{\pi/6}^{\pi/3} (\sec^2 x + \cos x) \, dx$

16. $\int_0^{\pi/3} (\sec x \tan x + \sin x) \, dx$

17. $\int_0^{\ln 4} e^{2x} \, dx$

18. $\int_0^{\ln 3} e^{-x} \, dx$

19. $\int_1^2 \sin^3 x \cos x \, dx$

20. $\int_{-3}^3 (1 + \cos x)^4 \sin x \, dx$

21. $\int_{0.1}^{0.2} \cos 3x \, dx$

22. $\int_0^{0.4} \sin 2x \, dx$

23. $\int_{-5}^5 (x^7 - 6x^3 + 4 \sin x + 2) \, dx$ (Be clever!)

24. $\int_{-1}^1 (\cos x + 10x^3 - \tan x) \, dx$ (Be very clever!)

25. $\int_{-1}^1 x^{-2} \, dx$ (Beware!)

26. $\int_{-2}^2 \sqrt{x} \, dx$ (Beware!)

For the functions in Problems 27–30, state whether the definite integral equals the area of the region bounded by the graph and the x-axis. Graphs may help.

27. $\int_3^6 (x^2 - 10x + 16) \, dx$

28. $\int_5^7 \cos x \, dx$

29. $\int_0^7 \sin \frac{\pi}{6} x \, dx$

30. $f(x) = \int_1^8 \left(\frac{1}{x^2} - \frac{1}{4} \right) dx$

For Problems 31–36, suppose that

$$\int_a^b f(x) \, dx = 7, \quad \int_a^b g(x) \, dx = 12, \quad \text{and}$$
$$\int_b^c g(x) \, dx = 13$$

Evaluate the given integral or state that it cannot be evaluated from the given information.

31. $\int_b^a f(x) \, dx$

32. $\int_a^b 4 f(x) \, dx$

33. $\int_a^c g(x) \, dx$

34. $\int_a^c f(x) \, dx$

35. $\int_a^c [f(x) + g(x)] \, dx$

36. $\int_a^b [f(x) + g(x)] \, dx$

For Problems 37 and 38, Figures 5-7h and 5-7i show the graph of the derivative, f', of a continuous function f. The graph of f is to contain the marked point. Recall that where $f'(x) = 0$, the tangent to the graph of f is horizontal. Also recall that $f(x) = \int f'(x) \, dx$, and thus $f(x)$ changes by an amount $\int_a^b f'(x) \, dx$ when x changes from a to b. Use this information and the fact that the definite integral represents an area to sketch the graph of f on a copy of the figure.

37.

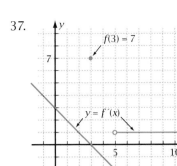

Figure 5-7h

38.

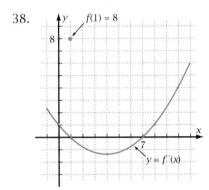

Figure 5-7i

39. The property of the upper bound of an integral states:

> If $f(x) < g(x)$ for all x in $[a, b]$, then
> $$\int_a^b f(x)\,dx < \int_a^b g(x)\,dx.$$

Write the converse of this statement and then show by example that the converse is false.

40. *"Plus C" Problem:* When you write an indefinite integral you always include "$+ C$," but you don't do this with a definite integral. Evaluate the integral

$$\int_1^4 x^2\,dx$$

by using the indefinite integral $\int x^2\,dx = \frac{1}{3}x^3 + C$. Then explain what happens to C and why you do not write "$+ C$" for a definite integral.

5-8   Definite Integrals Applied to Area and Other Problems

You have learned that a definite integral has a physical meaning as the product of x and y and a graphical meaning as the area of a region under a graph. In this section you will put these two ideas together to arrive at a systematic method for applying definite integrals.

OBJECTIVE   Given a problem in which a quantity y varies with x, learn a systematic way to write a definite integral for the product of y and x and evaluate the integral by using the fundamental theorem of calculus.

Suppose you are driving 60 ft/s (about 40 mi/h). You speed up to pass a truck. Assume that t seconds after you start accelerating, your velocity is given by

$$v = 60 + 6t^{1/2}$$

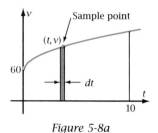

Figure 5-8a

You can solve the "big" problem of finding the distance you have gone from $t = 0$ s to $t = 10$ s by dividing it into little problems. Divide the time interval $[0, 10]$ into subintervals, each of width dt units. Figure 5-8a shows a representative subinterval and a narrow strip of width dt. Pick a sample point (t, v) on the graph within the subinterval. If dt is small, the velocity throughout the subinterval will be almost the same as it is at the sample point. Let y be the number of feet your car has gone. The distance dy it goes in the time interval dt is approximately

$$dy = v\,dt$$
$$dy = (60 + 6t^{1/2})\,dt$$

The total distance is approximately equal to the sum of the distances dy. Thus

$$y \approx \sum dy = \sum (60 + 6t^{1/2})\,dt$$

This sum is a Riemann sum. Its limit as dt approaches zero is a definite integral, so the distance traveled from $t = 0$ to $t = 10$ is

$$y = \int_0^{10} (60 + 6t^{1/2})\,dt \qquad \text{The definite integral is the exact distance.}$$

$$= (60t + 4t^{3/2})\Big|_0^{10} = 600 + 4(10^{3/2}) - 0 - 0$$

$$= 726.49... \approx 726 \text{ ft}$$

The thought process in the previous problem can be applied to any problem involving a definite integral. The process is summarized in this box.

TECHNIQUE: *Application of a Definite Integral*

To find the product of $P = y$ times x, where $y = f(x)$ and x goes from a to b, do the following.

1. Divide the interval $[a, b]$ into subintervals. Draw a strip of width dx, corresponding to one representative subinterval.

2. Pick a sample point (x, y) on the graph in that subinterval.

3. Write dP as a function of x and dx, using the fact that y is essentially constant throughout the subinterval.

4. Add the dP's and take the limit; that is, integrate.

Area Between Two Graphs

You can interpret a definite integral graphically as the area of a region under the graph of a function, that is, between the graph and the x-axis. Example 1 shows you how to find the area of a region *between* two graphs.

▶ **EXAMPLE 1**   Find algebraically (by the fundamental theorem of calculus) the area of the "triangular" region in Quadrant I that is bounded by the graphs of $y = 4 - x^2$ and $y = 4x - x^2$ and the y-axis.

Solution   Sketch the graphs of the two functions as shown in Figure 5-8b. Identify the correct "triangular" region and draw a thin vertical strip of width dx. A sample point at x on the x-axis determines a sample point on each graph, (x, y_1) and (x, y_2). Mark these sample points as shown in the figure. Because the strip is narrow, its length at any point is not much different from its length at the sample points, $y_1 - y_2$. So the area of the strip is approximately $(y_1 - y_2)\, dx$. This approximation becomes exact as dx approaches zero. Let dA be this area.

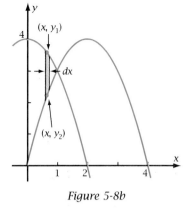

Figure 5-8b

$$dA = (y_1 - y_2)\, dx$$
$$= (4 - x^2 - 4x + x^2)\, dx$$
$$= (4 - 4x)\, dx$$

Be sure you subtract (larger y-value) minus (smaller y-value) so that the height of the rectangle will be positive. Before you can integrate, you must find the limits of integration. The first strip is at the y-axis, so the lower limit of integration is $x = 0$. The last strip is at the point where the two graphs intersect. If you cannot tell this point by inspection ($x = 1$, in this case), you can set the two y-values equal to each other and solve for x. The result is the upper limit of integration.

$$4 - x^2 = 4x - x^2 \Rightarrow 4 = 4x \Rightarrow x = 1$$
$$A \approx \sum dA = \sum (4 - 4x)\, dx \qquad \text{Total area is approximately the sum of the } dA\text{'s.}$$
$$A = \int_0^1 (4 - 4x)\, dx \qquad \text{A definite integral is a limit of a Riemann sum.}$$
$$= 4x - 2x^2 \Big|_0^1 \qquad \text{Use the fundamental theorem.}$$
$$= 4 - 2 - 0 + 0 = 2 \qquad\qquad\qquad\qquad\qquad\qquad\blacktriangleleft$$

Sometimes vertical slicing makes the problem hard to analyze. Figure 5-8c shows the region bounded by the graphs of $x = 3 + 2y - y^2$ and $x + y = -1$.

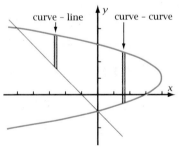

Awkward to slice vertically

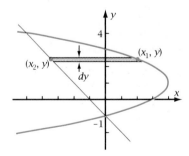

Appropriate to slice horizontally

Figure 5-8c

The length of the vertical strip is given by different rules for different parts of the domain. If x is negative, the length of the strip is given by

curve − line

For positive values of x, the length is given by

curve − curve

This difficulty is circumvented by slicing horizontally, as in Example 2.

▶ **EXAMPLE 2** For the region in Figure 5-8c bounded by the graphs of $x = 3 + 2y - y^2$ and $x + y = -1$, write an integral for the area and evaluate it numerically. Evaluate the integral exactly by the fundamental theorem and compare your answers.

Solution See the graph on the right in Figure 5-8c. Pick sample points (x_1, y) and (x_2, y). Then

$$dA = (\text{curve} - \text{line})\, dy = (x_1 - x_2)\, dy$$
$$= [3 + 2y - y^2 - (-y - 1)]\, dy = (4 + 3y - y^2)\, dy$$

The graphs intersect where the two x-values are equal.

$$3 + 2y - y^2 = -y - 1$$
$$0 = y^2 - 3y - 4 \Rightarrow (y + 1)(y - 4) = 0 \Rightarrow y = -1 \text{ or } y = 4$$

The total area is found by adding the dA's and taking the limit (that is, by integrating). Be sure the limits of integration are values of y.

$$A = \int_{-1}^{4} (4 + 3y - y^2)\, dy \qquad \text{Write an integral equal to the area.}$$
$$= 20.83333\ldots \qquad \text{Use numerical integration.}$$

You can also find the answer using the fundamental theorem.

$$A = 4y + \tfrac{3}{2}y^2 - \tfrac{1}{3}y^3 \Big|_{-1}^{4}$$
$$= 16 + 24 - \tfrac{64}{3} - (-4) - \tfrac{3}{2} + \tfrac{-1}{3}$$
$$= 20\tfrac{5}{6}, \text{ which equals } 20.8333\ldots$$ ◀

Problem Set 5-8

Q5. Sketch $f'(x)$ for Figure 5-8d.

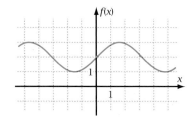

Figure 5-8d

Q6. Is $g(x) = 2 + |x - 1|$ continuous at $x = 1$?

Q7. Is $f(x) = \sin x$ increasing or decreasing at $x = 7$?

Q8. What hypothesis of Rolle's theorem is not a hypothesis of the mean value theorem?

Q9. How fast are you going at time $t = 9$ s if your displacement is $d(t) = 100t^{1.5}$ ft?

Q10. In Problem Q9, what is your acceleration at time $t = 9$ s?

---

1. *Displacement Problem 1:* Suppose you take a long trip. You are impatient to arrive at your destination, so you gradually let your velocity increase according to the equation

$$v = 55 + 12t^{0.6}$$

where v is in miles per hour and t is in hours (Figure 5-8e). Your displacement equals (velocity) · (time), but the velocity varies.

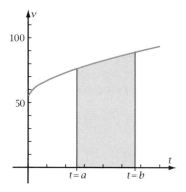

Figure 5-8e

a. On a copy of the graph, draw a narrow vertical strip of width dt in the interval from $t = a$ to $t = b$. Show a sample point (t, v) on the graph within the strip. Approximately what displacement, dy, does the car go in the time interval dt?

b. The total distance the car goes in the time interval $[a, b]$ is the limit of the sum of the dy-values from $t = a$ to $t = b$. That is, y is the integral of dy from $t = a$ to $t = b$. Write an integral and evaluate it to find how far the car goes from $t = 0$ to $t = 1$. Write and evaluate another integral to find out how far the car goes from $t = 1$ to $t = 2$.

c. The total displacement of the car in the first two hours is the integral of dy from $t = 0$ to $t = 2$. Evaluate this integral. Show that it equals the sum of the two integrals in part b, illustrating the sum of two integrals with the same integrand property.

d. To find out how long it takes to complete the entire 300-mile trip, you can set the integral of dy from $t = 0$ to $t = b$ equal to 300 and solve for b. Find this time.

e. How fast are you going at the end of the 300-mile trip?

2. *Displacement Problem 2:* As you approach a traffic light you take your foot off the accelerator. As your car slows down, its velocity is given by

$$v = 15e^{-0.1t}$$

where v is in meters per second t seconds after you start slowing. Sketch the graph of v for the first 20 s after you start slowing. Draw a vertical strip of width dt and show a sample point (t, v) on the graph within the strip. Write an equation for dy, the approximate distance you travel in time dt. Find, in terms of e, the exact distance you travel in the first 20 s as you are slowing down. Find a decimal approximation for this distance.

3. *Area Problem 1:* Figure 5-8f shows the region under the graph of $y = 10e^{0.2x}$ from $x = 0$ to $x = 2$.

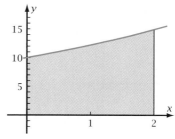

Figure 5-8f

a. On a copy of the graph, draw a narrow vertical strip of width dx. Show a sample point (x, y) on the graph within the strip.

b. Write an equation for dA, the approximate area of the strip, in terms of x, y, and dx. Transform dA so it involves x and dx only.

c. The area of the region is the limit of the sum of the values of dA as width dx approaches zero. That is, the area equals the definite integral of y with respect to x from $x = 0$ to $x = 2$. Find this area exactly, in terms of e, using the fundamental theorem.

d. Find a decimal approximation for your answer in part c. Perform a geometric calculation that shows that the answer is reasonable.

4. *Area Problem 2:* Sketch the region under the graph of $y = 6x - x^2$ that lies between the two x-intercepts. Draw a narrow vertical strip of the region of width dx, and show a sample point (x, y) on the graph within the strip. Write an equation for the differential of the area, dA (the altitude of the strip times its width). Then find the exact area of the region by adding the areas of the strips and taking the limit (that is, by integrating). Show that the area of this parabolic region is two-thirds the area of the circumscribed rectangle.

5. *Area Problem 3:* Figure 5-8g shows the region bounded by the graphs of

$$y_1 = x^2 - 2x - 2 \quad \text{and} \quad y_2 = x + 2$$

A thin strip of width dx is drawn, with sample points (x, y_1) and (x, y_2) on the two graphs within the strip.

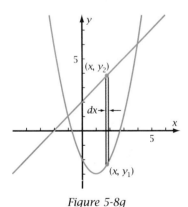

Figure 5-8g

a. Write an equation for dA, the area of the strip, assuming it to be a rectangle. Explain why dA is close to the actual area of the strip and why this approximation to the area gets closer to the actual area as dx approaches zero.

b. Find the exact area using the fundamental theorem. You can find the limits of integration by setting y_1 equal to y_2 and solving for x to find where the graphs intersect.

c. Check your answer by finding the midpoint Riemann sum R_{100}.

6. *Area Problem 4:* Sketch the "triangular" region in Quadrant I bounded by the y-axis and the graphs of

$$y_1 = \cos x \quad \text{and} \quad y_2 = \sin x$$

Show that the graphs intersect at $x = \pi/4$. Draw a thin vertical strip of width dx between sample points (x, y_1) and (x, y_2) on the two graphs within the strip. Write an equation for dA, the approximate area of the strip. Find the exact area of the region using the fundamental theorem.

7. *Work Problem 1:* The amount of work needed to stretch a spring equals the force exerted on the spring times the displacement of the end of the spring from one position to another. By **Hooke's law,** this force is proportional to the displacement from the rest position. Suppose that, for a particular spring, the force is given by

$$F = 0.6x$$

where F is in pounds and x is in inches (Figure 5-8h).

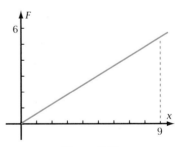

Figure 5-8h

a. On a copy of the graph, draw a narrow vertical strip of width dx in the region under the graph. Show a sample point (x, F) on the graph within the strip.

b. The amount of work, dW, done in stretching the spring by the displacement, dx, is approximately equal to $F\,dx$. Write dW in terms of x and dx. Then integrate to find the work done by stretching the spring from $x = 0$ to $x = 9$. (The integral can be interpreted as "Add up the dW's and take the limit as dx approaches zero.")

c. Confirm your answer to part b using simple geometry.

d. Based on the way you find dW, what are the units of work in this problem?

8. *Work Problem 2:* Calvin drags a heavy box 10 ft across the floor. He figures that the force F, in pounds, he exerts while the box is in motion is given by

$$F = 50 \cos \tfrac{\pi}{20} x$$

where x is the number of feet the box has been displaced from its starting point. Sketch the graph of F. Draw a thin vertical strip of width dx, and show a sample point (x, F) on the graph within the strip. Recalling that work is defined as force times displacement, write an equation for dW, the work done in moving the box dx feet. Then integrate to find the amount of work Calvin did in moving the box 10 ft. Check your answer by finding an appropriate Riemann sum.

9. *Degree-Days Problem:* A quantity used to measure the expense of air-conditioning a building is the degree-day. If the outside temperature is 20 degrees above that inside for two days, then there have been 40 degree-days. Usually the temperature difference varies. Suppose the temperature difference, T, is given by

$$T = 20 - 12 \cos 2\pi(x - 0.1)$$

where x is the time in days (Figure 5-8i).

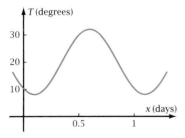

Figure 5-8i

a. On a copy of the graph, sketch a narrow vertical strip of width dx. Show a sample point (x, T) on the graph within the strip.

b. The number of degree-days, dD, during the time interval dx is approximately the value of T at the sample point times the length of time, dx. Write dD in terms of x and dx.

c. To find the total number of degree-days, D, from $x = 0$ to $x = 0.5$, add all the dD's and find the limit. That is, find the definite integral from $x = 0$ to $x = 0.5$.

d. The answer to part c is the number of degree-days from midnight to noon. Find the number of degree-days from noon to the next midnight. What is the total number of degree-days from one midnight till the next?

10. *Heat Problem:* The amount of heat needed to warm a substance equals the heat capacity of the substance times the number of degrees by which it is warmed. From experimental data it is found that the heat capacity for 1 lb mol of steam (a pound mole equals about 18 lb of steam) at normal atmospheric pressure is given approximately by the formula

$$C = -0.016T^3 + 0.678T^2 + 7.45T + 796$$

where T is in hundreds of degrees Fahrenheit and C is in Btu per 100°. (A Btu, or British thermal unit, is the amount of heat needed to warm 1 lb of water by 1°F.)

a. Plot the graph of C and sketch the result. Draw a narrow vertical strip of width dT for the region under the graph, and show a sample point (T, C) on the graph within the strip. Write an equation for dH, the amount of heat needed to warm one mole of steam by dT degrees. Then integrate using the fundamental theorem to find the number of Btu to warm one mole of steam from 1000° to 3000°.

b. The amount of heat needed to warm 2000 moles of steam is $2000C$. Find the amount of heat needed to warm 2000 moles of steam from 1000° to 3000°. What property of definite integrals lets you find this answer quickly?

11. *Total Cost Problem:* A mining company wants to dig a silver mine horizontally into the side of a mountain. The price per meter for digging the mine shaft gets longer because it is more expensive to bring out the dirt and the rock and to shore up the tunnel. Suppose the price, P, in dollars per meter for digging at a point x, in meters from the entrance, is given by

$$P = 100 + 0.06x^2$$

A silver mine in 1892

a. On a graph of P, draw a narrow strip of width dx. Show a sample point (x, P) on the graph within the strip.

b. Write an equation for dC, the number of dollars it costs to dig dx meters. Write an equation for C, the total cost of digging from $x = 0$ to $x = b$.

c. How much would it cost to dig from 0 m to 100 m? From 100 m to 200 m? From 0 m to 200 m? Show that the property of a sum of two integrals with the same integrand applies to the three answers.

12. *Golf Course Problem:* Suppose you are constructing a golf course in Scorpion Gulch. The plot for one of the putting greens is shown in Figure 5-8j. To estimate the area of the green, you draw parallel lines 10 ft apart and measure their lengths in feet. From the information given in the table, figure out about how many square feet of grass sod you must purchase to cover the green. Describe how you found the answer. Why can't you use the fundamental theorem of calculus to find this area?

x	Width
20	0
30	38
40	50
50	62
60	60
70	55
80	51
90	30
100	3

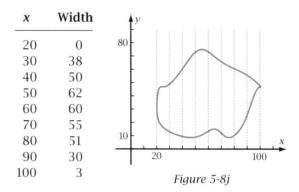

Figure 5-8j

For Problems 13–26, sketch the region bounded by the graph(s), write an integral for the area, and calculate the area exactly by the fundamental theorem.

13. $y = -x^2 + 6x - 5$ and the x-axis

14. $y = x^2 - x - 6$ and the x-axis

15. $x = (y - 1)(y - 4)$ and the y-axis

16. $x = 5 + 4y - y^2$ and the y-axis

17. $y = x^2 - 2x - 2$ and $y = x + 2$

18. $y = -2x + 7$ and $y = x^2 - 4x - 1$

19. $y = 0.5x^2 + 2x$ and $y = -x^2 + 2x + 6$

20. $y = 0.2x^2 + 3$ and $y = x^2 - 4x + 3$

21. $y = 2e^{0.2x}$ and $y = \cos x$, between $x = 0$ and $x = 5$

22. $y = \sec^2 x$ and $y = e^{2x}$, in Quadrant I, for $x \le 1$

23. $y = x + 3$ and $x = -y^2 + 6y - 7$

24. $y = -2x + 11$ and $x = 0.25y^2 - 0.5y - 0.75$

25. $y = x^3 - 4x$ and $y = 3x^2 - 4x - 4$ (The grapher will help you find the region.)

26. $y = x^{2/3}$ and $y = (x + 1)^{1/2} + 1$ (The grapher will help you find the region and locate the intersection points.)

27. Wanda Wye needs to find dA for the region bounded by $y = x^2$ and $y = x$ (Figure 5-8k). She wants to know whether to use

   $$(\text{line} - \text{curve}) \quad \text{or} \quad (\text{curve} - \text{line})$$

   Explain to Wanda how she can always choose the correct subtraction. Also tell her what her answer will be if she chooses the wrong one.

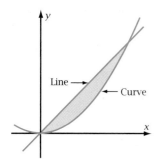

Figure 5-8k

28. Peter Doubt must find the area of the region shown in Figure 5-8l.

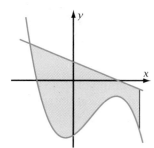

Figure 5-8l

   a. Explain to Peter why horizontal slicing is not appropriate for this problem.

   b. Peter is worried because some of the region is below the x-axis. Explain why, with proper slicing, each value of dA will be positive and thus he need not worry.

29. *Parabolic Region Problem:* Prove that the area of a parabolic region as shown in Figure 5-8m is always 2/3 the area of the circumscribed rectangle. Show the significance of what you have proved by using the result to find quickly the area of the parabolic region bounded by the graph of $y = 67 - 0.6x^2$ and the horizontal line $y = 7$.

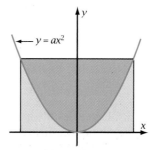

Figure 5-8m

30. *Sinusoidal Region Problem:* Show that the area under one arch of the sinusoid $y = \sin x$ (Figure 5-8n) is a rational number by finding this number. Demonstrate the significance of this fact by quickly finding the area under one arch of $y = 7 \cos 5x$.

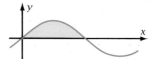

Figure 5-8n

31. *Area Check Problem:* Find the area of the region bounded by the graph of $y = x^2$ and the line $y = x + 6$ exactly, using the fundamental theorem. Then calculate midpoint Riemann sums with $n = 10$, $n = 100$, and $n = 1000$ increments. Show that the Riemann sums approach the exact answer as the number of increments becomes large.

32. *Curve Sketching Review Problem:* Given $t(x) = x + \sin x$, plot the graph of t using a window with an x-range of about 0 to 4π. Sketch the result. Then calculate the x-coordinates of all points at which the tangent to the graph is horizontal. Does $t'(x)$ change signs at any of these points? Show any such points on your graph.

5-9   Volume of a Solid by Plane Slicing

In Section 5-8, you set up definite integrals by slicing a region into narrow strips. In this section you will extend the technique to find volumes of solids that have curved boundaries. The idea is to slice the solid into thin, flat *slabs;* find the volume of a representative slab by multiplying its area by its thickness; then add the volumes and take the limit as the thickness approaches zero.

> **OBJECTIVE**   Given a solid whose cross-sectional area varies along its length, find its volume by slicing it into slabs and performing the appropriate calculus, and show that your answer is reasonable.

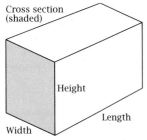

Figure 5-9a

The volume of a solid of uniform cross-sectional area is given by the formula

Volume = (area of cross section)(length of solid)

Figure 5-9a shows how this applies to a rectangular solid. The volume is given by (width)(height)(length), and (width)(height) is the cross-sectional area.

If a solid has some other shape, the volume still equals (area)(length), but the area may *vary.* Examples 1 and 2 show you how to deal with this situation.

▶ **EXAMPLE 1**   (*Rotate a region about the x-axis.*) The region under the graph of $y = x^{1/3}$ from $x = 0$ to $x = 8$ is rotated about the x-axis to form a solid. Write an integral for the volume of the solid and evaluate it to find the volume. Assume that x and y are in centimeters. Show that your answer is reasonable.

Solution   First draw a picture of the region (Figure 5-9b, left). Next slice the region into strips as though you planned to find the area. The idea is to slice perpendicular to the axis about which the region will rotate. Show a representative strip as in Figure 5-9b. Pick a sample point (x, y) that is on the graph of $y = x^{1/3}$.

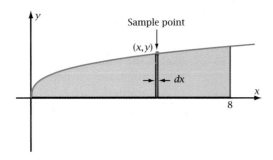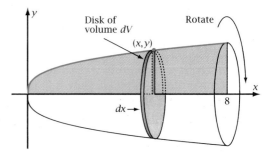

Draw the region.
Slice a strip perpendicular to the axis of rotation.
Pick a sample point (x, y).

Rotate the region and the strip to form a solid. The rotating strip forms a flat disk of volume dV.

Figure 5-9b

Chapter 5: Definite and Indefinite Integrals

Then draw what the solid looks like as the region rotates about the x-axis (Figure 5-9b, right). As the region turns, the rotating strip generates a **disk**. This disk can be thought of as a short, fat cylinder whose "altitude" is dx and whose cross section at any value of x is close to what it is at the sample point (x, y). The volume of the disk, dV, is

$$dV = \text{(cross-sectional area at the sample point)(altitude, } dx)$$

Because the radius of the disk is y, its volume will be

$$dV = \pi y^2 \, dx = \pi (x^{1/3})^2 \, dx = \pi x^{2/3} \, dx$$

The solid itself can be considered to be made up of a stack of these disks, so the approximate volume of the solid is equal to the sum of the disk volumes.

$$V \approx \sum dV = \sum \pi x^{2/3} \, dx$$

The exact volume is the limit of this Riemann sum, that is, the definite integral.

$$V = \int_0^8 \pi x^{2/3} \, dx \qquad \text{Volume is equal to the definite integral.}$$

$$= \pi \cdot \frac{3}{5} x^{5/3} \Big|_0^8 = \frac{3}{5}(32)\pi - \frac{3}{5}(0)\pi = 19.2\pi \qquad \text{Evaluate using the fundamental theorem.}$$

$$= 60.318... \approx 60.3 \text{ cm}^3 \qquad \text{Write a real-world answer, suitably rounded, with units.}$$

Checks:
Volume of circumscribed cylinder would be $\pi(2^2)(8) = 32\pi > 19.2\pi$. ✓
Volume of inscribed cone would be $(\frac{1}{3})\pi(2^2)(8) = 10.66...\pi < 19.2\pi$. ✓
Riemann sum approximation: $V \approx 19.2\pi$ ✓ ◀

▶ **EXAMPLE 2**   (*Rotate a region about the y-axis.*) The region in Quadrant I bounded by the parabola $y = 4 - x^2$ is rotated about the y-axis to form a solid paraboloid. Find the volume of the paraboloid if x and y are in inches. Show that your answer is reasonable.

Solution   First sketch the region as shown on the left in Figure 5-9c. Imagine the region sliced into strips perpendicular to the axis of rotation (the y-axis, in this case).

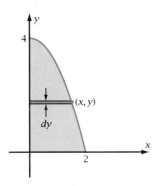

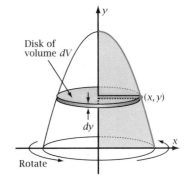

Draw the region.
Slice a strip perpendicular
to the axis of rotation.

Rotate the region and the strip
to form a solid. The rotating
strip forms a flat disk.

Figure 5-9c

Draw one such strip of width dy, and show a sample point (x, y) on the graph within the strip. As the region rotates to form the solid, the strips rotate to form disks. Because the strips are narrow, the cross-sectional area anywhere in the disk is not much different from the cross-sectional area at the sample point. Let dV be the volume of the disk shown.

$$dV = \pi x^2 \, dy$$
Volume = (area)(length); the radius is x this time.

$$= \pi(4 - y) \, dy$$
Get dV in terms of one variable.

$$\therefore V = \int_0^4 \pi(4 - y) \, dy = \pi\left(4y - \frac{1}{2}y^2\right)\Big|_0^4$$
Add the volumes of the slices and take the limit (integrate).

$$= \pi(16 - 8) - \pi(0 - 0) = 8\pi = 25.132\ldots \approx 25.1 \text{ in.}^3$$

Checks:
Volume of circumscribed cylinder would be $\pi(2^2)(4) = 16\pi > 8\pi$. ✓
Volume of inscribed cone would be $(1/3)\pi(2^2)(4) = 5.33\ldots\, \pi < 8\pi$. ✓
Riemann sum approximation: $V \approx 8\pi = 8\pi$ ✓ ◀

▶ **EXAMPLE 3**

(*Rotate a region bounded by two curves.*) Let R be the region bounded by the graphs of $y_1 = 6e^{-0.2x}$ and $y_2 = \sqrt{x}$ and by the vertical lines $x = 1$ and $x = 4$. Find the volume of the solid generated when R is rotated about the x-axis. Assume that x and y are in feet. Show that your answer is reasonable.

Solution

Sketch region R and slice it into strips perpendicular to the axis of rotation (x-axis). Show a representative strip with two sample points, one on each graph (Figure 5-9d, left). As R rotates, the strip will trace out a disk with a hole in the middle (a **washer**, for those of you familiar with nuts and bolts). The volume dV

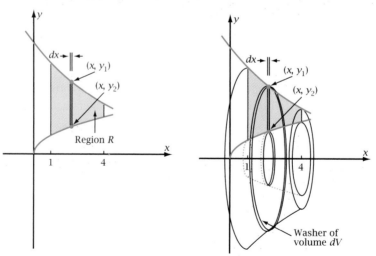

Draw the region. Slice a strip perpendicular to the axis of rotation. Pick *two* sample points, (x, y_1) and (x, y_2).

Rotate the region and the strip to form a solid. The rotating strip forms a flat washer of volume dV.

Figure 5-9d

of the washer will be the volume of the outer disk minus the volume of the inner disk (Figure 5-9d, right).

$$dV = \pi y_1^2\, dx - \pi y_2^2\, dx \qquad\qquad \text{Volume of outer disk minus volume of inner disk.}$$

$$= \pi(y_1^2 - y_2^2)\, dx \qquad\qquad \text{Start here if you're brave!}$$

$$= \pi(36e^{-0.4x} - x)\, dx \qquad\qquad \text{Substitute for } y_1 \text{ and } y_2 \text{ and square.}$$

$$\therefore V = \pi \int_1^4 (36e^{-0.4x} - x)\, dx \qquad \text{Add the volumes of the slices and take the limit (integrate).}$$

$$= \pi\left(\frac{36}{-0.4}e^{-0.4x} - \frac{1}{2}x^2\right)\Bigg|_1^4 \qquad \text{Use the fundamental theorem.}$$

$$= \pi(-90e^{-1.6} - 8) - \pi(-90e^{-0.4} - 0.5)$$

$$= 34.658\ldots\,\pi = 108.881\ldots \approx 108.9\ \text{ft}^3$$

Checks:
Volume of outer cylinder would be $\pi(6e^{-0.2})^2(4-1) = 72.39\ldots\,\pi$. ✓
Volume of inner cylinder would be $\pi(1)^2(4-1) = 3\pi$. ✓
∴ volume of solid is bounded above by $69.39\ldots\,\pi \geq 34.658\ldots\,\pi$. ✓ ◀

In Examples 1–3, the figure was generated by rotating a region. Such a figure is given the name **solid of revolution**. As the strip rotated, it generated a disk or washer. The same disk or washer would be generated if the solid were generated first, then sliced with planes perpendicular to the axis of rotation. As a result, the technique you have been using is called **finding volumes by plane slices**.

Once you realize this fact, you can find volumes of solids that are *not* generated by rotation. All you have to do is find the cross-sectional area in terms of the displacement perpendicular to the cross section. Example 4 shows you how.

▶ **EXAMPLE 4**   (*Plane slices of a noncircular solid*) A 2-in.-by-2-in.-by-4-in. wooden block is carved into the shape shown in Figure 5-9e. The graph of $y = \sqrt{4 - x}$ is drawn on the back face of the block. Then wood is shaved off the front and top faces in such a way that the remaining solid has square cross sections perpendicular to the x-axis. Find the volume of the solid. Show that your answer is reasonable.

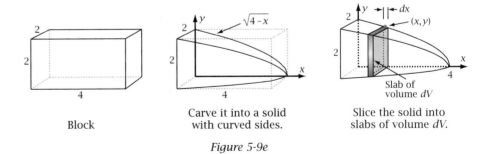

Block   Carve it into a solid   Slice the solid into
              with curved sides.   slabs of volume dV.

Figure 5-9e

Solution   The diagram at right in Figure 5-9e shows a **slab** formed by slicing the solid with planes perpendicular to the x-axis. Each such slab has length approximately equal to the y-value at the sample point shown. Thus the cross-sectional area of the slice is approximately equal to y^2 because the cross section is a square.

Using the fact that volume equals cross-sectional area times length (or thickness, in this case), you can write an equation for dV.

$$dV = y^2\,dx = (4 - x)\,dx \qquad \text{Do the geometry and algebra.}$$

$$\therefore V = \int_0^4 (4 - x)\,dx = 4x - \frac{1}{2}x^2 \Big|_0^4 = 16 - 8 - 0 + 0 = 8 \text{ in.}^3 \qquad \text{Do the calculus.}$$

Checks:
Volume of block is $(2)(2)(4) = 16 > 8.$ ✓
Volume of inscribed pyramid would be $(1/3)(2)(2)(4) = 5.33\ldots < 8.$ ✓
Riemann sum approximation: $V \approx 8 = 8$ ✓ ◀

All the previous examples involve the same basic reasoning, as summarized here.

TECHNIQUE: Volume of a Solid by Plane Slicing

- Cut the solid into flat slices, formed either by strips in a rotated region or by planes passed through the solid. Get disks, washers, or slabs whose volumes can be found in terms of the solid's cross section at sample point(s) (x, y).

- Use geometry to get dV in terms of the sample point(s).

- Use algebra to get dV in terms of one variable.

- Use calculus to add up all the dV's and take the limit (that is, integrate).

- Check your answer to make sure it's reasonable.

Problem Set 5-9

Quick Review

Q1. Integrate: $\int (x^2 + x + 1)\,dx$

Q2. Integrate: $\int x^{3/4}\,dx$

Q3. Differentiate: $y = x^{2/3}$

Q4. Integrate: $\int e^{-3x}\,dx$

Q5. Integrate: $\int \csc x \cot x\,dx$

Q6. Differentiate: $y = \ln 5x$

Q7. Write the definition of definite integral.

Q8. "…then there is a value of $x = c$ in (a, b) such that $f'(c) = [f(b) - f(a)]/(b - a)$" is the conclusion for the —?— theorem.

Q9. Sketch the graph of a function that is continuous at $x = 4$ but not differentiable there.

Q10. Sketch the graph of $y = 2^{-x}$.

---

1. *Paraboloid Volume Problem:* The region in Quadrant I under the graph of $y = 9 - x^2$ is rotated about the y-axis to form a solid paraboloid (Figure 5-9f). The figure shows

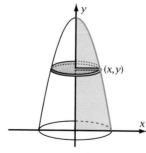

Figure 5-9f

a strip formed by slicing the region perpendicular to the axis of rotation and the disk formed as this strip rotates. Assume that x and y are in feet.

a. Write an equation for dV, the volume of the disk, in terms of the sample point (x, y) and the thickness of the disk, dy. Transform dV so it is in terms of y and dy.

b. The exact volume of the solid is the limit of the sum of the dV-values as dy approaches zero. That is, the volume equals the integral of dV between the proper limits of integration. Find the volume using the fundamental theorem.

c. Check your answer to part b by finding R_{100}, the midpoint Riemann sum with 100 increments.

d. Show that the volume of this paraboloid is exactly half the volume of the circumscribed cylinder.

2. *Cone Volume Problem:* A solid cone is formed by rotating about the y-axis the triangle in Quadrant I bounded by the line $y = 10 - 2x$ and the two axes (Figure 5-9g). The figure shows a strip formed by slicing the triangle perpendicular to the axis of rotation and the disk formed as this strip rotates. Assume that x and y are in centimeters. Find the volume of the cone by writing dV in terms of y and dy and then integrating between suitable limits. Show that your answer agrees with the formula for the volume of a cone you recall from geometry, namely, volume equals one-third the volume of the circumscribed cylinder.

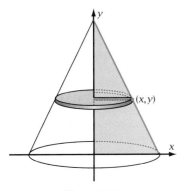

Figure 5-9g

3. *Volume Rotating About the x-axis Problem 1:* Figure 5-9h shows the solid formed by rotating about the x-axis the region under the graph of $y = 3e^{-0.2x}$ from $x = 0$ to $x = 5$.

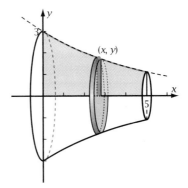

Figure 5-9h

a. Find dV, the volume of the disk generated by rotating the vertical strip of the region about the x-axis. Transform dV, if necessary, so it is in terms of x and dx.

b. Integrate to find the exact volume of the region. Then find a decimal approximation for this volume. Compare your algebraic answer to a numerical answer found by a suitable Riemann sum.

c. How do you decide whether to slice a region horizontally or vertically when you are finding the volume of a solid of revolution?

4. *Volume Rotating About the x-axis Problem 2:* Let R be the region under the graph of $y = 4x - x^2$ from $x = 1$ to $x = 4$. Find the exact volume of the solid formed by rotating R about the x-axis. Compare your algebraic answer to a numerical answer found by an appropriate Riemann sum.

5. Let R be the region under the graph of $y = x^{1.5}$ from $x = 1$ to $x = 9$. Find the exact volume of the solid generated by rotating R about the x-axis.

6. Let R be the region in Quadrant I bounded by the graphs of $y = \ln x$ and $y = 1$. Find the exact volume of the solid generated by rotating R about the y-axis.

7. Let R be the region bounded by the y-axis, the lines $y = 1$ and $y = 8$, and the curve $y = x^{3/4}$. Find the exact volume of the solid generated by rotating R about the y-axis.

8. *Washer Slices Problem:* Figure 5-9i shows the solid formed by rotating about the y-axis the region that is bounded by the graphs of $y = x^4$ and $y = 8x$. Find the exact volume of the solid using the fundamental theorem of calculus. Assume that x and y are in inches. Check your answer numerically by finding an appropriate Riemann sum.

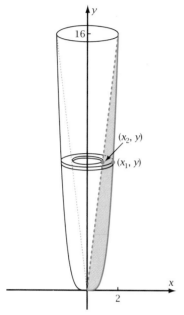

Figure 5-9i

9. *Exponential Horn Problem:* A horn for a public address system is to be made with the inside cross sections increasing exponentially with distance from the speaker. The horn will have the shape of the solid formed when the region bounded by the graphs of $y = e^{0.4x}$ and $y = x + 1$ from $x = 0$ to $x = 3$ is rotated about the x-axis (Figure 5-9j). Find the volume of the material used to make this speaker. Assume that x and y are in feet. Check your answer numerically by finding an appropriate Riemann sum.

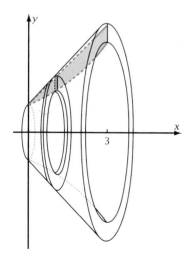

Figure 5-9j

10. The region between $x = 0$ and $x = 8$ and bounded by the graphs of $y = x^{1/3}$ and $y = 10e^{-0.1x}$ is rotated about the x-axis to form a solid. Find its exact volume.

11. The region bounded by the graphs of $y = 4 - x$ and $y = 4 - x^2$ is rotated about the y-axis to form a solid. Find its exact volume.

12. *Paraboloid Volume Formula Problem:* Prove that the volume of a paraboloid is always one-half the volume of the circumscribed cylinder. To do this, realize that any paraboloid is congruent to a paraboloid generated by rotating a parabola of the general equation $y = ax^2$ about the y-axis.

13. *Riemann Sum Limit Problem:* The region in Quadrant I bounded by the graphs of $y = 0.3x^{1.5}$, $x = 4$, and the x-axis, is rotated about the x-axis to form a solid.

    a. Find the volume of the solid by performing the appropriate calculus.

    b. Find three midpoint Riemann sums, M_{10}, M_{100}, and M_{1000}, for the volume of the solid. Show that these sums get closer to the exact value as the number of increments increases.

14. *Different Axis Problem I:* The region in Quadrant I under the graph of $y = 4 - x^2$ is rotated about the line $x = 3$ to form a solid (Figure 5-9k). Sketch the washer formed as the

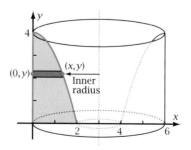

Figure 5-9k

Chapter 5: Definite and Indefinite Integrals

horizontal slice shown rotates. What are the inner and outer radii of the washer? Find the volume of the solid.

15. *Different Axis Problem II:* The region in Quadrant I under the graph of $y = 4 - x^2$ shown in Figure 5-9k is rotated about the line $y = -5$ to form another solid. Find its volume.

16. Let R be the region in the first quadrant bounded by the graphs of $y = x^{1/5}$ and $y = x^2$. Region R is the base of a solid whose cross sections perpendicular to the x-axis are squares, as shown in Figure 5-9l. Find the volume of this solid.

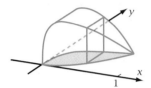

Figure 5-9l

17. *Pyramid Problem:* A pyramid has a square base 8 cm by 8 cm and altitude 15 cm (Figure 5-9m). Each cross section perpendicular to the y-axis is a square. On a sketch of the pyramid, draw a slab of thickness dy formed by slicing the solid perpendicular to the y-axis. Show a sample point (x, y) in the slab on the line segment connecting the vertex of the pyramid to the positive x-axis. Find dV, the volume of this slab, and use it to find the exact volume of the pyramid. Show that the volume is one-third the volume of the circumscribed rectangular box

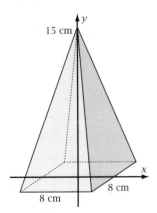

Figure 5-9m

with the same base and altitude. In what way is the volume of a pyramid analogous to the volume of a cone (see Problem 2)?

18. *Horn Problem:* Figure 5-9n shows a horn-shaped solid formed in such a way that a plane perpendicular to the x-axis cuts a circular cross section. Each circle has its center on the graph of $y = 0.2x^2$ and a radius ending on the graph of $y = 0.16x^2 + 1$. Find the volume of the solid if x and y are in centimeters.

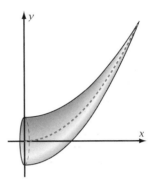

Figure 5-9n

19. *Triangular Cross Section Problem:* Region R under the graph of $y = x^{0.6}$ from $x = 0$ to $x = 4$ forms the back side of a solid (Figure 5-9o). Cross sections of the solid perpendicular to the x-axis are isosceles right triangles with their right angle on the x-axis.

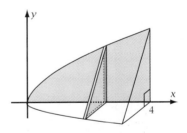

Figure 5-9o

a. The figure shows a slab formed by slicing the solid perpendicular to the x-axis. The width of the slab is dx, and the lengths of the two legs of the triangle are equal to y at the sample point. Recalling that the volume of the slab, dV, equals its area times its thickness, write an equation for dV and transform it so it is in terms of x and dx.

b. Find the volume of the solid algebraically using the fundamental theorem. Check your answer using an appropriate Riemann sum.

c. State what the volume of the solid would be if the cross sections were squares instead of right triangles.

20. Let R be the region bounded by the graphs of $y = e^x$, $y = 3$, and $x = 0$. Region R is the base of a solid. For this solid, each cross section perpendicular to the x-axis is a rectangle whose height is 4 times the length of its base. Find the volume of this solid.

21. Let R be the region in Quadrant I bounded by the graphs of $y = x^2$ and $y = 2 - x^2$. Region R is the base of a solid whose cross sections perpendicular to the x-axis are equilateral triangles. Find the volume of this solid.

22. Let R be the region in Quadrant I bounded by the graphs of $y = \ln x$ and $y = 1$. Region R is the base of a solid. For this solid, each cross section perpendicular to the y-axis is a rectangle whose height is half the length of its base. Find the volume of this solid.

23. *Wedge Problem:* Figure 5-9p shows a cylindrical log of radius 6 in. A wedge is cut from the log by sawing halfway through it perpendicular to its central axis, then sawing diagonally from a point 3 in. above the first cut.

a. Write the equation of the line that runs up the top surface of the wedge (in the xy-plane).

b. Write the equation of the circle in the xz-plane that forms the boundary for the bottom surface of the wedge.

c. Slice the wedge into slabs using planes perpendicular to the x-axis. Find an equation for the volume dV of a representative slab. Use your result to find the volume of the wedge.

24. *Generalized Wedge Problem:* Find a formula for the volume of a wedge of altitude h cut to the central axis of a log of radius r, as in Figure 5-9p, where one cut is perpendicular to the axis of the log.

25. *Cone Volume Formula Proof Problem:* Prove that the volume of a right circular cone of base radius r and altitude h is given by

$$V = \tfrac{1}{3}\pi r^2 h$$

You might find it helpful to draw the cone in a Cartesian coordinate system. If you put either the vertex of the cone or the center of its base at the origin, a slanted element of the cone will be a line segment whose equation you can find in terms of x and y.

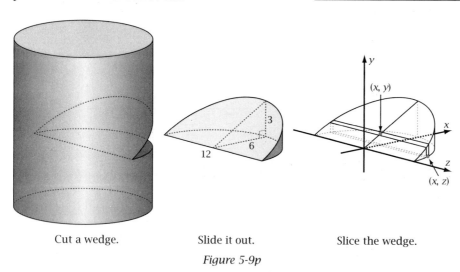

Cut a wedge.   Slide it out.   Slice the wedge.

Figure 5-9p

26. *Sphere Volume Problem:* Figure 5-9q shows a sphere of radius 10 cm.

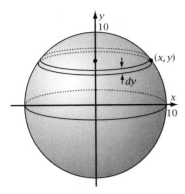

Figure 5-9q

a. Find the volume of the sphere using calculus.

b. Show that the answer you got in part a agrees with the formula you learned in geometry.

27. *General Volume of a Sphere Problem:* Using calculus, derive the formula

$$V = \frac{4}{3}\pi r^3$$

for the volume of a sphere of radius r.

28. *Volume of an Ellipsoid Problem:* Figure 5-9r shows the ellipsoid

$$\left(\frac{x}{a}\right)^2 + \left(\frac{y}{b}\right)^2 + \left(\frac{z}{c}\right)^2 = 1$$

Find the volume of the ellipsoid in terms of a, b, and c, the radii along the three axes. To get dV, show that every cross section perpendicular to the x-axis is an ellipse similar in proportions to the ellipse in the yz-plane. Use the fact that the area of an ellipse is given by πuv, where u and v are the radii along the major and minor axes.

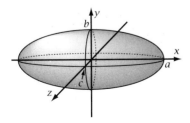

Figure 5-9r

29. *Highway Cut Problem:* Figure 5-9s shows a cut through a hill that is to be made for a new highway. Each vertical cross section of the cut perpendicular to the roadway is an isosceles trapezoid whose sides make angles of measure 52° with the horizontal. The roadway at the bottom of the trapezoid is 50 yd wide. The cut is 600 yd long from its beginning to its end.

x	y	x	y
0	0	330	47
30	5	360	49
60	9	390	51
90	15	420	46
120	28	450	39
150	37	480	37
180	42	510	31
210	42	540	22
240	48	570	14
270	53	600	0
300	52		

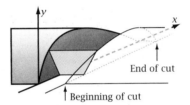

Highway cut through hill

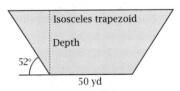

Vertical cross section

Figure 5-9s

You are to find how many cubic yards of earth must be removed in order to construct the cut. To estimate this volume, survey crews have measured the depth, y, of the cut at various distances, x, from the beginning. The table shows these depths. Write an integral involving y and dx that represents the volume of earth to be removed. Then evaluate the integral using a suitable numerical technique. If earth removal costs $12.00 per cubic yard, about how much will it cost to make the cut?

5-10 Definite Integrals Numerically by Grapher and by Simpson's Rule

Your grapher has a built-in function for evaluating definite integrals numerically. In this section you will see how to use this feature to evaluate integrals for which finding the antiderivative is difficult or impossible. You will also learn a numerical technique, called *Simpson's rule*, that gives more accurate answers than the trapezoidal rule or Riemann sums for definite integrals of functions that are specified only by tables of data.

OBJECTIVE   Use your grapher's built-in numerical integration feature or Simpson's rule to find approximate values of definite integrals.

Integrals by Grapher

The integral $\int_{0.3}^{4} \frac{\sin x}{x} \, dx$ is an example of the **sine-integral function**. You cannot evaluate it directly by the fundamental theorem because no elementary transcendental function is the antiderivative of $(\sin x)/(x)$. Example 1 shows you how to evaluate this integral numerically using the built-in integration feature of a typical grapher.

▶ **EXAMPLE 1**   Evaluate: $\int_{0.3}^{4} \frac{\sin x}{x} \, dx$

Solution   The numerical integration feature on a typical grapher might look like this.

$$\text{fnInt}((\sin x)/x, x, 0.3, 4)$$

The integrand appears first, followed by a comma. The name of the variable of integration appears next, x in this case. The third and fourth numbers are the lower and upper limits of integration, respectively. These four quantities are in the same order you would use in reading "the integral of $(\sin x)/x$, with respect to x, from 0.3 to 4."

When this command is executed, the result is 1.459699.... .

Some graphers have a numerical integrator that you can use directly from the graphing screen. The equation is stored on the y-menu. Then the graph is plotted, using a friendly window containing the two limits of integration. Upon accessing the integral command on a menu such as CALCULATE, the grapher allows you to trace to the lower limit, press ENTER, trace to the upper limit, and press ENTER.

The grapher will display the result and may shade the region on the graph, as shown in Figure 5-10a. The result, shown on the screen, is 1.4596991.... .   ◀

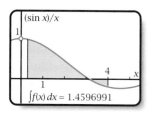

Figure 5-10a

Integrals by Simpson's Rule

The graph on the left in Figure 5-10b is the graph of function f approximated by line segments so you can approximate the area of the region by trapezoids. The graph on the right shows the same function approximated by segments of quadratic function graphs so you can approximate the area by parabolic regions. The parabolic segments are curved, so you would expect them to fit the graph more closely and thus give a more accurate estimate of the integral.

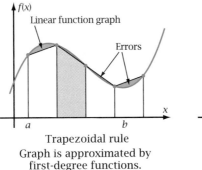
Trapezoidal rule
Graph is approximated by
first-degree functions.

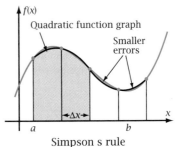
Simpson s rule
Graph is approximated by
second-degree functions.

Figure 5-10b

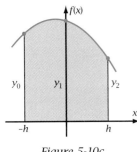

Figure 5-10c

It is possible to find the area of a parabolic region in terms of three y-values, one at the left, one at the middle, and one at the right of the region. Figure 5-10c shows the region under the parabola

$$y = ax^2 + bx + c$$

from $x = -h$ to $x = h$. The area of the region is

$$A = \int_{-h}^{h}(ax^2 + bx + c)\,dx$$

$$= \tfrac{2}{3}ah^3 + 2ch \qquad \text{Use the fundamental theorem and combine like terms.}$$

$$= \tfrac{1}{3}(h)(2ah^2 + 6c) \qquad \text{Factor (cleverly). The area is in terms of } h.$$

Set this result aside while you substitute $-h$, 0, and h for x in the equation $y = ax^2 + bx + c$.

$$y_0 = ah^2 - bh + c$$
$$y_1 = \qquad\quad\ + c$$
$$y_2 = ah^2 + bh + c$$

Adding the first and third equations gives

$$y_0 + y_2 = 2ah^2 + 2c$$

which is interesting because the quantity $(2ah^2 + 6c)$ appears in the equation for A by integration. Substituting $(y_0 + y_2)$ for the quantity $(2ah^2 + 2c)$ and substituting $4y_1$ for the remaining "$+ 4c$" gives

$$A = \tfrac{1}{3}h(y_0 + y_2 + 4y_1)$$
$$A = \tfrac{1}{3}h(y_0 + 4y_1 + y_2) \qquad \text{Area in terms of three } y\text{-values.}$$

Thus the area of a parabolic region may be found by summing the first y-value, four times the middle value, and the last value and then multiplying by one-third of the spacing between the x-values. Using this property, you can approximate the area of a region under a graph by parabolas, using only the y-values at regularly spaced points. Suppose integral I is given by

$$I = \int_a^b f(x)\, dx$$

Let $x_0, x_1, x_2, \ldots, x_n$ be values of x spaced Δx units apart, where $x_0 = a$ and $x_n = b$. Let $y_0, y_1, y_2, \ldots, y_n$ be the corresponding values of $f(x)$ (Figure 5-10d). It takes three points to determine a parabola, so group the strips in pairs and draw parabolic arcs, as shown in the figure. Integral I is equal to the area of the region under the graph, which is approximately equal to the sum of the areas of the parabolic regions.

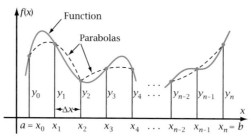

Figure 5-10d

Using the areas of the parabolic regions in terms of y, you can write

$$I \approx \tfrac{1}{3}h(y_0 + 4y_1 + y_2) + \tfrac{1}{3}h(y_2 + 4y_3 + y_4)$$
$$+ \tfrac{1}{3}h(y_4 + 4y_5 + y_6) + \cdots + \tfrac{1}{3}h(y_{n-2} + 4y_{n-1} + y_n)$$

Factoring out $(1/3)(h)$ from each term, combining like terms inside the parentheses, and replacing h with Δx gives the formula for **Simpson's rule**.

PROPERTY: Simpson's Rule

If the interval $[a, b]$ is divided into an even number, n, of subintervals of equal width Δx, then the integral of $f(x)$ from $x = a$ to $x = b$ is approximately equal to

$$\tfrac{1}{3}(\Delta x)(y_0 + 4y_1 + 2y_2 + 4y_3 + 2y_4 + \cdots + 2y_{n-2} + 4y_{n-1} + y_n)$$

Note: To use Simpson's rule, you must have an *even* number of increments, which means there must be an *odd* number of data points.

You apply Simpson's rule by multiplying each y-value by the appropriate weighting factor, adding them up, then multiplying the result by one-third of the spacing between x-values.

▶ **EXAMPLE 2**   The data in the table express y as a function of x. Use Simpson's rule with as many increments as possible to evaluate

$$\int_{30}^{70} y\, dx$$

x	y
30	83
35	79
40	74
45	68
50	61
55	49
60	37
65	31
70	33

Solution   Assume the function has a smooth graph that can be approximated by parabolic segments. For a small number of data points, such as those in this example, write the weighting factor by each point. Then multiply the factor by the y-value and add the results. The table shows a reasonable way to present your solution.

x	y	Factor	$y \times$ Factor
30	83	1	83
35	79	4	316
40	74	2	148
45	68	4	272
50	61	2	122
55	49	4	196
60	37	2	74
65	31	4	124
70	33	1	33
		Sum =	1368

Integral $\approx \frac{5}{3}(1368) = 2280$ ◀

You need only write out the y-values and the appropriate weighting factors. Accumulate the rest on your grapher. For larger numbers of data points, write a program that multiplies, accumulates, and performs the final calculation.

Problem Set 5-10

Quick Review  *5 min*

Q1. Evaluate: $\int (x^2 + 1)\, dx$

Q2. Evaluate: $\int_{-3}^{3} (x^2 + 1)\, dx$

Q3. Evaluate: $\int \sec^2 x\, dx$

Q4. Evaluate: $\dfrac{d}{dx}(\sec^2 x)$

Q5. Sketch a graph of velocity versus time for a moving object.

Q6. Show a strip of width dt on the graph you sketched in Problem Q5.

Q7. Show a sample point (t, v) on the graph within the strip in Problem Q6.

Q8. Write the displacement of the object in Problem Q5 for the time interval dt.

Q9. Write an expression for the exact displacement of the object in Problem Q5 from $t = a$ to $t = b$.

Q10. If the Riemann sum R_n is greater than or equal to $\int_a^b f(x)\,dx$ for a given function f, then R_n is called —?—.

    A. An upper sum

    B. A lower sum

    C. A midpoint sum

    D. A left sum

    E. A right sum

For Problems 1–4, evaluate the integral using your grapher's built-in integration feature.

1. $\int_{0.3}^{1.4} \cos x\,dx$

2. $\int_1^4 (x^2 - 3x + 5)\,dx$

3. $\int_0^3 2^x\,dx$

4. $\int_{0.1}^{1.4} \tan x\,dx$

5. Check your answer to Problem 1 algebraically using the fundamental theorem. How close to the exact value is the numerical value you found by your grapher?

6. Check your answer to Problem 2 algebraically using the fundamental theorem. How close to the exact value is the numerical value you found by your grapher?

7. *Volume Problem 1:* The region under one arch of the graph of $y = \sin x$ is rotated about the x-axis to form a football-shaped solid. Sketch the solid. Write an integral for the volume of the solid and evaluate it numerically using your grapher's built-in integration feature. Explain why you are not able at present to evaluate the integral algebraically.

8. *Volume Problem 2:* The region under the graph of $y = \ln x$ from $x = 2$ to $x = 5$ is rotated about the x-axis to form a solid. Sketch the solid. Write an integral for the volume of the solid and evaluate it numerically using your grapher's built-in integration feature. Explain why you are not able at present to evaluate the integral algebraically.

9. *Sine-Integral Function Problem:* The sine-integral function, Si x, is defined as

$$\text{Si } x = \int_0^x \frac{\sin t}{t}\,dt$$

  a. Use the grapher's integration feature to plot the graph of Si x from about $x = -20$ to $x = 20$, using a window with a y-range of -2 to 2. Sketch the result.

  b. The function $(\sin x)/(x)$ takes on the indeterminate form $0/0$ at $x = 0$. What limit does $(\sin x)/(x)$ seem to approach as x approaches zero?

  c. The National Bureau of Standards *Handbook of Mathematical Functions* lists Si 0.6 as 0.5881288096. How close to this number is your grapher's value?

  d. Does Si x seem to approach a limit as x approaches infinity? If so, what number do you conjecture this limit is? If not, explain why.

  e. Compare the graphs of Si x and $f(x) = (\sin x)/(x)$. How do you know from the graphs that one is the derivative of the other?

10. *Error Function Problem:* If you measure statistics on a large population, such as test scores or people's heights, the numbers are often **normally distributed**. That is, most of the data points are close to the mean, and fewer are far away from the mean. In statistics courses, you will learn that for a normal (or **Gaussian**) distribution the relative frequency with which a particular data point occurs is given by

$$y = \frac{1}{\sqrt{\pi}} e^{-t^2}$$

where t is the number of **standard deviations** the data point is from the mean and e is the base of the natural logarithm (approximately

2.718, as you will learn in Chapter 6). The area under the curve from $t = -x$ to $t = x$ is equal to the fraction of the population within x standard deviations of the mean (Figure 5-10e). This area is called the **error function of** x (erf x).

$$\text{erf } x = \frac{2}{\sqrt{\pi}} \int_0^x e^{-t^2} \, dt$$

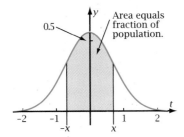

Figure 5-10e

a. What property of definite integrals explains the fact that the coefficient in the integral is $2/\sqrt{\pi}$ but the coefficient in the equation for y is only $1/\sqrt{\pi}$?

b. Plot the graph of erf x from $x = 0$ to $x = 2$. Use your grapher's built-in function e^x. Sketch the result.

c. If x is enough standard deviations from the mean, virtually all the data should be within that number of standard deviations. How do the values of erf x confirm this fact?

d. The National Bureau of Standards *Handbook of Mathematical Functions* lists erf 0.5 as 0.5204998778. How close to this number is your grapher's value?

e. Compare the graphs of erf x and $f(x) = (2/\sqrt{\pi})e^{-x^2}$. How do you know from the graphs that one is the derivative of the other?

11. *Velocity Problem:* People who sail ships at sea use dead reckoning to calculate the distance a ship has gone. (The term *dead reckoning* comes from "ded-reckoning," which is short for "deduced reckoning.") Suppose a ship is maneuvering by changing speed rapidly. The table shows its speed at 2-min intervals. (A knot—abbreviated kn—is a nautical mile per hour. A nautical mile is about 2000 yd.)

Time (min)	Speed (kn)
0	33
2	25
4	27
6	13
8	21
10	5
12	9

a. Use Simpson's rule to find the distance traveled in the 12-min time interval.

b. Find the distance using the trapezoidal rule.

c. Which result should be closer to the actual distance? Explain.

12. *Spleen Mass Problem:* Figure 5-10f is a CAT scan that shows a horizontal cross section of an eleven-year-old girl's body. The spleen is located at the right of the cross section (the girl's left). Doctors want to know whether the spleen has a mass that is within the normal range, 150 g to 200 g. Using the information contained in this and other CAT scans taken at 0.8-cm intervals up and down the girl's body, the doctors measure the spleen's cross-sectional areas.

D (cm)	A (cm²)	D (cm)	A (cm²)
0	6.8	4.8	38.4
0.8	6.8	5.6	33.9
1.6	20.1	6.4	15.8
2.4	25.3	7.2	6.1
3.2	29.5	8.0	2.3
4.0	34.6		

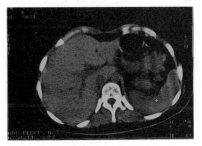

Figure 5-10f

a. The volume of an object equals its cross-sectional area times its thickness. If the cross section's area varies, the volume is the integral of the area with respect to the thickness. Use Simpson's rule to estimate the volume of the girl's spleen.

b. The density of the spleen is about the same as that of water, namely, 1.0 g/cm³. Is the girl's spleen within the normal mass range?

13. *Tensile Strength Test Problem:* The tensile strength of a metal bar is measured by the amount of force needed to stretch the bar until it breaks (Figure 5-10g). At first the force varies directly with the distance stretched. After a certain point, the bar begins to deform (to yield) and the force goes down. The amount of work done in breaking the bar is equal to the force times the displacement (the distance stretched). Suppose measurements yield the following data.

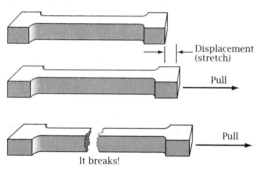

Figure 5-10g

Inches	Pounds	Inches	Pounds
0.00	0	0.30	290
0.05	120	0.35	280
0.10	240	0.40	270
0.15	360	0.45	270
0.20	370	0.50	190 (breaks)
0.25	330		

a. Draw a scatter plot showing force versus displacement.

b. Work equals force times displacement. If the force varies, you must find the work by integration. Use Simpson's rule to find the inch-pounds of work expended in breaking the bar.

14. *Heat Capacity Problem:* The amount of heat, C, needed to warm a mole (about 18 lb) of steam

by 1°F is called the *molal heat capacity* of the steam. The units of C are (Btu/lb mole)/°F, where a Btu (British thermal unit) is the amount of heat needed to warm a pound of water by 1°F. The amount of heat needed to warm the steam by x degrees would be Cx. However, as Figure 5-10h shows, C varies with temperature. Input for the graph comes from data published in Hougen and Watson's *Chemical Process Principles Charts* (John Wiley, 1946). Before this type of information was available by computer, engineers used such graphs for actual computations in the design of steam equipment. Read the value of C correct to the nearest 0.02 unit (last digit 0, 2, 4, 6, or 8) for each 500° from 500° through 4500°. Then use Simpson's rule to find the number of Btu needed to warm a pound mole of steam from 500° to 4500°.

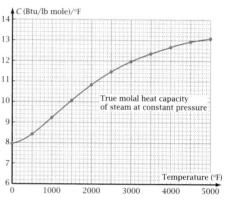

Figure 5-10h

15. *Simpson's Rule Versus Trapezoidal Rule Problem:* Simpson's rule is expected to give more accurate results than the trapezoidal rule. Let I be the integral

$$I = \int_0^\pi \sin x \, dx$$

a. Explain why you should expect Simpson's rule to give a more accurate result than the trapezoidal rule.

b. Find S_4 and T_4, the Simpson's rule and trapezoidal rule approximations, respectively, for I. Then find the exact value of I algebraically using the fundamental theorem. Is S_4 really closer than T_4 to the exact value?

16. *Simpson's Rule Program:* Write two programs to evaluate integrals using Simpson's rule. One program should read y-values from a table, and the other should calculate y-values from an equation in the Y= menu. The input should be a and b, the limits of integration, and n, the number of increments (an even number). A time-efficient way to write the program is to have the calculator compute two intervals at a time, using the starting y-value, four times the middle y-value, and the ending y-value for each pair of increments. Test the program that uses data on Example 2. Test the program that uses an equation on integral I in Problem 15, using ten increments. You should get 2.00010951... for the result.

17. *Spleen Mass Problem, Revisited:* Use your Simpson's rule program from Problem 16 to find the mass of the girl's spleen from the data in Problem 12.

18. *Error Function Problem, Revisited:* Use your Simpson's rule program from Problem 16 to find the approximations S_{50} and S_{100} for erf 0.5 (see Problem 10). Are both approximations close to the tabulated value, erf $0.5 = 0.5204998778$? Is S_{100} much closer than S_{50} to this value?

19. *Integral of the Reciprocal Function Problem:* Let $f(x) = \int_1^x \frac{1}{t} dt$.

    a. Sketch the graph of the integrand, $y = 1/t$. Sketch the region represented by the integral. Explain why the area of this region is a function of x.

    b. You can write the **reciprocal function**, $y = 1/t$, as $y = t^{-1}$. Show that you can't find the indefinite integral $\int t^{-1} dt$ using the power formula by showing what happens when you try this.

    c. Plot the graph of function f using your grapher's numerical integration feature. To avoid division by zero, use a window with an x-range of 0.001 to 10 and a y-range of -3 to 3. Sketch the graph. What familiar function does the graph resemble? Why is $f(x)$ negative for values of x less than 1, even though $y = 1/t$ is positive?

    d. Evaluate $f(2), f(3)$, and $f(6)$. What relationship seems to exist between the three numbers? What kind of function has this property?

5-11   Chapter Review and Test

In previous chapters you learned about definite integrals and antiderivatives. In this chapter you learned about the symbols and formal definitions for these concepts. The mean value theorem and Rolle's theorem led to the fundamental theorem of calculus, which allows you to find definite integrals exactly, by algebra, rather than simply numerically or graphically. This connection between the antiderivative and the definite integral explains why the word *integral* is also used for the antiderivative. You learned a method for applying definite integrals to applications such as the volume of a solid by slicing a region under a graph, picking a sample point, forming a Riemann sum, and then integrating. Finally, you learned two more ways to evaluate definite integrals numerically: your grapher's built-in integration feature and Simpson's rule.

Review Problems

R0. Update your journal with what you've learned. Include such things as

- The one most important thing you have learned in your study of Chapter 5
- Which boxes you have been working on in the "define, understand, do, apply" table
- The difference between definite integral and indefinite integral
- The fundamental theorem of calculus, how it is proved, and what it is useful for
- Application of definite integrals to real-world problems involving a product of variables
- Simpson's rule and your grapher's numerical integration feature
- Any ideas about calculus still unclear to you

R1. *Displacement from Velocity Problem:* A car traveling at 10 m/s starts accelerating at time $t = 0$ s. Its velocity, in meters per second, (Figure 5-11a) is given by the function

$$v(t) = 10 + 6x^{0.5}$$

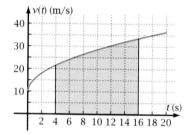

Figure 5-11a

a. The distance the car travels from $t = 4$ to $t = 16$ is the definite integral of $v(t)$ with respect to t from $t = 4$ to $t = 16$. Estimate this integral by dividing the region into three vertical strips and finding T_3, the trapezoidal rule sum. Show the procedure you use to calculate this sum. Explain why T_3 underestimates the integral.

b. Estimate the integral again, this time summing the areas of three rectangles to approximate the areas of the strips, and using the values of $v(t)$ at the midpoints of

the three intervals as the altitudes of the rectangles (that is, $v(6)$, $v(10)$, and $v(14)$). Is this *Riemann sum* close to the trapezoidal rule sum?

c. Use your program to find T_{50} and T_{100} for the integral. Based on the limit these sums seem to be approaching, make a conjecture about the exact value of the integral.

d. Let $g(t)$ be an indefinite integral (antiderivative) of $v(t)$. Evaluate the quantity $g(16) - g(4)$. What interesting thing do you notice about your result?

R2. a. For $f(x) = \sin \pi x$, find an equation for the linear function, $\ell(x)$, that best fits $f(x)$ at the point where $x = 1$. Plot $y_1 = f(x)$ and $y_2 = \ell(x)$ on the same screen and sketch the result. Then zoom in on the point $(1, \sin \pi)$. How does what you observe confirm that the graph of f is *locally linear* at that point? What error is introduced in using the linear function to estimate $f(1.1)$? To estimate $f(1.001)$?

b. Find dy.

i. $y = \csc^5 2x$

ii. $y = \dfrac{x^5}{5} - \dfrac{x^{-3}}{3}$

iii. $y = (7 - 3x)^4$

iv. $y = 5e^{-0.3x}$

v. $y = \ln (2x)^4$

c. Find the general equation for the antiderivative, y.

i. $dy = \sec x \tan x \, dx$

ii. $dy = (3x + 7)^5 \, dx$

iii. $dy = 5 \, dx$

iv. $dy = 0.2e^{-0.2x} \, dx$

v. $dy = 6^x \, dx$

d. Let $y = (2x + 5)^{1/2}$.

i. Find dy in terms of dx.

ii. Find dy if $x = 10$ and $dx = 0.3$.

iii. Find Δy if $x = 10$ and $dx = 0.3$.

iv. Show that dy is approximately equal to Δy.

R3. a. Write the definition of indefinite integral.

    b. Evaluate the indefinite integral.

       i. $\int 12x^{2/3}\,dx$

      ii. $\int \sin^6 x \cos x\,dx$

     iii. $\int (x^2 - 8x + 3)\,dx$

     iv. $\int 12e^{3x}\,dx$

      v. $\int 7^x\,dx$

R4. a. Write the definition of integrability.

    b. Write the definition of definite integral.

    c. Use the methods specified in parts i–iv to evaluate approximately $\int_{0.2}^{1.4} \sec x\,dx$.

       i. The upper Riemann sum with six increments

      ii. The lower Riemann sum with six increments

     iii. The midpoint Riemann sum with six increments

     iv. The trapezoidal rule with six increments

    d. Draw a diagram that shows the meaning of each sum in part c.

    e. Draw a sketch that shows how a sample point for an upper Riemann sum for some other function could be somewhere other than at the left end, right end, or midpoint of a given subinterval.

R5. a. What is the difference between the hypothesis of a theorem and the conclusion?

    b. A long pendulum swings slowly back and forth. Its displacement, $d(t)$ meters, from one wall of the museum in which it hangs is given by

$$d(t) = 20 + 3 \sin \tfrac{\pi}{4} t$$

    where t is time in seconds. The average velocity from $t = 0$ to $t = 2$ is equal to $[d(2) - d(0)]$ meters divided by 2 s. Find the time in the interval $(0, 2)$ at which the instantaneous velocity equals the average velocity.

    c. Find an interval $[a, b]$ on which the hypotheses of Rolle's theorem are satisfied for the function $g(x) = x^{4/3} - 4x^{1/3}$. Then find the value of x in (a, b) at which the conclusion of the theorem is true. Show that g is not differentiable at one of the two endpoints of $[a, b]$, and explain why this fact

is consistent with the hypotheses of the theorem.

    d. Explain why the piecewise function f in the top graph in Figure 5-11b is continuous on the interval $[2, 7]$, even though there are discontinuities at $x = 2$ and $x = 7$. Explain why the piecewise function g in the bottom graph in Figure 5-11b is *not* continuous on the interval $[2, 7]$. Sketch the two functions and show why the conclusion of the mean value theorem is true for function f on $[2, 7]$ but *not* true for function g, even though the two graphs look almost the same.

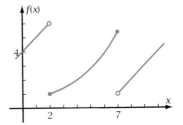

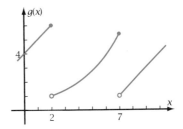

Figure 5-11b

    e. In the proof of the mean value theorem for $f(x)$ on $[a, b]$, two new functions were defined, g and h. Tell how they were defined and how, as a consequence, you can use Rolle's theorem as a lemma to prove the mean value theorem.

    f. Let $f(x) = 3 + 5 \cos 8\pi x$. Plot the graph on your grapher. Quickly find all the points in the interval $(0, 1)$ at which the conclusion of Rolle's theorem is true.

    g. What can you conclude about the values of $r(x)$ and $s(x)$ if $r'(x) = s'(x)$ for all values of x in an interval?

R6. a. Find a Riemann sum, R_3, for $\int_1^4 x^{1.5}\,dx$ by choosing the sample points in a special way: Let $g(x) = \int x^{1.5}\,dx$, the indefinite integral. Let c_1, c_2, and c_3 be the values of x in the three subintervals $(1, 2)$, $(2, 3)$, and $(3, 4)$ at which the conclusion of the mean value theorem is true for the integral, function g. Use these numbers as sample points for R_3 for the given integral. If the sample points are picked in this way, what can you conclude about the answer you get for the Riemann sum?

b. Evaluate $\int_{-1}^{3}(10 - x^2)\,dx$ exactly using the fundamental theorem of calculus.

c. Check your answer to part b by using the trapezoidal rule with a large number of increments. Show that the value is close to the exact answer.

d. Find the midpoint Riemann sums M_{10}, M_{100}, and M_{1000} for the integral in part b. Show that the greater the number of terms in the Riemann sum, the closer the sum gets to the actual value of the integral.

R7. a. Evaluate these integrals using the fundamental theorem of calculus. Check your results by using a Riemann sum or the trapezoidal rule.

    i. $\int_{1}^{5} x^{-2}\,dx$

    ii. $\int_{3}^{4}(x^2 + 3)^5(x\,dx)$

    iii. $\int_{0}^{\pi}(\sin x - 5)\,dx$

    iv. $\int_{0}^{\ln 5} 4e^{2x}\,dx$

    v. $\int_{1}^{4} 3^x\,dx$

b. Sketch a graph of the integral in part a.iii to show why you get a negative answer.

c. Evaluate $\int_{-10}^{10}(4\sin x + 6x^7 - 8x^3 + 4)\,dx$ quickly.

d. Sketch a graph to illustrate the property $\int_{a}^{b}f(x)\,dx = \int_{a}^{c}f(x)\,dx + \int_{c}^{b}f(x)\,dx$.

R8. a. *Displacement Problem:* Suppose that as a rocket rises from its launch pad its upward velocity, v, in feet per second, is given as a function of time, t, in seconds, by

$$v = 150t^{0.5}$$

Find the displacement, y, of the rocket from $t = 0$ to $t = 9$ by sketching the graph of v, showing a representative strip of width dt and a sample point (t, v), writing the differential dy, and integrating. Confirm the

property of the sum of two integrals with the same integrand by finding the displacement from $t = 0$ to $t = 4$ and from $t = 4$ to $t = 9$.

b. Figure 5-11c shows the region in Quadrant I bounded below by the graph of $y = \ln x$ and above by the line $y = \ln 4$. Slice a narrow strip in the region parallel to the x-axis. Find the differential of area, dA, in terms of a sample point (x, y) on the graph of $y = \ln x$. Transform dA so it is in terms of y and dy and then integrate algebraically to find the area of the region exactly, in terms of e and $\ln$ if necessary.

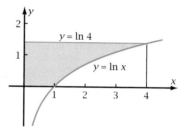

Figure 5-11c

R9. a. The region under the graph of $y = e^{0.2x}$ from $x = 0$ to $x = 4$ is rotated about the x-axis to form a solid. Find its volume.

b. The region in Quadrant I bounded by the graphs of $y = x^{0.25}$ and $y = x$ is rotated about the y-axis to form a solid. Find its volume.

c. *Oblique Cone Problem:* Figure 5-11d shows an oblique circular cone with base at $y = 0$. Each cross section perpendicular to the y-axis is a circle, with diameter extending from the graph of $y = x + 2$ to the graph of $y = 3x - 6$. Find the volume of the cone. Is its volume larger or smaller than the volume of a right circular cone of the same altitude, 6, and base radius, 2?

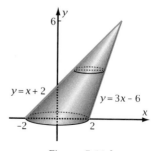

Figure 5-11d

R10. a. You can't yet evaluate $\int_1^{10} \log x \, dx$ using the fundamental theorem because you haven't learned an algebraic way to find the indefinite integral. Use your grapher's built-in integration feature to find the value of this integral. Sketch the graph of the integrand and use the graph to explain why your answer is reasonable.

b. *Worth of Land Problem:* Figure 5-11e shows a tract of land bordered by a highway along the y-axis, a dirt road along the x-axis, and a river whose path is given by the equation $y = 4 - 0.2x^2$, where x and y are in hundreds of meters.

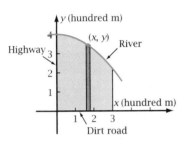

Figure 5-11e

The tract is 300 m deep along the dirt road. The value of the land is constant in any strip parallel to the highway and increases as you move away from the highway, with the value given by $v = 1000 + 50x$ dollars per 10,000 m^2 at the sample point (x, y). Find dW, the worth of a strip. Write an integral that equals the worth of the entire tract, and find the worth using your grapher's built-in numerical integration feature.

c. *Displacement by Simpson's Rule Problem:* The table shows the velocity, $v(t)$, in meters per minute, of a moving object. Write an integral to represent the displacement of the object from its position at time $t = 3$ min to its position at time $t = 5$ min. Show the steps involved in evaluating the integral by Simpson's rule. You may check your answer by your grapher, using the Simpson's rule program. Explain why you would expect Simpson's rule to give a more accurate answer for the displacement than would the trapezoidal rule.

t	v(t)	t	v(t)
3.0	29	4.2	28
3.2	41	4.4	20
3.4	50	4.6	11
3.6	51	4.8	25
3.8	44	5.0	39
4.0	33		

Concept Problems

C1. In this problem you will investigate the function $f(x) = \int_0^x \dfrac{1}{t^2 + 1} \, dt$.

a. On your grapher, plot $y_1 = f(x)$. Use a window with an x-range of −6 to 6 and with equal scales on the two axes. Sketch the result.

b. On the same screen, plot the lines $y_2 = \pi/2$ and $y_3 = -\pi/2$ and the curve $y_4 = \tan x$. For y_4, use Boolean variables to restrict the domain of x to the open interval $(-\pi/2, \pi/2)$. How are the two lines related to the graph of y_1? How is the graph of y_4 related to the graph of y_1?

c. Based on your graphs, make a conjecture about what function $f(x)$ really is. What algebraic evidence indicates that your conjecture is correct?

C2. *Mean Value Theorem for Quadratic Functions:* Let f be the general quadratic function, $f(x) = ax^2 + bx + c$. Given any interval $[d, e]$, show that the point $x = k$ in (d, e) at which the conclusion of the mean value theorem is true is the midpoint of that interval.

C3. *Sum of the Squares and Cubes Problem:* In Section 5-4, Problem 14, you learned that you can find the exact value of $\int_0^3 x^2 dx$ without calculus using the sum of the squares of the whole numbers,

$$S(n) = 0^2 + 1^2 + 2^2 + 3^2 + \cdots + n^2$$

a. $S(n)$ turns out to be a cubic function of n. That is,

$$S(n) = an^3 + bn^2 + cn + d$$

where $a, b, c,$ and d stand for constants. By summing the integers, find $S(0), S(1), S(2),$ and $S(3)$. Use the results to evaluate these four constants. Factor the resulting cubic equation as much as possible.

b. Show that your equation from part a gives the correct values for $S(4)$ and $S(5)$. Then use your equation to calculate $S(1000)$ without actually summing the integers.

c. Prove by mathematical induction that your equation for $S(n)$ works for any positive integer n.

d. The sum of the cubes of the whole numbers,

$$S(n) = 0^3 + 1^3 + 3^3 + 3^3 + \cdots + n^3$$

is a *quartic* (fourth-degree) function of n. Use the technique of part a to find an equation for $S(n)$. Validate your answer by showing that the equation gives the correct result for $S(5)$ and $S(6)$.

C4. *Radio Wave Integral Problem:* AM radio signals are transmitted by sending a high-frequency wave whose amplitude varies in the pattern of the sound being carried. (The amplitude is modulated, hence the abbreviation AM.) Figure 5-11f shows

$$f(x) = 4 \sin x \sin 10x$$

where $f(x)$ is the strength of the signal at any instant, x, in time. The sound represented by $y = 4 \sin x$, with a frequency of 1 cycle per 2π x-units, is being "carried" by the signal $y = \sin 10x$, with a frequency of 10 cycles per 2π x-units.

a. Using the fundamental theorem of calculus, evaluate the integral

$$\int_0^4 4 \sin x \sin 10x \, dx$$

To do this, you can first transform the product of sinusoids to a sum using the trigonometric property

$$2 \sin A \sin B = -\cos(A + B) + \cos(A - B)$$

Show that the integral is exactly equal to zero. Then explain how this fact relates graphically to the shaded regions in Figure 5-11f.

b. Show that the integral

$$\int_0^4 4 \sin x \sin nx \, dx$$

is equal to zero for any integer $n > 1$. Record in your journal that this kind of integral is related to *Fourier series,* which you will study later in your mathematical career.

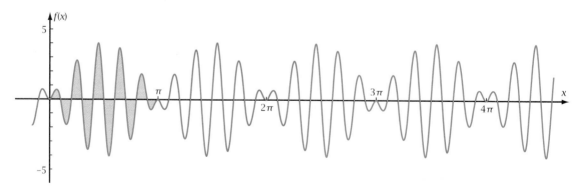

Figure 5-11f

C5. *Riemann Sums with Unequal Increments:*
Figure 5-11g shows the graph of $f(x) = 1.2^x$.
The interval $[1, 9]$ is partitioned into
subintervals of unequal width. The largest

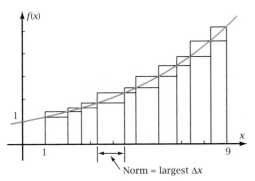

Figure 5-11g

value of Δx is called the **norm** of the partition,
written $\|P\|$. Upper and lower sums, U_n and L_n,
are shown.

a. Why is the difference $U_n - L_n$ no greater
than $\|P\|(1.2^9 - 1.2^1)$?

b. Suppose the number of subintervals is
allowed to approach infinity in such a way
that the limit of $\|P\|$ is zero. Use the
observation you made in part a to conclude
that f is integrable on $[1, 9]$.

c. Prove that $g(x) = 1/x$ is integrable on the
interval $[1, 4]$.

d. Could the reasoning of this problem be used
to prove that the function $h(x) = \sin x$ is
integrable on the interval $[0, 3]$? State why or
why not.

Chapter Test

PART 1: No calculators allowed (T1–T13)

T1. Write the definition of indefinite integral.

T2. Write the definition of definite integral.

T3. State the fundamental theorem of calculus.

T4. Figure 5-11h shows the graph of function f.
Explain why the hypotheses of the mean value
theorem are satisfied for f on the interval
$[3, 8]$, even though the graph has a cusp at
$x = 3$ and a vertical tangent at $x = 8$.

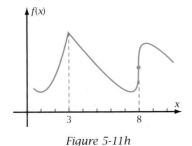

Figure 5-11h

T5. Show that you understand the conclusion of
the mean value theorem by appropriately
sketching on a copy of Figure 5-11h.

T6. Sketch a graph clearly showing that you
understand the hypotheses and the conclusion
of Rolle's theorem.

T7. Show that you know what an upper Riemann
sum is by sketching on a copy of Figure 5-11i
the sum U_5 for the integral from 3 to 8 of
$f(x)\,dx$.

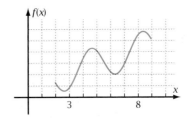

Figure 5-11i

T8. Find dy: $y = e^{\sin x}$

T9. Find $\int 0.1^x dx$.

T10. Find $\int (4x^3 + 13)^5 (x^2\,dx)$.

T11. Find $\int_1^4 x^2 dx$.

T12. Find $\int_2^{-2} (12x^3 + 10x^2)dx$.

T13. Figure 5-11j shows the solid formed by rotating about the *y*-axis the region in Quadrant I bounded by the graphs of two functions, y_1 and y_2. Write an integral for the volume of the solid if the region is sliced perpendicular to the axis of rotation.

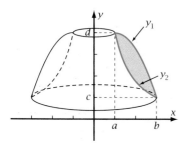

Figure 5-11j

PART 2: Graphing calculators allowed (T14–T18)

T14. Plot the graph of $y = x^3$. Write the equation of the linear function that best fits the graph at $x = 1$, and plot it on the same screen. Zoom in on the point on the graph where $x = 1$. What words describe what you observe as you zoom closer and closer to this point?

T15. For $\int_0^3 12e^{0.25x}\,dx$

a. Find the exact value of the integral using the fundamental theorem. Evaluate the answer in decimal form and store it in your grapher.

b. Find the midpoint Riemann sum M_{50}, the trapezoidal rule sum T_{50}, and the Simpson's rule sum S_{50}. Show that the error in M_{50} is only half (in absolute value) the error in T_{50}. Show that the error in S_{50} is much smaller than the error in either of the other sums.

T16. Figure 5-11k shows a solid that is to be carved out of a block of wood. The back vertical face of the solid is the region in the *xy*-plane under the graph of $y = \cos x$. Each cross section of the solid perpendicular to the *x*-axis is an isosceles right triangle with legs of length *y*, in feet.

a. On a copy of the figure, show a slab of thickness dx with the triangle as its cross section and a sample point (x, y) on the graph, in the slab.

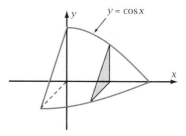

Figure 5-11k

b. Find dV, the volume of the slab. Transform it so it is in terms of *x* and dx. Then write an integral for the volume of the entire solid.

c. Explain why you cannot evaluate the integral in part b algebraically using the integration techniques you have learned so far. Find the volume numerically, using the built-in integration feature of your grapher.

T17. Let $f(x) = 0.3x^2$ and let $g(x) = \int f(x)\,dx$.

a. Integrate to find an equation for $g(x)$. Set the constant of integration, *C*, equal to zero.

b. On graph paper, plot graphs of *f* and *g* in the domain $[0, 4]$. Plot the graphs on different sets of axes, but use the same scale for both graphs.

c. Draw the secant line on the graph of *g* from $(1, g(1))$ to $(4, g(4))$. Find its slope. Then find the point $x = c$ in the interval $(1, 4)$ at which the conclusion of the mean value theorem is true for function *g*. Draw a line through the point $(c, g(c))$ parallel to the secant line and thus show that it is tangent to the graph of *g* at that point.

d. Using the value of *c* you calculated in part c, find the point $(c, f(c))$ on the graph of *f*. Draw a rectangle whose base is the segment from $x = 1$ to $x = 4$ and whose altitude is $f(c)$. How does the graph show that the area of this rectangle is equal to the value of the definite integral of $f(x)\,dx$ from $x = 1$ to $x = 4$?

T18. What did you learn as a result of this test that you did not know before?

The Calculus of Exponential and Logarithmic Functions

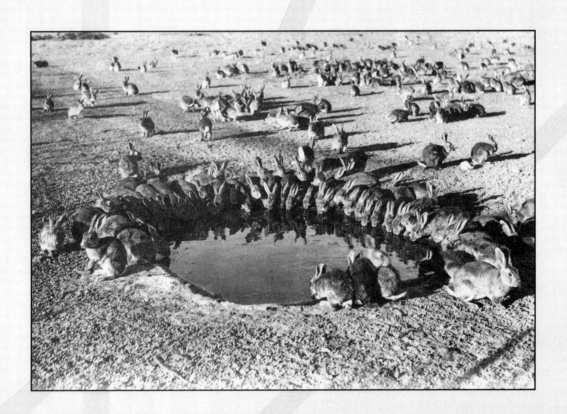

Rabbits introduced into Australia in the mid-1800s had no natural enemies, and their population grew unchecked. The rate of increase was proportional to the number of rabbits—the more there were, the faster the population grew. You can model this kind of population growth by exponential functions, with the help of their inverses, the logarithmic functions. As you have learned, these functions, which have variable exponents, behave differently from power functions, which have constant exponents.

Mathematical Overview

Exponential functions, in which the variable is an exponent, model population growth. In this chapter you will see how these functions arise naturally from situations where the rate of change of a dependent variable is directly proportional to the value of that variable. You will learn this information in four ways.

Graphically

The icon at the top of each even-numbered page of this chapter shows the natural logarithm function and its inverse, the base-*e* exponential function.

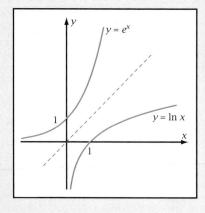

Numerically

x	e^x	$\ln x$
1	2.718...	0
2	7.389...	0.693...
3	20.085...	1.098...
⋮	⋮	⋮

Algebraically

$\ln x = \int_1^x \frac{1}{t}\, dt$, the definition of natural logarithm

Verbally

I didn't realize that you can define a function to be the definite integral of another function. The advantage of doing this is that you know immediately the derivative of that new function. I was surprised that the integral of 1/t from t = 1 to t = x is the natural logarithm function. I need to remember that derivatives and integrals of exponential functions are not the same as for powers.

6-1 Integral of the Reciprocal Function: A Population Growth Problem

You have learned how to differentiate and integrate power functions and exponential functions. However, there is one power function, the reciprocal function, for which the power rule doesn't work.

$$\int \frac{1}{x}\,dx = \int x^{-1}\,dx \text{ by the power rule would be } \tfrac{1}{0}x^0 + C, \text{which is undefined.}$$

In this section you will develop some background to see how to evaluate this integral.

> **OBJECTIVE** Work a problem in which the integral of the reciprocal function arises naturally.

Exploratory Problem Set 6-1

Population Problem: If the population, P, of a community is growing at an instantaneous rate of 5% of the population per year, then the derivative of the population is given by the **differential equation**

$$\frac{dP}{dt} = 0.05P$$

Suppose that the population, P, is 1000 at time $t = 0$ years, and grows to $P = N$ at the end of 10 years, where you must find the value of N. By **separating the variables** and then integrating, you get

$$\frac{1}{P}\,dP = 0.05\,dt \qquad \text{Get } P \text{ on one side and } t \text{ on the other side.}$$

$$\int_{1000}^{N} \frac{1}{P}\,dP = \int_{0}^{10} 0.05\,dt \qquad \begin{array}{l}P = 1000 \text{ when } t = 0, \\ \text{and } P = N \text{ when } t = 10.\end{array}$$

1. Explain why you cannot find the integral of the reciprocal function on the left using the power rule for integrals. Evaluate the integral numerically for $N = 1000, 1500, 2000, 2500, 500,$ and 100. Use these values to sketch the graph of the integral versus N. What kind of function has a graph of this shape?

2. Show that the integral on the right equals 0.5, which is between two of the values of the integral on the left that you found in Problem 1. Use the solver feature of your grapher to find the value of N that makes the integral equal exactly 0.5.

3. The value of N in Problem 2 is $P(10)$, the population at 10 years. Find $P(20)$ and $P(0)$. Use these three points to sketch what you think the graph of P looks like. What kind of function has a graph of this shape?

4. Find $\ln N - \ln 1000$, where N is the value you found in Problem 2. What interesting thing do you notice about the answer?

6-2 Antiderivative of the Reciprocal Function and Another Form of the Fundamental Theorem

In the last section you saw that you cannot find the integral of the reciprocal function by using the power rule for integrals. The power rule leads to division by zero.

$$\int \frac{1}{x}\,dx = \int x^{-1}\,dx \qquad \text{would be} \qquad \tfrac{1}{0}x^0 + C$$

In this section you will see that this integral turns out to be the natural logarithm function. To show this, you will learn the *derivative of an integral* form of the fundamental theorem of calculus. As a result, you will find that there is another definition of the natural logarithm function besides the inverse of the exponential function. You will also see some theoretical advantages of this new definition.

OBJECTIVE   Find out how to differentiate a function defined as a definite integral between a fixed lower limit and a variable upper limit, and use the technique to integrate the reciprocal function.

Fundamental Theorem—Derivative of an Integral Form

Suppose a definite integral has a variable for its upper limit of integration, for instance

$$\int_1^x \sin t\,dt$$

Evaluating the integral gives

$$-\cos t\,\Big|_1^x = -\cos x + \cos 1$$
$$= -\cos x + 0.5403\ldots$$

The answer is an expression involving the upper limit of integration, x. Thus, the integral is a function of x. Figure 6-2a shows what is happening. The integral equals the area of the region under the graph of $y = \sin t$ from $t = 1$ to $t = x$. Clearly (as mathematicians like to say!) the area is a function of the value you pick for x. Let $g(x)$ stand for this function.

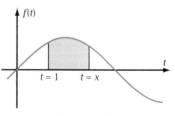

Figure 6-2a

The interesting thing is what results when you find the derivative of g.

$$g(x) = \int_1^x \sin t\,dt = -\cos x + 0.5403\ldots$$
$$\therefore\ g'(x) = \sin x$$

The answer is the integrand, evaluated at the upper limit of integration, x. Here's why this result happens. Because $g(x)$ equals the area of the region, $g'(x)$ is the rate of change of this area. Its value, $\sin x$, is equal to the height of the region at the boundary where the change is taking place. Figure 6-2b shows this situation for three values of x.

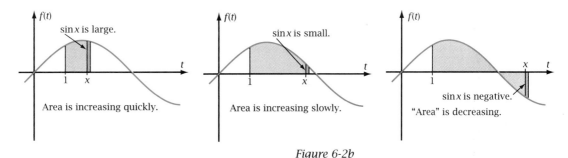

Figure 6-2b

The taller the region is, the faster its area increases as x changes. Think of painting horizontally with a brush whose width is the altitude of the region at x. The wider the paint brush, the faster the region gets painted for each inch the brush moves. If $\sin x$ is negative, the "area" decreases because the integrand is negative. This property is called the **fundamental theorem of calculus: derivative of an integral form**. It is stated here.

PROPERTY: *Fundamental Theorem of Calculus—Derivative of an Integral Form*

If $g(x) = \int_a^x f(t)\,dt$, where a stands for a constant, and f is continuous in the neighborhood of a, then $g'(x) = f(x)$.

▶ **EXAMPLE 1**   Prove the derivative of an integral form of the fundamental theorem of calculus.

Proof   You can prove this form of the fundamental theorem algebraically, using as a lemma the form of the theorem you already know.

Let $g(x) = \int_a^x f(t)\,dt$.

Let h be an antiderivative of f. That is, let $h(x) = \int f(x)\,dx$.

$\therefore g(x) = h(x) - h(a)$ Fundamental theorem (first form).

$\therefore g'(x) = h'(x) - 0$ Derivative of a constant is zero.

$\therefore g'(x) = f(x)$, Q.E.D.   Definition of indefinite integral. ◀

New Definition and Derivative of ln x

Let g be the function defined by the definite integral

$$g(x) = \int_1^x \frac{1}{t}\,dt$$

Figure 6-2c shows the integrand as a function of t, and the region whose area equals $g(x)$. Figure 6-2d shows the graph of $y_1 = g(x)$ as a function of x. The graph is zero at $x = 1$ because the region degenerates to a line segment. The graph increases as x increases because $1/t$ in Figure 6-2c is positive, but it increases at a decreasing rate because $1/t$ is getting closer to zero. The graph of g is negative for $x < 1$ because the values of dt in the Riemann sums are negative.

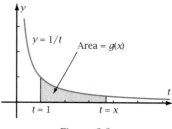

Figure 6-2c   *Figure 6-2d*

The graph of g looks remarkably like a logarithmic function. In fact, if you plot $y_2 = \ln x$ on the same screen, the graphs appear to be identical! This fact serves as a starting point for a way of developing logarithmic and exponential functions that has some theoretical advantages over what you have done before.

DEFINITION: The Natural Logarithm Function

$$\ln x = \int_1^x \frac{1}{t}\, dt$$

where x is a positive number.

From this definition of natural logarithm, the derivative follows immediately from the new form of the fundamental theorem.

$$\ln x = \int_1^x \frac{1}{t}\, dt$$

$$\therefore \ln' x = \frac{1}{x} \qquad \text{Evaluate the integrand, } 1/t, \text{ at the upper limit of integration, } x.$$

PROPERTY: Derivative of ln x

$$\frac{d}{dx}(\ln x) = \frac{1}{x}$$

To find derivatives involving ln, all you have to remember is that the derivative of $\ln x$ is the reciprocal function, $1/x$. You must also observe such rules as the chain rule and the product, power, and quotient rules.

▶ **EXAMPLE 2**   If $y = \ln(7x^5)$, find dy/dx.

Solution

$$y = \ln(7x^5)$$

$$\frac{dy}{dx} = \frac{1}{7x^5}(35x^4)$$ Find the derivative of ln (argument), then use the chain rule.

$$= \frac{5}{x}$$ Simplify.   ◀

In Section 6-3, you will find out why this answer is so simple.

▶ **EXAMPLE 3**   If $y = \csc(\ln x)$, find dy/dx.

Solution

$$y = \csc(\ln x)$$

$$\frac{dy}{dx} = -\csc(\ln x)\cot(\ln x) \cdot \frac{1}{x}$$ ◀

Observe that Example 3 is a straight application of the derivative of cosecant, a method you learned earlier, followed by the application of the chain rule on the inside function, $\ln x$.

▶ **EXAMPLE 4**   If $f(x) = x^3 \ln x$, find an equation for $f'(x)$. Then show graphically and numerically that your answer is correct.

Solution

$$f(x) = x^3 \ln x$$

$$f'(x) = 3x^2 \ln x + x^3 \cdot \frac{1}{x} = x^2(3\ln x + 1)$$ Equation for derivative.

Graphical Check: First plot.

$$y_1 = x^3 \ln x$$
$$y_2 = \text{numerical derivative of } y_1$$ Use regular style.
$$y_3 = x^2(3\ln x + 1)$$ Use thick style.

Figure 6-2e shows the three graphs. The graph of y_3 overlays the graph of y_2.

Numerical Check: Use your grapher's TABLE feature to generate values of these three functions.

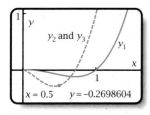

$x = 0.5$ $y = -0.2698604$

Figure 6-2e

x	y_1	y_2	y_3
0.5	−0.0866...	−0.2698...	−0.2698...
1.0	0	1	1.0000...
1.5	1.3684...	4.9868...	4.9868...
2.0	5.5451...	12.3177...	12.3177...
2.5	14.3170...	23.4304...	23.4304...

The numerical and algebraic derivatives give essentially the same values.   ◀

Integral of the Reciprocal Function

Because $\ln' x = 1/x$, the integral $\int (1/x)\,dx$ equals $\ln x + C$. But this works only for positive values of the variable. The function $\ln x$ is undefined for negative values of x because finding $\ln x$ requires you to integrate across a discontinuity in the graph (Figure 6-2f). However, it is possible to find $\int (1/x)\,dx$ if x is a negative number.

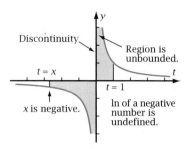

Figure 6-2f

Let $x = -u$. If x is negative, then u is a positive number. The differential, dx, is

$$dx = d(-u) = -du$$

$$\therefore \int \frac{1}{x}\,dx = \int \frac{1}{-u}(-du) \qquad \text{Substitute for } x \text{ and } dx.$$

$$= \int \frac{1}{u}\,du \qquad \text{By algebra.}$$

$$= \ln u + C \qquad \text{The integral has the same form as } \int (1/x)\,dx, \text{ and } u \text{ is a positive number.}$$

$$= \ln(-x) + C \qquad \text{Reverse substitution.}$$

Putting this result together with the original result gives the piecewise function

$$\int \frac{1}{x}\,dx = \begin{cases} \ln x + C \text{ if } x \text{ is positive} \\ \ln(-x) + C \text{ if } x \text{ is negative} \end{cases}$$

You can combine the two pieces with the aid of the absolute value function, as follows.

PROPERTY: *Integral of the Reciprocal Function*

$$\int \frac{1}{u}\,du = \ln|u| + C$$

The variable u, rather than x, has been used here to indicate that you can find the integral of the reciprocal of a function, u, in this way as long as the rest of the integrand is du, the *differential of the denominator.*

▶ **EXAMPLE 5**   Integrate: $\displaystyle \int \frac{\sec^2 5x}{\tan 5x}\,dx$

Solution

$$\int \frac{\sec^2 5x \, dx}{\tan 5x}$$

$$= \frac{1}{5} \int \frac{1}{\tan 5x} \cdot 5 \sec^2 5x \, dx$$ Write the fraction as a reciprocal. Multiply the other factor by 5 to make it equal to the differential of the denominator.

$$= \frac{1}{5} \ln |\tan 5x| + C$$ Integrate the reciprocal function.   ◀

Example 6 shows a case in which absolute value comes into action. The argument is often negative when the integral has negative limits of integration.

▶ **EXAMPLE 6**   Use the fundamental theorem to evaluate exactly. Check numerically.

$$\int_{-4}^{-5} \frac{x^2 \, dx}{1 + x^3}$$

Solution

$$\int_{-4}^{-5} \frac{x^2 \, dx}{1 + x^3}$$

$$= \frac{1}{3} \int_{-4}^{-5} \frac{1}{1 + x^3} \cdot 3x^2 \, dx$$ Make the numerator equal to the differential of the denominator.

$$= \frac{1}{3} \ln |1 + x^3| \Big|_{-4}^{-5}$$ Integrate the reciprocal function.

$$= \frac{1}{3} \ln |-124| - \frac{1}{3} \ln |-63| = \frac{1}{3} \ln 124 - \frac{1}{3} \ln 63$$ Exact answer.

$$= 0.225715613\ldots$$ Decimal value of exact answer.

Using your grapher's numerical integration feature gives 0.225715613..., or an answer very close to this, which checks!   ◀

Note that the answer is a positive number. Figure 6-2g reminds you that a definite integral is a limit of a Riemann sum. The terms have the form $f(x) \, dx$. The integrand function is negative, and so are the dx's because -5 is less than -4. If both $f(x)$ and dx are negative, the terms in the Riemann sum are positive.

Example 7 shows how to differentiate an integral when the upper limit of integration is a function of x rather than x itself.

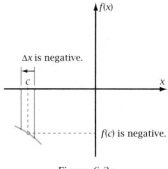

Figure 6-2g

▶ **EXAMPLE 7**   Find $f'(x): f(x) = \int_2^{5x} \sec t \, dt$

Solution

Let $g(x) = \int \sec x \, dx$.

Then $g'(x) = \sec x$.   Definition of indefinite integral (antiderivative).

$\therefore f(x) = g(5x) - g(2)$ Fundamental theorem, $g(b) - g(a)$ form.

$\therefore f'(x) = g'(5x) \cdot 5 - 0$ Chain rule and derivative of a constant.

$= 5 \sec 5x$ ◀

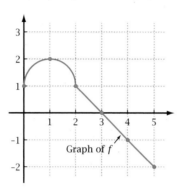

Once you see the pattern that appears in Example 7, you can write down the answer quickly, in one step. The sec 5x is the integrand, evaluated at the upper limit of integration. This is the result you would expect from the fundamental theorem in its derivative of the integral form. The 5 is the derivative of 5x, which is the inside function.

▶ **EXAMPLE 8**   Figure 6-2h shows the graph of function f.

a. Let $g(x) = \int_1^x f(t)\,dt$. Sketch the graph of function g.

b. Let $h(x) = \int_1^{(x/3)+1} f(t)\,dt$. Find $h'(3)$.

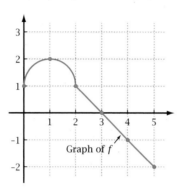

Figure 6-2h

Solution   a. By finding areas, $g(1) = 0$, $g(2) \approx 1.8$, $g(3) \approx 2.3$, $g(4) \approx 1.8$, $g(5) \approx 0.3$, and $g(0) \approx -1.8$. Plot these points and connect them, as shown in Figure 6-2i.

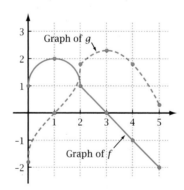

Figure 6-2i

b.
$$h(x) = \int_1^{(x/3)+1} f(t)\,dt$$

$$h'(x) = f(x/3 + 1) \cdot \tfrac{1}{3} \qquad \text{Fundamental theorem and chain rule.}$$

$$h'(3) = f(3/3 + 1) \cdot \tfrac{1}{3} = f(2) \cdot \tfrac{1}{3} = 1 \cdot \tfrac{1}{3} = \tfrac{1}{3} \qquad \text{Substitute for } x. \text{ Read } f(2) \text{ from graph.}$$ ◀

Problem Set 6-2

Quick Review  5 min

Q1. Integrate: $\int x^{-0.3}\, dx$

Q2. Integrate: $\int_0^3 x^2\, dx$

Q3. Differentiate: $f(x) = \cos^2 x$

Q4. $\lim_{x \to 100} \cos x = \cos 100$, so cos is —?— at $x = 100$.

Q5. If $f(x) = x^3$, then $f'(2) = 12$. Thus f is —?— at $x = 2$.

Q6. Find y': $y = \sin^{-1} x$

Q7. Find y': $y = \csc x$

Q8. $\sum f(x)\,\Delta x$ is called a(n) —?—.

Q9. $\int f(x)\, dx$ is called a(n) —?—.

Q10. $\log 3 + \log 4 = \log$ —?—.

---

0. *Look Ahead Problem 1:* Look at the derivatives and the integrals in Problem Set 6-6. Make a list, by problem number, of those you currently know how to do.

For Problems 1–26, find the derivative.

1. $y = \ln 7x$

2. $y = \ln 4x$

3. $f(x) = \ln x^5$

4. $f(x) = \ln x^3$

5. $h(x) = 6 \ln x^{-2}$

6. $g(x) = 13 \ln x^{-5}$

7. $r(t) = \ln 3t + \ln 4t + \ln 5t$

8. $v(z) = \ln 6z + \ln 7z + \ln 8z$

9. $y = (\ln 6x)(\ln 4x)$

10. $z = (\ln 2x)(\ln 9x)$

11. $y = \dfrac{\ln 11x}{\ln 3x}$

12. $y = \dfrac{\ln 9x}{\ln 6x}$

13. $p = (\sin x)(\ln x)$

14. $m = (\cos x)(\ln x)$

15. $y = \cos(\ln x)$

16. $y = \sin(\ln x)$

17. $y = \ln(\cos x)$ (Surprise?)

18. $y = \ln(\sin x)$ (Surprise?)

19. $T(x) = \tan(\ln x)$

20. $S(x) = \sec(\ln x)$

21. $y = (3x + 5)^{-1}$

22. $y = (x^3 - 2)^{-1}$

23. $y = x^4 \ln 3x$

24. $y = x^7 \ln 5x$

25. $y = \ln(1/x)$

26. $y = \ln(1/x)^4$

For Problems 27–46, integrate.

27. $\int (7/x)\, dx$

28. $\int (5/x)\, dx$

29. $\int \dfrac{1}{3x}\, dx$

30. $\int \dfrac{1}{8x}\, dx$

31. $\int \dfrac{x^2}{x^3 + 5}\, dx$

32. $\int \dfrac{x^5}{x^6 - 4}\, dx$

33. $\int \dfrac{x^5\, dx}{9 - x^6}$

34. $\int \dfrac{x^3\, dx}{10 - x^4}$

35. $\int \dfrac{\sec x \tan x\, dx}{1 + \sec x}$

36. $\int \dfrac{\sec^2 x\, dx}{1 + \tan x}$

37. $\int \dfrac{\cos x\, dx}{\sin x}$

38. $\int \dfrac{\sin x\, dx}{\cos x}$

39. $\int_{0.5}^{4} (1/w) \, dw$

40. $\int_{0.1}^{10} (1/v) \, dv$

41. $\int_{-0.1}^{-3} (1/x) \, dx$

42. $\int_{-0.2}^{-4} (1/x) \, dx$

43. $\int_{4}^{9} \dfrac{x^{1/2} \, dx}{1 + x^{3/2}}$

44. $\int_{1}^{8} \dfrac{x^{-1/3} \, dx}{2 + x^{2/3}}$

45. $\int (\ln x)^5 \dfrac{dx}{x}$ (Be clever!)

46. $\int \dfrac{\ln x}{x} \, dx$ (Be very clever!)

For Problems 47–54, find the derivative.

47. $f(x) = \int_{2}^{x} \cos 3t \, dt$

48. $f(x) = \int_{5}^{x} (t^2 + 10t - 17) \, dt$

49. $\dfrac{d}{dx} \left(\int_{2}^{x} \tan^3 t \, dt \right)$

50. $\dfrac{d}{dx} \left(\int_{-1}^{x} 2^t \, dt \right)$

51. $f(x) = \int_{1}^{x^2} 3^t \, dt$

52. $g(x) = \int_{0}^{\cos x} \sqrt{t} \, dt$

53. $h(x) = \int_{0}^{3x-5} \sqrt{1 + t^2} \, dt$

54. $p(x) = \int_{-1}^{x^3} (t^4 + 1)^7 \, dt$

55. Evaluate $\int_{1}^{3} (5/x) \, dx$ by using the fundamental theorem in its $g(b) - g(a)$ form. Then verify your answer numerically. Indicate which numerical method you used.

56. *Look Ahead Problem 1 Follow-Up:* In Problem 0, you were asked to look at Problem Set 6-6 and indicate which problems you knew how to do. Go back and list the problems in Problem Set 6-6 that you know how to do now but that you didn't know how to do before you worked on Problems 1–54 in this problem set.

57. Figure 6-2j shows the graph of function f.

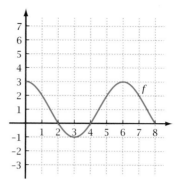

Figure 6-2j

a. Let $g(x) = \int_{1}^{x} f(t) \, dt$. On a copy of Figure 6-2j, sketch the graph of function g.

b. Let $h(x) = \int_{1}^{x^2 - 1} f(t) \, dt$. Find $h'(2)$.

58. Figure 6-2k shows the graph of function f.

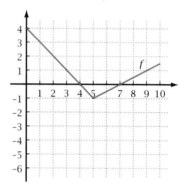

Figure 6-2k

a. Let $g(x) = \int_{2}^{x} f(t) \, dt$. On a copy of Figure 6-2k, sketch the graph of function g.

b. Let $h(x) = \int_{2}^{2x} f(t) \, dt$. Find $h'(4)$.

59. *Population Problem:* In the population problem of Problem Set 6-1, you evaluated

$$\int_{1000}^{N} \frac{1}{P} \, dP$$

where P stands for population as a function of time, t. Use what you have learned in this section to evaluate this integral using the fundamental theorem of calculus. You should get an answer in terms of N. Use the result to solve for N, the number of people when $t = 10$ yr, in the equation

$$\int_{1000}^{N} \frac{1}{P} \, dP = \int_{0}^{10} 0.05 \, dt$$

60. *Tire Pump Work Problem:* Figure 6-2l shows a bicycle tire pump. To compress the inside air, you exert a force F, in pounds, on the movable piston by pushing the pump handle. The outside air exerts a force of 30 lb, so the total force on the piston is $F + 30$. By Boyle's law, this total force varies inversely with h, in inches, the distance between the top of the pump base and the bottom of the movable piston. Consequently, the general equation is

$$F + 30 = \frac{k}{h}$$

where k is a constant of proportionality.

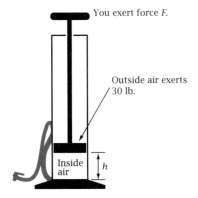

Figure 6-2l

a. Assume that the inside air is not compressed when $h = 20$ in., so that $F = 0$ when $h = 20$. Find the proportionality constant, k, and write the particular equation expressing F as a function of h.

b. In a sketch, show the region under the graph of F between $h = 10$ and $h = 20$.

c. The amount of work done in compressing the air is the product of the force exerted on the piston and the distance the piston moves. Explain why this work can be found using a definite integral.

d. Calculate the work done by compressing the air from $h = 20$ to $h = 10$. What is the mathematical reason why your answer is negative?

e. The units of work in this problem are inch-pounds (in.-lb). Why is this name appropriate?

61. *Radio Dial Derivative Problem:* Figure 6-2m shows an old AM radio dial. As you can see, the distances between numbers decrease as the frequency increases. If you study the theory behind the tuning of radios, you will learn that the distance from the left end of the dial to a particular frequency varies logarithmically with the frequency. That is,

$$d(f) = a + b \ln f$$

where $d(f)$ is the number of centimeters from the number 53 to the frequency number f on the dial, and where a and b stand for constants.

a. Solve for the constants a and b to find the particular equation for this logarithmic function.

b. Use the equation you found in part a to make a table of values of $d(f)$ for each frequency shown in Figure 6-2m. Then measure their

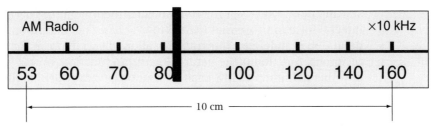

Figure 6-2m

distances with a ruler to the nearest 0.1 cm. If your calculated and measured answers do not agree, go back and fix your errors.

c. Write an equation for $d'(f)$. In the table from part b, put a new column that shows the instantaneous rates of change of distance with respect to frequency.

d. The numbers on the dial in Figure 6-2m are given in tens of kilohertz. (One hertz equals one cycle per second.) What are the units of $d'(f)$?

e. Do the values of $d'(f)$ increase or decrease as f increases? Explain how this fact is consistent with the way the numbers are spaced on the dial.

62. *Properties of ln Problem:* In this problem you will explore some properties of ln that you will prove in Section 6-3.

a. *ln of a product:* Evaluate $\ln 2$, $\ln 3$, and $\ln 6$. How are the three values related to one another? In general, what does $\ln(ab)$ equal in terms of $\ln a$ and $\ln b$?

b. *ln of a quotient:* Evaluate $\ln(10/2)$, $\ln 10$, and $\ln 2$. How are the three values related to one another? In general, what does $\ln(a/b)$ equal in terms of $\ln a$ and $\ln b$?

c. *ln of a power:* Evaluate $\ln(2^{10})$ and $\ln 2$. How are the two values related to each other? In general, what does $\ln(a^b)$ equal in terms of $\ln a$ and $\ln b$?

d. *Change-of-base:* Find $(\ln 2)/(\log 2)$. Find $(\ln 3)/(\log 3)$. What do you notice? The common logarithm function is $\log x = \log_{10} x$, and the natural logarithm function is $\ln x = \log_e x$. Find $\ln 10$. Find $1/(\log e)$. What do you notice? Give an example that shows

$$\log_{10} x = \frac{\log_e x}{\log_e 10}$$

63. *Journal Problem:* Update your journal with what you've learned since the last entry. Include such things as

• The one most important thing you have learned since your last journal entry

• The difference between the graph of $y = \ln x$ and the graph of $y = 1/t$, from which $\ln x$ is defined

• The second form of the fundamental theorem of calculus, as the derivative of a definite integral (You might give an example such as $g(x) = \int_1^x \sin t\, dt$, where you actually evaluate the integral, then show that $g'(x) = \sin x$.)

• The algebraic proof of the fundamental theorem in its second form

• The graphical interpretation of the fundamental theorem, second form, as the rate at which the area of a region changes

• Evidence (numerical, graphical, and algebraic) you have encountered so far that indicates that ln really is a logarithm

6-3 The Uniqueness Theorem and Properties of Logarithmic Functions

In Section 6-2, you defined $\ln x$ as a definite integral. With this definition as a starting point, you used the fundamental theorem in the derivative of an integral form to prove that $(d/dx)(\ln x) = 1/x$. This definition, coupled with the definition of indefinite integral, also allowed you to prove that the integral of the reciprocal function is the natural logarithm function. In this section you will see how this definition of $\ln x$ enables you to prove by calculus the properties of logarithms you studied in previous courses. To do this, you will learn the *uniqueness theorem for derivatives*, which says that two functions with identical derivatives and with a point in common are actually the same function.

OBJECTIVE   Learn the uniqueness theorem for derivatives, and use it to prove that $\ln x$, defined as an integral, is a logarithmic function and has the properties of logarithms. Use these properties to differentiate logarithmic functions with any acceptable base.

The Uniqueness Theorem for Derivatives

Figure 6-3a gives graphical evidence that if functions f and g have derivatives $f'(x) = g'(x)$ for all values of x in a given domain, then the function values $f(x)$ and $g(x)$ differ at most by a constant. If $f(a) = g(a)$ for some value $x = a$ in the domain, then the constant would be zero, and the function values would be identical for all x in the domain. The name *uniqueness theorem* is chosen because f and g are really only *one* ("unique") function. The property is stated and proven by contradiction here.

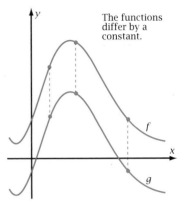

Figure 6-3a

PROPERTY: *The Uniqueness Theorem for Derivatives*

If: **1.** $f'(x) = g'(x)$ for all values of x in the domain, and

   2. $f(a) = g(a)$ for one value, $x = a$, in the domain,

then $f(x) = g(x)$ for all values of x in the domain.

Verbally: If two functions have the same derivative everywhere and they also have a point in common, then they are the same function.

Proof (by contradiction): Assume that the conclusion is false. Then there is a number $x = b$ in the domain for which $f(b) \neq g(b)$ (Figure 6-3b, left side). Let h be the difference function, $h(x) = f(x) - g(x)$ (Figure 6-3b, right side).

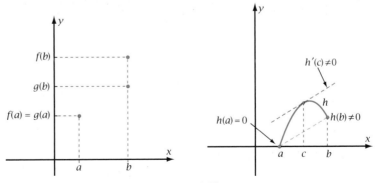

Figure 6-3b

Because $f(a) = g(a)$ and $f(b) \neq g(b)$, it follows that $h(a) = 0$ and $h(b) \neq 0$.

Thus, the secant line through the points $(a, h(a))$ and $(b, h(b))$ will have a slope not equal to zero.

Because both f and g are given to be differentiable for all x in the domain, h is also differentiable, and so the mean value theorem applies to function h on $[a, b]$.

Therefore, there is a number $x = c$ in (a, b) such that $h'(c)$ equals the slope of the secant line. Thus, $h'(c) \neq 0$.

But $h'(x) = f'(x) - g'(x)$. Because $f'(x)$ is given to be equal to $g'(x)$ for all x in the domain, $h'(x) = 0$ for all x. Thus, $h'(c)$ does equal zero.

This is a contradiction. Therefore, the assumption is false, and $f(x)$ does equal $g(x)$ for all x in the domain, Q.E.D.

Logarithm Properties of ln

Four properties of logarithms for natural logs are given in this box.

PROPERTY: Logarithm Properties of ln

(a and b are positive, and r is any real number.)

Product: $\ln(ab) = \ln a + \ln b$

Quotient: $\ln(a/b) = \ln a - \ln b$

Power: $\ln(a^r) = r \ln a$

Intercept: $\ln 1 = 0$

The intercept property is true because $\ln 1 = \int_1^1 (1/t)\, dt = 0$ by the properties of definite integrals. Example 1 shows how you can use the uniqueness theorem to prove the product property of ln. You will prove the other properties in Problem Set 6-3.

▶ **EXAMPLE 1**   Prove that $\ln(ab) = \ln a + \ln b$ for all $a > 0$ and $b > 0$.

Proof   Let b stand for a positive constant, and replace a with the variable x.

Let $f(x) = \ln(xb)$, and let $g(x) = \ln x + \ln b$.

Then $f'(x) = \dfrac{1}{xb} \cdot b = \dfrac{1}{x}$ for all $x > 0$, and $g'(x) = \dfrac{1}{x} + 0 = \dfrac{1}{x}$ for all $x > 0$.

Substituting 1 for x gives
$f(1) = \ln b$ and $g(1) = \ln 1 + \ln b = 0 + \ln b = \ln b$.

Thus, $f'(x) = g'(x)$ for all $x > 0$ and $f(1) = g(1)$. So, by the uniqueness theorem for derivatives, $f(x) = g(x)$ for all $x > 0$.

That is, $\ln(xb) = \ln x + \ln b$ for any positive number x and any positive number b. Replacing x with a gives

$\ln(ab) = \ln a + \ln b$ for all $a > 0$ and all $b > 0$, Q.E.D. ◀

Proof that ln x Is a Logarithm

The uniqueness theorem forms a lemma that you can use to prove that $\ln x$ is really a logarithmic function. Recall from Chapter 3 that the derivative of $\log_e x$ is $1/x$.

▶ **EXAMPLE 2** Prove that $\ln x = \log_e x$ for all $x > 0$.

Proof Let $f(x) = \ln x = \displaystyle\int_1^x \frac{1}{t}\, dt$, and let $g(x) = \log_e x$.

Then $f'(x) = 1/x$ for all $x > 0$ by the fundamental theorem in the derivative of the integral form, and $g'(x) = 1/x$ for all $x > 0$ by the proof in Chapter 3.

$\therefore f'(x) = g'(x)$ for all $x > 0$

But $f(1) = \ln 1 = \displaystyle\int_1^1 \frac{1}{t}\, dt = 0$ by the properties of definite integrals,

and $g(1) = \log_e 1 = 0$ by the properties of logarithms ($e^0 = 1$).

$\therefore f(1) = g(1)$

$\therefore f(x) = g(x)$ for all $x > 0$ by the uniqueness theorem

So $\ln x = \log_e x$ for all $x > 0$, Q.E.D. ◀

Derivatives of Base-*b* Logarithmic Functions

The new problem of differentiating $y = \log_b x$ can be transformed into an old problem using the *change-of-base property*. Logarithms to any given base are directly proportional to logs to any other base. In other words, $\log_b x = k \log_a x$. To prove this, recall the algebraic definition of logarithm.

ALGEBRAIC DEFINITION: Logarithm

$a = \log_b c$ if and only if $b^a = c$ where $b > 0$, $b \ne 1$, and $c > 0$

Verbally: A logarithm is an exponent.

Memory aid: For $y = \log_2 x$,

- 2 is the base. ($\log_2$ is read "log base 2.")

- y is the exponent, because a logarithm is an exponent.

- x is the answer to the expression 2^y.

To derive the change-of-base property, start with

$$y = \log_b x$$
$$b^y = x \qquad \text{By the algebraic definition of logarithm.}$$
$$\ln b^y = \ln x \qquad \text{Take the ln of both sides.}$$
$$y \cdot \ln b = \ln x \qquad \text{Log of a power property.}$$
$$y = \frac{\ln x}{\ln b}$$
$$\therefore \log_b x = \frac{\ln x}{\ln b} \qquad \text{Substitute for } y.$$

The property works for any other base logarithm ($b > 0, b \neq 1$), as well as for natural logarithms. Note that the logarithms in the numerator and denominator on the right side have the same base, e.

PROPERTY: Change-of-Base Property for Logarithms

In general: $\log_b x = \dfrac{\log_a x}{\log_a b}$

In particular: $\log_b x = \dfrac{\log_e x}{\log_e b} = \dfrac{\ln x}{\ln b} = \dfrac{1}{\ln b} \cdot \ln x$

▶ **EXAMPLE 3**

Find an equation for $f'(x)$ if $f(x) = \log_{10} x$. Check the formula by evaluating $f'(2)$ and showing that the line at the point $(2, \log_{10} 2)$, with slope $f'(2)$, is tangent to the graph.

Solution

$$f(x) = \log_{10} x = \frac{1}{\ln 10} \cdot \ln x \qquad \text{Use the change-of-base property.}$$

$$f'(x) = \frac{1}{\ln 10} \cdot \frac{1}{x} = \frac{1}{x \ln 10} \qquad \text{Derivative of a constant times a function.}$$

The line through the point $(2, \log_{10} 2)$, which equals $(2, 0.3010...)$, has slope

$$f'(2) = \frac{1}{\ln 10} \cdot \frac{1}{2} \approx 0.217$$

Thus, the line's equation is

$$y - 0.3010 \approx 0.217(x - 2) \qquad \text{or} \qquad y \approx 0.217x - 0.133$$

Figure 6-3c shows the graph of $f(x)$ and the line. The line really is tangent to the graph, Q.E.D. ◀

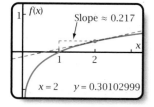

Figure 6-3c

▶ **EXAMPLE 4**

If $f(x) = \log_4 3x$, find $f'(x)$ algebraically and find an approximation for $f'(5)$. Show that your answer is reasonable by plotting f on your grapher and by showing that $f'(5)$ has a sign that agrees with the slope of the graph.

Solution

$$f(x) = \log_4 3x = \frac{\ln 3x}{\ln 4} = \frac{1}{\ln 4} \cdot \ln 3x$$

$$f'(x) = \frac{1}{\ln 4} \cdot \left(\frac{1}{3x} \cdot 3\right)$$

$$= \frac{1}{x \ln 4}$$

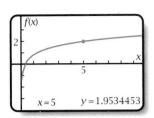

$x = 5$ $y = 1.9534453$

Figure 6-3d

By calculator, $f'(5) = 0.144269\ldots$.

To plot the graph, enter the transformed equation involving ln, shown above. As Figure 6-3d shows, $f(x)$ is increasing slowly at $x = 5$, which agrees with the small positive value of the derivative.   ◀

▶ **EXAMPLE 5**   Find $\dfrac{d}{dx}\left(\ln \dfrac{x^6}{\sin^2 x}\right)$.

Solution   First, use the properties of ln to transform the quotient and powers into sums that are more easily differentiated.

$$\frac{d}{dx}\left(\ln \frac{x^6}{\sin^2 x}\right) = \frac{d}{dx}[\ln x^6 - \ln(\sin^2 x)] \qquad \text{ln of a quotient.}$$

$$= \frac{d}{dx}[6\ln x - 2\ln(\sin x)] \qquad \text{ln of a power, twice.}$$

$$= \frac{6}{x} - \frac{2}{\sin x} \cdot \cos x \qquad \text{Derivative of ln, and the chain rule.}$$

$$= 6/x - 2\cot x \qquad \text{Optional simplification.} \qquad ◀$$

Conclusion

You have shown that the function defined by the integral

$$\ln x = \int_1^x \frac{1}{t}\, dt$$

has the graph of a logarithmic function, including an x-intercept of 1, and you have proved that $\ln x$, defined this way, is equivalent to $\log_e x$ and has the logarithm of a product property. In the following problem set you will prove by calculus that $\ln x$ has the other properties of logarithms.

Problem Set 6-3

Quick Review

Q1. Differentiate: $y = \tan^{-1} x$

Q2. Integrate: $\int (4x + 1)^{-1}\, dx$

Q3. Find: $\lim_{x\to 0}(\sin x)/(x)$

Q4. $\log 12 - \log 48 = \log$ —?—

Q5. $\log 7 + \log 5 = \log$ —?—

Q6. $3\log 2 = \log$ —?—

Q7. Sketch the graph of $\int_2^x y\,dx$ for the function shown in Figure 6-3e.

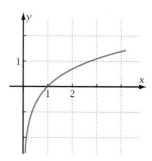

Figure 6-3e

Q8. Write one hypothesis of the mean value theorem.

Q9. Write the other hypothesis of the mean value theorem.

Q10. If $y = x^3$, then the value of dy when $x = 2$ and $dx = 0.1$ is —?—.

　　A. 12　　B. 1.2　　C. 0.8　　D. 0.6　　E. 0.4

---

0. *Look Ahead Problem 2:* Look at the derivatives and the integrals in Problem Set 6-6. Make a list, by problem number, of those you presently know how to do.

For Problems 1–6, show that the properties of ln are true by evaluating both sides of the given equation.

1. $\ln 24 = \ln 6 + \ln 4$

2. $\ln 35 = \ln 5 + \ln 7$

3. $\ln \frac{2001}{667} = \ln 2001 - \ln 667$

4. $\ln \frac{1001}{77} = \ln 1001 - \ln 77$

5. $\ln(1776^3) = 3\ln 1776$

6. $\ln(1066^4) = 4\ln 1066$

For Problems 7 and 8, prove the properties. Do this without looking at the proofs in the text. If you get stuck, look at the text proof just long enough to get going again.

7. Prove the uniqueness theorem for derivatives.

8. Prove that $\ln x = \log_e x$ for all $x > 0$.

9. Prove that $\ln(a/b) = \ln a - \ln b$ for all $a > 0$ and $b > 0$.

10. Prove that $\ln(a^b) = b\ln a$ for all $a > 0$ and for all b.

11. Prove that $\ln(a/b) = \ln a - \ln b$ again, using the property given in Problem 10 as a lemma.

12. Prove by counterexample that $\ln(a + b)$ does not equal $\ln a + \ln b$.

13. Write the definition of $\ln x$ as a definite integral.

14. Write the change-of-base property for logarithms.

For Problems 15 and 16, find an equation for the derivative of the given function and show numerically or graphically that the equation gives a reasonable value for the derivative at the given value of x.

15. $f(x) = \log_3 x$,　$x = 5$

16. $f(x) = \log_{0.8} x$,　$x = 4$

For Problems 17–24, find an equation for the derivative of the given function.

17. $g(x) = 8\ln(x^5)$

18. $h(x) = 10\ln(x^{0.4})$

19. $T(x) = \log_5(\sin x)$

20. $R(x) = \log_4(\sec x)$

21. $p(x) = (\ln x)(\log_5 x)$

22. $q(x) = \dfrac{\log_9 x}{\log_3 x}$

23. $f(x) = \ln \dfrac{x^3}{\sin x}$

24. $f(x) = \ln(x^4 \tan x)$

25. Find $\dfrac{d}{dx}(\ln x^{3x})$

26. Find $\dfrac{d}{dx}(\ln 5^{\sec x})$

27. *Lava Flow Problem:* Velocities are measured in miles per hour. (When a velocity is very low, people sometimes prefer to think of how many hours it takes for the lava to flow a mile.) Lava flowing down the side of a volcano flows more slowly as it cools. Assume that the distance, y, in miles, from the crater to the tip of the flowing lava is given by

$$y = 7 \cdot (2 - 0.9^x)$$

where x is the number of hours since the lava started flowing (Figure 6-3f).

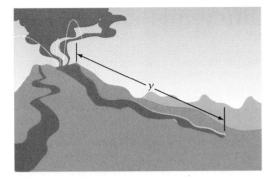

Figure 6-3f

Volcanologists in Hawaii stand on the fresh crust of recently hardened lava.

a. Find an equation for dy/dx. Use the equation to find out how fast the tip of the lava is moving when $x = 0, 1, 5$, and 10 h. Is the lava speeding up or slowing down as time passes?

b. Transform the equation $y = 7 \cdot (2 - 0.9^x)$ (from part a) so that x is in terms of y by taking the log of both sides, using some appropriate base for the logs.

c. Differentiate the equation given in part b with respect to y to find an equation for dx/dy. Calculate dx/dy when $y = 10$ mi. What are the units of dx/dy?

d. Calculate dx/dy for the value of y when $x = 10$ h.

e. You might think that dy/dx and dx/dy are reciprocals of each other. Based on your answers above, in what way is this reasoning true and in what way is it not true?

28. *Compound Interest Problem:* If interest on a savings account is compounded continuously, and the interest is such that the annual percentage rate (APR) equal 6%, then M, the amount of money after t years, is given by the exponential function $M = 1000 \times 1.06^t$. You can solve this equation for t in terms of M.

$$t = \log_{1.06}\left(\frac{M}{1000}\right)$$

a. Show the transformations needed to get t in terms of M.

b. Write an equation for dt/dM.

c. Evaluate dt/dM when $M = 1000$. What are the units of dt/dM? What real-world quantity does dt/dM represent?

d. Does dt/dM increase or decrease as M increases? How do you interpret the real-world meaning of your answer?

29. *Base e for Natural Logarithms Problem:* Figure 6-3g shows the graph of $y = \ln x$ and the horizontal line $y = 1$. Because $\log_b b = 1$ for any permissible base b, the value of x where the two graphs cross must be the base of the ln function. By finding this intersection graphically, confirm that e is the base of the natural logarithm function.

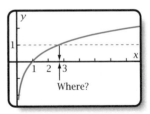

Figure 6-3g

30. *Journal Problem:* Update your journal with what you've learned about the natural and base-b logarithmic functions in this section.

6-4 The Number e, Exponential Functions, and Logarithmic Differentiation

You have studied the natural logarithm function starting with its definition as a definite integral. In this section you will define the **natural exponential function** as the inverse of the natural logarithm function. In doing so you will learn some properties of exponential functions that you can prove by calculus.

OBJECTIVE Derive properties of $y = e^x$ starting with its definition as the inverse of $y = \ln x$. Use the results to define b^x, and differentiate exponential functions by first taking the natural logarithm.

The Number e

In Chapter 3, you found that you can use e as a base for the exponential whose derivative is the same as the function. That is,

> If $f(x) = e^x$, then $f'(x) = e^x$.

There is a way to define e directly. Suppose that you evaluate $(1 + n)^{1/n}$ as n gets closer to zero. The base gets closer to 1, suggesting the limit may be 1, because 1 to any real power is 1. However, as n approaches zero from the positive side the exponent becomes very large because the reciprocal of a positive number close to zero is very large, and raising a number greater than 1 to a very large power gives a very large answer. And, as n approaches zero from the negative side, the base is less than 1, and raising a number less than 1 to a very large negative power also gives a very large answer. Figure 6-4a shows that the limit is e, which is more than 1 but is finite.

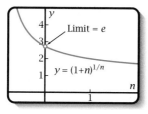

Figure 6-4a

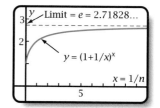

Figure 6-4b

The number e is defined to be this limit. The equivalent definition,

$$\lim_{x \to \infty} (1 + 1/x)^x$$

gives the same answer, e, as shown by graph in Figure 6-4b. In Problem 62 of Problem Set 6-4 you will see how this expression arises naturally when you use the definition of derivative to differentiate a base-b logarithmic function.

DEFINITION: *e*

$$e = \lim_{n \to 0} (1 + n)^{1/n} = \lim_{n \to \infty} (1 + 1/n)^n$$

Numerically: $e = 2.7182818284...$ (a nonrepeating decimal)

Note that in each of these cases, e takes the form 1^{∞}. Like $0/0$, this form is also **indeterminate**. You can't determine what the limit is just by looking at the expression.

Note also that although the digits ...1828... in the decimal part of e seem to be repeating, the repeat happens only once. Like π, the number e is a **transcendental** number. Not only is it irrational (a non-ending, nonrepeating decimal), but you cannot express it using only a finite number of the operations of algebra ($+$, $-$, $\times$, $\div$, and roots) performed on integers. It "transcends," or "goes beyond" these operations.

Inverse Relationship between e^x and $\ln x$

Recall from Section 3-9 that $y = e^x$ and $y = \ln x$ are inverses. Also, recall that for functions that are inverses of each other, $f(f^{-1}(x)) = x$ and $f^{-1}(f(x)) = x$. So,

$$\ln(e^x) = x \qquad \text{and} \qquad e^{\ln x} = x$$

▶ **EXAMPLE 1**   Differentiate: $y = \cos(\ln e^{7x})$

Solution

$$y = \cos(\ln e^{7x}) \Rightarrow y = \cos 7x \qquad \text{Function of an inverse function property.}$$
$$\therefore y' = -7 \sin 7x \qquad\qquad\qquad\qquad\qquad \blacktriangleleft$$

General Exponential Function: $y = b^x$

You can express an exponential expression with any other base in terms of e^x. For example, to express 8^x in terms of e^x, first write 8 as a power of e.

Let $8 = e^k$, where k stands for a positive constant.

$$\begin{aligned}
\text{Then } \ln 8 &= \ln e^k & &\text{Take ln of both sides.} \\
&= k(\ln e) & &\text{Reason?} \\
&= k & &\text{Reason?} \\
\therefore 8 &= e^{\ln 8} & &\text{Substitute ln 8 for } k.
\end{aligned}$$

This equation is an example of the inverse relationship

$$e^{\ln x} = x$$

This relationship leads to a definition of exponentials with bases other than e.

$$\begin{aligned}
b^x &= (e^{\ln b})^x & &\text{Replace } b \text{ with } e^{\ln b}. \\
&= e^{x \ln b} & &\text{Multiply the exponents.}
\end{aligned}$$

$$b^x = e^{x \ln b}$$

The definition of base-b exponential functions allows you to differentiate and integrate these functions based on the chain rule. Example 2 shows how to do this.

▶ **EXAMPLE 2**   a. Find $\dfrac{d}{dx}(8^x)$.   b. Find $\int 8^x dx$.

Solution   a.

$$\frac{d}{dx}(8^x) = \frac{d}{dx}(e^{x \ln 8}) \qquad \text{Definition of } 8^x.$$

$$= e^{x \ln 8} \cdot \ln 8 \qquad \begin{array}{l}\text{Derivative of base-}e \text{ exponential}\\ \text{function and the chain rule.}\end{array}$$

$$= 8^x \ln 8 \qquad \text{Definition of } 8^x.$$

b.

$$\int 8^x dx = \int e^{x \ln 8} dx \qquad \text{Definition of } 8^x.$$

$$= \frac{1}{\ln 8} \int e^{x \ln 8}(\ln 8 \, dx) \qquad \begin{array}{l}\text{Transform to get the differential}\\ \text{of the inside function.}\end{array}$$

$$= \frac{1}{\ln 8} e^{x \ln 8} + C \qquad \int e^u \, du = e^u + C$$

$$= \frac{1}{\ln 8} 8^x + C$$

◀

The formulas for derivatives and integrals of exponential functions that you learned earlier and proved in this chapter are summarized here for your reference.

Algebraically: For any positive constant $b \neq 1$ (to avoid division by zero),

$$\frac{d}{dx}(b^x) = b^x \ln b \qquad \int b^u \, du = \frac{b^u}{\ln b} + C$$

Verbally: To differentiate, multiply b^x by $\ln b$.

To integrate, divide b^u by $\ln b$ (and add C).

Special case: $\dfrac{d}{dx}(e^x) = e^x \qquad \int e^u \, du = e^u + C$

Note: The reason for using u and du in the integration formulas is to show that anything that is not a part of the exponential must be the differential of the inside function (that is, the differential of the exponent).

Logarithmic Differentiation

You can also evaluate the derivative of $y = 8^x$ using a procedure called **logarithmic differentiation**.

$$y = 8^x$$

$$\ln y = \ln 8^x \qquad \text{Take the ln of both sides.}$$

$$\ln y = x \ln 8 \qquad \text{ln of a power.}$$

$$\frac{1}{y} y' = 1(\ln 8) \qquad \text{Differentiate implicitly with respect to } x.$$

$$y' = y (\ln 8) \qquad \text{Solve for } y'.$$

$$\therefore \ y' = 8^x (\ln 8) \qquad \text{Substitute for } y.$$

This method is especially useful when differentiating functions where the base itself is a variable, as in Example 3.

▶ **EXAMPLE 3**   Find $f'(x)$: $f(x) = (x^3 + 4)^{\cos x}$

Solution

$$f(x) = (x^3 + 4)^{\cos x}$$

$$\ln f(x) = \ln (x^3 + 4)^{\cos x} \qquad \text{Take ln of both sides.}$$

$$\ln f(x) = \cos x \, [\ln (x^3 + 4)] \qquad \text{ln of a power.}$$

$$\frac{1}{f(x)} f'(x) = -\sin x \, [\ln (x^3 + 4)] + \cos x \left[\frac{3x^2}{x^3 + 4} \right] \qquad \begin{array}{l}\text{Differentiate implicitly} \\ \text{on the left. Derivative of} \\ \text{a product on the right.}\end{array}$$

$$f'(x) = (x^3 + 4)^{\cos x} \left(-\sin x \, [\ln (x^3 + 4)] + \cos x \left[\frac{3x^2}{x^3 + 4} \right] \right) \qquad ◀$$

The properties of logarithms allow you to turn a power into a product, which you know how to differentiate. The other properties let you transform products and quotients to sums and differences, which are even easier to differentiate. Example 4 shows you how this is done.

▶ **EXAMPLE 4**   If $f(x) = \dfrac{(3x + 7)^5}{\sqrt[3]{x + 2}}$, find $f'(x)$.

Solution

$$\ln f(x) = \ln \frac{(3x + 7)^5}{\sqrt[3]{x + 2}} \qquad \text{Take ln of both sides.}$$

$$\ln f(x) = \ln (3x + 7)^5 - \ln \sqrt[3]{x + 2} \qquad \text{ln of a quotient.}$$

$$\ln f(x) = 5 \ln (3x + 7) - \frac{1}{3} \ln (x + 2) \qquad \text{ln of a power.}$$

$$\frac{1}{f(x)} \cdot f'(x) = \frac{15}{3x + 7} - \frac{1/3}{x + 2} \qquad \begin{array}{l}\text{Differentiate implicitly and} \\ \text{use the chain rule.}\end{array}$$

$$f'(x) = \frac{(3x + 7)^5}{\sqrt[3]{x + 2}} \cdot \left(\frac{15}{3x + 7} - \frac{1/3}{x + 2} \right) \qquad ◀$$

Problem Set 6-4

Q1. Differentiate: $y = \ln 7x^3$

Q2. Integrate: $\int (5x)^{-3}\, dx$

Q3. Integrate: $\int (5x)^{-1}\, dx$

Q4. Differentiate: $y = \cos^{-1} x$

Q5. $\dfrac{d}{dx}\left(\displaystyle\int_3^{5x} \tan t\, dt \right) = -?-$

Q6. $\log_{17} 23 = \dfrac{\ln -?-}{\ln -?-}$

Q7. $\ln 12 + \ln 3 = \ln -?-$

Q8. $3 \ln 2 = \ln -?-$

For Problems Q9 and Q10, let $g(x) = \int_1^x f(t)\, dt$, where $f(t)$ is shown in Figure 6-4c.

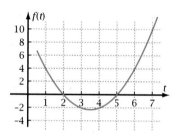

Figure 6-4c

Q9. $g(7)$ is closest to

   A. 2   B. 4   C. 6   D. 8   E. 10

Q10. $g'(7)$ is closest to $-?-$.

   A. 2   B. 4   C. 6   D. 8   E. 10

---

0. *Look Ahead Problem 3:* Look at the derivatives and the integrals in Problem Set 6-6. Make a list, by problem number, of those you presently know how to do.

1. *Rabbit Population Problem:* When rabbits were introduced to Australia in the mid-1800s, they had no natural enemies. As a result, their population grew exponentially with time. The general equation of the exponential function R, where $R(t)$ is the number of rabbits and t is time in years, is

$$R(t) = ae^{kt}$$

a. Suppose there were 60,000 rabbits in 1865, when $t = 0$, and that the population had grown to 2,400,000 by 1867. Substitute these values of t and R to get two equations involving the constants a and k. Use these equations to find values of a and k, then write the particular equation expressing $R(t)$ as a function of t.

b. How many rabbits does your model predict there would have been in 1870?

c. According to your model, when was the first pair of rabbits introduced into Australia?

2. *Depreciation Problem:* The value of a major purchase, such as a car, depreciates (decreases) each year because the purchase gets older. Assume that the value of Otto Price's car is given by

$$v(t) = 20,000\, e^{-0.1t}$$

where v is the value, in dollars, of the car at time t, in years after it was built.

a. How much was it worth when it was built?

b. How much did it depreciate during its eleventh year (from $t = 10$ to $t = 11$)?

c. What is the instantaneous rate of change (in dollars per year) of the value at $t = 10$? Why is this answer different from the answer you found in part b?

d. When will the value have dropped to $5,000?

3. *Compound Interest Problem:* In many real-world situations that are driven by internal forces, the variables are often related by an

exponential function. For instance, the more money there is in a savings account, the faster the amount in the account grows (Figure 6-4d).

If $1000 is placed in a savings account in which the interest is compounded continuously, and the interest is enough to make the annual percentage rate (APR) equal 6%, then the amount of money, m, in dollars, in the account at time t, in years after it is invested, is

$$m(t) = 1000(1.06)^t$$

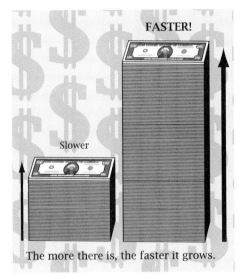

FASTER!

Slower

The more there is, the faster it grows.

Figure 6-4d

a. Find an equation for the derivative, $m'(t)$. At what rate is the amount growing at the instant $t = 0$ yr? At $t = 5$ yr? At $t = 10$ yr? What are the units of these rates?

b. Find the amount of money in the account at $t = 0$ yr, $t = 5$ yr, and $t = 10$ yr. Does the rate of increase seem to be getting larger as the amount increases?

c. Show that the rate of the increase is directly proportional to the amount of money present. One way to do this is to show that $m'(t)/m(t)$ is constant.

d. Show that the account earns exactly $60 the first year. Then explain why the rate of increase at time $t = 0$ is less than $60/yr.

4. *Door-Closer Problem:* In Section 1-1, you explored a problem in which a door was pushed open. As the automatic door-closer slowed the door down, the number of degrees,

$d(t)$, the door was open after time t, in seconds, was given as

$$d(t) = 200t \cdot 2^{-t}, \qquad 0 \le t \le 7$$

Use what you have learned about derivatives of exponential functions to analyze the motion of the door. For instance, how fast is it opening at $t = 1$ and at $t = 2$? At what time is it open the widest, and what is the derivative at that time? A graph might help.

5. *Definition of e Problem:* Write the two forms of the definition of e as a limit, as shown in this section. Explain why these two limits lead to *indeterminate forms*. Show numerically that the values of the expressions in the two forms of the definition get closer to 2.7182818284.... (Note that if you use large enough values of the exponent, the calculator may round the base to 1, making the limit seem to be the incorrect value 1.)

For Problems 6–47, find the derivative. Simplify your answer.

6. $y = 17e^{-5x}$

7. $y = 667e^{-3x}$

8. $h(x) = x^3 e^x$

9. $g(x) = x^{-6} e^x$

10. $r(t) = e^t \sin t$

11. $s(t) = e^t \tan t$

12. $u = 3e^x e^{-x}$

13. $v = e^{-4x} e^{4x}$

14. $y = \dfrac{e^x}{\ln x}$

15. $y = \dfrac{\ln x}{e^x}$

16. $y = 4e^{\sec x}$

17. $y = 7e^{\cos x}$

18. $y = 3 \ln e^{2x}$

19. $y = 4 \ln e^{5x}$

20. $y = (\ln e^{3x})(\ln e^{4x})$

21. $y = (\ln e^{-2x})(\ln e^{5x})$

22. $g(x) = 4e^{\ln x}$

23. $h(x) = 6e^{\ln 7x}$

24. $y = e^x + e^{-x}$

25. $y = e^x - e^{-x}$

26. $y = e^{5x^3}$

27. $y = 8e^{x^5}$

28. $f(x) = 0.4^{2x}$

29. $f(x) = 10^{-0.2x}$

30. $g(x) = 4(7^x)$

31. $h(x) = 1000(1.03^x)$

32. $c(x) = x^5 \cdot 3^x$

33. $m(x) = 5^x \cdot x^7$

34. $y = (\ln x)^{0.7x}$

35. $y = x^{\ln x}$

36. $y = (\csc 5x)^{2x}$

37. $y = (\cos 2x)^{3x}$

38. $y = \ln \dfrac{5x + 2}{7x - 8}$

39. $y = \ln[(4x - 7)(x + 10)]$

40. $y = (2x + 5)^3 \cdot \sqrt{4x - 1}$
    (Use logarithmic differentiation.)

41. $y = \dfrac{(10 + 3x)^{10}}{(4x - 5x)^3}$
    (Use logarithmic differentiation.)

42. $\dfrac{d}{dx}\left(\displaystyle\int_3^x 10^t \, dt\right)$

43. $\dfrac{d}{dx}\left(\displaystyle\int_3^x \ln t \, dt\right)$

44. $\dfrac{d}{dx}\left(\displaystyle\int_5^{4x} \log_2 t \, dt\right)$

45. $\dfrac{d}{dx}\left[\displaystyle\int_{6.3}^{x^2} \ln(\cos t) \, dt\right]$

46. $\dfrac{d^2}{dx^2}(\ln x^5)$

47. $\dfrac{d^2}{dx^2}(e^{7x})$

For Problems 48–59, find the indefinite integral. Simplify the answer.

48. $\displaystyle\int e^{5x} \, dx$

49. $\displaystyle\int e^{7x} \, dx$

50. $\displaystyle\int 7^{2x} \, dx$

51. $\displaystyle\int 1.05^x \, dx$

52. $\displaystyle\int 6e^x \, dx$

53. $\displaystyle\int e^{0.2x} \, dx$

54. $\displaystyle\int e^{\sin x} \cos x \, dx$

55. $\displaystyle\int e^{\tan x} \sec^2 x \, dx$

56. $\displaystyle\int e^{3 \ln x} \, dx$

57. $\displaystyle\int 60 e^{\ln 5x} \, dx$

58. $\displaystyle\int (1 + e^{2x})^{50} e^{2x} \, dx$

59. $\displaystyle\int (1 - e^{4x})^{100} e^{4x} \, dx$

For Problems 60 and 61, evaluate the definite integral using the fundamental theorem. Show by numerical integration that your answer is correct.

60. $\displaystyle\int_0^2 (e^x - e^{-x}) \, dx$

61. $\displaystyle\int_{-1}^2 (e^x + e^{-x}) \, dx$

62. *Derivative of Base-b Logarithm Function from the Definition of Derivative:* In this problem you will see how the definition of e as

$$e = \lim_{n \to 0} (1 + n)^{1/n}$$

arises naturally if you differentiate $f(x) = \log_b x$ starting with the definition of derivative. Give reasons for the indicated steps.

$f(x) = \log_b x$ Given.

$f'(x) = \lim\limits_{h \to 0} \dfrac{\log_b(x + h) - \log_b x}{h}$ Reason?

$= \lim\limits_{h \to 0} \dfrac{\log_b \dfrac{x + h}{x}}{h}$ Reason?

$= \lim\limits_{h \to 0} \left[\dfrac{1}{h} \cdot \log_b\left(1 + \dfrac{h}{x}\right)\right]$ Reasons?

$$= \lim_{h \to 0} \left[\frac{1}{x} \cdot \frac{x}{h} \cdot \log_b \left(1 + \frac{h}{x} \right) \right]$$

Multiply by 1 in the form x/x and rearrange.

$$= \frac{1}{x} \cdot \lim_{h \to 0} \left[\frac{x}{h} \cdot \log_b \left(1 + \frac{h}{x} \right) \right]$$

Why is this the limit of a "constant" times a function?

$$= \frac{1}{x} \cdot \lim_{h \to 0} \left[\log_b \left(1 + \frac{h}{x} \right)^{x/h} \right]$$

Reason?

$$= \frac{1}{x} \cdot \log_b \left[\lim_{h \to 0} \left(1 + \frac{h}{x} \right)^{x/h} \right]$$

Assuming limit of log equals log of limit.

$$= \frac{1}{x} \cdot \log_b e$$

Why is this the definition of e?

63. *Look Ahead Problem 3 Follow-Up:* In Problem 0, you were asked to look at Problem Set 6-6 and indicate which problems you knew how to do. Go back and make another list of the problems in Problem Set 6-6 that you know how to do now but that you didn't know how to do before you worked on this problem set.

64. *Journal Problem:* Update your journal with what you've learned about the exponential function in this section. In particular, note what you now know that you did not yet know when you studied exponential functions in Chapter 3.

6-5 Limits of Indeterminate Forms: l'Hospital's Rule

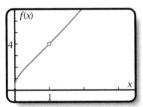

Figure 6-5a

When you use one of the formulas for finding derivatives, you are really using shortcuts to find the limit of the indeterminate form 0/0. You can use these formulas to evaluate other limits of the form 0/0. Figure 6-5a shows the graph of

$$f(x) = \frac{x^2 + 2x - 3}{\ln x}$$

If you try to evaluate $f(1)$, you get

$$f(1) = \frac{1 + 2 - 3}{\ln 1} = \frac{0}{0}$$

The graph suggests that the limit of $f(x)$ as x approaches 1 is 4.

The technique you will use to evaluate such limits is called **l'Hospital's rule**, named after G. F. A. de l'Hospital (1661–1704), although it was probably known earlier by the Bernoulli brothers. This French name is pronounced "lo-pee-tal′." (It is sometimes spelled l'Hôpital, with a circumflex placed over the letter *o*.) You will also learn how to use l'Hospital's rule to evaluate other indeterminate forms such as ∞ / ∞, 1^{∞}, 0^0, ∞^0, and $\infty - \infty$.

OBJECTIVE   Given an expression with an indeterminate form, find its limit using l'Hospital's rule.

L'Hospital's rule is easy to use but tricky to derive. Therefore, you will start out by seeing how it works, then get some insight into why it works. To find the limit of a fraction that has the form 0/0 or ∞ / ∞, you will take the derivatives of the numerator and the denominator, then find the limit. For instance, if $g(x)/h(x) = (x^2 + 2x - 3)/(\ln x)$ (given above),

$$\lim_{x \to 1} \frac{g(x)}{h(x)} = \lim_{x \to 1} \frac{g'(x)}{h'(x)} = \lim_{x \to 1} \frac{2x + 2}{1/x} = \frac{4}{1} = 4$$

which agrees with Figure 6-5a. Here is a formal statement of l'Hospital's rule.

PROPERTY: l'Hospital's Rule

If $f(x) = \dfrac{g(x)}{h(x)}$ and if $\lim\limits_{x \to c} g(x) = \lim\limits_{x \to c} h(x) = 0$,

then $\lim\limits_{x \to c} f(x) = \lim\limits_{x \to c} \dfrac{g'(x)}{h'(x)}$, provided the latter limit exists.

Corollaries of this rule lead to the same conclusion if $x \to \infty$ or if both $g(x)$ and $h(x)$ approach infinity.

Background Item: Limit-Function Interchange for Continuous Functions

If you take the limit of a continuous function that has another function inside, such as

$$\lim_{x \to c} \sin(\tan x)$$

it is possible to interchange the limit and the outside function,

$$\sin\left(\lim_{x \to c} \tan x\right)$$

Consider a simpler case, for which it's easy to see why continuity is sufficient for this interchange. The definition of continuity states that if g is continuous at $x = c$, then

$$\lim_{x \to c} g(x) = g(c)$$

But c is the limit of x as x approaches c. Replacing c with the limit gives

$$\lim_{x \to c} g(x) = g\left(\lim_{x \to c} x\right)$$

which shows that the limit and the outside function have been interchanged.

PROPERTY: Limit-Function Interchange for Continuous Functions

For the function $f(x) = g(h(x))$, if $h(x)$ has a limit, L, as x approaches c and if g is continuous at L, then $\lim\limits_{x \to c} g(h(x)) = g\left(\lim\limits_{x \to c} h(x)\right)$.

Here's why l'Hospital's rule works. Because $g(x)$ and $h(x)$ both approach zero as x approaches c, $g(c)$ and $h(c)$ either equal zero or can be defined to equal zero by removing a removable discontinuity. You can transform the fraction for $f(x)$ to a ratio of difference quotients by subtracting $g(c)$ and $h(c)$, which both equal zero, and by multiplying by clever forms of 1.

$$f(x) = \frac{g(x)}{h(x)} = \frac{g(x) - g(c)}{h(x) - h(c)} = \frac{\dfrac{g(x) - g(c)}{x - c}}{\dfrac{h(x) - h(c)}{x - c}}$$

$$\therefore \ \lim_{x \to c} f(x) = \frac{\displaystyle\lim_{x \to c} \frac{g(x) - g(c)}{x - c}}{\displaystyle\lim_{x \to c} \frac{h(x) - h(c)}{x - c}} = \frac{g'(c)}{h'(c)}$$

Limit of a quotient.
Definition of derivative.

If the derivatives of f and g are also continuous at $x = c$, you can write

$$g'(c) = g'(\lim_{x \to c} x) = \lim_{x \to c} g'(x) \qquad \text{and} \qquad h'(c) = h'(\lim_{x \to c} x) = \lim_{x \to c} h'(x)$$

Therefore,

$$\lim_{x \to c} f(x) = \frac{\lim_{x \to c} g'(x)}{\lim_{x \to c} h'(x)} = \lim_{x \to c} \frac{g'(x)}{h'(x)} \qquad \text{Q.E.D.}$$

where the last step is justified by the limit of a quotient property used "backwards." A formal proof of l'Hospital's rule must avoid the difficulty that $[h(x) - h(c)]$ might be zero somewhere other than at $x = c$. This proof and the proofs of the corollaries are not shown here because they would distract you from what you're learning. A graphical derivation of l'Hospital's rule is presented in Problem 34 of Problem Set 6-5.

Example 1 gives you a reasonable format to use when you apply l'Hospital's rule. The function is the one given at the beginning of this section.

▶ **EXAMPLE 1**   Find $L = \lim\limits_{x \to 1} \dfrac{x^2 + 2x - 3}{\ln x}$.

Solution

$$\lim_{x \to 1} \frac{x^2 + 2x - 3}{\ln x} \ \to \ \frac{0}{0}$$

L'Hospital's rule applies because the limit has the form 0/0.

$$= \lim_{x \to 1} \frac{2x + 2}{1/x} \ \to \ \frac{4}{1}$$

L'Hospital's rule is no longer needed because the limit is no longer indeterminate.

$$= 4$$

◀

Example 2 shows how to use l'Hospital's rule for an indeterminate form other than 0/0. It also shows a case where l'Hospital's rule is used more than once.

▶ **EXAMPLE 2**   Evaluate $\lim\limits_{x \to \infty} x^2 e^{-x}$.

Solution   As x goes to infinity, x^2 gets infinitely large and e^{-x} goes to zero. A graph of $y = x^2 e^{-x}$ suggests that the expression goes to zero as x becomes infinite (Figure 6-5b). To show this by l'Hospital's rule, first transform the expression into a fraction.

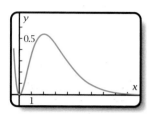

Figure 6-5b

$$\lim_{x \to \infty} x^2 e^{-x} \to \infty \cdot 0 \qquad \text{L'Hospital's rule doesn't apply yet.}$$

$$= \lim_{x \to \infty} \frac{x^2}{e^x} \to \frac{\infty}{\infty} \qquad \text{L'Hospital's rule does apply now. Find the derivative of the numerator and the denominator.}$$

$$= \lim_{x \to \infty} \frac{2x}{e^x} \to \frac{\infty}{\infty} \qquad \text{L'Hospital's rule applies again.}$$

$$= \lim_{x \to \infty} \frac{2}{e^x} \to \frac{2}{\infty} \qquad \text{L'Hospital's rule is no longer needed.}$$

$$= 0 \qquad \text{(finite)/(infinite)} \to 0 \qquad \blacktriangleleft$$

Indeterminate Exponential Forms

If you raise a number greater than 1 to a large power, the result is very large. A positive number less than 1 raised to a large power is close to zero. For instance,

$$\lim_{x \to \infty} 1.01^x = \infty \qquad \text{and} \qquad \lim_{x \to \infty} 0.99^x = 0$$

If an expression approaches 1^∞, the answer is indeterminate. You saw such a case with the definition of e, which is the limit of $(1 + 1/x)^x$ as x approaches infinity. Expressions that take on the form ∞^0 and 0^0 are also indeterminate. Example 3 shows you how to evaluate an indeterminate form with a variable base and exponent. As in logarithmic differentiation, shown earlier in this chapter, the secret is to take the log of the expression. Then you can transform the result to a fraction and apply l'Hospital's rule.

▶ **EXAMPLE 3**   Evaluate $\lim\limits_{x \to 1} x^{1/(1-x)}$.

Solution   The function $f(x) = x^{1/(1-x)}$ takes on the indeterminate form 1^∞ at $x = 1$. The graph of f (Figure 6-5c) shows a removable discontinuity at $x = 1$ and shows that the limit of $f(x)$ as x approaches 1 is a number less than 0.5. You can find the limit by using l'Hospital's rule after taking the log.

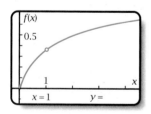

Figure 6-5c

$$\text{Let } L = \lim_{x \to 1} x^{1/(1-x)}.$$

$$\text{Then } \ln L = \ln \left[\lim_{x \to 1} x^{1/(1-x)} \right] = \lim_{x \to 1} [\ln x^{1/(1-x)}] \qquad \text{Reverse ln and lim.}$$

$$= \lim_{x \to 1} \left[\frac{1}{1 - x} \cdot \ln x \right] = \lim_{x \to 1} \frac{\ln x}{1 - x} \to \frac{0}{0} \qquad \text{L'Hospital's rule applies now.}$$

$$= \lim_{x \to 1} \frac{1/x}{-1} = -1 \qquad \text{Find the derivative of the numerator and the denominator.}$$

$$\therefore L = e^{-1} = 0.367879\ldots \qquad \ln L = -1 \Rightarrow L = e^{-1}$$

The solution to Example 3 agrees with the graph in Figure 6-5c.   ◀

Problem Set 6-5

Q1. $e \approx$ —?— (as a decimal)

Q2. $e =$ —?— (as a limit)

Q3. $\ln e =$ —?—

Q4. $\ln(\exp x) =$ —?—

Q5. $e^{\ln x} =$ —?—

Q6. If $\log_b x = \ln x$, then $b =$ —?—.

Q7. $\log_b x =$ —?— (in terms of the function ln)

Q8. If $f(x) = e^x$, then $f'(x) =$ —?—.

Q9. $\int e^{-x}\, dx =$ —?—

Q10. If $f(x) = \int_1^{\tan x} \sin t\, dt$, then $f'(x) =$ —?—.

    A. $\sin x$

    B. $\sin x + C$

    C. $\sin(\tan x)$

    D. $\sin(\tan x) - \sin 1$

    E. $\sin(\tan x)\sec^2 x$

For Problems 1 and 2, estimate graphically the limit of $f(x)$ as x approaches zero. Sketch the graph. Then confirm your estimate using l'Hospital's rule.

1. $\displaystyle\lim_{x \to 0} \frac{2\sin 5x}{3x}$

2. $\displaystyle\lim_{x \to 0} \frac{4\tan 3x}{5x}$

For Problems 3–30, find the indicated limit. Use l'Hospital's rule if necessary.

3. $\displaystyle\lim_{x \to 0} \frac{\tan x}{x}$

4. $\displaystyle\lim_{x \to 0} \frac{\sin x}{x}$

5. $\displaystyle\lim_{x \to 0} \frac{1 - \cos x}{x^2}$

6. $\displaystyle\lim_{x \to 0} \frac{x^2}{\cos 3x - 1}$

7. $\displaystyle\lim_{x \to 0^+} \frac{\sin x}{x^2}$

8. $\displaystyle\lim_{x \to 0} \frac{1 - \cos x}{x + x^2}$

9. $\displaystyle\lim_{x \to 0^+} \frac{\ln x}{1/x}$

10. $\displaystyle\lim_{x \to 0} \frac{e^{3x}}{x^2}$

11. $\displaystyle\lim_{x \to 1} \frac{e^x - e}{5\ln x}$

12. $\displaystyle\lim_{x \to 1} \frac{\ln x - x + 1}{x^2 - 2x + 1}$

13. $\displaystyle\lim_{x \to 2} \frac{3x + 5}{\cos x}$

14. $\displaystyle\lim_{x \to 2} \frac{\tan x}{x - 2}$

15. $\displaystyle\lim_{x \to \infty} \frac{e^x}{x^2}$

16. $\displaystyle\lim_{x \to \infty} \frac{x^3}{e^x}$

17. $\displaystyle\lim_{x \to \infty} \frac{3x + 17}{4x - 11}$

18. $\displaystyle\lim_{x \to \infty} \frac{2 - 7x}{3 + 5x}$

19. $\displaystyle\lim_{x \to \infty} \frac{x^3 - 5x^2 + 13x - 21}{4x^3 + 9x^2 - 11x - 17}$

20. $\displaystyle\lim_{x \to \infty} \frac{3x^5 + 2}{7x^5 - 8}$

21. $\displaystyle\lim_{x \to 0^+} x^x$

22. $\displaystyle\lim_{x \to 0^+} (\sin x)^{\sin x}$

23. $\displaystyle\lim_{x \to \pi/2^-} (\sin x)^{\tan x}$

24. $\displaystyle\lim_{x \to 1^+} x^{1/(x-1)}$

25. $\displaystyle\lim_{x \to \infty} (1 + ax)^{1/x}$ (where a = positive constant)

26. $\displaystyle\lim_{x \to 0} (1 + ax)^{1/x}$ (where a = constant)

27. $\displaystyle\lim_{x \to 0^+} x^{3/(\ln x)}$

28. $\displaystyle\lim_{x \to 0^+} (7x)^{5/(\ln x)}$

29. $\lim\limits_{x \to 0} \left(\dfrac{1}{x} - \dfrac{1}{e^x - 1} \right)$

30. $\lim\limits_{x \to 0} \left(\dfrac{1}{x} - \dfrac{1}{\sin x} \right)$

31. *Infinity Minus Infinity Problem:* Let $f(x) = \sec^2 \frac{\pi}{2}x - \tan^2 \frac{\pi}{2}x$. Because both $\sec(\pi/2)$ and $\tan(\pi/2)$ are infinite, $f(x)$ takes on the indeterminate form $\infty - \infty$ as x approaches 1. Naive thinking might lead you to suspect that $\infty - \infty$ is zero because the difference between two equal numbers is zero. But ∞ is not a number. Plot the graph of f. Sketch the result, showing what happens at $x = 1, 3, 5, \ldots$. Explain the graph based on what you recall from trigonometry.

32. *L'Hospital's Surprise Problem!* Try to evaluate $\lim\limits_{x \to \pi/2} \dfrac{\sec x}{\tan x}$ using l'Hospital's rule. What happens? Find the limit by using some other method.

33. *Zero to the Zero Problem:* Often, the indeterminate form 0^0 equals 1. For instance, the expression $(\sin x)^{\sin x}$ approaches 1 as x approaches 0. But a function of the form

    $$f(x) = x^{k/(\ln x)}$$

    (where k stands for a constant) that approaches 0^0 does not, in general, approach 1. Apply l'Hospital's rule appropriately to ascertain the limit of $f(x)$ as x approaches zero. On your grapher, investigate the graph of $f(x)$. Explain your results.

34. *L'Hospital's Rule, Graphically:* In this problem you will investigate

    $$f(x) = \dfrac{g(x)}{h(x)} = \dfrac{0.3x^2 - 2.7}{0.2x^2 - 2x + 4.2}$$

    which approaches $0/0$ as x approaches 3. You will see l'Hospital's rule graphically.

    a. Confirm that $g(3) = h(3) = 0$.

    b. Figure 6-5d shows the graphs of g and h, along with the tangent lines at $x = 3$. Find equations of the tangent lines. State your answers in terms of $(x - 3)$.

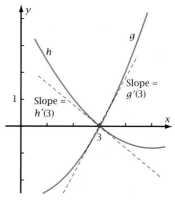

Figure 6-5d

c. Because g and h are differentiable at $x = 3$, they have local linearity in a neighborhood of $x = 3$. Thus, the ratio $g(x)/h(x)$ is approximately equal to the ratio of the two linear functions you found in part b. Show that this ratio is equal to $g'(3)/h'(3)$.

d. Explain the connection between the conclusion you made in part c and l'Hospital's rule. Explain why the conclusion might not be true if either $g(3)$ or $h(3)$ were not equal to zero.

e. Plot the graph of f. Sketch the result, showing its behavior at $x = 3$.

35. *Continuous Compounding of Interest Problem:* Suppose you deposit $1000 in a savings account that is earning interest at 6% per year, compounded annually (once a year). At the end of the first year, it will earn $(0.06)(1000)$, or $60, so there will be $1060 in your account. You can easily find this number by multiplying the original $1000 by 1.06, which is $(1 + \text{interest rate})$. At the end of each subsequent year, you multiply the amount in your account at the beginning of that year again by 1.06, as shown in this table.

Year	Total at End of Year
0	1000
1	$1000(1.06)$
2	$1000(1.06)^2$
3	$1000(1.06)^3$
⋮	⋮
t	$1000(1.06)^t$

a. If the interest is compounded semiannually (twice a year), your account gets half the interest rate for twice as many time periods. If $m(t)$ is the number of dollars in the account after time t, in years, explain why

$$m(t) = 1000\left(1 + \frac{0.06}{2}\right)^{2t}$$

b. Write an equation for $m(t)$ if the interest is compounded n times a year. Then find the limit of this equation as n approaches infinity to find $m(t)$ if the interest is compounded continuously. Treat t as a constant, because it is n that varies as you find the limit.

c. How much more money would you have with continuous compounding than you would have with annual compounding after 5 yr? After 20 yr? After 50 yr?

d. Quick! Write an equation for $m(t)$ if the interest is 7% per year compounded continuously.

36. *Order of Magnitude of a Function Problem:* Let L be the limit of $f(x)/g(x)$ as x approaches infinity. If L is infinite, then f is said to be of a **higher order of magnitude** than g. If $L = 0$, then f is said to be of a **lower order of magnitude** than g. If L is a finite nonzero number, then f and g are said to have the **same order of magnitude**.

a. Rank each kind of function according to its order of magnitude.

  i. Power function, $f(x) = x^n$, where n is a positive constant

  ii. Logarithmic function, $g(x) = \ln x$

  iii. Exponential function, $h(x) = e^x$

b. Quick! Without using l'Hospital's rule, evaluate the following limits.

  i. $\displaystyle\lim_{x \to \infty} \frac{\ln 3x}{x^5}$

  ii. $\displaystyle\lim_{x \to \infty} \frac{x^{100}}{e^{0.01x}}$

  iii. $\displaystyle\lim_{x \to \infty} \frac{e^{0.3x}}{100 \ln x}$

  iv. $\displaystyle\lim_{x \to \infty} \frac{\sqrt{x}}{x}$

  v. $\displaystyle\lim_{x \to \infty} \frac{e^x}{e^{0.2x}}$

37. *Journal Problem:* In your journal, write something about various indeterminate forms. Include examples of functions that approach the following forms.

Types of Indeterminate Form

Limits that take the following forms can equal different numbers at different times and thus cannot be found simply by looking at the form:

$$\frac{0}{0}, \quad \frac{\infty}{\infty}, \quad \infty \cdot 0, \quad \infty - \infty, \quad 1^\infty, \quad \infty^0, \quad \text{and} \quad 0^0$$

Try to pick examples for which the answer is not obvious. For instance, pick a function that approaches 0/0 but for which the limit does not equal 1. Show how other indeterminate forms can be algebraically transformed to 0/0 or to ∞/∞ so that you can use l'Hospital's rule.

6-6   Derivative and Integral Practice for Transcendental Functions

You have learned how to differentiate the elementary transcendental functions—trigonometric and inverse trigonometric, exponential and logarithmic—and how to integrate some of these. In this section you will learn how to integrate the remaining four trigonometric functions (sec, csc, tan, and cot). In Chapter 9, you will learn integration by parts, which will enable you to integrate the remaining elementary transcendental functions—logarithmic and inverse trigonometric.

OBJECTIVE   Differentiate and integrate algebraically functions involving logs and exponentials quickly and correctly so that you can concentrate on the applications in the following chapters.

Integrals of tan, cot, sec, and csc

You can integrate the tangent function by first transforming it to sine and cosine, using the quotient properties from trigonometry.

$$\int \tan x \, dx = \int \frac{\sin x}{\cos x} \, dx$$

$$= -\int \frac{1}{\cos x} (-\sin x \, dx) \qquad \text{Transform to the integral of the reciprocal function.}$$

$$= -\ln |\cos x| + C$$

$$= +\ln \left| \frac{1}{\cos x} \right| + C \qquad \ln n = -\ln (1/n)$$

$$= \ln |\sec x| + C$$

The cotangent function is integrated the same way. The integrals of secant and cosecant are trickier! A clever transformation is required to turn the integrand into the reciprocal function. The key to the transformation is that the derivative of $\sec x$ is $\sec x \tan x$ and the derivative of $\tan x$ is $\sec^2 x$. Here's how it works.

$$\int \sec x \, dx = \int \sec x \cdot \frac{\sec x + \tan x}{\sec x + \tan x} \, dx \qquad \text{Multiply by a "clever" form of 1.}$$

$$= \int \frac{1}{\sec x + \tan x} \cdot (\sec^2 x + \sec x \tan x) \, dx \qquad \text{Write as the reciprocal function.}$$

$$= \ln |\sec x + \tan x| + C \qquad (\sec^2 x + \sec x \tan x) \, dx \text{ is the differential of the denominator.}$$

The formulas for $\int \cot x \, dx$ and for $\int \csc x \, dx$ are derived similarly. These integrals are listed, along with sine and cosine, in this box.

PROPERTIES: *Integrals of the Six Trigonometric Functions*

$\int \sin x \, dx = -\cos x + C$

$\int \cos x \, dx = \sin x + C$

$\int \tan x \, dx = -\ln |\cos x| + C = \ln |\sec x| + C$

$\int \cot x \, dx = \ln |\sin x| + C = -\ln |\csc x| + C$

$\int \sec x \, dx = \ln |\sec x + \tan x| + C$

$\int \csc x \, dx = -\ln |\csc x + \cot x| + C = \ln |\csc x - \cot x| + C$

The next two boxes give properties of logs and exponentials that will help you do calculus algebraically.

PROPERTIES: *Natural Logs and Exponentials*

Definition of the Natural Logarithm Function:

$$\ln x = \int_1^x \frac{1}{t}\, dt \quad \text{(where } x \text{ is a positive number)}$$

Calculus of the Natural Logarithm Function:

$$\frac{d}{dx}(\ln x) = \frac{1}{x} \qquad \int \ln x\, dx \quad \text{(to be introduced in Chapter 9)}$$

Integral of the Reciprocal Function (from the definition):

$$\int \frac{1}{u}\, du = \ln |u| + C$$

Logarithm Properties of ln:

Product: $\ln(ab) = \ln a + \ln b$

Quotient: $\ln(a/b) = \ln a - \ln b$

Power: $\ln(a^b) = b \ln a$

Intercept: $\ln 1 = 0$

Calculus of the Natural Exponential Function:

$$\frac{d}{dx}(e^x) = e^x \qquad \int e^x dx = e^x + C$$

Inverse Properties of Log and Exponential Functions:

$$\ln(e^x) = x \quad \text{and} \quad e^{\ln x} = x$$

Function Notation for Exponential Functions:

$$\exp(x) = e^x$$

Definition of *e*:

$$e = \lim_{n \to 0}(1 + n)^{1/n} = \lim_{n \to \infty}(1 + 1/n)^n$$

$$e = 2.7182818284... \quad \text{(a transcendental number, a nonrepeating decimal)}$$

PROPERTIES: Base-b Logs and Exponentials

Equivalence of Natural Logs and Base-*e* Logs:

$$\ln x = \log_e x \text{ for all } x > 0$$

Calculus of Base-*b* Logs:

$$\frac{d}{dx}(\log_b x) = \frac{1}{\ln b} \cdot \frac{1}{x} \qquad \int \log_b x \, dx \quad \text{(to be introduced in Chapter 9)}$$

Calculus of Base-*b* Exponential Functions:

$$\frac{d}{dx}(b^x) = (\ln b)b^x \qquad \int b^x \, dx = \frac{1}{\ln b}b^x + C$$

Change-of-Base Property for Logarithms:

$$\log_a x = \frac{\log_b x}{\log_b a} = \frac{\ln x}{\ln a}$$

Problem Set 6-6

For Problems 1–56, find y'. Work the problems in the order in which they appear, rather than just the odds or just the evens.

1. $y = \ln(3x + 4)$

2. $y = \ln(3x^5)$

3. $y = \ln(e^{3x})$

4. $y = \ln(\sin 4x)$

5. $y = \ln(\cos^5 x)$

6. $y = \ln(e^5)$

7. $y = \ln[\cos(\tan x)]$

8. $y = \ln\sqrt{x^2 - 2x + 3}$

9. $y = \cos(\ln x)$

10. $y = \sin x \cdot \ln x$

11. $y = e^{7x}$

12. $y = e^{x^3}$

13. $y = e^{5\ln x}$

14. $y = e^{\cos x}$

15. $y = \cos(e^x)$

16. $y = (\cos^3 x)(e^{3x})$

17. $y = e^{x^5}$

18. $y = e^{e^x}$

19. $\sin y = e^x$

20. $y = e^x \cdot \ln x$

21. $y = \int_1^x \frac{1}{t} \, dt$

22. $\tan y = e^x$

23. $y = \ln(e^{\ln x})$

24. $y = 2^x$

25. $y = e^{x \ln 2}$

26. $y = e^{2 \ln x}$

27. $y = x^2$

28. $y = e^{x \ln x}$

29. $y = x^x$

30. $y = x \ln x - x$

31. $y = e^x(x - 1)$

32. $y = \frac{1}{2}(e^x + e^{-x})$

33. $y = \frac{1}{2}(e^x - e^{-x})$

34. $y = \dfrac{e^x}{1 + e^x}$

35. $y = 5^x$

36. $y = \log_5 x$

37. $y = x^{-7} \log_2 x$

38. $y = 2^{-x} \cos x$

39. $y = e^{-2x} \ln 5x$

40. $y = \dfrac{7^x}{\ln 7}$

41. $y = \dfrac{\log_3 x}{\log_3 e}$

42. $y = \dfrac{\log_{10} x}{\log_{10} e}$

43. $y = (\log_8 x)(\ln 8)$

44. $y = (\log_4 x)^{10}$

45. $y = \log_5 x^7$

46. $y = \tan e^x$

47. $y = e^{\sin x}$

48. $y = \ln \csc x$

49. $y = 3^5$

50. $y = \ln(\cos^2 x + \sin^2 x)$

51. $y = \sin x$

52. $y = \sin^{-1} x$

53. $y = \csc x$

54. $y = \tan^{-1} x$

55. $y = \tan x$

56. $y = \cot x$

For Problems 57–80, integrate. Work the problems in the order in which they appear, rather than just the odds or just the evens.

57. $\int e^{4x}\, dx$

58. $\int e^4\, dx$

59. $\int x^3 e^{x^4}\, dx$

60. $\int \cos x \cdot e^{\sin x}\, dx$

61. $\int \dfrac{(\ln x)^5}{x}\, dx$

62. $\int 5^x\, dx$

63. $\int e^{x \ln 5}\, dx$

64. $\int \dfrac{1}{2}(e^x + e^{-x})\, dx$

65. $\int_1^x \dfrac{1}{t}\, dt$

66. $\int e^{-x}\, dx$

67. $\int 2^x\, dx$

68. $\int (x^{-0.2} + 3^x)\, dx$

69. $\int \dfrac{3}{x}\, dx$

70. $\int_1^2 4^x\, dx$

71. $\int (\ln x)^9 \dfrac{1}{x}\, dx$

72. $\int \cos x\, dx$

73. $\int e^{\ln x}\, dx$

74. $\int \ln(e^{3x})\, dx$

75. $\int 0\, dx$

76. $\int \cos x \sec x\, dx$

77. $\int \sec 2x\, dx$

78. $\int \tan 3x\, dx$

79. $\int \cot 4x\, dx$

80. $\int \csc 5x\, dx$

For Problems 81–90, find the limit of the given expression.

81. $\lim\limits_{x \to 0} \dfrac{1 - \cos x}{x}$

82. $\lim\limits_{x \to 0} \dfrac{x}{1 - \cos x}$

83. $\lim\limits_{x \to \pi/2} \dfrac{x}{1 - \cos x}$

84. $\lim\limits_{x \to \pi} \dfrac{x}{1 + \cos x}$

85. $\displaystyle\lim_{x \to 0} \frac{5x - \sin 5x}{x^3}$

86. $\displaystyle\lim_{x \to \infty} \left(1 + \frac{0.03}{x}\right)^x$

87. $\displaystyle\lim_{x \to \infty}(1 + 0.03x)^{1/x}$

88. $\displaystyle\lim_{x \to \infty} \frac{2^x}{x^2}$

89. $\displaystyle\lim_{x \to 2}(0.5x)^{3/(2-x)}$

90. $\displaystyle\lim_{x \to 0}\left(\frac{1}{e^{3x} - 1} - \frac{1}{3x}\right)$

6-7   Chapter Review and Test

In this chapter you extended your knowledge of exponential and logarithmic functions. The integral of the reciprocal problem arose naturally from a population growth problem. The power rule does not work for $\int x^{-1}\, dx$, but you learned another form of the fundamental theorem that allows you to evaluate this integral by a clever definition of the natural logarithmic function. Using the uniqueness theorem, you proved that $\ln x$ defined this way has the properties of logarithms. Putting all this together enabled you to define b^x so that the answer is a real number for any real exponent, rational or irrational. Finally, you learned l'Hospital's rule, with which you can use derivatives to evaluate limits of indeterminate forms such as $0/0$ or ∞/∞.

Review Problems

R0. Update your journal with what you've learned since your last entry. Include such things as

- The one most important thing you've learned in studying Chapter 6

- Which boxes you've been working on in the "define, understand, do, apply" table

- Your ability to work calculus algebraically, not just numerically, on logs and exponentials

- New techniques and properties you've learned

- Any ideas about logs and exponentials you need to ask about before the test on Chapter 6

R1. a. If money in a savings account earns interest compounded continuously, the rate of change of the amount in the account is directly proportional to the amount of money there. Suppose that for a particular account, $dM/dt = 0.06M$, where M is the number of dollars in the account and t is the number of years the money has been

earning interest. Separate the variables so that M is on one side and t is on the other. If $100 is in the account at time $t = 0$, show that when $t = 5$ yr,

$$\int_{100}^{x} M^{-1}\, dM = \int_{0}^{5} 0.06\, dt = 0.3$$

where x is the number of dollars in the account after five years.

b. Use your grapher's numerical integration and solver features to find, approximately, the value of x for which the left integral equals 0.3.

c. To the nearest cent, how much interest will the account have earned when $t = 5$ yr?

R2. a. Explain why you cannot evaluate $\int x^{-1}\, dx$ using the power rule for integrals.

b. State the fundamental theorem of calculus in the derivative of an integral form. Write the definition of $\ln x$ as a definite integral, and use the fundamental theorem in this form to write the derivative of $\ln x$.

c. Differentiate.

   i. $y = (\ln 5x)^3$

   ii. $f(x) = \ln x^9$

   iii. $y = \csc(\ln x)$

   iv. $g(x) = \int_1^{x^2} \csc t \, dt$

d. Integrate.

   i. $\displaystyle\int \frac{\sec x \tan x}{\sec x} \, dx$

   ii. $\displaystyle\int_{-2}^{-3} \frac{10}{x} \, dx$

   iii. $\displaystyle\int x^2(x^3 - 4)^{-1} \, dx$

e. Let $h(x) = \int_2^x f(t) \, dt$ where the graph of f is shown in Figure 6-7a. On a copy of this figure, sketch the graph of $h(x)$ in the domain $[1, 11]$.

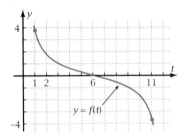

Figure 6-7a

f. *Memory Retention Problem:* Paula Tishan prides herself on being able to remember names. The number of names she can remember at an event is a logarithmic function of the number of people she meets and her particular equation is

$$y = 1 - 101 \ln 101 + 101 \ln(100 + x)$$

where y is the number of names she remembers out of x, the number of people she meets. The graph is shown in Figure 6-7b.

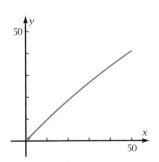

Figure 6-7b

i. How many names can Paula remember if she meets 100 people? If she meets just 1 person? What percent of the people she meets do these two numbers represent?

ii. At what rate does she remember names if she has met 100 people? If she has met just 1 person?

iii. What is the greatest number of names Paula is likely to remember without forgetting any? What assumptions did you make to come up with this answer?

R3. a. i. State the algebraic definition of $y = \log_b x$.

   ii. State the definition of $\ln x$ as a definite integral.

   iii. State the uniqueness theorem for derivatives.

   iv. Use the uniqueness theorem to prove that $\ln x = \log_e x$.

   v. Use the uniqueness theorem to prove the property of the ln of a power.

  b. i. Write the definition of e as a limit.

   ii. Write an equation expressing $\log_b x$ in terms of $\ln x$.

  c. Differentiate.

   i. $y = \log_4 x$

   ii. $f(x) = \log_2(\cos x)$

   iii. $y = \log_5 9^x$

R4. a. Sketch the graph.

   i. $y = e^x$

   ii. $y = e^{-x}$

   iii. $y = \ln x$

  b. Differentiate.

   i. $f(x) = x^{1.4} e^{5x}$

   ii. $g(x) = \sin e^{-2x}$

   iii. $\dfrac{d}{dx}(e^{\ln x})$

   iv. $y = 100^x$

   v. $f(x) = 3.7 \cdot 10^{0.2x}$

   vi. $r(t) = t^{\tan t}$

  c. Differentiate logarithmically:
$y = (5x - 7)^3(3x + 1)^5$

d. Integrate.

i. $\int 10e^{-2x}\,dx$

ii. $\int e^{\cos x}\sin x\,dx$

iii. $\int_{-2}^{2} e^{-0.1x}\,dx$

iv. $\int 10^{0.2x}\,dx$

e. *Chemotherapy Problem:* When a patient receives chemotherapy, the concentration, $C(t)$, in parts per million, of chemical in the blood decreases exponentially with time t, in days since the treatment. Assume that

$$C(t) = 150e^{-0.16t}$$

(See Figure 6-7c.)

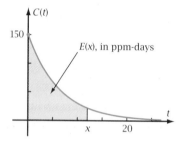

Figure 6-7c

i. You can express the amount of exposure to the treatment, $E(x)$, after $t = x$ days as the product of the concentration and the number of days. Explain why a definite integral must be used to calculate the exposure.

ii. Write an equation for $E(x)$. What is the patient's exposure in 5 days? In 10 days? Does there seem to be a limit to the exposure as x becomes very large? If so, what is the limit? If not, why not?

iii. Quick! Write an equation for $E'(x)$. At what rate is $E(x)$ changing when $x = 5$? When $x = 10$?

f. *Vitamin C Problem:* When you take vitamin C, its concentration, $C(t)$, in parts per million, in your bloodstream rises rapidly, then drops off gradually. Assume that if you take a 500-mg tablet, the concentration is given by

$$C(t) = 200t \times 0.6^{t}$$

where t is time in hours since you took the tablet. Figure 6-7d shows the graph of C.

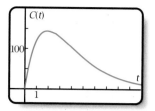

Figure 6-7d

i. Approximately what is the highest concentration, and when does it occur?

ii. How fast is the concentration changing when $t = 1$? When $t = 5$? How do you interpret the signs of these rates?

iii. For how long a period of time will the concentration of the vitamin C remain above 50 ppm?

iv. If you take the vitamin C with a cola drink, the vitamin C decomposes more rapidly. Assume that the base in the equation changes from 0.6 to 0.3. What effect will this change have on the highest concentration and when it occurs? What effect will this change have on the length of time the concentration remains above 50 ppm?

R5. Evaluate the limits.

a. $\displaystyle\lim_{x\to\infty} \frac{2x^2 - 3}{7 - 5x^2}$

b. $\displaystyle\lim_{x\to 0} \frac{x^2 - \cos x + 1}{e^x - x - 1}$

c. $\displaystyle\lim_{x\to\infty} x^3 e^{-x}$

d. $\displaystyle\lim_{x\to 1} x^{\tan(\pi x/2)}$

e. $\displaystyle\lim_{x\to 2} 3x^4$

f. $\displaystyle\lim_{x\to\pi/2} (\tan^2 x - \sec^2 x)$

g. Write as many indeterminate forms as you can think of.

R6. a. Differentiate.

i. $y = \ln(\sin^4 7x)$

ii. $y = x^{-3} e^{2x}$

iii. $y = \cos(2^x)$

iv. $y = \log_3(x^4)$

b. Integrate.

i. $\int e^{-1.7x}\,dx$

ii. $\int 2^{\sec x}(\sec x \tan x \, dx)$

iii. $\int (5 + \sin x)^{-1} \cos x \, dx$

iv. $\int_1^5 \dfrac{1}{z} \, dz$

c. Find each limit.

i. $\displaystyle\lim_{x \to 0} \dfrac{\tan 3x}{x^2}$

ii. $\displaystyle\lim_{x \to \infty} \left(1 - \dfrac{3}{x}\right)^x$

Concept Problems

C1. *Derivation of the Memory Equation Problem:* In Problem R2f, the number of names, y, that Paula remembered as a function of people met, x, was

$$y = 1 - 101 \ln 101 + 101 \ln(100 + x)$$

Suppose that, in general, the number of names a person remembers is

$$y = a + b \ln(x + c)$$

State why the following conditions are reasonable for y and y'. Calculate the constants a, b, and c so that these conditions will be met.

$$y = 1 \text{ when } x = 1$$
$$y' = 1 \text{ when } x = 1$$
$$y = 80 \text{ when } x = 100$$

C2. *Integral of ln Problem:* In Chapter 9, you will learn how to antidifferentiate $y = \ln x$. In this problem you will discover what this antiderivative equals by examining graphs and a table of values. Figure 6-7e shows

$$y_1 = \ln x$$
$$y_2 = \int \ln x \, dx, \text{ with } C = 0$$

(You will learn how to calculate y_2 in Chapter 9.) The table shows values of x, y_1, and y_2. From the tables and the graph, see if you can find an equation for y_2. Describe the methods you tried and state whether or not these methods helped you solve the problem.

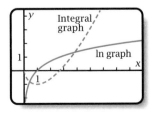

Figure 6-7e

x	$\ln x$	$\int \ln x \, dx$
0.5	$-0.6931471\ldots$	$-0.8465735\ldots$
1.0	0	-1
2.0	$0.6931471\ldots$	$-0.6137056\ldots$
3.0	$1.0986122\ldots$	$0.2958368\ldots$
4.0	$1.3862943\ldots$	$1.5451774\ldots$
5.0	$1.6094379\ldots$	$3.0471895\ldots$
6.0	$1.7917594\ldots$	$4.7505568\ldots$
10.0	$2.3025850\ldots$	$13.0258509\ldots$

C3. *Continued Exponentiation Function Problem:* Let $g(x) = x^x$, where both the base and the exponent are variable. Find $g'(x)$. Then suppose that the number of x's in the exponent is also variable. Specifically, define the **continued exponentiation** function, cont(x), as follows:

$$\text{cont}(x) = x^{x^{x^{x^{\cdot^{\cdot^{\cdot}}}}}}$$

where there is x number of x's in the exponent. For instance,

$$\text{cont}(3) = 3^{3^{3^3}} = 3^{3^{27}} = 3^{7,625,597,484,987}$$

which has over 3.6 billion digits. Find a way to define cont(x) for non-integer values of x. Try to do so in such a way that the resulting function is well defined, continuous, and differentiable for all positive values of x, including the integer values of x.

C4. *Every Real Number Is the ln of Some Positive Number:* Figure 6-7f (on the next page) shows the graph of $f(x) = \ln x$. The graph seems to be increasing, but slowly. In this problem you will prove that there is no horizontal asymptote and that the range of the ln function is all real numbers.

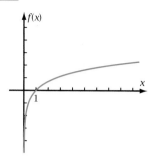

Figure 6-7f

a. Prove by contradiction that ln is unbounded above. That is, suppose that there is a positive number M such that $\ln x \le M$ for all values of $x > 0$. Then pick a clever value of x and show that you get an answer greater than M for $\ln x$.

b. Prove (quickly!) that ln is unbounded below.

c. Prove (quickly!) that ln is continuous for all positive values of x.

d. Prove that for any two numbers a and b, if k is between $\ln a$ and $\ln b$, then there is a number c between a and b such that $\ln c = k$. Sketching a graph and using the intermediate value theorem are helpful here.

e. Use the connections among parts a–d to prove that any real number, k, is the natural log of some positive number. That is, the range of ln is {real numbers}.

f. Use the fact that $\ln x$ and e^x are inverses of each other to show that the domain of $y = e^x$ is the set of all real numbers and that its range is the set of positive numbers.

C5. *Derivative of an Integral with Variable Upper and Lower Limits:* You have learned how to differentiate an integral such as $g(x) = \int_1^x \sin t \, dt$ between a fixed lower limit and a variable upper limit. In this problem you will see what happens if both limits of integration are variable.

a. Find $g'(x)$ if $g(x) = \int_{x^2}^4 \sin t \, dt$.

b. Find $g'(x)$ if $g(x) = \int_{x^2}^{\tan x} \sin t \, dt$.

c. Write a generalization: If $g(x) = \int_{u(x)}^{v(x)} f(t) \, dt$, then $g'(x) = \,\text{—?—}$. Include this generalization in your journal.

C6. What does $\displaystyle\int \frac{d(\text{cabin})}{\text{cabin}}$ equal?

Chapter Test

T1. Write the definition of $\ln x$ as a definite integral.

T2. Write the definition of e as a limit.

T3. State the fundamental theorem of calculus in the derivative of an integral form.

T4. State the uniqueness theorem for derivatives.

T5. Use the uniqueness theorem to prove that $\ln x = \log_e x$ for all positive numbers x.

T6. Let $f(x) = \ln(x^3 e^x)$. Find $f'(x)$ in the following two ways. Show that the two answers are equivalent.

a. Without first simplifying the equation for $f(x)$

b. First simplifying the equation for $f(x)$ by using ln properties

For Problems T7–T11, find an equation for the derivative, then simplify.

T7. $y = e^{2x} \ln x^3$

T8. $v = \ln(\cos 10x)$

T9. $f(x) = (\log_2 4x)^7$

T10. $t(x) = \ln(\cos^2 x + \sin^2 x)$

T11. $p(x) = \int_1^{\ln x} e^t \sin t \, dt$

For Problems T12–T15, evaluate the integral.

T12. $\int e^{5x} \, dx$

T13. $\int (\ln x)^6 \dfrac{dx}{x}$

T14. $\int \sec 5x \, dx$

T15. $\int_0^2 5^x \, dx$ (algebraically)

T16. Find: $\lim\limits_{x \to \infty} \dfrac{5 - 3x}{\ln 4x}$

T17. Find: $\lim\limits_{x \to \pi/2^-} (\tan x)^{\cot x}$

T18. Figure 6-7g shows the graph of function f.

    a. Let $g(x) = \int_2^x f(t)\,dt$. On a copy of Figure 6-7g, sketch the graph of function g.

    b. Let $h(x) = \int_2^{3x-5} f(t)\,dt$. Find $h'(3)$.

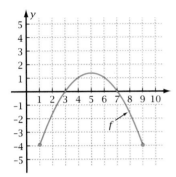

Figure 6-7g

PART 2: Graphing calculators allowed (T19–T23)

T19. Use the definition of ln as a definite integral to evaluate ln 1.8 approximately by using a midpoint Riemann sum M_4. Show the steps you used to evaluate this sum. Show that your answer is close to the calculator's value of ln 1.8.

T20. Let $g(x) = \int_2^{x^2} \sin t\,dt$. Find $g'(x)$ two ways: (1) by evaluating the integral using the fundamental theorem in the $g(b) - g(a)$ form, then differentiating the answer, and (2) directly by using the fundamental theorem in the derivative of the integral form. Show that the two answers are equivalent.

T21. *Proof of the Uniqueness Theorem:* The uniqueness theorem for derivatives states that if $f(a) = g(a)$ for some number $x = a$, and if $f'(x) = g'(x)$ for all values of x, then $f(x) = g(x)$ for all values of x. In the proof of the theorem,

you assume that there is a number $x = b$ for which $f(b)$ does not equal $g(b)$. Show how this assumption leads to a contradiction of the mean value theorem.

T22. *Force and Work Problem:* If you pull a box across the floor, you must exert a certain force. The amount of force needed increases with distance if the bottom of the box becomes damaged as it moves. Assume that the force needed to move a particular box is given by

$$F(x) = 60e^{0.1x}$$

where $F(x)$ is the number of pounds that must be exerted when the box has moved distance x, in feet (Figure 6-7h).

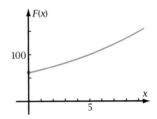

Figure 6-7h

    a. At what rate is the force changing when $x = 5$? When $x = 10$?

    b. Recall that work (foot-pounds) equals force times distance moved. Explain why a definite integral is used to find the work done moving the box to $x = 5$ from $x = 0$.

    c. Write an integral for the work done in moving the box from $x = 0$ to $x = 5$. Evaluate the integral by using the appropriate form of the fundamental theorem to find a "mathematical-world" answer (exact). Then write a real-world answer, rounded appropriately.

T23. What did you learn as a result of taking this test that you did not know before?

6-8   Cumulative Review: Chapters 1–6

The problems in this section constitute a "semester exam" that will help you check your mastery of the concepts you've studied so far. Another cumulative review appears at the end of Chapter 7.

Problem Set 6-8

1. The **derivative** of a function at a point is its instantaneous rate of change at that point. For the function

    $f(x) = 2^x$

    show that you can find a derivative numerically by calculating $f'(3)$, using a **symmetric difference quotient** with $\Delta x = 0.1$.

2. A **definite integral** is a product of x and y, where y is allowed to vary with x. Show that you can calculate a definite integral graphically by estimating the integral of $g(x)$, shown in Figure 6-8a, from $x = 10$ to $x = 50$.

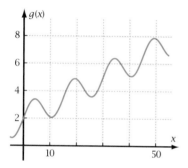

Figure 6-8a

3. Derivatives and definite integrals are defined precisely by using the concept of limit. Write the **definition of limit**.

4. Intuitively, a limit is a y-value that $f(x)$ stays close to when x is close to a given number c. Show that you understand the symbols for, and the meaning of, limit by sketching the graph of one function for which the following is true.

    $$\lim_{x \to 3^-} f(x) = 4 \quad \text{and} \quad \lim_{x \to 3^+} f(x) = -\infty$$

5. Limits are the basis for the formal **definition of derivative**. Write this definition.

6. Show that you can operate with the definition of derivative by using it to show that if $f(x) = x^3$, then $f'(x) = 3x^2$.

7. Properties such as the one used in Problem 6 allow you to calculate derivatives *algebraically*, so that you get exact answers. Find $f'(5)$ for the

function in Problem 6. Then find symmetric difference quotients for $f'(5)$ by using $\Delta x = 0.01$ and $\Delta x = 0.001$. Show that the difference quotients get closer to the exact answer as Δx approaches 0.

8. When **composite functions** are involved, you must remember the **chain rule**. Find the exact value of $f'(7)$ for

    $$f(x) = \sqrt{3x - 5}$$

9. Derivatives can be interpreted *graphically*. Show that you understand this graphical interpretation by constructing an appropriate line on a copy of the graph in Figure 6-8b for the function in Problem 8.

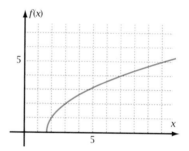

Figure 6-8b

10. You've learned to differentiate **products** and **quotients**, and to find **second derivatives**. Evaluate:

    a. y' if $y = e^{2x} \cos 3x$

    b. $q'(x)$ if $q(x) = \dfrac{\ln x}{\tan x}$

    c. $\dfrac{d^2}{dx^2} (5^x)$

11. **Piecewise functions** can sometimes be differentiable at a point where two pieces meet. For the function f, use one-sided limits to find the values of a and b that make f differentiable at $x = 2$.

    $$f(x) = \begin{cases} ax^2 + 1, & \text{if } x < 2 \\ -x^2 + 6x + b, & \text{if } x \geq 2 \end{cases}$$

    Plot the graph using Boolean variables to restrict the domain, and sketch the result.

12. Definite integrals can be calculated *numerically* by using **Riemann sums**. Show that you understand what a Riemann sum is by finding an upper sum, using $n = 6$ subintervals, for

$$\int_1^4 x^2 \, dx$$

13. The *definition* of **definite integral** involves the limit of a Riemann sum. For the integral in Problem 12, find midpoint Riemann sums with $n = 10$ and $n = 100$ increments. What limit do these sums seem to be approaching?

14. **Indefinite integrals** are **antiderivatives**. Evaluate:

a. $\int \cos^5 x \sin x \, dx$

b. $\int \dfrac{1}{x} \, dx$

c. $\int \tan x \, dx$

d. $\int \sec x \, dx$

e. $\int (3x - 5)^{1/2} \, dx$

15. The **fundamental theorem of calculus** gives an *algebraic* way to calculate definite integrals exactly using indefinite integrals. Use the fundamental theorem to evaluate

$$\int_1^4 x^2 \, dx$$

from Problem 12. Show that your answer is the number you conjectured in Problem 13 for the limit of the Riemann sums.

16. The fundamental theorem is proved using the **mean value theorem** as a lemma. State the mean value theorem. Draw a graph that clearly shows that you understand its conclusion.

17. Much of calculus involves learning how to work problems *algebraically* that you have learned to do graphically or numerically. Use **implicit differentiation** to find dy/dx if

$$y = x^{9/7}$$

thereby showing how the power rule for the derivative of functions with integer exponents is extended to functions with non-integer exponents.

18. Explain why the power rule for derivatives never gives x^{-1} as the answer to a differentiation problem.

19. The **fundamental theorem** in its other form lets you take the derivative of a function defined by a definite integral. Find $f'(x)$ if

$$f(x) = \int_1^{\tan x} \cos 3t \, dt$$

20. Show how the fundamental theorem in its second form lets you write a function whose derivative is x^{-1}.

21. The function you should have written in Problem 20 is the **natural logarithm function**. Use the **uniqueness theorem for derivatives** to show that this function has the property of the log of a power. That is, show that

$$\ln x^a = a \ln x$$

for any constant a and for all values of $x > 0$.

22. Using the **parametric chain rule**, you can find dy/dx for functions such as

$$x = 5 \cos t$$
$$y = 3 \sin t$$

Write a formula for dy/dx in terms of t.

23. The ellipse in Figure 6-8c has the parametric equations given in Problem 22. Find dy/dx if $t = 2$. On a copy of Figure 6-8c, show *graphically* that your answer is reasonable.

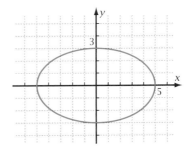

Figure 6-8c

24. You can apply derivatives to **real-world** problems. Suppose that a car's position is

$$y = \tan^{-1} t$$

where y is in feet and t is in seconds. The **velocity** is the instantaneous rate of change of position, and the **acceleration** is the instantaneous rate of change of velocity. Find an equation for the velocity and an equation for the acceleration, both as functions of time.

25. Derivatives can also be applied to problems from the **mathematical world**. For instance, derivatives can be used to calculate limits by using **l'Hospital's rule**. Find

$$\lim_{x \to 0} \frac{e^{3x} - 1}{\sin 5x}$$

26. In the differentiation of the base-b logarithm function, this limit appears:

$$L = \lim_{n \to 0}(1 + n)^{1/n}$$

By appropriate use of l'Hospital's rule, show that this limit equals e, the **base of natural logarithms**.

27. You can use derivatives to find **related rates** of moving objects. Figure 6-8d shows Calvin moving along Broadway in his car. He approaches its intersection with East Castano at 30 ft/s, and Phoebe, moving along East Castano, pulls away from the intersection at 40 ft/s. When Calvin is 200 ft from the intersection and Phoebe is 100 ft from the intersection, what is the instantaneous rate of change of z, the distance between Calvin and Phoebe? Is the distance z increasing or decreasing?

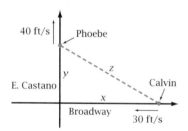

Figure 6-8d

28. **Simpson's rule** can be used to find definite integrals numerically if the integrand is specified only by a table of data. Use Simpson's rule to find the integral of $f(x)$ from $x = 2$ to $x = 5$.

x	$f(x)$
2.0	100
2.5	150
3.0	170
3.5	185
4.0	190
4.5	220
5.0	300

29. You can use integrals to find the **volume** of a solid object that has a variable cross-sectional area. The solid cone in Figure 6-8e is formed by rotating about the x-axis the region under the line $y = (r/h)x$ from $x = 0$ to $x = h$ (the height of the cone). Find the volume dV of the disk-shaped slice of the solid shown in terms of the sample point (x, y) and the differential dx. Then integrate to find the volume. Show that your answer is equivalent to the geometric formula for the volume of a cone,

$$V = \frac{1}{3}\pi r^2 h$$

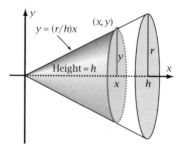

Figure 6-8e

30. It is important for you to be able to write about mathematics. Has writing in your journal helped you understand calculus better? If so, give an example. If not, why not?

The Calculus of Growth and Decay

Baobab trees, which grow in various parts of southern Africa, can live for thousands of years. You can find the age of one of these trees by measuring carbon-14, absorbed when the tree grew. The rate of decay of carbon-14 is proportional to the amount remaining. Integrating the differential equation that expresses this fact shows that the amount remaining is an exponential function of time.

Mathematical Overview

If you know the rate at which a population grows, you can use antiderivatives to find the population as a function of time. In Chapter 7, you will learn ways to solve differential equations for population growth and other related real-world phenomena. You will solve these differential equations in four ways.

Graphically   The icon at the top of each even-numbered page of this chapter shows three particular solutions of the same differential equation. This graph also shows the slope field for the differential equation.

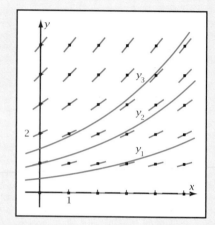

Numerically

x	y_1	y_2	y_3
0	0.5	1.0	1.5
1	0.64...	1.28...	1.92...
2	0.82...	2.64...	2.47...
3	1.05...	2.11...	3.17...
⋮	⋮	⋮	⋮

Algebraically $\dfrac{dy}{dx} = 0.25y \Rightarrow y = Ce^{0.25x}$, a differential equation

Verbally   *I learned that the constant of integration is of vital importance in the solution of differential equations. Different values of C give different particular solutions. So I must remember +C!*

7-1 Direct Proportion Property of Exponential Functions

In Chapters 1–6, you learned the heart of calculus. You now know precise definitions and techniques for calculating limits, derivatives, indefinite integrals, and definite integrals. In this chapter you will solve *differential equations*, which express the rate at which a function grows. The function can represent population, money in a bank, water in a tub, radioactive atoms, or other quantities. A *slope field*, shown in the graph in this chapter's Mathematical Overview, helps you solve complicated differential equations graphically. *Euler's method* allows you to solve them numerically. Antiderivatives let you solve them algebraically. The experience you gain in this chapter will equip you to apply these concepts intelligently when they arise in your study of such fields as biology, economics, physics, chemistry, engineering, medicine, history, and law.

OBJECTIVE Discover, on your own or with your study group, a property of exponential functions by working a real-world problem.

Exploratory Problem Set 7-1

1. Suppose the number of dollars, $D(t)$, in a savings account after time t, in years, is

    $D(t) = 500(1.06^t)$

    Figure 7-1a shows the graph of function D. Calculate the number of dollars at $t = 0$ yr, $t = 10$ yr, and $t = 20$ yr.

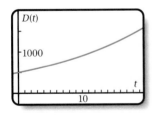

Figure 7-1a

2. For Problem 1, calculate $D'(0)$, $D'(10)$, and $D'(20)$. What are the units of $D'(t)$? Does the rate increase, decrease, or stay the same as the amount in the account increases?

3. For the account described in Problem 1, let $R(t)$ be the instantaneous rate of change of money in dollars per year *per dollar in the account*. Calculate $R(0)$, $R(10)$, and $R(20)$.

4. If you multiply the values of $R(t)$ you found in Problem 3 by 100, you get the interest rates for the savings account expressed as percentages. Does the interest rate go up, go down, or stay the same as the amount of money in the account increases?

5. Recall that if y is directly proportional to x, then $y = kx$, where k stands for a constant (called the constant of proportionality). Show that the following property is true.

    > **Property: Direct Proportion Property of Exponential Functions**
    >
    > If f is an exponential function, $f(x) = a \cdot b^x$, where a and b are positive constants, then $f'(x)$ is directly proportional to $f(x)$.

6. Just for fun, see if you can prove the converse of the property given in Problem 5. That is, prove that if $f'(x)$ is directly proportional to $f(x)$, then f is an exponential function of x.

7-2 Exponential Growth and Decay

At the beginning of Chapter 6, you encountered a population growth problem in which the rate of change of the population, dP/dt, was directly proportional to that population. An equation such as $dP/dt = kP$ is called a **differential equation**. Finding an equation for P as a function of t is called *solving the differential equation*. In this section you will learn an efficient way to solve this sort of differential equation.

OBJECTIVE Given a real-world situation in which the rate of change of y with respect to x is directly proportional to y, write and solve a differential equation and use the resulting solution as a mathematical model to make predictions and interpretations of that real-world situation.

▶ **EXAMPLE 1** *Population Problem:* The population of the little town of Scorpion Gulch is now 1000 people. The population is presently growing at about 5% per year. Write a differential equation that expresses this fact. Solve it to find an equation that expresses population as a function of time.

Solution Let P be the number of people after time t, in years after the present. The differential equation is

$$\frac{dP}{dt} = 0.05P$$

The growth rate is dP/dt; 5% of the population is 0.05 times the population.

$$\frac{dP}{P} = 0.05\,dt$$

Use algebra to separate the variables on opposite sides of the differential equation.

$$\int \frac{dP}{P} = \int 0.05\,dt$$

Integrate both sides of the differential equation.

$$\ln |P| = 0.05t + C$$

Evaluate the integral.

$$e^{\ln |P|} = e^{0.05t + C}$$

Exponentiate both sides of the integrated equation.

$$|P| = e^{0.05t} \cdot e^{C}$$

Exponential of an ln on the left, product of powers with equal bases on the right.

$$P = \pm e^{C} \cdot e^{0.05t}$$

$|P| = \pm P$

$$P = C_1 e^{0.05t}$$

e^{C} is a positive constant. Let $C_1 = \pm e^{C}$. This is the general solution.

$$1000 = C_1 e^{(0.05)(0)} = C_1$$

Substitute 0 for t and 1000 for P.

$$\therefore P = 1000\, e^{0.05t}$$

Substitute 1000 for C_1. This gives the particular solution. ◀

This differential equation was solved by **separating the variables**. The **general solution** represents a **family** of functions (Figure 7-2a), each with a different constant of integration. The population of 1000 at $t = 0$ is called an **initial condition**, or sometimes a **boundary condition**. The solution of a differential equation that meets a given initial condition is called a **particular solution**. Figure 7-2a shows the particular solution from Example 1. It also shows two other particular solutions, with $C_1 = 500$ and $C_1 = 1500$.

Example 2 shows how you can use a differential equation to solve another real-world situation. You can measure the air pressure in a car or bicycle tire in pounds per square inch (psi). If the pressure in each of the four tires of a 3000-lb vehicle is 30 psi, the tires will flatten enough on the bottoms to have "footprints" totaling 100 in². Lower tire pressure will cause the tires to deform more so that pressure times area still equals 3000 lb.

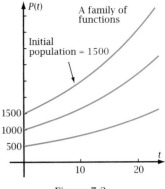

Figure 7-2a

▶ **EXAMPLE 2**   *Punctured Tire Problem:* You have just run over a nail. As air leaks out of your tire, the rate of change of the air pressure inside the tire is directly proportional to that pressure.

    a. Write a differential equation that states this fact. Evaluate the proportionality constant if, at time zero, the pressure is 35 psi and decreasing at 0.28 psi/min.

    b. Solve the differential equation subject to the initial condition given in part a.

    c. Sketch the graph of the function. Show its behavior a long time after the tire is punctured.

    d. What will the pressure be 10 min after the tire was punctured?

    e. The car is safe to drive as long as the tire pressure is 12 psi or greater. For how long after the puncture will the car be safe to drive?

Solution   a. Let p = pressure, in psi.
      Let t = time since the puncture, in minutes.

$$\frac{dp}{dt} = kp$$ Rate of change of pressure is directly proportional to pressure.

$$-0.28 = k(35)$$ Substitute for dp/dt and p. Because p is decreasing, dp/dt is negative.

$$-0.008 = k$$

$$\therefore \frac{dp}{dt} = -0.008p$$

b.

$$\frac{dp}{p} = -0.008\,dt$$ Separate the variables.

$$\int \frac{dp}{p} = \int -0.008\,dt$$ Integrate both sides.

$$\ln|p| = -0.008t + C_1$$ Use C_1 here so that you can use the simpler symbol C later on.

$$e^{\ln|p|} = e^{-0.008t+C_1}$$

$$|p| = e^{C_1}e^{-0.008t}$$

$$p = Ce^{-0.008t}$$ Let $C = \pm e^{C_1}$ so that you can use p instead of $|p|$.

$$35 = Ce^{-0.008(0)} \Rightarrow 35 = C$$ Substitute the initial condition.

$$\therefore p = 35e^{-0.008t}$$ Write the particular solution.

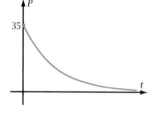

Figure 7-2b

c. The graph is shown in Figure 7-2b. The t-axis is a horizontal asymptote. You should be able to sketch the graph of a decreasing exponential function such as this one without plotting it on your grapher first.

d. $p = 35e^{-0.008(10)} = 32.30907\ldots$

The pressure 10 min after the tire is punctured will be about 32.3 psi.

e.

$$12 = 35e^{-0.008t}$$ Substitute 12 for p.

$$\frac{12}{35} = e^{-0.008t}$$

$$\ln\frac{12}{35} = -0.008t$$

$$t = \frac{\ln\frac{12}{35}}{-0.008} = 133.805\ldots$$

The car will be safe to drive for about 134 min, or slightly less than $2\frac{1}{4}$ h. ◀

In Problem Set 7-2, you will work more problems in which differential equations lead to exponential functions.

Problem Set 7-2

Quick Review

Q1. Sketch: $y = e^x$

Q2. Sketch: $y = e^{-x}$

Q3. Sketch: $y = \ln x$

Q4. Sketch: $y = x^2$

Q5. Sketch: $y = x^3$

Q6. Sketch: $y = 1/x$

Q7. Sketch: $y = x$

Q8. Sketch: $y = 3$

Q9. Sketch: $x = 4$

Q10. Sketch: $y = 3 - x$

1. *Bacteria Problem:* Bacteria in a lab culture (Figure 7-2c) grow in such a way that the instantaneous rate of change of the bacteria population is directly proportional to the number of bacteria present.

   a. Write a differential equation that expresses this relationship. Separate the variables and integrate the equation, solving for the number of bacteria as a function of time.

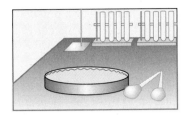

Figure 7-2c

b. Suppose that initially there are 5 million bacteria. Three hours later, the number has grown to 7 million. Write the particular equation that expresses the number of millions of bacteria as a function of the number of hours.

c. Sketch the graph of bacteria versus time.

d. What will the bacteria population be one full day after the first measurement?

e. When will the population reach 1 billion (1000 million)?

2. *Nitrogen-17 Problem:* When a water-cooled nuclear power plant is in operation, oxygen in the water is transmuted to nitrogen-17 (Figure 7-2d). After the reactor is shut down, the radiation from the nitrogen-17 decreases in such a way that the rate of change in the radiation level is directly proportional to the radiation level.

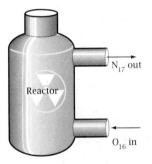

Figure 7-2d

a. Write a differential equation that expresses the rate of change in the radiation level in terms of the radiation level. Solve this equation to find an equation that expresses the radiation level in terms of time.

b. Suppose that when the reactor is first shut down, the radiation level is 3×10^{17} units. After 60 s the level has dropped to 5.6×10^{13} units. Write the particular equation.

c. Sketch the graph of radiation level versus time.

d. It is safe to enter the reactor compartment when the radiation level has dropped to 7×10^{-3} units. Will it be safe to enter the reactor compartment 5 min after the reactor has been shut down? Justify your answer.

3. *Chemical Reaction Problem:* Suppose that a rare substance called calculus foeride reacts in such a way that the rate of change in the amount of foeride left unreacted is directly proportional to that amount.

a. Write a differential equation that expresses this relationship. Integrate it to find an equation that expresses amount in terms of time. Use the initial conditions that the amount is 50 mg when $t = 0$ min, and 30 mg when $t = 20$ min.

b. Sketch the graph of amount versus time.

c. How much foeride remains an hour after the reaction starts?

d. When will the amount of foeride equal 0.007 mg?

4. *Car Trade-In Problem:* Major purchases, like cars, depreciate in value. That is, as time passes, their value decreases. A reasonable mathematical model for the value of an object that depreciates assumes that the instantaneous rate of change of the object's value is directly proportional to the value.

a. Write a differential equation that says that the rate of change of a car's trade-in value is directly proportional to that trade-in value. Integrate the equation and express the trade-in value as a function of time.

b. Suppose you own a car whose trade-in value is presently $4200. Three months ago its trade-in value was $4700. Find the particular equation that expresses the trade-in value as a function of time since the car was worth $4200.

c. Plot the graph of trade-in value versus time. Sketch the result.

d. What will the trade-in value be one year after the time the car was worth $4700?

e. You plan to get rid of the car when its trade-in value drops to $1200. When will this be?

f. At the time your car was worth $4700, it was 31 months old. What was its trade-in value when it was new?

g. The purchase price of the car when it was new was $16,000. How do you explain the difference between this number and your answer to part f?

5. *Biological Half-Life Problem:* While working, you accidentally inhale some mildly poisonous fumes. Twenty hours later you still feel a bit woozy, so you finally report to the medical facility where you are sent by your employer. From blood samples, the doctor measures a poison concentration of 0.00372 mg/mL and tells you to come back in 8 hours. On the second visit, she measures a concentration of 0.00219 mg/mL.

Let t be the number of hours that have elapsed since you first visited the doctor and let C be the concentration of poison in your blood, in milligrams per milliliter. From biology, you realize that the instantaneous rate of change of C with respect to t is directly proportional to C.

a. Write a differential equation that relates these two variables.

b. Solve the differential equation subject to the initial conditions specified. Express C as a function of t.

c. The doctor says that you could have suffered serious damage if the concentration of poison had ever been 0.015 mg/mL. Based on your mathematical model from part a, was the concentration ever that high? Justify your answer.

d. Plot the graph of this function. Sketch the function, showing the values in part c.

e. The **biological half-life** of a poison is the length of time it takes for its concentration to drop to half of its present value. Find the biological half-life of this poison.

6. *Carbon-14 Dating Problem:* Carbon-14 is an isotope of carbon that forms when radiation from the Sun strikes ordinary carbon dioxide in the atmosphere. Thus, plants such as trees, which get their carbon dioxide from the atmosphere, contain small amounts of carbon-14. Once a particular part of a plant stops growing, no more new carbon-14 is absorbed by that part. The carbon-14 in that part decays slowly, transmuting into nitrogen-14. Let P be the percentage of carbon-14 that remains in a part of a tree that grew a number of years ago, t.

a. The instantaneous rate of change of P with respect to t is directly proportional to P. Use this fact to write a differential equation that relates these two variables.

b. Solve the differential equation for P in terms of t. Use the fact that the half-life of carbon-14 is 5750 yr. That is, if $P = 100$ when $t = 0$, then $P = 50$ when $t = 5750$.

c. The oldest living trees in the world are the bristlecone pines in the White Mountains of California. Scientists have counted 4000 growth rings in the trunk of one of these trees, meaning that the innermost ring grew 4000 years ago. What percentage of the original carbon-14 would you expect to find remaining in this innermost ring?

d. A piece of wood claimed to have come from Noah's Ark is found to have 48.37% of its carbon-14 remaining. It has been suggested that the Great Flood occurred in 4004 B.C.E. Is the wood old enough to come from Noah's Ark? Explain.

e. Plot the graph of P versus t for times from 0 years through at least 20,000 years. Use your grapher's TRACE feature to demonstrate that your answers to parts c–e are correct by showing that they lie on this graph. Sketch the results.

7. *Compound Interest Problem I:* Banks compound interest on savings continuously, meaning that the instant the interest is earned, it also starts to earn interest. So the instantaneous rate at which the money in your account changes is directly proportional to the size of your account. As a result, your money, M, increases at a rate proportional to the amount of money in the account (Figure 7-2e),

$$\frac{dM}{dt} = kM$$

where M is in dollars, t is in years, and k is a proportionality constant.

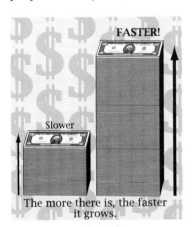

FASTER!

Slower

The more there is, the faster it grows.

Figure 7-2e

Based on what you have learned so far in calculus, determine how M varies with t. To find the value of k for a particular savings account, use the fact that if $100 is invested at an interest rate of 7% per year, then M is increasing at a rate of $7 per year at the instant $M = 100$. Once you have found a function that expresses M in terms of t, investigate the effects of keeping various amounts in the

account for various times at various interest rates. For instance, which option gives you more money in the long run: investing twice the amount of money, leaving the money twice as long, or finding an interest rate twice as high?

8. *Compound Interest Problem II:* If interest in a savings account is compounded at discrete intervals rather than continuously, then the amount of money, M, in the account is

$$M = M_0\left(1 + \frac{k}{n}\right)^{nt}$$

where M is the number of dollars at time t, in years after the investment was made, M_0 is the number of dollars invested when $t = 0$, k is the interest rate expressed as a decimal, and n is the number of times per year the interest is compounded. The ideas behind this equation are shown in Problem 35 of Problem Set 6-5. Compare the amount, M, you will have after a specified time if the money is compounded yearly, quarterly (four times a year), monthly, and daily. Compare these amounts with what you would get if the interest were compounded continuously, as in Problem 7. What conclusions can you make about the relative effects of higher interest rate versus more frequent compounding of interest? See if you can show that the function from Problem 7 for continuous compounding is a logical consequence of taking the limit of the compound interest formula in this problem as n approaches infinity.

9. *Negative Initial Condition Problem:* Suppose that the differential equation

$$\frac{dy}{dx} = 0.3y$$

has the initial condition $y = -4$ when $x = 0$. Show the steps you used to solve this differential equation. Make sure to show the step where you remove the absolute value of y. Sketch the graph of the particular solution.

10. *Initial Condition Not $x = 0$ Problem:* Suppose that the differential equation

$$\frac{dy}{dx} = -0.2y$$

has the initial condition $y = 30$ when $x = 7$. Show the steps you used to solve this differential equation. Make sure you show how you found the constant of integration.

11. *Generalization Problem:* In this section you worked problems in which the rate of change of y was directly proportional to y. Solving these differential equations always led to an exponential function. Good mathematicians are quick to spot possible generalizations that shorten the problem-solving process. Prove the following property.

> **Theorem: Converse of the Direct Proportion Property of Exponential Functions**
>
> If $\dfrac{dy}{dx} = ky$, where k stands for a constant, then $y = y_0 e^{kx}$, where y_0 is the value of y when $x = 0$.

7-3 Other Differential Equations for Real-World Applications

In Section 7-2, you worked real-world problems in which dy/dx was directly proportional to y. In this section you will work real-world problems in which the derivative has a more complicated property than being directly proportional to y. Some of the resulting functions will be exponential and others will not.

OBJECTIVE

Given the relationship between a function and its rate of change, write a differential equation, solve it to find an equation for the function, and use the function as a mathematical model.

▶ **EXAMPLE 1**

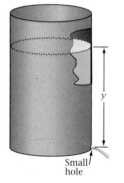

Small hole

Figure 7-3a

Tin Can Leakage Problem: Suppose you fill a tall (topless) tin can with water, then punch a hole near the bottom with an ice pick (Figure 7-3a). The water leaks quickly at first, then more slowly as the depth of the water decreases. In engineering or physics, you will learn that the rate at which water leaks out is directly proportional to the square root of its depth. Suppose that at time $t = 0$ min, the depth is 12 cm, and dy/dt is -3 cm/min.

a. Write a differential equation that states that the instantaneous rate of change of y with respect to t is directly proportional to the square root of y. Find the proportionality constant.

b. Solve the differential equation to find y as a function of t. Use the given information to find the particular solution. What type of function is this?

c. Plot the graph of y as a function of t. Sketch the graph. Consider the domain of t in which the function gives reasonable answers.

d. Solve the equation from part b algebraically to find the time it takes the can to drain. Compare your answer with the time it would take the can to drain at a constant rate of -3 cm/min.

Solution

a. $\dfrac{dy}{dt} = ky^{1/2}$ k is the proportionality constant.
The 1/2 power is equivalent to the square root.

At $t = 0$, $y = 12$ and $\dfrac{dy}{dt} = -3$.

dy/dt is negative because y is decreasing as t increases.

$\therefore -3 = k(12^{1/2}) \Rightarrow k = -3(12^{-1/2})$

b.
$$y^{-1/2} \, dy = k \, dt$$

Separate the variables. It's simpler to write k instead of $-3(12^{-1/2})$.

$$\int y^{-1/2} \, dy = \int k \, dt$$

$$2y^{1/2} = kt + C$$

$$2(12^{1/2}) = k \cdot 0 + C = C$$

Substitute the initial condition $y = 12$ when $t = 0$.

$$y = \tfrac{1}{4}(kt + C)^2 = \tfrac{1}{4}(k^2 t^2 + 2kCt + C^2)$$

Perform the algebra before you substitute for k and C.

$$y = \tfrac{3}{16}t^2 - 3t + 12$$

Use your pencil and paper to see how to get this!

This is a *quadratic* function.

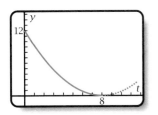

Figure 7-3b

c. Figure 7-3b shows the graph of this function. At time $t = 8$, the can is empty. Beyond that time the model indicates that the can is filling back up. Before $t = 0$, the can was not draining. Thus, the domain in which the mathematical model gives reasonable answers is $0 \le t \le 8$. The part of the graph beyond $t = 8$ is dotted to show the quadratic nature of the function.

d.
$$0 = \tfrac{3}{16}t^2 - 3t + 12$$

$$0 = \tfrac{3}{16}(t^2 - 16t + 64) \Rightarrow 0 = \tfrac{3}{16}(t - 8)^2 \Rightarrow t = 8$$

The can takes 8 min to drain, which is twice as long as it would take at the original rate of -3 cm/min. ◀

Example 2 shows what can happen if a population is growing at a constant rate because of one influence and decaying at another rate because of a second influence. The "population" in this case is water in a lake behind a dam.

▶ **EXAMPLE 2**   *Dam Leakage Problem:* A new dam is constructed across Scorpion Gulch (Figure 7-3c). The engineers need to predict the volume of water in the lake formed by the dam as a function of time. At time $t = 0$ days, the water starts flowing in at a fixed rate F, in ft³/h. Unfortunately, as the water level rises, some leaks out. The leakage rate, L, in ft³/h, is directly proportional to the volume of water, W, in ft³, present in the lake. Thus, the instantaneous rate of change of W is equal to $F - L$.

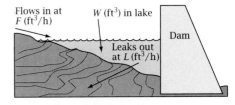

Figure 7-3c

a. What does L equal in terms of W? Write a differential equation that expresses dW/dt in terms of F, W, and t.

b. Solve for W in terms of t, using the initial condition $W = 0$ when $t = 0$.

c. The engineers know that water is flowing in at $F = 5000 \text{ ft}^3/\text{h}$. Based on geological considerations, the proportionality constant in the leakage equation is assumed to be $0.04/\text{h}$. Write the equation for W, substituting these quantities.

d. Predict the volume of water in the lake after 10 h, 20 h, and 30 h. After these intervals, how much water has flowed in and how much has leaked out?

e. When will there be $100,000 \text{ ft}^3$ of water in the lake?

f. Find the limit of W as t approaches infinity. State the real-world meaning of this number.

g. Draw the graph of W versus t, showing the asymptote.

Engineers at the Hoover Dam can control the flow of the Colorado River. The water that the Hoover Dam backs up forms Lake Mead.

Solution

In a problem this complicated, it helps to write what letters are being used and whether they stand for variables or constants.

$$W = \text{no. of ft}^3 \text{ of water in the lake (dependent variable)}$$
$$t = \text{no. of hours since water started flowing (independent variable)}$$
$$F = \text{no. of ft}^3/\text{h the water flows in (a constant)}$$
$$L = \text{no. of ft}^3/\text{h the water leaks out (a variable)}$$

a. Proceed by putting together the information asked for in the problem.

$L = kW$ Meaning of "directly proportional." k stands for the constant of proportionality.

$dW/dt = F - L$ dW/dt is the instantaneous rate of change of W.
$\therefore dW/dt = F - kW$

b. Separating the variables appears to be tricky. Recall that F and k are constants.

$$\frac{dW}{F - kW} = dt$$ Multiply by dt. Divide by $F - kW$.

$$\int \frac{dW}{F - kW} = \int dt$$ Integrate both sides.

The differential of the denominator, $d(F - kW)$, equals $-k\,dW$. Make the numerator equal to $-k\,dW$ by multiplying and dividing by $-k$.

$$\frac{1}{-k} \int \frac{-k\,dW}{F - kW} = \int dt$$

$$-\frac{1}{k} \ln |F - kW| = t + C$$ Integral of the reciprocal function.

$$\ln |F - kW| = -kt - kC$$ Isolate the ln term.

$$|F - kW| = e^{-kt-kC} = e^{-kt} \cdot e^{-kC}$$ Exponentiate both sides. Write the right side as a product.

$$F - kW = \pm e^{-kC} \cdot e^{-kt} = C_1 e^{-kt}$$ $|F - KW| = \pm(F - KW)$, so let $C_1 = \pm e^{kC}$.

$$W = \frac{1}{k}(F - C_1 e^{-kt})$$ Solve for W to get the general solution.

Substituting the initial condition $W = 0$ when $t = 0$,

$$0 = \frac{1}{k}(F - C_1 e^0)$$

$$0 = \frac{1}{k}(F - C_1), \text{ which implies that } C_1 = F$$

$$\therefore W = \frac{1}{k}(F - Fe^{-kt})$$

$$W = \frac{F}{k}(1 - e^{-kt})$$

c. Substituting 5000 for F and 0.04 for k gives

$$W = 125{,}000(1 - e^{-0.04t}) \qquad \text{This is the particular solution.}$$

d. Try using your grapher's TRACE or TABLE feature to find values of W. Round to some reasonable value, such as to the nearest cubic foot. The inflow values are calculated by multiplying 5000 by t. The values of leakage are found by subtraction.

t	W	Inflow	Leakage
10	41210	50000	8790
20	68834	100000	31166
30	87351	150000	62649

e. Substituting 100,000 for W and using the appropriate algebra gives

$$100000 = 125000(1 - e^{-0.04t})$$

$$0.8 = 1 - e^{-0.04t} \Rightarrow e^{-0.04t} = 0.2$$

$$-0.04t = \ln 0.2 \qquad \text{Take ln of both sides.}$$

$$t = \frac{\ln 0.2}{-0.04} = 40.2359...$$

So it will take a bit more than 40 h for the lake to fill up to 100,000 ft^3.

f. $$\lim_{t \to \infty} W = \lim_{t \to \infty} 125000(1 - e^{-0.04t})$$

$$= 125000(1 - 0)$$

$$= 125000$$

In the long run, the volume of water in the lake approaches 125,000 ft^3.

Note that as t approaches infinity, $e^{-0.04t}$ has the form $1/\infty$, and therefore approaches zero.

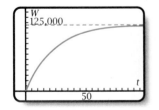

Figure 7-3d

g. The graph is shown in Figure 7-3d. ◄

▶ **EXAMPLE 3** The lake in Example 2 is now filling with water. The actual volume of water at time $t = 10$ h is exactly 40,000 ft^3. The flow rate is still 5000 ft^3/h, as predicted. Use this information to find a more precise value of the leakage constant k.

Solution   Substituting $F = 5000$ and the ordered pair $(t, W) = (10, 40000)$ gives

$$40000 = \frac{5000}{k}(1 - e^{-10k})$$

You cannot solve this equation analytically for k because k appears both algebraically (by division) and transcendentally (as an exponent). Fortunately, your grapher allows you to evaluate k as precisely as you like. You might first divide both sides by 5000 to make the numbers more manageable.

$$8 = \frac{1}{k}(1 - e^{-10k})$$

Then use your grapher's solver or intersect feature to find the value of k. The result is

$$k = 0.046421\ldots$$

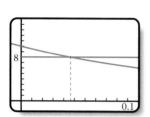

Figure 7-3e

You can also find k by plotting and tracing (Figure 7-3e). It is close to the 0.04 you assumed in part c of Example 2.   ◀

Problem Set 7-3

Q1. If $dy/dx = ky$, then $y = $ —?—.

Q2. If $dy/dx = kx$, then $y = $ —?—.

Q3. If $dy/dx = k$, then $y = $ —?—.

Q4. If $dy/dx = \sin x$, then $y = $ —?—.

Q5. If $y = \sin^{-1} x$, then $dy/dx = $ —?—.

Q6. $\ln(e^{5\cos x}) = $ —?—.

Q7. $e^{\ln \tan x} = $ —?—.

Q8. Sketch the graph of y (Figure 7-3f) if $y(1) = 0$.

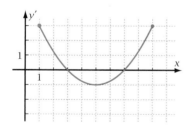

Figure 7-3f

Q9. What does it mean for f to be integrable on $[a, b]$?

Q10. If $\int_1^4 v(t)\, dt = 17$ and $\int_1^7 v(t)\, dt = 33$, then $\int_4^7 v(t)\, dt = $ —?—.

A. $33 + 17$ B. $33 - 17$ C. $17 - 33$
D. $33/17$ E. $17/33$

1. *Sweepstakes Problem I:* You have just won a national sweepstakes! Your award is an income of \$100 a day for the rest of your life! You decide to put the money into a fireproof filing cabinet (Figure 7-3g) and let it accumulate. But temptation sets in, and you start spending the money at rate S, in dollars per day.

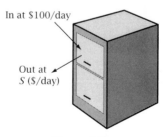

In at \$100/day

Out at
S (\$/day)

Figure 7-3g

a. Let M be the number of dollars you have in the filing cabinet and t be the number of days that you've been receiving the money. Assuming that the rates are continuous, write a differential equation that expresses dM/dt in terms of S.

b. Your spending rate, S, is directly proportional to the amount of money, M. Write an equation that expresses this fact, then substitute the result into the differential equation.

c. Separate the variables and integrate the differential equation in part b to get an equation for M in terms of t. For the initial condition, realize that $M = 0$ when $t = 0$.

d. Suppose that each day you spend 2% of the money in your filing cabinet, so the proportionality constant in the equation for S is 0.02. Substitute this value into the equation you found in part c to get M explicitly in terms of t.

e. Plot the graph of M versus t, and sketch the result.

f. How much money will you have in the filing cabinet after 30 days, 60 days, and 90 days? How much has come in? How much have you spent?

g. How much money do you have in the filing cabinet after one year? At what rate is the amount increasing at this time?

h. What is the limit of M as t approaches infinity?

2. *Sweepstakes Problem II:* You've won a national sweepstakes that will give you an income of $100 a day for the rest of your life! You put the money into a savings account at a bank (Figure 7-3h), where it earns interest at a rate directly proportional to the amount, M, in the account. Assume that the $100 per day rate is continuous, so that dM/dt equals $100 + kM$, where k is a proportionality constant. Solve this differential equation subject to the initial condition that you had no money in the account at $t = 0$ days. Find the proportionality constant if the interest rate is 0.02% (*not* 2%!) per day, or roughly 7% per year. Transform the solution so that M is in terms of t. Use your result to explore how M varies with t. (A graph might help.) Consider such information as how much of M and of dM/dt comes from the $100

per day and how much comes from interest after various numbers of days. What is the limit of M as t approaches infinity?

Figure 7-3h

3. *Electrical Circuit Problem:* When you turn on the switch in an electric circuit, a constant voltage (electrical "pressure"), E, is applied instantaneously to the circuit. This voltage causes an electrical current to flow through the circuit. The current is $I = 0$ A (ampere) when the switch is turned on at time $t = 0$ s. The part of this voltage that goes into overcoming the electrical resistance of the circuit is directly proportional to the current, I. The proportionality constant, R, is called the **resistance** of the circuit. The rest of the voltage is used to get the current moving through the circuit in the first place and varies directly with the instantaneous rate of change of the current with respect to time. The constant for this proportionality, L, is called the **inductance** of the circuit. Figure 7-3i shows an electric circuit diagram.

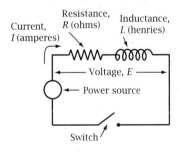

Figure 7-3i

a. Write a differential equation stating that E is the sum of the resistive voltage and the inductive voltage.

b. Solve this differential equation subject to the initial condition that $I = 0$ when $t = 0$. Write the resulting equation with I as a function of t.

c. Suppose that the circuit has a resistance of 10 Ω (ohms) and an inductance of 20 H (henries). If the circuit is connected to a normal 110-V (volt) outlet, write the particular equation and plot the graph. Sketch the result and show any asymptotes.

d. Predict the current for these times:

   i. 1 s after the switch is turned on

   ii. 10 s after the switch is turned on

   iii. At a steady state, after many seconds

e. At what time, t, will the current reach 95% of its steady-state value?

4. *Newton's Law of Cooling Problem:* When you turn on an electric heater, such as a burner on a stove (Figure 7-3j), its temperature increases rapidly at first, then more slowly, and finally approaches a constant high temperature. As the burner warms up, heat supplied by the electricity goes to two places.

   i. Storage in the heater materials, thus warming the heater

   ii. Losses to the room

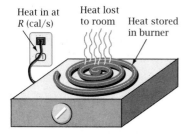

Figure 7-3j

Assume that heat is supplied at a constant rate, R. The rate at which heat is stored is directly proportional to the rate of change of temperature. Let T be the number of degrees above room temperature. Let t be the elapsed time, in seconds, since heat was applied. Then the storage rate is $C(dT/dt)$. The proportionality constant, C, (calories per degree), is called the **heat capacity** of the heater materials. According to Newton's law of

cooling, the rate at which heat is lost to the room is directly proportional to T. The (positive) proportionality constant, h, is called the **heat transfer coefficient**.

a. The rate at which heat is supplied to the heater is equal to the sum of the storage rate and the loss rate. Write a differential equation that expresses this fact.

b. Separate the variables and integrate the differential equation. Transform the answer so that temperature, T, is in terms of time, t. Use the initial condition that $T = 0$ when $t = 0$.

c. Suppose that heat is supplied at a rate $R = 50$ cal/s. Assume that the heat capacity is $C = 2$ cal/deg and that the heat transfer coefficient is $h = 0.04$ (cal/s)/deg. Substitute these values to get T in terms of t alone.

d. Plot the graph of T versus t. Sketch the result.

e. Predict T at times of 10, 20, 50, 100, and 200 s after the heater is turned on.

f. Find the limit of T as t approaches infinity. This is called the steady-state temperature.

g. How long does it take the heater to reach 99% of its steady-state temperature?

5. *Hot Tub Problem:* Figure 7-3k shows a cylindrical hot tub 8 ft in diameter and 4 ft deep. At time $t = 0$ min, the drain is opened and water flows out. The rate at which it flows is proportional to the square root of the depth, y, in feet. Because the tub has vertical sides, the rate is also proportional to the square root of the volume, V, in cubic feet, of water left.

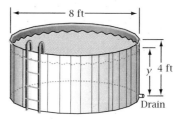

Figure 7-3k

a. Write a differential equation for the rate at which water flows from the tub. That is, write an equation for dV/dt in terms of V.

b. Separate the variables and integrate the differential equation you wrote in part a. Transform the result so that V is expressed explicitly in terms of t. Tell how V varies with t.

c. Suppose that the tub initially (when $t = 0$) contains 196 ft^3 of water and that when the drain is first opened the water flows out at 28 ft^3/min (when $t = 0$, $dV/dt = -28$). Find the particular solution of the differential equation that fits these initial conditions.

d. Naive thinking suggests that the tub will be empty after 7 min since it contains 196 ft^3 and the water is flowing out at 28 ft^3/min. Show that this conclusion is false, and justify your answer.

e. Does this mathematical model predict a time when the tub is completely empty, or does the volume, V, approach zero asymptotically? If there is a time, state it.

f. Draw a graph of V versus t in a suitable domain.

g. See Problem C4 in Section 7-7 to see what happens if a hose is left running in the hot tub while it drains.

6. *Burette Experiment:* In this problem you will simulate Problem 5 and Example 1. Obtain a burette (see Figure 7-3l) from a chemistry lab. Fill the burette with water, then open the stopcock so that water runs out fairly slowly. Record the level of water in the burette at various times as it drains. Plot volume versus time on graph paper. Does the volume seem to vary quadratically with time, as it did in Example 1? Find the best-fitting quadratic function for the data. Discuss the implications of the fact that the volume, V, read on the burette equals zero before the depth, y, of the water equals zero.

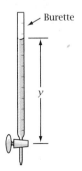

Figure 7-3l

7. *Differential Equation Generalization Problem:* The solutions of $dy/dx = ky^n$ are functions with different behaviors, depending on the value of the (constant) exponent n. If $n = 1$, then y varies exponentially with x. If $n = 0.5$, as in the *Hot Tub Problem,* then y varies quadratically with x. In this problem you will explore the graphs of various solutions of this equation.

a. Write the solution of the equation for $n = 1$. Let $k = 1$ and the constant of integration $C = -3$. Graph the solution and sketch the graph.

b. Solve the equation for $n = 0.5$. Let $k = 1$ and $C = -3$, as in part a. Graph the solution.

c. Show that if $n = -1$, then y is a square root function of x, and if $n = -2$, then y is a cube root function of x. Plot both graphs, using $k = 1$ and $C = -3$, as in part a.

d. Show that if $n > 1$, then there is a vertical asymptote at $x = -C/k$. Plot two graphs that show the difference in behavior for $n = 2$ and for $n = 3$. Use $k = 1$ and $C = -3$, as in part a.

e. What type of function is y when $n = 0$? Graph this function, using $k = 1$ and $C = -3$, as in parts a–d.

8. *Advertising Project:* A company that makes soft drinks introduces a new product. The company's salespeople want to predict B, the number of bottles per day they will sell as a function of t, the number of days since the product was introduced. One of the parameters is the amount per day spent on advertising. Here are some assumptions the salespeople make about future sales.

• The dependent variable is B; the independent variable is t, in days (Figure 7-3m).

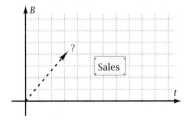

Figure 7-3m

- They spend a fixed amount, M, in dollars per day, to advertise this soft drink.
- Part of M, an amount proportional to B, is used to maintain present users of this soft drink (Figure 7-3n).

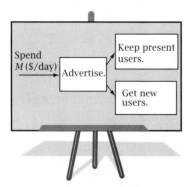

Figure 7-3n

- The rate of change of B, dB/dt, is directly proportional to the rest of M.
- Advertising costs need to be $80 per day to maintain sales of 1000 bottles per day.
- Due to advance publicity, dB/dt will be 500 bottles per day when $t = 0$, independent of M.

Use what you've learned in this section to find an equation for B as a function of t. Then show the effect of spending various amounts, M, on advertising. (Calculations and graphs would be convincing.) You might include such information as whether sales will continue to go up without bound or will eventually level off. To make an impression on management, assume a certain price per bottle and indicate how long it will take the product to start making a profit.

9. *Water Heater Project:* You have been hired by a water-heater manufacturer to determine some characteristics of a new line of water heaters (Figure 7-3o). Specifically, they want to know how long the new heater takes to warm up a tank of cold water to various temperatures. Once the tank reaches its upper storage temperature, the thermostat turns off the heat and the water starts to cool. So they also want to know how long it will be before the thermostat turns the heat on again.

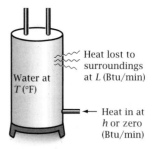

Figure 7-3o

Here's what you learn from the engineering and design departments.

- Heat is supplied at a constant rate of 1200 Btu (British thermal units) per minute.
- Heat is lost to the surroundings at a rate, L, proportional to the difference between the heater temperature and the room temperature. That is,

$$L = h(T - 70)$$

where L is loss rate in Btu/min and T is water temperature. The room temperature is $70°$F, and h is a proportionality constant called the heat transfer coefficient.
- The new heater warms water at a rate, dT/dt, proportional to $(1200 - L)$.
- The water would warm up at $3°$F/min if there were zero losses to the surroundings.
- In 10 min the heater warms the water to $96°$F from the room temperature of $70°$F.

Use this information to derive an equation that expresses temperature, T, in terms of the number of minutes, t, since the water heater was turned on. Use this equation to find the information the manufacturer is seeking (see the beginning of this problem). For instance, you might investigate how long it takes to warm water to $140°$F, to $160°$F, and to $180°$F. You can find out how long it takes, when the heat is off, for the water to cool from $160°$F to, say, $155°$F (when the heat turns on again). Impress your boss by pointing out any inadequacies in the proposed design of the heater and by suggesting which parameters might be changed to improve the design.

10. *Vapor Pressure Project:* The vapor pressure, P, in millimeters of mercury (mm Hg), of a liquid

or a solid (Figure 7-3p) increases as the temperature increases. The rate of change of the vapor pressure, dP/dT, is directly proportional to P and inversely proportional to the square of the Kelvin temperature, T. In physical chemistry you will learn that this relationship is called the Clausius-Clapeyron equation.

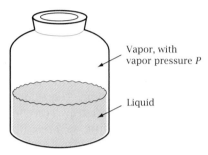

Vapor, with vapor pressure P

Liquid

Figure 7-3p

a. Write a differential equation that expresses dP/dT in terms of P and T. Integrate the equation, then solve for P in terms of T.

b. The table shows the vapor pressure for naphthalene (mothballs, $C_{10}H_8$) from an old edition of Lange's *Handbook of Chemistry*. Use the data for 293 K (20°C) and 343 K (70°C) to find the two constants in the equation you wrote in part a. You may solve the system of simultaneous equations either in their logarithmic form or in their exponential form, whichever is more

convenient. Don't be afraid of large numbers! And *don't* round them off!

°C	K	mm Hg
10	283	0.021
20	293	0.054
30	303	0.133
40	313	0.320
50	323	0.815
60	333	1.83
70	343	3.95
80	353	7.4 (melting point)
90	363	12.6
100	373	18.5
110	383	27.3
200	473	496.5

c. How well does your function fit the actual data? Does the same equation fit well above the melting point? If so, give information to support your conclusion. If not, find an equation that fits better above the melting point. Do any other types of functions available on your grapher seem to fit the data better than the function from the Clausius-Clapeyron equation?

d. Predict the boiling point of naphthalene, which is the temperature at which the vapor pressure equals atmospheric pressure, or 760 mm Hg.

e. What extensions can you think of for this project?

7-4 Graphical Solution of Differential Equations by Using Slope Fields

The function $y = e^{0.3x}$ is a particular solution of the differential equation $dy/dx = 0.3y$. The left graph in Figure 7-4a, on the next page, shows this solution, with a tangent line through the point $(2, e^{0.6})$. In the right graph, the curve and most of the tangent line have been deleted, leaving only a short segment of the tangent, centered at the point $(2, e^{0.6})$. You can also draw this segment without ever solving the differential equation. Its slope is $0.3e^{0.6} \approx 0.55$, the number you get by substituting $e^{0.6}$ for y in the original differential equation.

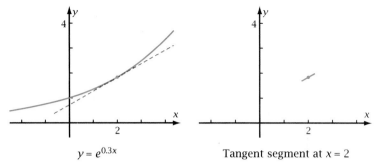

$y = e^{0.3x}$ Tangent segment at $x = 2$

Figure 7-4a

The left graph in Figure 7-4b shows what happens if you draw a short segment of slope $0.3y$ at every grid point (point with integer coordinates) on the plane. The result is called a **slope field** or sometimes a **direction field**. The right graph in Figure 7-4b shows the solution $y = e^{0.3x}$ (from Figure 7-4a) drawn on the slope field. The line segments show the direction of the graph. As a result, you can draw the graph of another particular solution simply by picking a starting point and going to the left and to the right "parallel" to the line segments. The dotted curves on the right graph of Figure 7-4b show three such graphs.

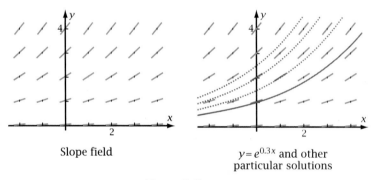

Slope field $y = e^{0.3x}$ and other
                                   particular solutions

Figure 7-4b

Slope fields are tedious to draw by hand and are best done by grapher. Once you get a slope field, however, you can graph any particular solution without ever solving the differential equation. As you can appreciate from difficulties you may have had integrating the differential equations in Section 7-3, such an approximate solution method is very welcome! In this section you will sketch the graphs by hand. However, you will compare some of the graphical solutions with exact, algebraic solutions. In Section 7-5, you will learn a numerical method for plotting such approximate graphs on the grapher itself.

OBJECTIVE   Given a slope field for a differential equation, graph an approximate particular solution by hand and, if possible, confirm the solution algebraically.

▶ *EXAMPLE 1*   Figure 7-4c shows the slope field for the differential equation

$$\frac{dy}{dx} = -\frac{0.36x}{y}$$

a. Use the differential equation to find the slope at the points (5, 2) and (−8, 9). Mark these points on the figure. Explain why the calculated slopes are reasonable.

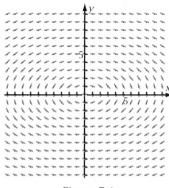

b. Start at the point (0, 6) and draw a graph that represents the particular solution of the differential equation that contains that point. Go both to the right and to the left. What geometric figure does the graph seem to be?

c. Start at the point (5, 2), from part a, and draw another particular solution of the differential equation. How is this solution related to that in part b?

Figure 7-4c

d. Solve the differential equation algebraically. Find the particular solution that contains the point (0, 6). Verify that the graph really is the figure indicated in part b.

Solution   a. At the point (5, 2), $\dfrac{dy}{dx} = -\dfrac{0.36(5)}{2} = -0.9$.

At the point $(-8, 9)$, $\dfrac{dy}{dx} = -\dfrac{0.36(-8)}{9} = 0.32$.

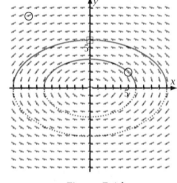

The circled points in Figure 7-4d show slopes of about −1 and 0.3, which agree with the calculations.

b. Figure 7-4d shows the graph. Start at the boxed point (0, 6). Where the graph goes between grid points, make its slope an average of the slopes shown. Don't try to head for the grid points themselves! The graph may not pass through these points. The graph appears to be a half-ellipse. The dotted curve below the x-axis shows the same elliptical pattern, but it is not part of the particular solution because a solution of a differential equation is a *function*.

Figure 7-4d

c. The solution that contains the point (5, 2) is the inner half-ellipse in Figure 7-4d. It's similar (has the same proportions) to the ellipse described in part b.

d. $y\,dy = -0.36x\,dx \Rightarrow \int y\,dy = -0.36\int x\,dx \Rightarrow 0.5y^2 = -0.18x^2 + C$

Substituting the point (0, 6) gives $0.5(36) = 0 + C \Rightarrow C = 18$.

$$\therefore\ 0.5y^2 = -0.18x^2 + 18 \Rightarrow 9x^2 + 25y^2 = 900$$

This is the equation of an ellipse centered at the origin, as shown in part b.   ◀

Figure 7-4e, on the next page, shows slope fields for three differential equations. For each slope field or its differential equation, three particular solutions are shown, along with the corresponding initial conditions. Note that the graph follows the pattern but usually goes between lattice points rather than through them. Note also that two of the solutions for the middle graph stop at the line $y = 0.5$. Following the pattern of the slope field would make the curve double back, giving two values of y for some x-values, resulting in a relation that is not a function.

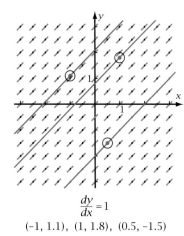

$$\frac{dy}{dx} = 1$$

$(-1, 1.1),\ (1, 1.8),\ (0.5, -1.5)$

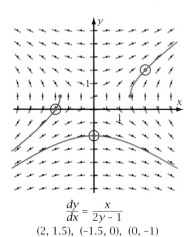

$$\frac{dy}{dx} = \frac{x}{2y - 1}$$

$(2, 1.5),\ (-1.5, 0),\ (0, -1)$

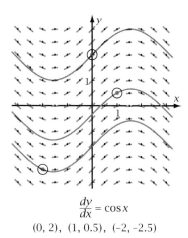

$$\frac{dy}{dx} = \cos x$$

$(0, 2),\ (1, 0.5),\ (-2, -2.5)$

Figure 7-4e

Problem Set 7-4

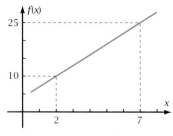
1. Figure 7-4g shows the slope field for the differential equation

$$\frac{dy}{dx} = \frac{x}{2y}$$

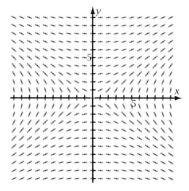

Figure 7-4g

Use a copy of this figure to answer these questions.

a. Show that you understand the meaning of slope field by first calculating dy/dx at the points (3, 5) and (−5, 1) and then showing that the results agree with the figure.

b. Sketch the graph of the particular solution of the differential equation that contains the point (1, 2). Draw on both sides of the y-axis. What geometric figure does the graph seem to be?

c. Sketch the graph of the particular solution that contains the point (5, 1). Draw on both sides of the x-axis.

d. Solve the differential equation algebraically. Find the particular solution that contains the point (5, 1). How well does your graphical solution from part b agree with the algebraic solution?

2. Figure 7-4h shows the slope field for the differential equation

$$\frac{dy}{dx} = 0.1y$$

Find the slopes at the points (3, 3) and (0, −2), and explain how you know that these agree with the slope field. On a copy of Figure 7-4h, sketch two particular solutions, one containing (3, 3) and the other containing (0, −2). From the two graphs, read to one decimal place the approximate values of y when $x = 6$. Then solve the differential equation algebraically and find the exact values of y for $x = 6$. How close do your graphical solutions come to the actual solutions?

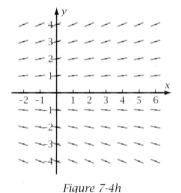

Figure 7-4h

3. Given the differential equation

$$\frac{dy}{dx} = -\frac{x}{2y}$$

a. Find the slope at the points (3, 2) and (1, 0). By finding slopes at other points, draw the slope field on dot paper using the window shown in Figure 7-4i. You can use symmetry to reduce the amount of computation.

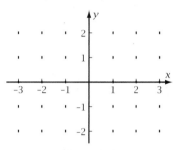

Figure 7-4i

b. On your graph from part a, sketch the particular solution that contains the point (2, 1) and the particular solution that contains the point (1, −1). What geometric figures do the graphs resemble?

c. Solve the differential equation algebraically using the initial condition (1, −1). By plotting this solution on your grapher, show that your sketch in part b is reasonable. Show algebraically that the graph is a half-ellipse.

4. Given the differential equation

$$\frac{dy}{dx} = x(1 - y)$$

a. Find the slope at the points (3, 1), (1, 2), and (0, −1). By finding slopes at other points, draw the slope field on dot paper using the window shown in Figure 7-4j.

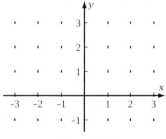

Figure 7-4j

b. On your graph from part a, sketch the particular solution that contains the point $(1, 2)$ and the particular solution that contains the point $(0, -1)$. What feature do both graphs have as x gets larger in the positive and negative directions?

c. Solve the differential equation algebraically using the initial condition $(0, -1)$. Plot this solution on your grapher to show that your sketch in part b is reasonable. Show algebraically that the graph has the feature you mentioned in part b as x increases in absolute value.

For Problems 5–8, sketch the solutions on copies of the slope field using the initial conditions given.

5. $(-1, 1), (1, -1)$

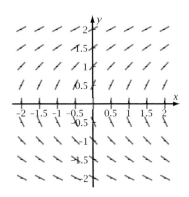

6. $(1, 1), (0.5, -1)$

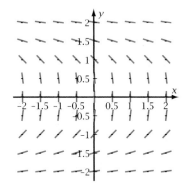

7. $(2, 2), (-1, 0), (-2, -2)$

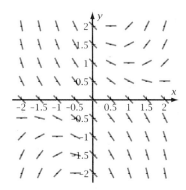

8. $(0.5, 1), (-2, 1), (-2, -2)$

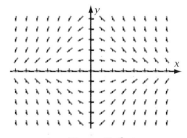

9. a. On a copy of the slope field shown in Figure 7-4k, sketch two particular solutions: one that contains the point $(3, 2)$ and one that contains the point $(1, -2)$.

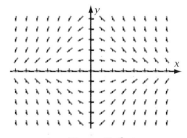

Figure 7-4k

b. In Quadrant I, the slope is always negative and gets steeper as x or y increases. The slope at the point $(1, 1)$ is about -0.2. Make a conjecture about a differential equation that could generate this slope field. Give evidence to support your conjecture.

10. *Dependence on Initial Conditions Problem:*
    Figure 7-4l shows the slope field for

$$\frac{dy}{dx} = 0.1x + 0.2y$$

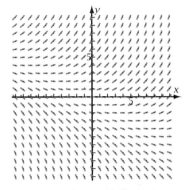

Figure 7-4l

a. On a copy of this figure, draw the particular solution that contains the point $(0, 2)$. Show the graph on both sides of the y-axis.

b. Show that the particular solution containing the point $(0, -5)$ exhibits a behavior different from the behavior in part a.

c. The solution to part a curves upward, and that to part b curves downward. It seems reasonable that somewhere between these two solutions there is one that has a straight-line graph. Draw this solution. Where does the graph cross the y-axis?

11. *Rabbit Population Overcrowding Problem:* In the population problems of Section 7-2, the rate of change of population was proportional to the population. In the real world, over-crowding limits the size of the population. One mathematical model, the **logistic equation**, says that dP/dt is proportional to the product of the population and a constant minus the population. Suppose that rabbits are introduced to a small uninhabited island in the Pacific. Naturalists find that the differential equation for population growth is

$$\frac{dP}{dt} = 0.038P(10.5 - P)$$

where P is in hundreds of rabbits and t is in months. Figure 7-4m shows the slope field.

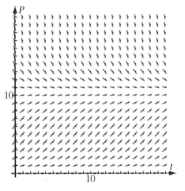

Figure 7-4m

a. Suppose that 200 rabbits arrive at time $t = 0$. On a copy of Figure 7-4m, graph the particular solution.

b. Draw another particular solution if the 200 rabbits are instead introduced at time $t = 4$. What are the differences and similarities in the population growth?

c. Draw a third particular solution if 1800 rabbits are introduced at time $t = 0$. With this initial condition, what is the major difference in population growth? What similarity does this scenario have to those in parts a and b?

d. Think of a real-world reason to explain the horizontal asymptote each graph approaches. Where does this asymptote appear in the differential equation?

12. *Terminal Velocity Problem:* A sky diver jumps from an airplane. During the free-fall stage, her speed increases at the acceleration of gravity, about 32.16 (ft/s)/s. But wind resistance causes a force that reduces the acceleration. The resistance force is proportional to the square of the velocity. Assume that the constant of proportionality is 0.0015 so that

$$\frac{dv}{dt} = 32.16 - 0.0015v^2$$

where v is her velocity, in feet per second, and t is the time she has been falling, in seconds. The slope field for this differential equation is shown in Figure 7-4n, on the next page.

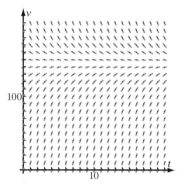

Figure 7-4n

a. What does the slope appear to be at the point (5, 120)? What does it actually equal? Explain any discrepancy between your two answers.

b. The sky diver starts at time $t = 0$ with zero initial velocity. On a copy of Figure 7-4n, sketch her velocity as a function of time.

c. Her velocity approaches an asymptote. What does this **terminal velocity** appear to equal? About how long does it take her to fall until she is essentially at this velocity?

d. A second sky diver starts 5 s later with zero initial velocity. Sketch the velocity-time graph. What similarities does this graph have to the graph you sketched in part b?

e. Suppose that the plane is descending steeply as a third diver jumps, giving him an initial downward velocity of 180 ft/s. Sketch this diver's velocity-time graph. How does it differ from the graphs you sketched in parts b and d?

f. The mathematical models for free fall in this problem and for population in Problem 11 have some similarities. Write a paragraph that discusses their similarities and differences. Do you find it remarkable that two different phenomena have similar mathematical models?

13. *Escape Velocity Problem:* If a spaceship has a high enough initial velocity, it can escape Earth's gravity and be free to go elsewhere in space. Otherwise, it will stop and fall back to Earth.

Eileen M. Collins, the first female space shuttle pilot (STS-63, 1995) and the first female shuttle commander (STS-93, 1999)

a. By Newton's second law of motion, the force, F, on the spaceship equals its mass, m, times the acceleration, a. By Newton's law of gravitation, F is also equal to mg/r^2, where g is the gravitational constant and r is the distance from the center of Earth to the spaceship. Give reasons for each step in these transformations.

$$ma = \frac{mg}{r^2}$$

$$\frac{dv}{dt} = \frac{g}{r^2}$$

$$\frac{dv}{dr} \cdot \frac{dr}{dt} = \frac{g}{r^2}$$

$$\frac{dv}{dr} \cdot v = \frac{g}{r^2}$$

$$\frac{dv}{dr} = \frac{g}{r^2 v}$$

b. If r is in earth-radii (1 earth-radius = 6380 km) and v is in kilometers per second, then

$$\frac{dv}{dr} = \frac{-62.44}{r^2 v}$$

The sign is negative because gravity acts opposite to the direction of motion. Figure 7-4o shows the slope field for this differential equation. Confirm that the differential equation produces the slopes shown at the points $(r, v) = (5, 2), (1, 10)$, and $(10, 4)$.

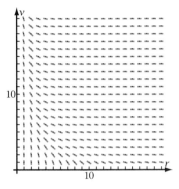

Figure 7-4o

c. If the spaceship starts at Earth's surface $(r = 1)$ with an initial velocity of $v = 10$ km/s, it will not escape Earth's gravity. On a copy of Figure 7-4o, sketch this particular solution. About how far from Earth's surface does the ship stop and begin to fall back?

d. On your copied figure, show that if the spaceship starts from Earth's surface with an initial velocity of 12 km/s, it will escape Earth's gravity. About how fast will it be going when it is far from Earth?

e. If the spaceship starts from Earth's surface with an initial velocity of 18 km/s, what will its velocity approach far from Earth? Does it lose as much speed starting at 18 km/s as it does starting at 12 km/s? How do you explain this observation?

f. Show that the spaceship will escape from Earth's gravity if it starts with an initial velocity of 10 km/s from a space platform in orbit that is 1 earth-radius above Earth's surface (that is, $r = 2$).

14. *Slope Fields on the Grapher:* On your grapher, generate the slope field for

$$\frac{dy}{dx} = 0.5y(1 - 0.15y)$$

as shown in Figure 7-4p. If your grapher doesn't have a built-in program to generate slope fields, obtain one or write one. Save this program because you will use it in Section 7-5.

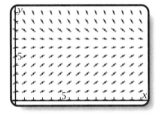

Figure 7-4p

7-5 Numerical Solution of Differential Equations by Using Euler's Method

In Section 7-4, you sketched approximate solutions of differential equations, using their slope fields. In this section you will learn a numerical method for calculating approximate y-values for a particular solution, and you'll plot the points either by hand or on your grapher. This method is called **Euler's method**, after Swiss mathematician Leonhard Euler (1707–1783). (Euler is pronounced "oi′-ler.")

OBJECTIVE   Given a differential equation and its slope field, calculate points on the graph iteratively by starting at one point and finding the next point by following the slope for a given x-distance.

Euler's method for solving a differential equation numerically is based on the fact that a differentiable function has local linearity at any given point. For instance, if $dy/dx = \cos xy$, you can calculate the slope at any point (x, y) and follow the linear function to another point Δx units away. If Δx is small, the new point will be close to the actual point on the graph. Figure 7-5a illustrates the procedure.

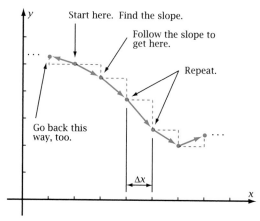

Figure 7-5a

Example 1 shows you a systematic way to use Euler's method.

▶ **EXAMPLE 1**   Figure 7-5b shows the slope field for the differential equation

$$\frac{dy}{dx} = 0.5(x + y)$$

a. Use Euler's method with $dx = 0.2$ to calculate approximate values of y for the particular solution containing $(0, 0.4)$ from $x = 0.2$ through $x = 1.2$.

b. Draw this approximate particular solution on a copy of Figure 7-5b.

c. Does the approximate solution overestimate the actual values of y or underestimate them? Explain.

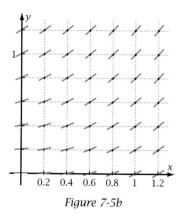

Figure 7-5b

Solution   a. $\dfrac{dy}{dx} = 0.5(x + y)$

$dy = 0.5(x + y)\,dx$ Get an equation for dy.

At $(0, 0.4)$, $dy = 0.5(0 + 0.4)(0.2) = 0.04$ Substitute $(0, 0.4)$ to calculate dy.

New $x = 0 + 0.2 = 0.2$, new $y \approx 0.4 + 0.04 = 0.44$ Add dx and dy to x and y.

At $(0.2, 0.44)$, $dy = 0.5(0.2 + 0.44)(0.2) = 0.064$ Repeat the procedure at $(0.2, 0.44)$.

New $x = 0.2 + 0.2 = 0.4$, new $y \approx 0.44 + 0.064 = 0.504$

At $(0.4, 0.504)$, $dy = 0.5(0.4 + 0.504)(0.2) = 0.0904$ Repeat the procedure at $(0.4, 0.504)$.

New $x = 0.4 + 0.2 = 0.6$, new $y \approx 0.504 + 0.0904 = 0.5944$

The table shows the results of repeating this procedure through $x = 1.2$.

x	y	dy	New y
0	0.4	0.04	0.44
0.2	0.44	0.064	0.504
0.4	0.504	0.0904	0.5944
0.6	0.5944	0.11944	0.71384
0.8	0.71384	0.151384	0.865224
1.0	0.865224	0.1865224	1.0517464
1.2	1.0517464	—	—

b. Figure 7-5c shows the points connected by line segments.

c. Figure 7-5c shows that the actual graph of y (shown dashed) is concave up. The Euler's method solution consists of tangent line segments, and tangent lines go on the *convex* side of the graph, so these tangent segments are *below* the actual graph. So solutions by Euler's method *underestimate* the actual values of y.

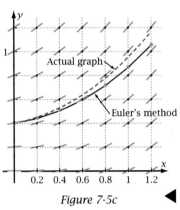

Figure 7-5c

The box summarizes Euler's method and how you determine whether the approximate numerical solution overestimates the actual solution or underestimates it.

TECHNIQUE: Euler's Method for Solving Differential Equations

- Solve the differential equation for dy in terms of x, y, and dx.
- Substitute values of x, y, and dx to calculate a value of dy.
- Find the approximate new value of y by adding dy to the old value of y.
- Repeat the procedure to find the next value of dy at the next value of x.

Note: The graph of the solution by Euler's method lies on the *convex* side of the actual graph. Thus, the Euler's solution *underestimates* the actual y-values if the convex side is downward and *overestimates* the actual y-values of the convex side is upward.

Problem Set 7-5

Quick Review 5 min

Q1. If dy/dx is directly proportional to y, then $dy/dx = $ —?—.

Q2. If $dy/dx = 3y$, then the general solution for y is —?—.

Q3. If $dy/dx = 0.1\,xy$, what is the slope of the slope-field line at the point $(6, 8)$?

Q4. If $y = Ce^{0.2x}$ and $y = 100$ when $x = 0$, then $C = $ —?—.

Q5. $\int dv/(1 - v) = $ —?—

Q6. $(d/dx)(\sec x) = $ —?—

Q7. Sketch the graph of y' for Figure 7-5d.

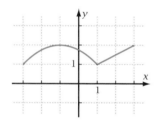

Figure 7-5d

Q8. Differentiate implicitly: $x^3 y^5 = x + y$

Q9. If $\lim_{x \to 4} f(x) = f(4)$, then f is —?— at $x = 4$.

Q10. If $f(x) = \int_1^x (3t + 5)^4\, dt$, then $f'(x) = $ —?—.

   A. $(3x + 5)^4$ B. $12(3x + 5)^3$

   C. $\frac{1}{5}(3x + 5)^5 + C$ D. $\frac{1}{5}(3x + 5)^5 - \frac{8}{5}$

   E. $\frac{1}{5}(3x + 5)^5 - \frac{8^5}{5}$

---

1. *How Euler's Method Works, Problem 1:*
   Figure 7-5e shows the slope field for the differential equation

   $$\frac{dy}{dx} = -\frac{x}{y}$$

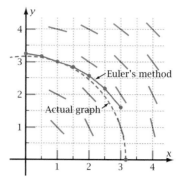

Figure 7-5e

Several points of the Euler's method solution containing the point $(1, 3)$ and with $dx = 0.5$ are shown in Figure 7-5e, along with the actual solution (dashed).

a. Show your calculations for the graph of y using Euler's method for $x = 1.5$ and $x = 2$, then make a table showing x and the Euler's method approximation of y for each 0.5 unit of x from 0 through 3. For values of x less than 1, use $dx = -0.5$. Do the values of y agree with those shown on the Euler graph? Explain how the pattern followed by the points tells you that the Euler's method solution overestimates the values of y, both for $x > 1$ and for $x < 1$. What happens to the size of the error as x gets farther away from 1?

b. Solve the differential equation algebraically. Show that the solution is a part of a circle, as shown in Figure 7-5e. Why does the particular solution stop at the x-axis? By how much does the Euler's method solution overestimate the actual value at $x = 3$?

2. *How Euler's Method Works, Problem 2:*
   Figure 7-5f shows the slope field for the differential equation

   $$\frac{dy}{dx} = \frac{x}{y}$$

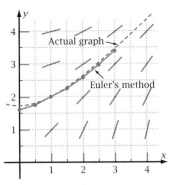

Figure 7-5f

Several points of the Euler's method solution containing the point (1, 2) and $dx = 0.5$ are shown in the figure, along with the actual solution (dashed).

a. Show the calculations by which you find the points on the Euler graph. For values of x less than 1, use $dx = -0.5$. Do the values of y agree with those shown on the graph? Explain how the pattern followed by the points tells you that the Euler's method solution underestimates the values of y, both for $x > 1$ and for $x < 1$. Why does the error at $x = 0$ seem to be larger in absolute value than the error at $x = 3$, even though 3 is farther from 1 than 0 is?

b. Solve the differential equation algebraically. Show that the solution is a part of a hyperbola, as shown in Figure 7-5f. By how much does the Euler's method solution underestimate the actual value at $x = 0$?

For Problems 3 and 4, the table shows values of the derivative dy/dx for a function at various values of x. Use the value of dx you find in going from one value of x to the next to calculate values of dy. Use these values of dy along with Euler's method to find approximate values of y for each value of x in the table, starting at the given value of y. Plot the points on graph paper. Is it possible to tell whether the last approximate value of y overestimates or underestimates the actual value of y?

3.

x	dy/dx	dy	y
2	3	___	1
2.2	5	___	___
2.4	4	___	___
2.6	1	___	___
2.8	−3	___	___
3	−6	___	___
3.2	−5	___	___
3.4	−3	___	___
3.6	−1	___	___
3.8	1	___	___
4	2	___	___

4.

x	dy/dx	dy	y
1	−3	___	2
1.3	−2	___	___
1.6	−1	___	___
1.9	0	___	___
2.2	1	___	___
2.5	2	___	___
2.8	3	___	___
3.1	4	___	___
3.4	5	___	___
3.7	6	___	___
3.9	7	___	___

5. *Numerical Program for Euler's Method:* Obtain or write a program for computing y-values by Euler's method. The program should allow you to enter the differential equation, say as y_1, in terms of both x and y. Then you should be able to input the initial condition (x, y) and the value of dx, such as the point (1, 3) and the number 0.5, respectively, for Problem 1. Test your program by using it to calculate the values of y for Problem 1.

6. *Graphical Program for Euler's Method:* Obtain or write a grapher program for plotting a particular solution of a differential equation by Euler's method. Adapt your program from Problem 5 if you prefer. You can enter the differential equation as y_1. The input should include the initial point and the value of Δx. As each point is calculated, the grapher should draw a line segment to it from the previous point. The program should allow the graph to be plotted to the right or left of the initial point, depending on the sign of Δx. The program should work in conjunction with the slope-field program of Section 7-4 so that you

superimpose the solution on the corresponding slope field. Test the program by using the information in Problem 1, and show that the graph resembles that in Figure 7-5e.

7. Figure 7-5g shows the slope field for the differential equation

$$\frac{dy}{dx} = -0.2xy$$

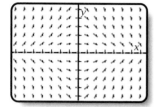

Figure 7-5g

a. Use your grapher programs to plot the slope field and the particular solution that contains the point (3, 2). Sketch the solution on a copy of Figure 7-5g.

b. Plot the particular solution that contains the point (1, −2). Sketch the solution on the copy.

8. Figure 7-5h shows the slope field for the differential equation

$$\frac{dy}{dx} = -0.1x + 0.2y$$

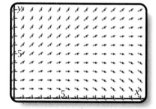

Figure 7-5h

a. Use your grapher programs to plot the slope field and the particular solution that contains the point (0, 2). Sketch the solution on a copy of Figure 7-5h.

b. Plot the particular solution that contains the point (0, 4). Sketch the solution on the copy.

c. The solution you plotted in part a curves downward, and the one in part b curves upward. It seems reasonable that

somewhere between these two solutions is a solution that has a straight-line graph. By experimenting on your grapher, find this particular solution. Record the initial point you used and sketch the solution on the copy.

9. *Escape Velocity Problem by Euler's Method:* In Problem 13 of Section 7-4, you learned that the acceleration (rate of change of velocity) of a spaceship moving away from Earth decreases due to Earth's gravity according to the differential equation

$$\frac{dv}{dr} = \frac{-62.44}{r^2 v}$$

where v is the spaceship's velocity, in km/s, and r is the distance from Earth's center, in earth-radii. The slope field for this differential equation is reproduced here (Figure 7-5i) for your use.

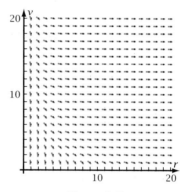

Figure 7-5i

a. Starting with an initial velocity of 12 km/s at Earth's surface ($r = 1$), use your Euler's method program with $dr = 0.6$ to show that the velocity appears to become negative, meaning that the spacecraft will stop and start to fall back to Earth.

b. Starting at (1, 12) as in part a, use $dr = 0.1$ to show that the velocity appears to be positive when $r = 20$ and the values of v seem to be leveling off.

c. Solve the differential equation algebraically and use the solution to find the actual velocity at $r = 20$. Explain why Euler's method using a relative large value of dr (as in part a) erroneously predicts that the spaceship will not escape Earth's gravity.

d. Show that if the spaceship starts from Earth's surface with an initial velocity less than the **escape velocity**, $v_e = \sqrt{2(62.44)}$ km/s, then the spacecraft will eventually reverse directions and start to fall back toward Earth. Show also that if the initial velocity is greater than v_e, then the velocity approaches a positive limit as r approaches infinity and the spacecraft never returns. What is the escape velocity in miles per hour?

10. *Terminal Velocity Problem by Euler's Method:* In Problem 12 of Section 7-4, you assumed that the vertical acceleration (rate of change of velocity) of a sky diver in free fall is given by the differential equation

$$\frac{dv}{dt} = 32.16 - 0.0015v^2$$

where v is the sky diver's velocity, in ft/s, and t is the time the sky diver has been falling, in seconds. The slope field for this differential equation is reproduced here (Figure 7-5j) for your use.

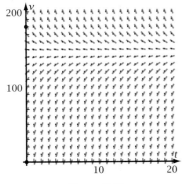

Figure 7-5j

a. The sky diver falls with an initial ($t = 0$) downward velocity of zero. Use your Euler's method program with $dt = 0.5$ to estimate her downward velocity at times 2, 4, 6, 8, 10, and 20 s. Are these underestimates or overestimates of her velocity? How can you tell?

b. On another occasion the plane is descending steeply, giving the diver an initial downward velocity of 180 ft/s at time $t = 0$. Use your Euler's method program to estimate this diver's downward velocity at times 2, 4, 6, 8, 10, and 20 s. Are these underestimates or

overestimates of her velocity? How can you tell?

c. When dv/dt reaches zero, a diver is at his or her **terminal velocity**. Calculate this terminal velocity. Store it, without rounding, for use in the next part of this problem.

d. When you study trigonometric substitution in Chapter 9, you will be able to show that the algebraic solution of this differential equation with initial condition (0, 0) as in part a is

$$v = \sqrt{\frac{32.16}{0.0015}} \cdot \frac{e^{0.439272...t} - 1}{e^{0.439272...t} + 1}$$

Calculate the actual values of v for the six times given in part a. True or false: "The farther the values of t are from the initial condition, the greater the error in Euler's approximation for v."

11. *Euler's Method for a Restricted Domain Problem:* Figure 7-5k shows the slope field for the differential equation

$$\frac{dy}{dx} = -\frac{9x}{25y}$$

The figure also shows the Euler's method solution with $dx = 0.1$ containing the point $(0, -3)$, and the algebraic solution (dashed) consisting of the half-ellipse with equation

$$y = -0.6\sqrt{25 - x^2}$$

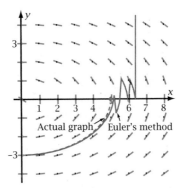

Figure 7-5k

a. Explain why the actual (algebraic) solution has a domain with $x \le 5$.

b. Without drawing the graphs, how can you tell from the slope field that the Euler's method value for y at $x = 4.9$ will underestimate the actual value? Show numerically that this Euler's method value actually does underestimate the actual value, but is reasonably close to it.

c. Find the Euler's method approximation for y at $x = 5.1$, 5.2, and 5.3. Based on Euler's method, explain why the predicted y-value jumps sharply, as shown in the figure, from the point at $x = 5.2$ to the point at $x = 5.3$. What does Euler's method predict for y at $x = 6.6$?

d. Why is it important to know the domain of a particular solution when you use Euler's method to find approximate solutions for a differential equation?

12. *Journal Problem:* Update your journal. Include such things as

- The one most important thing you have learned since your last entry

- How slope fields and numerical methods can be used to solve differential equations without finding an algebraic solution

- How much faith you would put into a computer-generated prediction of the U.S. population for the year 2050

- What you now better understand about differential equations

- Any technique or concept about differential equations that you're still unclear about

---

7-6 The Logistic Function, and Predator-Prey Population Problems

If a population, such as animals or people, has plenty of food and plenty of room, the population tends to grow at a rate proportional to the size of the population. You have seen that if $dy/dx = ky$, then y varies exponentially with x.

In this section you will learn about the *logistic function*, which accounts for the decreasing growth rate due to overcrowding. You will also explore what happens to populations when two species in an environment exist together as predator and prey. The two populations can rise and fall cyclically depending on abundance or scarcity of the food species (the prey), and the relative numbers of predators that prey on the species.

OBJECTIVE   Use slope fields and Euler's method to solve differential equations that model population growth in a restricted environment, and in instances where one species preys on another.

The Logistic Function

Suppose that a particular lake can sustain a maximum of 8000 fish (the **carrying capacity** of the lake) and that a number of fish, y, in thousands, live in the lake at time x, in years. If y is small compared to 8, dy/dx is expected to be proportional to y. It's also reasonable to assume that dy/dx is proportional to $(8 - y)/(8)$, the fraction of the maximum population at time x. Due to both of these phenomena,

$$\frac{dy}{dx} = ky \cdot \frac{8 - y}{8}$$

This is an example of the **logistic differential equation**. Examples 1, 2, and 3 show you how to solve the logistic differential equation graphically, numerically, and algebraically, and also show some conclusions that you can reach from the result.

▶ **EXAMPLE 1**   The number of fish, y, in thousands, in a particular lake at time x, in years, is given by the differential equation

$$\frac{dy}{dx} = 0.8y \cdot \frac{8 - y}{8}$$

a. On a copy of the slope field in Figure 7-6a, sketch the particular solution for the initial condition of 1000 fish in the lake at $x = 0$ years. Describe what happens to the fish population as time goes by.

b. From the differential equation, show that the rate of increase in the fish population is greatest when $y = 4$, halfway between the two horizontal asymptotes.

c. Suppose that at time $x = 4$ yr, the lake is restocked with fish, bringing the total population up to 11,000. On the same copy of Figure 7-6a, sketch the particular solution subject to this initial condition. Describe what happens to the fish population as time goes by. Surprising?

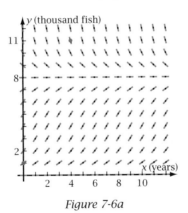

Figure 7-6a

Solution   a. Figure 7-6b, on the next page, shows that the population rises more rapidly at first, then levels off, approaching a horizontal asymptote at $y = 8$, corresponding to the maximum sustainable population of 8000 fish.

b.
$$\frac{dy}{dx} = 0.8(y) \cdot \frac{8-y}{8} = 0.1y(8-y) = -0.1y^2 + 0.8y \qquad \text{Transform the product to a sum.}$$

$$\frac{d}{dy}\left(\frac{dy}{dx}\right) = -0.2y + 0.8 \qquad \text{Differentiate } with \text{ respect to } y.$$

$$-0.2y + 0.8 = 0 \Rightarrow y = 4 \qquad \text{Set the derivative equal to 0 and solve for } y.$$

For $y < 4$, the derivative is positive, so dy/dx is increasing.

For $y > 4$, the derivative is negative, so dy/x is decreasing.

So the maximum value of dy/dx occurs if $y = 4$. (This means that the point of inflection is at $y = 4$.)

c. Figure 7-6b shows that when the population starts above the maximum sustainable value of 8000, some of the fish die out due to overcrowding. The population *decreases* rapidly at first, leveling off and approaching the same horizontal asymptote at $y = 8$. This is surprising because you might think that adding more fish would be good for the survival of the population.

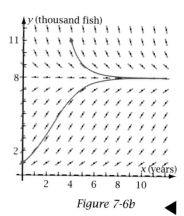

Figure 7-6b ◄

▶ **EXAMPLE 2** Solve the differential equation in Example 1, part a, numerically, using Euler's method with $dx = 0.1$. Write the value of y and the number of fish for $x = 2, 4, 6, 8,$ and 10 years. Do these values agree with the graph in Figure 7-6b? Do these values overestimate the actual fish population or underestimate it? Explain.

Solution Using the Euler's method program gives the values in the table. These agree with the graph. Between $x = 0$ and $x \approx 2.5$, the graph is concave up (convex side down) so the Euler's method tangents are below the graph, making the Euler's method values underestimate the actual values. For higher values of x, the graph is concave down (convex side up) so the Euler's method values are overestimates, compensating at least in part for the earlier underestimation.

x	y	Number of Fish
2	3.2536...	3254
4	6.1975...	6198
6	7.5769...	7577
8	7.9162...	7916
10	7.9840...	7984

◄

▶ **EXAMPLE 3** a. Solve the logistic equation in Example 1, part a, algebraically for y explicitly as a function of x.

b. Use the result of part a to show that the Euler's method approximation at $x = 2$ actually is an underestimation of the actual value.

350

c. As shown in Example 1, the rate of change in the fish population is the greatest when $y = 4$, halfway between the two horizontal asymptotes. Find the value of x at which the rate of change is the greatest.

Solution

a.
$$\frac{dy}{dx} = 0.8(y) \cdot \frac{8 - y}{8}$$

$$\int \frac{8}{y(8 - y)} dy = 0.8 \int dx \qquad \text{Separate the variables and integrate.}$$

To evaluate the integral on the left, resolve the integrand into **partial fractions**. This technique is presented in detail in Section 9-7 and is summarized here for your information.

Let $\dfrac{8}{y(8 - y)} = \dfrac{A}{y} + \dfrac{B}{8 - y}$ where A and B stand for constants.

$$\frac{8}{y(8 - y)} = \frac{A(8 - y) + By}{y(8 - y)} \qquad \text{Add the fractions on the right.}$$

$$= \frac{8A - Ay + By}{y(8 - y)}$$

$$= \frac{8A + (-A + B)y}{y(8 - y)} \qquad \text{Combine "like terms" in the numerator.}$$

$8 + 0y = 8A + (-A + B)y \qquad$ Equate the numerators. (*Note:* $8 = 8 + 0y$).

$8 = 8A$ and $0 = -A + B \qquad$ Equate the constant terms and the y-coefficients.

$A = 1$, so $0 = -1 + B \Rightarrow B = 1$

$\therefore \dfrac{8}{y(8 - y)} = \dfrac{1}{y} + \dfrac{1}{8 - y} \qquad$ Substitute for A and B.

$\therefore \displaystyle\int \left(\frac{1}{y} + \frac{1}{8 - y} \right) dy = 0.8 \int dx \quad$ Substitute into the differential equation.

$\ln |y| - \ln |8 - y| = 0.8x + C \qquad$ Integrate a *sum* of two *reciprocal* functions. (Why the minus between terms?)

$\ln \left| \dfrac{8 - y}{y} \right| = -0.8x - C \qquad$ Multiply both sides by -1 to make the next step easier. Use the ln of a quotient property.

$\left| \dfrac{8 - y}{y} \right| = e^{-0.8x - C} = e^{-C} \cdot e^{-0.8x} \qquad$ Exponentiate both sides.

$\dfrac{8 - y}{y} = C_1 e^{-0.8x} \qquad$ Remove the absolute value signs. $C_1 = \pm e^{-C}$.

$\dfrac{8 - 1}{1} = C_1 e^0 \Rightarrow C_1 = 7 \qquad$ Substitute the initial condition to find C_1.

$\dfrac{8 - y}{y} = 7e^{-0.8x}$

$8 - y = 7ye^{-0.8x}$

$(1 + 7e^{-0.8x})y = 8 \qquad$ Factor out y from the y-terms.

$y = \dfrac{8}{1 + 7e^{-0.8x}} \qquad$ Divide by the coefficient of y.

b. $\qquad x = 2: y = \dfrac{8}{1 + 7e^{-1.6}} = 3.3149...$

The Euler's method value of 3.2536... is an underestimate of this value, as predicted.

c. $\qquad 4 = \dfrac{8}{1 + 7e^{-0.8x}} \qquad$ Set $y = 4$ and solve numerically or algebraically for x.

$x \approx 2.4323...$, which agrees with the graph. ◀

The general form of the logistic differential equation follows from Examples 1, 2, and 3.

DEFINITIONS: The Logistic Differential Equation and Logistic Function

The **logistic differential equation** is

$$\frac{dy}{dx} = ky \cdot \frac{M - y}{M} \quad \text{or} \quad \frac{dy}{dx} = \frac{k}{M}(y)(y - M)$$

The constant M, called the **carrying capacity** of the system being modeled, is the maximum sustainable value of y as x increases. The constant k is a proportionality constant.

The **logistic function**

$$y = \frac{M}{1 + ae^{-kx}}$$

is the solution of the logistic differential equation, where the constant a is determined by the initial condition.

Predator-Prey Population Problems

In Problems 9–23 of Problem Set 7-6, you will investigate a situation in which foxes prey on rabbits, and find that under certain assumptions about population growth rates, a cyclical behavior occurs, in which the population of one species increases while the population of the other decreases, then vice versa.

Problem Set 7-6

Quick Review

Q1. $\int_a^b f(x)\,dx = \lim\limits_{n \to \infty} \Sigma f(c)\Delta x$ is a brief statement of the —?—.

Q2. $\int_a^b f(x)\,dx = g(b) - g(a)$ is a brief statement of the —?—.

Q3. $\int f(x)\,dx = g(x)$ if and only if $f(x) = g'(x)$ is a statement of the —?—.

Q4. "... then there is a point c in (a, b) such that $f(c) = k$" is the conclusion of —?—.

Q5. "... then there is a point c in (a, b) such that $f'(c) = 0$" is the conclusion of —?—.

Q6. "... then there is a point c in (a, b) such that $f'(c) = \dfrac{f(b) - f(a)}{b - a}$," is the conclusion of —?—.

Q7. "... then $f'(x) = g'(h(x)) \cdot h'(x)$" is the conclusion of —?—.

Q8. $f(x) = \cos x + C$ is the —?— solution of a differential equation.

Q9. $f(x) = \cos x + 5$ is a(n) —?— solution of a differential equation.

Q10. $f(0) = 6$ is a(n) —?— condition for the differential equation in Problem Q9.

---

1. *Bacteria Problem:* Harry and Hermione start a culture of bacteria in a laboratory dish by introducing 3 million bacteria at time $t = 0$ hours. The number of bacteria increases rapidly at first, then levels off. The two lab partners estimate that the dish can support a maximum of 30 million bacteria. They let B stand for the size of the bacteria population, in millions, at time t, in hours, and assume that the rate of change of B is given by the logistic differential equation

$$\frac{dB}{dt} = 0.21B \cdot \frac{30 - B}{30}$$

The slope field for this differential equation is shown in Figure 7-6c.

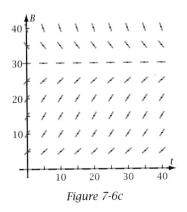

Figure 7-6c

   a. Explain the real-world influence on the rate of bacteria growth of the two factors B and $(30 - B)/30$ in the differential equation. Explain why it is reasonable that dB/dt is positive for $0 < B < 30$ and negative for $B > 30$.

   b. On a copy of Figure 7-6c, sketch the graph of the particular solution subject to the given initial condition of 3 million bacteria. Also, sketch the graph of the particular solution if Harry and Hermione try to speed up the process by adding enough bacteria at time $t = 10$ to make a total of 40 million bacteria. What is the major difference in the behavior of the population for the two different initial conditions?

   c. Use Euler's method with $dt = 0.5$ and initial condition $(0, 3)$ to estimate B at times $t = 10$, 20, 30, and 40 h. Mark these points on your graph from part b. Does your graph agree reasonably well with these values?

   d. Show the steps you used to solve the differential equation algebraically, subject to the initial condition that $B = 3$ when $t = 0$. Write B explicitly as a function of t. Use the particular solution to find out how close the Euler's method estimate of B is to the precise predicted value of B when $t = 20$.

   e. Use the differential equation to show that the greatest rate of increase in the bacteria population occurs when B is halfway between the horizontal asymptotes at $B = 0$ and $B = 30$. Find the value of t at this "point of inflection."

2. *Subdivision Building Problem:* A real estate developer opens up a small subdivision of 120 lots on which to build houses. She builds houses on five of the lots, then opens the subdivision for others to build. The total number of houses increases slowly at first, then faster, and finally slows down as the last few lots are built on.

   a. Explain why a logistic function is a reasonable mathematical model for the number of houses as a function of time.

   b. The developer estimates that the proportionality constant is 0.9, so that the differential equation is

$$\frac{dy}{dx} = 0.9y \cdot \frac{120 - y}{120}$$

This development in Davis, California, utilizes solar panels installed on the roof of each building.

where y is the number of houses that have been built after time x, in years. Figure 7-6d shows the slope field for this differential equation. Show the steps you used to separate the variables and integrate this differential equation subject to the initial condition that $y = 5$ when $x = 0$. Write the particular solution for y explicitly as a function of x.

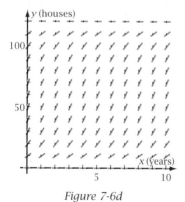

Figure 7-6d

c. Find the time when 70% of the lots have been built on, at which time the developer

no longer needs to participate in the homeowners' association. Find the time at which you expect only one unbuilt lot to remain.

d. Demonstrate how you can use the original differential equation to show that the point of inflection in the graph of y versus x is at $y = 60$, halfway between the asymptotes at $y = 0$ and $y = 120$. Explain the real-world significance of this point of inflection.

3. *Merchandise Sales Problem:* When a new product is brought onto the market, sales typically increase slowly at first, then increase more rapidly as more people learn about the product, and finally taper off as there are fewer new customers who want to buy the product. Suppose that Ajax Studios releases a new CD by popular singer Nellie Wilson. The studio assumes that a logistic function is a reasonable mathematical model. They know that the general equation is

$$\frac{dy}{dx} = ky \cdot \frac{M - y}{M}$$

where y is the number of CDs, in thousands, sold after a time x, in days. But k and M are unknown constants. At a particular time designated as $x = 0$ days, they find that $y = 10$ (10,000 CDs sold) and that sales are increasing at 500 ($\frac{1}{2}$ of 1000) CDs per day. When y has reached 24, sales are increasing at 1100 per day.

a. Use the given information to find values of the constants k and M in the differential equation. Based on the values of these constants, how many of Nellie's CDs does Ajax eventually expect to sell?

b. Plot the slope field for this differential equation using a computer algebra system or dynamic geometry system such as The Geometer's Sketchpad (or at least using the low-resolution graphics of your hand-held grapher). Does the slope field confirm the carrying capacity you found in part a?

c. Find the particular solution for y explicitly as a function of x subject to the initial condition $y = 10$ when $x = 0$. Plot this solution on the slope field of part b. Does the graph follow the slope lines?

d. Use the particular solution to find out the total sales when $x = 50$ and when $x = 51$. How many CDs does Ajax expect to sell on this 51st day?

e. Merchandisers look for the point of inflection in sales graphs such as this so that they will know when sales start to slip and it is time to start advertising. How many CDs will Ajax have sold at the point of inflection? At what time should they plan to start an ad campaign?

4. *General Solution of the Logistic Differential Equation:* Start with the general logistic differential equation and show the steps in separating the variables and integrating to prove that

$$\text{If } \frac{dy}{dx} = ky \cdot \frac{M - y}{M}, \text{ then } y = \frac{M}{1 + ae^{-kx}}$$

where the constant a is determined by the constant of integration.

5. *Snail Darter Endangered Species Problem:* The snail darter is a small fish found naturally only in certain waterways. Suppose that a particular waterway currently has 190 snail darters living in it. Based on known data, researchers believe that in this waterway the maximum sustainable population is 1000 darters and the minimum sustainable population is 200. In a reasonable mathematical model for the snail darter population, F, dF/dt is proportional to the fraction of the minimum number times the fraction of the maximum number. That is,

$$\frac{dF}{dt} = 130 \cdot \frac{F - 200}{200} \cdot \frac{1000 - F}{1000}$$

where F is the number of snail darters and t is time in years from the present, and the proportionality constant is estimated to be 130. Figure 7-6e shows the slope field for this differential equation.

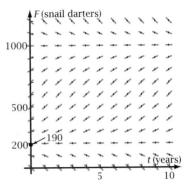

Figure 7-6e

a. The 190 snail darters present initially is below the minimum sustainable population! Use Euler's method with $dt = 0.1$ to show that the darter population is predicted to decrease, eventually becoming extinct. At approximately what time does the model predict extinction? On a copy of Figure 7-6e, sketch the population as a function of time.

b. Concerned about the threat of extinction, the National Wildlife Federation restocks the waterway at time $t = 3$ years with enough snail darters to bring F up to 1200. On a copy of the slope field, sketch the graph of the darter population under this condition, and make a prediction about what will happen.

c. Suppose that at time $t = 0$, enough snail darters are introduced to bring the population up to $F = 300$. Sketch the graph of F under this condition, and make a prediction about what will happen.

d. Solve the differential equation algebraically for F as a function of t, using the initial conditions of part c. To transform this new problem into a more familiar problem, let $F = y + 200$. Then dF/dt is the same as dy/dt, and the differential equation becomes

$$\frac{dy}{dt} = 130 \cdot \frac{y}{200} \cdot \frac{800 - y}{1000}$$

Solve this differential equation algebraically for y, then find F by suitable algebra. Plot the graph of F as a function of t on the slope field. Does your graphical solution from part c agree with the precise algebraic solution?

6. *Rumor-Spreading Experiment:* Number off the members of your class. Pick person 1 to be a person who will start a rumor, and have him or her stand. By calculator, choose two random integers to decide which two people will be told the rumor. Those people stand. Then choose two more random integers for each person standing, and have the people with these numbers stand. At each iteration, x, record N, the number of people standing. Continue until all class members are standing. Then perform logistic regression on your grapher to find an equation for N as a function of x. Plot the equation and the data on the same screen. Does the graph confirm that a logistic function is a reasonable mathematical model in this case?

7. *U.S. Population Project:* The following table shows the U.S. population, in millions, from 1940 through 1990. In this problem you will use these data to make a mathematical model for predicting the population in future years and for seeing how far back the model fits previous years.

a. For the years 1950, 1960, 1970, and 1980, find symmetric difference quotients, $\Delta P/\Delta t$, where P is population in millions and t is time in years since 1940. (Why can't you do this for 1940 and 1990?)

Year	Population
1940	131.7
1950	151.4
1960	179.3
1970	203.2
1980	226.5
1990	248.7

b. For each year given in part a, find $\Delta P/\Delta t$ as a fraction of P. That is, find $(\Delta P/\Delta t)/P$.

c. It is reasonable to assume that the rate of growth of a population in a fixed region such as the United States (as a fraction of the size of that population) is some function of the population. For instance, when the population gets too large, its growth rate slows because of overcrowding. Find the function (linear, logarithmic, exponential, or power) that best fits the values of $(\Delta P/\Delta t)/P$ as a function of P. Justify your answer.

d. Assume that dP/dt obeys the same equation as $\Delta P/\Delta t$. Write a differential equation based on your answer to part c. Transform the equation so that dP/dt is by itself on one side. The result is a logistic equation.

e. Plot a slope field. Use $-50 \le t \le 100$ and $0 \le P \le 500$. Print the slope field or sketch it on a copy of Figure 7-6f.

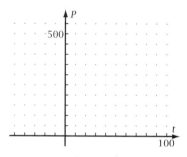

Figure 7-6f

f. Make a table of population predicted by Euler's method for each 10 years from $t = -50$ through $t = 100$. Use the population in 1940 ($t = 0$) as the initial condition. Use steps of $\Delta t = 1$ yr. Plot the points on the graph you drew in part e and connect them with a smooth curve.

g. According to this mathematical model, what will be the ultimate population of the United States? How does this number appear in the differential equation and on the slope field?

h. Plot the populations for the six years given in the table on your graph from part e. Does the population really seem to follow the solution by Euler's method?

i. Write a paragraph describing how well the predicted populations in part f agree with the actual populations from 1940 through 1990.

j. Consult some reference material to find the results of censuses dating back through 1900. How well do your predicted values compare with the actual ones? How can you explain any large discrepancies between predicted and actual values?

k. Suppose that in the year 2010, 200 million people immigrate to the United States. Predict the population for the next 40 years. What does the logistic-equation model say about the population growth under this condition?

8. *Algebraic Solution of the Logistic Equation:* As you have seen, it is possible to solve the logistic differential equation like that in Problem 7 algebraically. Suppose that

$$\frac{dy}{dx} = 3y(10 - y)$$

Separating the variables and integrating gives

$$\int \frac{1}{y(10 - y)} \, dy = 3 \int dx$$

In this problem you will learn how to integrate on the left side. Then you will solve the logistic equation in Problem 7 algebraically.

a. The fraction in the left integral can be split into partial fractions like this:

$$\frac{1}{y(10 - y)} = \frac{A}{y} + \frac{B}{10 - y}$$

where A and B stand for constants. Using suitable algebra, find the values of A and B.

b. Integrate the differential equation. Show that you can transform the integrated equation into

$$y = \frac{10}{1 + ke^{-30x}}$$

where k is a constant related to the constant of integration.

c. Solve the logistic equation from part d of Problem 7 algebraically. Transform the answer so that population is in terms of time. Use the initial condition $P = 131.7$ in 1940 (when $t = 0$) to evaluate k.

d. Use the algebraic solution you found in part c to predict the population in 1950, 1960, 1970, 1980, and 1990. How well do the approximate solutions found by Euler's method in part f of Problem 7 compare with these exact solutions? How well do the exact solutions compare with the actual population in these years? Write a paragraph that describes your observations about how well different mathematical models agree with each other and about how well they fit data from the real world.

For Problems 9–23, Ona Nyland moves to an uninhabited island. Being lonely for company, she imports some pet rabbits. The rabbits multiply and become a nuisance! So she imports some foxes to control the rabbit population.

9. Let R be the number of rabbits, in hundreds, and F be the number of foxes at any given time, t. If there are no foxes, the rabbit population grows at a rate proportional to the population. That is, dR/dt equals $k_1 R$, where k_1 is a positive constant. Show that the rabbit population grows exponentially with time under this condition.

10. If there are no rabbits for the foxes to eat, the fox population decreases at a rate proportional to the population. That is, dF/dt equals $-k_2 F$, where k_2 is a positive constant. Show that the fox population decreases exponentially under this condition.

11. Assume that the foxes eat rabbits at a rate proportional to the number of encounters between foxes and rabbits. This rate is proportional to the product of the number of rabbits and foxes. (If there are twice as many rabbits, there are twice as many encounters, and vice versa.) So, there are two rates operating on each population: The rabbit population is decreasing at a rate $k_3 RF$ at the same time it is increasing at $k_1 R$. The fox population is increasing at a rate $k_4 RF$ and decreasing at $k_2 F$. Write differential equations for dR/dt and for dF/dt under these conditions.

12. Use the chain rule to write a differential equation for dF/dR. What happens to dt?

13. Assume that the four constants in the differential equation are such that

$$\frac{dF}{dR} = \frac{-F + 0.025RF}{R - 0.04RF}$$

If $R = 70$ and $F = 15$, calculate dF/dR.

14. Figure 7-6g shows the slope field for this differential equation. On a copy of the figure, show the initial condition given in Problem 13. Then show the relative populations of rabbits and foxes as time progresses. How do you tell from the differential equations you wrote for Problem 11 whether to start going to the right or to the left?

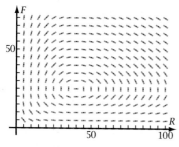

Figure 7-6g

15. How would you describe the behavior of the rabbit and fox populations?

16. Is there a fixed point at which both the rabbit and the fox populations do not change? Explain.

17. The logistic equation of Problem 7 shows that, because of overcrowding, the rate of change in the population is decreased by an amount proportional to the square of the population. Assume that

$$\frac{dR}{dt} = R - 0.04RF - 0.01R^2$$

Calculate dF/dR (not dR/dt!) at $R = 70$ and $F = 15$ under this condition.

18. The slope field in Figure 7-6h is for dF/dR, which you calculated in Problem 17. On a copy of the figure, use the initial condition in Problem 17 to sketch the predicted populations.

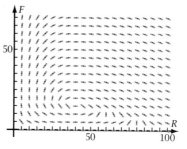

Figure 7-6h

19. How does the graph in Problem 18 differ from that in Problem 14? How does overcrowding by rabbits affect the ultimate rabbit population? The ultimate fox population?

20. Ona tries to reduce the rabbit population by allowing hunters to come to the island. She allows the hunters to take 1000 rabbits per unit of time, so dR/dt is decreased by an additional 10. Calculate dF/dR at the point $(R, F) = (70, 15)$ under these conditions.

21. The slope field in Figure 7-6i is for the differential equation

$$\frac{dF}{dR} = \frac{-F + 0.025RF}{R - 0.04RF - 0.01R^2 - 10}$$

On a copy of Figure 7-6i, trace the predicted populations under these conditions, starting at the point $(70, 15)$.

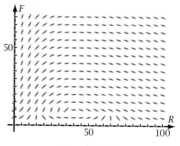

Figure 7-6i

22. Describe what happens to the populations of rabbits and foxes under these conditions.

23. Worried about the fate of the foxes in Problem 21, Ona imports 15 more of them. Starting at the point $(70, 30)$, trace the populations. According to this mathematical model, what is the effect of importing more foxes? Surprising?

7-7 Chapter Review and Test

In this chapter you have seen that by knowing the rate at which a population is changing, you can write an equation for the derivative of the population. You can solve this differential equation numerically by Euler's method, graphically by slope field, or exactly by algebraic integration.

Review Problems

R0. Update your journal with what you've learned since your last entry. Include such things as

- The one most important thing you have learned in studying Chapter 7
- Which boxes you've been working on in the "define, understand, do, apply" table
- The proportion property of exponential functions and their derivatives, and this property's converse
- The fact that you can find a function equation from the rate of change of the function
- How to solve differential equations graphically and numerically
- Any ideas about calculus that you're still unclear about

R1. *Punctured Tire Problem:* You've run over a nail! The pressure, $P(t)$, in pounds per square inch (psi), of the air remaining in your tire is given by

$$P(t) = 35(0.98^t)$$

where t is the number of seconds since the tire was punctured. Calculate $P(0)$, $P(10)$, and $P(20)$. Show by example that although $P'(t)$ decreases as t increases, the ratio $P'(t)/P(t)$ stays constant. Prove in general that $P'(t)/P(t)$ is constant.

R2. *Ramjet Problem:* A ramjet (Figure 7-7a) is a relatively simple jet engine. The faster the plane goes, the more air is "rammed" into the engine and the more power the engine generates.

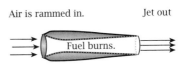

Air is rammed in.　　Jet out

Fuel burns.

Figure 7-7a

a. Assume that the rate at which the plane's speed changes is directly proportional to its speed. Write a differential equation that expresses this assumption.

b. Solve the differential equation. Show the integration step and describe what happens to the absolute value sign.

c. Evaluate the constants in the equation if the plane is flying at 400 mi/h at time $t = 0$ s and 500 mi/h at time $t = 40$ s.

d. When will the plane reach the speed of sound, 750 mi/h?

The X-43A hypersonic research aircraft, powered by a ramjet engine, can reach speeds up to 10 times the speed of sound (Mach 10).

R3. a. Find the general solution of the differential equation $dy/dx = 6y^{1/2}$.

b. Find the particular solution of the equation in part a that contains the point (3, 25).

c. Plot the graph of the solution you found in part b. Sketch the result.

d. Find dy/dx for this differential equation when $x = 2$. On your graph, show that your answer is reasonable.

e. *Memory Retention Problem:* Paula Lopez starts her campaign for election to state senate. She meets people at a rate of about 100 per day, and she tries to remember as many names as possible. She finds that after seven full days, she remembers the names of 600 of the 700 people she met.

   i. Assume that the rate of change in the number of names she remembers, dN/dt, equals 100 minus an amount that is directly proportional to N. Write a differential equation that expresses this assumption, and solve the equation subject to the initial condition that she knew no names when $t = 0$.

   ii. How many names should Paula remember after 30 days?

   iii. Does your mathematical model predict that her brain will "saturate" after a long time, or does it predict that she can remember unlimited numbers of names?

   iv. After how many days of campaigning will Paula be able to remember the names of only 30 of the people she meets that day?

R4. Figure 7-7b shows the slope field for

$$\frac{dy}{dx} = -\frac{20}{xy} + 0.05y$$

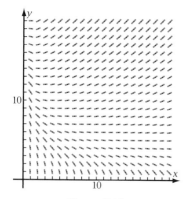

Figure 7-7b

a. Calculate the slope at the points $(2, 5)$ and $(10, 16)$. Show that these slopes agree with the graph.

b. On a copy of Figure 7-7b, draw the particular solutions that contain $(1, 8)$ and $(1, 12)$. Describe the major difference in the behavior of the two graphs.

c. Does the particular solution that contains $(1, 10)$ behave like that containing $(1, 12)$ or that containing $(1, 8)$? Justify your answer.

R5. a. For the differential equation given in Problem R4, use Euler's method to calculate values of y for the particular solution that contains $(1, 9)$. Use $\Delta x = 1$. Where does the graph seem to cross the x-axis? On a copy of Figure 7-7b, plot the points you calculated.

b. Use Euler's method, as you did in part a, but with an increment of $\Delta x = 0.1$. Record the y-value for each integer value of x that shows in Figure 7-7b. Plot these points on your copy of Figure 7-7b.

c. Write a few sentences commenting on the accuracy of Euler's method far away from the initial point when you use a relatively large value of Δx.

d. At what value of x would the graph described in part c cross the x-axis?

R6. *Beaver Logistic Function Problem:* The beaver population, y, in hundreds of beavers, in a certain area is presently 1200 and is changing at rate dy/dx, in hundred beavers per year, given by

$$\frac{dy}{dx} = 0.6y \cdot \frac{9 - y}{9}$$

Figure 7-7c shows the slope field for the logistic differential equation.

a. On a copy of Figure 7-7c, sketch the particular solution subject to the initial condition $y = 12$ when $x = 0$ years. How do you interpret the fact that the beaver population is decreasing? Use Euler's method with $dx = 0.1$ to estimate the number of beavers at time $x = 3$.

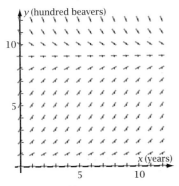

Figure 7-7c

b. At time $x = 3$ years, a flood washes away most of the remaining beavers, leaving only 100. On the copy of the slope field, sketch the graph of the beaver population subject to the initial condition that $y = 2$ when $x = 3$. Explain what happens to the population under these conditions.

c. Find the algebraic solution of the differential equation using the initial condition given in part b. At what time after $x = 3$ is the population increasing most rapidly?

Predator-Prey Problem: Space explorers visiting a planet in a nearby star system discover a population of 600 humanlike beings called Xaltos living by preying on a herd of 7000 creatures that bear a remarkable resemblance to yaks. They find that the differential equation that relates the two populations is

$$\frac{dy}{dx} = \frac{-0.5(x - 6)}{(y - 7)}$$

where x is the number of hundreds of Xaltos and y is the number of thousands of yaks.

Figure 7-7d shows the slope field for this differential equation.

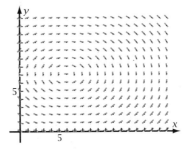

Figure 7-7d

d. The numerator of the fraction in the differential equation is dy/dt and the denominator is dx/dt. Explain why the two populations presently seem to be in equilibrium with each other.

e. Suppose that 300 more Xaltos move into the community. Starting at the point (9, 7), draw the particular solution of the differential equation on a copy of Figure 7-7d. Explain why the graph goes clockwise from this initial point. Describe what happens to the two populations as time goes on.

f. Instead of 300, suppose that 1300 more Xaltos move into the community. Draw the particular solution, using this initial condition. What dire circumstance befalls the populations under this condition? Surprising?

g. What if only 900 more Xaltos move in instead of the 1300 described in part c? Would the same fate befall the populations? Justify your answer.

Concept Problems

C1. *Differential Equations Leading to Polynomial Functions:* You have shown that if dy/dx is directly proportional to y, then y is an exponential function of x. In this problem you will investigate similar differential equations that lead to other types of functions.

a. If dy/dx is directly proportional to $y^{1/2}$, show that y is a quadratic function of x.

b. Make a conjecture about what differential equation would make y a cubic function of x.

c. Verify or refute your conjecture by solving the differential equation. If your conjecture is wrong, make other conjectures until you find one that is correct.

d. Once you succeed in part c, you should be able to write a differential equation whose solution is any specified degree. Demonstrate that you see the pattern by writing and solving a differential equation whose solution is an eighth-degree function.

*C2. *Film Festival Problem:* In this chapter you assumed a certain behavior for the derivative of a function. Then you integrated to find an equation for the function. In this problem you will reverse the procedure. You will use measured values of a function, then find the derivative to make use of the mathematical model. In order to make money for trips to contests, the math club at Chelmsdale High plans to rent some videocassettes and present an all-night Halloween film festival in the school gym. The club members want to predict how much money they could make from such a project and to set the admission price so that they make the greatest amount of money.

The club conducts a survey of the entire student body, concluding with the question "What is the most you would pay to attend the festival?" Here are the results.

Maximum Price ($)	Number of People
2.00	100
2.50	40
3.00	60
4.00	120
4.50	20
5.50	40
6.00	80

a. Make a chart that shows the total number of people likely to attend as a function of the admission price.

b. Plot the data you charted in part a. What type of function might be a reasonable mathematical model for people in terms of dollars? Fit an equation of this function to the data.

c. The amount of money club members expect to make is the product of ticket price and the number of tickets purchased. Write an equation that expresses amount of money as a function of ticket price.

d. What price should club members charge to make the greatest amount of money? Justify your answer.

e. Why would club members expect to make less money if they charged more than the price you determined in part d? Why would they expect to make less money if they charged less than the price in part d?

C3. *Gompertz Growth Curve Problem:* Another function with a sigmoid (S-shaped) graph sometimes used for population growth is the Gompertz function, whose general equation is

$$g(t) = ae^{-ce^{-kt}}$$

where $g(t)$ is the population at time t, and a, c, and k are positive constants. The graphs of these functions look somewhat like Figure 7-7e. In this problem you will investigate effects of the constants, maximum growth rates, and limiting population values.

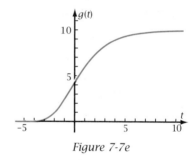

Figure 7-7e

a. Let $a = 10$, $c = 0.8$, and $k = 0.5$ so that the equation is

$$g(t) = 10e^{-0.8e^{-0.5t}}$$

Plot the graph of this particular Gompertz function. Confirm that it looks like the graph in Figure 7-7e. What does the limit of $g(t)$ appear to be as t approaches infinity? Confirm your answer by taking the limit in the equation. If $g(t)$ represents population growth, what is the significance of this limit in the real world?

b. Find the equation of the particular Gompertz function that fits the U.S. population figures for 1960, 1970, and 1980, given in the table on the next page.

---

* Adapted from data by Landy Godbold, as cited by Dan Teague.

Year	U.S. Population (Millions)
1960	179
1970	203
1980	226

Let $t = 0$ in 1970. To deal with the exponential constants, you may take the ln of both sides of the equation twice. By clever use of algebra, you can get two equations that involve only the constant a. Then you can use your grapher to calculate the value of a. Plot the graph of the function. At what value does the population seem to level off?

c. Suppose that the 1980 data point is 227 million instead of 226 million. How does this change affect the predicted ultimate U.S. population? Does the Gompertz equation seem to be fairly sensitive to slight changes in initial conditions?

C4. *Hot Tub Problem, Continued:* In Problem 5 of Problem Set 7-3, you wrote a differential equation for the volume of water remaining in a hot tub as it drained. That equation is

$$\frac{dV}{dt} = -2V^{1/2}$$

where V is the volume of water that remains t minutes after the drain is opened. By solving the differential equation, you found that the 196 ft^3 of water initially in the tub drained in 14 min. Now suppose that while the drain is open, water flows in at a rate of F ft^3/min. Explore the effect of such an inflow on the remaining amount as a function of time.

Chapter Test

PART 1: No calculators allowed (T1–T8)

T1. Write a differential equation stating that the instantaneous rate of change of y with respect to x is directly proportional to the value of y.

T2. What does it mean to *solve* a differential equation?

T3. What is the difference between the *general* solution of a differential equation and a *particular* solution?

T4. On a copy of Figure 7-7f, sketch the particular solution of the differential equation with slope field shown subject to the initial condition $y = -4$ when $x = 0$.

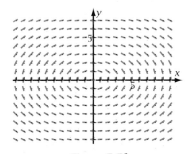

Figure 7-7f

T5. Why will Euler's method for solving differential equations give an underestimate of the values of y for the particular solution in Problem T4?

T6. If the constant M is the maximum sustainable value of a dependent variable y, write the general logistic differential equation for dy/dx.

T7. Show that you know how to solve a differential equation algebraically by solving

$$\frac{dy}{dx} = 0.4y$$

for y explicitly as a function of x if the point $(0, -5)$ is on the graph.

T8. Separate the variables and integrate this differential equation. You may leave the answer as an implicit relation (that is, you don't need to solve for y explicitly in terms of x).

$$\frac{dy}{dx} = 12y^{1/2}$$

T9. *Phoebe's Space Leak Problem:* Phoebe is returning to Earth in her spaceship when she detects an oxygen tank leak. She knows that the rate of change of pressure is directly proportional to the pressure of the remaining oxygen.

a. Write a differential equation that expresses this fact and solve it subject to the initial condition that pressure is 3000 psi (pounds per square inch) at time $t = 0$ when Phoebe discovers the leak.

b. Five hours after she discovers the leak, the pressure has dropped to 2300 psi. At that time, Phoebe is still 20 h away from Earth. Will she make it home before the pressure drops to 800 psi? Justify your answer.

T10. *Swimming Pool Chlorination Problem:* Suppose that a pool is filled with chlorine-free water. The chlorinator is turned on, dissolving chlorine in the pool at a rate of 30 g/h. But chlorine also escapes to the atmosphere at a rate proportional to the amount dissolved in the water. For this particular pool, the escape rate is 13 g/h when the amount dissolved is 100 g.

a. Write a differential equation that expresses this information and solve it to express the amount of chlorine, in grams, in the pool as a function of the number of hours the chlorinator has been running. Be clever in finding an initial condition!

b. How long will it take for the chlorine content to build up to the desired 200 g?

T11. *Water Lily Problem:* Phoebe plans to plant water lilies in a small lake in the tropics. She realizes that the lilies will multiply, eventually filling the lake. Consulting a Web site on the Internet, she figures that the maximum number of lilies her lake can sustain is 1600. She assumes that a logistic differential equation is a reasonable mathematical model, and figures that the differential equation is

$$\frac{dy}{dx} = 0.5y \cdot \frac{16 - y}{16}$$

where y is in hundreds of lilies and x is in months. The slope field is shown in Figure 7-7g.

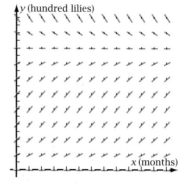

Figure 7-7g

a. Use what you have learned about the relationship between the logistic differential equation and its solution to find the particular solution if Phoebe plants 200 lilies ($y = 2$) at time $x = 0$ months.

b. Demonstrate that you understand how Euler's method works by showing the steps in estimating y at $x = 0.2$ with $dx = 0.1$. Show numerically that this Euler's solution is an underestimate, as expected, because of the concavity of the graph.

c. Phoebe could speed up the process by planting 400 lilies, or she could wait awhile until the original number grows to 400. How long would she have to wait?

d. On a copy of Figure 7-7g, show what would happen if Phoebe is impatient and plants 2000 lilies at time $x = 0$.

T12. *Coyote and Roadrunner Problem:* Coyotes are reputed to prey on roadrunners. Figure 7-7h shows the slope field for the differential equation

$$\frac{dR}{dC} = \frac{dR/dt}{dC/dt}$$

where C is the number of coyotes at a particular time t and R is the corresponding number of roadrunners.

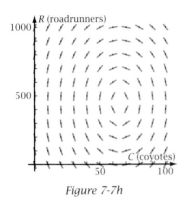

Figure 7-7h

a. Starting at the point $(C, R) = (80, 700)$, which way will the graph go, to the left or to the right? How can you tell? On a copy of Figure 7-7h, sketch the particular solution subject to this initial condition.

b. You have learned that a solution of a differential equation is a *function* whose derivative appears in the equation. If you continue the pattern in the slope field, there will be places where there are two different values of R for the same value of C. Explain why this situation *is* satisfactory in this problem.

T13. Write a paragraph about the most important thing you learned as a result of studying this chapter.

7-8   Cumulative Review: Chapters 1–7

In your study of calculus so far, you have learned that calculus involves four major concepts, studied by four techniques. You should be able to do four major things with the concepts.

Concepts	Techniques	Be able to
Limits	Graphical	Define them.
Derivatives	Numerical	Understand them.
Indefinite integrals	Algebraic	Do them.
Definite integrals	Verbal	Apply them.

Two of these concepts, derivatives and definite integrals, are used to work problems involving the rate of change of a function (derivatives), and the product of x and y for a function in which y depends on x (definite integrals). Both derivatives and definite integrals are founded on the concept of limit. Indefinite integrals, which are simply antiderivatives, provide an amazing link between derivatives and definite integrals via the fundamental theorem of calculus.

The following problems constitute a "semester exam" in which you will demonstrate your mastery of these concepts as you have studied them so far.

Problem Set 7-8

Rocket Problems: Princess Leia is traveling in her rocket ship. At time $t = 0$ min, she fires her rocket engine. The ship speeds up for a while, then slows down as the planet Alderaan's gravity takes its effect. The graph of her velocity, $v(t)$, in miles per minute, as a function of t, in minutes, is shown in

Figure 7-8a, on the next page. In Problems 1–16, you will analyze Leia's motion.

1. On a copy of Figure 7-8a, draw a narrow vertical strip of width dt. Show a sample point $(t, v(t))$ on the graph within the strip. What physical quantity does $v(t)\,dt$ represent?

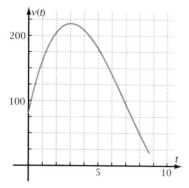

Figure 7-8a

2. If you take the sum $\sum v(t)\,dt$ from $t = 0$ to $t = 8$, what calculus concept equals the limit of this sum as dt approaches zero?

3. Leia figures that her velocity is given by

$$v(t) = t^3 - 21t^2 + 100t + 80$$

Use the fundamental theorem of calculus to find the distance she travels from $t = 0$ to $t = 8$.

4. Calculate midpoint Riemann sums with $n = 100$ and $n = 1000$ increments. How do the results confirm that the fundamental theorem gives the correct answer for the integral even though it has nothing to do either with Riemann sums or with limits?

5. On a copy of Figure 7-8a, draw a representation of an upper sum with $n = 8$ increments.

6. Explain why, for an integrable function, any Riemann sum is squeezed to the same limit as the upper and lower sums as the widths of the increments approach zero.

7. Write the definition of definite integral. State the fundamental theorem of calculus. Be sure to tell which is which.

8. Calculate the integral in Problem 3 numerically by using your grapher's integrate feature. Calculate the integral again graphically by counting squares. Compare the answers with the exact value.

9. Use symmetric difference quotients with $\Delta t = 0.1$ min and $\Delta t = 0.01$ min to estimate the rate of change of Leia's velocity when $t = 4$ min.

10. Write the definition of derivative.

11. For most types of functions there is a way to find the derivative algebraically. Use the appropriate method to find the exact rate of change of Leia's velocity when $t = 4$.

12. At $t = 4$, was Leia speeding up or slowing down? Justify your answer.

13. On a copy of Figure 7-8a, draw a line at the point $(4, v(4))$ with slope $v'(4)$. Show how you constructed the line. How is the line related to the graph?

14. What is the physical name of the instantaneous rate of change of velocity?

15. Leia's maximum velocity seems to occur at $t = 3$. Use derivatives appropriately to find out whether the maximum occurs when t is exactly 3 s.

16. Find an equation for $v''(t)$, the second derivative of $v(t)$ with respect to t.

Related Rates Problems: Suppose that you have constructed the diagram shown in Figure 7-8b on a computer screen using a dynamic geometry utility such as The Geometer's Sketchpad. The curve has equation $y = 6e^{-0.5x}$. A vertical line y units long connects the x-axis to a point on the graph, and a diagonal line z units long connects this point on the graph to the origin.

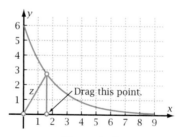

Figure 7-8b

17. Suppose that you drag the point on the x-axis at a constant rate of 0.3 unit per second. Find dy/dt when $x = 2$. Is y increasing or decreasing at this instant? At what rate?

18. Find dz/dt when $x = 2$. Is z increasing or decreasing at this instant? At what rate?

Compound Interest Problems: When money is left in a savings account that compounds interest continuously, the instantaneous rate at which the money increases, dm/dt, is directly proportional to m, the amount in the account at that instant.

19. Write a differential equation that expresses this property.

20. Solve the differential equation in Problem 19 for m as a function of t, and show the steps that you use.

21. In one word, how does m vary with t?

22. The solution in Problem 20 is called the —?— solution of the differential equation.

23. Find the particular solution in Problem 20 if m is $10,000 at $t = 0$ and $10,900 at $t = 1$.

24. In Problem 23, the amount of money in the account grew by $900 in one year. True or false: The amount of money will grow by $9000 in 10 yr. Justify your answer.

Discrete Data Problems: The techniques of calculus were invented for dealing with continuous functions. These techniques can also be applied to functions specified by a table of data. This table gives values of y for various values of x.

x	y
30	74
32	77
34	83
36	88
38	90
40	91
42	89

25. Use Simpson's rule to estimate $\int_{30}^{42} y\, dx$.

26. Estimate dy/dx if $x = 36$. Show how you got your answer.

Mean Value Theorem Problems: The proof of the fundamental theorem is based on the mean value theorem. This theorem is a corollary of Rolle's theorem.

27. State Rolle's theorem.

28. Sketch a graph that illustrates the conclusion of the mean value theorem.

Graphing Problems: Calculus is useful for analyzing the behavior of graphs of functions.

29. Figure 7-8c shows the graph of function f. On a copy of this figure, sketch the derivative graph, f'.

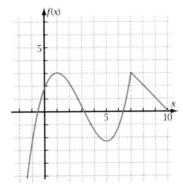

Figure 7-8c

30. The function
$$f(x) = 2^x - \frac{|x - 1|}{x - 1}$$

has a discontinuity at $x = 1$. Sketch the graph. What type of discontinuity is it?

31. The function $g(x) = x^{1/3}(x - 1)$ has $g(0) = 0$. Show that $g'(0)$ is undefined. Show what the graph of g looks like in a neighborhood of $x = 0$. You may use your grapher's cube root function, if it has one.

Area and Volume Problems: Figure 7-8d shows the graphs of $y = e^{0.2x}$ and $y = 0.6x$.

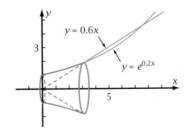

Figure 7-8d

32. Find numerically the two intersections of the graphs shown in the figure. Store these, without rounding. Find the area of the narrow region bounded by these two graphs between the two intersections.

33. The region bounded by the two graphs in Problem 32 from $x = 0$ to the first intersection is rotated about the x-axis to form a solid. Find the volume of this solid.

Differential Equation Problems: Figure 7-8e shows the slope field for the differential equation

$$\frac{dy}{dx} = 0.25\frac{x}{y}$$

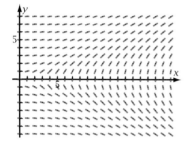

Figure 7-8e

34. On a copy of Figure 7-8e, sketch the particular solutions that contain the points $(0, 3)$ and $(10, 4)$.

35. The two solutions in Problem 34 share a common asymptote. Sketch the asymptote. State an initial condition that would give the asymptote as the graph of the solution.

36. Solve the differential equation by separating the variables and integrating. Find the equation of the particular solution that contains the point $(10, 4)$.

37. Use the function in Problem 36 to calculate the exact value of y when $x = 10.5$.

38. Demonstrate that you understand the idea behind Euler's method by calculating the first point to the right of the point $(10, 4)$ in Problem 36, with $\Delta x = 0.5$. How does this value compare with the exact value in Problem 37?

Algebraic Techniques Problems: You have learned algebraic techniques for differentiating, antidifferentiating, and calculating limits.

39. Find: $\dfrac{d}{dx}(\sin^{-1} x^3)$

40. Find $\dfrac{dy}{dx}$ if $x = \ln(\cos t)$ and $y = \sec t$.

41. Find: $\displaystyle\int \frac{dx}{4 - 3x}$

42. Find $h'(x)$ if $h(x) = 5^x$.

43. Find: $\displaystyle\lim_{x \to 0} \frac{\sin 5x + \cos 3x - 5x - 1}{x^2}$

44. Plot the graph of the fraction given in Problem 43. Sketch the result. Show how the graph confirms your answer to Problem 43.

Journal Problems: Take a moment and think about what you've learned and what you're still unsure about. In your journal:

45. Write what you think is the one most important thing you have learned so far as a result of taking calculus.

46. Write one thing in calculus that you are still not sure about.

The Calculus of Plane and Solid Figures

A cable hanging under its own weight forms a curve called a catenary. A cable supporting a uniform horizontal load, such as the cables in the Brooklyn Bridge, forms a parabola. By slicing such graphs into short segments, you can find the differential of arc length. Integrating this differential allows engineers to compute the exact length of hanging cables and chains, important information for constructing bridges.

Mathematical Overview

In Chapter 8, you will learn how definite integrals help you find exact area, volume, and length by slicing an object into small pieces, then adding and taking the limit. You will also use derivatives to find maxima, minima, and other interesting features of geometric figures. You will explore geometric figures in four ways.

Graphically   The icon at the top of each even-numbered page of this chapter shows an object for which you can find length, area, volume, and points of inflection.

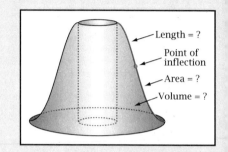

Length = ?
Point of inflection
Area = ?
Volume = ?

Numerically

x	$f'(x)$	$f(x)$
1.8	0.72	13.931
1.9	0.33	13.984
2.0	0	14 ← Max.
2.1	−0.27	13.987
2.2	−0.48	13.949
⋮	⋮	⋮

Algebraically $V = \pi \int_a^b (x_2^2 - x_1^2)\, dy$, volume by slicing into washers

Verbally   *I think the most important thing I learned is that I can use the same technique to find length and surface area that I used to find volume and plane area. To do this, I draw a figure that shows a representative slice of the object, pick a sample point within the slice, find the differential of the quantity I'm trying to find, then add up the differentials and take the limit, which means integrate.*

8-1 Cubic Functions and Their Derivatives

Recall that the graph of a quadratic function, $f(x) = ax^2 + bx + c$, is always a parabola. The graph of a cubic function, $f(x) = ax^3 + bx^2 + cx + d$, is called a **cubic parabola**. To begin your application of calculus to geometric figures, you will learn about the second derivative, which tells the rate at which the (first) derivative changes. From the second derivative you can learn something about the curvature of a graph and whether the graph curves upward or downward.

OBJECTIVE   Work alone or with your study group to explore the graphs of various cubic functions and to make connections between a function's graph and its derivatives.

Figure 8-1a shows the graphs of three cubic parabolas. They have different shapes depending on the relative sizes of the coefficients $a, b,$ and c. (The constant d affects only the vertical placement of the graph, not its shape.) Sometimes they have two distinct vertices, sometimes none at all. In Exploratory Problem Set 8-1, you will accomplish the objective of this section.

Exploratory Problem Set 8-1

1. In Figure 8-1a,

$$f(x) = x^3 - 6x^2 + 9x + 3$$
$$g(x) = x^3 - 6x^2 + 15x - 9$$
$$h(x) = x^3 - 6x^2 + 12x - 3$$

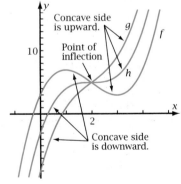

Figure 8-1a

For each function, find an equation for the derivative. Plot the function and its derivative on the same screen. Then list as many connections as you can find between the function graph and the derivative graph. Sketches will help.

2. What connection can you see between the graph of the derivative of a function and whether the function has two distinct vertex points (high or low points)?

3. The **second derivative** of a function is the derivative of the (first) derivative. For instance, $f''(x)$ (read "f double prime of x") is equal to $6x - 12$. Find equations for the second derivatives $g''(x)$ and $h''(x)$. What do you notice?

4. Figure 8-1a illustrates a curve that is **concave up** for certain points and **concave down** for others. What do you notice about the sign of the second derivative and the direction of the concave side of the graph?

5. A graph has a *point of inflection* where it changes from concave down to concave up or vice versa. State two ways you can use derivatives to locate a point of inflection.

8-2   Critical Points and Points of Inflection

If a moving object comes to a stop, several things can happen. It can remain stopped, start off again in the same direction, or start off again in some different direction. When a car stops or reverses direction, its velocity goes through zero (hopefully!). When a baseball is hit by a bat, its velocity changes abruptly and is undefined at the instant of contact. Figure 8-2a shows how displacement, d, and velocity, v (derivative), vary with time, x.

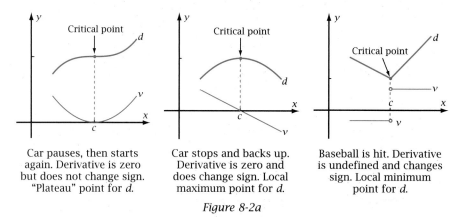

Car pauses, then starts again. Derivative is zero but does not change sign. "Plateau" point for d.

Car stops and backs up. Derivative is zero and does change sign. Local maximum point for d.

Baseball is hit. Derivative is undefined and changes sign. Local minimum point for d.

Figure 8-2a

A point where the derivative is zero or undefined is called a **critical point**, a term that comes from "crisis." (When one reaches a crisis, things stop and can then go in different directions.) "Critical point" is sometimes used for the point on the x-axis and sometimes for the point on the graph itself. You must decide from the context which is meant.

The y-value at a critical point can be a **local maximum** or a **local minimum** (Figure 8-2a, center and right). The word *local* indicates that $f(c)$ is the maximum or minimum of $f(x)$ when x is kept in a neighborhood (locality) of c. The **global maximum** and **global minimum** are the largest and smallest of the local maxima and minima, respectively. (Maxima and minima are the plural forms.) A critical point with zero derivative but no maximum or minimum (Figure 8-2a, left) is called a **plateau point**. (*Relative* and *absolute* are often used in place of local and global when expressing maxima and minima.)

There are connections between the derivative of a function and the behavior of its graph at a critical point. For instance, if the derivative changes from positive to negative (Figure 8-2a, center), there is a maximum point in the graph of the function. As you saw in Section 8-1, the second derivative of a function tells which direction the concave side of the graph points. A **point of inflection**, or **inflection point**, occurs where the concavity changes direction.

OBJECTIVE   From information about the first and second derivatives of a function, decide whether the y-value is a local maximum or minimum at a critical point and whether the graph has a point of inflection, then use this information to sketch the graph or find the equation of the function.

▶ **EXAMPLE 1**   For the function graphed in Figure 8-2b, sketch a number-line graph for f' and a number-line graph for f'' that shows the sign of each derivative in a neighborhood of the critical point at $x = 2$. On the number-line graphs, indicate whether $f(2)$ is a local maximum or a local minimum and whether the graph has a point of inflection at $x = 2$.

Solution   Sketch a number-line graph for f' and another one for f''. Each number-line graph needs three regions: one for x, one for the derivative, and one for $f(x)$. Figure 8-2c shows a convenient way to sketch them. The abbreviation "e.p." signifies an endpoint of the domain.

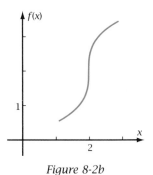

Figure 8-2b

The graph is vertical at $x = 2$, so $f'(2)$ is infinite. Insert the infinity symbol, ∞, in the $f'(x)$ region above $x = 2$, and draw a vertical arrow above it in the $f(x)$ region.

The graph of f slopes up on both sides of $x = 2$. Since the derivative is positive when the function is increasing, put a plus sign in the $f'(x)$ region on both sides of $x = 2$. Show upward sloping arrows in the $f(x)$ region above the plus signs. There is not a maximum or minimum value of $f(x)$ at $x = 2$, so you don't need to write any words in that region.

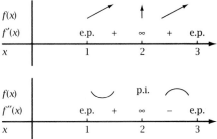

Figure 8-2c

The graph is concave up for $x < 2$ and concave down for $x > 2$. Because a positive second derivative indicates concave up and a negative second derivative indicates concave down, put a plus sign in the $f''(x)$ region to the left of $x = 2$ and a minus sign to the right. Draw arcs in the $f(x)$ region to indicate the direction of concavity of the f graph. The concavity changes (from up to down) at $x = 2$, so the graph has a point of inflection there. Write "p.i." in the $f(x)$ region above $x = 2$.   ◀

A Note on Concavity and Curvature

The word *concave* comes from the Latin cavus, meaning "hollow." (So do "cave" and "cavity.") If the second derivative is positive, the first derivative is increasing. Figure 8-2d shows why the concave side is upward in this case and vice versa. As shown in Figure 8-2e, the larger the absolute value of $f''(x)$, the more sharply the graph curves. However, as you will learn in Section 10-6,

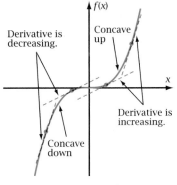

Figure 8-2d

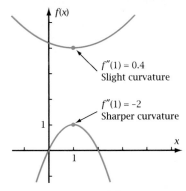

Figure 8-2e

the curvature also depends on the slope of the graph. For a given value of $f''(x)$, the steeper the slope, the less the curvature.

In Example 2, you will reverse the procedure of Example 1 and construct the graph of a function from the number lines for its first and second derivatives.

▶ **EXAMPLE 2** Figure 8-2f shows number-line graphs for the first and second derivatives of a continuous function f. Use this information to sketch the graph of f if $f(4) = 0$. Describe the behavior of the function at critical points.

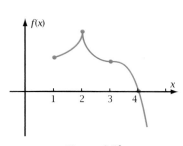

Figure 8-2f

Solution Sketch the number-line graphs. Add arrows and arcs in the $f(x)$ region to show the slope and concavity in the intervals between critical points (Figure 8-2g). Add words to describe what features the graph will have at the critical points of f and f'. Sketch a continuous function (no asymptotes) that has the prescribed features and crosses the x-axis at $x = 4$ (Figure 8-2h). The graph you draw may be somewhat different, but it must have the features shown on the number-line graphs in Figure 8-2g.

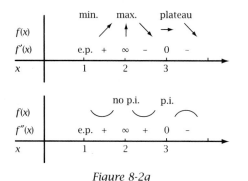

Figure 8-2g

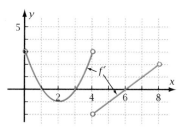

Figure 8-2h

Example 3 shows you how to sketch the graph of a function if the actual graph of the derivative function is given, not just the number-line information.

▶ **EXAMPLE 3** Figure 8-2i shows the graph of the derivative of a continuous, piecewise function f defined on the closed interval $x \in [0, 8]$. On a copy of the figure, sketch the graph of f, given the initial condition that $f(1) = 5$. Put a dot at the approximate location of each critical point and each point of inflection.

Figure 8-2i

Solution Figure 8-2j shows that

• $f(1) = 5$, the initial condition, is a local maximum because $f'(x)$ changes from positive to negative as x increases through 1.

- $f(0)$ is an endpoint local minimum because $f(x)$ is increasing between 0 and 1. By counting squares, $f(0) \approx 5 - 1.3 = 3.7$. It is defined because the domain is $[0, 8]$.

- $f(2) \approx 5 - 0.7 = 4.3$ is a point of inflection because $f'(x)$ has a local minimum at $x = 2$.

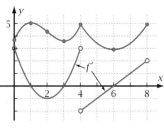

Figure 8-2j

- $f(3) \approx 4.3 - 0.7 = 3.6$ is a local minimum because $f'(x)$ changes from negative to positive at $x = 3$.

- $f(4) \approx 3.6 + 1.3 = 4.9$ exists because f is specified to be continuous, and is a local maximum because $f'(x)$ changes from positive to negative at $x = 4$. There is a cusp or corner (see the note on page 379) at $x = 4$ because the direction of the graph changes abruptly (discontinuous derivative).

- $f(6) \approx 4.9 - 2.0 = 2.9$ is a local minimum because $f'(x)$ changes from negative to positive at $x = 6$.

- $f(8) \approx 2.9 + 2.0 = 4.9$ is an endpoint local maximum because $f(x)$ is increasing from $x = 6$ to $x = 8$. ◀

In Example 4, you are given both the equation for the function and an accurate graph. You will be asked to find critical features algebraically, and some of them may be hard to see.

▶ **EXAMPLE 4**   Figure 8-2k shows the graph of $f(x) = x^{4/3} + 4x^{1/3}$.

a. Sketch number-line graphs for f' and f'' that show features that appear clearly on the graph.

b. Find equations for $f'(x)$ and $f''(x)$. Show algebraically that the critical points you drew in part a are correct. Fix any errors.

c. Write x- and y-coordinates of all maxima, minima, and points of inflection.

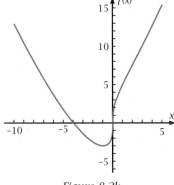

Figure 8-2k

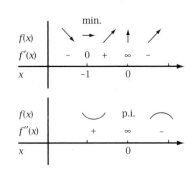

Figure 8-2l

Solution   a. Figure 8-2l shows the two number-line graphs. Note that $f'(x)$ is zero at $x = -1$ and infinite at $x = 0$. The graph is concave up for $x < 0$ and appears to be concave down for $x > 0$.

b. $f'(x) = \frac{4}{3}x^{1/3} + \frac{4}{3}x^{-2/3} = \frac{4}{3}x^{-2/3}(x + 1)$ Factor out the power of x with the smaller exponent.

$f''(x) = \frac{4}{9}x^{-2/3} - \frac{8}{9}x^{-5/3} = \frac{4}{9}x^{-5/3}(x - 2)$

Critical points occur where either $f'(x) = 0$ or $f'(x)$ is undefined.

$f'(x) = 0 \iff \frac{4}{3}x^{-2/3}(x + 1) = 0$

$x^{-2/3} = 0$ or $x + 1 = 0$ A product is zero if and only if one of its factors is zero.

$\therefore x = -1$ $x^{-2/3} = 1/x^{2/3}$, which cannot equal zero. So the other factor must be zero.

$f'(x)$ is undefined $\iff x = 0$ $0^{-2/3} = 1/0^{2/3} = 1/0$, which is infinite.

$\therefore$ critical points occur at $x = 0$ and $x = -1$, as observed in part a.

Inflection points can occur where f' has critical points; that is, $f''(x)$ is zero or undefined.

$f''(x) = 0 \iff \frac{4}{9}x^{-5/3}(x - 2) = 0$

$x^{-5/3} = 0$ or $x - 2 = 0$

$\therefore x = 2$

$f''(x)$ is undefined $\iff x = 0$

For there to be a point of inflection, $f''(x)$ must change sign.

At $x = 2$, the factor $(x - 2)$ in $f''(x)$ changes sign. At $x = 0$, the factor $(4/9)x^{-5/3}$ changes sign. (Any power of a positive number is positive. If x is negative, then $x^{-5/3}$ is negative. The cube root of a negative number is negative, the fifth power of that answer is also negative, and the reciprocal (negative exponent) of that negative answer is still negative.)

So inflection points are at $x = 0$ and $x = 2$.

The point at $x = 2$ did not show up in the original number-line graph for f'' in part a, so add this feature to your sketch, as shown in Figure 8-2m.

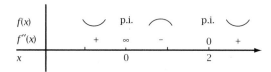

Figure 8-2m

c. To find the y-coordinates of the maxima and minima, substitute the x-values from part b into the $f(x)$ equation.

$f(-1) = (-1)^{4/3} + 4(-1)^{1/3} = 1 - 4 = -3$

$f(0) = 0$

$f(2) = (2)^{4/3} + 4(2)^{1/3} = 7.559...$

The local and global minima of $f(x)$ are both -3 at $x = -1$.

Points of inflection are at $(0, 0)$ and at $(2, 7.559...)$.

There are no local or global maxima, because $f(x)$ approaches ∞ as x approaches $\pm\infty$.   ◀

Sometimes you will find critical points from just the equation for a function. Example 5 shows how to do this graphically and numerically and how to confirm the results algebraically.

▶ **EXAMPLE 5**   Let $f(x) = -x^3 + 4x^2 + 5x + 20$, with domain $x \in [-2.5, 5]$.

    a. Plot the graph. Estimate the x- and y-coordinates of all local maxima or minima and of all points of inflection. State the global maximum and minimum.

    b. Write equations for $f'(x)$ and $f''(x)$. Use them to find, either numerically or algebraically, the precise values of the x-coordinates in part a.

    c. Show that the second derivative is *negative* at the local maximum point and *positive* at the local minimum point. Explain the graphical meaning of these facts.

    d. Explain why there are no other critical points or points of inflection.

Solution

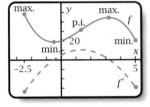

Figure 8-2n

    a. Figure 8-2n shows the graph in the given domain. Using your grapher's maximum and minimum features to find the high points and low point, and TRACE to find points of inflection, you get

Local minima of 20 at the endpoint $x = 5$, and about 18.625 at $x \approx -0.5$

Global minimum at about 18.625

Local maxima of 48.125 at the endpoint $x = -2.5$, and about 44.192 at $x \approx 3.2$

Global maximum at about 48.125

Point of inflection at approximately $(1.3, 31)$

    b. $f'(x) = -3x^2 + 8x + 5$

       $f''(x) = -6x + 8$

The graph of f' is shown on the same screen as f in Figure 8-2n. To locate the critical points precisely, either use your grapher's solver feature to find numerically where $f'(x) = 0$, or use the quadratic formula. Thus,

$$x = \frac{-8 \pm \sqrt{64 - 4(-3)(5)}}{2(-3)}$$
$$= -0.5225... \text{ or } 3.1892...$$

Both of these confirm the estimates in part a.

To find the point of inflection precisely, set $f''(x) = 0$ and solve. So,

$$-6x + 8 = 0 \iff x = \tfrac{4}{3}$$

which confirms the estimate of $x \approx 1.3$ in part a.

c. $f''(-0.5225...) = 11.1355...$, which is positive.

$f''(3.1892...) = -11.1355...$, which is negative.

The graph is *concave up* at a low point and *concave down* at a high point.

d. Because $f'(x)$ is quadratic, it can have at most two zeros, both of which were found in part b. Since $f''(x)$ is linear, it has exactly one zero, which was also found in part b. Therefore, there are no more critical points or points of inflection. ◀

Notes on the Second Derivative

From Example 5 you saw that a point of inflection occurs where the derivative stops increasing and starts decreasing, and vice versa. You can conclude that a point of inflection occurs where the *derivative* function has a local maximum or local minimum. So the *derivative* of the derivative (the *second* derivative), if it is defined, will be zero at a point of inflection.

Also from Example 5 you can see an algebraic way to determine whether a critical point is a high point or a low point. If the graph has a zero first derivative (horizontal tangent), then there is a *high* point if the second derivative is negative (concave down) or a *low* point if the second derivative is positive (concave up). This fact is called the **second derivative test** for maxima and minima. It is illustrated in Figure 8-2o and summarized in the box on page 380. The figure also shows that the test fails to distinguish among maximum, minimum, and plateau points if the second derivative is *zero* where the first derivative is zero.

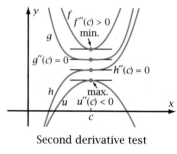

Second derivative test

Figure 8-2o

Figure 8-2p and the accompanying boxes present the definitions and properties of this section.

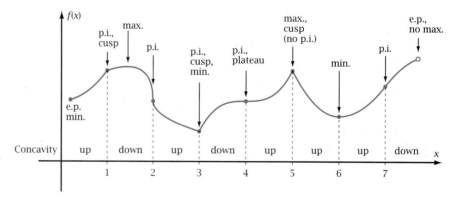

Figure 8-2p

Note that the figure shows *local* maxima and minima. To find the *global* maximum and minimum, you must check each local one to find out which is the greatest and which is the least.

Note on Cusps and Corners

The cusp at $x = 5$ in Figure 8-2p has the property that the slope becomes infinite as x approaches 5 from both directions, and so the graph has a vertical tangent line. At $x = 3$, there is a step change in the first derivative (an abrupt change in direction of the graph) but the slopes approach different values from the positive and negative sides. The name **corner** is sometimes used for such a point.

DEFINITIONS: *Critical Points and Related Features*

- A **critical point** on a graph occurs at $x = c$ if and only $f(c)$ is defined and $f'(c)$ either is zero or is undefined.

- $f(c)$ is a **local maximum** (or relative maximum) of $f(x)$ if and only if $f(c) \geq f(x)$ for all x in a neighborhood of c (that is, in an open interval containing c).

- $f(c)$ is a **local minimum** (or relative minimum) of $f(x)$ if and only if $f(c) \leq f(x)$ for all x in a neighborhood of c.

- $f(c)$ is the **global maximum** (or absolute maximum) of $f(x)$ if and only if $f(c) \geq f(x)$ for all x in the domain of f.

- $f(c)$ is the **global minimum** (or absolute minimum) of $f(x)$ if and only if $f(c) \leq f(x)$ for all x in the domain of f.

- The graph of f is **concave up** at $x = c$ if and only if for all x in a neighborhood of c, the graph lies *above* the tangent line at the point $(c, f(c))$. The graph of f is **concave down** at $x = c$ if and only if for all x in a neighborhood of c, the graph lies *below* the tangent line at $x = c$. The value of $f''(c)$ is called the **concavity** of the graph of f at $x = c$.

- The point $(c, f(c))$ is a **point of inflection**, or **inflection point**, if and only if $f''(x)$ changes sign at $x = c$. (Old spelling: inflexion, meaning "not bent.")

- The point $(c, f(c))$ is a **cusp** if and only if f' is discontinuous at $x = c$. If the concavity changes at $x = c$, but the secant lines on both sides of c do not approach a common tangent line, the term **corner** is sometimes used instead of cusp.

- The point $(c, f(c))$ is a **plateau point** if and only if $f'(c) = 0$, but $f'(x)$ does not change sign at $x = c$.

PROPERTIES: Maximum, Minimum, and Point of Inflection

- If $f'(x)$ goes from positive to negative at $x = c$ and f is continuous at $x = c$, then $f(c)$ is a local maximum.

- If $f'(x)$ goes from negative to positive at $x = c$, and f is continuous at $x = c$, then $f(c)$ is a local minimum.

- If $f''(c)$ is positive, then the graph of f is concave up at $x = c$.

- If $f''(c)$ is negative, then the graph of f is concave down at $x = c$.

- If $f''(x)$ changes sign at $x = c$ and f is continuous at $x = c$, then $(c, f(c))$ is a point of inflection (by definition).

- The **second derivative test**: If $f'(c) = 0$ and $f''(c)$ is positive (graph concave up), then $f(c)$ is a local minimum. If $f'(c) = 0$ and $f''(c)$ is negative (graph concave down), then $f(c)$ is a local maximum. If $f'(c) = 0$ and $f''(c) = 0$, then the second derivative test fails to distinguish among maximum, minimum, and plateau points.

- A maximum or minimum point (but not a point of inflection) can occur at an endpoint of the domain of a function.

Problem Set 8-2

Quick Review

Q1. Sketch: $y = x^2$

Q2. Sketch: $y = x^3$

Q3. Sketch: $y = \cos x$

Q4. Sketch: $y = \sin^{-1} x$

Q5. Sketch: $y = e^{-x}$

Q6. Sketch: $y = \ln x$

Q7. Sketch: $y = \tan x$

Q8. Sketch: $y = x$

Q9. Sketch: $y = 1/x$

Q10. Sketch: $x = 2$

For Problems 1–10, sketch number-line graphs for f' and f'' that show what happens to the value and to the sign of each derivative in a neighborhood of $x = 2$.

1.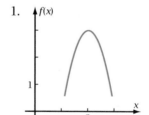

2.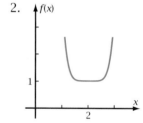

3.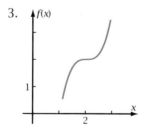

4.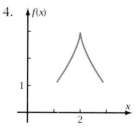

5.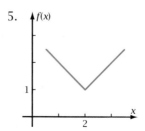

6.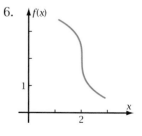

7.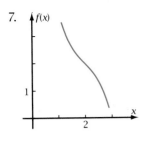

8.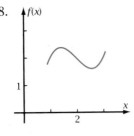

9.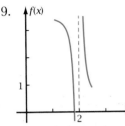

10.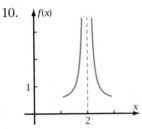

15.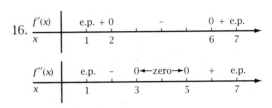

16.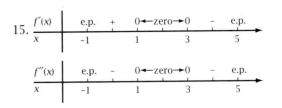

For Problems 11–16, on a copy of the number-line graphs, mark information about the behavior of the graph of f. Sketch a graph of a continuous function f consistent with the information about the derivatives.

For Problems 17–20, the graph of $y = f'(x)$, the derivative of a continuous function f, is given. On a copy of the graph, sketch the graph of the parent function $y = f(x)$ in the indicated domain, subject to the given initial condition.

17. Initial condition: $f(2) = -2$
    Domain: $x \in [1, 5]$

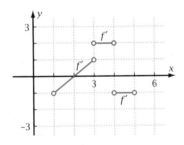

18. Initial condition: $f(1) = 3$
    Domain: $x \in [1, 9]$

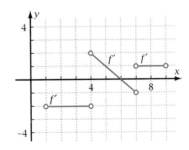

11.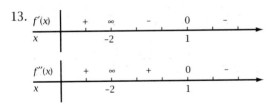

12.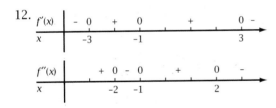

13.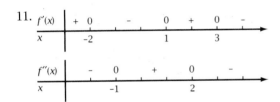

14.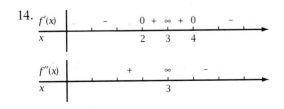

19. Initial condition: $f(2) = 4$
    Domain: $x \in [0, 8]$

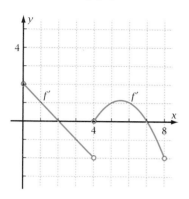

20. Initial condition: $f(0) = 2$
    Domain: $x \in [-2, 4]$

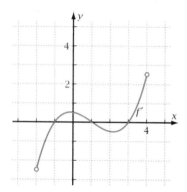

For Problems 21–26, show that a critical point occurs at $x = 2$, and use the second derivative test to determine algebraically whether this critical point is a relative maximum or a relative minimum. If the test fails, use the sign of the first derivative to decide. Plot the graph and sketch it in a neighborhood of $x = 2$, thus confirming your conclusion.

21. $f(x) = 3e^x - xe^x$

22. $f(x) = -\sin \frac{\pi}{4} x$

23. $f(x) = (2 - x)^2 + 1$

24. $f(x) = -(x - 2)^2 + 1$

25. $f(x) = (x - 2)^3 + 1$

26. $f(x) = (2 - x)^4 + 1$

27. Let $f(x) = 6x^5 - 10x^3$ (Figure 8-2q).

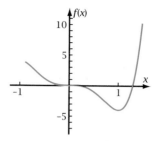

Figure 8-2q

a. Use derivatives to find the x-coordinates for all critical points of f and f'.

b. Explain why there are critical points in part a that do not show up on this graph.

c. Explain why there is no maximum or minimum point at $x = 0$, even though $f'(0)$ equals zero.

28. Let $f(x) = 0.1x^4 - 3.2x + 7$ (Figure 8-2r).

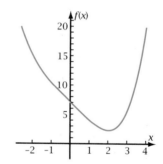

Figure 8-2r

a. Use derivatives to find the x-coordinates of all critical points of f and f'.

b. Explain why there is no point of inflection at $x = 0$, even though $f''(0)$ equals zero.

c. Under what conditions for $f'(x)$ and $f''(x)$ can a graph be "straight" without being horizontal?

29. Let $f(x) = xe^{-x}$ (Figure 8-2s).

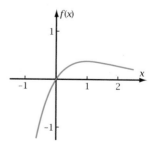

Figure 8-2s

a. Use derivatives to find the x-coordinates of all critical points of f and f'.

b. How can you tell that there is a point of inflection even though it does not show up on the graph?

c. Does the graph cross the x-axis at any point other than $(0, 0)$? Justify your answer.

30. Let $f(x) = x^2 \ln x$ (Figure 8-2t).

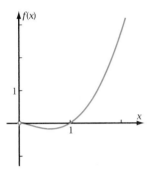

Figure 8-2t

a. Use derivatives to find the x-coordinates of all critical points of f and f'.

b. Show that the limit of $f(x)$ is zero as x approaches zero from the right, but not from the left. L'Hospital's rule will help.

c. Are there any critical points that do not show up on the graph? If so, where, and what type? If not, explain how you know.

31. Let $f(x) = x^{5/3} + 5x^{2/3}$ (Figure 8-2u).

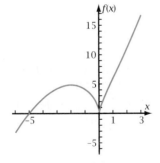

Figure 8-2u

a. Use derivatives to find the x-coordinates of all critical points of f and f'.

b. Explain why there is a tangent line at the cusp, even though $f'(x)$ is undefined there.

c. Is there a point of inflection at the cusp? Is there a point of inflection anywhere else?

32. Let $f(x) = x^{1.2} - 3x^{0.2}$ (Figure 8-2v).

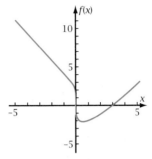

Figure 8-2v

a. Use derivatives to find the x-coordinates of all critical points of f and f'.

b. The tangent is vertical at $x = 0$. How do you know that there aren't several different values of y at $x = 0$?

c. Is the graph straight or curved when x is less than -2? If it is curved, which way is the concave side directed?

For Problems 33–36,

a. Plot the graph. Using TRACE, and the maximum and minimum features of your grapher, find graphically the approximate x- and y-coordinates of all local maxima, minima, and points of inflection. Find the global maximum and global minimum.

b. Write equations for $f'(x)$ and $f''(x)$. Use them to find numerically or algebraically the precise values of the x-coordinates in part a.

c. Show how the second derivative test applies to the leftmost interior critical point.

d. Explain how you know that there are no other critical points or points of inflection.

33. $f(x) = -x^3 + 5x^2 - 6x + 7$

34. $f(x) = x^3 - 7x^2 + 9x + 10$

35. $f(x) = 3x^4 + 8x^3 - 6x^2 - 24x + 37$, for $x \in [-3, 2]$

36. $f(x) = (x - 1)^5 + 4$, for $x \in [-1, 3]$

37. *Point of Inflection of a Cubic Function:* The general equation of a quadratic function is

$$y = ax^2 + bx + c \qquad \text{where } a \neq 0$$

Recall from algebra that the "middle" of the graph of a quadratic function (that is, the vertex) is at $x = -b/(2a)$. The "middle" of the graph of a cubic function is at its point of inflection. Prove that if $f(x) = ax^3 + bx^2 + cx + d$, where $a \neq 0$, then the point of inflection is located at $x = -b/(3a)$.

38. *Maximum and Minimum Points of a Cubic Function:* The maximum and minimum points of a cubic function are located symmetrically on either side of the point of inflection. Prove that this is true in general for the cubic function $f(x) = ax^3 + bx^2 + cx + d$. In terms of the coefficients a, b, c, and d, how far on either side of the point of inflection do the maximum and minimum points occur?

For Problems 39 and 40, find the particular equation of the cubic function described. Use your grapher to confirm your solutions.

39. Local maximum at the point $(5, 10)$ and point of inflection at $(3, 2)$

40. Local maximum at the point $(-1, 61)$ and point of inflection at $(2, 7)$

41. *Concavity Concept Problem:* Figure 8-2w shows the graph of $f(x) = x^3$. Tangent lines are drawn where $x = -0.8, -0.5, 0.5,$ and 0.8.

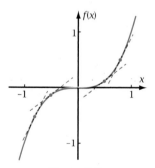

Figure 8-2w

a. Calculate the slope for each given tangent point.

b. What happens to the slope as x increases from -0.8 to -0.5? As x increases from 0.5 to 0.8? How do the values of the second derivative confirm these findings?

c. On which side of the tangent line does the graph of a function lie if the graph is concave up at the point of tangency?

42. *Naive Graphing Problem:* Ima Evian plots the graph of $y = x^3$, using $x = -1, 0,$ and 1 (Figure 8-2x). From these three points she concludes that the graph is a straight line. Explain to Ima how she can use derivatives to avoid making this false conclusion.

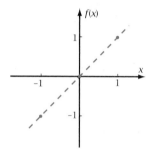

Figure 8-2x

43. *Connection Between a Zero First Derivative and the Graph:* If $f'(c) = 0$, the only thing you know for sure about the graph of f is that a horizontal tangent occurs at $x = c$. Sketch a graph that shows these behaviors in a neighborhood of $x = c$.

a. $f(x)$ stops increasing and starts decreasing as x increases through c.

b. $f(x)$ stops decreasing and starts increasing as x increases through c.

c. $f(x)$ stops increasing but starts increasing again as x increases through c.

d. $f(x)$ stops decreasing but starts decreasing again as x increases through c.

e. $f(x)$ stays locally constant as x increases through c.

44. *Infinite Curvature Problem:* Show that the graph of

$$f(x) = 10(x - 1)^{4/3} + 2$$

is defined and differentiable at $x = 1$, but that the second derivative is infinite there. Explore the behavior of $f(x)$ close to $x = 1$ by zooming in on that point on the graph or by constructing a table of values. Describe what you discover.

45. *Exponential and Polynomial Function Look-Alike Problem:* Figure 8-2y shows the graphs of

$$f(x) = e^{0.06x} \text{ and}$$
$$g(x) = 1 + 0.06x + 0.0018x^2 + 0.000036x^3$$

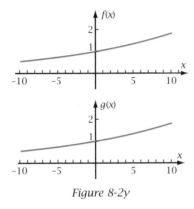

Figure 8-2y

Show that although the graphs look similar and f and g have equal function values and equal first, second and **third derivatives** at $x = 0$, they are *not* identical functions.

46. *A Pathological Function:* Consider the piecewise function

$$f(x) = \begin{cases} (x-1)^3 \sin \dfrac{1}{x-1} + 2, & \text{if } x \neq 1 \\ 2, & \text{if } x = 1 \end{cases}$$

Plot the graph of f using a window with $[0, 2]$ for x and $[1.99, 2.01]$ for y. Show that f, as defined, is continuous and has a zero derivative at $x = 1$, even though the graph makes an infinite number of cycles as x approaches 1 from either side.

47. *Journal Problem:* Update your journal with what you've learned since the last entry. Include such things as

- The one most important thing you've learned since your last entry
- The relationships between the signs of the first and second derivatives and the behavior of the graph of the function itself
- How the first and second derivatives can be used to locate maxima, minima, and points of inflection algebraically
- How your understanding of derivatives has improved
- Any techniques or ideas about the behavior of graphs that are still unclear

---

8-3 Maxima and Minima in Plane and Solid Figures

In Section 8-2, you found maximum and minimum values of a function where the equation was already given. The problems of this section require you first to find an equation for the area, volume, or perimeter of a geometric figure, then to use the now-familiar techniques to find extreme values. As a result you will be able to investigate real-world situations such as how the canning industry saves money by packaging the maximum volume of product with the minimum amount of metal.

OBJECTIVE Given a plane or solid figure, find the maximum or minimum perimeter, area, or volume.

▶ **EXAMPLE 1**

Suppose you need to build a rectangular corral along the riverbank. Three sides of the corral will be fenced with barbed wire. The river forms the fourth side of the corral (Figure 8-3a). The total length of fencing available is 1000 ft. What is the maximum area the corral could have? How should the fence be built to enclose this maximum area? Justify your answers.

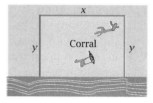

Figure 8-3a

Solution

The first thing to note is that the problem asks you to maximize the area. So you need an equation for area as a function of one or more variables. Letting A stand for area and x and y stand for the length of fence parallel to and perpendicular to the river, respectively, you can write

$$A = xy$$

Next you must find A in terms of one variable. Because there is a total of 1000 ft of fencing, you can write an equation relating x and y.

$$x + 2y = 1000 \implies x = 1000 - 2y, \text{ where } y \in [0, 500] \qquad \text{If } x = 0, \text{ then } y = 500.$$

$$\therefore A = (1000 - 2y)(y) = 1000y - 2y^2$$

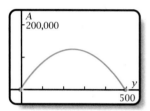

Figure 8-3b

As Figure 8-3b shows, the graph of A versus y is a parabola opening downward, with its maximum point halfway between the two y-intercepts. Since these intercepts are $y = 0$ and $y = 500$, and these points are the ends of the domain, the maximum point is at

$$y = 250$$

Confirming this fact by derivatives,

$$A' = 1000 - 4y$$
$$A' = 0 \text{ if and only if}$$
$$1000 - 4y = 0 \iff y = 250$$
$$\therefore x = 1000 - 2(250) = 500$$
$$\therefore A = (500)(250) = 125{,}000$$

You should build the corral 250 ft perpendicular to the river and 500 ft parallel to the river. The corral can have a maximum area of 125,000 ft². ◀

▶ **EXAMPLE 2**

In Figure 8-3c, the part of the parabola $y = 4 - x^2$ from $x = 0$ to $x = 2$ is rotated about the y-axis to form a surface. A cone is inscribed in the resulting paraboloid with its vertex at the origin and its base touching the parabola. At what radius and height does the maximum volume occur? What is this maximum volume? Justify your answer.

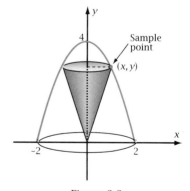

Figure 8-3c

Solution

Pick a sample point (x, y) on the parabola in the first quadrant where the cone touches it. Since you are asked to maximize volume, find an equation for volume V in terms of x and y.

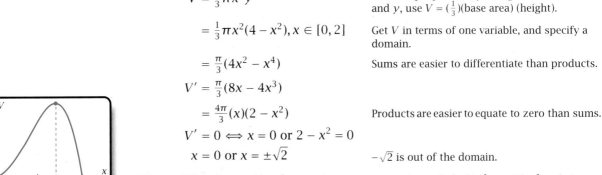

$$V = \frac{1}{3}\pi x^2 y \qquad \text{Pick sample point } (x, y). \text{ To get } V \text{ in terms of } x \text{ and } y, \text{ use } V = (\frac{1}{3})(\text{base area})(\text{height}).$$

$$= \frac{1}{3}\pi x^2 (4 - x^2), x \in [0, 2] \qquad \text{Get } V \text{ in terms of one variable, and specify a domain.}$$

$$= \frac{\pi}{3}(4x^2 - x^4) \qquad \text{Sums are easier to differentiate than products.}$$

$$V' = \frac{\pi}{3}(8x - 4x^3)$$

$$= \frac{4\pi}{3}(x)(2 - x^2) \qquad \text{Products are easier to equate to zero than sums.}$$

$$V' = 0 \iff x = 0 \text{ or } 2 - x^2 = 0$$

$$x = 0 \text{ or } x = \pm\sqrt{2} \qquad -\sqrt{2} \text{ is out of the domain.}$$

Figure 8-3d shows that the maximum comes at $x \approx 1.4$. At the critical point $x = 0$, the volume is a minimum. At the other endpoint, $x = 2$, the volume is also minimum.

The maximum volume is at $x = \sqrt{2}$. The volume at this point is

$$V = \frac{\pi}{3}(2)(4 - 2)$$

$$= 4\pi/3 = 4.18879... \qquad \blacktriangleleft$$

There are key steps in Examples 1 and 2 that will help you succeed in max-min problems. These steps are listed in this box.

Figure 8-3d (caption under left figure)

TECHNIQUE: Analysis of Maximum-Minimum Problems

1. Make a sketch if one isn't already drawn.

2. Write an equation for the variable you are trying to maximize or minimize.

3. Get the equation in terms of one variable and specify a domain.

4. Find an approximate maximum or minimum by grapher.

5. Find the exact maximum or minimum by seeing where the derivative is zero or infinite. Check any endpoints of the domain.

6. Answer the question by writing what was asked for in the problem statement.

Problem Set 8-3

Quick Review

Q1. Differentiate: $y = (3x + 5)^{-1}$

Q2. Integrate: $\int (x + 6)^{-1} dx$

Q3. Differentiate: $y = x^{-2/3}$

Q4. Integrate: $\int x^{-2/3} dx$

Q5. Integrate: $\int x^{-2} dx$

Q6. Integrate: $\int x^0 dx$

Q7. $\int \cot x \, dx = \text{—?—}$

Q8. Sketch the graph: $y = x^{1/3}$.

Q9. Sketch the graph of y'' for the cubic function in Figure 8-3e.

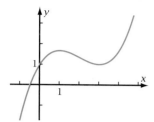

Figure 8-3e

Q10. $(d/dx)(\cos^{-1} x) = -?-$

   A. $\sin^{-1} x$ B. $-\sin^{-1} x$ C. $-(\cos x)^{-2}$

   D. $-\dfrac{1}{\sqrt{x^2 - 1}}$ E. $\dfrac{1}{\sqrt{x^2 - 1}}$

---

1. *Divided Stock Pen Problem:* Suppose you are building a rectangular stock pen (Figure 8-3f) using 600 ft of fencing. You will use part of this fencing to build a fence across the middle of the rectangle (see the diagram). Find the length and width of the rectangle that give the maximum total area. Justify your answer.

Figure 8-3f

2. *Motel Problem:* A six-room motel is to be built with the floor plan shown in Figure 8-3g. Each room will have 350 ft^2 of floor space.

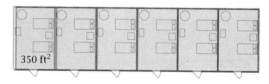

Figure 8-3g

  a. What dimensions should be used for the rooms in order to minimize the total length of the walls? Justify your answer.

  b. How would your answer to part a change if the motel had ten rooms? If it had just three rooms?

3. *Two-Field Problem:* Ella Mentary has 600 ft of fencing to enclose two fields. One field will be a rectangle twice as long as it is wide, and the other will be a square (Figure 8-3h). The square field must contain at least 100 ft^2. The rectangular one must contain at least 800 ft^2.

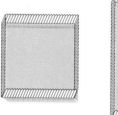

Figure 8-3h

  a. If x is the width of the rectangular field, what is the domain of x?

  b. Plot the graph of the total area contained in the two fields as a function of x.

  c. What is the greatest area that can be contained in the two fields? Justify your answer.

4. *Two-Corral Problem:* You work on Bill Spender's Ranch. Bill tells you to build a circular fence around the lake and to use the remainder of your 1000 yards of fencing to build a square corral (Figure 8-3i). To keep the fence out of the water, the diameter of the circular enclosure must be at least 50 yards.

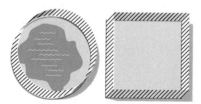

Figure 8-3i

  a. If you must use all 1000 yards of fencing, how can you build the fences to enclose the minimum total area? Justify your answer.

  b. What would you tell Bill if he asked you to build the fences to enclose the maximum total area?

5. *Open Box I:* A rectangular box with a square base and no top (Figure 8-3j) is to be constructed using a total of 120 cm^2 of cardboard.

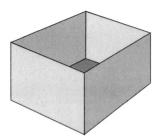

Figure 8-3j

a. Find the dimensions of the box of maximum volume.

b. Make a conjecture about the depth of the maximum-volume box in relation to the base length if the box has a fixed surface area.

6. *Open Box II (Project):* For this project you will investigate the volume of a box. Form the box, which will not have a top, by cutting out squares from the four corners of a 20-by-12-unit piece of graph paper (Figure 8-3k) and folding up the edges.

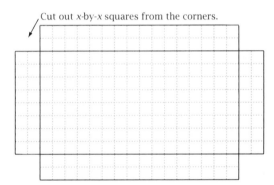

Cut out x-by-x squares from the corners.

Figure 8-3k

a. Each group should pick a different value of x, such as $1, 2, 3, 4, \ldots$, then cut out the squares from the graph paper, and fold and tape it to form a box. What is the largest possible value of x?

b. Calculate the volume of each box. Which integer value of x gives the largest volume?

c. Find the value of x that gives the maximum volume. What is this volume?

d. Construct a box of maximum volume.

7. *Open Box III:* You are building a glass fish tank that will hold 72 ft³ of water. You want its base and sides to be rectangular and the top, of course, to be open. You want to construct the tank so that its width is 5 ft but the length and depth are variable. Building materials for the tank cost $10 per square foot for the base and $5 per square foot for the sides. What is the cost of the least expensive tank? Justify your answer.

8. *Open Box IV (Project):* Figure 8-3l shows an open-top box with rectangular base x-by-y units and rectangular sides. The depth of the box is z units.

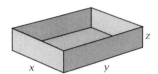

Figure 8-3l

a. Hold the depth constant. Show that the maximum volume of the box for a given amount of material occurs when $x = y$.

b. Set $y = x$, but let both vary as the depth varies. Find the values of x and z that give the maximum volume for a given amount of material.

c. In what ratio are the values of x and z for the maximum-volume box in part b? Is the maximum-volume box tall and narrow or short and wide? Based on geometry, why is your answer reasonable?

9. *Shortest-Distance Problem:* In Figure 8-3m, what point on the graph of $y = e^x$ is closest to the origin? Justify your answer.

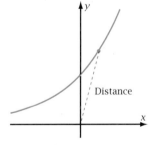

Figure 8-3m

10. *Track and Field Problem:* A track of perimeter 400 m is to be laid out on the practice field (Figure 8-3n). Each semicircular end must have a radius of at least 20 m, and each straight section must be at least 100 m. How should the

track be laid out so that it encompasses the least area? Justify your answer.

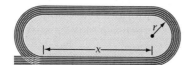

Figure 8-3n

11. *Ladder Problem:* A ladder is to reach over a fence 8 ft high to a wall that is 1 ft behind the fence (Figure 8-3o). What is the length of the shortest ladder that you can use? Justify your answer.

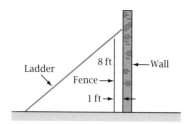

Figure 8-3o

12. *Ladder in the Hall Problem:* A nonfolding ladder is to be taken around a corner where two hallways intersect at right angles (Figure 8-3p). One hall is 7 ft wide, and the other is 5 ft wide. What is the maximum length the ladder can be so that it will pass around such a corner, given that you must carry the ladder parallel to the floor?

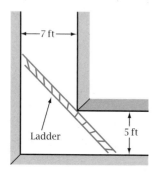

Figure 8-3p

13. *Rotated Rectangle Problem:* A rectangle of perimeter 1200 mm is rotated in space using one of its legs as the axis (Figure 8-3q). The volume enclosed by the resulting cylinder depends on the proportions of the rectangle. Find the dimensions of the rectangle that maximize the cylinder's volume.

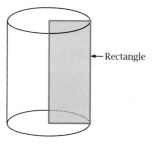

Figure 8-3q

14. *Rotated Rectangle Generalization Problem:* The rectangle of maximum area for a given perimeter P is a square. Does rotating a square about one of its sides (as in Problem 13) produce the maximum-volume cylinder? If so, prove it. If not, what proportions do produce the maximum-volume cylinder?

15. *Tin Can Problem:* A popular size of tin can with "normal" proportions has diameter 7.3 cm and height 10.6 cm (Figure 8-3r).

Figure 8-3r

a. What is its volume?

b. The volume is to be kept the same, but the proportions are to be changed. Write an equation expressing total surface of the can (lateral surface plus two ends) as a function of radius and height. Transform the equation so that the volume is in terms of radius alone.

c. Find the radius and height of the can that minimize its surface area. Is the can tall and narrow or short and wide? What is the ratio of diameter to height? Justify your answers.

d. Does the normal can use close to the minimum amount of metal? What percentage of the metal in the normal can could you save by using cans with the minimum dimensions?

e. If the United States uses 20 million of these cans a day and the metal in a normal can is worth $0.06, how much money could be saved in a year by using minimum-area cans?

16. *Tin Can Generalization Project:* The tin can of minimum cost in Problem 15 is not necessarily the one with minimum surface area. In this problem you will investigate the effects of wasted metal in the manufacturing process and of overlapping metal in the seams.

   a. Assume that the metal for the ends of the can in Problem 15 costs k times as much per square centimeter as the metal for the cylindrical walls. Find the value of k that makes the minimum-cost can have the proportions of the normal can. Is it reasonable in the real world for the ends to cost this much more (or less) per square centimeter than the walls? Explain.

   b. Assume that the ends of the normal tin can in Problem 15 are cut from squares and that the remaining metal from the squares is wasted. What value of k in part a minimizes the cost of the normal can under this assumption? Is the can that uses the minimum amount of metal under this assumption closer to the proportions of the normal can or farther away?

   c. The specifications require that the ends of the can be made from metal disks that overhang by 0.6 cm all the way around. This provides enough overlap to fabricate the constructed can's top and bottom joints. There must also be an extra 0.5 cm of metal in the can's circumference for the overlap in the vertical seam. How do these specifications affect the dimensions of the minimum-area can in Problem 15?

17. *Cup Problem:* You have been hired by the Yankee Cup Company. They currently make a cylindrical paper cup of diameter 5 cm and height 7 cm. Your job is to find ways to save paper by making cups that hold the same amount of liquid but have different proportions.

   a. Find the dimensions of a same-volume cylindrical cup that uses a minimum amount of paper.

   b. What is the ratio of diameter to height for the minimum-area cup?

c. Paper costs $2.00 per square meter. Yankee makes 300 million of this type of cup per year. Write a proposal to your boss telling her whether you think it would be worthwhile to change the dimensions of Yankee's cup to those of the minimum cup. Be sure to show that the area of the proposed cup really is a minimum.

d. Show that in general if a cup of given volume V has minimum total surface area, then the radius is equal to the height.

18. *Duct Problem:* A duct made of sheet metal connects one rectangular opening in an air-conditioning system to another rectangular opening (Figure 8-3s). The rectangle on the left is at $x = 0$ in. and the one on the right is at $x = 100$ in. Cross sections perpendicular to the x-axis are rectangles of width $z = 30 + 0.2x$ and $y = 40 - 0.2x$.

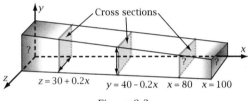

Figure 8-3s

   a. Find the areas of the two end rectangles.

   b. What is the cross-sectional area of the duct at $x = 80$?

   c. What x-values give the maximum and the minimum cross-sectional areas?

19. *Rectangle in Sinusoid Problem:* A rectangle is inscribed in the region bounded by one arch of the graph of $y = \cos x$ and the x-axis (Figure 8-3t). What value of x gives the maximum area? What is the maximum area?

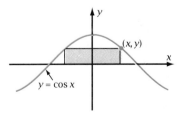

Figure 8-3t

20. *Building Problem:* Tom O'Shea plans to build a new hardware store. He buys a rectangular lot

that is 50 ft by 200 ft, the 50-ft dimension being along the street. The store will have 4000 ft^2 of floor space. Construction costs $100 per linear foot for the part of the store along the street but only $80 per linear foot for the parts along the sides and back. What dimensions of the store will minimize construction costs? Justify your answer.

21. *Triangle under Cotangent Problem:* A right triangle has a vertex at the origin and a leg along the x-axis. Its other vertex touches the graph of $y = \cot x$, as shown in Figure 8-3u.

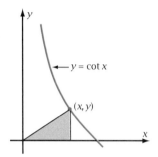

Figure 8-3u

a. As the right angle approaches the origin, the height of the triangle approaches infinity, and the base length approaches zero. Find the limit of the area as the right angle approaches the origin.

b. What is the maximum area the triangle can have if the domain is the half-open interval $(0, \pi/2]$? Justify your answer.

22. *Triangle under Exponential Curve Problem:* A right triangle has one leg on the x-axis. The vertex at the right end of that leg is at the point $(3, 0)$. The other vertex touches the graph of $y = e^x$. The entire triangle is to lie in the first quadrant. Find the maximum area of this triangle. Justify your answer.

23. *Rectangle in Parabola Problem:* A rectangle is inscribed in the region bounded by the x-axis and the parabola $y = 9 - x^2$ (Figure 8-3v).

a. Find the length and width of the rectangle of greatest area. Justify your answer.

b. Find the length and width of the rectangle of greatest perimeter. Justify your answer.

c. Does the rectangle of greatest area have the greatest perimeter?

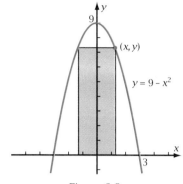

Figure 8-3v

24. *Cylinder in Paraboloid Problem:* In Figure 8-3w, the parabola $y = 9 - x^2$ is rotated about the y-axis to form a paraboloid. A cylinder is coaxially inscribed in the paraboloid.

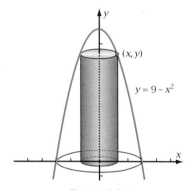

Figure 8-3w

a. Find the radius and height of the cylinder of maximum volume. Justify your answer.

b. Find the radius and height of the cylinder of maximum lateral area.

c. Find the radius and height of the cylinder of maximum total area (including the ends). Justify your answer.

d. Does the maximum-volume cylinder have the same dimensions as either of the maximum-area cylinders in part b or c?

e. Does rotating the maximum-area rectangle in Problem 23a produce the maximum-volume cylinder in this problem?

f. If the cylinder of maximum volume is inscribed in the paraboloid formed by rotating the parabola $y = a^2 - x^2$ about the y-axis, does the ratio

(cylinder radius) : (paraboloid radius)

depend in any way on the length of the paraboloid? (That is, does it depend on the value of the constant a?) Justify your answer.

25. *Cylinder in Sphere Problem:* A cylinder is to be inscribed in a sphere of radius 10 cm (Figure 8-3x). The bottom and top bases of the cylinder are to touch the surface of the sphere. The volume of the cylinder will depend on whether it is tall and narrow or short and wide.

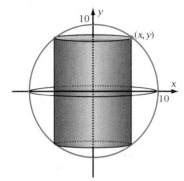

Figure 8-3x

a. Let (x, y) be the coordinates of a point on the circle, as shown. Write an equation for the volume of the cylinder in terms of x and y.

b. What radius and height of the cylinder will give it the maximum volume? What is this volume? Justify your answer.

c. How are the radius and height of the maximum-volume cylinder related to each other? How is the maximum cylinder volume related to the volume of the sphere?

26. *Conical Nose Cone Problem:* In the design of a missile nose cone, it is important to minimize the surface area exposed to the atmosphere. For aerodynamic reasons, the cone must be long and slender. Suppose that a right-circular nose cone is to contain a volume of 5π ft^3. Find the radius and height of the nose cone of minimum lateral surface area, subject to the restriction that the height must be at least twice the radius of the base. (The differentiation will be easier if you minimize the square of the area.)

27. *Cylinder in Cone Problem:* In Figure 8-3y, a cone of height 7 cm and base radius 5 cm has a cylinder inscribed in it, with the base of the cylinder contained in the base of the cone.

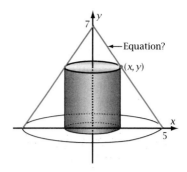

Figure 8-3y

a. Find the radius of the cylinder of maximum lateral area (sides only).

b. Find the radius of the cylinder of maximum total area (including the top and bottom bases). Justify your answer.

28. *General Cylinder in Cone Problem:* A given cone has a cylinder inscribed in it, with its base contained in the base of the cone.

a. How should the radius of the cone and cylinder be related for the lateral surface of the cylinder to maximize its area? Justify your answer.

b. Find the radius of the cylinder that gives the maximum total area.

c. If the cone is short and fat, the maximum total area occurs where the height of the cylinder drops to zero and all the material is used in the top and bottom bases. How must the radius and height of the cone be related for this phenomenon to happen?

29. *Elliptical Nose Cone Problem:* The nose of a new cargo plane is to be a half-ellipsoid of diameter 8 m and length 9 m (Figure 8-3z). The nose swings open so that a cylindrical cargo container can be placed inside. What is the greatest volume this container could hold?

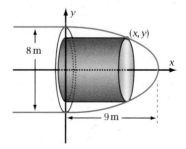

Figure 8-3z

What are the radius and height of this largest container? Justify your answer.

30. *Submarine Pressure Hull Project:* According to a new design, the forward end of a submarine hull is to be constructed in the shape of a paraboloid 16 m long and 8 m in diameter (Figure 8-3aa). Since this is a doubly curved surface, it is hard to bend thick steel plates into this shape. So the paraboloid is to be made of relatively thin steel, and the pressure hull built inside as a cylinder (a singly curved surface). A frustum of a cone is also a singly curved surface, which would be about as easy to make and which might be able to contain more volume (Figure 8-3bb). How much more volume could be contained in the maximum-volume frustum than in the maximum-volume cylinder? Some things you will need to find are the equation of this particular parabola and the equation for the volume of a frustum of a cone.

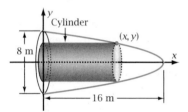

Figure 8-3aa

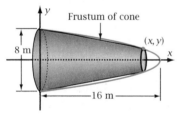

Figure 8-3bb

31. *Local Maximum Property Problem:* The definition of local maximum is as follows: $f(c)$ is a local maximum of f on the interval (a, b) if and only if $f(c) \geq f(x)$ for all values of x in (a, b). Figure 8-3cc illustrates this definition.

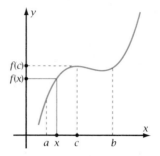

Figure 8-3cc

a. Prove that if $f(c)$ is a local maximum of f on (a, b) and f is differentiable at $x = c$ in (a, b), then $f'(c) = 0$. (Consider the sign of the difference quotient when x is to the left and to the right of c, then take left and right limits.)

b. Explain why the property in part a would be false without the hypothesis that f is differentiable at $x = c$.

c. Explain why the converse of the property in part a is false.

32. *Corral with Short Wall Project:* Millie Watt is installing an electric fence around a rectangular corral along a wall. Part or all of the wall forms all or part of one side of the corral. The total length of fence (excluding the wall) is to be 1000 ft. Find the maximum area that she can enclose if

a. The wall is 600 ft long (Figure 8-3dd, left).

b. The wall is 400 ft long (Figure 8-3dd, center).

c. The wall is 200 ft long (Figure 8-3dd, right).

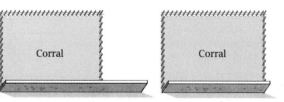

Figure 8-3dd

33. *Journal Problem:* Update your journal with what you've learned in this section. Include such things as

• The one most important thing you've learned since your last entry

• The critical features of the graph of a function that its two derivatives tell you

• How you solve real-world problems that involve a maximum or minimum value

• Any techniques or ideas about extreme-value problems that are still unclear to you

8-4 Volume of a Solid of Revolution by Cylindrical Shells

In Chapter 5, you learned how to find the volume of a solid of revolution by slicing the rotated region perpendicular to the axis of rotation. Figure 8-4a shows the region under the graph of $y = 4x - x^2$ from $x = 0$ to $x = 3$ that is to be rotated about the y-axis to form a solid. Slicing perpendicular to the axis of rotation causes the length of the strip to be a *piecewise* function of y, (line − curve) from $y = 0$ to $y = 3$ and (curve − curve) from there up. Slicing *parallel* to the axis of rotation avoids this difficulty.

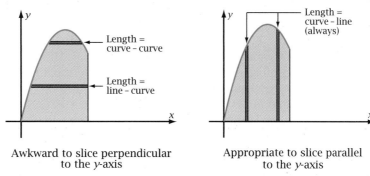

Length = curve − curve

Length = line − curve

Length = curve − line (always)

Awkward to slice perpendicular to the y-axis

Appropriate to slice parallel to the y-axis

Figure 8-4a

As the region rotates, each strip parallel to the axis of rotation generates a **cylindrical shell**, as shown in Figure 8-4b. In this section you will find the volume of the solid by integrating dV, the volume of a typical shell.

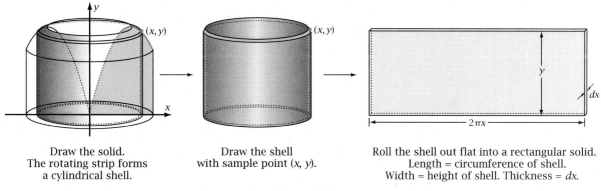

(x, y)

(x, y)

y

dx

$2\pi x$

Draw the solid. The rotating strip forms a cylindrical shell.

Draw the shell with sample point (x, y).

Roll the shell out flat into a rectangular solid. Length = circumference of shell. Width = height of shell. Thickness = dx.

Figure 8-4b

OBJECTIVE   Find the volume of a solid of revolution by slicing it into cylindrical shells.

The shells in Figure 8-4b resemble tin cans without ends. Since a shell is thin, you can find its volume dV by cutting down its side and rolling it out flat (Figure 8-4b, right). The resulting rectangular solid has the following approximate dimensions.

Length: Circumference of the shell at the sample point ($2\pi x$, in this case)

Width: Height of the shell at the sample point (y, in this case)

Thickness: Width of the strip (dx, in this case)

Consequently, the volume of the shell is given by this property.

PROPERTY: *Differential of Volume for Cylindrical Shells*

$$dV = (\text{circumference})(\text{height})(\text{thickness})$$

The rings of a tree are like cylindrical shells that you can use to measure the volume of the tree.

The volume of the solid will be approximately equal to the sum of the volumes of the shells (Figure 8-4c). The exact volume will be the limit of this sum—that is, the definite integral. The innermost shell is at $x = 0$, and the outermost is at $x = 3$. Thus, the limits of integration will be from 0 to 3. (The part of the solid from $x = -3$ to $x = 0$ is simply the image of the region being rotated, not the region itself.) Example 1 shows the details of calculating the volume of this solid.

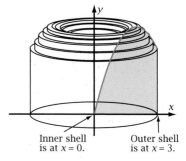

Inner shell is at $x = 0$.   Outer shell is at $x = 3$.

Figure 8-4c

▶ **EXAMPLE 1**   The region under the graph of $y = 4x - x^2$ from $x = 0$ to $x = 3$ is rotated about the y-axis to form a solid. Find the volume of the solid by slicing into cylindrical shells. Use the fundamental theorem to obtain the exact answer. Show that your answer is reasonable.

Solution   The volume of a representative cylindrical shell is

$$dV = (\text{circumference})(\text{height})(\text{thickness})$$

Recall the rolled-out shell in Figure 8-4b.

$$= (2\pi x)(y)(dx)$$

Substituting $4x - x^2$ for y gives

$$dV = 2\pi x(4x - x^2)\,dx = 2\pi(4x^2 - x^3)\,dx$$

The volume is found by adding all the dV's and taking the limit (integrating).

$$V = \int_0^3 2\pi(4x^2 - x^3)\,dx$$

$$= 2\pi\left(\tfrac{4}{3}x^3 - \tfrac{1}{4}x^4\right)\Big|_0^3$$

$$= 2\pi\left(36 - \tfrac{81}{4} - 0 + 0\right)$$

$$= 31.5\pi = 98.96\ldots$$

Checks:

Volume of circumscribed cylinder: $\pi(3^2)(4) = 36\pi > 31.5\pi$ ✓

Numerical integration: Integral$= 31.5\pi$ ✓ ◀

In case you are wondering whether the distortion of the shell as you roll it out flat causes the final answer, 31.5π, to be inaccurate, the answer is no. As Δx approaches zero, so do the inaccuracies in the shell approximation. In Problem C4 of Section 11-7, you will learn that if the approximate value of dV differs from the exact value by nothing more than *infinitesimals of higher order*, for instance $(dx)(dy)$, then the integral will give the exact volume.

You can use cylindrical shells to find volumes when these conditions are encountered.

- The rotation is not around the y-axis.

- The axis of rotation is not a bound of integration.

- Both ends of the shell's height are variable.

Example 2 shows how this can be done.

▶ **EXAMPLE 2** Let R be the region bounded below by the graph of $y = x^{1/2}$, above by the graph of $y = 2$, and on the left by the graph of $y = x$. Find the volume of the solid generated when R is rotated about the line $y = -1$. Assume that x and y are in feet. Show that your answer is reasonable.

Solution First draw the region, as in the left diagram of Figure 8-4d. Find the points of intersection of the graphs. Slice parallel to the axis of rotation. Mark the two resulting sample points as (x_1, y) and (x_2, y). Then rotate the region about the line $y = -1$. As shown in the center diagram of Figure 8-4d, it helps to draw only the back half of the solid. (Otherwise, the diagram becomes so cluttered it is hard to tell which lines are which.) Roll out the shell and find dV.

$$dV = (\text{circumference})(\text{height})(\text{thickness})$$

$$= 2\pi(y + 1)(x_1 - x_2)\,dy$$

Height is always larger value minus smaller value. Radius is the difference between the y-values, $y - (-1) = y + 1$.

For the curve $y = x^{1/2}$, solve to get $x_1 = y^2$.

For the line $y = x$, "solve" to get $x_2 = y$.

$$\therefore dV = 2\pi(y + 1)(y^2 - y)\,dy$$

$$\therefore V = \int_1^2 2\pi(y + 1)(y^2 - y)\,dy \qquad$$ Innermost shell is at $y = 1$; outermost shell is at $y = 2$.

$$= 4.5\pi \qquad\qquad\qquad$$ Integrate numerically or algebraically.

$$= 14.1371\ldots \approx 14.1 \text{ ft}^3$$

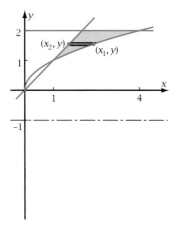

Draw the region. Slice parallel to the axis of rotation. Show two sample points.

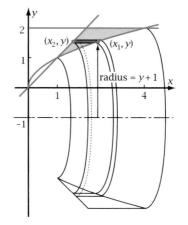

Rotate about the x-axis. Show only the back half of the solid. The rotating strip generates the cylindrical shell.

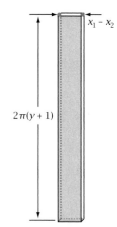

Roll out the (whole) shell into a flat rectangular solid.

Figure 8-4d

Check:

Outer cylinder − inner cylinder $= \pi(3^2)(3) - \pi(2^2)(3) = 15\pi > 4.5\pi$ ✓ ◀

Problem Set 8-4

Q1. Sketch the graph: $y = x^2$

Q2. Sketch the graph: $y = -x^2$

Q3. Sketch the graph: $y = x^{-2}$

Q4. Sketch the graph: $y = 2^x$

Q5. Sketch the graph: $y = 2^{-x}$

Q6. Sketch the graph: $y = 2x$

Q7. Sketch the graph: $y = \ln x$

Q8. Sketch the graph of a continuous function whose derivative is shown in Figure 8-4e.

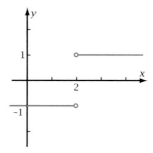

Figure 8-4e

Q9. $\int \sec^2 x\,dx = \text{—?—}$

Q10. $y = x^3 - 3x$ has a local minimum at $x = \text{—?—}$.

　　A. 0　　B. 1　　C. $\sqrt{3}$　　D. -1　　E. $-\sqrt{3}$

1. Figure 8-4f shows the solid formed by rotating about the y-axis the region in Quadrant I under the graph of $y = 4 - x^2$.

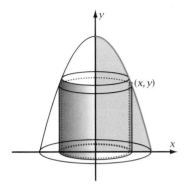

Figure 8-4f

a. Find the volume dV of a cylindrical shell. Transform dV so that it is in terms of one variable.

b. Find the exact volume of the solid by using the fundamental theorem.

c. Find the volume again by plane slicing. Use planes perpendicular to the y-axis to form slabs of thickness dy. Show that you get the same answer as in part b.

2. Figure 8-4g shows the solid formed by rotating about the x-axis the region under the graph of $y = x^{2/3}$ from $x = 0$ to $x = 8$.

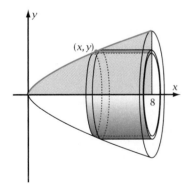

Figure 8-4g

a. What is the height of the cylindrical shell in terms of the sample point (x, y)?

b. Find the volume dV of a cylindrical shell. Transform dV so that it is in terms of one variable.

c. Find the exact volume of the solid using the fundamental theorem.

d. Find the volume again, by plane slicing. Is the answer the same as in part c?

For Problems 3–18, find the volume of the solid by slicing into cylindrical shells. You may use numerical integration. Use familiar geometric relationships to show geometrically that your answer is reasonable.

3. Rotate about the y-axis the region under the graph of $y = -x^2 + 4x + 3$ from $x = 1$ to $x = 4$.

4. Rotate about the y-axis the region under the graph of $y = x^2 - 8x + 17$ from $x = 2$ to $x = 5$.

5. Rotate about the x-axis the region bounded by the y-axis and the graph of $x = -y^2 + 6y - 5$.

6. Rotate about the x-axis the region bounded by the y-axis and the graph of $x = y^2 - 10y + 24$.

7. Rotate about the y-axis the region above the graph of $y = x^3$ that is bounded by the lines $x = 1$ and $y = 8$ (Figure 8-4h).

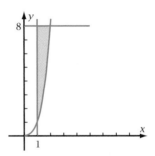

Figure 8-4h

8. Rotate about the y-axis the region in Quadrant I above the graph of $y = 1/x$ that is bounded by the lines $y = 4$ and $x = 3$.

9. Rotate about the x-axis the region in Quadrant I above the graph of $y = 1/x^2$ that is bounded by the lines $x = 5$ and $y = 4$ (Figure 8-4i).

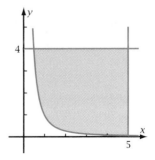

Figure 8-4i

10. Rotate about the x-axis the region in Quadrant I below the graph of $y = x^{2/3}$, above the line $y = 1$ and bounded by the line $x = 8$.

11. Rotate about the y-axis the region bounded by the graph of $y = x^2 - 6x + 7$ and the line $x - y = -1$ (Figure 8-4j).

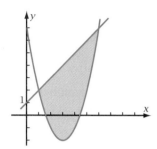

Figure 8-4j

12. Rotate about the x-axis the region in Quadrant I bounded by the graph of $y = x^{1/3}$ and the line $y = 0.5x - 2$.

13. Rotate about the line $x = 5$ the region under the graph of $y = x^{3/2}$ from $x = 1$ to $x = 4$ (Figure 8-4k).

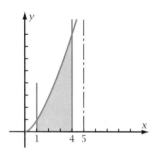

Figure 8-4k

14. Rotate about the line $x = 3$ the region under the graph of $y = x^{-2}$ from $x = 1$ to $x = 2$.

15. Rotate about the line $x = 4$ the region bounded by the graph of $y = x^4$ and the line $y = 5x + 6$.

16. Rotate about the line $x = -1$ the region bounded by the graph of $y = \sqrt{x}$ and the lines $x + y = 6$ and $x = 1$.

17. Rotate about the line $x = -2$ the region bounded by the graphs of $y = -x^2 + 4x + 1$

and $y = 1.4^x$ (Figure 8-4l). You will need to find one of the intersections numerically.

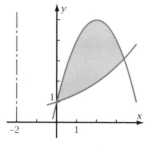

Figure 8-4l

18. Rotate about the line $y = -1$ the region in Figure 8-4l from Problem 17. Explain why it is not appropriate to find the volume of this figure by cylindrical shells.

For Problems 19 and 20, find the volume of the solid by slicing into plane slabs, thus verifying the answer obtained by cylindrical shells.

19. Use the solid given in Problem 7.

20. Use the solid given in Problem 8.

21. *Limit of Riemann Sum Problem:* The region under the graph of $y = x^{1/3}$ from $x = 0$ to $x = 8$ is rotated about the x-axis to form a solid. Find the volume exactly by slicing into cylindrical shells and using the fundamental theorem. Then find three midpoint Riemann sums for the integral, using $n = 8$, $n = 100$, and $n = 1000$ increments. Show that the Riemann sums approach the exact answer as n increases.

22. *Unknown Integral Problem:* Figure 8-4m shows the region under $y = \sin x$ from $x = 0$ to $x = 2$, rotated about the y-axis to form a solid.

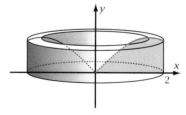

Figure 8-4m

a. Write an integral for the volume of this solid using cylindrical shells. Evaluate the integral numerically.

b. Explain why you cannot evaluate the integral by the fundamental theorem using the techniques you have learned so far.

23. *Parametric Curve Problem:* Figure 8-4n shows the ellipse with parametric equations

$$x = 5 \cos t$$
$$y = 3 \sin t$$

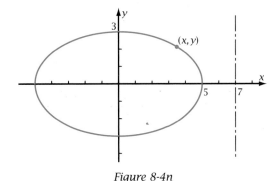

Figure 8-4n

a. Slice the region horizontally, then rotate it about the x-axis to form an ellipsoid. Find the volume of the ellipsoid by first writing dV in terms of the parameter t.

b. Slice the region vertically, then rotate it about the x-axis to form the same ellipsoid. Show that you get the same volume.

c. Find the volume of the solid generated by rotating the ellipse about the line $x = 7$.

24. *Journal Problem:* Update your journal with what you've learned since your last entry. Include such things as

• The one most important thing you have learned in this section

• The basic concept from geometry that is used to find volumes by calculus

• The similarities of slicing into disks, washers, and other plane slices

• The difference between plane slicing and cylindrical shells

• Any techniques or ideas about finding volumes that are still unclear to you

---

8-5   Length of a Plane Curve—Arc Length

At the beginning of this chapter you were introduced to the geometry of plane and solid figures, such as the bell-shaped solid shown in Figure 8-5a. You can now find the volume of a solid by plane slices or cylindrical shells and the area of a plane region. In the next two sections you will find the length of a curved line and the area of a curved surface in space.

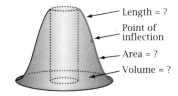

Figure 8-5a

OBJECTIVE   Given the equation for a plane curve, find its length approximately by calculating and summing the lengths of the chords, or exactly by calculus.

▶ **EXAMPLE 1** Find approximately the length of the parabola $y = x^2$ from $x = -1$ to $x = 2$ (Figure 8-5b, left) by dividing it into three chords.

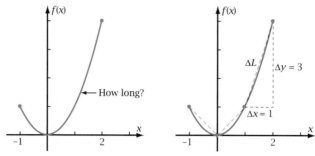

Figure 8-5b

Solution The right diagram in Figure 8-5b shows three chords drawn to the graph. The sum of the lengths of the chords is approximately the length of the graph. By the Pythagorean theorem,

$$L \approx \sqrt{1^2 + 1^2} + \sqrt{1^2 + 1^2} + \sqrt{1^2 + 3^2} = 5.990704... \approx 5.99 \text{ units}$$ ◀

In general, the length of any one chord will be

$$\Delta L = \sqrt{\Delta x^2 + \Delta y^2} \qquad \sqrt{\Delta x^2 + \Delta y^2} \text{ means } \sqrt{(\Delta x)^2 + (\Delta y)^2}.$$

Using smaller values of Δx (that is, a greater number n of chords), the following is true.

$n = 30$: $L \approx 6.12417269...$ By the program of Problem 33 in
$n = 100$: $L \approx 6.12558677...$ Problem Set 8-5.
$n = 1000$: $L \approx 6.12572522...$ The values are approaching a limit!

The limit of the sums of the chord lengths equals the exact length of the curve.

PROPERTY: Length of a Plane Curve (Arc Length)

A curve between two points in the xy-plane has length

$$L = \lim_{\Delta x \to 0, \Delta y \to 0} \sum \sqrt{\Delta x^2 + \Delta y^2}$$

provided that this limit exists.

You can find the limit of a chord length sum exactly by transforming it to a Riemann sum,

$$\sum g(c) \Delta x$$

The first thing to do is make the factor Δx appear. Although Δx^2 is not a factor of both terms in the expression $\Delta x^2 + \Delta y^2$, you can still factor it out.

$$\Delta x^2 + \Delta y^2 = \left[1 + \frac{\Delta y^2}{\Delta x^2}\right] \Delta x^2 = \left[1 + \left(\frac{\Delta y}{\Delta x}\right)^2\right] \Delta x^2$$

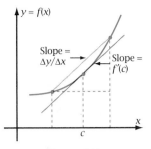

Figure 8-5c

So you can write a chord length sum

$$\sum \sqrt{\left[1 + \left(\frac{\Delta y}{\Delta x}\right)^2\right] \Delta x^2} = \sum \sqrt{1 + \left(\frac{\Delta y}{\Delta x}\right)^2} \, \Delta x$$

The remaining radicand contains $\Delta y / \Delta x$, which you should recognize as the difference quotient that approaches $f'(x)$ as Δx approaches zero. In fact, if f is a differentiable function, the mean value theorem tells you that there is a number $x = c$ within the interval where $f'(c)$ is exactly equal to $\Delta y / \Delta x$ (Figure 8-5c). So you can write a chord length sum

$$\sum \sqrt{1 + f'(c)^2} \, \Delta x$$

where the sample points, $x = c$, are chosen at a point in each subinterval where the conclusion of the mean value theorem is true. The length L of the curve is thus the limit of a Riemann sum and hence a definite integral.

$$L = \int_a^b \sqrt{1 + f'(x)^2} \, dx \qquad \text{The } exact \text{ length of the curve!}$$

You can write the quantity $dL = \sqrt{1 + f'(x)^2} \, dx$, the differential of curve length (often called "arc length"), in a form that is easier to remember and use. Recalling that $f'(x) = dy/dx$, and that the differentials dy and dx can be written as separate quantities, you can write

$$dL = \sqrt{1 + \left(\frac{dy}{dx}\right)^2} \, dx$$

or, more simply,

$$dL = \sqrt{dx^2 + dy^2} \qquad \text{Differential of arc length.}$$

It's easy to remember dL in this last form because it looks like the Pythagorean theorem. There are also some algebraic advantages of this form, as you will see in later examples.

▶ **EXAMPLE 2**   Write an integral to find the exact length of the curve in Example 1 and evaluate it.

Solution
$$y = x^2 \Rightarrow dy = 2x \, dx$$
$$\therefore dL = \sqrt{dx^2 + (2x \, dx)^2} = \sqrt{1 + 4x^2} \, dx$$
$$\therefore L = \int_{-1}^{2} \sqrt{1 + 4x^2} \, dx$$

When you study trigonometric substitution in Chapter 9, you will be able to evaluate integrals like this using the fundamental theorem. Numerical integration gives

$$L = 6.1257266\ldots \qquad \blacktriangleleft$$

You can write equations for more complex curves in parametric form. Example 3 shows how this can be done.

▶ **EXAMPLE 3**   The ellipse in Figure 8-5d has parametric equations

$$x = 6 + 5 \cos t$$
$$y = 4 + 3 \sin t$$

Write an integral equal to the length of the graph and evaluate it numerically. Check that your answer is reasonable.

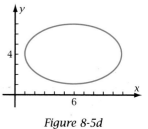

Figure 8-5d

Solution
$$x = 6 + 5 \cos t \Rightarrow dx = -5 \sin t \, dt$$
$$y = 4 + 3 \sin t \Rightarrow dy = 3 \cos t \, dt$$
$$dL = \sqrt{dx^2 + dy^2}$$
$$= \sqrt{(-5 \sin t \, dt)^2 + (3 \cos t \, dt)^2}$$
$$= \sqrt{25 \sin^2 t + 9 \cos^2 t} \, dt \qquad \text{Explain the "}dt\text{."}$$
$$L = \int_0^{2\pi} \sqrt{25 \sin^2 t + 9 \cos^2 t} \, dt$$
$$\approx 25.52699...$$

Check: A circle of radius 4 has length $2\pi \cdot 4 = 25.13274...$. ✓   ◀

The indefinite integral in Example 3 cannot be evaluated using any of the elementary functions. It is called an *elliptic integral*, which you will study in later courses.

Occasionally a curve length problem involves an integral that you can evaluate by the fundamental theorem. Example 4 shows one instance.

▶ **EXAMPLE 4**   Plot the graph of $y = \frac{2}{3}x^{3/2}$. Find exactly the length from $x = 0$ to $x = 9$.

Solution   The graph is shown in Figure 8-5e. It starts at the origin and rises gently to the point (9, 18).

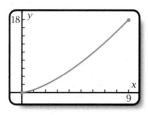

Figure 8-5e

$$dy = x^{1/2} \, dx$$
$$dL = \sqrt{dx^2 + dy^2}$$
$$= \sqrt{dx^2 + (x^{1/2} \, dx)^2}$$
$$= \sqrt{1 + x} \, dx$$
$$L = \int_0^9 \sqrt{1 + x} \, dx = \int_0^9 (1 + x)^{1/2} \, dx$$
$$= \frac{2}{3}(1 + x)^{3/2} \Big|_0^9 = \frac{2}{3}(10^{3/2}) - \frac{2}{3}(1^{3/2})$$
$$= \frac{2}{3}(10^{3/2} - 1) \qquad \text{Exact length.}$$
$$= 20.41518... \qquad \text{Approximation for exact length.}$$ ◀

Problem Set 8-5

Quick Review 5 min

Q1. Sketch: $y = x^2$

Q2. Show the region under $y = x^2$ from $x = 1$ to $x = 4$.

Q3. Write an integral for the area of the region in Problem Q2.

Q4. Find the antiderivative in Problem Q3.

Q5. Evaluate the integral in Problem Q4.

Q6. Sketch the solid generated by rotating the region in Problem Q2 about the y-axis.

Q7. Write an integral for the volume of the solid in Problem Q6.

Q8. Find the antiderivative in Problem Q7.

Q9. Evaluate the integral in Problem Q8.

Q10. Figure 8-5f illustrates the —?— theorem.

    A. Fundamental   B. Parallelism

    C. Average slope   D. Derivative

    E. Mean value

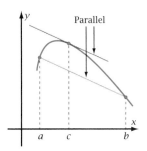

Figure 8-5f

For Problems 1–4,

    a. Sketch the graph for x in the given interval.

    b. Find its approximate length using five chords with equal values of Δx.

    c. Find its length more precisely using a definite integral evaluated numerically.

    1. $y = e^x$ $x \in [0, 2]$

    2. $y = 2^x$ $x \in [0, 3]$

    3. $y = \tan x$ $x \in [0, 1.5]$

    4. $y = \sec x$ $x \in [0, 1.5]$

For Problems 5–16,

    a. Plot the graph for x in the given interval. Sketch the result.

    b. Find the approximate length using a definite integral evaluated numerically.

    c. Show that your answer is reasonable.

    5. $y = x^2 - 5x + 3$ $x \in [1, 6]$

    6. $y = 4x - x^2$ $x \in [0, 4]$

    7. $y = 16 - x^4$ $x \in [-1, 2]$

    8. $y = x^3 - 9x^2 + 5x + 50$ $x \in [-1, 9]$

    9. $y = (\ln x)^2$ $x \in [0.1, e]$

    10. $y = x \sin x$ $x \in [0, 4\pi]$

    11. $y = \tan x$ $x \in [0, 1.5]$

    12. $y = \sec x$ $x \in [0, 1.5]$

    13. Astroid: $t \in [0, 2\pi]$
        $x = 5 \cos^3 t$
        $y = 5 \sin^3 t$

    14. Cardioid: $t \in [0, 2\pi]$
        $x = 5(2 \cos t - \cos 2t)$
        $y = 5(2 \sin t - \sin 2t)$

    15. Epicycloid: $t \in [0, 2\pi]$
        $x = 5 \cos t - \cos 5t$
        $y = 5 \sin t - \sin 5t$

    16. Involute of a circle: $t \in [0, 4\pi]$
        $x = \cos t + t \sin t$
        $y = \sin t - t \cos t$

For Problems 17–20,

    a. Plot the graph for x in the given interval. Sketch the result.

    b. Find the exact length using a definite integral evaluated by the fundamental theorem.

    c. Show that your answer is reasonable. (For Problem 19, find a common denominator under the radical sign.)

    17. $y = 4x^{3/2}$ $x \in [0, 4]$

    18. $y = \dfrac{x^3}{12} + \dfrac{1}{x}$ $x \in [1, 2]$

19. $y = 3x^{2/3} + 5 \qquad x \in [1, 8]$

20. $\frac{1}{3}(x^2 + 2)^{3/2} \qquad x \in [0, 3]$

21. *Golden Gate Bridge Problem:* The photograph shows the Golden Gate Bridge across San Francisco Bay in California. The center span of the bridge is about 4200 ft long. The suspension cables hang in parabolic arcs from towers about 750 ft above the water's surface. These cables come as close as 220 ft to the water at the center of the span. Use this information to write an equation of the particular quadratic function that expresses the distance of the cables from the water as a function of the horizontal displacement from center span. Use the equation to calculate the length of the parabolic cable.

22. *Chain Problem:* When a chain hangs under its own weight, its shape is a **catenary** (from the Latin *catina*, meaning "chain"). Figure 8-5g shows a catenary with its vertex on the *y*-axis. Its equation is

$$y = 0.2(e^x + e^{-x})$$

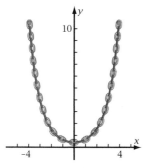

Figure 8-5g

where *x* and *y* are in feet. Find the length of this chain from $x = -4$ to $x = 4$. How does this length compare with that of a parabola,

$$y = ax^2 + c$$

which has the same vertex and endpoints?

23. *Stadium Problem:* Figure 8-5h shows the seating area for a sports stadium. The ellipses have these parametric equations.

Outer Ellipse:	Inner Ellipse:
$x = 120 \cos t$	$x = 100 \cos t$
$y = 100 \sin t$	$y = 50 \sin t$

Both *x* and *y* are in meters. Find the lengths of the boundaries between the outer ellipse and the parking lot, and between the inner ellipse and the playing field.

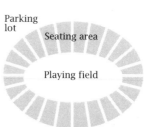

Figure 8-5h

24. *Parabola Surprise Problem!:* A parabola has parametric equations

$$x = 8 \cos 2t$$
$$y = 5 \sin t$$

Find the length from $t = 0$ to $t = 2\pi$. Why does the answer seem unreasonably high?

25. *Implicit Relation Problem I:* Use the fundamental theorem to find exactly the length of the graph of $9x^2 = 4y^3$ between the points $(0, 0)$ and $(2\sqrt{3}, 3)$. Consider *y* to be the independent variable.

26. *Implicit Relation Problem II:* Use the fundamental theorem to find exactly the length of the semicubical parabola $x^2 = y^3$ between the points $(-1, 1)$ and $(8, 4)$. Consider *y* to be the independent variable. You will have to break the graph into two branches (sketch a graph).

27. *Spiral Problem:* Figure 8-5i shows the spiral whose parametric equations are

$$x = \frac{t}{\pi} \cos t$$

$$y = \frac{t}{\pi} \sin t$$

What range of t generates the part of the spiral shown in the figure? Find the length of the spiral by the fundamental theorem if you can, or by numerical methods.

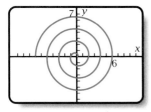

Figure 8-5i

28. *Length of a Circle Problem:* Write parametric equations for a circle of radius r centered at the origin. Then use appropriate algebra, trigonometry, and calculus to prove the familiar circumference formula $C = 2\pi r$.

29. *Sinusoid Length Investigation Problem:* Write an integral for the length of one cycle of the sinusoid of constant amplitude A,

$$y = A \sin x$$

Find lengths of the sinusoid for various values of A. From the results of your work, try to reach a conclusion about how the length varies with A. For instance, does doubling A double the length?

30. *Ellipse Length Investigation Problem:* Write an integral for the length of the ellipse

$$x = \cos t$$

$$y = A \sin t$$

Find lengths of the ellipse for various values of A. From the results of your work, try to reach a conclusion about how the length varies with A. For instance, does doubling A double the length?

31. *Fatal Error Problem:* Mae wants to find the length of $y = (x - 2)^{-1}$ from $x = 1$ to $x = 3$. She partitions $[1, 3]$ into five equal subintervals and gets 18.2774... for the length. Explain to Mae why her approach to the problem has a fatal error.

32. *Mistake Problem:* Amos finds the length of the curve $y = \sin 2\pi x$ from $x = 0$ to $x = 10$ by dividing the interval $[0, 10]$ into 5 subintervals of equal length. He gets an answer of exactly 10. Feeling he has made a mistake, he tries again with 20 subintervals and gets the same answer, 10. Show Amos that he really did make a mistake. Show him how he can get a very accurate answer using only 5 subintervals.

33. *Program for Arc Length by Brute Force:* Write a program that calculates the approximate length of a curve by summing the lengths of the chords. Store the equation for the function in the y = menu. Your program should allow you to input the lower and upper bounds of the domain and the number of increments to be used. The output should be the approximate length of the curve. To make the program more entertaining to run, you might have it display the increment number and the current sum of the lengths at each pass through the loop. You can assume that your program is working correctly if it gives 6.12417269... for the length of $y = x^2$ from $x = -1$ to $x = 2$ (Example 1) with $n = 30$ increments.

8-6   Area of a Surface of Revolution

Suppose that the graph of a function $y = f(x)$ is rotated about the x-axis. The result will be a surface in space (Figure 8-6a, left, on the next page). You are to find the area of the surface.

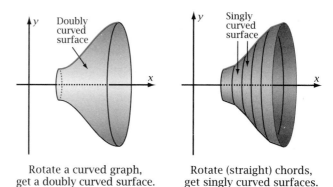

Rotate a curved graph,
get a doubly curved surface.

Rotate (straight) chords,
get singly curved surfaces.

Figure 8-6a

A graph curved in one direction that rotates in another direction forms a **doubly curved** surface. Like a map of Earth, a doubly curved surface cannot be flattened out. But if you draw chords on the graph as you did for finding arc length, the rotating chords generate singly curved **frustums of cones**, as shown in the right diagram in Figure 8-6a. Flattening the frustums (Figure 8-6b) allows you to find their areas geometrically.

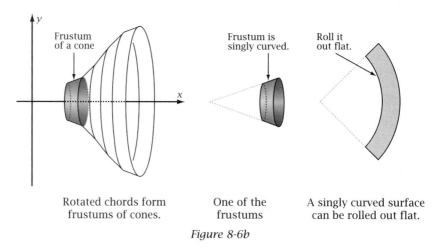

Rotated chords form
frustums of cones.

One of the
frustums

A singly curved surface
can be rolled out flat.

Figure 8-6b

OBJECTIVE Find the area of a surface of revolution by slicing the surface into frustums of cones.

The surface area of a cone is $S = \pi R L$, where R is the base radius and L is the slant height (Figure 8-6c). The area of a frustum is the area of the large cone minus the area of the small one; that is,

$$S = \pi R L - \pi r l$$

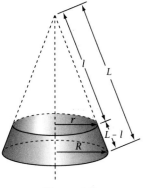

Figure 8-6c

where R and L are for the large cone and r and l are for the small one. By clever algebra (which you will be asked to use in Problem 26), you can transform this equation to

$$S = 2\pi \left(\frac{R + r}{2} \right) (L - l)$$

The quantity $(R + r)/2$ is the average of the two radii. The quantity $(L - l)$ is the slant height of the frustum. It is the same as dL in the arc length problems of Section 8-5. So the differential of surface area, dS, is

$$dS = 2\pi(\text{average radius})(\text{slant height}) = 2\pi(\text{average radius})\, dL$$

Note that 2π(average radius) equals the distance traveled by the midpoint of the chord as the chord rotates.

PROPERTY: Area of a Surface of Revolution

If y is a differentiable function of x, then the area of the surface formed by rotating the graph of the function about an axis is

$$S = \int_a^b (\text{circumference})\, dL = 2\pi \int_a^b (\text{radius})\, dL$$

where $dL = dx^2 + dy^2$ and a and b are the x- or y-coordinates of the endpoints of the graph.

The radius must be found from information about the surface.

▶ **EXAMPLE 1** The graph of $y = \sin x$ from $x = 1$ to $x = 3$ is rotated about various axes to form surfaces. Find the area of the surface if the graph is rotated about

    a. The y-axis

    b. The line $y = 2$

Show that your answers are reasonable.

Solution a. Figure 8-6d shows the surface for the graph rotated about the y-axis.

$$dy = \cos x\, dx$$

$$\sqrt{dx^2 + \cos^2 x\, dx^2} = \sqrt{1 + \cos^2 x}\, dx$$

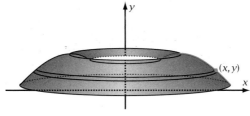

Figure 8-6d

Radius $= x$, so

$$dS = 2\pi x\sqrt{1 + \cos^2 x}\, dx$$

$$\therefore S = 2\pi \int_1^3 x\sqrt{1 + \cos^2 x}\, dx$$

By numerical integration,

$$S \approx 9.5111282...\pi = 29.88009...$$

As a check on this answer, consider the area of a flat washer of radii 1 and 3 (Figure 8-6e). Its area is

$$\pi(3^2 - 1^2) = 25.132...$$

So the 29.88... answer is reasonable.

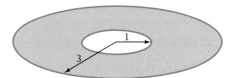

Figure 8-6e

b. Figure 8-6f shows the surface for rotation about $y = 2$. Note that even though the graph is rotated about a horizontal axis instead of a vertical one (and therefore must use a different radius), dL can still be expressed in terms of dx.

$$\text{radius} = 2 - y = 2 - \sin x$$

So the surface area is

$$S = 2\pi \int_1^3 (2 - \sin x)\sqrt{1 + \cos^2 x}\, dx$$

By numerical integration,

$$S \approx 5.836945...\pi = 18.337304...$$

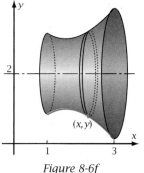

Figure 8-6f

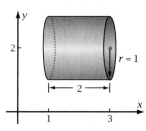

Figure 8-6g

To be reasonable, the answer should be a bit more than a cylinder of altitude 2 and radius 1 (Figure 8-6g). That area is

$$2\pi(1^2)(2) = 12.566...$$

so the answer is reasonable.

◀

Problem Set 8-6

Q1. If $y = x^3$, then dL (arc length) $= -?-$.

Q2. If $y = \tan x$, then $dL = -?-$.

Q3. $\int \sin^5 x \cos x\, dx = -?-$

Q4. $\int_1^5 x^3\, dx = -?-$

Q5. If $y = x\, e^x$, then $y' = -?-$.

Q6. The maximum of $y = x^2 - 8x + 14$ on the interval $[1, 6]$ is $-?-$.

Q7. Write the definition of derivative.

Q8. Give the physical meaning of derivative.

Q9. $\int \sec 2x\, dx = -?-$

Q10. If $\lim U_n = \lim L_n$ for function f, then f is $-?-$.

    A. Differentiable   B. Continuous

    C. Constant   D. Integrable

    E. Squeezed

1. *Paraboloid Problem:* A paraboloid is formed by rotating about the y-axis the graph of $y = 0.5x^2$ from $x = 0$ to $x = 3$.

   a. Write an integral for the area of the paraboloid. Evaluate it numerically.

   b. Show that your answer is reasonable by comparing it with suitable geometric figures.

   c. The indefinite integral in part a is relatively easy to evaluate. Do so, and thus find the exact area. Show that your answer in part a is close to the exact answer.

2. *Rotated Sinusoid Problem:* One arch of the graph of $y = \sin x$ is rotated about the x-axis to form a football-shaped surface.

   a. Sketch the surface.

   b. Write an integral equal to the area of the surface. Evaluate it numerically.

   c. Show that your answer is reasonable by comparing it with suitable geometric figures.

3. *In Curved Surface, Problem I:* The graph of $y = \ln x$ from $x = 1$ to $x = 3$ is rotated about the x-axis to form a surface. Find the area of the surface.

4. *In Curved Surface, Problem II:* The graph of $y = \ln x$ from $x = 1$ to $x = 3$ is rotated about the y-axis to form a surface. Find the area of the surface.

5. *Reciprocal Curved Surface Problem I:* The graph of $y = 1/x$ from $x = 0.5$ to $x = 2$ is rotated about the y-axis to form a surface. Find its area.

6. *Reciprocal Curved Surface Problem II:* The graph of $y = 1/x$ from $x = 0.5$ to $x = 2$ is rotated about the x-axis to form a surface. Find its area. How does this answer compare with that in Problem 5?

7. *Cubic Paraboloid Problem I:* The cubic paraboloid $y = x^3$ from $x = 0$ to $x = 2$ is rotated about the y-axis to form a cuplike surface. Find the area of the surface.

8. *Cubic Paraboloid Problem II:* The part of the cubic parabola

$$y = -x^3 + 5x^2 - 8x + 6$$

in Quadrant I is rotated about the y-axis to form a surface. Find the area of the surface.

For Problems 9–16, write an integral equal to the area of the surface. Evaluate it exactly, using the fundamental theorem. Find a decimal approximation for the exact area.

9. $y = \sqrt{x}$, from $x = 0$ to $x = 1$, about the x-axis

10. $y = x^3$, from $x = 1$ to $x = 2$, about the x-axis

11. $y = \dfrac{x^4}{8} + \dfrac{x^{-2}}{4}$, from $x = 1$ to $x = 2$, about the x-axis

12. $y = x^2$, from $x = 0$ to $x = 2$, about the y-axis

13. $y = \frac{1}{3}(x^2 + 2)^{3/2}$, from $x = 0$ to $x = 3$, about the y-axis

14. $y = 2x^{1/3}$, from $x = 1$ to $x = 8$, about the y-axis

15. $y = \dfrac{x^3}{3} + \dfrac{1}{4x}$, from $x = 1$ to $x = 3$, about the line $y = -1$

16. $y = \dfrac{x^3}{3} + \dfrac{1}{4x}$, from $x = 1$ to $x = 3$, about the line $x = 4$

17. *Sphere Zone Problem:* The circle with equation $x^2 + y^2 = 25$ is rotated about the x-axis to form a sphere (Figure 8-6h).

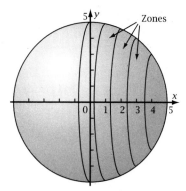

Figure 8-6h

a. Slice the sphere with planes perpendicular to the x-axis. Write the differential of surface area, dS, in terms of x.

b. Find the area of the zone between
   i. $x = 0$ and $x = 1$
   ii. $x = 1$ and $x = 2$
   iii. $x = 2$ and $x = 3$
   iv. $x = 3$ and $x = 4$
   v. $x = 4$ and $x = 5$

c. As you progress from the center of a sphere toward a pole, you would expect the areas of zones of equal height to decrease because their radii are decreasing, but also to increase because their arc lengths are increasing (Figure 8-6i). From the results of part b, which of these two opposing features seems to predominate in the case of a sphere?

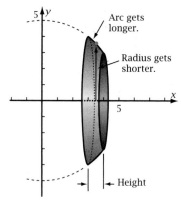

Figure 8-6i

18. *Sphere Total Area Formula Problem:* Prove that the surface area of a sphere of radius r is given by $S = 4\pi r^2$.

19. *Sphere Volume and Surface Problem:* You can find the volume of a sphere by slicing it into **spherical shells** (Figure 8-6j). If the shell is thin, its volume is approximately equal to its surface area times its thickness. The approximation becomes exact as the thickness of the shell approaches zero. Use the area formula in Problem 18 to derive the volume formula for a sphere,

$$V = \tfrac{4}{3}\pi r^3$$

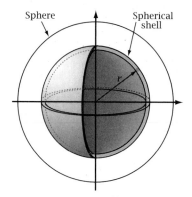

Figure 8-6j

20. *Sphere Rate of Change of Volume Problem:* Prove that the instantaneous rate of change of the volume of a sphere with respect to its radius is equal to the sphere's surface area.

21. *Paraboloid Surface Area Problem:* Figure 8-6k shows the paraboloid formed by rotating about the y-axis the graph of a parabola $y = ax^2$. Derive a formula for the surface area of a

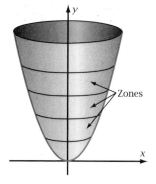

Figure 8-6k

paraboloid in terms of its base radius r and the constant a in the equation.

22. *Zone of a Paraboloid Problem:* Zones of equal altitude on a sphere have equal areas (Problem 17). Is this property also true for a paraboloid (Figure 8-6k)? If so, support your conclusion with appropriate evidence. If not, does the area increase or decrease as you move away from the vertex?

23. *Ellipsoid Problem:* The ellipse with x-radius 5 and y-radius 3 and parametric equations

$$x = 5 \cos t$$
$$y = 3 \sin t$$

is rotated about the x-axis to form an ellipsoid (a football-shaped surface). Write an integral for the surface area of the ellipsoid and evaluate it numerically. Show why the Cartesian equation $(x/5)^2 + (y/3)^2 = 1$ for the same ellipsoid would be difficult to use because of what happens to dL at the end of the ellipsoid, that is, at $x = 5$.

24. *Cooling Tower Problem:* Cooling towers for some power plants are made in the shape of hyperboloids of one sheet (see the photograph). This shape is chosen because it uses all straight reinforcing rods. A framework is constructed, then concrete is applied to form a relatively thin shell that is quite strong, yet has no structure inside to get in the way. In this problem you will find the area of such a cooling tower.

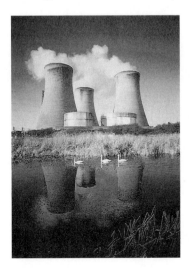

The cooling tower shown in Figure 8-6l is formed by rotating about the y-axis the hyperbola with the parametric equations

$$x = 35 \sec t$$
$$y = 100 + 80 \tan t$$

where x and y are in feet.

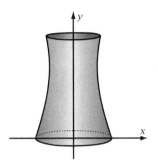

Figure 8-6l

a. The hyperbola starts at $y = 0$. What is the radius of the hyperboloid at its bottom?

b. The hyperbola stops where $t = 0.5$. What is the radius at the top of the hyperboloid? How tall is the cooling tower?

c. What is the radius of the cooling tower at its narrowest? How high up is this narrowest point located?

d. Find the surface area of the hyperboloid.

e. The walls of the cooling tower are to be 4 in. thick. Approximately how many cubic yards of concrete will be needed to build the tower?

25. *Lateral Area of a Cone Problem:* Figure 8-6m shows a cone of radius R and slant height L. The cone is a singly curved surface, so you can cut it and roll it out into a plane surface that is a sector of a circle. Show that the area of the lateral surface of a cone is

$$S = \pi R L$$

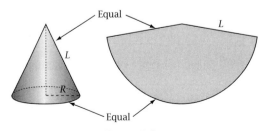

Figure 8-6m

26. **Lateral Area of a Frustum Problem:** Figure 8-6n shows that a frustum of a cone is a difference

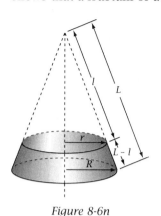

Figure 8-6n

between two similar cones, one of radius and slant height R and L, the other of r and l. By the properties of similar triangles,

$$\frac{R}{L} = \frac{r}{l}$$

Use this fact and clever algebra to transform the area of the frustum so that it is in terms of the average radius and the frustum slant height,

$$S = \pi RL - \pi rl = 2\pi \left(\frac{R+r}{2}\right)(L-l)$$

8-7 Lengths and Areas for Polar Coordinates

You have seen how to find lengths of curves specified by parametric equations and by regular Cartesian equations. In this section you will find lengths and areas when the curve is specified by polar coordinates. In polar coordinates, the position of a point is given by the displacement from the origin (the **pole**) and the angle with the positive x-axis (the **polar axis**).

Suppose that an object is located at point (x, y) in the Cartesian plane (Figure 8-7a). Let r (for "radius") be the directed distance from the pole to the object. Let θ be the directed angle from the polar axis to a ray from the pole through the point (x, y). Then the ordered pair (r, θ) contains polar coordinates of (x, y). Note that in (r, θ) the variable r is the dependent variable, not the independent one. Figure 8-7a also shows how you can plot (r, θ) if r is negative. In this case, θ is an angle to the ray opposite the ray through (x, y).

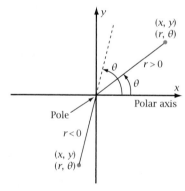

Figure 8-7a

Suppose that the polar coordinates of a moving object are given by

$$r = 5 + 4\cos\theta$$

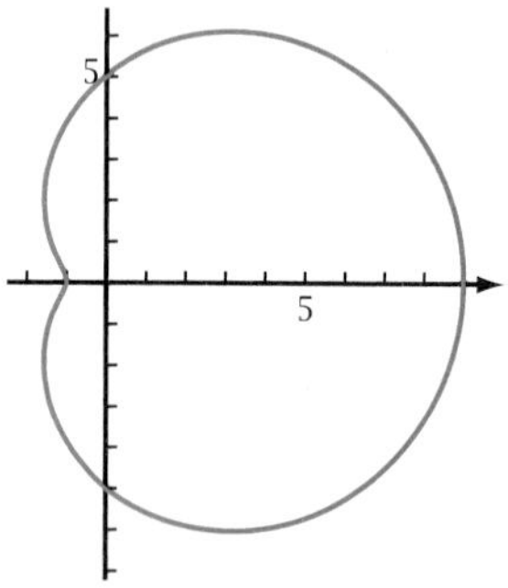

By picking values of θ, you can calculate and plot by hand or by grapher the corresponding values of r. The polar graph in this case (Figure 8-7b) is a **limaçon**, French for "snail." (The cedilla under the c makes its pronunciation "s.")

Figure 8-7b

> **OBJECTIVE**
>
> Given the equation of a polar function, find the area of a region bounded by the graph and find the length of the graph.

Area

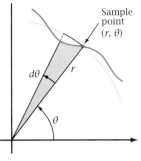

Sample point (r, θ)

Figure 8-7c

Figure 8-7c shows a wedge-shaped region between a polar curve and the pole, swept out as θ increases by a small amount $d\theta$. The point (r, θ) on the graph can be used as a sample point for a Riemann sum. The area of the region is approximately the area of a circular sector of radius r and central angle $d\theta$. The sector has area $d\theta/(2\pi)$ of a whole circle. Let dA be the sector's area.

$$\therefore \, dA = \pi r^2 \cdot \frac{d\theta}{2\pi} = \frac{1}{2} r^2 d\theta$$

The area of an entire region swept out as θ increases from a to b is found by summing the areas of the sectors between a and b and taking the limit as $d\theta$ approaches zero (that is, integrating).

DEFINITION: Area of a Region in Polar Coordinates

The area A of the region swept out between the graph of $r = f(\theta)$ and the pole as θ increases from a to b is given by

$$A = \lim_{\Delta\theta \to 0} \sum \tfrac{1}{2} r^2 \Delta\theta = \int_a^b \tfrac{1}{2} r^2 d\theta$$

▶ **EXAMPLE 1**   Find the area of the region enclosed by the limaçon $r = 5 + 4 \cos \theta$.

Solution   The graph is shown in Figure 8-7b. You can plot it with your grapher in polar mode. If you start at $\theta = 0$, the graph makes a complete cycle and closes at $\theta = 2\pi$.

$$dA = \tfrac{1}{2}(5 + 4 \cos \theta)^2 \, d\theta \qquad \text{Find the area of a sector, } dA.$$

The entire limaçon is generated as θ increases from 0 to 2π.

$$\therefore \, A = \tfrac{1}{2} \int_0^{2\pi} (5 + 4 \cos \theta)^2 d\theta \qquad \text{Add the sector areas and take the limit (that is, integrate).}$$

$$= 103.672\ldots \text{ square units} \qquad \text{By numerical integration.}$$

Note: Because the limaçon is symmetrical, you can integrate from 0 to π and double the answer.

As a rough check, the limaçon is somewhat larger than a circle of diameter 10 units. The area of the circle is $25\pi = 78.5\ldots$. So 103.6... is reasonable for the limaçon.   ◀

You cannot use the integral of a power property in Example 1 because you cannot make $d\theta$ the differential of the inside function. In Section 9-5, you will learn a way to evaluate this integral algebraically using the fundamental theorem. You will find that the exact answer is 33π. If you divide the unrounded numerical answer, 103.672..., by π you should get 33.

Example 2 shows you how to find the area swept out as a polar curve is generated if r becomes negative somewhere, or if part of the region overlaps another part.

▶ **EXAMPLE 2**   Figure 8-7d shows the limaçon $r = 1 + 3 \sin \theta$.

a. Find the area of the region inside the inner loop.

b. Find the area of the region between the outer loop and the inner loop.

Solution   a. Figure 8-7d shows a wedge-shaped slice of the region and a sample point (r, θ) on the inner loop of the graph.

$$dA = \tfrac{1}{2}r^2\, d\theta = \tfrac{1}{2}(1 + 3 \sin \theta)^2$$

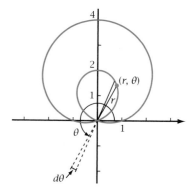

Figure 8-7d

By plotting and tracing, you will find that the inner loop corresponds to values of θ roughly between π and 2π. You will also find that r is negative, as shown in Figure 8-7d. However, dA will be positive for positive values of $d\theta$ because the r is squared. To find the limits of integration precisely, set r equal to zero and solve numerically or algebraically for θ.

$$1 + 3 \sin \theta = 0 \Rightarrow \sin \theta = -\tfrac{1}{3}$$
$$\Rightarrow \theta = -0.3398... + 2\pi n \text{ or } \pi - (-0.3398...) + 2\pi n$$

By using $n = 1$ in the first equation and $n = 0$ in the second equation, you can find values of θ in the desired range.

$$\theta = 3.4814... \text{ or } 5.9433...$$

For convenience, store these values as a and b in your grapher.

$$A = \int_a^b \tfrac{1}{2}(1 + 3 \sin \theta)^2 d\theta \approx 2.527636... \qquad \text{By numerical integration.}$$

As a rough check, the inner loop has an area slightly smaller than that of a circle of diameter 2. See Figure 8-7d. That area is $\pi \cdot 1^2 = 3.141...$.

b. The outer loop begins and ends where the inner loop ends and begins. The appropriate values of the limits of integration as found in part a are $a = -0.3398...$ and $b = 3.4814...$. Store these values in your grapher and repeat the numerical integration.

$$A = \int_a^b \tfrac{1}{2}(1 + 3 \sin \theta)^2 d\theta \approx 14.751123...$$

It's important to realize that this is the area of the entire outer loop, between the graph and the pole, swept out as θ increases from $-0.33...$ to $3.48...$. The area of the region between the two loops is the difference between this and the area of the inner loop.

$$A \approx 14.751123... - 2.527636... = 12.223487...$$

◀

Arc Length

Figure 8-7e shows a part of a polar curve traced out as θ increases by $d\theta$. An arc of a circle drawn at (r, θ) would have length $r\, d\theta$ since θ is measured in radians. The length dL is close to the length of the hypotenuse of a right triangle of legs $r\, d\theta$ and dr. By the Pythagorean theorem,

$$dL = \sqrt{dr^2 + (r\, d\theta)^2}$$

Factoring out $d\theta^2$ and taking its square root,

$$dL = \sqrt{\left(\frac{dr}{d\theta}\right)^2 + r^2}\; d\theta$$

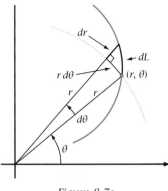

Figure 8-7e

The length of the entire path traced as θ increases from a to b is found by summing the dL's and finding the limit (that is, integrating).

DEFINITION: Length of a Curve in Polar Coordinates

The length L traced out along the polar curve $r = f(\theta)$ as θ increases from a to b is given by

$$L = \lim_{\Delta\theta \to 0} = \sum dL = \int_a^b \sqrt{dr^2 + (r\, d\theta)^2} = \int_a^b \sqrt{(dr/d\theta)^2 + r^2}\; d\theta$$

▶ **EXAMPLE 3**   Find the length of the limaçon $r = 1 + 3\sin\theta$ in Example 2 (Figure 8-7d).

Solution

$$\frac{dr}{d\theta} = 3\cos\theta$$

$$\therefore dL = \sqrt{(3\cos\theta)^2 + (1 + 3\sin\theta)^2}\; d\theta$$

By plotting the graph and tracing, you will find that the graph starts repeating itself after θ has increased by 2π radians. So, convenient limits of integration are 0 to 2π. A smaller interval will not generate the entire graph. A larger interval will count parts of the graph more than once.

$$L = \int_0^{2\pi} \sqrt{(3\cos\theta)^2 + (1 + 3\sin\theta)^2}\; d\theta$$

$$L \approx 19.3768\ldots \text{ units} \qquad\qquad \text{By numerical integration.}$$

The inner and outer loops of the limaçon are close to circles of diameters 2 and 4, respectively. The circumferences of these circles sum to $2\pi \cdot 1 + 2\pi \cdot 2 = 18.84\ldots$. So 19.3... is reasonable for the length of the limaçon.   ◀

Problem Set 8-7

Q1. Differentiate: $f(x) = 5x^3 - 7x^2 + 4x - 11$

Q2. Differentiate: $g(x) = (4x - 9)^3$

Q3. Differentiate: $h(x) = \sin^3 x$

Q4. Differentiate: $u(x) = \sec 3x$

Q5. Differentiate: $v(x) = e^{-x}$

Q6. Differentiate: $r(x) = 1/x$

Q7. Integrate: $\int (1/x)\,dx$

Q8. Integrate: $\int x\,dx$

Q9. Integrate: $\int 3\,dx$

Q10. Integrate: $\int dx$

1. Figure 8-7f shows the polar graph of the circle $r = 10 \sin \theta$, with diameter 10.

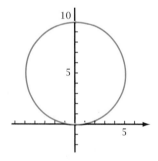

Figure 8-7f

a. Find the area of the region swept out between the graph and the pole in one revolution as θ increases from 0 to 2π.

b. Why is the answer in part a twice the area of the region inside the circle? Why don't you get a negative value for the area integral as θ increases from π to 2π, even though r is negative for these values of θ?

2. Find the length of the circle $r = 10 \sin \theta$ in Figure 8-7f that is traced out as θ makes one revolution, increasing from 0 to 2π radians. Why is the answer twice the circumference of the circle? Why do you suppose the polar coordinate length formula is phrased dynamically, in terms of the length "traced out," rather than statically in terms of the length "of" the curve?

For Problems 3–10,

a. Plot the graph, thus confirming the one given here.

b. Find the area of the region enclosed by the graph.

c. Find the length of the graph.

3. The limaçon $r = 4 + 3 \sin \theta$

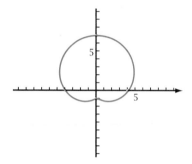

4. The limaçon $r = 5 - 3 \cos \theta$

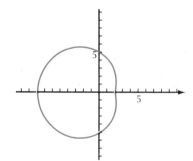

5. The curve $r = 7 + 3 \cos 2\theta$

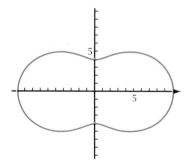

6. The four-leaved rose $r = 8\cos 2\theta$

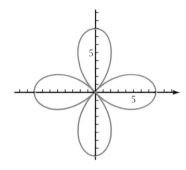

7. The cardioid $r = 5 + 5\cos\theta$

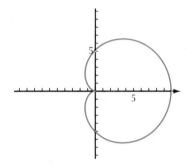

8. The ellipse $r = \dfrac{10}{3 - 2\cos\theta}$

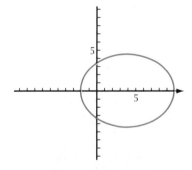

9. The *three-leaved rose* $r = \sin 3\theta$ (Be careful of the range of θ-values!)

10. The *cissoid of Diocles* $r = 4\sec\theta - 4\cos\theta$, and the lines $\theta = -1$ and $\theta = 1$

11. Figure 8-7g shows the *lemniscate of Bernoulli* with polar equation

$$r = \sqrt{49\cos 2\theta}$$

What range of values of θ causes the right loop to be generated? What is the total area of the region enclosed by both loops?

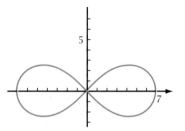

Figure 8-7g

12. The polar graph of $r = \csc\theta + 4$ is a *conchoid of Nicomedes*. The graph is unbounded but contains a closed loop. Find the area of the region inside the loop.

13. Figure 8-7h shows these polar graphs.

Cardioid: $r = 4 + 4\cos\theta$

Circle: $r = 10\cos\theta$

Where do the graphs intersect? What is the area of the region outside the cardioid and inside the circle?

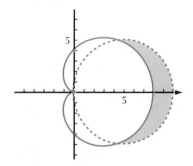

Figure 8-7h

14. Find the area of the region that is inside the circle $r = 5$ and outside the cardioid $r = 5 - 5\cos\theta$.

15. Figure 8-7i shows the Archimedean spiral $r = 0.5\theta$.

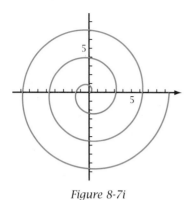

Figure 8-7i

a. Find the length of the part of the spiral shown.

b. Find the area of the region in Quadrant I that lies between the outermost branch of the spiral and the next branch in toward the pole.

16. For the limaçon $r = 4 + 6\cos\theta$, find the area of the region that lies between the inner and outer loops.

17. *Column Scroll Problem:* The spiral design at the top of Ionic columns in ancient Greek architecture is an example of a *lituus* (pronounced "lit'-you-us"). Plot the lituus

$$r = 5\theta^{-1/2}$$

traced out as θ increases from 0 to 6π.

This column at Villa Barbaro in Maser, Italy, was built in the 1500s.

a. Find the length from $\theta = \pi/2$ to $\theta = 6\pi$.

b. Let $\theta = 1$ radian. Sketch an arc of a circle centered at the pole, from the polar axis to the point $(r, 1)$ on the lituus. Find the area of the sector of the circle corresponding to this arc. Repeat the calculation for $\theta = 2$ and $\theta = 3$. What seems to be true about these areas?

18. *Line Problem:* Show that the graph of $r = \sec\theta$ is a line. Find the length of the segment from $\theta = 0$ to $\theta = 1.5$ using the calculus of polar coordinates. Then confirm that your answer is correct by appropriate geometry.

19. *LP Record Project:* In this project you will calculate the length of the groove on an old $33\frac{1}{3}$ rpm record. Obtain such a record. Figure out a way to measure the number of grooves per centimeter in the radial direction. Then find a polar equation for the spiral formed by the grooves. By integrating, calculate the length of the groove from the outer one to the inner one. Perform a quick calculation to show that your answer is reasonable.

20. *Kepler's Law Project:* Figure 8-7j shows the path of a satellite in an elliptical orbit. Earth is at the pole (one focus of the ellipse). The polar equation of the ellipse is

$$r = \frac{100}{3 - 2\cos\theta}$$

where θ is in radians and r is in thousands of miles. In this problem you will investigate the speed of the satellite at various places.

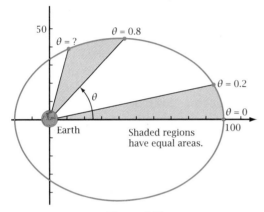

Figure 8-7j

a. Find the area of the elliptical sector from $\theta = 0$ to $\theta = 0.2$.

b. German astonomer Johannes Kepler (1571–1630) observed that an object in orbit sweeps out sectors of equal area in equal times. This fact is known as **Kepler's second law** of planetary motion. (His first law states that the orbit is an ellipse where the object being orbited is located at one focus.) If the sector in Figure 8-7j starting at $\theta = 0.8$ has area equal to the one in part a, at what value of θ does the sector end?

c. **Kepler's third law** states that the period of an orbiting object is related to its distance from the object being orbited. If a is half the major axis of the orbit, then the period P is

$$P = ka^{1.5}$$

Find the value of the constant k using data for the moon. The moon is about $a = 240{,}000$ mi from Earth and has a period of about 655 h ($27.3\,\mathrm{d} \cdot 24\,\mathrm{h/d}$).

d. What is the period of the satellite in Figure 8-7j?

e. How many hours does it take the satellite to travel from $\theta = 0$ to $\theta = 0.2$? How many hours does it take to travel from $\theta = 0.8$ to the value of θ in part b?

f. How many miles does the satellite travel on its elliptical path between $\theta = 0$ and $\theta = 0.2$? How many miles does it travel between $\theta = 0.8$ and the value of θ in part b?

g. Find the average speed (distance/time) of the satellite for each of the two arcs in part f.

h. See whether you can explain physically why a satellite would move faster when it is closer to Earth than it does when it is farther away.

21. *The Derivative dy/dx for Polar Coordinates Problem:* Figure 8-7k shows the polar graph of the spiral

$$r = \theta$$

superimposed on a Cartesian xy-plane. A tangent line is plotted at the point $(r, 7)$.

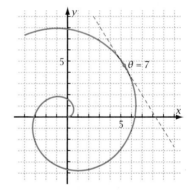

Figure 8-7k

a. Estimate the slope of the tangent line.

b. Polar and Cartesian coordinates are related by the parametric equations

$$x = r\cos\theta$$
$$y = r\sin\theta$$

By appropriate use of the parametric chain rule, find an equation for dy/dx and use it to show algebraically that the slope you found in part a is correct.

22. *Project—The Angle Between the Radius and the Tangent Line:* A remarkably simple relationship exists between $dr/d\theta$ in polar coordinates and the angle ψ (Greek letter psi, pronounced "sigh") measured counterclockwise from the radius to the tangent line (Figure 8-7l). In this project you will derive and apply this relationship.

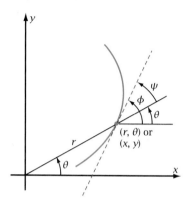

Figure 8-7l

a. Explain why $\tan \theta = y/x$.

b. Let ϕ (Greek letter phi, "fee" or "fye") be the angle from the positive horizontal direction to the tangent line (Figure 8-7l). The slope of the tangent line is dy/dx. Explain why

$$\tan \phi = \frac{dy/d\theta}{dx/d\theta}$$

c. Recall from trigonometry that

$$\tan(A - B) = \frac{\tan A - \tan B}{1 + \tan A \tan B}$$

Use this property, the results of parts a and b, and appropriate algebra to show that

$$\tan \psi = \frac{x(dy/d\theta) - y(dx/d\theta)}{x(dx/d\theta) + y(dy/d\theta)}$$

d. Use the fact that $x = r \cos \theta$ and $y = r \sin \theta$ to show that the numerator in part c equals r^2.

e. Use the fact that $r^2 = x^2 + y^2$, and its derivative, to show that $r\dfrac{dr}{d\theta}$ equals the

denominator in part c, and thus the following property holds:

Property: The Angle ψ in Polar Coordinates

If ψ is the angle measured counterclockwise from the radius to the tangent line of a polar graph, then

$$\tan \psi = \frac{r}{dr/d\theta} = \frac{r}{r'}$$

f. Figure 8-7m shows the cardioid

$$r = a - a \cos \theta$$

where a stands for a nonzero constant. Prove that for this cardioid the angle ψ is always equal to $\frac{1}{2}\theta$. The half-argument properties for tangent, which you may recall from trigonometry, are helpful.

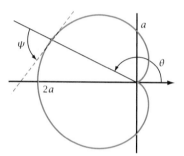

Figure 8-7m

g. Figure 8-7n shows a cross section through a chambered nautilus shell. The spiral has the property that the angle ψ is a constant. Use

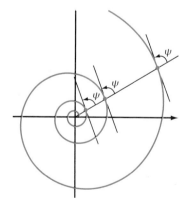

Equiangular spiral. ψ is constant.

Figure 8-7n

this fact and the property in part e to find the general equation of this **equiangular spiral**. Choose two values of θ on the photograph and measure the corresponding values of r. (You may arbitrarily pick *any* point on the spiral to be $\theta = 0$. Treat points closer to the center as $\theta < 0$ and points farther from the center as $\theta > 0$.) Use these values as initial conditions to find the particular equation for the spiral. Confirm that your equation is correct by plotting it, tracing to another value of θ, and showing that the point is actually on the photographed spiral. Use the constants in your equation to calculate the value of ψ. Measure a copy of the shell to show that your calculated value of ψ is correct.

8-8   Chapter Review and Test

In this chapter you have seen applications of derivatives and integrals to geometrical problems. The rate of change of the area or volume of a figure describes how fast these quantities change as a given dimension changes. Maximum or minimum areas or volumes occur where the rate of change equals zero. Areas, volumes, and curved lengths can be calculated by slicing a figure into small pieces, adding the pieces, and taking the limit. The resulting limits of Riemann sums are equal to definite integrals. You have evaluated these integrals numerically or by the fundamental theorem if you were able to find the indefinite integral.

Review Problems

R0. Update your journal with what you've learned since your last entry. Include such things as

- The one most important thing you have learned in studying Chapter 8
- How the volume, length, and surface area of a geometric figure are found
- How volume, length, and surface area are found in polar coordinates or with parametric functions
- Which boxes you have been working on in the "define, understand, do, apply" table

R1. Three cubic functions have equations

$$f(x) = x^3 - 9x^2 + 30x - 10$$
$$g(x) = x^3 - 9x^2 + 27x - 10$$
$$h(x) = x^3 - 9x^2 + 24x - 10$$

a. Plot the graphs on the same screen. Sketch the results.

b. Write equations for the first and second derivatives of each function.

c. Which function has two distinct values of x at which the first derivative is zero? What are these values of x? What features occur at these points?

d. Which function has a horizontal tangent line somewhere but no local maximum or minimum points?

e. Each function has a point of inflection. Show that the second derivative is zero at each of the points of inflection.

R2. a. For the function in Figure 8-8a, sketch a number-line graph for f' and for f'' that shows the sign of each derivative in a neighborhood of the critical point at $x = 2$. Indicate on the number lines whether there is a local maximum, a local minimum, or a point of inflection at $x = 2$.

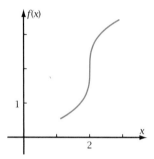

Figure 8-8a

b. Sketch the graph of a function whose derivatives have the features given in Figure 8-8b.

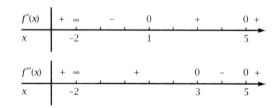

Figure 8-8b

c. Figure 8-8c shows the graph of

$$f(x) = x^{2/3} - x$$

i. Write equations for $f'(x)$ and $f''(x)$.

ii. The graph appears to slope downward for all x. Does $f(x)$ have any local maxima or minima? If so, where? If not, explain how you can tell.

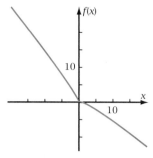

Figure 8-8c

iii. The graph appears to be concave down for all x. Are there any points of inflection? If so, where? If not, explain how you can tell.

iv. Write the global maximum and minimum values of $f(x)$ for x in the closed interval $[0, 5]$.

d. For $f(x) = x^2 e^{-x}$, find the maxima, minima, and points of inflection and sketch the graph.

R3. a. *Storage Battery Problem:* A normal automobile battery has six cells divided by walls. For a particular battery, each cell must have an area of 10 in.^{2}, looking down from the top, as shown in Figure 8-8d. What dimensions of the battery will give the minimum total wall length (including outsides)? A typical battery is 9 in. by 6.7 in. (that is, 1.5-in. cell width). Does minimum wall length seem to be a consideration in battery design?

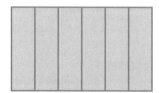

Figure 8-8d

b. *Cylinder in Cubic Paraboloid Problem:* A rectangle is inscribed in the region in Quadrant I under the cubic parabola $y = 8 - x^3$. Two sides of the rectangle are on the x- and y-axes, and the opposite corner of the rectangle touches the graph. The figure is rotated about the y-axis. The curve generates a cubic paraboloid, and the rectangle generates a cylinder. What rectangle dimensions give the largest-volume cylinder?

R4. a. The region in Quadrant I bounded by the graphs of $y = x^{1/3}$ and $y = x^2$ is rotated about the x-axis to form a solid. Find the volume of the solid by cylindrical shells.

   b. Find the volume of the solid in part a by plane slices. Show that the answer is the same.

   c. *Various Axes Problem:* The region bounded by the parabola $y = x^2$ and the line $y = 4$ is rotated to form various solids. Find the volume of the solid for the following axes of rotation:

   i. y-axis

   ii. x-axis

   iii. Line $y = 5$

   iv. Line $x = 3$

R5. a. Write an integral equal to the length of the parabola $y = x^2$ between $x = -1$ and $x = 2$. Evaluate the integral numerically.

   b. Find exactly the length of the graph of $y = x^{3/2}$ from $x = 0$ to $x = 9$ using the fundamental theorem. Find a decimal approximation for the answer. Check your answer by employing suitable geometry.

   c. Find the length of the following spiral from $t = 0$ to $t = 4$.

   $$x = t \cos \pi t$$
   $$y = t \sin \pi t$$

R6. a. Find exactly, by the fundamental theorem, the area of the surface formed by rotating

about the y-axis the graph of $y = x^{1/3}$ from $x = 0$ to $x = 8$. Find a decimal approximation for the answer. Check your answer by using geometry.

   b. The graph of $y = \tan x$ from $x = 0$ to $x = 1$ is rotated about the line $y = -1$ to form a surface. Write an integral for the area of the surface. Evaluate it numerically.

   c. The spiral in Problem R5c is rotated about the y-axis to form a "sea shell." Find its surface area.

R7. Figure 8-8e shows the spiral

$$r = \theta$$

from $\theta = 0$ to $\theta = 5\pi/2$.

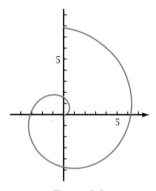

Figure 8-8e

   a. Find the length of this part of the spiral.

   b. Find the area of the region in Quadrant I that is outside the first cycle and inside the second cycle of the spiral.

Concept Problems

C1. *Oil Viscosity Problem:* The viscosity (resistance to flow) of normal motor oil decreases as the oil warms up. "All-weather" motor oils retain about the same viscosity throughout the range of operating temperatures. Suppose that the viscosity of 10W-40 oil is given by

$$\mu = 130 - 12T + 15T^2 - 4T^3 \quad \text{for } 0 \le T \le 3$$

where μ (Greek letter *mu*, pronounced "mew" or "moo") is the viscosity in centipoise and T is the temperature in hundreds of degrees.

   a. At what temperature in this domain will the maximum viscosity occur?

   b. What is the minimum viscosity in this domain? Justify your answer.

C2. *"Straight Point" Problem:* Show that the graph of $f(x) = (x - 1)^4 + x$ has a zero second derivative at $x = 1$ but does not have a point of inflection there. Sketch what the graph will look like in the vicinity of $x = 1$. Describe what is true about the graph at $x = 1$.

C3. *Infinite Derivative Problem:* The functions $f(x) = x^{2/3}$ and $g(x) = x^{-2/3}$ both have infinite first derivatives at $x = 0$, but the behavior of each is quite different there. Sketch a graph showing the difference.

C4. *Chapter Logo Problem:* The icon at the top of each even-numbered page of this chapter shows a solid formed by rotating about the line $x = 4$ the part of the graph

$$y = 3 + 5\left[0.5 + 0.5\cos\left(\frac{\pi}{3}(x - 5)\right)\right]^2$$

from $x = 5$ to $x = 7.5$. A cylindrical hole 1 unit in radius is coaxial with the solid. Figure 8-8f shows the coordinate system in which this diagram was drawn.

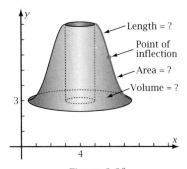

Figure 8-8f

a. Find the length of the segment of graph that was rotated.

b. Find the x-coordinate of the point of inflection.

c. Find the area of the doubly curved surface of the solid.

d. Find the volume of the solid.

C5. *Area by Planimeter Project:* You have learned how to calculate the area of a region algebraically using the fundamental theorem and also numerically. There is a mechanical device called a planimeter (compensating polar planimeter) that finds the area geometrically

from a drawing of the region. In this problem you will borrow a planimeter and use it to find the area of Brazil from a map.

a. Make a copy of the map of Brazil in Figure 8-8g. Be sure to include the scale because some copy machines shrink the picture.

Figure 8-8g

b. Set up the planimeter with the tracer point at a convenient starting point on the map. Set the dial to zero. Then trace around the boundary until you return to the starting point.

c. Read the final setting on the dial. The planimeter may have a vernier scale for reading tenths of a unit.

d. Measure the scale on the map to find out how many miles correspond to 1 cm. Be clever! Then find out how many square miles correspond to 1 cm². Finally, calculate the area of Brazil to as many significant digits as the data justify.

e. Find out from the planimeter's instruction manual the theoretical basis on which the instrument works. Write a paragraph or two describing what you learned.

f. Check an almanac or encyclopedia to see how accurate your measurement is.

C6. *Hole in the Cylinder Project:* A cylinder of uranium 10 cm in diameter has a hole 6 cm in diameter drilled through it (Figure 8-8h). The axis of the hole intersects the axis of the cylinder at right angles. Find the volume of the uranium drilled out. Find the value of the uranium drilled out assuming that uranium costs $200 a gram. Be resourceful to find out the density of uranium.

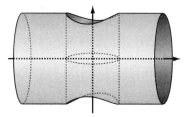

Figure 8-8h

C7. *Three-Hole Project:* A cube 2 cm on each edge has three mutually perpendicular holes drilled through its faces (Figure 8-8i). Each hole has diameter 2 cm and so extends to the edge of the cube face to which it is parallel. Find the volume of the solid remaining after the three holes are drilled.

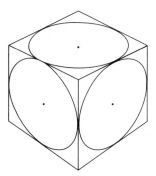

Figure 8-8i

Chapter Test

PART 1: No calculators allowed (T1–T5)

T1. Figure 8-8j shows critical values of $f'(x)$ and $f''(x)$ for a continuous function f, as well as the signs of the derivatives in the intervals between these points. Sketch a possible graph of f using the initial condition that $f(0) = 3$.

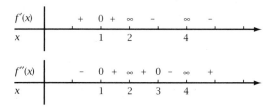

Figure 8-8j

T2. Figure 8-8k shows the graph of $y = f'(x)$, the *derivative* of a continuous function. Sketch the graph of $y = f(x)$ if $f(0) = 4$. Put a dot at the approximate location of each critical point and each point of inflection.

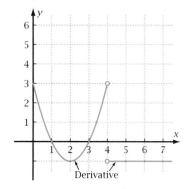

Figure 8-8k

T3. Figure 8-8l shows a rectangular field with one side along a river. A fence of total length 1000 ft encloses the remaining three sides. Find the values of x and y that give the field maximum area. Show how you know that the area is a maximum rather than a minimum.

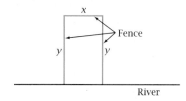

Figure 8-8l

T4. Figure 8-8m shows the solid formed by rotating about the y-axis the region in Quadrant I bounded by the graphs of two functions y_1 and y_2. Write an integral for the volume of the solid if the region is sliced *parallel* to the axis of rotation.

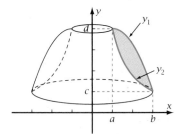

Figure 8-8m

T5. Write these differentials:

a. Area, dA, in polar coordinates

b. Arc length, dL, for a plane curve in polar coordinates

c. Arc length, dL, for a plane curve in Cartesian coordinates in terms of dx and dy

d. Area, dS, for a surface of revolution around the line $x = 1$ in terms of x or y, and dL

PART 2: Graphing calculators allowed (T6–T15)

T6. Figure 8-8n shows the graph of

$$f(x) = x^3 - 7.8x^2 + 20.25x - 13$$

Ascertain whether the graph has a relative maximum and a relative minimum, a horizontal tangent at the point of inflection, or simply a point of inflection with no horizontal tangent. Justify your answer.

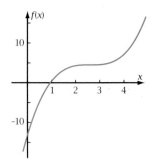

Figure 8-8n

Problems T7–T11 are concerned with the region R shown in Figure 8-8o. Region R is bounded by the graph of $y = x^3$ from $x = 0$ to $x = 2$, the y-axis, and the line $y = 8$.

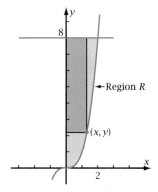

Figure 8-8o

T7. Write a definite integral to find the length of the graph of $y = x^3$ from $x = 0$ to $x = 2$. Evaluate the integral to find the length.

T8. Write a definite integral equal to the area of the surface generated by rotating the segment of graph in Problem T7 about the y-axis. Evaluate the integral.

T9. A rectangular region is inscribed in region R as shown in Figure 8-8o. As R rotates about the y-axis, the rectangular region generates a cylinder. Find exactly the maximum volume the cylinder could have. Justify your answer.

T10. Using slices of R parallel to the y-axis, write an integral equal to the volume of the solid formed by rotating R about the y-axis. Evaluate the integral algebraically using the fundamental theorem.

T11. Suppose that a cylinder is circumscribed about the solid in Problem T10. What fraction of the volume of this cylinder is the volume of the solid?

T12. For the ellipse with parametric functions

$$x = 5 \cos t$$
$$y = 2 \sin t$$

a. Plot the graph and sketch it.

b. Find the length of the ellipse.

c. Show that the volume of the ellipsoid formed by rotating the ellipse about the x-axis is

$$V = \frac{4}{3}\pi \, (x\text{-radius}) \, (y\text{-radius})^2$$

For Problems T13 and T14, use the spiral $r = 5e^{0.1\theta}$ shown in Figure 8-8p.

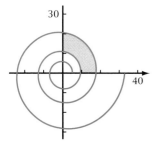

Figure 8-8p

T13. Find the length of the part of the spiral shown.

T14. Find the area of the region in Quadrant I that is outside the second revolution of the spiral and inside the third revolution.

T15. What did you learn as a result of taking this test that you did not know before?

Algebraic Calculus Techniques for the Elementary Functions

The Gateway Arch in St. Louis is built in the form of a catenary, the same shape a chain forms when it hangs under its own weight. In this shape the stresses act along the length of the arch and cause no bending. The equation of a catenary involves the hyperbolic functions, which have properties similar to the circular functions of trigonometry.

Mathematical Overview

In Chapter 9, you will learn ways to integrate each of the elementary functions and their inverses. The elementary functions are

- Algebraic
- Trigonometric
- Exponential
- Logarithmic
- Hyperbolic

You will do the integration in four ways.

Graphically   The icon at the top of each even-numbered page of this chapter shows the graphical meaning of the integration by parts formula, which is used to integrate products.

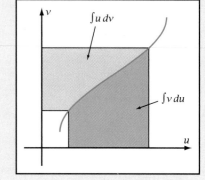

Numerically

x	$\ln x$	$\int_1^x \ln t\, dt$
1	0	0
2	0.693...	0.386...
3	1.098...	1.295...
4	1.386...	2.545...
5	1.609...	4.047...
⋮	⋮	⋮

Algebraically $\int u\, dv = uv - \int v\, du$, the integration by parts formula

Verbally   *The most fundamental method of integration seems to be integration by parts. With it I can integrate products of functions. I can also use it to find algebraic integrals for the inverse trigonometric functions.*

9-1 Introduction to the Integral of a Product of Two Functions

Suppose you are to find the volume of the solid formed by rotating the region under $y = \cos x$ about the y-axis (Figure 9-1a). The value of dV, the differential of volume, is

$$dV = 2\pi x \cdot y \cdot dx = 2\pi x \cos x \, dx$$

Thus, the volume is

$$V = 2\pi \int_0^{\pi/2} x \cos x \, dx$$

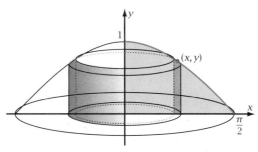

Figure 9-1a

The integrand, $x \cos x$, involves a *product* of two functions. So far you have been able to evaluate such integrals only by approximate numerical methods because you usually could not find the antiderivative of a product.

In this chapter you will learn algebraic techniques for integrating the so-called *elementary functions*. These are the **algebraic functions**, involving no operations other than addition, subtraction, multiplication, division, and roots (rational exponential powers); and the **elementary transcendental functions**, which are the trigonometric and inverse trigonometric functions, logarithmic and exponential functions, and *hyperbolic* functions. The hyperbolic functions are defined in terms of exponential functions but have properties similar to those of the trigonometric functions.

Some of the techniques you will learn were essential before the advent of the computer made numerical integration easily available. They are now interesting mostly for historical reasons and because they give you insight into how to approach a problem. Your instructor will guide you to the techniques that are important for your course. You will see how algebraically "brave" one had to be to learn calculus in the days "BC" ("before calculators"). But for each technique you learn, you will get the thrill of knowing, "I can do it on my own, without a calculator!"

OBJECTIVE   On your own or with your study group, find the indefinite integral $\int x \cos x \, dx$, and use the result to find the exact volume of the solid in Figure 9-1a using the fundamental theorem.

Exploratory Problem Set 9-1

1. Find the approximate volume of the solid in Figure 9-1a by numerical integration.

2. Let $f(x) = x \sin x$. Use the derivative of a product formula to find an equation for $f'(x)$. You should find that $x \cos x$, the integrand in Problem 1, is one of the terms.

3. Multiply both sides of the equation for $f'(x)$ in Problem 2 by dx. Then integrate both sides. (That's easy! You simply write an integral sign ($\int$) in front of each term!)

4. The integral $\int x \cos x \, dx$ should be one term in the equation of Problem 3. Use suitable algebra to isolate this integral. Evaluate the integral on the other side of the equation. Recall what $\int f'(x) \, dx$ equals!

5. Use the result of Problem 4 to find the exact volume of the solid in Figure 9-1a.

6. Find a decimal approximation for the exact volume in Problem 5. How close did the approximation in Problem 1 come to this exact volume?

7. The technique of this exercise is called *integration by parts*. Why do you suppose this name is used? How do you suppose the function $f(x) = x \sin x$ was chosen in Problem 2?

9-2   Integration by Parts—A Way to Integrate Products

Figure 9-2a shows the solid formed by rotating about the y-axis the region under the graph of $y = \cos x$. The volume of this solid is given by

$$V = 2\pi \int_0^{\pi/2} x \cos x \, dx$$

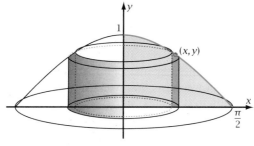

Figure 9-2a

In Section 9-1, you saw that you can integrate this product of functions algebraically. In this section you will learn why the integration by parts technique works.

OBJECTIVE   Given an integral involving a product, evaluate it algebraically using integration by parts.

To integrate a product of two functions, it helps to start with the formula for the differential of a product. If $y = uv$, where u and v are differentiable functions of x, then

$$dy = du \cdot v + u \cdot dv$$

Commuting the du and v in the right side of the equation and integrating both sides gives

$$\int dy = \int v\,du + \int u\,dv$$

You can isolate either integral on the right side and write in terms of the remaining quantities. For instance,

$$\int u\,dv = \int dy - \int v\,du$$

Because $\int dy = y$ (ignoring, for the time being, $+ C$) and $y = uv$, you can write

$$\int u\,dv = uv - \int v\,du$$

For $\int x \cos x\, dx$, let $u = x$ and let $dv = \cos x\, dx$. Then the following holds.

$$du = dx \text{ and } v = \int \cos x\, dx = \sin x + C \qquad \begin{array}{l} \text{Differentiate } u \text{ and} \\ \text{integrate } dv. \end{array}$$

$$\therefore \int x \cos x\, dx = x(\sin x + C) - \int (\sin x + C)\, dx \qquad \begin{array}{l} \text{Substitute for } u, v, \text{ and} \\ du \text{ in the equation} \\ \int u\,dv = uv - \int v\,du. \end{array}$$

$$= x \sin x + Cx + \cos x - Cx + C_1 \qquad \begin{array}{l} \text{A new constant of} \\ \text{integration comes from} \\ \text{the second integral.} \end{array}$$

$$= x \sin x + \cos x + C_1 \qquad \begin{array}{l} \text{The old constant of} \\ \text{integration cancels out!} \end{array}$$

You can always write the integral of a product of two functions as

$$\int (\text{one function})(\text{differential of another function})$$

Associating the integrand into two factors leads to the name *integration by parts*. It succeeds if the new integral, $\int v\,du$, is simpler to integrate than the original integral, $\int u\,dv$.

TECHNIQUE: Integration by Parts

A way to integrate a product is to write it in the form

$$\int (\text{one function})(\text{differential of another function})$$

If u and v are differentiable functions of x, then

$$\int u\,dv = uv - \int v\,du$$

You should memorize the integration by parts formula, because you will use it often. Example 1 shows a way to use this formula.

▶ **EXAMPLE 1**   Evaluate the integral: $\int 5xe^{3x}\,dx$

Solution   Write u and dv to the right, out of the way, as shown.

$$\int 5xe^{3x}\,dx \qquad \text{Let } u = 5x, \quad dv = e^{3x}\,dx$$

$$\therefore du = 5\,dx, \quad v = \int e^{3x}\,dx \qquad \qquad \text{Differentiate } u \text{ and integrate } dv \text{ to find } du \text{ and } v.$$

$$= \tfrac{1}{3}e^{3x} + C$$

$$= 5x\left(\tfrac{1}{3}e^{3x} + C\right) - \int\left(\tfrac{1}{3}e^{3x} + C\right)5\,dx \qquad \text{Use the integration by parts formula, } \int u\,dv = uv - \int v\,du.$$

$$= \tfrac{5}{3}xe^{3x} + 5Cx - \int \tfrac{5}{3}e^{3x}\,dx - \int 5C\,dx$$

$$= \tfrac{5}{3}xe^{3x} + 5Cx - \tfrac{5}{9}e^{3x} - 5Cx + C_1$$

$$= \tfrac{5}{3}xe^{3x} - \tfrac{5}{9}e^{3x} + C_1 \qquad\qquad\qquad\qquad\qquad\qquad ◀$$

Again, the original constant C in integrating $\int dv$ conveniently drops out. The C_1 comes from the last integral. In Problem 46 of Problem Set 9-3, you will prove that this is always the case. So you don't need to worry about putting in the $+C$ until the last integral disappears. Example 2 shows that you may have to integrate by parts more than once.

▶ **EXAMPLE 2**   Evaluate the integral: $\int x^2 \cos 4x\,dx$

Solution
$$\int x^2 \cos 4x\,dx \qquad\qquad u = x^2 \qquad\qquad dv = \cos 4x\,dx$$

$$du = 2x\,dx \qquad\qquad v = \tfrac{1}{4}\sin 4x$$

$$= x^2 \cdot \tfrac{1}{4}\sin 4x - \int\left(\tfrac{1}{4}\sin 4x\right)(2x\,dx) \qquad \text{Use } \int u\,dv = uv - \int v\,du.$$
Note that the integral still involves a product of two functions.

$$= \tfrac{1}{4}x^2 \sin 4x - \int (2x)\left(\tfrac{1}{4}\sin 4x\,dx\right) \qquad \text{Associate the } dx \text{ with the sine factor, as it was originally.}$$

$$u = 2x \qquad\qquad dv = \tfrac{1}{4}\sin 4x\,dx$$

$$du = 2\,dx \qquad\qquad v = -\tfrac{1}{16}\cos 4x$$

$$= \tfrac{1}{4}x^2 \sin 4x - \left[-\tfrac{1}{8}x \cos 4x - \int\left(-\tfrac{1}{16}\cos 4x\right)(2)\,dx\right] \qquad \text{Use integration by parts again.}$$

$$= \tfrac{1}{4}x^2 \sin 4x + \tfrac{1}{8}x \cos 4x - \tfrac{1}{32}\sin 4x + C \qquad\qquad ◀$$

Integration by parts is successful in this example because, at both steps, the second integral, $\int v\,du$, is less complex than $\int u\,dv$ at the start of the step. You could also have chosen the terms

$$u = \cos 4x \qquad\qquad dv = x^2\,dx$$

$$du = -4\sin 4x\,dx \qquad v = \tfrac{1}{3}x^3$$

However, the new integral, $-\frac{4}{3}\int x^3 \sin 4x\,dx$, would have been more complicated than the original integral. The techniques in this box will help you decide how to split an integral of a product into appropriate parts.

TECHNIQUE: *Choosing the Parts in Integration by Parts*

To evaluate $\int u\,dv$, these two criteria should be met.

Primary criterion: dv must be something you can integrate.

Secondary criterion: u should, if possible, be something that gets simpler (or at least not much more complicated) when it is differentiated.

Problem Set 9-2

Quick Review

Q1. Differentiate: $y = x\tan x$

Q2. Integrate: $\int x^{10}\,dx$

Q3. Sketch: $y = e^{-x}$

Q4. Integrate: $\int \cos 3x\,dx$

Q5. Differentiate: $y = \cos 5x \sin 5x$

Q6. Sketch: $y = 2/x$

Q7. $r(x) = \int t(x)\,dx$ if and only if —?—.

Q8. Definition: $f'(x) = $ —?—

Q9. If $f(6.2) = 13$, $f(6.5) = 19$, and $f(6.8) = 24$, then $f'(6.5) \approx$ —?—.

Q10. If region R (Figure 9-2b) is rotated about the line $x = c$, the volume of the solid is —?—.

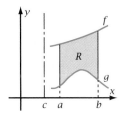

Figure 9-2b

A. $\pi \int_a^b [f(x) - g(x)]^2\,dx$

B. $\pi \int_a^b [(f(x))^2 - (g(x))^2]\,dx$

C. $2\pi \int_a^b [f(x) - g(x)](x - c)\,dx$

D. $2\pi \int_a^b [f(x) - g(x)](c - x)\,dx$

E. $2\pi \int_a^b x[f(x) - g(x)]\,dx$

For Problems 1–10, integrate by parts.

1. $\int x\sin x\,dx$

2. $\int x\cos 3x\,dx$

3. $\int xe^{4x}\,dx$

4. $\int 6x\,e^{-3x}\,dx$

5. $\int (x + 4)e^{-5x}\,dx$

6. $\int (x + 7)e^{2x}\,dx$

7. $\int x^3 \ln x\,dx$

8. $\int x^5 \ln 3x\,dx$

9. $\int x^2\,e^x\,dx$

10. $\int x^2 \sin x\,dx$

11. *Integral of the Natural Logarithm Problem:* You can evaluate the integral $\int \ln x\,dx$ by parts, but you must be clever to make the choice of parts! Find $\int \ln x\,dx$.

9-3 Rapid Repeated Integration by Parts

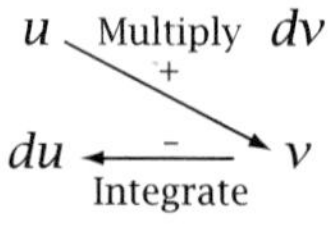

Figure 9-3a

A pattern exists that can help you remember the integration by parts formula. Write u and dv on one line. Below them write du and v. The pattern is "Multiply diagonally down, then integrate across the bottom." The arrows in Figure 9-3a remind you of this pattern. The plus and minus signs on the arrows say to add uv and subtract $\int v\, du$.

This pattern is particularly useful when you must integrate by parts several times.

Use the pattern of Figure 9-3a to simplify repeated integration by parts.

Here's the way Example 2 of Section 9-2 was done.

$$\int x^2 \cos 4x\, dx \qquad\qquad u = x^2 \qquad dv = \cos 4x\, dx$$
$$du = 2x\, dx \qquad v = \tfrac{1}{4}\sin 4x$$
$$= x^2 \cdot \tfrac{1}{4}\sin 4x - \int \left(\tfrac{1}{4}\sin 4x\right)(2x\, dx)$$
$$= \tfrac{1}{4}x^2 \sin 4x - \int (2x)\left(\tfrac{1}{4}\sin 4x\, dx\right) \qquad u = 2x \qquad dv = \tfrac{1}{4}\sin 4x\, dx$$
$$du = 2\, dx \qquad v = -\tfrac{1}{16}\cos 4x$$
$$= \tfrac{1}{4}x^2 \sin 4x - \left[-\tfrac{1}{8}x \cos 4x - \int \left(-\tfrac{1}{16}\cos 4x\right)(2)\, dx\right]$$
$$= \tfrac{1}{4}x^2 \sin 4x + \tfrac{1}{8}x \cos 4x - \tfrac{1}{32}\sin 4x + C$$

Note that the function to be differentiated appears in the left column and the function to be integrated appears in the right column. For instance, $2x$ appears as part of du in the first step and again as u in the next step. If you head the left column "u" and the right column "dv," and leave out dx and other redundant information, you can shorten the work this way:

$$\int x^2 \cos 4x\, dx$$

$$= \tfrac{1}{4}x^2 \sin 4x + \tfrac{1}{8}x \cos 4x - \tfrac{1}{32}\sin 4x + C$$

u		dv
x^2	$+$	$\cos 4x$
$2x$	$-$	$\dfrac{1}{4}\sin 4x$
2	$+$	$-\dfrac{1}{16}\cos 4x$
0	$-$	$-\dfrac{1}{64}\sin 4x$

The second diagonal arrow has a minus sign because the minus sign from the first $\int v\, du$ carries over. The third diagonal arrow has a plus sign because multiplying two negatives from the step before gives a positive. If you put in the third step (fourth line), the last integral is $\int 0\, dx$, which equals C, the constant of integration.

> **TECHNIQUE: Rapid Repeated Integration by Parts**
> - Choose parts u and dv. Differentiate the u's and integrate the dv's.
> - Multiply down each diagonal.
> - Integrate once across the bottom.
> - Use alternating signs shown on the arrows.
>
> If you get 0 in the u-column, the remaining integral will be $\int 0 \, dx$, which equals C.

Examples 1–4 show some special cases and how to handle them.

Make the Original Integral Reappear

When you perform repeated integration by parts, the original integral may reappear.

▶ **EXAMPLE 1**   Evaluate the integral: $\int e^{6x} \cos 4x \, dx$

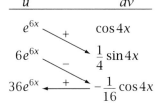

Solution   Set up columns for u and dv.

$$\int e^{6x} \cos 4x \, dx$$
$$= \tfrac{1}{4} e^{6x} \sin 4x + \tfrac{6}{16} e^{6x} \cos 4x - \tfrac{36}{16} \int e^{6x} \cos 4x \, dx$$

The integral on the right is the same as the original one on the left but has a different coefficient. Adding $\tfrac{36}{16} \int e^{6x} \cos 4x \, dx$ to both sides of the equation gives

$$\tfrac{52}{16} \int e^{6x} \cos 4x \, dx = \tfrac{1}{4} e^{6x} \sin 4x + \tfrac{6}{16} e^{6x} \cos 4x + C$$

You must display the $+ C$ on the right side of the equation because no indefinite integral remains there. Multiplying both sides by $\tfrac{16}{52}$ and simplifying gives

$$\int e^{6x} \cos 4x \, dx = \tfrac{1}{13} e^{6x} \sin 4x + \tfrac{3}{26} e^{6x} \cos 4x + C_1 \qquad C_1 = \tfrac{16}{52} C. \qquad ◀$$

If you have doubts about the validity of what has been done, you can check the answer by differentiation.

$$y = \tfrac{1}{13} e^{6x} \sin 4x + \tfrac{3}{26} e^{6x} \cos 4x + C_1$$
$$y' = \tfrac{6}{13} e^{6x} \sin 4x + \tfrac{4}{13} e^{6x} \cos 4x + \tfrac{9}{13} e^{6x} \cos 4x - \tfrac{6}{13} e^{6x} \sin 4x$$
$$y' = e^{6x} \cos 4x \qquad \text{The original integrand.}$$

Use Trigonometric Properties to Make the Original Integral Reappear

In Example 1, you found that the original integral reappeared on the right side of the equation. Sometimes you can use properties of the trigonometric functions to make this happen. Example 2 shows you how. Note the clever choice of u and dv!

▶ **EXAMPLE 2**   Evaluate the integral: $\int \sin^2 x \, dx$

Solution

$$\int \sin^2 x \, dx$$
$$= -\sin x \cos x + \int \cos^2 x \, dx$$
$$= -\sin x \cos x + \int (1 - \sin^2 x) \, dx \qquad \text{By the Pythagorean properties.}$$
$$= -\sin x \cos x + \int 1 \, dx - \int \sin^2 x \, dx \qquad \int \sin^2 x \, dx \text{ reappears!}$$
$$\therefore 2\int \sin^2 x \, dx = -\sin x \cos x + x + C \qquad \text{Do the indicated algebra.}$$
$$\int \sin^2 x \, dx = -\tfrac{1}{2} \sin x \cos x + \tfrac{1}{2}x + C_1$$

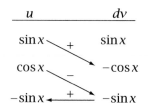

It would have been possible to make $\int \sin^2 x \, dx$ reappear by carrying the integration by parts one more step. Something unfortunate happens in this case, however, as shown here.

$$\int \sin^2 x \, dx$$

$$= -\sin x \cos x + \cos x \sin x + \int \sin^2 x \, dx$$

This time the original integral appears on the right side but with 1 as its coefficient. When you subtract $\int \sin^2 x \, dx$ from both sides and simplify, you end up with

$$0 = 0$$

which is true, but not very helpful! Stopping one step earlier, as in Example 2, avoids this difficulty.

Reassociate Factors Between Steps

The table form of repeated integration by parts relies on the fact that u and dv stay separate at each step. Sometimes it is necessary to reassociate some of the factors of u with dv, or vice versa, before taking the next step. You can still use the table format. Example 3 shows you how.

▶ **EXAMPLE 3**   Evaluate the integral: $\int x^3 e^{x^2} \, dx$

Solution   You can integrate the quantity $x \, e^{x^2} \, dx$ like this:

$$\int x \, e^{x^2} \, dx = \tfrac{1}{2}\int e^{x^2}(2x \, dx) = \tfrac{1}{2}e^{x^2} + C$$

The technique is to make $xe^{x^2}\,dx$ show up in the dv column.

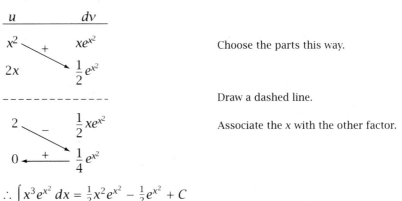

u	dv
x^2	xe^{x^2}
$2x$	$\frac{1}{2}e^{x^2}$

Choose the parts this way.

Draw a dashed line.

2	$\frac{1}{2}xe^{x^2}$
0	$\frac{1}{4}e^{x^2}$

Associate the x with the other factor.

$$\therefore \int x^3 e^{x^2}\,dx = \tfrac{1}{2}x^2 e^{x^2} - \tfrac{1}{2}e^{x^2} + C$$

The dashed line across the two columns indicates that the factors above the line have been reassociated to the form shown below the line. The arrows indicate which of the terms are actually multiplied to give the answer.

Integral of the Natural Logarithm Function

In Problem 11 of Problem Set 9-2, you were asked to integrate $\ln x$. To do this, you must be clever in selecting the parts. Example 4 shows you how.

▶ **EXAMPLE 4**   Evaluate the integrals: $\int \ln x\,dx$

Solution

$$\int \ln x\,dx$$
$$= x\ln x - \int \frac{1}{x}(x)\,dx$$
$$= x\ln x - x + C$$

u	dv
$\ln x$	1
$1/x$	x

Let $dv = 1\,dx$

You can integrate $(1/x)(x) = 1$ ◀

PROPERTY: Integral of the Natural Logarithm Function

$$\int \ln x\,dx = x\ln x - x + C$$

In Problem Set 9-3, you will practice rapid repeated integration by parts. You will justify that the $+C$ can be left out in the integration of dv. You will also practice integrating other familiar functions.

Problem Set 9-3

Quick Review   5 min

Q1. Integrate: $\int r^5\,dr$

Q2. Differentiate: $g(m) = m\sin 2m$

Q3. Integrate: $\int \sec^2 x\,dx$

Q4. Integrate: $\int (x^3 + 11)^5 (x^2\,dx)$

Q5. Integrate: $\int (x^3 + 11)\,dx$

Q6. Find: $\lim\limits_{x \to 0} \left(\dfrac{\sin x}{x} \right)$

Q7. Find: $\lim\limits_{x \to 0} \left(\dfrac{\sin x}{2x} \right)$

Q8. If region R (Figure 9-3b) is rotated about the x-axis, the volume of the solid is —?—.

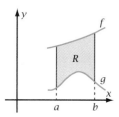

Figure 9-3b

Q9. Sketch the paraboloid formed by rotating $y = x^2$ from $x = 0$ to 2 about the y-axis.

Q10. Differentiate: $f(t) = te^t$

   A. e^t B. $te^t + e^t$ C. $e^t + t$
   D. $t^2e^t + t$ E. $e^t + 1$

For Problems 1–20, evaluate the integral.

1. $\int x^3 e^{2x}\, dx$

2. $\int x^5 e^{-x}\, dx$

3. $\int x^4 \sin x\, dx$

4. $\int x^2 \cos x\, dx$

5. $\int x^5 \cos 2x\, dx$

6. $\int x^3 \sin 5x\, dx$

7. $\int e^x \sin x\, dx$

8. $\int e^x \cos x\, dx$

9. $\int e^{3x} \cos 5x\, dx$

10. $\int e^{4x} \sin 2x\, dx$

11. $\int x^7 \ln 3x\, dx$

12. $\int x^5 \ln 6x\, dx$

13. $\int x^4 \ln 7\, dx$

14. $\int e^{7x} \cos 5\, dx$

15. $\int \sin^5 x \cos x\, dx$

16. $\int x\,(3 - x^2)^{2/3}\, dx$

17. $\int x^3 (x + 5)^{1/2}\, dx$

18. $\int x^2 \sqrt{2 - x}\, dx$

19. $\int \ln x^5\, dx$

20. $\int e^{\ln 7x}\, dx$

For Problems 21–32, evaluate the integral. You may reassociate factors between steps or use the trigonometric functions or logarithm properties in the integration by parts.

21. $\int x^5 e^{x^2}\, dx$

22. $\int x^5 e^{x^3}\, dx$

23. $\int x\,(\ln x)^3\, dx$

24. $\int x^3\,(\ln x)^2\, dx$

25. $\int x^3\,(x^2 + 1)^4\, dx$

26. $\int x^3 \sqrt{x^2 - 3}\, dx$

27. $\int \cos^2 x\, dx$

28. $\int \sin^2 0.4x\, dx$

29. $\int \sec^3 x\, dx$

30. $\int \sec^2 x \tan x\, dx$ (Be clever!)

31. $\int \log_3 x\, dx$

32. $\int \log_{10} x\, dx$

For Problems 33–38, write the antiderivative.

33. $\int \sin x\, dx$

34. $\int \cos x\, dx$

35. $\int \csc x\, dx$

36. $\int \sec x\, dx$

37. $\int \tan x\, dx$

38. $\int \cot x\, dx$

39. Wanda evaluates $\int x^2 \cos x \, dx$, letting $u = x^2$ and $dv = \cos x \, dx$. She gets

$$x^2 \sin x - \int 2x \sin x \, dx$$

For the second integral, she lets $u = \sin x$ and $dv = 2x \, dx$. Show Wanda why her second choice of u and dv is inappropriate.

40. Amos evaluates $\int x^2 \cos x \, dx$ by parts, letting $u = \cos x$ and $dv = x^2 \, dx$. Show Amos that although he can integrate his choice for dv, it is a mistake to choose the parts as he did.

41. If you evaluate $\int e^x \sin x \, dx$, the original integral reappears after two integrations by parts. It will also reappear after four integrations by parts. Show why it would be unproductive to evaluate the integral this way.

42. You can integrate the integral $\int \cos^2 x \, dx$ by clever use of trigonometric properties, as well as by parts. Substitute $\frac{1}{2}(1 + \cos 2x)$ for $\cos^2 x$ and integrate. Compare this answer with what you obtain using integration by parts, and show that the two answers are equivalent.

43. *Area Problem:* Sketch the graph of $y = xe^{-x}$ from $x = 0$ to $x = 3$. Where does the function have its high point in the interval $[0, 3]$? Use the fundamental theorem to find algebraically the area of the region under the graph from $x = 0$ to $x = 3$.

44. *Unbounded Region Area Problem:* Figure 9-3c shows the region under the graph of $y = 12x^2 e^{-x}$ from $x = 0$ to $x = b$. Find an equation for this area in terms of b. Then find the limit of the area as b approaches infinity. Does the area approach a finite number, or does it increase without bound as b increases? Justify your answer.

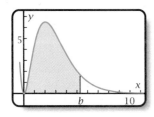

Figure 9-3c

45. *Volume Problem:* The region under the graph of $y = \ln x$ from $x = 1$ to $x = 5$ is rotated about the

x-axis to form a solid. Use the fundamental theorem to find its volume exactly.

46. *Proof Problem:* In setting up an integration by parts problem, you select dv to equal something useful and integrate to find v. Prove that whatever number you pick for the constant of integration at this point will cancel out later in the integration by parts process.

47. *Areas and Integration by Parts:* Figure 9-3d shows the graph of function v plotted against function u. As u goes from a to b, v goes from c to d. Show that the integration by parts formula can be interpreted in terms of areas on this diagram. (This diagram is the same as the icon at the top of each even-numbered page in this chapter.)

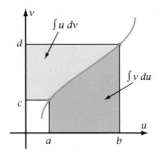

Figure 9-3d

48. *Integral of ln Generalization Problem:* Derive a formula for $\int \ln ax \, dx$, where a stands for a nonzero constant.

49. *Introduction to Reduction Formulas Problem:* For $\int \sin^7 x \, dx$, integrate once by parts. Use the Pythagorean properties in an appropriate manner to write the remaining integral as two integrals, one involving $\sin^5 x$ and the other involving $\sin^7 x$. Then use algebra to combine the two integrals involving $\sin^7 x$, and thus express $\int \sin^7 x \, dx$ in terms of $\int \sin^5 x \, dx$. Use the resulting pattern repeatedly to finish evaluating $\int \sin^7 x \, dx$. The pattern leads to a reduction formula, as you will learn in the next section.

50. *Journal Problem:* Update your journal with techniques and concepts you've learned since the last entry. In particular, describe integration by parts, the kind of integral it is used for, and the rapid way in which you can accomplish it.

9-4 Reduction Formulas and Computer Algebra Systems

In this section you will develop formulas that allow you to algebraically integrate powers of the trigonometric functions. You can use repeated integration by parts for integrals such as

$$\int \sin^6 x \, dx$$

The choice for dv is $\sin x \, dx$ because you can integrate it and the result is not more complicated. But the other part, $u = \sin^5 x$, gets more complex when you differentiate it.

u	dv
$\sin^5 x$	$\sin x$
$5 \sin^4 x \cos x$	$-\cos x$

So you simply put the integral back together and hope for the best!

$$\int \sin^6 x \, dx$$
$$= -\sin^5 x \cos x + 5 \int \sin^4 x \cos^2 x \, dx$$
$$= -\sin^5 x \cos x + 5 \int \sin^4 x \, (1 - \sin^2 x) \, dx \qquad \text{Use the Pythagorean property to return to sines.}$$
$$= -\sin^5 x \cos x + 5 \int \sin^4 x \, dx - 5 \int \sin^6 x \, dx$$

The original integral appears on the right side with a coefficient of -5. Adding $5 \int \sin^6 x \, dx$ to the first and last members of the above equation (and using the transitive property to ignore the two members in the middle) gives

$$6 \int \sin^6 x \, dx = -\sin^5 x \cos x + 5 \int \sin^4 x \, dx$$

Dividing by 6 gives

$$\int \sin^6 x \, dx = -\tfrac{1}{6} \sin^5 x \cos x + \tfrac{5}{6} \int \sin^4 x \, dx$$

The integral $\int \sin^6 x \, dx$ has been replaced by an expression in terms of $\int \sin^4 x \, dx$. The new integral has the same form as the original but is "reduced" in complexity. By repeating this integration where n is the exponent instead of 6, you can find an equation expressing the integral with any nonzero exponent in terms of an integral with that exponent reduced by 2.

$$\int \sin^n x \, dx = -\frac{1}{n} \sin^{n-1} x \cos x + \frac{n-1}{n} \int \sin^{n-2} x \, dx$$

An equation such as this one is called a **reduction formula**. You can use it again on the new integral. For the previous example, the work would look like this. (Note that you can use ditto marks, as shown, to avoid rewriting the first terms so many times.)

$$\int \sin^6 x \, dx$$

$$= -\frac{1}{6} \sin^5 x \cos x + \frac{5}{6} \int \sin^4 x \, dx$$

$$= {}'' + \frac{5}{6} \left(-\frac{1}{4} \sin^3 x \cos x + \frac{3}{4} \int \sin^2 x \, dx \right)$$

$$= {}'' - \frac{5}{24} \sin^3 x \cos x + \frac{5}{8} \int \sin^2 x \, dx$$

$$= {}'' - {}'' + \frac{5}{8} \left(-\frac{1}{2} \sin^1 x \cos x + \frac{1}{2} \int \sin^0 x \, dx \right)$$

$$= {}'' - {}'' - \frac{5}{16} \sin x \cos x + \frac{5}{16} \int dx$$

$$= -\frac{1}{6} \sin^5 x \cos x - \frac{5}{24} \sin^3 x \cos x - \frac{5}{16} \sin x \cos x + \frac{5}{16} x + C$$

DEFINITION: Reduction Formula

A reduction formula is an equation expressing an integral of a particular kind in terms of an integral of exactly the same type, but of reduced complexity.

OBJECTIVE

Given an integral involving powers of functions, derive a reduction formula, and given a reduction formula, use it to evaluate an indefinite integral either by pencil and paper or by computer algebra systems (CAS).

The formulas for integration of the six trigonometric functions (recall from Section 6-6) are repeated here to refresh your memory.

PROPERTIES: Integrals of the Trigonometric Functions

$$\int \sin x \, dx = -\cos x + C \qquad\qquad \int \cos x \, dx = \sin x + C$$

$$\int \tan x \, dx = \ln|\sec x| + C \qquad\qquad \int \cot x \, dx = -\ln|\csc x| + C$$
$$\qquad\qquad = -\ln|\cos x| + C \qquad\qquad\qquad\quad = \ln|\sin x| + C$$

$$\int \sec x \, dx = \ln|\sec x + \tan x| + C \qquad \int \csc x \, dx = -\ln|\csc x + \cot x| + C$$
$$\qquad\qquad = -\ln|\sec x - \tan x| + C \qquad\qquad\quad = \ln|\csc x - \cot x| + C$$

The work at the beginning of this section shows that you can derive a reduction formula by choosing a particular case, integrating, and looking for a pattern in the answer. However, to be perfectly sure the result is correct, start with a general integral of the type in question, using a letter such as n to stand for the (constant) exponent. Example 1 shows you how to derive the integral of $\sec^n x \, dx$.

▶ **EXAMPLE 1**
a. Derive a reduction formula for $\int \sec^n x \, dx$ that can be used if $n \geq 2$.

b. Use the reduction formula to evaluate $\int \sec^5 x \, dx$.

c. Check your answer to part b by CAS.

Solution
a. The best choice for dv is $\sec^2 x \, dx$, because you can integrate it easily and its integral, $\tan x$, is simpler than the original integrand. The work looks like this.

u		dv
$\sec^{n-2} x$	$\searrow +$	$\sec^2 x$
$(n-2)\sec^{n-3} x \sec x \tan x$	$\xleftarrow{\quad - \quad}$	$\tan x$

$$\int \sec^n x \, dx$$

$$= \sec^{n-2} x \tan x - (n-2)\int \sec^{n-2} x \tan^2 x \, dx$$

$$= \sec^{n-2} x \tan x - (n-2)\int \sec^{n-2} x (\sec^2 x - 1) \, dx$$

$$= \sec^{n-2} x \tan x - (n-2)\int (\sec^n x - \sec^{n-2} x) \, dx$$

$$= \sec^{n-2} x \tan x - (n-2)\int \sec^n x \, dx + (n-2)\int \sec^{n-2} x \, dx$$

The desired integral now appears in the second term of the equation, with $-(n-2)$ as its coefficient. Adding $(n-2)\int \sec^n x \, dx$ to the first and last members (and eliminating the middle members by transitivity) gives

$$(n-1)\int \sec^n x \, dx = \sec^{n-2} x \tan x + (n-2)\int \sec^{n-2} x \, dx$$

Dividing both members by $(n-1)$ produces the desired reduction formula.

$$\int \sec^n x \, dx = \frac{1}{n-1} \sec^{n-2} x \tan x + \frac{n-2}{n-1}\int \sec^{n-2} x \, dx$$

b.
$$\int \sec^5 x \, dx$$

$$= \tfrac{1}{4} \sec^3 x \tan x + \tfrac{3}{4}\int \sec^3 x \, dx$$

$$= \tfrac{1}{4} \sec^3 x \tan x + \tfrac{3}{4}\left(\tfrac{1}{2}\sec x \tan x + \tfrac{1}{2}\int \sec x \, dx\right)$$

$$= \tfrac{1}{4} \sec^3 x \tan x + \tfrac{3}{8}\sec x \tan x + \tfrac{3}{8}\int \sec x \, dx$$

$$= \tfrac{1}{4} \sec^3 x \tan x + \tfrac{3}{8}\sec x \tan x + \tfrac{3}{8}\ln|\sec x + \tan x| + C$$

c. A symbol-manipulating CAS may present the answer in a slightly different form. For instance, a popular handheld device presents the answer this way.

$$\frac{3 \cdot (\cos(x))^4 \cdot \ln(|\cos(x)|) - 3 \cdot (\cos(x))^4 \cdot \ln(|\sin(x)-1|) + \sin(x) \cdot [3 \cdot (\cos(x))^2 + 2]}{8 \cdot (\cos(x))^4}$$

◀

Sometimes you can derive a reduction formula without integrating by parts, as Example 2 shows.

▶ **EXAMPLE 2**   Derive a reduction formula for $\int \cot^n x \, dx$ that you can use if $n \geq 2$.

Solution

$$\int \cot^n x \, dx$$

$$= \int \cot^{n-2} x \, (\cot^2 x \, dx)$$

$$= \int \cot^{n-2} x \, (\csc^2 x - 1) \, dx \qquad \text{Use the Pythagorean properties to transform to cosecant.}$$

$$= \int \cot^{n-2} x \csc^2 x \, dx - \int \cot^{n-2} x \, dx$$

$$= -\frac{1}{n-1} \cot^{n-1} x - \int \cot^{n-2} x \, dx \qquad \text{Integral of a power function.}$$

The last expression has the integral of cotangent with an exponent 2 less than the original integral and is thus a reduction formula. ◀

▶ **EXAMPLE 3**   Use the reduction formula in Example 2 to evaluate $\int \cot^5 x \, dx$.

Solution

$$\int \cot^5 x \, dx$$

$$= -\tfrac{1}{4} \cot^4 x - \int \cot^3 x \, dx$$

$$= -\tfrac{1}{4} \cot^4 x - \left(-\tfrac{1}{2} \cot^2 x - \int \cot x \, dx \right)$$

$$= -\tfrac{1}{4} \cot^4 x + \tfrac{1}{2} \cot^2 x + \ln |\sin x| + C$$ ◀

You will derive these reduction formulas in Problem Set 9-4.

REDUCTION FORMULAS: Integrals of Powers of Trigonometric Functions

(For reference only. Do not try to memorize these!)

$$\int \sin^n x \, dx = -\frac{1}{n} \sin^{n-1} x \cos x + \frac{n-1}{n} \int \sin^{n-2} x \, dx, \text{ for } n \geq 2$$

$$\int \cos^n x \, dx = \frac{1}{n} \cos^{n-1} x \sin x + \frac{n-1}{n} \int \cos^{n-2} x \, dx, \text{ for } n \geq 2$$

$$\int \tan^n x \, dx = \frac{1}{n-1} \tan^{n-1} x - \int \tan^{n-2} x \, dx, \text{ for } n \geq 2$$

$$\int \cot^n x \, dx = -\frac{1}{n-1} \cot^{n-1} x - \int \cot^{n-2} x \, dx, \text{ for } n \geq 2$$

$$\int \sec^n x \, dx = \frac{1}{n-1} \sec^{n-2} x \tan x + \frac{n-2}{n-1} \int \sec^{n-2} x \, dx, \text{ for } n \geq 2$$

$$\int \csc^n x \, dx = -\frac{1}{n-1} \csc^{n-2} x \cot x + \frac{n-2}{n-1} \int \csc^{n-2} x \, dx, \text{ for } n \geq 2$$

Don't try to memorize the reduction formulas! You are almost bound to make a mistake, and there is nothing quite as useless as a wrong formula. Fortunately you will not have to make a career out of evaluating reduction formulas. Your purpose here is to see how the computer comes up with an answer, as in Example 1 when it integrates a power of a trigonometric function.

Problem Set 9-4

Quick Review

Q1. In integration by parts, $\int u\, dv = -?-$.

Q2. Sketch: $y = 3 \cos x$

Q3. Sketch: $y = \cos 2x$

Q4. Differentiate: $y = x \ln 5x$

Q5. Integrate: $\int \sin^5 x \cos x\, dx$

Q6. Integrate: $\int dx/x$

Q7. Sketch the graph of a function that is continuous at point $(3, 1)$, but not differentiable there.

Q8. Find: $(d/dx)(\tan^{-1} x)$

Q9. Integrate: $\int \sec x\, dx$

Q10. Integrate: $\int_1^3 e^{2x}\, dx$

A. $2e^4$ B. $\dfrac{1}{2}e^4$ C. $e^6 - e^2$

D. $\dfrac{1}{2}(e^6 - e^2)$ E. e^4

For Problems 1–6, take the first step in integration by parts or use appropriate trigonometry to write the given integral in terms of an integral with a reduced power of the same function.

1. $\int \sin^9 x\, dx$

2. $\int \cos^{10} x\, dx$

3. $\int \cot^{12} x\, dx$

4. $\int \tan^{20} x\, dx$

5. $\int \sec^{13} x\, dx$

6. $\int \csc^{100} x\, dx$

For Problems 7–12, derive the reduction formula in the box on page 447 using n as the exponent rather than using a particular constant, as in Problems 1–6.

7. $\int \cos^n x\, dx, \quad n \neq 0$

8. $\int \sin^n x\, dx, \quad n \neq 0$

9. $\int \tan^n x\, dx, \quad n \neq 0$

10. $\int \cot^n x\, dx, \quad n \neq 0$

11. $\int \csc^n x\, dx, \quad n \neq 0$

12. $\int \sec^n x\, dx, \quad n \neq 0$

For Problems 13–18, integrate by

    a. Using pencil and paper and the appropriate reduction formula.

    b. Using computer algebra systems.

13. $\int \sin^5 x\, dx$

14. $\int \cos^5 x\, dx$

15. $\int \cot^6 x\, dx$

16. $\int \tan^7 x\, dx$

17. $\int \sec^4 x\, dx$

18. $\int \csc^4 x\, dx$

19. *Cosine Area Problem:* Figure 9-4a shows the graphs of

$$y = \cos x$$
$$y = \cos^3 x$$
$$y = \cos^5 x$$

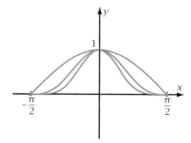

Figure 9-4a

    a. Which graph goes with which function?

    b. By numerical integration, find the approximate area of the region under each graph from $x = -\pi/2$ to $\pi/2$.

c. Find exactly each area in part b by the fundamental theorem. Use the reduction formulas.

d. Based on the graphs, explain why the areas you calculated are reasonable.

e. Plot the graph of $y = \cos^{100} x$. Sketch the result.

f. As the exponent n gets larger, the graph of $y = \cos^n x$ gets "narrower." Does the limit of the area as n approaches infinity seem to be zero? Explain.

20. *Integral of $\cos^5 x$ Another Way:* You can write the integral $\int \cos^5 x \, dx$

$$\int \cos^5 x \, dx = \int (\cos^4 x) \cos x \, dx$$

You can convert the factor $\cos^4 x$ to powers of sine by appropriate use of the Pythagorean properties of trigonometric functions. The result will be three integrals that you will be able to evaluate by the fundamental theorem. Find the area of the region under the graph of $y = \cos^5 x$ from $x = -\pi/2$ to $\pi/2$ using this technique for integration. Compare your answer with that in Problem 19, part c.

21. *Integral of Secant Cubed Problem:* The integral $\int \sec^3 x \, dx$ often appears in applied mathematics, as you will see in the next few sections. Use an appropriate technique to evaluate the integral. Then see if you can figure out a way to remember the answer.

22. *Reduction Formula for $\int \sin^n ax \, dx$:* Derive a reduction formula for

$$\int \sin^n ax \, dx, \ n \geq 2, \ a \neq 0$$

Then use the reduction formula to evaluate $\int \sin^5 3x \, dx$.

23. Prove:
$$\int \sin^3 ax \, dx = -\frac{1}{3a}(\cos ax)(\sin^2 ax + 2) + C$$
$(a \neq 0)$

24. Prove:
$$\int \cos^3 ax \, dx = \frac{1}{3a}(\sin ax)(\cos^2 ax + 2) + C$$
$(a \neq 0)$

9-5 Integrating Special Powers of Trigonometric Functions

In Section 9-4, you found algebraically the indefinite integrals of any positive integer power of any trigonometric function. In this section you will learn how to integrate some special powers of these functions without having to resort to reduction formulas.

OBJECTIVE Integrate odd powers of sine or cosine, even powers of secant or cosecant, and squares of sine or cosine without having to use the reduction formulas.

Odd Powers of Sine and Cosine

▶ *EXAMPLE 1* Evaluate the integral: $\int \sin^7 x \, dx$

Solution   The key to the technique is associating one $\sin x$ factor with dx, then transforming the remaining (even) number of sines into cosines using the Pythagorean properties.

$$\int \sin^7 x \, dx$$

$$= \int \sin^6 x \, (\sin x \, dx)$$

$$= \int (\sin^2 x)^3 (\sin x \, dx)$$

$$= \int (1 - \cos^2 x)^3 (\sin x \, dx)$$

$$= \int (1 - 3\cos^2 x + 3\cos^4 x - \cos^6 x)(\sin x \, dx)$$

$$= \int \sin x \, dx - 3\int \cos^2 x \sin x \, dx + 3\int \cos^4 x \sin x \, dx - \int \cos^6 x \sin x \, dx$$

$$= -\cos x + \cos^3 x - \tfrac{3}{5}\cos^5 x + \tfrac{1}{7}\cos^7 x + C \qquad \blacktriangleleft$$

Each integral in the next-to-last line of Example 1 has the form of the integral of a power, $\int u^n \, du$. So you end up integrating power functions rather than trigonometric functions. This technique will work for odd powers of sine or cosine because associating one of the factors with dx leaves an even number of sines or cosines to be transformed into the cofunction.

Squares of Sine and Cosine

In Section 8-7, you found the area inside the limaçon $r = 5 + 4\cos\theta$ (Figure 9-5a) is

$$A = \tfrac{1}{2}\int_0^{2\pi}(5 + 4\cos\theta)^2 \, d\theta$$

Expanding the binomial power gives

$$A = \tfrac{1}{2}\int_0^{2\pi}(25 + 40\cos\theta + 16\cos^2\theta) \, d\theta$$

The last term in the integral has $\cos^2\theta$. The integrals $\int \sin^2 x \, dx$ and $\int \cos^2 x \, dx$ occur frequently enough to make it worthwhile to learn an algebraic shortcut. The double-argument property for cosine is

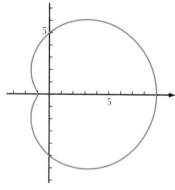

Figure 9-5a

$$\cos 2x = \cos^2 x - \sin^2 x$$

You can also write the right side either entirely in terms of cosine or entirely in terms of sine.

$$\cos 2x = 2\cos^2 x - 1$$
$$\cos 2x = 1 - 2\sin^2 x$$

Performing algebra on these two equations gives the property at the top of the next page.

These two equations allow you to transform $\int \sin^2 x \, dx$ and $\int \cos^2 x \, dx$ so that the integrand is linear in $\cos 2x$.

$$\int \cos^2 x \, dx = \tfrac{1}{2}\int (1 + \cos 2x) \, dx = \tfrac{1}{2}x + \tfrac{1}{4}\sin 2x + C$$

$$\int \sin^2 x \, dx = \tfrac{1}{2}\int (1 - \cos 2x) \, dx = \tfrac{1}{2}x - \tfrac{1}{4}\sin 2x + C$$

▶ **EXAMPLE 2** Evaluate the integral: $\int \sin^2 8x \, dx$

Solution

$$\int \sin^2 8x \, dx$$

$$= \tfrac{1}{2}\int (1 - \cos 16x) \, dx$$

$$= \tfrac{1}{2}x - \tfrac{1}{32}\sin 16x + C \qquad \blacktriangleleft$$

Even Powers of Secant and Cosecant

You can adapt the technique for integrating odd powers of sine and cosine to even powers of secant and cosecant. Example 3 shows how.

▶ **EXAMPLE 3** Evaluate the integral: $\int \sec^8 5x \, dx$

Solution

$$\int \sec^8 5x \, dx$$

$$= \int \sec^6 5x \,(\sec^2 5x \, dx)$$

$$= \int (\sec^2 5x)^3 (\sec^2 5x \, dx)$$

$$= \int (\tan^2 5x + 1)^3 (\sec^2 5x \, dx)$$

$$= \int (\tan^6 5x + 3\tan^4 5x + 3\tan^2 5x + 1)(\sec^2 5x \, dx)$$

$$= \int \tan^6 5x \,\sec^2 5x \, dx + 3\int \tan^4 5x \,\sec^2 5x \, dx + 3\int \tan^2 5x \,\sec^2 5x \, dx$$

$$\quad + \int \sec^2 5x \, dx$$

$$= \tfrac{1}{35}\tan^7 5x + \tfrac{3}{25}\tan^5 5x + \tfrac{1}{5}\tan^3 5x + \tfrac{1}{5}\tan 5x + C \qquad \blacktriangleleft$$

The advantage of the techniques in Examples 2 and 3 is that you don't have to remember the reduction formulas. The disadvantages are that you must remember certain trigonometric properties, the binomial formula for expanding powers of binomials, and which powers of which functions you can integrate this way. Problem Set 9-5 gives you some opportunities for practice. You will also find the exact area of the limaçon in Figure 9-5a.

Problem Set 9-5

Quick Review *5 min*

Q1. Find $f'(1)$: $f(x) = x^3 - 7x$

Q2. Find $g'(2)$: $g(x) = \ln x$

Q3. Find $h'(3)$: $h(x) = (2x - 7)^6$

Q4. Find $t'(4)$: $t(x) = \sin \frac{\pi}{12}x$

Q5. Find $p'(5)$: $p(x) = xe^x$

Q6. Solve: $x^{1/3} = 8$

Q7. Sketch the graph: $(x/3)^2 + (y/5)^2 = 1$

Q8. $\int u\, dv = uv - \int v\, du$ is the —?— formula.

Q9. For Figure 9-5b, sketch the (continuous) antiderivative, $f(x)$, that contains point (2, 1).

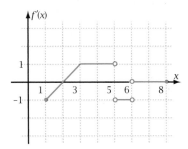

Figure 9-5b

Q10. $\int [f(x)]^n\, dx = g(x) + \int [f(x)]^{n-1}\, dx$ is called a(n) —?— formula.

A. Power expansion

B. Double argument

C. Integration by parts

D. Power integration

E. Reduction

For Problems 1–30, evaluate the integral.

1. $\int \sin^5 x\, dx$

2. $\int \cos^7 x\, dx$

3. $\int \cos^7 9x\, dx$

4. $\int \sin^3 10x\, dx$

5. $\int \sin^4 3x \cos 3x\, dx$

6. $\int \cos^8 7x \sin 7x\, dx$

7. $\int \cos^6 8x \sin^3 8x\, dx$

8. $\int \sin^4 2x \cos^3 2x\, dx$

9. $\int \sin^5 x \cos^2 x\, dx$

10. $\int \cos^3 x \sin^2 x\, dx$

11. $\int \cos^2 x\, dx$

12. $\int \sin^2 x\, dx$

13. $\int \sin^2 5x\, dx$

14. $\int \cos^2 6x\, dx$

15. $\int \sec^4 x\, dx$

16. $\int \csc^6 x\, dx$

17. $\int \csc^8 6x\, dx$

18. $\int \sec^4 100x\, dx$

19. $\int \tan^{10} x \sec^2 x\, dx$

20. $\int \cot^8 x \csc^2 x\, dx$

21. $\int \sec^{10} x \tan x\, dx$

22. $\int \csc^8 x \cot x\, dx$

23. $\int \sec^{10} 20\, dx$

24. $\int \csc^8 12\, dx$

25. $\int (\cos^2 x - \sin^2 x)\, dx$

26. $\int (\cos^2 x + \sin^2 x)\, dx$

27. $\int (\sin x)^{-2}\, dx$

28. $\int (\cos 3x)^{-2}\, dx$

29. $\int \sec^3 x\, dx$

30. $\int \csc^3 x\, dx$

31. *Area Problem I:* Figure 9-5c shows the region bounded by the graph of

$$y = \cos 5x \sin 3x$$

from $x = 0$ to $x = 2\pi$.

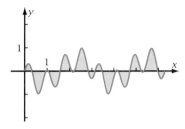

Figure 9-5c

a. Using integration by parts, find the indefinite integral $\int \cos 5x \sin 3x \, dx$.

b. Show that the region has just as much area below the x-axis as it has above.

32. *Area Problem II:* Let $f(x) = \sin^3 x$.

a. Plot the graph from $x = 0$ to $x = \pi$. Sketch the result.

b. Find the exact area of the region under the graph of f from $x = 0$ to $x = \pi$, using the fundamental theorem and the integration techniques of this section.

c. Verify your answer to part b by integrating numerically.

d. Quick! Find the integral of $f(x)$ from $x = -\pi$ to $x = \pi$. State what property allows you to answer this question so quickly.

33. *Volume Problem I:* One arch of the graph of $y = \sin x$ is rotated about the x-axis to form a football-shaped solid (Figure 9-5d). Find its volume. Use the fundamental theorem.

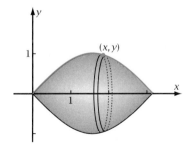

Figure 9-5d

34. *Volume Problem II:* The region under the graph of $y = \sec^2 x$ from $x = 0$ to $x = 1$ is rotated about the line $y = -3$ to form a solid.

a. Find the exact volume of the solid.

b. Find the exact volume of the solid if the region is rotated about the line $x = -3$.

35. *Limaçon Area Problem:* Figure 9-5e shows the limaçon

$$r = 5 + 4 \cos \theta$$

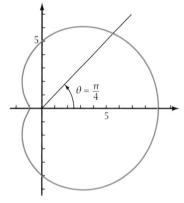

Figure 9-5e

which appears at the beginning of this section. By numerical integration the area of the region from $\theta = 0$ to $\theta = \pi/4$ is 29.101205... square units. Find this area exactly, using the fundamental theorem. Write the answer using radicals and π, if necessary. Show that this exact answer, when evaluated, gives the same answer as obtained numerically.

36. *Cardioid Area Problem:* Figure 9-5f shows the general cardioid

$$r = a(1 + \cos \theta)$$

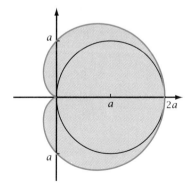

Figure 9-5f

where a is a constant. A circle of radius a is inscribed in the cardioid. By the fundamental theorem, find the exact area inside the cardioid in terms of a. Explain how this area corresponds to the area of the circle.

37. *Journal Problem:* Update your journal with what you've learned. You should include such things as

- The one most important thing you've learned since your last journal entry

- The basis behind integration by parts, and what kind of function you can integrate that way

- The reason you might want to go to the trouble of using the fundamental theorem to integrate, rather that simply using a Riemann sum or the trapezoidal rule

- Any techniques or ideas about derivatives that are still unclear to you

9-6 Integration by Trigonometric Substitution

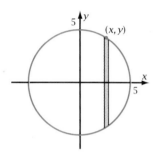

Figure 9-6a

Recall from geometry that the area of a circle is $A = \pi r^2$, where r is the radius. You can derive this formula by calculus if you know how to integrate certain square root functions.

The circle in Figure 9-6a has equation

$$x^2 + y^2 = 25$$

Draw a vertical strip. Pick a sample point in the strip on the upper half of the circle. The area of any one strip is $2y\,dx$. The upper half of the circle has equation

$$y = \sqrt{25 - x^2}$$

So the area of the entire circle is

$$A = 2\int_{-5}^{5} \sqrt{25 - x^2}\,dx$$

You cannot integrate the indefinite integral $\int \sqrt{25 - x^2}\,dx$ as a power because dx is not the differential of the inside function $25 - x^2$. But there is an algebraic way to evaluate the integral.

Your thought process might go something like this:

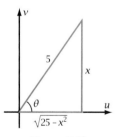

Figure 9-6b

- Hmm ... This looks Pythagorean—like somebody was trying to find the third side of a right triangle.

- Draw a right triangle as in Figure 9-6b, placing the angle θ in standard position in a uv-coordinate system.

- The radicand is $25 - x^2$. So the hypotenuse must be 5 and one leg must be x.

- Label the vertical leg x (to avoid a minus sign in dx later) and the other leg $\sqrt{25 - x^2}$.

$$\therefore \frac{x}{5} = \sin\theta \Rightarrow x = 5\sin\theta \Rightarrow dx = 5\cos\theta\, d\theta \qquad \text{Use trigonometric ratios.}$$

$$\text{and } \frac{\sqrt{25-x^2}}{5} = \cos\theta \Rightarrow \sqrt{25-x^2} = 5\cos\theta \qquad \text{Use trigonometric ratios.}$$

$$\therefore \int \sqrt{25-x^2}\, dx$$

$$= \int (5\cos\theta)(5\cos\theta\, d\theta) \qquad \text{Use trigonometric substitution.}$$

$$= 25\int \cos^2\theta\, d\theta$$

$$= \frac{25}{2}\int (1+\cos 2\theta)\, d\theta \qquad \text{Double-argument properties.}$$

$$= \frac{25}{2}\theta + \frac{25}{4}\sin 2\theta + C \qquad \text{Evaluate the integral.}$$

$$= \frac{25}{2}\theta + \frac{25}{2}\sin\theta\cos\theta + C \qquad \text{Double-argument properties.}$$

$$= \frac{25}{2}\sin^{-1}\frac{x}{5} + \frac{25}{2}\cdot\frac{x}{5}\cdot\frac{\sqrt{25-x^2}}{5} + C \qquad \text{Use reverse substitution. See triangle in Figure 9-6b.}$$

$$= \frac{25}{2}\sin^{-1}\frac{x}{5} + \frac{x}{2}\sqrt{25-x^2} + C \qquad \text{Answer.}$$

Multiplying by 2 from the original integral and evaluating from $x = -5$ to $x = 5$ gives

$$A = 2\int_{-5}^{5}\sqrt{25-x^2}\, dx = 25\sin^{-1}\frac{x}{5} + x\sqrt{25-x^2}\,\Big|_{-5}^{5}$$

$$= 25\sin^{-1}1 + 5\sqrt{25-25} - 25\sin^{-1}(-1) - 5\sqrt{25-25}$$

$$= 25\cdot\frac{\pi}{2} + 0 - 25\cdot\frac{-\pi}{2} - 0 = 25\pi$$

which agrees with the answer from the area formula, $A = \pi \cdot 5^2$.

The substitution rationalizes the integrand by taking advantage of the Pythagorean properties of trigonometric functions. The technique works for square roots of quadratics. Your success in using this **trigonometric substitution** depends on drawing the triangle, then deciding what to call the legs and hypotenuse. After the substitution has been made, you must also be able to integrate the trigonometric functions that appear.

OBJECTIVE

Rationalize an integrand containing the square root of a quadratic binomial by using trigonometric substitution, then perform the integration.

In the preceding work, the radical contained $(\text{constant})^2 - x^2$. The situation is different if the radical contains $x^2 - (\text{constant})^2$. Example 1 shows you what happens.

▶ **EXAMPLE 1**   Evaluate the integral: $\displaystyle\int \frac{dx}{\sqrt{x^2 - 9}}$

Solution   Draw the triangle. This time x is the hypotenuse and 3 is a leg. Putting 3 on the horizontal leg allows you to use secant instead of cosecant (Figure 9-6c), thus avoiding a minus sign when you find dx.

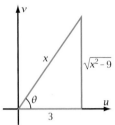

Figure 9-6c

$$\frac{x}{3} = \sec \theta \Rightarrow x = 3 \sec \theta \Rightarrow dx = 3 \sec \theta \tan \theta \, d\theta$$

and $\displaystyle\frac{\sqrt{x^2 - 9}}{3} = \tan \theta \Rightarrow \sqrt{x^2 - 9} = 3 \tan \theta$

$$\therefore \int \frac{dx}{\sqrt{x^2 - 9}}$$

$$= \int \frac{3 \sec \theta \tan \theta \, d\theta}{3 \tan \theta} \qquad\qquad \text{Substitute.}$$

$$= \int \sec \theta \, d\theta$$

$$= \ln |\sec \theta + \tan \theta| + C$$

$$= \ln \left| \frac{x}{3} + \frac{\sqrt{x^2 - 9}}{3} \right| + C \qquad\qquad \text{Use reverse substitution.}$$

$$= \ln |x + \sqrt{x^2 - 9}| - \ln 3 + C \qquad\qquad \text{Log of a quotient.}$$

$$= \ln |x + \sqrt{x^2 - 9}| + C_1 \qquad\qquad \ln 3 \text{ is constant.} \qquad ◀$$

Example 2 shows what to do if the sign in the quadratic is a plus instead of a minus. It also shows that you can use trigonometric substitution even if there is no radical and if the constant is not a square.

▶ **EXAMPLE 2**   Evaluate the integral: $\displaystyle\int \frac{dx}{x^2 + 37}$

Solution   Draw a triangle (Figure 9-6d). The sign between terms is a plus sign, so both x and $\sqrt{37}$ are legs. By putting x on the vertical leg, you can use $\tan \theta$ instead of $\cot \theta$. The hypotenuse is $\sqrt{x^2 + 37}$.

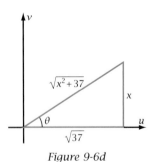

Figure 9-6d

$$\frac{x}{\sqrt{37}} = \tan \theta \Rightarrow x = \sqrt{37} \tan \theta \Rightarrow dx = \sqrt{37} \sec^2 \theta \, d\theta$$

and $\displaystyle\frac{\sqrt{x^2 + 37}}{\sqrt{37}} = \sec \theta \Rightarrow \sqrt{x^2 + 37} = \sqrt{37} \sec \theta$

$$\therefore \int \frac{dx}{x^2 + 37} = \int \frac{\sqrt{37} \sec^2 \theta \, d\theta}{37 \sec^2 \theta}$$

$$= \frac{1}{\sqrt{37}} \int d\theta = \frac{1}{\sqrt{37}} \theta + C$$

$$= \frac{1}{\sqrt{37}} \tan^{-1} \frac{x}{\sqrt{37}} + C \qquad \text{Because } x/\sqrt{37} = \tan \theta, \ \theta = \tan^{-1}(x/\sqrt{37}). \qquad ◀$$

Definite Integrals by Trigonometric Substitution

The reverse substitution in Examples 1 and 2 is done to revert to x, as in the original integral. For a definite integral, you can avoid the reverse substitution if you change the limits of integration to θ instead of x. Example 3 shows you this can be done with the integral for the area of the circle at the beginning of this section.

▶ **EXAMPLE 3**

Find the area of the zone of the circle

$$x^2 + y^2 = 25$$

between $x = -2$ and $x = 3$ (Figure 9-6e).

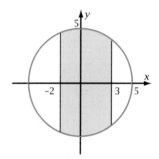

Figure 9-6e

Solution

This problem is similar to the problem at the beginning of this section except that the limits of integration are different. The area is

$$A = 2\int_{-2}^{3} \sqrt{25 - x^2}\, dx$$

Let $x = 5\sin\theta$. Then $dx = 5\cos\theta\, d\theta$, $\sqrt{25 - x^2} = 5\cos\theta$, and $\theta = \sin^{-1}\frac{x}{5}$.

If $x = 3$, then $\theta = \sin^{-1}\frac{3}{5} = \sin^{-1} 0.6$. If $x = -2$, then $\theta = \sin^{-1}\frac{-2}{5} = \sin^{-1}(-0.4)$.

$$\therefore A = 2\int_{\sin^{-1}(-0.4)}^{\sin^{-1} 0.6} (5\cos\theta)(5\cos\theta\, d\theta) \qquad \text{Substitute } \theta \text{ limits.}$$

$$= 25\,\theta + \tfrac{25}{2}\sin 2\theta \,\Big|_{\sin^{-1}(-0.4)}^{\sin^{-1} 0.6} \qquad \text{See the problem at the beginning of this section.}$$

$$= 25\sin^{-1} 0.6 + \tfrac{25}{2}\sin(2\sin^{-1} 0.6) - 25\sin^{-1}(-0.4) - \tfrac{25}{2}\sin[2\sin^{-1}(-0.4)]$$

$$= 47.5406002\ldots$$

By numerical integration, $A \approx 47.5406\ldots$, which is close to $47.5406002\ldots$. ◀

In Problem 35 of Problem Set 9-6, you will see what happens in trigonometric substitution if x is negative, and thus θ is not in Quadrant I.

Problem Set 9-6

Quick Review

Q1. Integrate: $\int \cos 3x\, dx$

Q2. Integrate: $\int \sin 4x\, dx$

Q3. Integrate: $\int \tan 5x\, dx$

Q4. Integrate: $\int \cot 6x\, dx$

Q5. Integrate: $\int \sec 7x\, dx$

Q6. Find: $(d/dx)(\tan 5x)$

Q7. Find y': $y = \sin 4x$

Q8. For Figure 9-6f, the maximum acceleration on interval $[a, b]$ is at time $t = $ —?—.

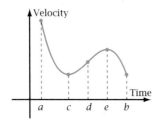

Figure 9-6f

Q9. State the fundamental theorem of calculus.

Q10. Write the definition of indefinite integral.

For Problems 1–16, evaluate the integral.

1. $\displaystyle\int \sqrt{49 - x^2}\, dx$

2. $\displaystyle\int \sqrt{100 - x^2}\, dx$

3. $\displaystyle\int \sqrt{x^2 + 16}\, dx$

4. $\displaystyle\int \sqrt{81 + x^2}\, dx$

5. $\displaystyle\int \sqrt{9x^2 - 1}\, dx$

6. $\displaystyle\int \sqrt{16x^2 - 1}\, dx$

7. $\displaystyle\int \frac{dx}{\sqrt{17 - x^2}}$

8. $\displaystyle\int \frac{dx}{\sqrt{13 - x^2}}$

9. $\displaystyle\int \frac{1}{\sqrt{x^2 + 1}}\, dx$

10. $\displaystyle\int \frac{1}{\sqrt{x^2 - 121}}\, dx$

11. $\displaystyle\int x^2 \sqrt{x^2 - 9}\, dx$

12. $\displaystyle\int x^2 \sqrt{9 - x^2}\, dx$

13. $\displaystyle\int (1 - x^2)^{3/2}\, dx$

14. $\displaystyle\int (x^2 - 81)^{3/2}\, dx$

15. $\displaystyle\int \frac{dx}{81 + x^2}$

16. $\displaystyle\int \frac{dx}{25x^2 + 1}$

Some integrals that you can do by trigonometric substitution you can also do other ways. For problems 17 and 18,

    a. Evaluate the integral by trigonometric substitution.

    b. Evaluate the integral again as a power. Show that the two answers are equivalent.

17. $\displaystyle\int \frac{x\, dx}{\sqrt{x^2 + 25}}$

18. $\displaystyle\int \frac{x\, dx}{\sqrt{x^2 - 49}}$

You can evaluate integrals such as in Problems 19 and 20 by trigonometric substitution. For instance, in Problem 19, the hypotenuse would be 3 and one leg would be $(x - 5)$. Evaluate the integral.

19. $\displaystyle\int \frac{dx}{\sqrt{9 - (x - 5)^2}}$

20. $\displaystyle\int \frac{dx}{\sqrt{36 - (x + 2)^2}}$

You can transform integrals such as in Problems 21 and 22 into ones like Problems 19 and 20 by first completing the square. In Problem 21, $x^2 + 8x - 20 = (x^2 + 8x + 16) - 36$, which equals $(x + 4)^2 - 36$. Evaluate the integral.

21. $\displaystyle\int \frac{dx}{\sqrt{x^2 + 8x - 20}}$

22. $\displaystyle\int \frac{dx}{\sqrt{x^2 - 14x + 50}}$

For Problems 23 and 24, evaluate the integral exactly using the fundamental theorem. Compare the answer with the one you get by numerical integration.

23. $\displaystyle\int_{-3}^{8} \sqrt{100 - x^2}\, dx$

24. $\displaystyle\int_{-1}^{4} \sqrt{x^2 + 25}\, dx$

25. *Arc Length of a Parabola Problem:* Use the fundamental theorem to find the exact length of the parabola $y = 3x^2$ from $x = 0$ to $x = 5$. Find a decimal approximation for the answer. Compare the decimal approximation with the answer by numerical integration.

26. *Area of an Ellipse Problem:*

    a. Use the fundamental theorem to find the exact area of the region bounded by the ellipse $9x^2 + 25y^2 = 225$ between $x = -3$ and $x = 4$ (Figure 9-6g). Compare a decimal approximation of this answer with the answer by numerical integration.

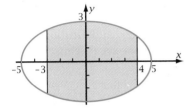

Figure 9-6g

b. Find the exact area of the entire ellipse. How is this area related to the 5 and 3, which are the x- and y-radii of the ellipse?

27. *Circle Area Formula Problem:* Derive by calculus the area formula, $A = \pi r^2$, for the circle

$$x^2 + y^2 = r^2$$

28. *Ellipse Area Formula Problem:* Derive by calculus the area formula for an ellipse. You may start with the general equation for an ellipse with x- and y-radii a and b,

$$\left(\frac{x}{a}\right)^2 + \left(\frac{y}{b}\right)^2 = 1$$

(Figure 9-6h). Show that the formula reduces to the area of a circle if $a = b$.

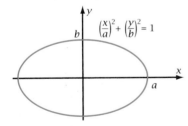

Figure 9-6h

29. *Ellipsoid Problem:* The ellipse in Problem 28 is rotated about the y-axis to form an ellipsoid. Find the volume inside the ellipsoid in terms of the constants a and b. What difference would there be in the answer if the ellipse had been rotated about the x-axis?

30. *Hyperbola Area Problem:* Use the fundamental theorem to find the exact area of the region

bounded above and below by the hyperbola

$$x^2 - y^2 = 9$$

from $x = 3$ to $x = 5$ (Figure 9-6i). Find an approximation for this answer, and compare it with the answer obtained by integrating numerically.

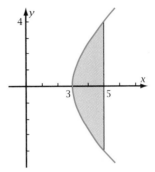

Figure 9-6i

31. *Hyperboloid Problem:* The region in Problem 30 is rotated about the y-axis to form a hollow solid. (The inside surface of the solid is a *hyperboloid of one sheet.*) Find the volume of the solid.

32. *Average Radius Problem:* You may define the **average radius** of a solid of rotation to be the distance $\bar{x}$ for which

Volume $= 2\pi\bar{x} \cdot A$

where a is the area of the region being rotated. Find the average radius of the hyperboloidal solid in Problem 31. Is the average radius more than, less than, or exactly halfway through the region in the x direction as you progress outward from the y-axis?

33. *Area of an Ellipse, Parametrically:* The ellipse in Figure 9-6h has parametric equations

$$x = a\cos t$$
$$y = b\sin t$$

Find the area of the ellipse directly from the parametric equations. Show that the answer is the same as in Problem 28. How does the integration technique used in this case compare with the trigonometric substitution method used in Problem 28?

34. *Length of a Spiral in Polar Coordinates:* In Problem 15 of Problem Set 8-7, you found the length of the spiral with polar equation

$$r = 0.5\theta$$

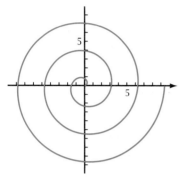

Figure 9-6j

The spiral is shown in Figure 9-6j. Now that you know how to integrate by trigonometric

substitution, you can find the exact length. Use the fundamental theorem to find the length of the part of the spiral shown. Compare this answer with the value you get by numerical integration.

35. *Trigonometric Substitution for Negative Values of x:* In trigonometric substitutions you let x/a equal $\sin\theta$, $\tan\theta$, or $\sec\theta$. Thus,

$$\theta = \sin^{-1}\frac{x}{a}, \quad \theta = \tan^{-1}\frac{x}{a}, \quad \text{or} \quad \theta = \sec^{-1}\frac{x}{a}$$

If x is negative, then θ is not in Quadrant I. If you restrict θ to the other quadrant in the range of the inverse trigonometric function, you find the same indefinite integral as if you naively assumed that θ is always in Quadrant I. Show that this is the case for each of these three trigonometric substitutions.

9-7 Integration of Rational Functions by Partial Fractions

In unrestrained population growth, the rate of change of a population is proportional to the number of people in that population. This happens because the more people there are, the more babies are born each year. In restrained population growth, there is a maximum population a region can sustain. So, the rate of population growth is also proportional to how close the number of people is to that maximum. For instance, if a region can sustain 10.5 million people, then a differential equation for population growth could be

$$\frac{dP}{dt} = 0.038P(10.5 - P)$$

where P is population in millions and 0.038 is the proportionality constant. You may already have seen this differential equation in connection with the *logistic equation* in Problem 11 of Section 7-4. Separating the variables and integrating gives

$$\int \frac{1}{P(10.5 - P)}\, dP = 0.038 \int dt$$

The integral on the left side contains a **rational algebraic function** of P. That is, the integrand can be written as (polynomial)/(polynomial). In this section you will learn an algebraic method to find the antiderivative on the left side of the equation. The method involves breaking the rational expression into a sum of relatively simple **partial fractions**, each of which is easy to integrate. In theory, at least, you can write many ratios of polynomials that have a numerator of degree lower than the denominator degree as

$$\frac{\text{Polynomial}}{\text{Polynomial}} = \frac{\text{constant}}{\text{linear}} + \frac{\text{constant}}{\text{linear}} + \cdots + \frac{\text{constant}}{\text{linear}}$$

OBJECTIVE   Find the integral of a rational algebraic function by first resolving the integrand into partial fractions.

▶ **EXAMPLE 1**   (Heaviside method) Evaluate the indefinite integral: $\displaystyle\int \frac{4x + 41}{x^2 + 3x - 10}\, dx$

Solution   The first step is factoring the denominator. The rational expression thus becomes

$$\frac{4x + 41}{(x + 5)(x - 2)}$$

Your thought process might go something like this: "Hmmm...It looks like someone has been adding fractions, where $(x + 5)(x - 2)$ is the common denominator!" So you write

$$\frac{4x + 41}{(x + 5)(x - 2)} = \frac{A}{x + 5} + \frac{B}{x - 2}$$

A clever way to isolate the constant A is to multiply both sides of the equation by $(x + 5)$.

$$(x + 5)\frac{4x + 41}{(x + 5)(x - 2)} = (x + 5)\frac{A}{x + 5} + (x + 5)\frac{B}{x - 2}$$

$$\frac{4x + 41}{x - 2} = A + (x + 5)\frac{B}{x - 2}$$

Substituting -5 for x in the transformed equation gives

$$\frac{4(-5) + 41}{-5 - 2} = A + (-5 + 5)\frac{B}{-5 - 2} = A + 0$$

$$\therefore \; -3 = A$$

Similarly, multiplying both sides by $(x - 2)$ isolates the constant B.

$$(x - 2)\frac{4x + 41}{(x + 5)(x - 2)} = (x - 2)\frac{A}{x + 5} + (x - 2)\frac{B}{x - 2}$$

$$\frac{4x + 41}{x + 5} = (x - 2)\frac{A}{x + 5} + B$$

Substituting 2 for x eliminates the A term and gives

$$\frac{4(2) + 41}{2 + 5} = (2 - 2)\frac{A}{2 + 5} + B = 0 + B$$

$$\therefore \; 7 = B$$

Substituting −3 for *A* and 7 for *B* and putting the integral back together gives

$$\int \frac{4x + 41}{x^2 + 3x - 10}\, dx = \int \left(\frac{-3}{x + 5} + \frac{7}{x - 2} \right) dx$$
$$= -3 \ln |x + 5| + 7 \ln |x - 2| + C \qquad \blacktriangleleft$$

The act of resolving into partial fractions, as shown in Example 1, is called the **Heaviside method** after Oliver Heaviside (1850–1925). You can shorten the method enough to do it in one step in your head! Here's how.

▶ **EXAMPLE 2**   (Heaviside shortcut) Integrate by resolving into partial fractions: $\int \dfrac{x - 2}{(x - 5)(x - 1)}\, dx$

Solution   Thought process:
Write the integral and the denominators of the partial fractions.

$$\int \frac{x - 2}{(x - 5)(x - 1)}\, dx = \int \left(\frac{}{x - 5} + \frac{}{x - 1} \right) dx$$

Tell yourself, "If *x* is 5, then (*x* − 5) equals zero." Cover up the (*x* − 5) with your finger, and substitute 5 into what is left.

$$\int \frac{x - 2}{\qquad (x - 1)}\, dx \qquad \text{Simplify: } \frac{5 - 2}{5 - 1} = \frac{3}{4}$$

The answer, 3/4, is the numerator for (*x* − 5). To find the numerator for (*x* − 1), repeat the process, but cover up the (*x* − 1) and substitute 1, the number that makes (*x* − 1) zero.

$$\int \frac{x - 2}{(x - 5)\qquad}\, dx \qquad \text{Simplify: } \frac{1 - 2}{1 - 5} = \frac{1}{4}$$

Fill in 3/4 and 1/4 where they belong. The entire process is just one step, like this:

$$\int \frac{x - 2}{(x - 5)(x - 1)}\, dx = \int \left(\frac{\frac{3}{4}}{x - 5} + \frac{\frac{1}{4}}{x - 1} \right) dx$$
$$= \tfrac{3}{4} \ln |x - 5| + \tfrac{1}{4} \ln |x - 1| + C \qquad \blacktriangleleft$$

In Problem Set 9-7, you will practice integrating by partial fractions. You will also find out what to do if the denominator has

- Unfactorable quadratic factors
- Repeated linear factors

462

Problem Set 9-7

Quick Review 5 min

Q1. Factor: $x^2 - 25$

Q2. Multiply: $(x - 3)(x + 5)$

Q3. Factor: $x^2 - 4x - 12$

Q4. Multiply: $(x + 7)^2$

Q5. Factor: $x^2 + 8x + 16$

Q6. Multiply: $(x - 8)(x + 8)$

Q7. If $f(x) = \ln x$, then $f^{-1}(x) = $ —?—.

Q8. Show that you cannot factor $x^2 + 50x + 1000$ using real numbers only.

Q9. Show that you cannot factor $x^2 + 36$ using real numbers only.

Q10. For the data in the table, using Simpson's rule, $\int_1^9 g(x)\, dx = $ —?—.

x	$g(x)$
1	10
3	15
5	16
7	14
9	13

  A. 136 B. 114 C. 113

  D. 110 E. 134

For Problems 1–10, integrate by first resolving the integrand into partial fractions.

1. $\int \dfrac{11x - 15}{x^2 - 3x + 2}\, dx$

2. $\int \dfrac{7x + 25}{x^2 - 7x - 8}\, dx$

3. $\int \dfrac{(5x - 11)\, dx}{x^2 - 2x - 8}$

4. $\int \dfrac{(3x - 12)\, dx}{x^2 - 5x - 50}$

5. $\int \dfrac{21\, dx}{x^2 + 7x + 10}$

6. $\int \dfrac{10x\, dx}{x^2 - 9x - 36}$

7. $\int \dfrac{9x^2 - 25x - 50}{(x + 1)(x - 7)(x + 2)}\, dx$

8. $\int \dfrac{7x^2 + 22x - 54}{(x - 2)(x + 4)(x - 1)}\, dx$

9. $\int \dfrac{4x^2 + 15x - 1}{x^3 + 2x^2 - 5x - 6}\, dx$

10. $\int \dfrac{-3x^2 + 22x - 31}{x^3 - 8x^2 + 19x - 12}\, dx$

Improper Algebraic Fractions:

If the numerator is of higher degree than the denominator, long division will reduce the integrand to a polynomial plus a fraction. For instance,

$$\frac{x^3 - 9x^2 + 24x - 17}{x^2 - 6x + 5} = x - 3 + \frac{x - 2}{x^2 - 6x + 5}$$

For Problems 11 and 12, evaluate the integral by first dividing.

11. $\int \dfrac{3x^3 + 2x^2 - 12x + 9}{x - 1}\, dx$

12. $\int \dfrac{x^3 - 7x^2 + 5x + 40}{x^2 - 2x - 8}\, dx$

Unfactorable Quadratics:

Heaviside's method does not work for integrals such as

$$\int \frac{7x^2 - 4x}{(x^2 + 1)(x - 2)}\, dx$$

that have an unfactorable quadratic in the denominator (unless you are willing to use imaginary numbers!). However, the quadratic term can have a linear numerator. In this case you can write

$$\frac{7x^2 - 4x}{(x^2 + 1)(x - 2)} = \frac{Ax + B}{x^2 + 1} + \frac{C}{x - 2}$$

$$= \frac{(Ax + B)(x - 2) + C(x^2 + 1)}{(x^2 + 1)(x - 2)}$$

$$= \frac{Ax^2 - 2Ax + Bx - 2B + Cx^2 + C}{(x^2 + 1)(x - 2)}$$

So $Ax^2 + Cx^2 = 7x^2$, $-2Ax + Bx = -4x$, and $-2B + C = 0$. Solving the system

$$A + C = 7$$
$$-2A + B = -4$$
$$-2B + C = 0$$

gives $A = 3$, $B = 2$, and $C = 4$. Therefore,

$$\frac{7x^2 - 4x}{(x^2 + 1)(x - 2)} = \frac{3x + 2}{x^2 + 1} + \frac{4}{x - 2}$$

For Problems 13 and 14, integrate by first resolving into partial fractions.

13. $\int \dfrac{4x^2 + 6x + 11}{(x^2 + 1)(x + 4)}\, dx$

14. $\int \dfrac{4x^2 - 15x - 1}{x^3 - 5x^2 + 3x + 1}\, dx$

Repeated Linear Factors:

If a power of a linear factor appears in the denominator, the fraction could have come from adding partial fractions with that power or any lower power. For instance, the integral

$$\int \frac{x^2 - 4x + 18}{(x + 4)(x - 1)^2}\, dx$$

can be written

$$\int \left[\frac{A}{x + 4} + \frac{B}{x - 1} + \frac{C + Dx}{(x - 1)^2} \right] dx$$

However, the numerator of the original fraction has only three coefficients, and the right side of the equation has four unknown constants. So one of the constants is arbitrary and can take on any value you decide. The smart move is to let $D = 0$ so that there will be three partial fractions that are as easy to integrate as possible.

For Problems 15 and 16, integrate by resolving the integrand into partial fractions.

15. $\int \dfrac{4x^2 + 18x + 6}{(x + 5)(x + 1)^2}\, dx$

16. $\int \dfrac{3x^2 - 53x + 245}{x^3 - 14x^2 + 49x}\, dx$

"Old Problem" New Problems: Sometimes a problem that seems to fit the pattern of a new problem

actually reduces to an old problem. For Problems 17 and 18, evaluate the integral with this idea in mind.

17. $\int \dfrac{dx}{x^3 - 6x^2 + 12x - 8}$

18. $\int \dfrac{1}{x^4 + 4x^3 + 6x^2 + 4x + 1}\, dx$

19. *Rumor Problem:* There are 1000 students attending Lowe High. One day 10 students arrive at school bearing the rumor that final exams will be canceled! On average, each student talks to other students at a rate of two students per hour, passing on the rumor to students, some of whom have already heard it and some of whom have not. Thus, the rate at which students hear the rumor for the first time is 2 times the number who have already heard it times the fraction of the students who have not yet heard it. If y is the number of students who have heard the rumor at time t, in hours since school started, then

$$\frac{dy}{dt} = 2y \cdot \frac{1000 - y}{1000}$$

a. Solve this differential equation algebraically, subject to the initial condition that $y = 10$ when school started at $t = 0$.

b. How many students had heard the rumor after the first hour? At lunchtime ($t = 4$)? At the end of the school day ($t = 8$)?

c. How many students had heard the rumor at the time it was spreading the fastest? At what time was this?

d. Figure 9-7a shows the slope field for the differential equation. Plot the solution in part a on a copy of the figure. How does the curve relate to the slope field?

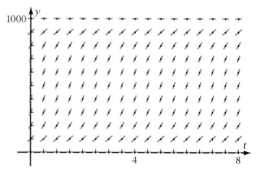

Figure 9-7a

20. *Epidemic Problem:* A new disease arrives at the town of Scorpion Gulch. When a person with the disease comes into contact with a person who has not yet had the disease, the uninfected person may or may not catch the disease. The disease is not fatal, but it persists with the infected person forever after. Of the N people who live there, let P be the number of them who have the disease after time t, in days.

   a. Suppose that each person contacts an average of three people per day. In terms of P and N, how many contacts will there be between infected and uninfected people per day?

   b. Suppose that the probability of passing on the disease to an uninfected person is only 10% at each contact. Explain why

   $$\frac{dP}{dt} = 0.3P\frac{N - P}{N}$$

   c. Solve the differential equation in part b for P in terms of t. Use the initial condition that $P = P_0$ at time $t = 0$.

   d. If 1000 people live in Scorpion Gulch and 10 people are infected at time $t = 0$, how many will be infected after 1 week?

   e. For the conditions in part d, how long will it be until 99% of the population is infected?

21. *Area Problem:* Find an equation for the area of the region under the graph of

   $$y = \frac{25}{x^2 + 3x - 4}$$

   from $x = 2$ to $x = b$, where b is a constant greater than 2 (Figure 9-7b). Let $b = 7$ and

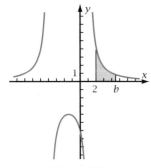

Figure 9-7b

check your answer by numerical integration. Does the area approach a finite limit as b approaches infinity? Justify your answer.

22. *Volume Problem:* The region in Problem 21 is rotated about the y-axis to form a solid. Find an equation in terms of the constant b for the volume of the solid. Check your answer by numerical integration using $b = 7$. Does the volume approach a finite limit as b approaches infinity? Justify your answer.

23. *Equivalent Answers Problem:* Evaluate this integral three ways.

   $$\int \frac{x - 3}{x^2 - 6x + 8}\,dx$$

   a. First resolve into partial fractions.

   b. Complete the square, then follow with trigonometric substitution.

   c. Evaluate directly, as the integral of the reciprocal function.

   d. Show that the three answers are equivalent.

24. *Logistic Curve Problem, Algebraically:* For unrestrained population growth, the rate of change of population is directly proportional to the population. That is, $dp/dt = kp$, where p is population, t is time, and k is a constant. One assumption for restrained growth is that there is a certain maximum size, m, for the population, and the rate goes to zero as the population approaches that size. Use this information to answer these questions.

   a. Show that the differential equation $dp/dt = kp(m - p)$ has the properties mentioned.

   b. At what value of p is the growth rate the greatest?

   c. Separate the variables, then solve the equation by integrating. If you have worked correctly, you can evaluate the integral on one side of the equation by partial fractions.

   d. Transform your answer so that p is explicitly in terms of t. Show that you can express it in

the form of the logistic equation,

$$p = p_0 \frac{1 + b}{1 + be^{-Kt}}$$

where p_0 is the population at time $t = 0$ and b is a constant.

e. Census figures for the United States are

    1960: 179.3 million
    1970: 203.2 million
    1980: 226.5 million
    1990: 248.7 million

Let t be time, in years, that has elapsed since 1960. Use these data as initial conditions to evaluate p_0, b, and k. Write the particular solution.

f. Predict the outcome of the 2000 census. How close does your answer come to the actual 2000 population, 281.4 million?

g. Based on the logistic model, what will be the ultimate U.S. population?

h. Is this mathematical model very sensitive to the initial conditions? For instance, suppose that the 1970 population had really been 204.2 million instead of 203.2 million. How much would this affect the predicted ultimate population?

Stanton and Irene Katchatag of Unalakeet, Alaska, were the first two people counted for the 2000 U.S. census.

9-8   Integrals of the Inverse Trigonometric Functions

The beginning of this chapter mentioned that there are three types of functions that, along with their inverses, come under the category **elementary transcendental functions**. They are

Trigonometric	Inverse trigonometric
Logarithmic	Inverse logarithmic (exponential)
Hyperbolic	Inverse hyperbolic

The name *transcendental* implies that the operations needed to calculate values of the functions "transcend," or go beyond, the operations of algebra (addition, subtraction, multiplication, division, and root extraction). You have already learned algebraic calculus techniques for the first four types of functions, except for integrating the inverse trigonometric functions. In this section you will learn to integrate the inverse trigonometric functions algebraically. In Section 9-9, you will explore the hyperbolic functions, which are related to both exponential and trigonometric functions.

OBJECTIVE   Integrate (antidifferentiate) each of the six inverse trigonometric functions.

Background: Definition and Derivatives of the Inverse Trigonometric Functions

In Section 4-5, you learned definitions of the inverse trigonometric functions and how to find algebraic formulas for their derivatives. These definitions and derivative formulas are repeated here for easy reference.

DEFINITIONS: *Inverse Trigonometric Functions (Principal Branches)*

$$y = \sin^{-1} x \text{ if and only if } \sin y = x \quad \text{and} \quad y \in \left[-\frac{\pi}{2}, \frac{\pi}{2}\right]$$

$$y = \cos^{-1} x \text{ if and only if } \cos y = x \quad \text{and} \quad y \in [0, \pi]$$

$$y = \tan^{-1} x \text{ if and only if } \tan y = x \quad \text{and} \quad y \in \left(-\frac{\pi}{2}, \frac{\pi}{2}\right)$$

$$y = \cot^{-1} x \text{ if and only if } \cot y = x \quad \text{and} \quad y \in (0, \pi)$$

$$y = \sec^{-1} x \text{ if and only if } \sec y = x \quad \text{and} \quad y \in [0, \pi], \text{ but } y \neq \frac{\pi}{2}$$

$$y = \csc^{-1} x \text{ if and only if } \csc y = x \quad \text{and} \quad y \in \left[-\frac{\pi}{2}, \frac{\pi}{2}\right], \text{ but } y \neq 0$$

PROPERTIES: *Derivatives of the Six Inverse Trigonometric Functions*

$$\frac{d}{dx}(\sin^{-1} x) = \frac{1}{\sqrt{1 - x^2}} \qquad \frac{d}{dx}(\cos^{-1} x) = -\frac{1}{\sqrt{1 - x^2}}$$

$$\frac{d}{dx}(\tan^{-1} x) = \frac{1}{1 + x^2} \qquad \frac{d}{dx}(\cot^{-1} x) = -\frac{1}{1 + x^2}$$

$$\frac{d}{dx}(\sec^{-1} x) = \frac{1}{|x|\sqrt{x^2 - 1}} \qquad \frac{d}{dx}(\csc^{-1} x) = -\frac{1}{|x|\sqrt{x^2 - 1}}$$

Note: Your grapher must be in the radian mode.

Memory Aid: The derivative of each co-inverse function is the opposite of the derivative of the corresponding inverse function because each co-inverse function is decreasing as x starts increasing from zero.

Integrals

The technique for integrating inverse trigonometric functions is (surprisingly!) integration by parts. Example 1 shows you how this is done.

▶ **EXAMPLE 1**   Integrate: $\int \tan^{-1} x \, dx$

Solution

$$\int \tan^{-1} x \, dx$$

$$= x \tan^{-1} x - \int \frac{x \, dx}{x^2 + 1}$$

$$= x \tan^{-1} x - \frac{1}{2} \ln |x^2 + 1| + C$$

$$
\begin{array}{c|c}
u & dv \\
\hline
\tan^{-1} x & 1 \\
\dfrac{1}{x^2 + 1} & x
\end{array}
$$

You can transform the last integral to the integral of the reciprocal function. The absolute value in the argument of ln is optional because $x^2 + 1$ is always positive. ◀

In Problem Set 9-8, you will derive algebraic formulas for the integrals of the other five inverse trigonometric functions. These formulas are listed in the box. They were more important in the first 300 years of calculus, before such technology as your grapher made numerical integration quick and easily accessible. Like climbing Mount Everest, these formulas are interesting more from the standpoint that they can be done rather than because they are of great practical use.

PROPERTIES: Algebraic Integrals of the Inverse Trigonometric Functions

$$\int \sin^{-1} x \, dx = x \sin^{-1} x + \sqrt{1 - x^2} + C$$

$$\int \cos^{-1} x \, dx = x \cos^{-1} x - \sqrt{1 - x^2} + C$$

$$\int \tan^{-1} x \, dx = x \tan^{-1} x - \tfrac{1}{2} \ln |x^2 + 1| + C = x \tan^{-1} x - \ln \sqrt{x^2 + 1} + C$$

$$\int \cot^{-1} x \, dx = x \cot^{-1} x + \tfrac{1}{2} \ln |x^2 + 1| + C = x \cot^{-1} x + \ln \sqrt{x^2 + 1} + C$$

$$\int \sec^{-1} x \, dx = x \sec^{-1} x - \operatorname{sgn} x \ln |x + \sqrt{x^2 - 1}| + C$$

$$\int \csc^{-1} x \, dx = x \csc^{-1} x + \operatorname{sgn} x \ln |x + \sqrt{x^2 - 1}| + C$$

Problem Set 9-8

Quick Review 5 min

Q1. To integrate a product, use —?—.

Q2. To integrate a rational function, use —?—.

Q3. To integrate $\int \sqrt{x^2 + 1} \, dx$, use the —?— trigonometric substitution.

Q4. To integrate $\int \sqrt{x^2 - 1} \, dx$, use the —?— trigonometric substitution.

Q5. To integrate $\int \sqrt{1 - x^2} \, dx$, use the —?— trigonometric substitution.

Q6. Integrate: $\int (x^2 + 1)^7 (x \, dx)$

Q7. If $f(x) = 3 + |x - 5|$, then the maximum of $f(x)$ on $[1, 6]$ is —?—.

Q8. If $f(x) = 3 + |x - 5|$, then the minimum of $f(x)$ on $[1, 6]$ is —?—.

Q9. If $f(x) = 3 + |x - 5|$, then $f'(5)$ is —?—.

Q10. If $h(x) = x^3 + x$, then the graph of h has a point of inflection at $x = $ —?—.

   A. $\sqrt{3}$ B. 0   C. $1 + \sqrt{3}$

   D. $\dfrac{1}{\sqrt{3}}$ E. $-\dfrac{1}{\sqrt{3}}$

For Problems 1–6, find the indefinite integral. Check your answer against those in the preceding box.

1. $\int \tan^{-1} x \, dx$

2. $\int \cot^{-1} x \, dx$

3. $\int \cos^{-1} x \, dx$

4. $\int \sin^{-1} x \, dx$

5. $\int \sec^{-1} x \, dx$

6. $\int \csc^{-1} x \, dx$

7. *Answer Verification Problem:* Evaluate the integral $\int_1^4 \tan^{-1} x \, dx$ algebraically. Find a decimal approximation for the answer. Then evaluate the integral numerically. How close does the numerical answer come to the exact algebraic answer?

8. *Simpson's Rule Review Problem:* Plot the graph of $y = \sec^{-1} x$ from $x = 1$ to $x = 3$. You may do this in parametric mode, with $x = 1/\cos t$ and $y = t$. Sketch the graph. Use Simpson's rule with $n = 10$ increments to find a numerical approximation for the area of the region under this graph from $x = 1$ to $x = 3$. Then find the exact area by integrating algebraically. How close does the answer using Simpson's rule come to the exact answer?

9. *Area Problem:* Figure 9-8a shows the region above the graph of $y = \sin^{-1} x$, below $y = \pi/2$, and to the right of the y-axis. Find the area of this region by using vertical slices. Find the area again, this time by using horizontal slices. Show that the two answers are equivalent.

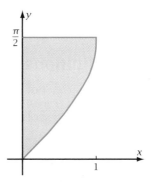

Figure 9-8a

10. *Volume Problem:* Figure 9-8b shows the region under the graph of $y = \tan^{-1} x$ from $x = 0$ to $x = 1$, rotated about the y-axis to form a solid. Use the fundamental theorem to find the exact volume of the solid. Show that your answer is reasonable by numerical integration and by comparing it with a suitable geometric figure.

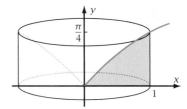

Figure 9-8b

9-9 Calculus of the Hyperbolic and Inverse Hyperbolic Functions

Figure 9-9a shows what a chain might look like suspended between two nails driven into the frame of a chalkboard. Although the shape resembles a parabola, it is actually a **catenary** from the Latin *catena*, meaning "chain." Its graph is the **hyperbolic cosine** function,

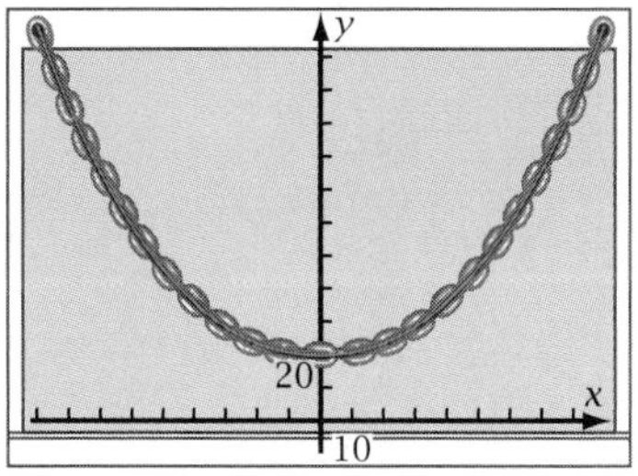

$$y = a + b \cosh cx$$

(pronounced "kosh," with a short "o"). As shown in Figure 9-9b, a parabola is more sharply curved at the vertex than a catenary is.

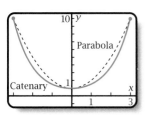

Figure 9-9b

Cosh and the related hyperbolic sine and tangent are important enough to be on most graphers. In this section you will see that the hyperbolic functions are related both to the natural exponential function and to the circular sine and

Figure 9-9a

cosine functions. You will see why a chain hangs in the shape of a hyperbolic cosine and why the functions are called hyperbolic.

Definitions of the Six Hyperbolic Functions

Let two functions, u and v, be defined as

$$u = \tfrac{1}{2}(e^x + e^{-x}) \qquad \text{and} \qquad v = \tfrac{1}{2}(e^x - e^{-x})$$

The u-graph is identical to $y = \cosh x$, as you could see by plotting $y = u$ and $y = \cosh x$ on the same screen (Figure 9-9c, left). The v-graph is identical to $y = \sinh x$ (pronounced "sinch of x," although the letter "c" does not appear in writing). The right diagram in Figure 9-9c shows $y = \sinh x$ and $y = v$. The dashed graphs are $y = 0.5e^x$ and $y = \pm 0.5e^{-x}$.

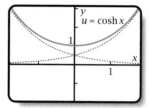 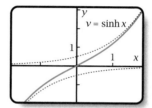

Figure 9-9c

The reason these functions are called "hyperbolic" becomes apparent if you eliminate x and get an equation with u and v alone. Squaring u and v, then subtracting, gives

$$u^2 = \tfrac{1}{4}(e^{2x} + 2 + e^{-2x}) \qquad \text{Think about why 2 is the middle term.}$$

$$v^2 = \tfrac{1}{4}(e^{2x} - 2 + e^{-2x})$$

$$\overline{u^2 - v^2 = 1} \qquad \text{A unit equilateral hyperbola in the } uv\text{-coordinate system.}$$

This is the equation of a hyperbola in a uv-coordinate system. Function u is the horizontal coordinate of a point on the hyperbola, and function v is the vertical coordinate. These coordinates have the same relationship to the unit equilateral hyperbola $u^2 - v^2 = 1$ as the "circular" functions cosine and sine have to the unit circle, $u^2 + v^2 = 1$ (Figure 9-9d).

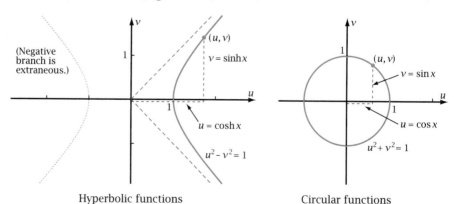

Hyperbolic functions   Circular functions

Figure 9-9d

The other four hyperbolic functions are defined in the same way as the other four circular (trigonometric) functions, namely, as reciprocals and quotients of the first two. The symbols come from adding an "h" to the corresponding circular function symbol. There are no widely accepted pronunciations for symbols such as $\tanh x$ and $\operatorname{csch} x$ other than "hyperbolic tangent of x," and so on.

DEFINITIONS: The Hyperbolic Functions

$$\sinh x = \tfrac{1}{2}(e^x - e^{-x}) \qquad\qquad \cosh x = \tfrac{1}{2}(e^x + e^{-x})$$

$$\tanh x = \frac{\sinh x}{\cosh x} \qquad\qquad\qquad \coth x = \frac{\cosh x}{\sinh x}$$

$$\operatorname{sech} x = \frac{1}{\cosh x} \qquad\qquad\qquad \operatorname{csch} x = \frac{1}{\sinh x}$$

From the fact that $u^2 - v^2 = 1$ and the previous definitions, the Pythagorean properties follow.

Pythagorean Properties of the Hyperbolic Functions

$$\cosh^2 x - \sinh^2 x = 1$$

$$1 - \tanh^2 x = \operatorname{sech}^2 x$$

$$\coth^2 x - 1 = \operatorname{csch}^2 x$$

OBJECTIVE Differentiate and integrate any of the six hyperbolic functions and their inverses, and use the hyperbolic cosine function as a mathematical model.

Derivatives of the Hyperbolic Functions

The derivatives of cosh and sinh have an interesting relationship.

$$\frac{d}{dx} \cosh x = \tfrac{1}{2}(e^x - e^{-x}) = \sinh x \quad \text{and} \quad \frac{d}{dx} \sinh x = \tfrac{1}{2}(e^x + e^{-x}) = \cosh x$$

This "cyclical" property of the derivatives is similar to that of the circular sine and cosine.

$$\sin' x = \cos x \qquad \text{and} \qquad \cos' x = -\sin x$$

You can find the derivatives of the other four hyperbolic functions by first transforming to $\sinh x$ and $\cosh x$. Example 1 shows you how.

▶ **EXAMPLE 1** Find the derivative: $y = \coth x$

Solution   By definition of coth,

$$y = \frac{\cosh x}{\sinh x}$$

Applying the derivative of a quotient property gives

$$y' = \frac{(\sinh x)(\sinh x) - (\cosh x)(\cosh x)}{\sinh^2 x}$$

$$= \frac{\sinh^2 x - \cosh^2 x}{\sinh^2 x}$$

$$= \frac{-1}{\sinh^2 x} \qquad \text{By the Pythagorean properties.}$$

$$y' = -\text{csch}^2 x \qquad\qquad\qquad\qquad\qquad \blacktriangleleft$$

The derivatives of the hyperbolic functions are summarized in this box. In Problems 37 and 38 of Problem Set 9-9, you will be asked to derive some of these derivatives.

PROPERTIES: *Derivatives of Hyperbolic Functions*

$\sinh' x = \cosh x$	$\cosh' x = \sinh x$
$\tanh' x = \text{sech}^2 x$	$\coth' x = -\text{csch}^2 x$
$\text{sech}' x = -\text{sech}\, x \tanh x$	$\text{csch}' x = -\text{csch}\, x \coth x$

Integrals of Hyperbolic Functions

The integrals of cosh and sinh come directly from the derivative formulas. You find the integrals of tanh and coth in much the same way you integrate tan and cot. Example 2 shows how to integrate coth. The integrals of sech and csch require clever substitutions you have not yet learned. In Problems C1 and C2 of Problem Set 9-13, you will see how to do this.

▶ *EXAMPLE 2*   Integrate: $\int \coth x \, dx$

Solution   The technique is to use the definition of coth. The resulting quotient has a numerator that is the derivative of the denominator. You wind up integrating the reciprocal function.

$$\int \coth x \, dx$$

$$= \int \frac{\cosh x}{\sinh x} \, dx$$

$$= \int \frac{1}{\sinh x} \cosh x \, dx$$

$$= \ln |\sinh x| + C \qquad\qquad\qquad\qquad \blacktriangleleft$$

The integrals of the six hyperbolic functions are listed in this box. In Problem Set 9-9, you will be asked to derive some of these integrals.

PROPERTIES: *Integrals of the Hyperbolic Functions*

$$\int \sinh x \, dx = \cosh x + C \qquad \int \cosh x \, dx = \sinh x + C$$

$$\int \tanh x \, dx = \ln(\cosh x) + C \qquad \int \coth x \, dx = \ln|\sinh x| + C$$

$$\int \operatorname{sech} x \, dx = \sin^{-1}(\tanh x) + C \qquad \int \operatorname{csch} x \, dx = \ln|\tanh(x/2)| + C$$

Inverse Hyperbolic Functions

Recall that the inverse of a function is the relation you obtain by interchanging the two variables. For instance, if $y = \sinh x$, then the inverse relation has the equation

$$x = \sinh y$$

The dependent variable y is called the **inverse hyperbolic sine of x**. The symbol often used is similar to that for inverse circular functions,

$$y = \sinh^{-1} x$$

This is read, "inverse hyperbolic sine of x," or simply "sinch inverse of x." The term **$y = \operatorname{argsinh} x$** is also used, because y is the "argument whose sinh is x."

Figure 9-9e shows the graphs of $y = \sinh x$ and $y = \sinh^{-1} x$. As is true with any function and its inverse, the two graphs are reflections of each other in the line $y = x$.

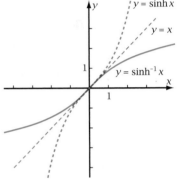

Figure 9-9e

Derivatives of the Inverse Hyperbolic Functions

To differentiate an inverse hyperbolic function, simply use the definition to transform back to the natural function. Then differentiate implicitly.

▶ **EXAMPLE 3** Find the derivative: $y = \cosh^{-1} x$, for $x > 1$

Solution

$y = \cosh^{-1} x$

$\cosh y = x$

$\sinh y \dfrac{dy}{dx} = 1$ dy/dx comes from the chain rule.

$\dfrac{dy}{dx} = \dfrac{1}{\sinh y}$ Divide each member by $\sinh y$.

$$\frac{dy}{dx} = \frac{1}{\sqrt{\cosh^2 y - 1}}$$ $\cosh^2 y - \sinh^2 y = 1$, so $\sinh^2 y = \cosh^2 y - 1$.

$$\frac{dy}{dx} = \frac{1}{\sqrt{x^2 - 1}}$$ $\cosh y = x$.   ◀

You can derive differentiation formulas for the other five inverse hyperbolic functions in a similar way, as you will do in Problem 38 of Problem Set 9-9. The six derivatives are summarized in this box.

PROPERTIES: *Inverse Hyperbolic Function Derivatives*

$$\frac{d}{dx}(\sinh^{-1} x) = \frac{1}{\sqrt{x^2 + 1}} \qquad\qquad \frac{d}{dx}(\cosh^{-1} x) = \frac{1}{\sqrt{x^2 - 1}}, \quad x > 1$$

$$\frac{d}{dx}(\tanh^{-1} x) = \frac{1}{1 - x^2}, \quad |x| < 1 \qquad \frac{d}{dx}(\coth^{-1} x) = \frac{1}{1 - x^2}, \quad |x| > 1$$

$$\frac{d}{dx}(\operatorname{sech}^{-1} x) = -\frac{1}{x\sqrt{1 - x^2}}, \quad x \text{ in } (0, 1) \quad \frac{d}{dx}(\operatorname{csch}^{-1} x) = -\frac{1}{|x|\sqrt{1 + x^2}}, \quad x \neq 0$$

Integrals of Inverse Hyperbolic Functions

You can find the integrals of the inverse hyperbolic functions by straightforward integration by parts. Example 4 shows you how.

▶ **EXAMPLE 4**   Integrate: $\int \sinh^{-1} x \, dx$

Solution

$$\int \sinh^{-1} x \, dx$$

$$= x \sinh^{-1} x - \int \frac{1}{\sqrt{x^2 + 1}} x \, dx$$

$$= x \sinh^{-1} x - \frac{1}{2}\int (x^2 + 1)^{-1/2}(2x \, dx)$$

$$= x \sinh^{-1} x - \frac{1}{2} \cdot \frac{2}{1}(x^2 + 1)^{1/2} + C \qquad \text{Give a reason.}$$

$$= x \sinh^{-1} x - (x^2 + 1)^{1/2} + C \qquad\qquad\qquad\qquad ◀$$

u	dv
$\sinh^{-1} x$	1
$\dfrac{1}{\sqrt{x^2 + 1}}$	x

The indefinite integrals of the inverse hyperbolic functions are listed in the box on page 475. You should understand that these properties exist, and you should know how to derive them if you are called upon to do so. However, unless you plan to make a career out of integrating inverse hyperbolic functions, there is no need to memorize them!

$$\int \sinh^{-1} x \, dx = x \sinh^{-1} x - (x^2 + 1)^{1/2} + C$$

$$\int \cosh^{-1} x \, dx = x \cosh^{-1} x - (x^2 - 1)^{1/2} + C, \quad x > 1$$

$$\int \tanh^{-1} x \, dx = x \tanh^{-1} x + \tfrac{1}{2} \ln |1 - x^2| + C, \quad |x| < 1$$

$$\int \coth^{-1} x \, dx = x \coth^{-1} x + \tfrac{1}{2} \ln |1 - x^2| + C, \quad |x| > 1$$

$$\int \operatorname{sech}^{-1} x \, dx = x \operatorname{sech}^{-1} x + \sin^{-1} x + C, \quad x \text{ in } (0, 1)$$

$$\int \operatorname{csch}^{-1} x \, dx = x \operatorname{csch}^{-1} x + \operatorname{sgn} x \sinh^{-1} x + C, \quad x \neq 0$$

Hyperbolic Cosine as a Mathematical Model

A chain hangs in the shape of the catenary

$$y = k \cosh \tfrac{1}{k} x + C$$

where k and C stand for constants. In Problem 25 of Problem Set 9-9, you will learn why this is true. Example 5 shows you how to derive the particular equation.

▶ **EXAMPLE 5**   A chain hangs from above a chalkboard (Figure 9-9f). Its ends are at point ($\pm 90, 120$), and its vertex is at point (0, 20), where x and y are in centimeters along and above the chalk tray, respectively.

   a. Find the particular equation of the catenary.

   b. How high is the chain above the chalk tray when $x = 50$?

   c. At what values of x is the chain 110 cm above the chalk tray?

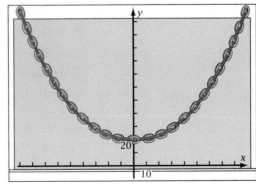

Figure 9-9f

Solution

a. First write the general equation of the catenary.

$$y = k \cosh \tfrac{1}{k}x + C$$

$$20 = k \cosh \tfrac{1}{k}(0) + C = k + C \qquad \text{Substitute (0, 20); } \cosh 0 = 1.$$

$$120 = k \cosh \tfrac{1}{k}(90) + C \qquad \text{Substitute (90, 120).}$$

$$100 = k \cosh \tfrac{90}{k} - k \qquad \text{Eliminate } C \text{ by subtracting equations.}$$

$$0 = k \cosh \tfrac{90}{k} - k - 100 \Rightarrow k = 51.780122\ldots \qquad \text{Use your grapher's solver feature.}$$

$$20 = 51.78\ldots + C \Rightarrow C = -31.78\ldots$$

$$\therefore y = 51.78\ldots \cosh \tfrac{1}{51.78\ldots}x - 31.78\ldots$$

b. Using the equation found in part a,

$$y = 51.78\ldots \cosh \tfrac{50}{51.78\ldots} - 31.78\ldots = 46.0755\ldots$$

The chain is about 46.1 cm above the chalk tray when $x = 50$.

c.
$$110 = 51.78\ldots \cosh \tfrac{1}{51.78\ldots}x - 31.78$$

$$\cosh \tfrac{1}{51.78\ldots}x = 2.7381\ldots$$

$$\tfrac{1}{51.78\ldots}x = \pm \cosh^{-1} 2.7381\ldots = \pm 1.66526\ldots$$

$$x = \pm 86.2278\ldots$$

The chain is about 110 cm above the chalk tray when $x \approx \pm 86.2$ cm. ◀

Problem Set 9-9

Quick Review

Q1. What trigonometric substitution should be used for $\int (x^2 + 5)^{3/2}\, dx$?

Q2. Integrate: $\int xe^x\, dx$

Q3. Integrate: $\int \sec^2 3x\, dx$

Q4. Integrate: $\int x^{-1/2}\, dx$

Q5. Integrate: $\int x^{-1}\, dx$

Q6. $\int \sin^n x\, dx = \dfrac{1}{n} \sin^{n-1} x \cos x + \dfrac{n-1}{n} \int \sin^{n-2} x\, dx$ is called a(n) —?—.

Q7. True or false: $\sec' x = \ln |\sec x + \tan x| + C$

Q8. Write the formula for dL, the differential of arc length.

Q9. If $f(x) = \sin^{-1} x$, then $f'(x) = $ —?—.

Q10. The appropriate integration method for $\int (x+5)/[(x-3)(x+2)]\, dx$ is —?—.

    A. Integration by parts

    B. Trigonometric substitution

    C. Partial fractions

    D. Reduction

    E. Rational integration

---

1. *Hyperbolic Function Graphing Problem:* Sketch the graphs of each of the six hyperbolic functions. Plot these on your grapher first to see what they look like. You may plot the graphs of coth, sech, and csch by taking advantage of their definitions to write their equations in terms of functions that appear on your grapher.

2. *Inverse Hyperbolic Function Graphing Problem:* Sketch the graphs of the inverses of the six hyperbolic functions. You can do this most easily in parametric mode on your grapher, letting $x = \cosh t$ and $y = t$, for example. For each inverse that is not a function, darken what you think would be the principal branch (just one value of y for each value of x), and write an inequality that restricts the range to specify this branch.

For Problems 3–22, evaluate the integral or differentiate.

3. $f(x) = \tanh^3 x$

4. $f(x) = 5 \operatorname{sech} 3x$

5. $\int \cosh^5 x \sinh x\, dx$

6. $\int (\sinh x)^{-3} \cosh x\, dx$

7. $g(x) = \operatorname{csch} x \sin x$

8. $g(x) = \tan x \tanh x$

9. $\int \operatorname{sech}^2 4x\, dx$

10. $\int \operatorname{sech} 7x \tanh 7x\, dx$

11. $h(x) = x^3 \coth x$

12. $h(x) = x^{2.5} \operatorname{csch} 4x$

13. $\int_1^3 \tanh x\, dx$

14. $\int_{-4}^4 \sinh x\, dx$

15. $q(x) = \dfrac{\sinh 5x}{\ln 3x}$

16. $r(x) = \dfrac{\cosh 6x}{\cos 3x}$

17. $\int_0^1 x \sinh x\, dx$

18. $\int_a^b x^2 \cosh x\, dx$

19. $y = 3 \sinh^{-1} 4x$

20. $y = 5 \tanh^{-1}(x^3)$

21. $\int \tanh^{-1} 5x\, dx$

22. $\int 4 \cosh^{-1} 6x\, dx$

Hyperbolic Substitution Problems: For Problems 23 and 24, integrate by hyperbolic substitution, using the fact that $\cosh^2 t = \sinh^2 t + 1$ and $\sinh^2 t = \cosh^2 t - 1$.

23. $\int \sqrt{x^2 + 9}\, dx$

24. $\int \sqrt{x^2 - 25}\, dx$

25. *Hanging Chain or Cable Problem:* If a chain (or a flexible cable that doesn't stretch) is hung between two supports, it takes the shape of a hyperbolic cosine curve,

$$y = \frac{h}{w} \cosh \frac{w}{h} x + C$$

where x and y are horizontal and vertical distances to a point on the chain, h is the horizontal tensile force exerted on the chain, and w is the weight of the chain per unit length. In this problem you will show why this is true. Figure 9-9g shows the graph of the chain in an xy-coordinate system. Any point on the chain experiences horizontal and vertical forces of h and v, respectively. Force h is constant and depends on how tightly the chain is pulled at its ends. Force v varies and equals the weight of the part of the chain below point (x, y). The resultant tension vector points along the graph.

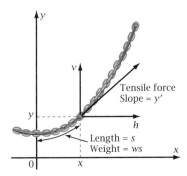

Figure 9-9g

a. Explain why the slope, $y' = dy/dx$, of the graph at point (x, y) is equal to v/h.

b. The weight of the chain from 0 to x equals the length, s, times the weight per unit length, w. Explain why this equation is true.

$$y' = \frac{w}{h} s$$

c. If you differentiate both sides of the equation in part b, you get the differential equation

$$d(y') = \frac{w}{h}\, ds$$

Using what you know about arc length, show that you can write this differential equation as

$$d(y') = \frac{w}{h}\left[1 + (y')^2\right]^{1/2} dx$$

d. Separating the variables in the equation in part c and integrating gives

$$\int [1 + (y')^2]^{-1/2}\, d(y') = \int \frac{w}{h}\, dx$$

Perform a hyperbolic substitution on the left side, letting $y' = \sinh t$, and integrate to find

$$\sinh^{-1} y' = \frac{w}{h}x + C$$

e. Use the fact that $y' = 0$ at the vertex, $x = 0$, to evaluate the constant of integration, C.

f. Based on the above work, show that

$$y' = \sinh \frac{w}{h}x$$

g. From part f, show that the equation of the hanging chain is as shown in this box.

Property: Equation of Hanging Chain or Cable

$$y = \frac{h}{w}\cosh \frac{w}{h}x + C$$

where w is the weight of chain per unit length,

h is the horizontal tensile force on chain (in units consistent with w),

x is the distance from the axis of symmetry to point (x, y) on the chain,

y is the vertical distance from the x-axis to point (x, y), and

C is a vertical distance determined by the position of the chain.

---

26. *Can You Duplicate This Graph?* Figure 9-9h shows the graph of a hyperbolic cosine function with general equation

$$y = k \cosh \frac{1}{k}x + C$$

a. Find the particular equation. Check your equation by showing that its graph agrees with points in the figure.

b. Calculate y if x is 20.

c. Calculate x if y is 4. On a copy of Figure 9-9h, show that your answer is consistent with the graph.

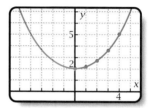

Figure 9-9h

d. Find the slope of the graph if x is 3. Show that a line of this slope through the point on the graph where $x = 3$ is tangent to the graph.

e. Find the area of the region under the graph from $x = -1$ to $x = 3$.

f. Find the length of the graph from $x = -1$ to $x = 3$.

27. *Power Line Problem:* An electrical power line is to be suspended between pylons 300 ft apart (Figure 9-9i). The cable weighs 0.8 lb/ft and will be connected to the pylons 110 ft above ground.

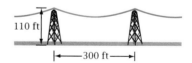

Figure 9-9i

a. Contractors plan to use a horizontal tensile force of $h = 400$ lb to hold the cable. Find the particular equation of the resulting catenary. Use the equation to calculate how close to the ground the cable will come.

b. How long will the cable in part a be? How much will it weigh?

c. For the cable in part a, where will the maximum total tensile force be, at the middle or at the ends? What will this maximum tension be equal to?

d. The power company decides that the cable must come no closer to the ground than 100 ft. How strong a horizontal force would be needed to achieve this clearance if the cables are still connected on 110-ft pylons?

28. *Hanging Chain Experiment:* Suspend a chain, about 6 to 10 feet in length between two convenient suspension points, such as nails driven into the upper frame of a chalkboard. Let the chain hang down fairly far, as shown in Figure 9-9f (page 475). Set up a coordinate system with the *y*-axis halfway between the two suspension points. The *x*-axis can be some convenient horizontal line, such as a chalk tray. In this experiment you are to derive an equation for the vertical distance from the *x*-axis to the chain and to check it by actual measurement.

   a. Measure, in centimeters, the *x*- and *y*-coordinates of the two suspension points and the vertex.

   b. Find the particular equation of the particular catenary that fits the three data points.

   c. Make a table of values of *y* versus *x* for each 10 cm, going both ways from *x* = 0 and ending at the two suspension points.

   d. Mark the chain links at the suspension points, then take down the chain. Plot the points you calculated in part c on the chalkboard. Find some way to make sure the *y*-distances are truly vertical. Then rehang the chain to see how well it fits the catenary.

e. Find the particular equation of the parabola (quadratic function) that fits the three measured data points. Plot this function and the catenary from part b on the same screen. Sketch the result, showing the difference between a catenary and a parabola.

f. Use your equation from part b to calculate the length of the chain between the two suspension points. Then stretch out the chain on the floor and measure it. How close does the calculated value come to the measured value?

29. *Bowl Problem:* The graph of $y = \sinh x$ from $x = 0$ to $x = 1$ is rotated about the *y*-axis to form a bowl (Figure 9-9j). Assume that *x* and *y* are in feet.

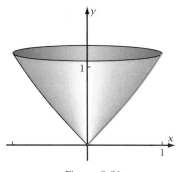

Figure 9-9j

   a. Find the surface area of the bowl.

   b. The bowl is to be silver-plated inside and out. The cost of plating is $57 per square foot. How much will plating cost?

   c. How much liquid could be held inside the bowl if it were filled to within a half-inch of the top?

30. *Gateway Arch Problem:* The Gateway to the West Arch in St. Louis, Missouri, is built in the shape of an inverted catenary (Figure 9-9k, on the next page). This shape was used because compression forces act tangentially to the structure, as do the tensile forces on a chain, thus avoiding bending of the stainless steel from which the arch is constructed. The outside of the arch is 630 ft wide at the base and 630 ft high. The inside of the arch is 520 ft wide and 612 ft high.

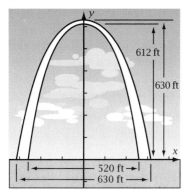

Figure 9-9k

a. Find particular equations of the inner and outer catenaries.

b. Verify that your equations are correct by plotting them on your grapher.

c. The stress created by wind blowing against the arch depends on the area of the region between the two graphs. Find this area.

d. A spider starts at the point where the left end of the outside of the arch meets the ground. It crawls all the way up, then down the other side to the ground. As it crawls, it leaves one strand of web. How long is the strand it leaves?

e. How steeply must the spider climb when it first begins?

f. José Vasquez wants to fly a plane underneath the arch. The plane has a (total) wingspan of 120 ft. Below what altitude, y, can José fly to ensure that each wing misses the inside of the arch by at least 50 ft horizontally?

31. *Derivative Verification Problem:* For $H(x) = \operatorname{csch} x$,

    a. Find $H'(1)$ exactly, using the differentiation formula.

    b. Find $H'(1)$ approximately, using the symmetric difference quotient with $\Delta x = 0.01$. By what percentage does the approximate answer differ from the exact answer?

32. *Integral Verification Problem:* Evaluate $\int_1^2 \operatorname{sech} x \, dx$ by the fundamental theorem, using the antiderivative formula for the hyperbolic secant. Then evaluate the integral approximately using numerical integration. How does the numerical answer compare with the exact answer?

33. *Integration by Parts Problem:* Evaluate $\int e^x \sinh 2x \, dx$ by parts. Then integrate again by first using the definition of sinh to transform to exponential form. Show that the two answers are equivalent. Which technique is easier?

34. *Integration Surprise Problem:* Try to integrate by parts.

$$\int e^x \sinh x \, dx$$

What causes integration by parts to fail in this case? Recalling the definition of sinh, find another way to evaluate the integrals, and do it.

35. *Derivations of the Pythagorean Properties of Hyperbolic Functions:*

    a. Starting with the definition of $\cosh x$ and $\sinh x$, prove that $\cosh^2 x - \sinh^2 x = 1$.

    b. Divide both members of the equation in part a by $\cosh^2 x$, and thus derive the property $1 - \tanh^2 x = \operatorname{sech}^2 x$.

    c. Derive the property $\coth^2 x - 1 = \operatorname{csch}^2 x$.

36. *Double-Argument Properties of Hyperbolic Functions:*

    a. Explain why $\sinh 2x = \frac{1}{2}(e^{2x} - e^{-2x})$.

    b. Derive the double-argument property $\sinh 2x = 2 \sinh x \cosh x$.

    c. Derive the double-argument property $\cosh 2x = \cosh^2 x + \sinh^2 x$.

    d. Derive the other form of the double-argument property, $\cosh 2x = 1 + 2 \sinh^2 x$.

    e. Derive the property $\sinh^2 x = \frac{1}{2}(\cosh 2x - 1)$.

37. *Hyperbolic Radian Problem:* In trigonometry you learn that the argument x, in radians, for the circular functions $\sin x$ or $\cos x$ equals an arc length on the unit circle (Figure 9-9l, left). In this problem you will show that the same is not true for x, in hyperbolic radians, for $\sinh x$ and $\cosh x$. You will show that the arguments in both circular and hyperbolic radians equal the area of a sector of the circular or hyperbolic region shown in Figure 9-9l.

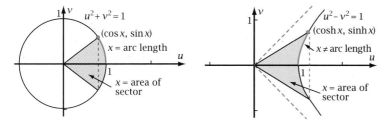

Figure 9-9l

a. Show that the unit circle, $u^2 + v^2 = 1$, between $u = \cos 2$ and $u = 1$ is 2 units long, but the hyperbola, $u^2 - v^2 = 1$, between $u = 1$ and $u = \cosh 2$ is greater than 2 units long.

b. Show that 2 is the area of the hyperbolic sector in Figure 9-9l with the point $(\cosh 2, \sinh 2)$ as its upper boundary and $(\cosh 2, -\sinh 2)$ as its lower boundary.

c. Show that x is the area of the circular sector with the point $(\cos x, \sin x)$ as its upper boundary and $(\cos x, -\sin x)$ as its lower boundary.

d. Show in general that x is the area of the hyperbolic sector with the point $(\cosh x, \sinh x)$ as its upper boundary and $(\cosh x, -\sinh x)$ as its lower boundary.

38. *Algebraic Derivatives of the Other Five Inverse Hyperbolic Functions:* Derive the differential formulas given in this section for these expressions.

a. $\dfrac{d}{dx}(\sinh^{-1} x)$

b. $\dfrac{d}{dx}(\tanh^{-1} x)$

c. $\dfrac{d}{dx}(\coth^{-1} x)$

d. $\dfrac{d}{dx}(\operatorname{sech}^{-1} x)$

e. $\dfrac{d}{dx}(\operatorname{csch}^{-1} x)$

9-10   Improper Integrals

Suppose you are driving along the highway at 80 ft/s (about 55 mi/h). At time $t = 0$ s, you take your foot off the accelerator and let the car start slowing down. Assume that your velocity is given by

$$v(t) = 80e^{-0.1t}$$

where $v(t)$ is in feet per second. According to this mathematical model, the velocity approaches zero as time increases but is never equal to zero. So you are never quite stopped. Would the distance you go approach a limiting value, or would it increase without bound? In this section you will learn how to answer such questions by evaluating *improper integrals*.

OBJECTIVE   Given an improper integral, determine whether it **converges** (that is, approaches a finite number as a limit). If it does, find the number to which it converges.

Figure 9-10a shows the velocity function previously mentioned. The distance the car goes between $t = 0$ and $t = b$ is equal to the area of the region under the graph. So,

$$\text{Distance} = \int_0^b 80e^{-0.1t}\, dt$$

$$= -800e^{-0.1t}\Big|_0^b$$

$$= -800e^{-0.1b} + 800$$

If $b = 10$ s, the distance is 505.6... ft. As b approaches infinity, the $-800e^{-0.2b}$ term approaches zero. Thus, the distance approaches 800 ft. The mathematical model shows you that the car never passes a point 800 ft from where you started slowing. The integral

$$\int_0^\infty 80e^{-0.1t}\, dt$$

is called an **improper integral** because one of its limits of integration is not finite. The integral *converges* to 800 because the integral from 0 to b approaches 800 as b approaches infinity. Suppose the velocity function had been $v(t) = 320(t + 4)^{-1}$. The graph (Figure 9-10b) looks almost the same. The velocity still approaches zero as time increases. So,

$$\text{Distance} = \int_0^b 320(t + 4)^{-1}\, dt$$

$$= 320\ln|t + 4|\Big|_0^b$$

$$= 320\ln|b + 4| - 320\ln 4$$

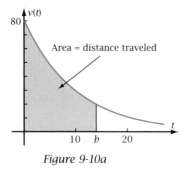

Figure 9-10a

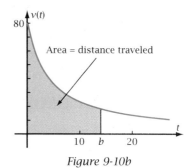

Figure 9-10b

As b approaches infinity, so does $\ln|b + 4|$. The integral *diverges*. Unlike the first mathematical model, this one shows that the car would go arbitrarily far from the starting point if you waited long enough!

A definite integral is improper if one of these holds true.

- The upper or lower limit of integration is infinite.

- The integrand is discontinuous for at least one value of x in the closed interval determined by the limits of integration.

DEFINITION: Improper Integrals

$$\int_a^\infty f(x)\,dx = \lim_{b \to \infty} \int_a^b f(x)\,dx$$

$$\int_{-\infty}^b f(x)\,dx = \lim_{a \to -\infty} \int_a^b f(x)\,dx$$

$$\int_a^b f(x)\,dx = \lim_{k \to c^+} \int_k^b f(x)\,dx + \lim_{k \to c^-} \int_a^k f(x)\,dx, \ f \text{ is discontinuous}$$

at $x = c$ in $[a, b]$

An improper integral converges to a certain number if each applicable limit shown is finite. Otherwise, the integral **diverges**.

Note: An improper integral with an infinite limit of integration always diverges if the integrand has a limit other than zero as the variable of integration approaches infinity.

▶ **EXAMPLE 1**   For the improper integral $\int_0^\infty x^2 e^{-x}\,dx$,

a. Graph the integrand and explain whether the integral might converge.

b. If the integral might converge, find out whether or not it does, and if so, to what limit it converges.

Solution   a. Figure 9-10c shows that the integral might converge because the integrand seems to approach zero as x gets very large.

b. Replace ∞ with b and let b approach infinity.

$$\int_0^\infty x^2 e^{-x}\,dx = \lim_{b \to \infty} \int_0^b x^2 e^{-x}\,dx$$

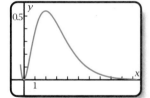

Figure 9-10c

Integrating by parts twice gives

$$\lim_{b \to \infty}(-x^2 e^{-x} - 2xe^{-x} - 2e^{-x})\Big|_0^b = \lim_{b \to \infty}(-b^2 e^{-b} - 2be^{-b} - 2e^{-b} + 0 + 0 + 2)$$

The limit of $b^2 e^{-b}$ can be found by two applications of l'Hospital's rule.

$$\lim_{b \to \infty} b^2 e^{-b} = \lim_{b \to \infty} \frac{b^2}{e^b} \to \frac{\infty}{\infty}$$ Write the expression as a quotient.

$$= \lim_{b \to \infty} \frac{2b}{e^b} \to \frac{\infty}{\infty}$$ Use l'Hospital's rule. (Take the derivative of numerator and denominator.)

$$= \lim_{b \to \infty} \frac{2}{e^b} \to \frac{2}{\infty}$$ Use l'Hospital's rule again.

$$= 0$$ Limit of the form (finite/infinite) is zero.

Similarly, the second and third terms in the limit each go to zero. Therefore, $\displaystyle\lim_{b \to \infty}(-b^2 e^{-b} - 2be^{-b} - 2e^{-b} + 0 + 0 + 2) = 2$.   ◀

▶ **EXAMPLE 2**   For the improper integral $\int_0^\infty x^{0.2}\,dx$,

a. Graph the integrand and explain whether the integral might converge.

b. If the integral might converge, find out whether or not it does, and if so, to what limit it converges.

Solution   a. The graph in Figure 9-10d indicates that the integral does not converge. The limit of the integrand, $x^{0.2}$, is not 0 as x approaches infinity, so the integral diverges.

b. Nothing remains to be done because the integral diverges.

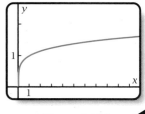

Figure 9-10d   ◀

▶ **EXAMPLE 3**   For the improper integral $\int_{-2}^{2} \dfrac{1}{x}\,dx$,

a. Graph the integrand and explain whether the integral might converge.

b. If the integral might converge, find out whether or not it does, and if so, to what limit it converges.

Solution   a. Figure 9-10e shows the graph of the function. There is a discontinuity at $x = 0$, so you must evaluate two integrals, one from -2 to b and the other from a to 2. You write

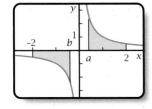

Figure 9-10e

$$\int_{-2}^{2} \frac{1}{x}\,dx = \lim_{b \to 0^-} \int_{-2}^{b} \frac{1}{x}\,dx + \lim_{a \to 0^+} \int_{a}^{2} \frac{1}{x}\,dx$$

At first glance, you might think that the integral converges to zero. If a and b are the same distance from the origin, then there is as much "negative" area below the x-axis as there is area above. For the integral to converge, however, both of the integrals shown must converge. Checking the first one,

$$\lim_{b \to 0^-} \int_{-2}^{b} \frac{1}{x}\,dx = \lim_{b \to 0^-} \ln |x| \Big|_{-2}^{b} = \lim_{b \to 0^-} (\ln |b| - \ln |-2|)$$

Because $\ln 0$ is infinite, this integral diverges. Thus, the original integral also diverges.

b. There is nothing to be done for part b because the integral diverges.   ◀

Problem Set 9-10

Q1. Sketch: $y = e^{-x}$

Q2. $\int \cosh x\,dx = $ —?—

Q3. $y = \cosh x \Rightarrow y' = $ —?—

Q4. $y = \cos x \Rightarrow y' = $ —?—

Q5. $\int \cos x\,dx = $ —?—

Q6. What function has a graph like Figure 9-10f?

Q7. What function has a graph like Figure 9-10g?

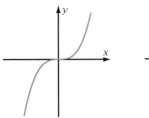

Figure 9-10f   Figure 9-10g

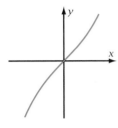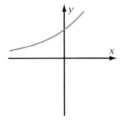

Figure 9-10h   Figure 9-10i

Q8. What function has a graph like Figure 9-10h?

Q9. What function has a graph like Figure 9-10i?

Q10. The maximum value of y if $y = -x^2 + 10x + 7$ is —?—.

    A. 32   B. 5   C. 7

    D. 0   E. 82

For Problems 1–20,

    a. Explain from the graph of the integrand whether the integral might converge.

    b. If the integral might converge, find out whether or not it does, and if so, the limit to which it converges.

1. $\int_2^\infty \dfrac{1}{x^2}\,dx$ 2. $\int_3^\infty \dfrac{1}{x^4}\,dx$

3. $\int_1^\infty \dfrac{1}{x}\,dx$ 4. $\int_0^1 \dfrac{1}{x}\,dx$

5. $\int_1^\infty \dfrac{1}{x^{0.2}}\,dx$ 6. $\int_1^\infty \dfrac{1}{x^{1.2}}\,dx$

7. $\int_0^1 \dfrac{1}{x^{0.2}}\,dx$ 8. $\int_0^1 \dfrac{1}{x^{1.2}}\,dx$

9. $\int_0^\infty \dfrac{dx}{1 + x^2}$ 10. $\int_0^\infty \dfrac{dx}{1 + x}$

11. $\int_0^1 \dfrac{dx}{x \ln x}$ 12. $\int_3^\infty \dfrac{dx}{x (\ln x)^2}$

13. $\int_2^\infty e^{-0.4x}\,dx$ 14. $\int_0^\infty e^{0.02x}\,dx$

15. $\int_{-1}^2 \sqrt{x}\,dx$ 16. $\int_1^7 (x - 3)^{-2/3}\,dx$

17. $\int_0^\infty xe^{-x}\,dx$ 18. $\int_0^3 (x - 1)^{-2}\,dx$

19. $\int_0^\infty \cos x\,dx$ 20. $\int_0^\infty \sin x\,dx$

21. *Divergence by Oscillation Problem:* The improper integrals in Problems 19 and 20 are said to diverge by oscillation. Explain why these words make sense. A graph may help.

22. *p-Integral Problem:* An integral of the form

$$I_p = \int_1^\infty \frac{1}{x^p}\,dx$$

where p stands for a constant, is called a **p-integral**. For some values of the exponent p, the integral converges and for others it doesn't.

Figure 9-10j shows an example for which the two graphs look practically identical but only one of the integrals converges. In this problem your objective is to find the values of p for which the p-integral converges and those for which it diverges.

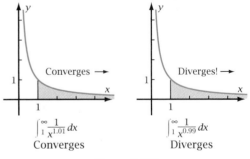

Figure 9-10j

    a. Show that I_p converges if $p = 1.001$ but not if $p = 0.999$.

    b. Does I_p converge if $p = 1$? Justify your answer.

    c. Complete the statement "I_p converges if p —?— and diverges if p —?—."

23. *Volume of an Unbounded Solid Problem:* Figure 9-10k (bottom of the page) shows the region under the graph of $y = 1/x$ from $x = 1$ to $x = b$.

   a. Does the region's area approach a finite limit as b approaches infinity? Explain.

   b. The region is rotated about the x-axis to form a solid. Does the volume of the solid approach a finite limit as $b \to \infty$? If so, what is the limit? If not, explain why.

   c. The region is rotated about the y-axis to form a different solid. Does this volume approach a finite limit as $b \to \infty$? If so, what is the limit? If not, explain why.

   d. True or false: "If a region has infinite area, then the solid formed by rotating that region about an axis has infinite volume."

24. *Infinite Paint Bucket Problem:* The graph of $y = -1/x$ from $x = 0$ to $x = 1$ is rotated about the y-axis to form an infinitely deep paint bucket (Figure 9-10l).

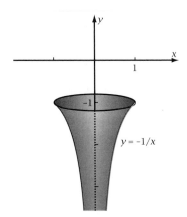

Figure 9-10l

Explain why a vertical cross section along the y-axis will have an infinite area and thus why the surface area of the bucket itself is infinite.

Then show that the bucket could be completely filled with a finite volume of paint, thus coating the infinite surface area. Surprising?

25. *The Gamma Function and Factorial Function:* In this problem you will explore

$$f(x) = \int_0^\infty t^x e^{-t}\, dt$$

where x is a constant with respect to the integration. Figure 9-10m shows the integrand for $x = 1$, $x = 2$, and $x = 3$.

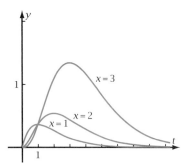

Figure 9-10m

   a. Find $f(1)$, $f(2)$, and $f(3)$ by evaluating the improper integral. Along the way you will have to show, for instance, that

$$\lim_{b \to \infty} b^3 e^{-b} = 0$$

   b. From the pattern you see in the answers to part a, make a conjecture about what $f(4)$, $f(5)$, and $f(6)$ are equal to.

   c. Integrate by parts once and thus show that

$$f(x) = x \cdot f(x - 1)$$

Use the answer to confirm your conjecture in part b.

   d. The result of the work in parts a–c forms a basis for the definition of the factorial

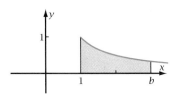
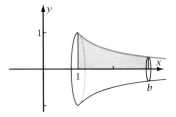
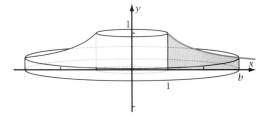

Figure 9-10k

function. Explain why this definition is consistent with the definition

$$x! = (x)(x-1)(x-2)\ldots(2)(1)$$

Definition: The Factorial Function (and the Gamma Function)

$$x! = \int_0^\infty t^x e^{-t}\, dt \qquad \text{The factorial function.}$$
$$\Gamma(x) = (x-1)! \qquad \text{The gamma function.}$$
(Γ is the uppercase Greek letter gamma.)

a. Confirm that the integral for 3! approaches 6 by integrating numerically from $t = 0$ to $t = b$ for some fairly large value of b. How large a value of b makes the integral come within 0.000001 of 6?

b. You can use the improper integral to define factorials for non-integer values of x. Write an integral equal to 0.5!. Evaluate it numerically, using the value of b from part e. How can you tell from the graphs in Figure 9-10m that your answer will be closer than 0.000,001 to the correct answer? How does your answer compare with the value in the National Bureau of Standards Handbook of Mathematical Functions, namely, $0.5! = 0.8862269255$?

c. Quick! Without further integration, calculate 1.5!, 2.5!, and 3.5!. (Remember part c.)

d. Show that $0! = 1$, as you probably learned in algebra.

e. Show that $(-1)!, (-2)!, (-3)!, \ldots$, are infinite but $(-0.5)!, (-1.5)!$, and $(-2.5)!$ are finite.

f. Show that the value of 0.5! in part f can be expressed, rather simply, in terms of π.

26. *Spaceship Work Problem:* A 1000-lb spaceship is to be sent to a distant location. The work required to get the spaceship away from Earth's gravity equals the force times the distance the spaceship is moved. But the force, F, which is 1000 lb at Earth's surface, decreases with the square of the distance from Earth's center,

$$F = \frac{1000}{r^2}$$

where r is the number of earth-radii. There is always some force no matter how far you travel from Earth, so additional work is always being

done. Does the amount of work increase without bound as r goes to infinity? Show how you arrive at your answer.

The Space Shuttle Endeavor (STS-47) blasting off on September 12, 1992

27. *Piecewise Continuity Problem:* Figure 9-10n shows the graph of

$$y = 2^x - \frac{|x-2|}{x-2}$$

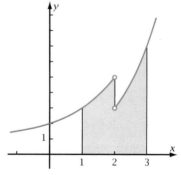

Figure 9-10n

Suppose that you are to evaluate

$$\int_1^3 \left(2^x - \frac{|x-2|}{x-2}\right) dx$$

Although the integrand is discontinuous on the closed interval [1, 3], there is only a step discontinuity at $x = 2$. The integrand is continuous everywhere else in [1, 3]. Such a function is said to be *piecewise-continuous* on the given interval. In this problem you will show that a piecewise-continuous function is integrable on the given interval.

Definition: Piecewise Continuity

Function f is **piecewise-continuous** on the interval $[a, b]$ if and only if there is a finite number of values of x in $[a, b]$ at which $f(x)$ is discontinuous, the discontinuities are either removable or step discontinuities, and f is continuous elsewhere on $[a, b]$.

a. Write the integral as the sum of two integrals, one from $x = 1$ to $x = 2$ and the other from $x = 2$ to $x = 3$.

b. Both integrals in part a are improper. Write each one using the correct limit terminology.

c. Show that both integrals in part b converge. Observe that the expression $|x - 2|/(x - 2)$ equals one constant to the left of $x = 2$ and a different constant to the right. Find the value to which the original integral converges.

d. Explain why this property is true.

Property: Integrability of Piecewise-Continuous Functions

If function f is piecewise-continuous on the interval $[a, b]$, then f is integrable on $[a, b]$.

e. True or false: "A function is integrable on the interval $[a, b]$ if and only if it is continuous on $[a, b]$." Justify your answer.

28. *Journal Problem:* Update your journal with what you've learned. You should include such things as

   • The one most important thing you've learned since the last journal entry

   • The big integration technique that allows you to integrate a product of two functions

   • Other integration techniques that involve substitutions and algebraic transformations

   • Hyperbolic functions

   • Improper integrals

   • Why the fundamental theorem, not numerical integration, is needed for improper integrals

   • Any techniques or ideas about the calculus of transcendental functions that are still unclear to you

---

9-11 Miscellaneous Integrals and Derivatives

By the time you finish this section you will have seen all of the classical algebraic techniques for performing calculus. These techniques were the only way you could do calculus before the advent of the computer made numerical methods easily implementable.

OBJECTIVE   Algebraically integrate or differentiate expressions containing the elementary functions.

488   Chapter 9: Algebraic Calculus Techniques for the Elementary Functions

TECHNIQUES: Differentiation

- Sum: $(u + v)' = u' + v'$

- Product: $(uv)' = u'v + uv'$
  Or use the logarithmic differentiation technique.

- Quotient: $(u/v)' = (u'v - uv')/v^2$
  Or use the logarithmic differentiation technique.

- Composite: $[f(u)]' = f'(u)u'$
  Chain rule.

- Implicit: $f(y) = g(x) \Rightarrow f'(y)y' = g'(x)$
  Use the chain rule, where y is the inside function.

- Power function: $(x^n)' = nx^{n-1}$

- Exponential function: $(n^x)' = (n^x)\ln n$
  Use the logarithmic differentiation technique.

- Logarithmic function: $(\log_b x)' = (1/x)(1/\ln b)$

- Logarithmic differentiation technique:
  $$y = f(x) \Rightarrow \ln y = \ln f(x) \Rightarrow (1/y)y' = [\ln f(x)]' \Rightarrow y' = y[\ln f(x)]'$$

- Trigonometric function:
  $\sin' x = \cos x; \quad \cos' x = -\sin x$
  For tan x, cot x, sec x, and csc x, write as sines and cosines and use the quotient rule.

- Inverse trigonometric function: Differentiate implicitly.

- Hyperbolic function: $\sinh' x = \cosh x, \quad \cosh' x = \sinh x$

- Inverse hyperbolic function: Differentiate implicitly.

TECHNIQUES: Indefinite Integration

- Known derivative: $\int f'(x)\,dx = f(x) + C$

- Sum: $\int (u + v)\,dx = \int u\,dx + \int v\,dx$

- Product: $\int u\,dv = uv - \int v\,du$
  Integrate by parts.

- Reciprocal function: $\int u^{-1}\,du = \ln|u| + C$

- Power function: $\int u^n\,du = u^{n+1}/(n+1) + C, \ n \neq -1$
  "u-substitution" method.

- Power of a function: $\int f^n(x)\,dx$
  Use a reduction formula.

- Square root of a quadratic: Integrate by trigonometric substitution; complete the square first, if necessary.

- Rational algebraic function: Convert to a sum by long division and by resolving into partial fractions.

- Inverse function [exponential (logarithmic), trigonometric, or hyperbolic]: Integrate by parts.

Problem Set 9-11

For Problems 1–100, differentiate the given function, or evaluate the given integral.

1. $y = \sec 3x \tan 3x$

2. $y = \sinh 5x \tanh 5x$

3. $\int x \cosh 4x \, dx$

4. $\int x \cos x \, dx$

5. $f(x) = (3x + 5)^{-1}$

6. $f(x) = (5 - 2x)^{-1}$

7. $\int (3x + 5)^{-1} \, dx$

8. $\int (5 - 2x)^{-1} \, dx$

9. $t(x) = \tan^5 4x$

10. $h(x) = \operatorname{sech}^3 7x$

11. $\int \sin^2 x \, dx$

12. $\int \cos^2 x \, dx$

13. $y = \dfrac{6x - 11}{x + 2}$

14. $y = \dfrac{5x + 9}{x - 4}$

15. $\int \dfrac{6x - 11}{x + 2} \, dx$

16. $\int \dfrac{5x + 9}{x - 4} \, dx$

17. $f(t) = \sqrt{1 + t^2}$

18. $g(t) = \sqrt{t^2 - 1}$

19. $\int \sqrt{1 + t^2} \, dt$

20. $\int \sqrt{t^2 - 1} \, dt$

21. $y = x^3 e^x$

22. $y = x^4 e^{-x}$

23. $\int x^3 e^x \, dx$

24. $\int x^4 e^{-x} \, dx$

25. $f(x) = \sin^{-1} x$

26. $g(x) = \tan^{-1} x$

27. $\int \sin^{-1} x \, dx$

28. $\int \tan^{-1} x \, dx$

29. $\int \dfrac{1}{x^2 + 4x - 5} \, dx$

30. $\int \dfrac{1}{x^2 - 6x - 7} \, dx$

31. $\int \dfrac{1}{\sqrt{x^2 + 4x - 5}} \, dx$

32. $\int \dfrac{1}{\sqrt{x^2 - 6x - 7}} \, dx$

33. $f(x) = \tanh x$

34. $f(x) = \coth x$

35. $\int \tanh x \, dx$

36. $\int \coth x \, dx$

37. $y = e^{2x} \cos 3x$

38. $y = e^{-3x} \cos 4x$

39. $\int e^{2x} \cos 3x \, dx$

40. $\int e^{-3x} \cos 4x \, dx$

41. $g(x) = x^3 \ln 5x$

42. $h(x) = x^2 \ln 8x$

43. $\int x^3 \ln 5x \, dx$

44. $\int x^2 \ln 8x \, dx$

45. $y = \dfrac{x}{(x + 2)(x + 3)(x + 4)}$

46. $y = \dfrac{x}{(x - 1)(x - 2)(x - 3)}$

47. $\int \dfrac{x}{(x + 2)(x + 3)(x + 4)} \, dx$

48. $\int \dfrac{x}{(x - 1)(x - 2)(x - 3)} \, dx$

49. $y = \cos^3 x \sin x$

50. $y = \sin^5 x \cos x$

51. $\int \cos^3 x \sin x \, dx$

490

52. $\int \sin^5 x \cos x \, dx$

53. $\int \cos^3 x \, dx$

54. $\int \sin^5 x \, dx$

55. $\int \cos^4 x \, dx$

56. $\int \sin^6 x \, dx$

57. $g(x) = (x^4 + 3)^3$

58. $f(x) = (x^3 - 1)^4$

59. $\int (x^4 + 3)^3 \, dx$

60. $\int (x^3 - 1)^4 \, dx$

61. $\int (x^4 + 3)^3 x^3 \, dx$

62. $\int (x^3 - 1)^4 x^2 \, dx$

63. $\int (x^4 + 3) \, dx$

64. $\int (x^3 - 1) \, dx$

65. $f(x) = \int_1^x (t^4 + 3)^3 \, dt$

66. $h(x) = \int_5^x (t^3 - 1)^4 \, dt$

67. $\int_1^2 x e^x \, dx$

68. $\int_0^2 x e^{-x} \, dx$

69. $r(x) = x e^x$

70. $s(x) = x e^{-x}$

71. $q(x) = \dfrac{\ln x + 2}{x}$

72. $r(x) = \dfrac{(\ln x)^3 + 4}{x}$

73. $\int \dfrac{\ln x + 2}{x} \, dx$

74. $\int \dfrac{(\ln x)^3 + 4}{x} \, dx$

75. $f(x) = e^{x^2}$

76. $f(x) = e^{x^3}$

77. $\int x e^{x^2} \, dx$

78. $\int x^2 e^{x^3} \, dx$

79. $\int x^3 e^{x^2} \, dx$

80. $\int x^5 e^{x^3} \, dx$

In Problems 81–100, a, b, c, d, and n stand for constants.

81. $\int e^{ax} \cos bx \, dx$

82. $\int e^{ax} \sin bx \, dx$

83. $\int \sin^2 cx \, dx$

84. $\int \cos^2 cx \, dx$

85. $f(x) = \dfrac{ax + b}{cx + d}$

86. $f(x) = (ax + b)^n$

87. $\int \dfrac{ax + b}{cx + d} \, dx$

88. $\int (ax + b)^n \, dx$

89. $\int (x^2 + a^2)^{-1/2} x \, dx$

90. $\int (a^2 - x^2)^{-1/2} x \, dx$

91. $\int (x^2 + a^2)^{-1/2} \, dx$

92. $\int (a^2 - x^2)^{-1/2} \, dx$

93. $f(x) = x^2 \sin ax$

94. $f(x) = x^2 \cos ax$

95. $\int x^2 \sin ax \, dx$

96. $\int x^2 \cos ax \, dx$

97. $\int \sinh ax \, dx$

98. $\int \cosh ax \, dx$

99. $\int \cos^{-1} ax \, dx$

100. $\int \sin^{-1} ax \, dx$

Historical Topic 1—Rationalizing Algebraic Substitutions

Before calculators and computers were readily available to perform numerical integration, it was important to be able to find algebraic formulas for as many integrals as possible. Users of mathematics spent much time searching for clever integration techniques as did students of mathematics in learning these techniques. Two such techniques are shown here and in Problem 107. You can transform integrals such as

$$I = \int \frac{1}{1 + \sqrt[3]{x}}\, dx$$

that have a radical in the denominator to a rational integrand by substituting a variable either for the radical or for the entire denominator. Here's how you would go about it. Let $u = 1 + \sqrt[3]{x}$. Then $(u - 1)^3 = x$, from which $dx = 3(u - 1)^2\, du$. Substituting u for the denominator and $3(u - 1)^2\, du$ for dx gives

$$I = 3\int \frac{(u - 1)^2}{u}\, du = 3\int \left(u - 2 + \frac{1}{u}\right) du$$

$$= \frac{3}{2} u^2 - 6u + 3 \ln |u| + C$$

$$= \frac{3}{2}(1 + \sqrt[3]{x})^2 - 6(1 + \sqrt[3]{x}) + 3 \ln |1 + \sqrt[3]{x}| + C$$

For Problems 101–106, evaluate the integrals by algebraic substitution.

101. $\displaystyle\int \frac{1}{1 + \sqrt{x}}\, dx$

102. $\displaystyle\int \frac{1}{1 - \sqrt{x}}\, dx$

103. $\displaystyle\int \frac{1}{1 + \sqrt[4]{x}}\, dx$

104. $\displaystyle\int \frac{1}{\sqrt{x} + \sqrt[3]{x}}\, dx$

105. $\displaystyle\int \frac{1}{\sqrt{e^x + 1}}\, dx$

106. $\displaystyle\int \frac{1}{\sqrt{e^x - 1}}\, dx$

Historical Topic 2—Rational Functions of $\sin x$ and $\cos x$ by $u = \tan(x/2)$

107. From trigonometry, recall the double-argument properties for cosine and sine,

$$\cos 2t = 2 \cos^2 t - 1 \quad \text{and} \quad \sin 2t = 2 \sin t \cos t$$

a. Explain how these properties justify these equations.

$$\cos x = 2 \cos^2 \tfrac{x}{2} - 1 \quad \text{and}$$
$$\sin x = 2 \sin \tfrac{x}{2} \cos \tfrac{x}{2}$$

b. Show that the equations in part a can be transformed to

$$\cos x = \frac{1 - \tan^2 \tfrac{x}{2}}{1 + \tan^2 \tfrac{x}{2}} \quad \text{and}$$

$$\sin x = \frac{2 \tan \tfrac{x}{2}}{1 + \tan^2 \tfrac{x}{2}}$$

c. Let $u = \tan(x/2)$. Show that $x = 2 \tan^{-1} u$, and thus that these properties are true.

> **Property: $u = \tan(x/2)$ Substitution**
>
> $$\cos x = \frac{1 - u^2}{1 + u^2}, \quad \sin x = \frac{2u}{1 + u^2}, \quad \text{and}$$
>
> $$dx = \frac{2}{1 + u^2}\, du$$

d. Let $u = \tan(x/2)$. Use the results of part c to show that this integral reduces to $\int du$.

$$\int \frac{1}{1 + \cos x}\, dx$$

e. Perform the integration in part d and then do the reverse substitution to show that the integral equals $\tan(x/2) + C$.

108. *Another Indefinite Integral of Secant:*

a. Transform the integral

$$\int \sec x\, dx$$

using the substitution $u = \tan(x/2)$ from Problem 107 to get

$$\int \frac{2}{1 - u^2}\, du$$

b. Perform the integration in part a. Show that the result is

$$\ln \left| \frac{1 + \tan \tfrac{x}{2}}{1 - \tan \tfrac{x}{2}} \right| + C$$

c. Recall from trigonometry

   i. $\tan(A + B) = \dfrac{\tan A + \tan B}{1 - \tan A \tan B}$

   ii. $\tan \tfrac{\pi}{4} = 1$

492

Use this information to show that
$$\int \sec x \, dx = \ln |\tan\left(\tfrac{\pi}{4} + \tfrac{x}{2}\right)| + C$$

d. Evaluate $\int_0^1 \sec x \, dx$ two ways:
  i. Using the result of part c
  ii. Using the more familiar integral formula
  Show that the answers are equivalent.

For Problems 109–111, use the substitution $u = \tan(x/2)$ to evaluate the integral.

109. $\displaystyle\int \frac{1}{1 - \cos x} \, dx$

110. $\displaystyle\int \frac{1}{1 + \sin x} \, dx$

111. $\displaystyle\int \frac{\cos x}{1 - \cos x} \, dx$

9-12   Integrals in Journal

In this chapter you have learned algebraic techniques by which you can integrate and differentiate the elementary transcendental functions. The integration techniques include

- Recognition of the integrand as the derivative of a familiar function
- Integral of the power function, $\int u^n \, du$, $n \neq -1$
- Integral of the reciprocal function, $\int u^{-1} \, du$
- Integration by parts
- Reduction formulas
- Trigonometric substitution
- Partial fractions
- Other substitutions

These techniques let you find the equation of a function whose derivative is given. They also allow you to use the fundamental theorem to find the exact value of a definite integral.

Although differentiating a function is relatively easy, the reverse process, integrating, can be like unscrambling eggs! Integrals that look almost the same, such as

$$\int (x^2 + 1)^{10} x \, dx \qquad \text{and} \qquad \int (x^2 + 1)^{10} \, dx$$

may require completely different techniques. An integrand, such as e^{-x^2} in

$$\int e^{-x^2} \, dx$$

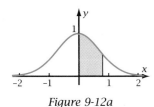

Figure 9-12a

may be an elementary function but not the derivative of any other elementary function. [This integral gives the area under the bell curve in statistics (Figure 9-12a).]

In this section you will record a short table of integrals in your journal. Constructing the table will bring together the various techniques of integration. The end product will give you a reference that you can use to recall various integrals and how you derived them. As a result, you will better be able to use publications, such as Chemical Rubber Company (CRC) tables, and to understand the output from symbol-manipulating computers.

Problem Set 9-12

1. *Table of Integrals Problem:* Record a short table of integrals in your journal. Arrange the table by the nature of the integrand, rather than by the technique used. Include integrals of the algebraic functions and each one of the elementary transcendental functions:

   - Power and reciprocal
   - Exponential and logarithmic

   - Circular and reverse circular
   - Hyperbolic and inverse hyperbolic

   You should include examples of other frequently occurring forms, such as rational function, square root of a quadratic, and power of a trigonometric function (especially $\int \sin^2 x \, dx$, $\int \sec^3 x \, dx$, and so on). For each entry, state or show how the formula is derived.

9-13   Chapter Review and Test

In this chapter you have learned to do algebraically the calculus of elementary transcendental functions—exponential and logarithmic, circular (trigonometric), hyperbolic, and their inverses.

Review Problems

R0. Update your journal with what you've learned since the last entry. You should include such things as

  - The one most important thing you have learned in studying Chapter 9
  - Which boxes you have been working on in the "define, understand, do, apply" table
  - Any techniques or ideas about calculus that are still unclear to you

R1. Let $f(x) = x \cos x$. Find $f'(x)$, observing the derivative of a product property. From the results, find an equation for the indefinite integral $\int x \sin x \, dx$. Check your work by using the equation to evaluate the definite integral

$$\int_1^4 x \sin x \, dx$$

and comparing it with the approximate answer you get by numerical integration.

R2. Integrate: $\int 5x \sin 2x \, dx$

R3. a. Integrate: $\int x^3 \cos 2x \, dx$
   b. Integrate: $\int e^{4x} \sin 3x \, dx$
   c. Integrate: $\int x (\ln x)^2 \, dx$
   d. The region under the graph of $y = x \ln x$ from $x = 1$ to $x = 2$ is rotated about the y-axis to form a solid. Find the volume of the solid.

R4. a. Integrate by parts once to express $\int \cos^{30} x \, dx$ in terms of an integral of a reduced power of $\cos x$.
   b. Use the appropriate reduction formula to evaluate $\int \sec^6 x \, dx$.
   c. Derive the reduction formula for $\int \tan^n x \, dx$.

R5. a. Integrate without reduction formula:
$\int \cos^5 x\, dx$

  b. Integrate without reduction formula:
$\int \sec^6 x\, dx$

  c. Integrate without reduction formula:
$\int \sin^2 7x\, dx$

  d. Integrate without reduction formula:
$\int \sec^3 x\, dx$

  e. Integrate without reduction formula:
$\int \tan^9 32\, dx$

  f. Find the exact area of the region inside the limaçon with polar equation $r = 9 + 8\sin\theta$ from $\theta = 0$ to $\theta = \pi/4$.

R6. a. Integrate: $\int \sqrt{x^2 - 49}\, dx$

  b. Integrate: $\int \sqrt{x^2 - 10x + 34}\, dx$

  c. Integrate: $\int \sqrt{1 - 0.25x^2}\, dx$

  d. Using the fundamental theorem, find the exact area of the zone of a circle of radius 5 between the lines 3 units and 4 units from the center (Figure 9-13a).

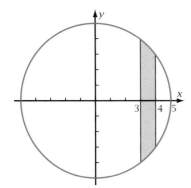

Figure 9-13a

R7. For parts a–d, integrate:

  a. $\displaystyle \int \frac{(6x + 1)\, dx}{x^2 - 3x - 4}$

  b. $\displaystyle \int \frac{5x^2 - 21x - 2}{(x - 1)(x + 2)(x - 3)}\, dx$

  c. $\displaystyle \int \frac{5x^2 + 3x + 45}{x^3 + 9x}\, dx$

  d. $\displaystyle \int \frac{5x^2 + 27x + 32}{x(x + 4)^2}\, dx$

e. *Differential Equation Problem:* Figure 9-13b shows the slope field for the differential equation

$$\frac{dy}{dx} = 0.1(y - 3)(y - 8)$$

Solve this differential equation subject to the initial condition that $y = 7$ when $x = 0$. On a copy of the slope field, plot the graph of your solution, thus showing that it is reasonable.

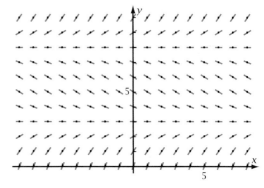

Figure 9-13b

R8. a. Sketch the graph: $y = \cos^{-1} x$

  b. Differentiate: $f(x) = \sec^{-1} 3x$

  c. Integrate: $\int \tan^{-1} 5x\, dx$

  d. Find the area of the region in Quadrant I bounded by the graph of $y = \cos^{-1} x$.

R9. a. Sketch the graph: $f(x) = \sinh x$

  b. Sketch the graph: $g(x) = \cosh^{-1} x$

  c. Differentiate: $h(x) = x^2 \operatorname{sech} x$

  d. Differentiate: $f(x) = \sinh^{-1} 5x$

  e. Integrate: $\int \tanh 3x\, dx$

  f. Integrate: $\int \cosh^{-1} 7x\, dx$

  g. Using the definitions of $\cosh x$ and $\sinh x$, prove that $\cosh^2 x - \sinh^2 x = 1$.

  h. Find a particular equation of the catenary with vertex $(0, 5)$ and point $(3, 7)$. Use the equation to predict the value of y if $x = 10$. Find the values of x if $y = 20$.

R10. a. Evaluate: $\int_3^{\infty} (x - 2)^{-1.2}\, dx$

  b. Evaluate: $\int_{\pi/2}^{0} \tan x\, dx$

  c. Evaluate: $\int_{-1}^{1} x^{-2/3}\, dx$

d. Evaluate: $\int_0^4 \left(\sqrt{x} - \dfrac{|x-1|}{x-1} \right) dx$

e. For what values of p does the p-integral $\int_1^\infty x^{-p} dx$ converge?

R11. a. Differentiate: $f(x) = x \sin^{-1} x$

b. Integrate: $\int x \sin^{-1} x \, dx$

c. Differentiate: $\tanh(e^x)$

d. Integrate: $\int (x^3 - x)^{-1} dx$

e. Differentiate: $f(x) = (1 - x^2)^{1/2}$

f. Integrate: $\int (1 - x^2)^{1/2} dx$

g. Differentiate: $g(x) = (\ln x)^2$

h. Integrate: $\int x \ln x \, dx$

R12. Explain why $\int (9 - x^2)^{-1/2} dx$ has an inverse sine in the answer but $\int (9 - x^2)^{-1/2} x \, dx$ does not.

Concept Problems

C1. *Integral of sech x Problem:* Derive the formula

$$\int \operatorname{sech} x \, dx = \sin^{-1}(\tanh x) + C$$

You can transform the integrand to a square root involving $\tanh x$ by using the Pythagorean property relating $\operatorname{sech} x$ and $\tanh x$. Then you can use a very clever trigonometric substitution to rationalize the radical. Confirm that the formula works by evaluating the integral on the interval $[0, 1]$, then checking by numerical integration.

C2. *Integral of csch x Problem:* Derive the formula

$$\int \operatorname{csch} x \, dx = \ln \left| \tanh \dfrac{x}{2} \right| + C$$

You can transform the integrand to functions of $\tanh(x/2)$ by first observing that

$$\sinh 2A = 2 \sinh A \cosh A$$

from which

$$\operatorname{csch} x = \dfrac{1}{2 \sinh \frac{x}{2} \cosh \frac{x}{2}}$$

Clever algebra, followed by an application of the Pythagorean properties, produces the desired result. Then substitute u for $\tanh(x/2)$. You will have to be clever again to figure out what to substitute for dx in terms of du. The resulting integral is remarkably simple! Confirm that the formula works by evaluating the integral on the interval $[1, 2]$, then checking by numerical integration.

C3. *Another Integral of csc x:* Derive the formula

$$\int \csc x \, dx = \ln \left| \tan \dfrac{x}{2} \right| + C$$

Confirm that the formula works by evaluating the integral on the interval $[0.5, 1]$, then checking by numerical integration.

C4. *Another Definition of π Problem:* Figure 9-13c shows the region under the graph of $y = (x^2 + 1)^{-1}$, extending to infinity in both directions. Show that the area of this infinitely long region is exactly equal to π. This fact is remarkable because the integrand has nothing to do with circles, yet the answer is the most fundamental number concerned with circles!

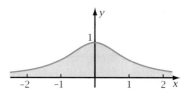

Figure 9-13c

C5. *Upper Bound Problem:* Figure 9-13d shows the graphs of $f(x) = \ln x$ and $g(x) = \tan^{-1} x$. As x gets larger, both graphs increase but are concave down. The inverse tangent graph approaches $\pi/2$. Prove that the graph of $f(x) = \ln x$ is unbounded above. You can do this by assuming that it is bounded above by some number M, then finding a contradiction by finding a value of x in terms of M for which $\ln x > M$.

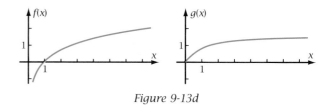

Figure 9-13d

Chapter Test

No calculators allowed (T1–T8)

For Problems T1–T6, evaluate the indefinite integral.

T1. $\int \sin^5 x \cos x \, dx$

T2. $\int x^3 \sinh 6x \, dx$

T3. $\int \cos^{-1} x \, dx$

T4. $\int \sec^3 x \, dx$

T5. $\int e^{2x} \cos 5x \, dx$

T6. $\int \ln 3x \, dx$

For Problems T7 and T8, differentiate.

T7. $f(x) = \operatorname{sech}^3(e^{5x})$

T8. $g(x) = \sin^{-1} x$

Graphing calculators allowed (T9–T15)

T9. Given $f(x) = \tanh^{-1} x$, find a formula for $f'(x)$ in terms of x by appropriate implicit differentiation. Demonstrate that the formula is correct by approximating $f'(0.6)$ using numerical differentiation.

T10. Find the particular equation of the form $y = k \cosh(1/k)x + C$ for the catenary containing vertex $(0, 1)$ and point $(5, 3)$.

T11. a. Integrate $\int \dfrac{x - 3}{x^2 - 6x + 5} \, dx$ three ways:

    i. By trigonometric substitution, after completing the square

    ii. By partial fractions

    iii. As the integral of the reciprocal function

    b. Show that all three answers are equivalent.

T12. Evaluate $\int \cos^2 x \, dx$ by appropriate use of the double-argument properties.

T13. a. Evaluate $\int \cos^5 x \, dx$ two ways:

    i. By transforming four of the cosines to sines, and integrating as powers of sine.

    ii. By using the reduction formula,

$$\int \cos^n x \, dx$$

$$= \frac{1}{n} \cos^{n-1} x \sin x + \frac{n-1}{n} \int \cos^{n-2} x \, dx$$

    b. Show that the two answers are equivalent.

T14. Evaluate the improper integral: $\int_0^\infty x e^{-0.1x} \, dx$

T15. What did you learn as a result of taking this test that you did not know before?

The Calculus of Motion—Averages, Extremes, and Vectors

The distance a spaceship travels equals velocity multiplied by time. But the velocity varies. Displacement is the integral of velocity, and velocity is the integral of acceleration. By measuring acceleration of the spaceship at frequent time intervals, you can calculate the displacement by numerical calculus methods.

Mathematical Overview

Chapter 10 extends your study of objects in motion. You will distinguish between concepts such as

- Distance versus displacement
- Acceleration versus velocity
- Maximum versus minimum
- Linear versus planar motion

You will do this in four ways.

Graphically   The icon at the top of each even-numbered page of this chapter shows the position vector, velocity vector, and acceleration vector for an object moving in a curved, planar path.

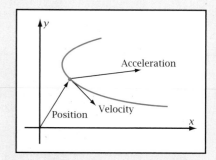

Numerically

Time	Acceleration	Velocity
0	1.3	20.0
2	1.7	23.0
4	2.2	26.9
6	2.1	31.2
8	1.5	34.8
⋮	⋮	⋮

Algebraically $\text{Distance} = \int_a^b |v(t)|\, dt$ $\text{Displacement} = \int_a^b v(t)\, dt$

Verbally   *Now I know the precise definition of the average value of a function. It is the integral of the function between two limits divided by the difference between those limits.*

10-1 Introduction to Distance and Displacement for Motion Along a Line

You have learned that velocity is the rate of change of position with respect to time. In this chapter you will concentrate on the distinction between distance (how far) and displacement (how far and in which direction). You will sharpen your understanding of the difference between speed (how fast) and velocity (how fast and in which direction). Once you have made these distinctions for motion in one dimension (along a line), you will use vectors to analyze motion in two dimensions (in a plane). Along the way you will find maximum, minimum, and average values of velocity and position functions.

OBJECTIVE Given an equation for the velocity of a moving object, find the distance traveled and the displacement from the starting point for a specified time interval.

Suppose that you drive 100 mi and then return 70 of those miles (Figure 10-1a). Although you have gone a **distance** of 170 mi, your **displacement**, which is measured from the starting point, is only 30 mi. Problem Set 10-1 will clarify the distinction between these two quantities. You may work on your own or with your study group.

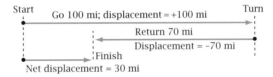

Total distance = 170 mi

Figure 10-1a

Exploratory Problem Set 10-1

Calvin's Swimming Problem: Calvin enters an endurance swimming contest. The objective is to swim upstream in a river for a period of 10 min. The river flows at 30 ft/min. Calvin jumps in and starts swimming upstream at 100 ft/min. Phoebe ascertains that as Calvin tires, his speed through the water decreases exponentially with time according to the equation $v_c = 100(0.8)^t$.

Thus, Calvin's net velocity (Figure 10-1b) is only

$$v = 100(0.8)^t - 30$$

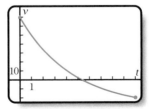

Figure 10-1b

1. At what time will Calvin's velocity become negative?

2. How far will Calvin travel upstream—that is, while his velocity is positive? How far will he travel back downstream (while his velocity is negative) until $t = 10$? What is the total distance Calvin will have traveled in the 10 min?

3. What will Calvin's displacement from the starting point be at the end of the 10 min? Will he be upstream or downstream of his starting point?

4. Write a definite integral that you can use to find the displacement after 10 min in one computation. Check it by evaluating the integral and comparing with Problem 3.

5. Write one definite integral that represents the total distance Calvin travels in the 10 min. A clever application of absolute value will help. Integrate numerically.

10-2 Distance, Displacement, and Acceleration for Linear Motion

Most real objects, such as cars and birds, travel in two or three dimensions. In Section 10-6, you will learn how to use vectors to analyze such motion. For the time being, consider only objects moving in a straight line. In Section 10-1, you saw the distinction between the distance a moving object travels and its displacement from the starting point. If the velocity is positive, the displacement is positive. If the velocity is negative, then the displacement is negative. In the latter case the distance traveled is the opposite of the displacement. You can combine the two ideas with the aid of the absolute value function.

PROPERTY: Distance and Displacement

$$\text{Displacement} = \int_a^b (\text{velocity})\, dt$$

$$\text{Distance} = \int_a^b |\text{velocity}|\, dt$$

OBJECTIVE   Given velocity or acceleration as a function of time for an object in linear motion, find the displacement at a given time and the distance traveled in a given time interval.

▶ **EXAMPLE 1**   A moving object has velocity $v(t) = t^2 - 7t + 10$, in feet per second, in the time interval $[1, 4]$.

    a. Find the time subintervals in which the velocity is positive and the subintervals in which it is negative.

    b. Find the distance the object travels in each of these subintervals.

c. Use a single integral to find the displacement in the time interval $[1, 4]$, and another single integral to find the distance traveled in this interval.

d. Show how you can find the answers to part c from the answers to parts a and b.

Solution

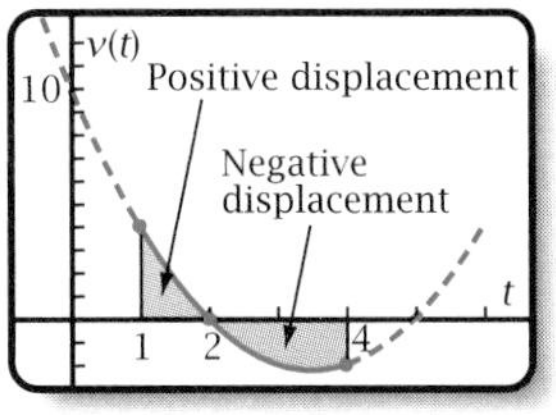

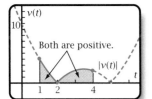

Figure 10-2a

a. The graph of v versus t (Figure 10-2a) shows that the velocity changes from positive to negative at $t = 2$. You can confirm this fact algebraically by solving

$$t^2 - 7t + 10 = 0 \Rightarrow t = 2 \text{ or } t = 5$$

Positive velocity: $[1, 2)$
Negative velocity: $(2, 4]$

b. For the time interval $[1, 2]$,

$$\text{displacement} = \int_1^2 (t^2 - 7t + 10)\, dt = 1.8333... = 1\tfrac{5}{6}$$
Numerically, or by fundamental theorem.

$$\therefore \text{ distance } = 1\tfrac{5}{6} \text{ ft}$$

For the time interval $[2, 4]$,

$$\text{displacement} = \int_2^4 (t^2 - 7t + 10)\, dt = -3.3333... = -3\tfrac{1}{3}$$

$$\therefore \text{ distance } = 3\tfrac{1}{3} \text{ ft}$$

c. For the time interval $[1, 4]$,

$$\text{displacement} = \int_1^4 (t^2 - 7t + 10)\, dt = -1.5$$

$$\text{distance} = \int_1^4 |t^2 - 7t + 10|\, dt$$
Use the absolute value function on your grapher.

$$= 5.1666... = 5\tfrac{1}{6} \text{ ft}$$
Use numerical integration (see note).

d. For displacement, $1\tfrac{5}{6} + \left(-3\tfrac{1}{3}\right) = -1.5$ (✔).

For distance, $1\tfrac{5}{6} + 3\tfrac{1}{3} = 5\tfrac{1}{6}$ (✔). ◀

Note: If you want to evaluate the integral in part c algebraically using the fundamental theorem, divide the interval $[1, 4]$ as in part b. The integrand has a cusp at $t = 2$ (Figure 10-2b) at which the absolute value function changes from $+(t^2 - 7t + 10)$ to $-(t^2 - 7t + 10)$. (Recall from algebra that the absolute value of a negative number is the opposite of that number.) The final answer is still the sum,

$$1\tfrac{5}{6} + \left(+3\tfrac{1}{3}\right) = 5\tfrac{1}{6} \text{ ft}$$

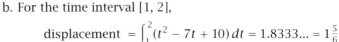

Figure 10-2b

Figure 10-2b also shows that the integrand for distance is always positive (or zero), meaning that the distance traveled is positive. This is true even though the object in Example 1 is displaced a negative amount from its starting point.

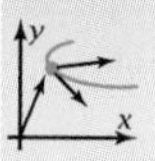

▶ **EXAMPLE 2**   A car accelerates for 24 s. Its acceleration in (mi/h)/s is measured every 3 s and is listed in the table.

Time (s)	Acceleration [(mi/h)/s]
0	1.3
3	1.7
6	2.2
9	2.1
12	1.5
15	0.3
18	−0.4
21	−1.1
24	−1.4

a. Plot the graph of acceleration versus time.

b. At time $t = 0$, the car was going 20 mi/h. Predict its velocity at each 3-s instant from 0 through 24.

c. Plot the graph of velocity versus time.

d. Approximately how far did the car go in this 24-s interval?

Solution   a. Figure 10-2c shows the graph of acceleration versus time.

b. Acceleration is the derivative of velocity with respect to time, so the velocity is the integral of acceleration.

$$v = \int a \, dt$$

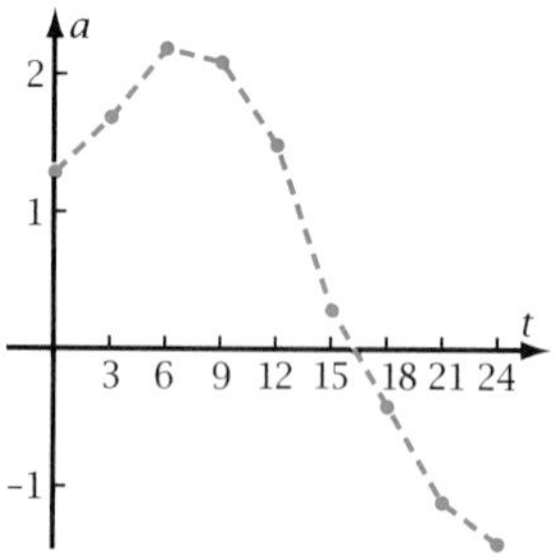

You can estimate the average acceleration for each 3-s interval. For interval 1,

$$\text{Average acceleration} = \tfrac{1}{2}(1.3 + 1.7) = 1.50$$

So, the change in velocity is $(1.50)(3) = 4.50$. The initial velocity is given to be 20, so the velocity at the end of the interval is about $20 + 4.50 = 24.50$.

The process is equivalent to integrating by the trapezoidal rule. You can do the calculations by extending the given table using a computer spreadsheet or the LIST feature on your grapher.

Figure 10-2c

Time (s)	Acceleration [(mi/h)/s]	Average Acceleration [(mi/h)/s]	Δv (mi/h)	Velocity (mi/h)
0	1.3	—	—	20 (given)
3	1.7	1.50	4.50	24.5
6	2.2	1.95	5.85	30.35
9	2.1	2.15	6.45	36.8
12	1.5	1.80	5.40	42.2
15	0.3	0.90	2.70	44.9
18	−0.4	−0.05	−0.15	44.75
21	−1.1	−0.75	−2.25	42.5
24	−1.4	−1.25	−3.75	38.75

504

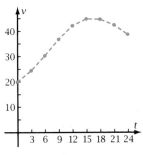

Figure 10-2d

c. Figure 10-2d shows the graph of velocity versus time.

d. You can find the displacement at the end of each 3-s interval the same way you found the velocity. Find an average velocity for the interval, multiply by Δt, and add the result to the displacement at the beginning of the interval. The displacement at $t = 0$ is zero because you are finding the displacement from the starting point. The only catch is that you must use 3/3600 h for Δt because v is in miles per hour. The calculations can be done by extending the table from part b and putting appropriate instructions into the spreadsheet.

Time (s)	Acceleration [(mi/h)/s]	Average Acceleration [(mi/h)/s]	Δv (mi/h)	Velocity (mi/h)	Average Velocity (mi/h)	Displacement (mi)
0	1.3	—	—	20	—	0.000
3	1.7	1.50	4.50	24.5	22.25	0.018...
6	2.2	1.95	5.85	30.35	27.425	0.041...
9	2.1	2.15	6.45	36.8	33.575	0.069...
12	1.5	1.80	5.40	42.2	39.5	0.102...
15	0.3	0.90	2.70	44.9	43.55	0.138...
18	−0.4	−0.05	−0.15	44.75	44.825	0.175...
21	−1.1	−0.75	−2.25	42.5	43.625	0.212...
24	−1.4	−1.25	−3.75	38.75	40.625	0.246...

The last column shows the total displacement from $t = 0$ to the end of the time interval. For the 24 s, the total displacement was about 0.246 mi, or about 1300 ft. ◀

Problem Set 10-2

Quick Review

Q1. You traveled 30 mi/h for 4 h. How far did you go?

Q2. You traveled 75 mi in 3 h. How fast did you go?

Q3. You traveled 50 mi at 40 mi/h. How long did it take?

Q4. Differentiate: $f(x) = \ln x$

Q5. Integrate: $\int \ln x \, dx$

Q6. Differentiate: $f(t) = \tan t$

Q7. Differentiate: $g(t) = \tanh t$

Q8. Integrate: $\int x^2 \, dx$

Q9. Integrate: $\int 2^x \, dx$

Q10. Differentiate: $h(x) = 2^x$

For Problems 1–4, an object moving in a straight line has velocity $v(t)$ in the given time interval.

a. Find the time subintervals in which the velocity is positive and those in which it is negative.

b. Find the distance the object travels in each of these subintervals.

c. Use a single integral to find the displacement in the given time interval, and use a single integral to find the distance traveled in this interval.

d. Show how you can find the answers to part c from the answers to parts a and b.

e. Find the acceleration of the object at the midpoint of the time interval.

1. $v(t) = t^2 - 10t + 16$ ft/s, from $t = 0$ to $t = 6$ s

2. $v(t) = \tan 0.2t$ cm/s, from $t = 10$ s to $t = 20$ s

3. $v(t) = \sec \frac{\pi}{24} t - 2$ km/h, from $t = 1$ s to $t = 11$ h

4. $v(t) = t^3 - 5t^2 + 8t - 6$ mi/min, from $t = 0$ to $t = 5$ min

For Problems 5–8, first find an equation for the velocity of a moving object from the equation for acceleration. Recall that acceleration is the derivative of velocity. Then find the displacement and distance traveled by the moving object in the given time interval.

5. $a(t) = t^{1/2}$ (ft/s)/s, $v(0) = -18$ ft/s, from $t = 0$ to $t = 16$ s

6. $a(t) = t^{-1}$ (cm/s)/s, $v(1) = 0$ cm/s, from $t = 0.4$ s to $t = 1.6$ s

7. $a(t) = 6 \sin t$ (km/h)/h, $v(0) = -9$ km/h, from $t = 0$ to $t = \pi$ h

8. $a(t) = \sinh t$ (mi/h)/h, $v(0) = -2$ mi/h, from $t = 0$ to $t = 5$ h

9. *Meg's Velocity Problem:* Meg accelerates her car, giving it a velocity of $v = t^{1/2} - 2$, in feet per second, at time t, in seconds after she started accelerating.

   a. Find the time(s) at which $v = 0$.

   b. Find her net displacement for the time interval [1, 9].

   c. Find the total distance she travels for the time interval [1, 9].

10. *Periodic Motion Problem:* The velocity of a moving object is given by $v = \sin 2t$, in centimeters per second.

    a. Find the distance the object travels from the time it starts ($t = 0$) to the first time it stops.

    b. Find the displacement of the object from its starting point and the total distance it travels from $t = 0$ to $t = 4.5\pi$. Be clever!

11. *Car on the Hill Problem:* Faye Ling's car runs out of gas as she drives up a long hill. She lets the car roll without putting on the brakes. As it slows down, stops, and starts rolling backward, the car's velocity up the hill is given by

    $$v = 60 - 2t$$

    where v is in feet per second and t is the number of seconds since the car ran out of gas.

a. What is the car's net displacement between $t = 10$ and $t = 40$?

b. What is the total distance the car rolls between $t = 10$ and $t = 40$?

12. *Rocket Problem:* If a rocket is fired straight up from Earth, it experiences acceleration from two sources:

    • Upward, a_u, in (meters per second) per second, due to the rocket engine

    • Downward, a_d, in (meters per second) per second, due to gravity

    The net acceleration is $a = a_u + a_d$. However, the upward acceleration has a discontinuity at the time the rocket engine stops. Suppose that the accelerations are given by

    $$a_u = \begin{cases} 40 \cos 0.015t, & 0 \le t \le 100 \\ 0, & t > 100 \end{cases}$$

    $$a_d = -9.8, \text{ for all } t$$

    a. Plot graphs of a and v versus t for the first 300 s.

    b. At what value of t does a become negative? At what value of t does v become negative?

    c. Find the displacement of the rocket at time $t = 300$ and the distance the rocket traveled between $t = 0$ and $t = 300$. What does the relationship between these two numbers tell you about what is happening in the real world at $t = 100$?

    d. How fast and in what direction is the rocket traveling at $t = 300$?

13. *Subway Problem:* A train accelerates as it leaves one subway station, then decelerates as it approaches the next one. Three calculus students take an accelerometer aboard the train. They measure the accelerations, a, given in the table in (miles per hour) per second, at the given values of t, in seconds.

    a. Calculate the velocity and displacement at the end of each time interval. Assume that the velocity was zero at time zero. You may use a computer spreadsheet.

    b. Show that the train has stopped at $t = 60$.

    c. How can the velocity be zero at $t = 0$ but the acceleration be positive at that time?

Time (s)	Acceleration [(mi/h)/s]
0	1.2
5	4.7
10	2.9
15	0.6
20	0
25	0
30	0
35	0
40	−0.4
45	−1.4
50	−3.8
55	−3.2
60	0

d. How can acceleration be zero from $t = 20$ to $t = 35$, while the velocity is not zero?

e. How far is it between the two stations?

14. *Spaceship Problem:* A spaceship is to be sent into orbit around Earth. You must determine whether the proposed design of the last-stage booster rocket will get the spaceship going fast enough and far enough so that it can orbit. Based on the way the fuel burns, the acceleration of the spaceship is predicted to be as shown in the table, where time is in seconds and acceleration is in miles per hour per second.

a. Initially the spaceship is 400 mi from the launchpad, going 6000 mi/h. Calculate the velocity and acceleration at the end of each time interval. You may use a spreadsheet.

Time (s)	Acceleration [(mi/h)/s]	
0	3	
10	14	
20	30	
30	36	
40	43	
50	42	
60	64	
70	78	
80	89	
90	6	
100	0	(Rocket burns out.)

b. Consulting the specifications, you find that when the last stage finishes firing (100 s, in this case) the spaceship must

• Be at least 1,000 mi from the launchpad

• Be moving at least 17,500 mi/h

Based on your work, conclude whether each of these specifications will be

• Definitely met

• Definitely not met

• Too close to say without more information

15. *Physics Formula Problem:* Elementary physics courses usually deal only with motion under a constant acceleration, such as motion under the influence of gravity. Under this condition, certain formulas relate acceleration, velocity, and displacement. These formulas are easily derived by calculus. Let a be the acceleration

(a constant). Let v_0 be the initial velocity (when time $t = 0$). Let s_0 be the initial displacement (again, when time $t = 0$). Derive the following formulas.

a. $v = v_0 + at$

b. $s = v_0 t + \frac{1}{2}at^2 + s_0$

16. *Elevator Project:* When a normal elevator starts going up, you feel a jerk until it gets up to speed. This is because the acceleration changes almost instantly from zero to some positive value. In this problem you will explore another way for an elevator's acceleration to be arranged such that the jerk is minimized.

   a. Suppose an elevator starts from rest ($v = 0$) at the bottom floor ($s = 0$ ft) and is given a constant acceleration of 2 ft/s² for 6 s. Thereafter, the elevator rises with constant velocity. Find the velocity and displacement of the elevator as functions of time. Sketch three graphs—acceleration, velocity, displacement—for times $t = 0$ to $t = 10$ s.

   b. How does the acceleration graph show that passengers on the elevator get a jerk at $t = 0$ and another jerk at $t = 6$?

   c. If the acceleration increases gradually to a maximum value, then decreases gradually to

zero, the jerks will be eliminated. Show that if the acceleration is given by

$$a = 2 - 2 \cos(\pi/3)t$$

for the first 6 s, then the elevator's acceleration has this property.

   d. Using the acceleration function in part c, find the velocity as a function of time.

   e. Sketch the graph of velocity as a function of time, if, as the elevator goes up, the velocity remains at the value it had at $t = 6$ s. How does this velocity graph differ from that in part a?

   f. How far does the elevator go while it is getting up to top speed?

   g. The elevator is to be slowed down the same way it was sped up, over a 6-s time period. If the elevator is to go all the way to the 50th story, 600 ft above the bottom floor, where should it start slowing down?

   h. How long does it take for the elevator to make the complete trip?

   i. If the elevator were to go up just one floor (12 ft), would the new acceleration and deceleration functions in parts c and f still provide a smooth ride? If so, how can you tell? If not, what functions could you use to smooth out the ride?

10-3   Average Value Problems in Motion and Elsewhere

Suppose that the velocity of a moving object is given by

$$v(t) = 12t - t^2$$

where t is in seconds and $v(t)$ is in feet per second. What would be meant by the average velocity in a time interval such as from $t = 2$ to $t = 11$? The equation distance = (rate)(time) is the basis for the answer. Dividing both sides by time gives rate = (distance)/(time). This concept is extended to velocity by using displacement instead of distance.

$$\text{Average velocity} = \frac{\text{displacement}}{\text{time}}$$

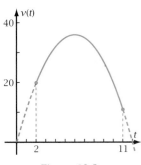

Figure 10-3a

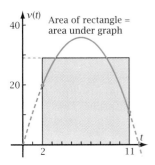

Figure 10-3b

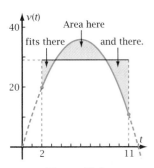

Figure 10-3c

Figure 10-3a shows the graph of $v(t)$. The total displacement s, in feet, is the integral of the velocity from 2 to 11.

$$s = \int_2^{11} (12t - t^2)\, dt$$
$$= 6t^2 - \tfrac{1}{3}t^3 \Big|_2^{11}$$
$$= 261$$

The time to travel the 261 ft is $(11 - 2)$, or 9 s. So, the average velocity is

$$v_{av} = \frac{261}{9} = 29$$

Plotting the average velocity on the $v(t)$-graph (Figure 10-3b) reveals several conclusions.

- The area of the rectangle with height equal to v_{av} and base 9 is also equal to the total displacement. So, the area of the rectangle equals the area under the $v(t)$-graph.

- Because the area of the rectangle equals the area of the region, there is just as much of the rectangular region above the v_{av} line as there is empty space below the line but above the rectangle. If you could move the region above the line into the spaces below the line, it would just fit (Figure 10-3c).

- The average velocity, v_{av}, is not equal to the average of the initial and final velocities. From the equation, $v(2) = 20$ and $v(11) = 11$. The average, 29 (which equals (displacement)/(time)), is higher than either one and is thus not equal to their average.

- Another object, starting at the same time and place but moving with a constant velocity equal to the average velocity, would finish at the same time and place as the first object.

The facts that average velocity equals (displacement)/(time) and that displacement is found by integrating the velocity lead to this general definition of average velocity.

DEFINITION: Average Velocity

If $v(t)$ is the velocity of a moving object as a function of time, then the average velocity from time $t = a$ to $t = b$ is

$$v_{av} = \frac{\int_a^b v(t)\, dt}{b - a}$$

That is, average velocity $= \dfrac{\text{displacement}}{\text{time}}$.

The reasoning used to define average velocity can be extended to define average value for any function. The definition of the average value of a function is shown at the top of the next page.

DEFINITION: *Average Value of a Function*

If function f is integrable on the interval $[a, b]$, then the average value of $y = f(x)$ on the interval $x = a$ to $x = b$ is

$$y_{av} = \frac{\int_a^b f(x)\,dx}{b - a}$$

OBJECTIVE Calculate the average value of a function given its equation.

Example 1 shows you how to find a value of x for which y equals the average value of the function.

▶ **EXAMPLE 1** For the function $v(t) = 12t - t^2$, graphed in Figure 10-3b (on the previous page), find a value $t = c$ for which $v(c)$ equals the average velocity, 29 ft/s.

Solution
$$29 = 12c - c^2$$
$$c^2 - 12c + 29 = 0$$
$$c = 3.3542\ldots \text{ or } 8.6457\ldots$$

At about 3.6 s and 8.6 s, the object is going 29 ft/s. ◀

Example 1 illustrates the **mean value theorem for definite integrals**, which states that for a continuous function, there will be at least one point at which the exact value of a function equals the average value of that function over a given interval.

PROPERTY: *Mean Value Theorem for Integrals*

If $y = f(x)$ is continuous on the closed interval $[a, b]$, then there is at least one point $x = c$ in $[a, b]$ for which

$$f(c) = y_{av} = \frac{\int_a^b f(x)\,dx}{b - a}$$

or, equivalently,

$$f(c)(b - a) = \int_a^b f(x)\,dx$$

The mean value theorem you learned in Chapter 5 is sometimes called the **mean value theorem for derivatives** to distinguish it from this new theorem.

Problem Set 10-3

Quick Review  5 min

Q1. What is your average speed if you go 40 mi in 0.8 h?

Q2. How far do you go in 3 min if your average speed is 600 mi/h?

Q3. How long does it take to go 10 mi at an average speed of 30 mi/h?

Q4. The first positive value of x at which $y = \cos x$ has a local maximum is —?—.

Q5. $y = e^x$ has a local maximum at what value of x?

Q6. $y = x^2 - 3x + 11$ has a local minimum at $x = $ —?—.

Q7. For $f(x) = (x - 5)^2$, the global maximum for $x \in [1, 3]$ is —?—.

Q8. If $f(x) = x^{0.8}$, then $f'(0) = $ —?—.

Q9. Name the theorem that states that under suitable conditions, a function's graph has a tangent line parallel to a given secant line.

Q10. The graph of $\left(\frac{x}{3}\right)^2 + \left(\frac{y}{5}\right)^2 = 1$ is a(n) —?—.

A. Circle   B. Hyperbola   C. Line

D. Ellipse   E. Parabola

For Problems 1–6,

    a. Find the average value of the function on the given interval.

    b. Sketch a graph showing the graphical interpretation of the average value.

    c. Find a point c in the given interval for which the conclusion of the mean value theorem for definite integrals is true.

1. $f(x) = x^3 - x + 5, \ x \in [1, 5]$

2. $f(x) = x^{1/2} - x + 7, \ x \in [1, 9]$

3. $g(x) = 3 \sin 0.2x, \ x \in [1, 7]$

4. $h(x) = \tan x, \ x \in [0.5, 1.5]$

5. $v(t) = \sqrt{t}, \ t \in [1, 9]$

6. $v(t) = 100(1 - e^{-t}), \ t \in [0, 3]$

For Problems 7–10, find a formula in terms of k for the average value of the given function on the interval $[0, k]$, where k is a positive constant (a stands for a constant).

7. $f(x) = ax^2$

8. $f(x) = ax^3$

9. $f(x) = ae^x$

10. $f(x) = \tan x, \ k < \frac{\pi}{2}$

11. *Average Velocity from Acceleration Problem:* Suppose you are driving 60 ft/s (about 40 mi/h) behind a truck. When you get the opportunity to pass, you step on the accelerator, giving the car an acceleration $a = 6/\sqrt{t}$, where a is in (feet per second) per second and t is in seconds. How fast are you going 25 s later when you have passed the truck? How far did you travel in that time? What was your average velocity for the 25-s interval?

12. *Ida's Speeding Ticket Problem:* Ida Livermore is rushing to take pizzas to her customers when she is stopped for speeding. Her ticket states that she was clocked at speeds up to 50 mi/h during a 4-min period, and that her fine will be $140 ($7 for each mile per hour over the 30-mi/h speed limit). Ida is good at calculus. She figures that her speed was a quadratic function of time over the 4-min period— 30 mi/h at the beginning and end, and peaking at 50 mi/h (Figure 10-3d). She argues she should be charged only for her average speed above 30 mi/h. How much less will her fine be if she wins her appeal?

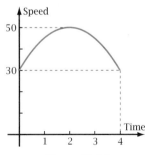

Figure 10-3d

13. *Average Velocity for Constant Acceleration Problem:* Prove that if an object moves with constant acceleration a, such as it does in ideal free fall, then its average velocity over a time interval is the average of the velocities at the beginning and end of the interval. (This result leads to one of the physics formulas you may have learned and that may have led you to a false conclusion about average velocity when the acceleration is not constant.)

14. *Average Velocity for Other Accelerations Problem:* Show by counterexample that if the acceleration of an object varies over a time interval, then the average velocity over that interval might not equal the average of the velocities at the beginning and end of the interval.

15. *Average Cost of Inventory Problem:* Merchants like to keep plenty of merchandise on hand to meet the needs of customers. But there is an expense associated with storing and stocking merchandise. Suppose that it costs a merchant $0.50 per month for each $100 worth of merchandise kept in inventory. Figure 10-3e shows the value of the inventory for a particular 30-day month.

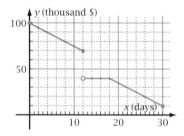

Figure 10-3e

a. Find the average value of the inventory, and use the answer to find the inventory cost for that month.

b. The function is discontinuous at $x = 12$ days. What do you suppose happened that day to cause the discontinuity? Is there a day $x = c$ at which the conclusion of the mean value theorem for integrals is true, even though the hypothesis of the theorem is not met for this piecewise function? Illustrate your answer on a copy of Figure 10-3e.

16. *Swimming Pool Average Depth Problem:* Figure 10-3f shows a vertical cross section along the length of a swimming pool. On a copy of the figure, draw a horizontal line at your estimate of the average depth of the water. Show on the graph that your answer is reasonable. How could you use this average depth to estimate the number of gallons of water the pool will hold?

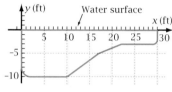

Figure 10-3f

17. *Average Temperature Problem:* Figure 10-3g shows the temperature recorded by a weather bureau station at various times on a day when a cold front arrived in the city at 11:00 a.m. Use the trapezoidal rule in an appropriate way to calculate the average temperature for that day. Compare this average temperature with the temperature found by averaging the high and low temperatures for the day. Give an advantage for each method of averaging.

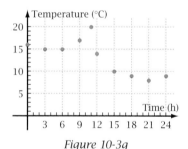

Figure 10-3g

18. *Average Vitamin C Amount Problem:* Calvin takes a 200-mg vitamin C tablet at 8:00 a.m. and another at 11:00 a.m. The quantity, y, in milligrams, of vitamin C remaining in his body after time x, in hours after 8:00 a.m., is given by the piecewise function shown in Figure 10-3h.

a. Between 8:00 and 11:00, $y = 200e^{-0.3x}$. How much vitamin C remains in Calvin's body at 11:00 a.m., before he takes the second pill? What was the average number of milligrams over that 3-h period?

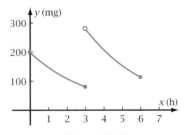

Figure 10-3h

b. At 11:00 a.m., when Calvin takes the next 200 mg of vitamin C, the value of y jumps immediately to $200 + k$, where k is the amount of the first dose remaining at 11:00. Between 11:00 a.m. and 2:00 p.m., $y = (200 + k)e^{-0.3(x-3)}$. Find the average number of milligrams over the 6-h period from 8:00 a.m. to 2:00 p.m.

c. The mean value theorem for integrals does not apply to y on the interval [0, 6] because the function is discontinuous at $x = 3$. Is the conclusion of the theorem true in spite of the discontinuity? Give numbers to support your conclusion.

19. *Average Voltage Problem:* For the normal alternating current supplied to houses, the voltage varies sinusoidally with time (Figure 10-3i), making 60 complete cycles each second. So,

$$v = A \sin 120\pi t$$

where v is in volts, t is in seconds, and A is the maximum voltage during a cycle. The average of the absolute value of the voltage is 110 volts. Use the fundamental theorem to find the average value of $y = |A \sin 120\pi t|$ from $t = 0$ to $t = 1/60$. Use the result to calculate the maximum voltage A if the average is 110 v. Show how you can find this number more easily using $y = \sin x$ and an appropriate interval of integration.

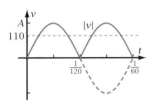

Figure 10-3i

20. *Root Mean Square Deviation Problem:* To measure how hilly a landscape is or how rough a machined surface is, people ask the question "On average, how far do points on the surface deviate from the mean level?" If you simply average the deviations, you will get zero. As Figure 10-3j shows, there is just as much area above the mean as there is below. One way to overcome this difficulty is to

• Square the deviations.

• Find the average of the squares.

• Take the square root of the average to get an answer with the same dimensions as the original deviations.

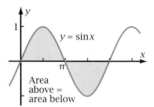

Figure 10-3j

The result is called the **root mean square deviation**. For instance, the roughness of a machined surface might be reported as "0.1 microinch, rms," where rms stands for root mean square and a microinch is one-millionth of an inch.

a. Suppose that the deviations from average are sinusoidal, as in Figure 10-3j. That is, $d = k \sin x$, where d is deviation, x is displacement along the surface, and k is a constant amplitude. Find the average of d^2 for one complete cycle. Use the result to calculate the rms deviation.

b. Plot the graph of $y = \sin^2 x$. Sketch the result. Show that the resulting graph is itself a sinusoid, and find its equation.

c. Show that you can determine the answer to part a graphically from part b, without having to use calculus.

d. Suppose a surface is lumpy, as in Figure 10-3k, and has the shape of the graph $y = |\sin x|$. Find y_{av}, the average value of y. Then find the rms deviation, using the fact that

$$\text{Deviation} = y - y_{av}$$

Based on your answer, would this surface be rougher or smoother than a sinusoidal surface with the same maximum distance between high points and low points, as in Figure 10-3k?

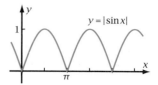

Figure 10-3k

10-4   Minimal Path Problems

Suppose you are swimming in the ocean. When you finish, you could swim straight to the place on the shoreline where you left your towel, or you could swim straight to the shoreline, then walk to the towel (Figure 10-4a). Swimming straight to the towel minimizes the distance you must go, but you swim slower than you walk, so this might increase your time. Heading for the closest point on the shoreline minimizes your time in the water, but increases your total distance. A third alternative, which might reduce your total time, is to swim to some point between the towel and the closest point on the shoreline, shortening the walk while increasing the swim by just a little.

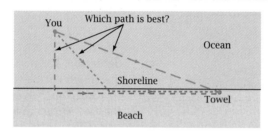

Figure 10-4a

Your objective in this section will be to analyze such problems to find the **minimal path**, in the case, the path which takes the least total time.

OBJECTIVE   Given a situation in which something goes from one place to another through two different media at different rates, find the path that minimizes a total time or cost.

▶ **EXAMPLE 1**   Suppose you are 200 yd from the beach and, in the minimum possible time, you want to get to the place where you left your towel. You can walk on the beach at 110 yd/min but can swim only 70 yd/min. Let x be the distance from the closest point on the beach to the point where you will make landfall (Figure 10-4b). What value of x minimizes your total time if your towel is

   a. 600 yd from the closest point
   b. 100 yd from the closest point

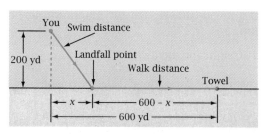

Figure 10-4b

Solution

a. Swim distance is $\sqrt{x^2 + 200^2}$.
   Walk distance is $(600 - x)$, where $x \le 600$.
   $\therefore$ total time will be

$$t = \frac{1}{70}\sqrt{x^2 + 200^2} + \frac{1}{110}(600 - x)$$

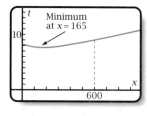

Figure 10-4c

To find the minimum total time, find the distance, x, at which t stops decreasing and starts increasing. You can do this graphically by plotting t as a function of x and tracing (Figure 10-4c). The minimum is somewhere between $x = 160$ and $x = 170$.

You can find the exact value algebraically by finding the value of x that makes the derivative of t equal to zero.

$$\frac{dt}{dx} = \frac{1}{140}(x^2 + 200^2)^{-1/2}(2x) - \frac{1}{110}$$

$$= \frac{x}{70}(x^2 + 200^2)^{-1/2} - \frac{1}{110}$$

$$\frac{dt}{dx} = 0 \qquad \text{if and only if}$$

$$\frac{x}{70}(x^2 + 200^2)^{-1/2} = \frac{1}{110}$$

$$70(x^2 + 200^2)^{1/2} = 110x \qquad \text{Take the reciprocal of both sides,}$$
$$\text{then multiply by } x.$$

$$4900(x^2 + 200^2) = 12100x^2$$

$$4900(200^2) = 7200x^2$$

$$x = \pm 164.9915... \qquad \text{+164.9... is in the desired range, thus}$$
$$\text{confirming the graphical solution.}$$

Head for a point about 165 yd from the closest point on the beach.

b. If the towel is only 100 yd from the closest point on the beach, the domain of x is [0, 100], and the equation for t is

$$t = \frac{1}{70}\sqrt{x^2 + 200^2} + \frac{1}{110}(100 - x)$$

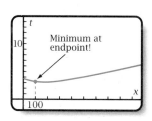

Figure 10-4d

Figure 10-4d shows the graph of this equation. Surprisingly, the minimum is at the same value, $x \approx 165$, as in part a. The 600 or 100 disappears when you differentiate, so the zero of the derivative is not affected by this number. However, 165 is out of the domain. It is beyond the towel! By the graph you can see that the minimum time comes at the endpoint, $x = 100$.

Head straight for the towel.   ◀

These problems are sometimes called "drowning swimmer" problems, because they are often phrased in such a way that a person on the beach must rescue a drowning swimmer and wants to take the path that reaches the swimmer in a minimum length of time. (The swimmer might drown if one were to take all that time to do the calculations!)

In Problem Set 10-4, you will work more minimal path problems. Then you will show that there is a remarkably simple way to find the minimal path if you do calculus algebraically on the general case instead of "brute-force" plotting on a particular case.

Problem Set 10-4

Quick Review   5 min

Q1. Solve: $\sqrt{x} = 9$

Q2. Differentiate: $y = \sqrt{100 - x^2}$

Q3. Integrate: $\int x\sqrt{100 - x^2}\,dx$

Q4. Differentiate: $y = \sin^{-1} 3x$

Q5. Integrate: $\int xe^{2x}\,dx$

Q6. Differentiate: $y = \tanh x$

Q7. Going 60 cm at 40 cm/h takes —?— h.

For Problems Q8–Q10, use the velocity-time function in Figure 10-4e:

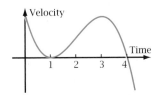

Figure 10-4e

Q8. At what time(s) is the moving object at rest?

Q9. At what time(s) does the moving object change directions?

Q10. The acceleration function has a local maximum at time(s) —?—.

    A. $t = 2$ B. $t = 0$ and $t = 3$ C. $t = 3$

    D. $t = 4$ E. $t = 0$

1. *Swim-and-Run Problem:* In a swim-and-run biathlon, Ann Athlete must get to a point on the other side of a 50-m-wide river, 100 m downstream from her starting point (Figure 10-4f). Ann can swim 2 m/s and run 5 m/s. Toward what point on the opposite side of the river should Ann swim to minimize her total time?

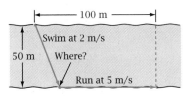

Figure 10-4f

2. *Scuba Diver Problem:* A scuba diver heads for a point on the ocean floor that is 30 m below the surface and 100 m horizontally from the point where she entered the water. She can move 13 m/min on the surface but only 12 m/min as she descends. How far from her entry point should she start descending to reach her destination in minimum time?

Chapter 10: The Calculus of Motion—Averages, Extremes, and Vectors

3. *Pipeline Problem:* Earl owns an oil lease. A new well in a field 300 m from a road is to be connected to storage tanks 1000 m down that same road (Figure 10-4g). Building pipeline across the field costs $50 per meter, whereas building it along the road costs only $40 per meter. How should the pipeline be laid out to minimize its total cost?

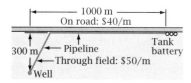

Figure 10-4g

4. *Elevated Walkway Problem:* Suppose you are building a walkway from the corner of one building to the corner of another building across the street and 400 ft down the block. It is 120 ft across the street (Figure 10-4h). Engineering studies show that the walkway will weigh 3000 lb/ft where it parallels the street and 4000 lb/ft where it crosses the street. How should you lay out the walkway to minimize its total weight?

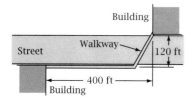

Figure 10-4h

5. *Minimal Path Discovery Problem:* In this problem you will explore a relationship between the minimal path and the two speeds or costs per meter.

   a. In the *Swim-and-Run Problem,* let θ be the (acute) angle between the slant path and a perpendicular to the river (Figure 10-4i, top). Show that the sine of this angle equals the ratio of the two speeds. That is, show that $\sin\theta = 2/5$.

   b. In the *Pipeline Problem,* let θ be the acute angle between the slant path and a

perpendicular to the road (Figure 10-4i, bottom). Show that the sine of this angle equals the ratio of the two costs per meter. That is, show that $\sin\theta = 40/50$.

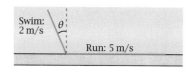

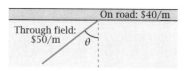

Figure 10-4i

6. *Minimal Path Generalization Problem:* A swimmer is at a distance p, in feet, from the beach. His towel is at the water's edge, at a distance k, in feet, along the beach (Figure 10-4j). He swims at an angle θ to a line perpendicular to the beach and reaches land at a distance x, in feet, from the point on the beach that was originally closest to him. He can swim at velocity s, in feet per minute, and walk at velocity w, in feet per minute, where $s < w$. Prove that his total time is a minimum if the sine of the angle the slant path makes with the perpendicular equals the ratio of the two speeds. That is, prove this property:

Property: Minimal Path

For the path shown in Figure 10-4j, if $\sin\theta = s/w$, then the total time taken is a minimum.

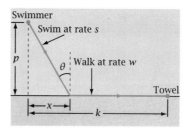

Figure 10-4j

7. *Scuba Diver Problem Revisited:* Work Problem 2 again, using the minimal path property of Problem 6. Which way to work the problem is easier? Give one reason why mathematicians find general solutions to typical problems.

8. *Elevated Walkway Problem Revisited:* Work Problem 4 again, using the minimal path property of Problem 6. Weight equals distance multiplied by cost per foot (rather than distance divided by velocity), so you will have to adapt the property appropriately. Which way to work the problem is easier? Give one reason why mathematicians find general solutions to typical problems.

9. *Pipeline Problem, Near Miss:* Suppose you present your boss with a solution to the *Pipeline Problem*, but to save some trees, the value of x must be a few meters from the optimum value. Will this fact make much difference in the total cost of the pipeline? Explain how you reach your conclusion.

10. *Calvin and Phoebe's Commuting Problem:* Calvin lives at the corner of Alamo and Heights Streets (Figure 10-4k). Phoebe lives on High Street, 500 ft from its intersection with Heights Street. Since they started dating, Calvin has found that he is spending a lot of time walking between the two houses. He seeks to minimize the time by cutting across the field to a point that is distance x, in feet, from the Heights-High intersection. He finds that he can walk 5 ft/s along the streets but only 3 ft/s across the rough, grassy field. What value of x minimizes Calvin's time getting to Phoebe's?

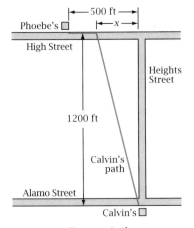

Figure 10-4k

11. *Robinson Crusoe Problem:* Robinson Crusoe is shipwrecked on a desert island. He builds a hut 70 yd from the shore. His wrecked ship is 120 yd from the shore, 300 yd down from the hut (Figure 10-4l). Crusoe makes many trips between the hut and the ship and wants to minimize the time each trip takes. He can walk 130 yd/min and pole his raft 50 yd/min. Where on the shoreline should he moor his raft so that he can make the trips in minimum time?

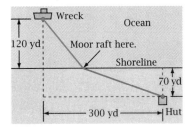

Figure 10-4l

12. *Robinson Crusoe Follow-Up Problem:* Let θ_1 and θ_2 be the angles between the two paths and a line perpendicular to the beach (Figure 10-4m) in the *Robinson Crusoe Problem.* Use the answer to the problem to calculate the measures of these angles. Then show that the ratio of the sines of the angles equals the ratio of the two speeds. That is, show

$$\frac{\sin \theta_1}{\sin \theta_2} = \frac{50}{130}$$

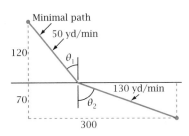

Figure 10-4m

13. *Robinson Crusoe Generalization Problem:* Figure 10-4n shows the general case of the *Robinson Crusoe Problem,* where the wreck, A, and the hut, B, are a and b units from the shore, respectively, and k units apart parallel to the shore. The velocities through the water and

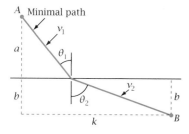

Figure 10-4n

on land are v_1 and v_2, respectively. Prove that the minimum time from wreck to hut is where

$$\frac{\sin \theta_1}{\sin \theta_2} = \frac{v_1}{v_2}$$

14. *Snell's Law of Refraction Problem:* About 350 years ago the Dutch physicist Willebrord Snell observed that when light passes from one substance into another, such as from air to water, the rays bend at the interface (Figure 10-4o). He found that the angles θ_1 and θ_2, which the incoming and outgoing rays make with a perpendicular to the interface, obey the rule

$$\frac{\sin \theta_1}{\sin \theta_2} = \frac{v_1}{v_2}$$

where v_1 and v_2 are the speeds of light in the two substances.

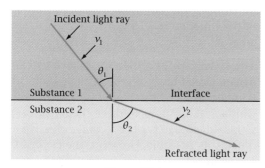

Figure 10-4o

a. In what ways does this real-world situation correspond to the *Robinson Crusoe Problems?*

b. What can you conclude about the time that light takes to travel from one point to another when it passes through different substances?

c. When you are above the surface of the water in a swimming pool, objects on the bottom always appear closer to the surface than they actually are. Explain how this can be true and what this fact tells you about the relative speeds of light in air and in water.

15. *Journal Problem:* Update your journal with what you've learned. You should include such things as

   • The one most important thing you've learned since your last journal entry

   • The big difference between displacement and distance

   • The meaning of average value of a function

   • The main idea behind minimal path problems

   • Any techniques or ideas about objects in motion that are still unclear

10-5 Maximum and Minimum Problems in Motion and Elsewhere

In Section 10-4, you found minimal paths, whereby one could travel between points in the least time or build structures between points for the least cost. In this section you will find maximum and minimum values in other phenomena.

OBJECTIVE   Given a situation in the real or mathematical world where a function is to be maximized or minimized, write an equation for the function and find the maximum or minimum values.

The technique is similar to that in the previous section and in the analysis of plane and solid figures from Section 8-3. Therefore, no specific examples are given in this section.

Problem Set 10-5

Q1. If $f(x) = x \sin x$, find $f'(x)$.

Q2. If $g(x) = x \ln x$, find $g''(x)$.

Q3. If $h(x) = xe^x$, find $\int h(x)\,dx$.

Q4. The minimal path problems of Section 10-4 are equivalent to what law from physics?

Q5. $\ln(\exp x) = \text{—?—}$

Q6. Sketch a paraboloid.

Q7. Sketch: $y = x^{1/3}$

Q8. $\int_a^b |velocity|\,d(time)$ = total distance or net displacement?

Q9. Who is credited with inventing calculus?

Q10. The sum of $1 + 2 + 3 + \cdots + 98 + 99 + 100$ is —?—.

    A. 303   B. 5,000   C. 5,050

    D. 5,151   E. 10,000

1. *Rocket Problem:* Jeff is out Sunday driving in his spaceship. As he approaches Mars, he fires his retro rockets. Starting 30 s later, his distance from Mars is given by

$$D = t + \frac{1}{t}$$

where D is in thousands of miles and t is in minutes. Plot the graphs of D and D' versus t. Sketch them on your paper. What are his maximum and minimum distances from Mars in the time interval $[0.5, 3]$? Justify your answers.

2. *Truck Problem:* Miles Moore owns a truck. His driver, Ouida Liver, regularly makes the 100-mi trip between Tedium and Ennui. Miles must pay Ouida \$20 per hour to drive the truck, so it is to his advantage for her to make the trip as quickly as possible. However, the cost of fuel varies directly with the square of the speed and is \$0.18 per mile at a speed of 30 mi/h. The truck can go as fast as 85 mi/h. Plot cost and derivative of cost versus speed. Sketch the graphs. What speed gives the minimum total cost for the 100-mi trip? (Ignore the cost of tickets, and so on, for going over the speed limit.) Justify your answer.

3. *Number Problem I:* Find the number that exceeds its square by the greatest amount. That is, find x if $x - x^2$ is to be as large as possible.

4. *Number Problem II:* Find the number greater than or equal to 2 that exceeds its square by the greatest amount.

5. *Fran's Optimal Study Time Problem:* Fran forgot to study for her calculus test until late Sunday night. She knows she will score zero if she doesn't study at all and that her potential score will be

$$S = \frac{100t}{t + 1}$$

if she studies for time t, in hours. She also realizes that the longer she studies, the more fatigued she will become. So, her actual score will be less than the potential score. Her "fatigue factor" is

$$F = \frac{9}{t + 9}$$

This is the number she must multiply by the potential score to find her actual grade, G. That is, $G = S \cdot F$.

a. Sketch the graphs of S, F, and G versus time, t.

b. What is the optimal number of hours for Fran to study? That is, how long should she study to maximize her estimated grade, G?

c. How many points less than the optimum will Fran expect to make if she studies

i. 1 hour more than the optimum

ii. 1 hour less than the optimum

6. *Motor Oil Viscosity Problem:* The viscosity (resistance to flow) of normal motor oil decreases as temperature goes up. All-temperature oils have roughly the same viscosity throughout their range of operating temperatures. Suppose that 10W-30 motor oil has viscosity

$$\mu = 130 - 12T + 15T^2 - 4T^3$$

where μ (Greek letter "mu") is the viscosity in centipoise, T is the temperature in hundreds of degrees Celsius, and the equation applies for temperatures from 0°C through 300°C.

a. Find the temperature in the domain at which the maximum viscosity occurs.

b. Find the minimum viscosity in the domain.

c. Suppose that the oil heats in such a way that $T = \sqrt{t}$, where t is time in minutes. At what rate is the viscosity changing when the temperature is 100°C?

7. *Cylinder-in-the-Cone Problem I:* A right circular cone has height 6 in. and base radius 10 in. A right circular cylinder is inscribed in the cone, coaxial with it.

a. Plot the graphs of volume and total surface area of the cylinder as a function of its radius. Sketch the graphs.

b. What radius and height of the cylinder give it the maximum volume? The maximum total surface area? Do the two maxima occur at the same radius?

8. *Cylinder-in-the-Cone Problem II:* A cone of radius 6 in. and height 18 in. has a cylinder inscribed in it. The cylinder's height starts at 0 in. and increases at 2 in./min.

a. When the altitude of the cylinder is 12 in., will its volume be increasing or decreasing? At what rate?

b. What will be the maximum volume of the cylinder in the time interval [0, 9]? Justify your answer.

c. What will be the maximum volume of the cylinder in the time interval [4, 6]?

9. *Quartic Parabola Tank Problem:* A water storage tank has the shape of the surface formed by rotating about the y-axis the graph of $y = x^4 + 5$, where x and y are in meters (Figure 10-5a). At what rate is the depth of the water changing when the water is 3 m deep in the tank and draining at 0.7 m³/min?

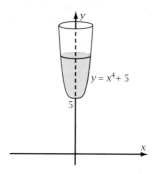

$y = x^4 + 5$

Figure 10-5a

10. *Cylinder in Paraboloid Problem:* A paraboloid is formed by rotating the graph of $y = 4 - x^2$ about the y-axis. A cylinder is inscribed in the paraboloid (Figure 10-5b).

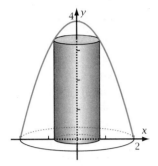

Figure 10-5b

a. If the radius of the cylinder is 1.5 units and is increasing at 0.3 units per second, is the volume of the cylinder increasing or decreasing? At what rate?

b. What is the maximum volume the cylinder could have? Justify your answer.

11. *Pig Sale Problem:* Ann's pig weighs 1000 lb and is gaining 15 lb/day. She could sell it for $900 at today's price of $0.90/lb, or she could wait until it gains some more weight and hope to get more than $900. Unfortunately, the price per pound is dropping at $0.01/lb each day. So, she must decide when is the best time to sell.

a. Write functions for weight and for price per pound in terms of the days after today. Then write a function for the total amount Ann will get for the pig.

b. Find the time when the derivative of total amount will be zero. Convince Ann that at this time the total amount is a maximum, not a minimum.

c. If Ann sells at the time in part b, how much will she get for the pig?

12. *Bridge Problem:* Suppose that you work for a construction company that has a contract to build a new bridge across Scorpion Gulch, downstream from the present bridge (Figure 10-5c). You collect information from various sources to decide exactly where to build the bridge. From the surveyors, you find that the width, in ft, of the gulch is

$$\text{Width} = 10(x^2 - 8x + 22)$$

where x is the number of miles downstream from the old bridge. The depth, in ft, of the water is

$$\text{Depth} = 20x + 10 \text{ ft}$$

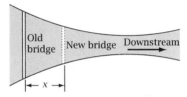

Figure 10-5c

a. Structural engineers specify that you can build the bridge in water as deep as 130 ft, but that it can be no more than 310 ft long. The city traffic department specifies that the bridge must be at least 1 mi downstream from the present bridge. What is the domain of x?

b. What are the shortest and longest lengths the bridge could be?

c. The cost of building the bridge is proportional to the product of the length of the bridge and the depth of the water. Where should the bridge be built to minimize the cost? Justify your answer.

d. Is the shortest bridge also the cheapest bridge? Explain.

10-6   Vector Functions for Motion in a Plane

Until now you have considered velocity and acceleration of an object moving back and forth in a line. The displacement, x, from some fixed point depends on time, t. In this section you will consider objects moving along a path in a plane.

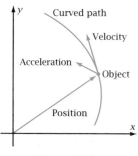

Figure 10-6a

Both x and y will depend on t. For such motion the velocity and acceleration may act at an angle to each other. Therefore, it is convenient to use vectors to represent the object's position, velocity, and acceleration.

Figure 10-6a shows the path a moving object might take in a plane. Its **velocity vector** points in the direction of motion and has **magnitude** (length) equal to the speed of the object. Its **acceleration vector**, acting at an angle to the path, changes both the object's speed and direction, thus pulling it into a curved path. You can represent the position of the object by a **position vector**, which tells the direction and distance from the origin to the object at any given time, t.

You can use parametric equations to write the position vector as a function of time. You can differentiate the resulting **vector function** to find velocity and acceleration vectors.

OBJECTIVE Given the equation of a vector function for the position of a moving object, find the first and second derivatives of position with respect to time, and interpret the way these vectors influence the motion of the object.

Background: Vectors

A **vector quantity** is a quantity that has both magnitude and direction. Quantities that have only magnitude are called **scalar quantities**. Volume, mass, time, distance (not displacement!), and money are scalar quantities. You can represent them by points on a scale such as a number line. (The word *scalar* comes from the Latin *scalaris*, meaning "like a ladder.")

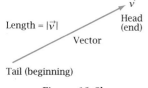

Figure 10-6b

You can represent vector quantities mathematically by directed line segments, called simply **vectors**. An overhead arrow such as $\vec{v}$ denotes a variable used for a vector (Figure 10-6b). The length of the line segment represents the magnitude of the vector quantity, also called its **absolute value** or its **norm**. The direction the segment points represents the direction in which the vector quantity acts.

Two vectors are **equal** if and only if they have the same magnitude and the same direction (Figure 10-6c, left). You are free to move vectors around from place to place, as long as you keep them pointing the same direction and don't change their lengths. The **opposite** of a vector, written $-\vec{v}$, is a vector of the same magnitude as $\vec{v}$ but pointing in the opposite direction (Figure 10-6c, middle). The **zero vector** is a vector of magnitude zero. It can be pointing in any direction!

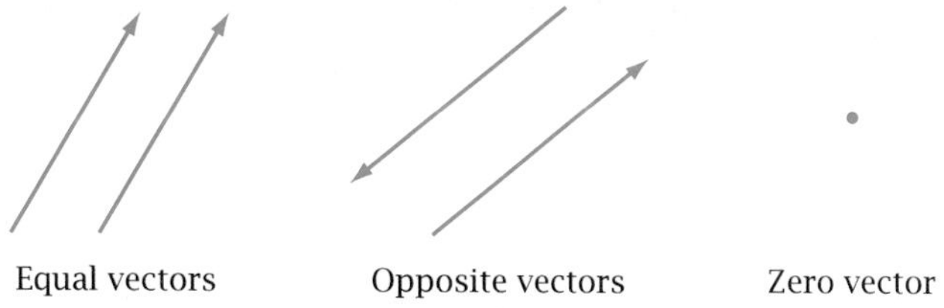

Equal vectors   Opposite vectors   Zero vector

Figure 10-6c

You add vectors by moving one of them so that its tail is at the head of the other (Figure 10-6d). The sum is the vector from the beginning of the first vector to the end of the second. You subtract vectors by adding the opposite of the second vector to the first one.

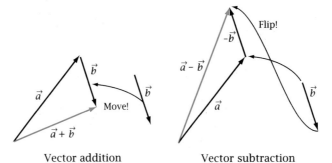

Vector addition Vector subtraction

Figure 10-6d

The kite surfer moves in a direction dictated by the force of the wind and the force of the current.

Vectors are most easily worked with in a coordinate system (Figure 10-6e). Any vector $\vec{v}$ is a sum of a vector in the x-direction and one in the y-direction (called **components** of the vector). If $\vec{i}$ and $\vec{j}$ are **unit vectors** in the x- and y-directions, respectively, then

$$\vec{v} = x\vec{i} + y\vec{j}$$

Other symbols are sometimes used for vectors because they are easier to type or easier to write. Some of these are shown in the box.

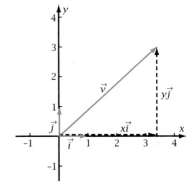

Figure 10-6e

TERMINOLOGY: Other Symbols for Vectors

Vector $\vec{v} = x\vec{i} + y\vec{j}$ may be written in these formats:

$\langle x, y \rangle$	Bracket format
(x, y)	Ordered-pair format, easy if used in vector context but may be confused with coordinates of a point
$\mathbf{v} = \mathbf{x} + \mathbf{y}$	Boldface format, easy to typeset but difficult to write

---

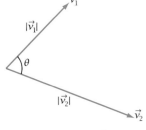

Figure 10-6f

The **dot product** (or *inner* product or *scalar* product) of two vectors is the number you get by placing the vectors tail-to-tail (Figure 10-6f), then multiplying their magnitudes and the cosine of the angle between the vectors; that is,

$$\vec{v}_1 \cdot \vec{v}_2 = |\vec{v}_1||\vec{v}_2| \cos \theta$$

In Problem 18 of Problem Set 10-6, you will show that the dot product is also equal to

$$\vec{v}_1 \cdot \vec{v}_2 = x_1 x_2 + y_1 y_2$$

where $\vec{v}_1 = x_1\vec{i} + y_1\vec{j}$ and $\vec{v}_2 = x_2\vec{i} + y_2\vec{j}$. The dot product is useful for finding the length of a "shadow" that one vector would cast on another. This **scalar projection**, p, of $\vec{v}_1$ on $\vec{v}_2$ is given by

$$p = |\vec{v}_1| \cos \theta$$

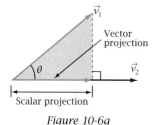

as shown in Figure 10-6g. Multiplying by a clever form of 1 makes a dot product appear on the right side of the equation.

$$p = \frac{|\vec{v}_1||\vec{v}_2| \cos \theta}{|\vec{v}_2|} = \frac{\vec{v}_1 \cdot \vec{v}_2}{|\vec{v}_2|}$$

Figure 10-6g

The **vector projection**, $\vec{p}$, of $\vec{v}_1$ on $\vec{v}_2$ is a vector in the direction of $\vec{v}_2$, with magnitude p. That is, $\vec{p} = p\vec{u}$ where $\vec{u}$ is a unit vector in the direction of $\vec{v}_2$, onto which $\vec{v}_1$ is projected.

Derivatives of a Position Vector Function—Velocity and Acceleration

The derivative of a scalar-valued function is the limit of $\Delta y / \Delta x$ as Δx approaches zero. Similarly the derivative of a vector function, $\vec{r}$, is

$$\frac{d\vec{r}}{dt} = \lim_{\Delta t \to 0} \frac{\Delta \vec{r}}{\Delta t}$$

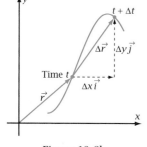

Figure 10-6h shows vector $\vec{r}$ to the position of a moving object at time t. Vector $\Delta \vec{r}$ is the vector from this position to a new position at time $t + \Delta t$. You can resolve vector $\Delta \vec{r}$ into horizontal and vertical components, $\Delta x\vec{i}$ and $\Delta y\vec{j}$. So,

$$\frac{d\vec{r}}{dt} = \lim_{\Delta t \to 0} \frac{\Delta x\vec{i} + \Delta y\vec{j}}{\Delta t}$$

Figure 10-6h

$$= \lim_{\Delta t \to 0} \left(\frac{\Delta x}{\Delta t}\vec{i} + \frac{\Delta y}{\Delta t}\vec{j} \right) \qquad \text{Δt distributes and can be associated with Δx and Δy.}$$

$$= \lim_{\Delta t \to 0} \frac{\Delta x}{\Delta t}\vec{i} + \lim_{\Delta t \to 0} \frac{\Delta y}{\Delta t}\vec{j} \qquad \text{The limit of a sum property applies to vectors.}$$

$$= \frac{dx}{dt}\vec{i} + \frac{dy}{dt}\vec{j} \qquad \text{Definition of derivative. (The unit vectors are constant.)}$$

Thus, if the components of a *position vector* are specified by functions of t, you can find $d\vec{r}/dt$ simply by differentiating each function and multiplying the answers by $\vec{i}$ and $\vec{j}$. As you will see in Example 1, the resulting vector is *tangent* to the path of the moving object. By the Pythagorean theorem, its length is

$$\left| \frac{d\vec{r}}{dt} \right| = \sqrt{\left(\frac{dx}{dt} \right)^2 + \left(\frac{dy}{dt} \right)^2} = \frac{\sqrt{dx^2 + dy^2}}{dt} = \frac{dL}{dt}$$

But dL/dt is the *speed* of the object along its curved path. Thus, $d\vec{r}/dt$ is the *velocity vector* for the moving object. You can find the *acceleration vector* by differentiating the velocity vector's components, the same way you can find the velocity from the displacement.

PROPERTIES: Velocity and Acceleration Vectors

If $\vec{r}(t) = x(t)\vec{i} + y(t)\vec{j}$ is the **position vector** for a moving object, then $\vec{v}(t) = \vec{r}\,'(t) = x'(t)\vec{i} + y'(t)\vec{j}$ is the **velocity vector**, which is tangent to the path of the object, in the direction of motion, and $\vec{a}(t) = \vec{v}\,'(t) = \vec{r}\,''(t) = x''(t)\vec{i} + y''(t)\vec{j}$ is the **acceleration vector**, which lies on the concave side of the path. The speed of the moving object equals $|\vec{v}(t)| = \sqrt{(dx/dt)^2 + (dy/dt)^2}$.

▶ **EXAMPLE 1**   Given the vector equation

$$\vec{r}(t) = (5\sin t)\,\vec{i} + (5\cos^2 t)\,\vec{j}$$

for the position, $\vec{r}(t)$, of a moving object, where distances are in feet and time t is in seconds,

    a. Plot the path of the object. Sketch the result. Then show the position vectors $\vec{r}(0)$, $\vec{r}(0.5)$, $\vec{r}(1)$, $\vec{r}(1.5)$, and $\vec{r}(2)$. Interpret the location of $\vec{r}(2)$.

    b. Calculate the difference vector $\Delta\vec{r} = \vec{r}(1) - \vec{r}(0.5)$ by subtracting the respective components. Sketch $\Delta\vec{r}$ with its tail at the head of $\vec{r}(0.5)$. Where is the head of $\Delta\vec{r}$?

    c. The **average velocity** vector of the object for interval $[0.5, t]$ is the difference quotient

$$\vec{v}_{\text{av}} = \frac{\vec{r}(t) - \vec{r}(0.5)}{t - 0.5}$$

       Find the average velocities for the time intervals $[0.5, 1]$ and $[0.5, 0.6]$. Sketch these vectors starting at the head of the position vector $\vec{r}(0.5)$.

    d. Find the (instantaneous) velocity vector $\vec{v}(0.5)$. Plot it starting at the end of $\vec{r}(0.5)$, thus showing that it is tangent to the path. How do the average velocity vectors in part c relate to $\vec{v}(0.5)$?

    e. Find the speed of the object at time $t = 0.5$.

Solution   a. With your grapher in parametric mode, plot

$$x = 5\sin t$$
$$y = 5\cos^2 t$$

If you use the grid-on format, it will be easier to sketch the graph on dot paper (Figure 10-6i). The position vectors are

$$\vec{r}(0) = 0\vec{i} + 5\vec{j}$$
$$\vec{r}(0.5) = 2.39...\vec{i} + 3.85...\vec{j}$$
$$\vec{r}(1) = 4.20...\vec{i} + 1.45...\vec{j}$$
$$\vec{r}(1.5) = 4.98...\vec{i} + 0.02...\vec{j}$$
$$\vec{r}(2) = 4.54...\vec{i} + 0.86...\vec{j}$$

The heads of the vectors lie on the path. (That's why they are called position vectors.) They progress clockwise from the vertical. At time $t = 2$, the object has started back in the counterclockwise direction, as you can see by tracing.

b. $$\Delta\vec{r} = \vec{r}(1) - \vec{r}(0.5)$$
$$= (5\sin 1 - 5\sin 0.5)\vec{i} + (5\cos^2 1 - 5\cos^2 0.5)\vec{j}$$
$$= 1.81...\vec{i} - 2.39...\vec{j}$$

Figure 10-6j shows this difference vector. If you start this vector at the head of $\vec{r}(0.5)$, then $\Delta\vec{r}$ ends at the head of $\vec{r}(1)$, as you can see by counting spaces on the dot paper.

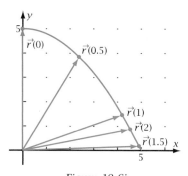

Figure 10-6i

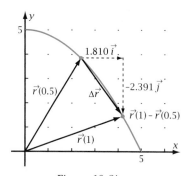

Figure 10-6j

c. $$t = 1: \vec{v}_{av} = \frac{\vec{r}(1) - \vec{r}(0.5)}{1 - 0.5}$$

$$= \frac{1.81...\vec{i} - 2.39...\vec{j}}{0.5}$$

$$= 3.62...\vec{i} - 4.78...\vec{j}$$

$$t = 0.6: \vec{v}_{av} = \frac{\vec{r}(0.6) - \vec{r}(0.5)}{0.6 - 0.5}$$

$$= \frac{0.426...\vec{i} - 0.444...\vec{j}}{0.1}$$

$$= 4.26...\vec{i} - 4.44...\vec{j}$$

Figure 10-6k shows these average vectors plotted on Figure 10-6j. You plot them by counting spaces on the dot paper.

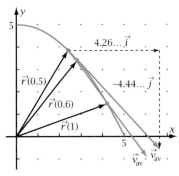

Figure 10-6k   Figure 10-6l

d. $\vec{v}(t) = \vec{r}\,'(t) = (5\cos t)\,\vec{i} + (-10\cos t\sin t)\,\vec{j}$
    $\therefore \vec{v}(0.5) = 4.38...\,\vec{i} - 4.20...\,\vec{j}$

Figure 10-6l shows that $\vec{v}$ is tangent to the path at the position $\vec{r}(0.5)$. The average velocity vectors approach $\vec{v}$ as a limit as Δt approaches zero.

e.   Speed $= |\vec{v}(0.5)|$
        $= \sqrt{(4.38...)^2 + (-4.20...)^2}$
        $= 6.079...$

Thus, the object is traveling at about 6.1 ft/s.   ◀

Example 2 shows how to find the acceleration vector and how it helps interpret the motion.

▶ **EXAMPLE 2**   For parts a–f, use the vector equation

$$\vec{r}(t) = (5\sin t)\,\vec{i} + (5\cos^2 t)\,\vec{j}$$

for the position of a moving object, as in Example 1.

  a. Find vector equations for the velocity vector, $\vec{v}(t)$, and the acceleration vector, $\vec{a}(t)$.

  b. On a graph of the path of the object, sketch the position vector $\vec{r}(4)$ at time $t = 4$. From the head of $\vec{r}(4)$, sketch the vectors $\vec{v}(4)$ and $\vec{a}(4)$.

  c. How fast is the object going at time $t = 4$?

  d. Compute $\vec{a}_t(4)$, the *tangential component* of the acceleration (parallel to the path).

  e. Is the object speeding up or slowing down at $t = 4$? How can you tell? At what rate is it speeding up or slowing down?

  f. Compute $\vec{a}_n(4)$, the *normal component* of acceleration (perpendicular to the path). Toward which side of the path does $\vec{a}_n(4)$ point? What effect does this component have on the motion of the object?

Solution

a. You can find the velocity as you did in Example 1. You can find the acceleration by differentiating the velocity vector.

$$\vec{v}(t) = \vec{r}\,'(t) = (5\cos t)\vec{i} + (-10\cos t \, \sin t)\vec{j}$$
$$= (5\cos t)\vec{i} + (-5\sin 2t)\vec{j}$$

From trigonometry,
$2\sin t \cos t = \sin 2t$.

$$\vec{a}(t) = \vec{v}\,'(t) = (-5\sin t)\vec{i} + (-10\cos 2t)\vec{j}$$

b. Substituting 4 for t gives

$$\vec{r}(4) = -3.78...\vec{i} + 2.13...\vec{j}$$
$$\vec{v}(4) = -3.26...\vec{i} - 4.94...\vec{j}$$
$$\vec{a}(4) = 3.78...\vec{i} + 1.45...\vec{j}$$

Figure 10-6m shows $\vec{v}(4)$ and $\vec{a}(4)$ drawn on a graph plotted as in Example 1. The vectors are drawn by counting spaces on the dot paper. Note that although $\vec{v}(4)$ is tangent to the path, the acceleration is at an angle to the path.

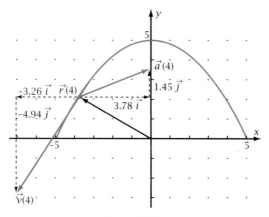

Figure 10-6m

c. Speed is equal to the absolute value of the velocity.

$$\text{Speed} = |\vec{v}(4)|$$
$$= \sqrt{(5\cos 4)^2 + (-5\sin 8)^2}$$
$$= 5.928...$$

Thus, the object is going about 5.93 ft/s.

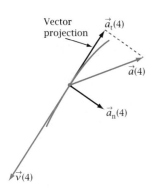

Figure 10-6n

d. The tangential acceleration, $\vec{a}_t(4)$, parallel to the path, is the vector projection of $\vec{a}(4)$ on $\vec{v}(4)$ (Figure 10-6n). First find the scalar projection, p.

$$p = \frac{\vec{a}(4) \cdot \vec{v}(4)}{|\vec{v}(4)|} = \frac{(-5\sin 4)(5\cos 4) + (-10\cos 8)(-5\sin 8)}{5.928...} = -3.2998...$$

To find a vector in the direction of $\vec{v}(4)$ with length 3.2998..., pointed the other way, multiply the scalar projection, $-3.2998...$, by a unit vector in the direction of $\vec{v}(4)$.

$$\vec{a}_t(4) = p \frac{\vec{v}(4)}{|\vec{v}(4)|}$$

$$= (-3.2998...)\frac{(5\cos 4)\vec{i} + (-5\sin 8)\vec{j}}{5.928...}$$

$$= 1.818...\vec{i} + 2.753...\vec{j}$$

e. The tangential acceleration points in the direction opposite to $\vec{v}(4)$, so the object is slowing down when $t = 4$. Both the obtuse angle between $\vec{v}(4)$ and $\vec{a}(4)$ (Figure 10-6n) and the negative value of p, the scalar projection of $\vec{a}(4)$ on $\vec{v}(4)$, reveal this fact. The rate at which the object is slowing equals the magnitude of $\vec{a}_t(4)$, namely, $|p|$, or 3.2998.... So the object is slowing down at about 3.3 (ft/s)/s when $t = 4$.

f. As Figure 10-6n shows, the normal acceleration, $\vec{a}_n(4)$, perpendicular to the path, equals the vector that gives $\vec{a}(4)$ when added to $\vec{a}_t(4)$. Therefore,

$$\vec{a}_n(4) = \vec{a}(4) - \vec{a}_t(4) = (3.78...\vec{i} + 1.45...\vec{j}) - (1.818...\vec{i} + 2.753...\vec{j})$$
$$= 1.96...\vec{i} - 1.29...\vec{j}$$

The normal component of acceleration will always point toward the concave side of the graph, as Figure 10-6n shows. It is responsible for pulling the object out of a straight-line path and into a curved path. ◀

From Example 2, you can reach these conclusions about acceleration vectors.

PROPERTY: *Components of the Acceleration Vector*

- The acceleration points toward the concave side of the path of the object.

- The scalar projection, p, of $\vec{a}$ on $\vec{v}$ is given by $p = \dfrac{\vec{a} \cdot \vec{v}}{|\vec{v}|}$.

- The **tangential component**, $\vec{a}_t$, of the acceleration is the vector projection of $\vec{a}$ on $\vec{v}$. It is a vector in the direction of $\vec{v}$ with magnitude $|p|$. It changes the speed of the object. The tangential component is given by

  $\vec{a}_t = p\vec{u}$

  where $\vec{u} = \vec{v}/|\vec{v}|$ is a unit vector in the direction of $\vec{v}$.

- $|\vec{a}_t| = |p|$ is the **rate of change of speed** of the object.

- If $p > 0$, the angle between $\vec{a}$ and $\vec{v}$ is acute and the object is speeding up. If $p < 0$, the angle between $\vec{a}$ and $\vec{v}$ is obtuse and the object is slowing down.

- The **normal component** of acceleration, $\vec{a}_n$, changes the direction of motion. Because $\vec{a} = \vec{a}_t + \vec{a}_n$, the normal component can be found by $\vec{a}_n = \vec{a} - \vec{a}_t$.

In Example 3, you will find the distance traveled by an object moving in a curved path.

▶ **EXAMPLE 3**   For the object in Example 1, $\vec{r}(t) = (5\sin t)\vec{i} + (5\cos^2 t)\vec{j}$. Find the distance traveled by the object in the time interval $[0, 4]$.

Solution   Recognize that a vector equation is the same as two parametric equations. The distance the object travels thus equals the arc length of a parametric curve (Section 8-5).

$$dL = \sqrt{x'(t)^2 + y'(t)^2}\,dt = \sqrt{(5\cos t)^2 + (-10\cos t \sin t)^2}\,dt$$

$$L = \int_0^4 \sqrt{(25 + 100\sin^2 t)\cos^2 t}\,dt$$

$$= 19.7246... \approx 19.72 \text{ ft} \qquad\text{By numerical integration.}$$

Check: A circle of radius 5 approximates the curve. A 4-radian arc of the circle has length $(4/2\pi)(2\pi \cdot 5) = 20$, which compares well with 19.72.   ◀

Problem Set 10-6

Q1. Integrate: $\int x \sin x\,dx$

Q2. Differentiate: $x^2 e^{3x}$

Q3. Integrate: $\int 2^x\,dx$

Q4. Evaluate: $\dfrac{5^{2001}}{5^{1998}}$

Q5. Find $\int x^3\,dx$ if the integral equals 11 when $x = 2$.

Q6. Find dy/dx if $x = e^{3t}$ and $y = \tan 6t$.

Q7. The function in Problem Q6 is called a(n) —?— function.

Q8. Integrate: $\int \ln x\,dx$

Q9. Simplify: $e^{2\ln x}$

Q10. In polar coordinates, the graph of $r = \theta$ is a(n) —?—.

    A. Ray   B. Circle   C. Parabola

    D. Line   E. Spiral

1. Figure 10-6o shows the path of an object moving counterclockwise. On a copy of the figure, sketch the position vector to the point shown on the path. Sketch a velocity vector and an acceleration vector, with tails at the head of the position vector if the object is slowing down.

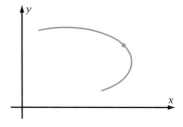

Figure 10-6o

2. Repeat Problem 1 if the object is speeding up.

3. A particle moves with position vector $\vec{r} = (e^t \cos t)\vec{i} + (e^t \sin t)\vec{j}$, where x and y are distances in centimeters and t is time in seconds.

   a. Find the particle's velocity vector at time $t = 1$. At this time, is its x-coordinate increasing or decreasing? How fast is the particle going?

   b. How far did the particle travel between $t = 0$ and $t = 1$? How far is the particle from the origin at that time?

   c. Find the acceleration vector at time $t = 1$.

4. Figure 10-6p shows the paths of two moving particles. For times $t \geq 0$ min, their position vectors (distances in meters) are given by

Particle 1: $\vec{r}_1 = (t - 2)\vec{i} + (t - 2)^2\,\vec{j}$
Particle 2: $\vec{r}_2 = (1.5t - 4)\vec{i} + (1.5t - 2)\,\vec{j}$

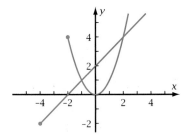

Figure 10-6p

a. Find the velocity and acceleration of each particle at time $t = 1$. How fast is each going at that time?

b. How far does Particle 1 go along its curved path between $t = 1$ and $t = 4$?

c. The paths of the particles cross each other at two points in the figure. Do the particles collide at either of these points? How can you tell?

5. *Parabolic Path Problem I:* An object moves along the parabolic path (Figure 10-6q)

$$\vec{r}(t) = (10 \sin 0.6t)\,\vec{i} + (4 \cos 1.2t)\,\vec{j}$$

where distance is in feet and time is in seconds.

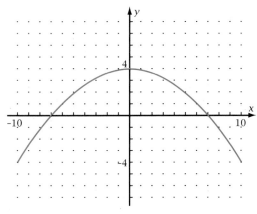

Figure 10-6q

a. Find equations for $\vec{v}(t)$ and $\vec{a}(t)$.

b. Calculate $\vec{r}(0.5)$, $\vec{v}(0.5)$, and $\vec{a}(0.5)$. On a copy of Figure 10-6q, plot $\vec{r}$ as a position

vector, and plot $\vec{v}$ and $\vec{a}$ with their tails at the head of $\vec{r}$. Explain why the three vectors are reasonable.

c. Based on the graphs of the vectors in part b, does the object seem to be speeding up or slowing down at time $t = 0.5$ s? How can you tell?

d. Verify your answer to part c by finding the tangential and normal components of $\vec{a}(0.5)$. Sketch these components on your sketch from part b, starting at the tail of $\vec{a}$.

e. At what rate is the object speeding up or slowing down at $t = 0.5$?

f. Calculate $\vec{r}(7)$, $\vec{v}(7)$, and $\vec{a}(7)$. Sketch these vectors on your sketch from part b. At time $t = 7$, does the object seem to be speeding up or slowing down?

g. Show that at time $t = 0$ the acceleration vector is perpendicular to the path. How do you interpret this fact in terms of motion of the object at $t = 0$?

6. *Parabolic Path Problem II:* An object moves along the parabolic path (Figure 10-6r)

$$\vec{r}(t) = (8 \cos 0.8t)\vec{i} + (6 \sin 0.4t)\vec{j}$$

where distance is in feet and time is in seconds.

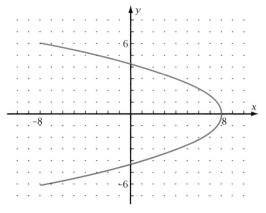

Figure 10-6r

a. Find equations for $\vec{v}(t)$ and $\vec{a}(t)$.

b. Calculate $\vec{r}(1)$, $\vec{v}(1)$, and $\vec{a}(1)$. On a copy of Figure 10-6r, plot $\vec{r}$ as a position vector, and plot $\vec{v}$ and $\vec{a}$ with their tails at the head of $\vec{r}$. Explain why the three vectors are reasonable.

c. Based on the graphs of the vectors in part b, does the object seem to be speeding up or slowing down at time $t = 1$? How can you tell?

d. Verify your answer to part c by finding the tangential and normal components of $\vec{a}(1)$. Sketch these components on your sketch from part b, starting at the tail of $\vec{a}$.

e. At what rate is the object speeding up or slowing down at $t = 1$?

f. Calculate $\vec{r}(10.5)$, $\vec{v}(10.5)$, and $\vec{a}(10.5)$. Sketch these vectors on your sketch from part b. At time $t = 10.5$, does the object seem to be speeding up or slowing down?

g. What is the first positive value of t at which the object is stopped? What is the acceleration vector at that time? Plot this vector on your sketch from part b. Surprising?

7. *Elliptical Path Problem:* An object moves along the elliptical path (Figure 10-6s)

$$\vec{r}(t) = (10 \cos \tfrac{\pi}{6} t)\vec{i} + (6 \sin \tfrac{\pi}{6} t)\vec{j}$$

where time is in seconds and distances are in feet.

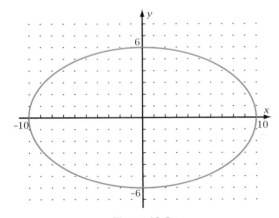

Figure 10-6s

a. On your grapher, plot the path followed by the heads of the velocity vectors if their tails are placed at the heads of the respective position vectors. You can use parametric mode to plot $\vec{r} + \vec{v}$. Sketch the result on a copy of Figure 10-6s.

b. Sketch the velocity vectors for each integer value of t from 0 through 12. You may find their beginnings by tracing the $\vec{r}$ vector and their ends by tracing $\vec{r} + \vec{v}$.

c. Prove that the heads of the velocity vectors in part b lie along an ellipse.

d. On your grapher, plot the path followed by the heads of the acceleration vectors when their tails are placed at the heads of the respective position vectors. Sketch the result on the copy of Figure 10-6s from part a.

e. Sketch the acceleration vectors as you did for the velocity vectors in part b. What seems to be true about the direction of each of these acceleration vectors?

8. *Spiral Path Problem:* An object moves on the spiral path (Figure 10-6t)

$$\vec{r}(t) = (0.5t \cos t)\vec{i} + (0.5t \sin t)\vec{j}$$

where distance is in miles and time is in hours.

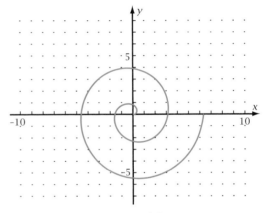

Figure 10-6t

a. Find vector equations for $\vec{v}(t)$ and $\vec{a}(t)$.

b. Find the position, velocity, and acceleration at $t = 8.5$ and $t = 12$. On a copy of Figure 10-6t, plot $\vec{r}(8.5)$ and $\vec{r}(12)$, thus showing that these vectors really do terminate on the path.

c. Plot $\vec{v}(8.5)$ and $\vec{a}(8.5)$, starting at the head of $\vec{r}(8.5)$. Do the same for $\vec{v}(12)$ and $\vec{a}(12)$. Explain why these velocity vectors have the proper relationships to the path.

d. At $t = 8.5$ and $t = 12$, is the object speeding up or slowing down? How can you tell?

e. Find the tangential and normal components of the acceleration vector at $t = 12$. Show these components on the copy of Figure 10-6t from part b.

f. How fast is the object going at $t = 12$? At what rate is the speed changing then?

g. On your grapher, plot the path followed by the heads of the acceleration vectors when they are placed with their tails at the heads of the respective position vectors. What graphical figure does the path appear to be? Prove algebraically that your conjecture is correct.

9. *Parabolic Path Problem III:* An object moves along the parabola $y = x^2$. At various times, t, the object is at various points, (x, y), where x and y are in centimeters and t is in seconds.

a. Write the position vector $\vec{r}(x)$ as a function of x alone (and the two unit vectors $\vec{i}$ and $\vec{j}$, of course!). Then find the velocity vector $\vec{v}(x)$ as a function of x and dx/dt.

b. Assume that the object moves in such a way that x decreases at a constant rate of 3 cm/s. Find $\vec{v}(2)$. How fast is the object moving when $x = 2$?

c. Sketch the graph of the parabola and draw $\vec{r}(2)$ and $\vec{v}(2)$ at the point $(2, 4)$. Explain why the graph of $\vec{v}(2)$ is reasonable.

d. Find the acceleration vector, $\vec{a}(x)$, and evaluate $\vec{a}(2)$. Sketch $\vec{a}(2)$ on your graph.

e. Find the tangential and normal components of acceleration at $x = 2$. Show these components on your graph. Based on the graphs, why are your answers reasonable?

f. When $x = 2$, is the object speeding up or slowing down? Justify your answer.

g. The object changes its motion and goes in such a way that its speed along its curved path is 5 cm/min. Write an expression in terms of x for dL, the differential of arc length along the curve. Find dx/dt when $x = 2$.

10. *Velocity Vector Limit Problem:* An object moves along one petal of a four-leafed rose (Figure 10-6u),

$$\vec{r} = (12 \sin t \cos 0.5t)\vec{i} + (12 \sin t \sin 0.5t)\vec{j}$$

On a copy of this figure, plot $\vec{r}(1)$. From the end of $\vec{r}(1)$, plot the average velocity difference quotient vectors,

$$\vec{q}(t) = \frac{\vec{r}(t) - \vec{r}(1)}{t - 1}$$

for $t = 2$, $t = 1.5$, and $t = 1.1$. Then plot the velocity vector, $\vec{v}(1)$. How does the velocity vector relate to the path of the object and to the average velocity vectors?

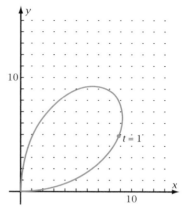

Figure 10-6u

11. Find the distance traveled by the object in Problem 5 from time $t = 0$ to $t = 2$.

12. Find the distance traveled by the object in Problem 7 in one complete cycle.

13. *Baseball Problem:* Saul Teen pitches a baseball. As it leaves his hand, it moves horizontally toward home plate at 130 ft/s (about 90 mi/h), so its velocity vector at time $t = 0$ s is

$$\vec{v}(0) = -130\vec{i} + 0\vec{j}$$

The minus sign is used because the distance from the plate is decreasing. As the ball moves, it drops vertically due to the acceleration of gravity. The vertical acceleration is 32 (ft/s)/s. Assuming that there is no loss of speed due to air resistance, its acceleration vector is

$$\vec{a}(t) = 0\vec{i} - 32\vec{j}$$

a. Write an equation for $\vec{v}(t)$, the velocity as a function of time.

b. When Saul releases the ball at time $t = 0$, it is at $y = 8$ ft above the playing field and $x = 60.5$ ft from the plate. Write the position vector, $\vec{r}(t) = x(t)\vec{i} + y(t)\vec{j}$.

c. Saul's sister, Phyllis, stands at the plate ready to hit the ball. How long does it take the ball to reach the plate at $x = 0$? As it passes over the plate, will it be in Phyllis's strike zone, between $y = 1.5$ ft and

$y = 4.5$ ft above the plate? Show how you reach your conclusion.

d. Saul pitches another time. Phyllis hits the ball, making it leave a point 3 ft above the plate at an angle of 15 deg to the horizontal, going at a speed of 200 ft/s. What are the initial horizontal and vertical velocities? Assuming that the horizontal velocity stays constant and the vertical velocity is affected by gravity, write the position vector $\vec{r}(t)$ as a function of the number of seconds since she hit the ball.

e. Will Phyllis make a home run with the hit in part d? The fence for which the ball is heading is 400 ft from the plate and 10 ft high. You may do this algebraically or by plotting the position graph in parametric mode and seeing where the ball is when $x = 400$.

14. *Sinusoidal Path Problem:* An object moves along the graph of $y = \sin x$ (Figure 10-6v), where x and y are in meters and t is in minutes. Its x-acceleration is 3 m/min², a constant. At time $t = 0$, the object is at the point $(0, 0)$ and has velocity equal to the zero vector.

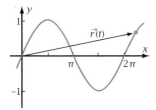

Figure 10-6v

a. Find $\vec{r}(t)$, the position vector as a function of time.

b. How fast is the object moving when it is at the point $(6, \sin 6)$?

15. *Figure Skating Problem:* One figure that roller skaters do in competition has a large loop and a small loop (Figure 10-6w). Specifications by the Roller Skating Rink Operations Association of America require the outer loop to be 240 cm from the origin where the loops cross and the inner loop to be 60 cm from the origin. The figure closely resembles a limaçon, with polar equation

$$d = a + b \cos t$$

where d is the directed distance from the origin at angle t, in radians, and a and b are constants.

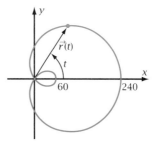

Figure 10-6w

a. At $t = 0$, $d = 240$. At $t = \pi$, $d = -60$ so that the point is on the positive x-axis. Find the particular equation of this limaçon.

b. Suppose that Annie skates with an angular velocity of 1 rad/s (radians per second). So, t is also her time in seconds, and her position vector is

$$\vec{r}(t) = (d \cos t)\vec{i} + (d \sin t)\vec{j}$$

where d is given by the equation in part a. Find Annie's velocity vector at time $t = 1$. How fast is she going at that time?

c. Find Annie's acceleration vector at $t = 1$. Find the tangential and normal components of the acceleration. Is she speeding up or slowing down at this time? At what rate?

16. *River Bend Problem:* A river meanders slowly across the plains (Figure 10-6x). A log floating on the river has position vector

$$\vec{r}(t) = (0.5t + \sin t)\vec{i} + (4\cos 0.5t)\vec{j}$$

where distances are in miles and time is in hours.

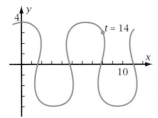

Figure 10-6x

a. When $t = 14$, what are the log's velocity and acceleration vectors? How fast is it going? What are its tangential and normal acceleration vectors? Is it speeding up or slowing down? At what rate?

b. How far does the log move along its curved path from $t = 0$ to $t = 14$? What is its average speed for this time interval?

17. *Roller Coaster Problem:* Assume that a roller coaster track is a prolate cycloid (Figure 10-6y) and that the position of a car on the track, in feet, at time t, in seconds, is

$$\vec{r}(t) = (5t - 12\sin t)\vec{i} + (15 + 12\cos t)\vec{j}$$

a. Write the velocity and acceleration vectors as functions of t.

b. Find the velocity and acceleration vectors at the point shown, where $t = 2.5$. Plot these vectors on a copy of Figure 10-6y, starting at the point on the graph.

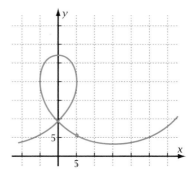

Figure 10-6y

c. Find the tangential and normal components of the acceleration vector at $t = 2.5$.

d. Analyze the motion of the roller coaster at $t = 2.5$. For instance, how can you tell that the velocity vector is reasonable? Is the normal component of acceleration reasonable? How fast is the roller coaster going? Is it speeding up or slowing down?

e. Show that the acceleration vector is straight down when the roller coaster is at a high point and straight up when it is at a low point.

f. How long is the track from one high point to the next?

18. *Dot Product Problem:* The dot product of two vectors is defined to be

$$\vec{v}_1 \cdot \vec{v}_2 = |\vec{v}_1||\vec{v}_2|\cos\theta$$

If $\vec{v}_1 = x_1\vec{i} + y_1\vec{j}$ and $\vec{v}_2 = x_2\vec{i} + y_2\vec{j}$, show, as in Figure 10-6z, that

$$\vec{v}_1 \cdot \vec{v}_2 = x_1x_2 + y_1y_2$$

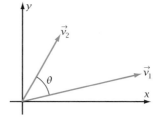

Figure 10-6z

Assume the distributive property, $\vec{a} \cdot (\vec{b} + \vec{c}) = \vec{a} \cdot \vec{b} + \vec{a} \cdot \vec{c}$. You will need to figure out what $\vec{i} \cdot \vec{i}$, $\vec{j} \cdot \vec{j}$ and $\vec{i} \cdot \vec{j}$ equal, based on the angles between these unit vectors.

19. *Three-Dimensional Vector Problem:*
A three-dimensional vector (Figure 10-6aa) can be resolved into three mutually perpendicular components. If $\vec{i}$, $\vec{j}$, and $\vec{k}$ are unit vectors in the *x*-, *y*-, and *z*-directions, respectively, then position vector $\vec{r}$ from the origin to the point (x, y, z) can be written

$$\vec{r} = x\vec{i} + y\vec{j} + z\vec{k}$$

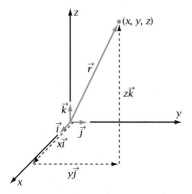

Figure 10-6aa

Suppose that a moving object's position is given by the vector function

$$\vec{r}(t) = (10\sin 0.8t)\vec{i} + (10\cos 0.6t)\vec{j} + (6t^{0.5})\vec{k}$$

Find the velocity and acceleration vectors at time $t = 1$. At that time, is the object speeding up or slowing down? Justify your answer.

20. *Curvature Project:* Figure 10-6bb shows an object moving with velocity $\vec{v}$ along a path. The velocity makes an angle ϕ (lowercase Greek letter "phi") with the positive *x*-axis. The curvature of the path is κ (lowercase Greek letter "kappa") defined to be the rate of change of ϕ with respect to distance, *s*, along the path.

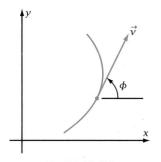

Figure 10-6bb

That is, κ is given by

$$\kappa = d\phi/ds$$

a. Explain why $d\phi/ds = (d\phi/dt)(dt/ds)$.
b. Explain why $\tan\phi = dy/dx$, which equals $(dy/dt)/(dx/dt)$.
c. Let x' and x'' be the first and second derivatives of *x* with respect to *t*, and similarly for *y*. Show that the formula shown in this box is true.

> **Calculation of Curvature**
>
> $$\kappa = \frac{d\phi}{ds} = \frac{x'y'' - x''y'}{|\vec{v}|^3}$$
>
> where the derivatives are taken with respect to *t*.

d. Figure 10-6cc shows the ellipse

$$x = 5\cos t$$
$$y = 3\sin t$$

Show that the maximum curvature is at each end of the major axis.

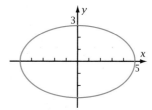

Figure 10-6cc

e. Show that the curvature of a circle,

$$x = r\cos t$$
$$y = r\sin t$$

is constant.

f. Show that the curvature of this line is zero.

$$x = 5\cos^2 t$$
$$y = 3\sin^2 t$$

g. The **radius of curvature** is defined to be the reciprocal of the curvature. Find the radius of curvature of the ellipse in part d at the right-most vertex $(5, 0)$.

h. On your grapher, plot the ellipse in Figure 10-6cc. Then plot a circle tangent to the ellipse at point (5, 0), on the concave side of the ellipse, with radius equal to the radius of curvature. Sketch the result. This

circle is called the **osculating circle** ("kissing" circle). Appropriately, it is the circle that best fits the curve at the point of tangency.

10-7 Chapter Review and Test

In this chapter you have studied applications of calculus to motion. You distinguished between distance traveled by a moving object and displacement from its starting point. You made precise the concept of average velocity and extended it to other average values. You extended your study of maximum and minimum values to problems involving motion and then to other similar problems. Finally, you applied the concepts to objects moving in a plane, using vectors as a tool.

Review Problems

R0. Update your journal with what you've learned since your last entry. You should include such things as

- The one most important thing you've learned in studying Chapter 10
- Which boxes you have been working on in the "define, understand, do, apply" table.
- The distinction among displacement, velocity, and acceleration
- How to find average rates
- How to use rates of change to find extreme values of functions
- How to analyze motion of objects moving in two dimensions
- Any techniques or ideas about calculus that are still unclear

R1. *Popeye and Olive Problem:* Olive Oyl is on a conveyor belt moving 3 ft/s toward the sawmill. At time $t = 0$, Popeye rescues her and starts running in the other direction along the conveyor belt. His velocity with respect to the ground, v, in ft/s, is given by

$$v = \sqrt{t} - 3$$

When does Popeye's velocity become positive? How far have he and Olive moved toward the

sawmill at this time? What is their net displacement from the starting point at $t = 25$? What total distance did they go from $t = 0$ to $t = 25$?

R2. a. The velocity of a moving object is given by $v(t) = 2^t - 8$ cm/min.

i. Graph velocity versus time. Sketch the result.

ii. Find the net displacement between $t = 1$ and $t = 4$.

iii. Find the total distance traveled between $t = 1$ and $t = 4$.

b. *Acceleration Data Problem:* An object initially going 30 ft/s has the following accelerations, in (feet per second) per second, measured at 5-s intervals.

Time	Acceleration
0	2
5	8
10	1
15	0
20	−10
25	−20

Find the estimated velocities at the ends of the time intervals. For each entry in the table, explain whether the object was speeding up, slowing down, or neither at that instant.

R3. a. *Average Velocity Problem:* An object moves with velocity $v(t) = \sin(\pi t/6)$. Find the average velocity on the time interval

   i. $[0, 3]$

   ii. $[3, 9]$

   iii. $[0, 12]$

   b. *Average Value Problem:* For the function $f(x) = 6x^2 - x^3$,

   i. Find the average value of $f(x)$ on the interval between the two x-intercepts.

   ii. Sketch a graph showing the graphical significance of this average value.

   iii. Show that the average value is not equal to the average of the two values of $f(x)$ at the endpoints of the interval.

R4. a. *Campus Cut-Across Problem:* Juana makes daily trips from the math building to the English building. She has three possible routes (Figure 10-7a):

   • Along the sidewalk all the way

   • Straight across the grass

   • Angle across to the other sidewalk

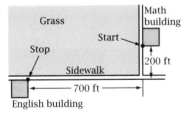

Figure 10-7a

She figures her speed is 6.2 ft/s on the sidewalk and 5.7 ft/s across the grass. Which route takes the least time? Explain.

   b. *Resort Island Causeway Problem:* Moe Tell owns a resort on the beach. He purchases an island 6 km offshore, 10 km down the beach (Figure 10-7b). So that his guests may drive to the island, he plans to build a causeway from the island to the beach, connecting to a road along the beach to the hotel. The road

will cost \$5,000 per kilometer, and the bridge will cost \$13,000 per kilometer. What is the minimum cost for the road and causeway system? How much money is saved by using the optimum path over what it would cost to build a causeway from the hotel straight to the island?

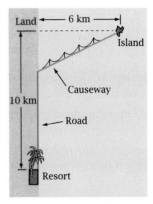

Figure 10-7b

R5. a. An object's acceleration is given by $a(t) = 6t - t^2$ in the interval $[0, 10]$. Find:

   i. The maximum and minimum accelerations for t in $[0, 10]$

   ii. The maximum and minimum velocities for t in $[0, 10]$, assuming $v(0) = 0$

   iii. The maximum and minimum displacements from the starting point for t in $[0, 10]$

   b. *Inflation Problem:* Dagmar lives in a country where the inflation rate is very high. She saves at a rate of 50 pillars (the currency in her country) a day. But the value of money is decreasing exponentially with time in such a way that at the end of 200 days a pillar will purchase only half of what it would at the beginning.

   i. Find an equation for the purchasing power of the money Dagmar has saved as a function of the number of days since she started saving.

   ii. On what day will Dagmar's accumulated savings have the maximum total purchasing power? Justify your answer.

R6. a. Draw a sketch showing how the velocity and acceleration vectors are related to each other and to the curved path of an object moving in a plane if

  i. The object is speeding up.

  ii. The object is slowing down.

b. An object moves along the hyperbola shown in Figure 10-7c. The position vector at any time t, in minutes, is given by

$$\vec{r} = (5\cosh t)\vec{i} + (3\sinh t)\vec{j}$$

  i. Find the position, velocity, and acceleration vectors for the object at $t = 1$.

  ii. Draw these vectors at appropriate places on a copy of the object's path.

  iii. At $t = 1$, how fast is the object moving? Is it speeding up or slowing down? At what rate?

  iv. How far does the object move between $t = 0$ and $t = 1$?

  v. Show that if the tails of the velocity vectors are placed at the respective points on the path, then their heads lie on one of the asymptotes of the hyperbola.

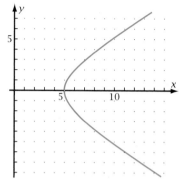

Figure 10-7c

Concept Problems

C1. *One-Problem Test on Linear Motion and Other Concepts:* A particle moves up and down the y-axis with velocity v, in feet per second, given by

$$v = t^3 - 7t^2 + 15t - 9$$

during the time interval $[0, 4]$. At time $t = 0$, its position is $y = 4$.

a. Sketch the velocity-time graph.

b. At what time(s) is the particle stopped?

c. At what time is the velocity the maximum? The minimum? Justify your answers.

d. At what time(s) does the velocity-time graph have a point of inflection?

e. What is happening to the particle at the point(s) of inflection?

f. Find the position, y, as a function of time.

g. Sketch the position-time graph.

h. At what time is y the maximum? The minimum? Justify your answers.

i. At what time(s) does the position-time graph have a point of inflection?

j. What is happening to the particle at the point(s) of inflection?

k. Is y ever negative? Explain.

l. What is the net displacement of the particle from $t = 0$ to $t = 4$?

m. How far does the particle travel from $t = 0$ to $t = 4$?

n. What is the average velocity from $t = 0$ to $t = 4$?

o. What is the average speed from $t = 0$ to $t = 4$?

C2. *New York to Los Angeles Problem:* What is the shortest possible time in which a person could get from New York to Los Angeles? If you ignore such things as getting to and from airports, the type of vehicle used, and so forth, the problem reduces to how much stress the human body can take from acceleration and deceleration (g forces). Recall that an acceleration of 1 g is the same as the acceleration due to gravity.

a. What have you heard from the media or elsewhere about the maximum g force the human body can withstand?

b. About how far is it from New York to Los Angeles?

c. You must be stopped both at the beginning of the trip and at the end. What, then, is the minimum length of time a person could take to get from New York to Los Angeles?

C3. *Spider and Clock Problem:* A spider sitting on a clock face at the "12" attaches one end of its web there 25 cm from the center of the clock. As the second hand passes by, it jumps onto it and starts crawling toward the center at a rate of 0.7 cm/s (Figure 10-7d). As the second hand turns, the spider spins more web. The length of this web depends on the number of seconds the spider has been crawling and can be calculated using the law of cosines. Find the rate of change of this length at the instant the spider has been crawling for 10 s.

Figure 10-7d

C4. *Submerging Cone Problem:* A cone of base radius 5 cm and height 12 cm is being lowered at 2 cm/min, vertex down, into a cylinder of radius 7 cm that has water 15 cm deep in it (Figure 10-7e).

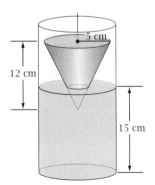

Figure 10-7e

As the cone dips into the water, the water level in the cylinder rises. Find the rate at which the level is rising when the vertex of the cone is

a. 10 cm from the bottom of the cylinder

b. 1 cm from the bottom of the cylinder

C5. *The Horse Race Theorem:* Sir Vey and Sir Mount run a horse race. They start at time $t = a$ at the same point. At the end of the race, time $t = b$, they are tied. Let $f(t)$ be Sir Vey's distance from the start and $g(t)$ be Sir Mount's distance from the start. Assuming that f and g are differentiable, prove that there was a time $t = c$ between a and b at which each knight was going exactly the same speed.

C6. *Hemispherical Railroad Problem:* A mountain has the shape of a perfect hemisphere with a base radius 1000 ft (unlikely in the real world, but it makes an interesting problem!). A railroad track is to be built to the top of the mountain. The train can't go straight up, so the track must spiral around the mountain (Figure 10-7f). The steeper the track spirals, the shorter it will be, but the slower the train will go. Suppose that the velocity of the train is given by

$$v = 30 - 60 \sin \theta$$

where v is in feet per second and θ is the (constant) angle the track makes with the horizontal. If the track is built in the optimal way, what is the minimum length of time the train could take to get to the top?

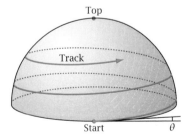

Figure 10-7f

Chapter Test

PART 1: No calculators allowed (T1–T4)

T1. For an object moving back and forth along a straight line, how can you tell from the velocity and acceleration whether the object is speeding up or slowing down?

T2. *Freight Elevator Problem:* A freight elevator moves up and down with velocity v, in feet per second, as shown in Figure 10-7g. Find the elevator's net displacement and the total distance it traveled in the 1-min period shown in the figure.

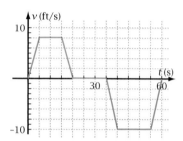

Figure 10-7g

T3. Estimate the average value of $f(x)$ on the interval $[1, 9]$ for the function graphed in Figure 10-7h. On a copy of the figure, show that you understand the graphical meaning of average value and also the conclusion of the mean value theorem for integrals.

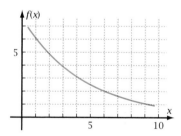

Figure 10-7h

T4. Figure 10-7i shows the path followed by an object moving in the xy-plane. Write the position vector for the point marked. On a copy of the figure, sketch a possible velocity vector and acceleration vector when the object is at this point, showing that the object is moving in the negative x-direction and slowing down.

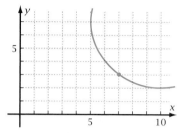

Figure 10-7i

PART 2: Graphing calculators allowed (T5–T20)

T5. *Truck Passing Problem:* You accelerate your car to pass a truck, giving your car velocity

$$v = \sqrt{t} + 60$$

where v is in feet per second and t is time in seconds since you started to pass. Find out how far you go in the 25 s it takes you to get around the truck.

T6. *Power Line Problem:* Ima Gardner builds a camp house that she wants to supply with electricity. The house is 3 mi from the road. The electrical contractor tells her it will cost $2520 ($360 per mile) to run the power line 7 mi along the highway to the point nearest the camp house and $2400 ($800 per mile) more to run it the 3 mi from the highway to the camp house. You believe that Ima could save money by making the line cut off from the highway before the 7-mi point and angle across to the house. How should the power line be run to minimize its total cost? How much could Ima save over the $4920 that the contractor proposes?

T7. For the function $f(x) = x^3 - 4x + 5$, find the maximum, the minimum, and the average values of the function on the interval $[1, 3]$. Sketch a graph showing the graphical significance of the average value.

T8. An object moving along a line has velocity $v(t) = 10(0.5 - 2^{-t})$ ft/s.

a. Find the distance it travels and its net displacement from the starting point for the time interval $[0, 2]$.

b. Find its acceleration at time $t = 0$.

c. At $t = 0$, was the particle speeding up or slowing down? Justify your answer.

T9. An object is moving at 50 cm/s at time $t = 0$. It has accelerations of 4, 6, 10, and 13 (cm/s)/s at $t = 0$, 7, 14, and 21, respectively. Approximately what was the object's average velocity for the 21-s time interval? About how far did the object go?

Figure 10-7j shows the path of an object moving with vector equation

$$\vec{r}(t) = (10 \cos 0.4t)\vec{i} + (10 \sin 0.6t)\vec{j}$$

where distance is in miles and time is in hours.

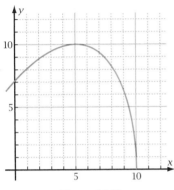

Figure 10-7j

T10. Find an equation for $\vec{v}(t)$.

T11. Find an equation for $\vec{a}(t)$.

T12. Find $\vec{r}(2)$. Make a copy of Figure 10-7j and draw $\vec{r}(2)$ on it.

T13. Find $\vec{v}(2)$. On the copy of Figure 10-7j, plot this vector starting at the object's position when $t = 2$. How is this vector related to the path of the object?

T14. Find $\vec{a}(2)$. On the copy of Figure 10-7j, plot this vector starting at the object's position when $t = 2$.

T15. Sketch the components of $\vec{a}(2)$, one of them directed tangentially to the path and the other normal to it.

T16. Based on the components of $\vec{a}(2)$, would you expect that the object is slowing down or speeding up when $t = 2$? How can you tell?

T17. At what rate is the object speeding up or slowing down when $t = 2$?

T18. Explain why the normal component of $\vec{a}(2)$ is pointing toward the concave side of the path.

T19. Find the distance the object travels between $t = 0$ and $t = 2$.

T20. What did you learn as a result of taking this test that you did not know before?

The Calculus of Variable-Factor Products

I-beams used in construction must be stiff so that they do not bend too much. The stiffness depends on the shape of the beam's cross section. Stiffness is measured by the second moment of area of the cross section, which is defined to be area times the square of the distance from the centroid of the cross section. Different parts of the cross section are at different distances from the centroid, so you can use definite integrals to compute the stiffness of a given type of beam.

Mathematical Overview

A definite integral enables you to find the product of x and y, where y varies. In Chapter 11, you will apply integrals to

- Work = force × displacement
- Force = pressure × area
- Mass = density × volume
- Moment = displacement × quantity

You will perform the applications in four ways.

Graphically   The icon at the top of each even-numbered page of this chapter reminds you that work equals force times displacement.

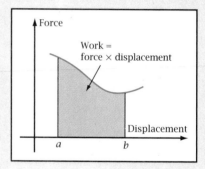

Numerically

Force	Displacement	Work
50	10	0
53	12	103
58	14	214
70	16	342
90	18	502
⋮	⋮	⋮

Algebraically $M_y = \int_a^b x \cdot dA$, the definition of moment of area

Verbally   *I calculated the balance point of a piece of cardboard by finding its centroid. Then I showed that I was right by cutting out the cardboard. It actually did balance on a pencil point placed at the calculated centroid!*

11-1 Review of Work—Force Times Displacement

In previous chapters you learned that the work done in moving an object from one place to another equals the force with which it is pushed or pulled times the displacement through which it moves. For instance, if you push a chair 7 ft across the floor with a force of 11 lb (Figure 11-1a), you do 77 ft-lb (**foot-pounds**) of work.

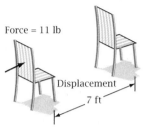

Force = 11 lb

Displacement

7 ft

Figure 11-1a

In Problem Set 11-1, you will refresh your memory about how to compute the work done if the force is variable. As you study this chapter you will see how the thought process you use for this one application can be used for many others. You will learn such things as how to find the balance point of a solid object and how to calculate volumes and masses of objects quickly, without actually evaluating any integrals.

OBJECTIVE   By yourself or with your study group, find the work done in moving a chair across the floor if the force you exert on it varies as you push.

Exploratory Problem Set 11-1

Chair Work Problem: Suppose that you push a chair across the floor with a force

$$F = 20xe^{-0.5x}$$

where F is the force in pounds and x is the distance in feet the chair has moved since you started pushing.

1. Figure 11-1b shows the graph of F. On a copy of this figure, draw a narrow vertical strip of width $\Delta x = 0.2$ centered at $x = 4$. Approximately what is the force at any value of x in this strip? Approximately how much work is done in moving the chair a distance Δx at $x = 4$?

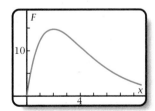

Figure 11-1b

2. Write an equation for dW, the work done in moving the chair a distance dx, in feet, when the chair is at point x, where the force is given by the previous equation.

3. Add all the dW's from $x = 0$ to $x = 7$. That is, find the definite integral of dW.

4. How much work was done in moving the chair from $x = 0$ to $x = 7$?

5. If you continue to push the chair with the force shown and it continues to move, what limit would the amount of work approach as x approaches infinity?

11-2 Work Done by a Variable Force

In ordinary English the word *work* is used in different contexts and with different meanings. For instance, you may feel that you did a lot of work on your calculus assignment last night. Physically, however, the word *work* has a precise definition. Work is the product of the *force* applied to an object and the *displacement* the object moves as a result of that force. For instance, if you push a chair 7 ft across the floor by exerting a force of 11 lb, you have done 77 ft-lb of work, as you saw in Section 11-1.

DEFINITION: Work

If an object moves a certain displacement as a result of being acted upon by a certain force, then the amount of work done is given by

$$\text{Work} = (\text{force})(\text{displacement})$$

In most real-world situations, the force does not remain constant as the object moves. By now you should realize that finding the work under these conditions is a job for definite integration!

OBJECTIVE Given a situation in which a varying force acts on an object, or where different parts of the object move through different displacements, calculate the amount of work done.

There are two ways to analyze a work problem.

- Move the whole object a small part of the displacement.
- Move a small part of the object the whole displacement.

Examples 1 and 2 show how this analysis can be done.

▶ **EXAMPLE 1** **Move the whole object a small part of the displacement**. A ship is at anchor in 80 ft of water. The anchor weighs 5000 lb, and the chain weighs 20 lb/ft (Figure 11-2a). The anchor is to be pulled up as the ship gets under way.

a. How much force must be exerted to lift the anchor as it comes aboard the ship? Write an equation expressing force in terms of the displacement, y, of the anchor from the ocean floor.

b. How much work must be done to raise the anchor the 80 ft from the ocean floor to the point where it comes aboard the ship?

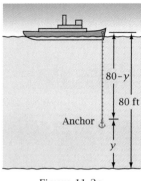

Figure 11-2a

Solution

a. When the anchor is on the ocean floor, and neglecting buoyancy, you must pull with enough force to lift the 5000-lb anchor and the 80-ft chain. Letting F stand for force,

$$F = (20)(80) + 5000 = 6600 \text{ lb}$$

At the ship, the only force needed is to lift the 5000-lb anchor.

$$F = 5000 \text{ lb}$$

In between, the force varies linearly with the length of the chain. If y is the displacement from the ocean floor to the anchor, then this length is equal to $(80 - y)$. Therefore,

$$F = 20(80 - y) + 5000$$
$$= 6600 - 20y$$

b. If the anchor is raised a small displacement, dy, the force would be essentially constant, the same as at some sample point in that particular subinterval. So the work, dW, done in lifting the anchor this small displacement would be

$$dW = F \, dy$$
$$= (6600 - 20y) \, dy$$

You can find the total amount of work, W, by adding the dW's, then taking the limit as dy approaches zero. You should recognize by now that this process is definite integration.

$$W = \int_0^{80} (6600 - 20y) \, dy$$
$$= 6600y - 10y^2 \Big|_0^{80} \qquad \text{You could integrate numerically.}$$
$$= 464{,}000 \text{ ft-lb} \qquad\qquad\qquad\qquad \blacktriangleleft$$

▶ **EXAMPLE 2** **Move a small part of the object the whole displacement.** A conical tank has top diameter 10 ft and height 15 ft (Figure 11-2b). It is filled to the top with liquid of density k, in pounds per cubic foot. A pump takes suction from the bottom of the tank and pumps the liquid to a level 8 ft above the top of the tank. Find the total amount of work done.

Solution The first thing to realize is that the amount of work done lifting any small volume of liquid is independent of the path the liquid takes. It depends only on how far the volume is

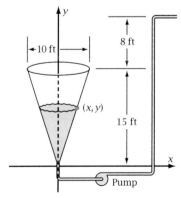

Figure 11-2b

displaced upward from where it starts to where it finishes. Liquid at a sample point (x, y) travels down through the tank, through the pump, and up to the discharge 8 ft above the top of the tank (and thus 23 ft above the bottom of the tank, where $y = 0$). So its net displacement is equal to $(23 - y)$.

If you "slice" the liquid horizontally, liquid at each point in the slice will be displaced essentially the same distance as that at the sample point. The force, dF, needed to lift the water the displacement $(23 - y)$ is equal to the weight of the slice, namely, $k\,dV$. Letting W stand for amount of work, the work done in lifting one slice is

$$dW = (k\,dV)(23 - y)$$

$$= k\pi x^2(23 - y)\,dy \qquad \text{Substitute } \pi x^2\,dy \text{ for } dV, \text{ and commute.}$$

The element of the cone where the sample point (x, y) is located is a line segment through the origin, containing the point $(5, 15)$. So its slope is $15/5 = 3$, and its equation is

$$y = 3x, \quad \text{or} \quad x = \frac{y}{3}$$

$$dW = k\pi\left(\frac{y}{3}\right)^2(23 - y)\,dy \qquad \text{Substitute } y/3 \text{ for } x.$$

$$= \frac{k\pi}{9}(23y^2 - y^3)\,dy$$

$$W = \frac{k\pi}{9}\int_0^{15}(23y^2 - y^3)\,dy \qquad \text{Add the } dW\text{'s and take the limit (that is, integrate).}$$

$$= \frac{k\pi}{9}\left(\frac{23}{3}y^3 - \frac{1}{4}y^4\right)\Big|_0^{15} \qquad \text{You could integrate numerically.}$$

$$= \frac{5875k\pi}{4} = 4614.214...\,k$$

If the liquid were water, with density $k = 62.4$ lb/ft^3, the total work would be about 287,927 ft-lb. ◀

Note: Work is a physical quantity equivalent to energy. For instance, the work done compressing a spring is stored in that spring as energy. Foot-pounds of work can be converted directly to joules or calories by the appropriate conversion factors. Although first moment of force, called **torque** (pronounced "tork"), also has the units (distance)(force), torque is not the same physical quantity as work. For this reason, torque is usually called pound-feet rather than foot-pounds (see Section 11-4).

Problem Set 11-2

Quick Review ⏱ *5 min*

Q1. What is the area under one arch of $y = \sin x$?

Q2. What is the area under $y = 4 - x^2$ from $x = -2$ to $x = 2$?

Q3. Find a velocity equation if the acceleration is $a = \tan t$.

Q4. Find the acceleration equation if the velocity is $v = \ln t$.

Q5. Name the theorem that allows definite integrals to be calculated by antiderivatives.

Q6. $\sum f(x)\,dx$ is a (n) —?— sum.

Q7. Name the technique for integrating $\int e^x \cos x\, dx$.

Q8. Name the technique for finding dy/dx if $x^3 y^5 = x \sin^2 y$.

Q9. Name the quick method for resolving an expression into partial fractions.

Q10. The average value of $y = \sin x$ for one complete cycle is —?—.

    A. 0   B. 1   C. 2   D. π E. 2π

---

1. *Leaking Bucket Problem:* Llara pulls a bucket of water up from the bottom of the well (Figure 11-2c). When she starts pulling, she exerts a force of 20 lb. But by the time she gets the bucket to the top, 50 ft up, enough water has leaked out so that she pulls with only 12 lb. Assume that both the rate she pulls the bucket and the rate the water leaks are constant, so the force she exerts decreases linearly with displacement from the bottom. How much work did Llara do in pulling the bucket out of the well?

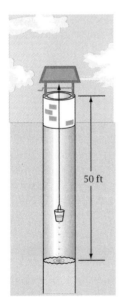

Figure 11-2c

2. *Spaceship Problem:* A spaceship on a launchpad weighs 30 tons (Figure 11-2d). By the time it

reaches an altitude of 70 mi, the spaceship weighs only 10 tons because it has used 20 tons of fuel. Assume that the weight of the spaceship decreases linearly with displacement above Earth.

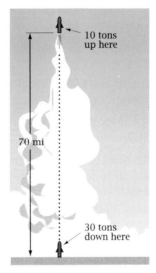

Figure 11-2d

a. How many mile-tons of work are done in lifting the spaceship to an altitude of 70 mi?

b. The rocket engines exert a constant thrust (that is, force) of 90 tons. How much work was done by the engines in lifting the spaceship to 70 mi? What do you think happens to the excess energy from part a?

3. *Spring Problem:* It takes work to compress a spring (work = force × displacement). However, the amount of force exerted while compressing the spring varies and is directly proportional to the displacement, s, the spring has been compressed (Figure 11-2e). This property is known as Hooke's law. Let k be the proportionality constant. Find the work required to compress a spring from $s = 0$ to $s = 10$.

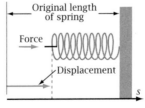

Figure 11-2e

4. *Table Moving Problem:* You push a table across the floor. At first, you push hard to get it moving, then you ease off as it starts to move. The force drops to zero at a displacement of 4 ft, where the table stops. Assume that the force, *F*, in pounds is given by

$$F = -x^3 + 6x^2 - 12x + 16$$

where *x* is the number of feet the table has been displaced.

a. Sketch the graph of *F* and show that it really does have these properties.

b. How much work is done pushing the table the 4 ft?

5. *Conical Reservoir Problem:* A conical reservoir 30 ft in diameter and 10 ft deep is filled to the top with water of density 62.4 lb/ft³. Find the work done in pumping all this water to a level of 7 ft above the top of the reservoir.

6. *Paraboloidal Tank Problem:* A tank is made in the shape of the paraboloid formed by rotating about the *y*-axis the graph of $y = x^2$ from $x = 0$ to $x = 4$ (Figure 11-2f). The tank is filled with benzene, an organic liquid with density 54.8 lb/ft³. Find the work done in pumping a full tank of benzene to a level of 10 ft above the top of the tank.

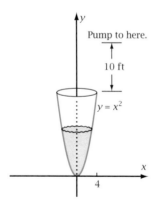

Figure 11-2f

7. *Spherical Water Tower Problem:* A spherical water tower 40 ft in diameter has its center 120 ft above the ground (Figure 11-2g). A pump at ground level fills the tank with water of density 62.4 lb/ft³.

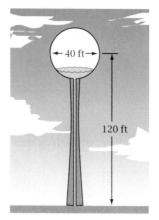

Figure 11-2g

a. How much work is done in filling the tank half full?

b. Quick! How much work is done in filling the tank completely? (Be careful!)

8. *Flooded Ship Problem:* A compartment in a ship is flooded to a depth of 16 ft with sea water of density 67 lb/ft³ (Figure 11-2h). The vertical bulkheads at both ends of the compartment have the shape of the region above the graph of

$$y = 0.0002x^4$$

where *x* and *y* are in feet. The compartment is 15 ft long. How much work must the bilge pumps do to pump all of the water over the side of the ship, which is 30 ft above the bottom?

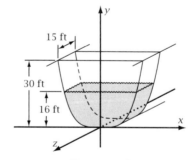

Figure 11-2h

9. *Carnot Cycle Problem:* An automobile engine works by burning gasoline in its cylinders. Assume that a cylinder in a particular engine has diameter 2 in. (Figure 11-2i). When the spark plug fires, the pressure inside the cylinder is 1000 psi (pounds per square inch),

and the volume is at its minimum, 1 in.^{3}. As the piston goes out, the hot gases expand adiabatically (that is, without losing heat to the surroundings). The pressure drops according to the equation

$$p = k_1 V^{-1.4}$$

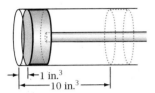

Figure 11-2i

where p is pressure, V is volume, and k_1 is a proportionality constant. When the piston is

farthest out and the volume is 10 in.^{3}, the exhaust valve opens and the pressure drops to atmospheric pressure 15 psi. As the piston comes back, the cool gases are compressed and the cycle is repeated. For compression, $p = k_2 V^{-1.4}$, where k_2 is a different proportionality constant.

a. Find the work done on the piston by the expanding hot gas.

b. Find the work done by the piston as it compresses the cool gas.

c. Find the net amount of work done. This is the amount of work that is available for moving the car.

d. How is "Carnot" pronounced? Who was Carnot?

11-3 Mass of a Variable-Density Object

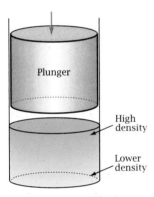

Figure 11-3a

The density of an object is defined to be the mass per unit volume. For instance, water has density 1 g/cm^3; iron, 7.86 g/cm^3; and uranium, 18.5 g/cm^3. You calculate density by dividing the mass of an object by its volume, therefore the mass is equal to density times volume.

$$\text{Mass} = (\text{density})(\text{volume})$$

The density of a real object may vary from point to point within the object. For example, the density of the materials composing Earth varies from about 12 g/cm^3 at the center of Earth to about 4 g/cm^3 at Earth's surface.

As another example, uranium oxide pellets, used as fuel in nuclear reactors, are made by compacting uranium oxide powder with a press (Figure 11-3a). The powder closer to the plunger in the press compacts to a density higher than the density of the powder that is farther away.

In this section you will explore ways of predicting the mass of an object if you know how its density behaves at various places.

OBJECTIVE Given a function specifying the density of an object at various places within that object, calculate the total mass of the object.

Example 1 shows how you can use techniques of definite integration that you know to calculate the total mass of a hypothetical object where the density varies. Your purpose is to apply these techniques to real-world problems.

▶ **EXAMPLE 1**   A solid is formed by rotating about the x-axis the region under the graph of $y = x^{1/3}$ from $x = 0$ to $x = 8$. Find the mass of the solid if the density, ρ (Greek letter "rho"),

> a. Varies axially (in the direction of the axis of rotation), being directly proportional to the square of the distance from the yz-plane
>
> b. Varies radially (in the direction of the radius), being directly proportional to the distance from the x-axis

Solution   a. Figure 11-3b shows the solid. A vertical slice of the rotated region generates a disk parallel to the yz-plane. So each point in the disk has essentially the same density as at the sample point (x, y). Letting m stand for mass and V stand for volume, the mass of a representative slice is defined as

$$dm = \rho \, dV$$

$$\rho = kx^2 \qquad\qquad \text{ρ is directly proportional to the square of x, the distance from the yz-plane.}$$

$$dV = \pi y^2 \, dx \qquad\qquad \text{By geometry, volume = (cross-sectional area)(length).}$$

$$\therefore dm = kx^2 \cdot \pi y^2 \, dx \qquad \text{Substitute for ρ and for dV.}$$

$$= k\pi x^2 (x^{1/3})^2 \, dx = k\pi x^{8/3} \, dx$$

$$\therefore m = \int_0^8 k\pi x^{8/3} \, dx \qquad\qquad \text{Add the dm's and find the limit. That is, integrate.}$$

$$= \tfrac{3}{11} k\pi x^{11/3} \Big|_0^8$$

$$= \tfrac{6144}{11} k\pi \qquad\qquad \text{(Find the actual mass by substituting for k and doing the arithmetic.)}$$

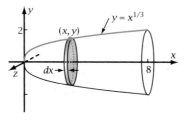

Figure 11-3b

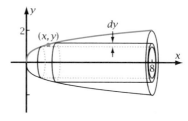

Figure 11-3c

b. The density varies radially (with the distance from the x-axis), so slicing the rotated region horizontally will produce a constant density within the slice. Rotating slices parallel to the axis of rotation produces cylindrical shells, as shown in Figure 11-3c. Again, picking a sample point on the curve, the mass of a representative shell is defined as

$$dm = \rho \, dV$$

$$= ky \cdot 2\pi y(8 - x) \, dy \qquad\qquad \rho = ky \text{ and}$$
$$dV = 2\pi y(8 - x)\, dy.$$
$$= 2k\pi y^2 (8 - y^3) \, dy = 2k\pi (8y^2 - y^5) \, dy$$

$$m = 2k\pi \int_0^2 (8y^2 - y^5)\, dy$$

Add the dm's and find the limit.
Note that the bounds are from 0 to 2.

$$= 2k\pi \left(\tfrac{8}{3}y^3 - \tfrac{1}{6}y^6 \right) \Big|_0^2$$

$$= 2k\pi \left(\tfrac{64}{3} - \tfrac{32}{3} - 0 + 0 \right) = \tfrac{64}{3} k\pi \qquad \blacktriangleleft$$

Note: Part b of the example shows the real reason for slicing objects into cylindrical shells. If the density varies radially, it will be constant (essentially) at all points in the shell. Slicing into plane slices would give a slice in which the density varies.

Problem Set 11-3

Quick Review

Q1. What is the volume of a cone inscribed in a 6-cm^3 cylinder?

Q2. What is the volume of a paraboloid inscribed in a 6-cm^3 cylinder?

Q3. Sketch the graph: $y = \sin x$

Q4. Sketch the graph: $y = \ln x$

Q5. Sketch the graph: $y = 2^x$

Q6. Sketch the graph: $y = x^2$

Q7. Density = $(-?-)/(-?-)$

Q8. Work = $(-?-)(-?-)$

Q9. If $y = \sin^{-1} x$, then $y' = -?-$.

Q10. y varies linearly with x. If $x = 0$, $y = 12$ and if $x = 2$, $y = 20$. If $x = 3$, then $y = -?-$.

    A. 18   B. 24   C. 26   D. 28   E. 40

1. The region bounded by the graph of $y = \ln x$, the x-axis, and the line $x = 3$ is rotated about the line $x = 0$ to form a solid. Find the mass of the solid if

  a. The density varies inversely with the distance from the axis of rotation

  b. The density varies linearly with y, being 5 if $y = 0$, and 7 if $y = 1$

2. The region bounded by the graph of $y = \sin x$ and the x-axis, between $x = 0$ and $x = \pi$, is

rotated about the y-axis to form a solid. The density of the solid varies directly with the distance from the axis of rotation. Find the mass of the solid.

3. The region under the graph of $y = 9 - x^2$ is rotated about the y-axis to form a solid. In parts a–c, find the mass of the solid if

  a. The density is a constant, k

  b. The density is constant in any thin horizontal slice but varies directly with the square of y in the y-direction

  c. The density does not vary in the y-direction but is directly proportional to the quantity $(1 + x)$, where x is the distance between the sample point and the axis of rotation

  d. Which of the solids in parts a, b, and c has the greatest mass? Assume that the constant k is the same in all three parts.

4. The region bounded by the graphs of $y = \sqrt{x}$ and $y = 0.5x$ is rotated about the x-axis to form a solid. Find the mass of the solid if

  a. The density varies axially, being directly proportional to the distance between the sample point and the yz-plane

  b. The density varies radially, being directly proportional to the square of the distance between the sample point and the axis of rotation

5. *Two Cone Problem:* Two cones have the same size—base radius 3 in. and height 6 in. (Figure 11-3d). Both have the same weight densities at their two ends—50 oz/in.^{3} and 80 oz/in.^{3}—but one has the higher density at the base and the other has the higher density at the vertex. In both, the density varies linearly with the distance from the plane of the base.

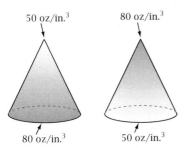

50 oz/in.^{3} 80 oz/in.^{3}

80 oz/in.^{3} 50 oz/in.^{3}

Figure 11-3d

a. Without performing any calculations, predict which cone has the higher mass. Explain your reasoning.

b. Confirm (or refute!) your prediction by calculating the mass of each cone.

6. *Two Cylinder Problem:* Two cylinders have the same shape—3-in. radius and 6-in. height (Figure 11-3e). One has density 50 oz/in.^{3} along the axis and 80 oz/in.^{3} at the walls. The other has density 80 oz/in.^{3} along the axis and 50 oz/in.^{3} at the walls. In both cylinders the density varies linearly with the distance from the axis.

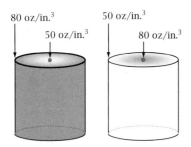

80 oz/in.^{3} 50 oz/in.^{3}

50 oz/in.^{3} 80 oz/in.^{3}

Figure 11-3e

a. Without performing any calculations, predict which cylinder has the higher mass. Explain your reasoning.

b. Confirm (or refute!) your prediction by calculating the mass of each cylinder.

7. The region bounded by the graphs of $y = 4 - 2x^2$, $y = 3 - x^2$, and $x = 0$ (Figure 11-3f) is rotated about the x-axis to form a solid. Both x and y are in centimeters. The density varies directly as the square of the distance from the yz-plane (that is, the base of the solid). Find the mass of the solid.

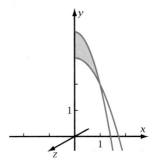

Figure 11-3f

8. The region in Problem 7 (Figure 11-3f) is rotated about the y-axis to form a different solid. The density decreases exponentially with distance from the y-axis, according to the equation $\rho = e^{-x}$. Find the mass of the solid.

9. *Uranium Fuel Pellet Problem:* Uranium oxide is used as a fuel in nuclear power plants that generate electricity. Powdered uranium oxide is compressed into pellets, as shown in Figure 11-3a on page 553. The compressing makes the pellets slightly denser at the top than they are at the bottom. Suppose that the cylindrical pellets have base diameter 1 cm and height 2 cm. Assume that the density is constant in the radial direction but varies with y, the distance from the bottom of the pellet (as shown in the table), being 9 g/cm^3 at the bottom and 10 g/cm^3 at the top. Predict the mass of the pellet, taking into consideration the variable density.

y (cm)	Density (g/cm^3)
2.0	10.0
1.6	9.9
1.2	9.8
0.8	9.6
0.4	9.4
0	9.0

10. The "triangular" region in the first quadrant, bounded by the graphs of $y = 4 - x^2$, $y = 4x - x^2$, and $x = 0$, is rotated about various axes to form various solids.

  a. Find the mass of the solid formed by rotating the region about the y-axis if the density varies directly with x, the distance from the y-axis.

  b. Find the mass of the solid formed by rotating the region about the x-axis if the density varies directly with x, the distance from the yz-plane.

  c. Find the mass of the solid in part a if the density varies directly with y, the distance from the xz-plane, instead of with x.

11. Find the mass of a sphere of radius r if

  a. The density varies directly with the distance from a plane through the center

  b. The density varies directly with the distance from one of its diameters

  c. The density varies directly with the distance from the center (use *spherical* shells.)

12. *Mass of Earth Problem:* The density of Earth is about $12\,\mathrm{g/cm^3}$ at its center and about $4\,\mathrm{g/cm^3}$ at its surface. Assume that the density varies linearly with the distance from the center. Find the mass of Earth.
  Useful information:

  • 1 mi = 5280 ft; 1 ft = 12 in.; 2.54 cm = 1 in.

  • Radius of Earth is about 3960 mi.

  • Slice into spherical shells.

13. The region under the graph of $y = e^x$ from $x = 0$ to $x = \pi/2$ is rotated about the y-axis to form a solid. The density is given by $\rho = \cos x$. Find the mass of the solid.

14. *Buckminster's Elliptical Dome Problem:* Architect Buckminster Fuller (1895–1983) once proposed that a dome should be built over Manhattan Island and an air-conditioning system be built to regulate the temperature, remove air pollution, and so forth. Your job is to find out how much air would be inside such a dome. Figure 11-3g shows what that dome might look like. Assume that the dome is a half-ellipsoid 8 mi long, 2 mi wide, and 1/2 mi high.

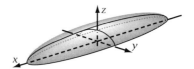

Figure 11-3g

  a. The three-dimensional equation for an ellipsoid is

$$\left(\frac{x}{a}\right)^2 + \left(\frac{y}{b}\right)^2 + \left(\frac{z}{c}\right)^2 = 1$$

  where a, b, and c are the x-, y-, and z-radii, respectively. Show that any cross section of the ellipsoid parallel to the xy-plane is an ellipse.

  b. The weight density of air at sea level ($z = 0$) is about $0.08\,\mathrm{lb/ft^3}$. But it decreases with altitude according to the equation $\rho = 0.08e^{-0.2z}$, where z is in miles. Find the mass of air inside the dome. How many tons is this?

  c. Suppose that you assume the density of the air is constant throughout the dome, $0.08\,\mathrm{lb/ft^3}$. How many more pounds of air would you have assumed are in the dome than are actually there? Surprising?

  d. The volume of a (whole) ellipsoid is $V = (4/3)\pi abc$. See if you can derive this formula by integrating the dV from this problem.

11-4   Moments, Centroids, Center of Mass, and the Theorem of Pappus

If you hold a meterstick at one end and hang a weight on it, the force caused by the weight twists the meterstick downward (Figure 11-4a). The farther from your hand you hang the weight, the more twisting there is. Doubling the displacement doubles the twisting for a given weight. And doubling the weight hung at the same displacement also doubles the twisting. The amount of twisting is the **torque**. The torque with respect to an axis through your hand is defined to be the amount of force created by the weight, multiplied by the displacement from your hand perpendicular to the direction of the force.

Torque = (force)(displacement)

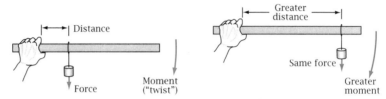

Figure 11-4a

Torque is just one example of a more general concept called **moment**. The word *moment* comes from the Latin *movere*, "to move," and *momentum*, "moving power." The moment of a physical quantity equals the magnitude of that quantity multiplied by some power of the displacement from a reference point, line, or plane to the place where the quantity is located.

DEFINITION: *nth Moment*

nth moment of quantity = (magnitude of quantity)(displacement)^n

---

The torque produced by the weight in Figure 11-4a is thus the *first* moment of *force* with respect to an *axis* through your hand. You can find any order moment of any quantity with respect to a point, line, or plane. Some moments have interesting meanings in the real world. Others are of interest in the mathematical world only. In this section you will explore first and second moments of mass, volume, arc length, and area. You will use the results to calculate the **center of mass** of a solid, the point where all of the mass could be concentrated to produce the same first moment, and the related **centroid**, which is a center of volume, length, or area. Calculus is used if parts of the object are at different displacements.

OBJECTIVE   Given the description of a solid or a plane region, find its first or second moment of volume, area, or mass with respect to a point, line, or plane, and its center of mass, volume, or area.

▶ **EXAMPLE 1**   A solid paraboloid has the shape of the solid formed by rotating about the y-axis the region in Quadrant I above the graph of $y = x^2$ and below the line $y = 9$ (Figure 11-4b).

  a. Find the volume of the solid if the dimensions are in centimeters.

  b. Find M_{xz}, the first moment of volume with respect to the xz-plane.

  c. Find $\overline{y}$ ("y bar"), the y-coordinate of the centroid, at which $\overline{y} \cdot$ volume $= M_{xz}$ from part b.

  d. At the centroid $(\overline{x}, \overline{y}, \overline{z})$, $\overline{x} \cdot$ volume $= M_{yz}$, and $\overline{z} \cdot$ volume $= M_{xy}$, where the moments are with respect to the yz- and xy-planes, respectively. Explain why both $\overline{x}$ and $\overline{z}$ are zero for this solid.

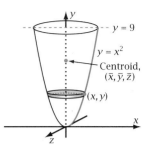

Figure 11-4b

Solution   a. Let V stand for volume. Slicing perpendicular to the y-axis gives

$$dV = \pi x^2 \, dy = \pi y \, dy$$

$$V = \int_0^9 \pi y \, dy$$

$$= \frac{\pi}{2} y^2 \Big|_0^9$$

$$= 40.5\pi = 127.234\ldots \text{ cm}^3$$

  b. Let M_{xz} stand for the moment with respect to the xz-plane. Slicing perpendicular to the y-axis, as in part a, makes the slice parallel to the xz-plane, so every point in the slice has approximately the same displacement from the xz-plane as the sample point (x, y).

$$dM_{xz} = y \, dV \qquad\qquad \text{Moment = (displacement)(volume).}$$

$$= y \cdot \pi y \, dy = \pi y^2 \, dy \qquad\qquad \text{Substitute for } dV.$$

$$M_{xz} = \int_0^9 \pi y^2 \, dy \qquad\qquad \text{Add the } dM\text{'s and take the limit (integrate).}$$

$$= \frac{\pi}{3} y^3 \Big|_0^9 = 243\pi = 763.407\ldots \text{ cm}^4 \qquad \text{The units are (cm)(cm}^3).$$

  c. $$\overline{y} \cdot 40.5\pi = 243\pi \qquad\qquad (\overline{y})(\text{volume}) = \text{moment.}$$

$$\overline{y} = \frac{243\pi}{40.5\pi} = 6 \text{ cm} \qquad\qquad \text{The units are (cm}^4)/(\text{cm}^3).$$

Note that the centroid is two-thirds of the way up from the vertex to the base. This fact is to be expected because the solid is wider near the base.

  d. The displacement is a directed distance. The solid is symmetrical with respect to the xy- and yz-planes, so there is just as much negative moment on one side of the plane as there is positive moment on the other side. The moments with respect to these planes are thus zero. So the x- and z-coordinates of the centroid are also zero, and the centroid lies on the y-axis.   ◀

Centroids, or geometrical centers, of plane regions and curves (center of area and center of arc length) are similarly defined. For instance, a region of area A in the xy-plane has a centroid at a displacement from the y-axis given by $\bar{x}A = M_y$. A curve of length L in the xy-plane has a centroid at a displacement from the x-axis given by $\bar{y}L = M_x$.

If a solid has a uniform density, its **center of mass**, or balance point, will be at the centroid. If the density of the solid is not uniform, the center of mass can be at a place other than the centroid. Example 2 shows how to calculate the center of mass when the density of the solid is not uniform.

After a firework explodes, its motion still follows its center of mass.

DEFINITION: Centroid

The centroid of a solid is the point $(\bar{x}, \bar{y}, \bar{z})$ at which

$$\bar{x}V = M_{yz}, \quad \bar{y}V = M_{xz}, \quad \text{and} \quad \bar{z}V = M_{xy}$$

where V is volume, and M_{yz}, M_{xz}, and M_{xy} are the first moments of volume of the solid with respect to the yz-, xz-, and xy-planes, respectively.

Note: To find the center of mass of a solid, $(\bar{x}, \bar{y}, \bar{z})$, replace volume, V, with mass, m. M_{yx}, M_{xz}, and M_{xy} then represent the first moments of mass rather than volume.

▶ **EXAMPLE 2** Suppose that the solid in Example 1 has a density that is constant radially but varies axially, being equal to $y^{1/2}$ g/cm³.

    a. Find the mass of the solid.

    b. Find the first moment of mass with respect to the xz-plane.

    c. Find the y-coordinate of the center of mass, the point $(\bar{x}, \bar{y}, \bar{z})$ for which $(\bar{y})(\text{mass}) = $ moment of mass with respect to the xz-plane. Show that the answer is reasonable.

Solution a. Figure 11-4c shows the solid, sliced horizontally as in Example 1. The density varies only axially and all points in the slice are about the same displacement from the xz-plane, therefore the mass, dm, of any slice is the density at the sample point times the volume of the slice.

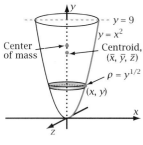

Figure 11-4c

$$dm = \rho \, dV = y^{1/2} \pi y \, dy = \pi y^{3/2} \, dy \qquad \text{From Example 1,}\ dV = \pi y \, dy.$$

$$m = \int_0^9 \pi y^{3/2} \, dy = \tfrac{2}{5} \pi y^{5/2} \Big|_0^9 = 97.2\pi \approx 305.4 \text{ g} \qquad \text{The units are } (\text{g/cm}^3)(\text{cm}^3).$$

    b. Let M_{xz} stand for the first moment of mass with respect to the xz-plane. All points in a slice are about the same displacement from this plane. So the moment of the slice, dM_{xz}, equals the displacement from the xz-plane to the sample point times the mass of the slice.

$$dM_{xz} = y \, dm = y\pi y^{3/2} \, dy = \pi y^{5/2} \, dy$$

$$M_{xz} = \int_0^9 \pi y^{5/2} \, dy = \frac{2}{7} \pi y^{7/2} \Big|_0^9 = \frac{4374\pi}{7} \approx 1963.0 \text{ cm-g} \qquad \text{The units are } (\text{cm})(\text{g}).$$

c. The y-coordinate of the center of mass, $\overline{y}$, is found using the fact that (displacement from center of mass)(total mass) = moment.

$$\overline{y} \cdot m = M_{xz}$$

$$\overline{y} = \frac{4374\pi/7}{97.2\pi} = 6\frac{3}{7} = 6.428... \text{ cm} \qquad \text{The units are } [(\text{cm})(g)]/(g).$$

The answer is reasonable because it is a little larger than 6 cm, the y-coordinate of the centroid. The density is greater near the base than at the top of the solid, so the center of mass is closer to the base than is the centroid.   ◀

Example 3 shows how to extend the concept of moment and centroid to a plane region.

▶ **EXAMPLE 3**   Let R be the region in Quadrant I bounded by the graph of $y = 3 \cos x$ and the two coordinate axes, where x and y are in inches.

    a. Find the first moment of area of R with respect to the y-axis.

    b. Find the first moment of area of R with respect to the x-axis.

    c. Find the centroid of R, the point $(\overline{x}, \overline{y})$ at which $(\overline{x})(\text{area})$ = first moment with respect to the y-axis and $(\overline{y})(\text{area})$ = first moment with respect to the x-axis.

Solution   a. Figure 11-4d shows the region R. Slicing parallel to the y-axis gives strips in which all points are about the same displacement from the y-axis. Use A for area and M_y for moment with respect to the y-axis. The moment, dM_y, of a strip of area dA is

$$dM_y = x\,dA = x(3\cos x)\,dx$$
$$= 3x \cos x\,dx$$
$$M_y = \int_0^{\pi/2} 3x \cos x\,dx$$
$$= 3x \sin x + 3 \cos x \Big|_0^{\pi/2} = \frac{3\pi}{2} + 0 - 0 - 3$$
$$= 1.71238... \text{ in.}^3 \qquad \text{Note that the units are in.}^3.$$

    b. To find M_x, the moment with respect to the x-axis, slice horizontally (Figure 11-4e). That way, each point in the strip will be about the same displacement from the x-axis.

$$dM_x = y\,dA = y(x\,dy)$$

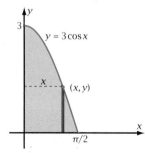

Figure 11-4d

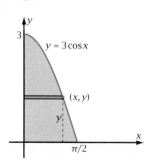

Figure 11-4e

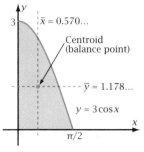

$\overline{x} = 0.570...$

Centroid
(balance point)

$\overline{y} = 1.178...$

$y = 3\cos x$

$\pi/2$

Figure 11-4f

You could put $\cos^{-1}(y/3)$ in place of x and integrate from 0 to 3. If you are integrating algebraically, it is easier to substitute $3\cos x$ for y and $-3\sin x\,dx$ for dy. Because x is $\pi/2$ when y is 0, and 0 when y is 3, the limits of integration are $\pi/2$ to 0.

$$M_x = \int_{\pi/2}^{0} (3\cos x)(x)(-3\sin x\,dx)$$

$$= -9\int_{\pi/2}^{0} x\cos x\sin x\,dx = -\frac{9}{2}\int_{\pi/2}^{0} x\sin 2x\,dx$$

$$= -\frac{9}{2}\left(-\frac{1}{2}x\cos 2x + \frac{1}{4}\sin 2x\right)\Big|_{\pi/2}^{0}$$

$$= -\frac{9}{2}(0 + 0 - \frac{\pi}{4} + 0) = \frac{9\pi}{8} = 3.534291... \approx 3.53 \text{ in.}^3$$

c. To find the centroid, you need to find the area of the region first. You can either integrate $3\cos x$ from 0 to $\pi/2$ or recall that the area under a half-arch of $y = \cos x$ equals 1 so that the total area is 3.

$$\overline{x}A = M_y$$

$$\overline{x} = \frac{3\pi/2 - 3}{3} = 0.570796... \approx 0.57 \text{ in.}$$

$$\overline{y}A = M_x$$

$$\overline{y} = \frac{9\pi/8}{3} = 1.178097... \approx 1.18 \text{ in.} \quad \blacktriangleleft$$

Figure 11-4f shows the location of the centroid, point $(0.57..., 1.178...)$.

The photograph shows the region in Example 3 drawn on cardboard. The cutout will balance on a pencil point placed at the centroid!

The last example for this section shows you the meaning of second moment.

▶ **EXAMPLE 4**   For the region R in Example 3,

    a. Find the second moment of area of the region with respect to the y-axis.

    b. Find $\overline{x}$, the x-coordinate of the center of second moment, which is the displacement from the y-axis for which

$$(\text{Displacement})^2(\text{area}) = \text{second moment}$$

Solution   a. Slice parallel to the y-axis, as shown in Figure 11-4d, so that points in a strip will be about the same displacement from that axis.

$$dM_y = x^2\,dA = x^2(3\cos x\,dx) \qquad \text{Second moment} = (\text{area})(\text{square of displacement}).$$

$$M_y = \int_{0}^{\pi/2} 3x^2\cos x\,dx = 1.402203... \text{ in.}^4 \qquad \text{Note that the units are in.}^4.$$

    b. To find the center of second moment, you must use the fact that displacement squared times area equals second moment.

$$(\overline{x})^2 A = M_y(3) = 1.402203...$$

$$\overline{x} = \sqrt{\frac{1.402203...}{3}} = 0.683667... \approx 0.68 \text{ in.} \qquad \text{Notice that the units come out inches.}$$

$\blacktriangleleft$

The stiffness of a beam is related to the second moment of area of a region that has the cross-sectional shape of the beam. The resistance of a uniform flat plate to being rotated about an axis is related to the second moment of area about that axis. The larger the second moment, the more difficult it is to start the region rotating, or stop it once it gets started. For this reason the second moment of area is sometimes called the **moment of inertia** of the region. The displacement to the center of second moment from the axis of reference is called the **radius of gyration**. The same terms apply to second moments of volume or mass of a solid figure.

Problem Set 11-4

1. *Paraboloid Problem:* The region in Quadrant I under the graph of $y = 9 - x^2$ is rotated about the y-axis to form a solid. Assume that dimensions are in centimeters.

   a. Find the volume of the solid.

   b. Find its first moment of volume with respect to the xz-plane.

   c. Find its centroid.

2. *Ellipsoid Problem:* A half-ellipsoid (Figure 11-4g) is formed by rotating about the x-axis the region in the first quadrant bounded by

$$\left(\frac{x}{12}\right)^2 + \left(\frac{y}{5}\right)^2 = 1 \text{ (dimensions in feet)}$$

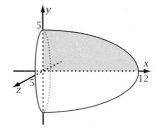

Figure 11-4g

   a. Confirm by appropriate integration that the volume is $\frac{2}{3}\pi(12)(5^2)$.

   b. Find the first moment of volume with respect to the yz-plane.

   c. Find the centroid.

3. *Paraboloid Mass Problem:* The paraboloid in Problem 1 has density, in grams per cubic centimeter, that is constant in the radial direction but directly proportional to the cube root of y in the axial direction.

   a. Find the mass of the solid.

   b. Find the first moment of mass of the solid with respect to the xz-plane.

   c. Find the center of mass.

   d. True or false: "The center of mass of a solid is always at the centroid."

4. *Ellipsoid Mass Problem:* The half-ellipsoid in Problem 2 has a weight density, in pounds per cubic foot, that is constant in the radial direction but varies directly with x in the axial direction.

   a. Find the weight of the solid.

   b. Find the first moment of weight of the solid with respect to the yz-plane.

   c. Find the center of weight (which is the same as the center of mass in this case).

   d. True or false: "The center of mass of a solid is always at the centroid."

5. *Exponential Region and Solid Problem:* Let R be the region under the graph of $y = e^x$ from $x = 0$ to $x = 2$ (Figure 11-4h).

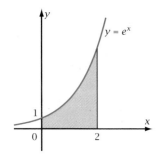

   Figure 11-4h

   a. Find the x-coordinate of the centroid of R.

   b. R is rotated about the x-axis to form a solid. Find the x-coordinate of the centroid.

   c. True or false: "The centroid of a region and the centroid of the solid formed by rotating that region about the x-axis have the same x-coordinate."

6. *Secant Curve Region Problem:* Let R be the region under the graph of $y = \sec x$ from $x = 0$ to $x = \pi/3$ (Figure 11-4i).

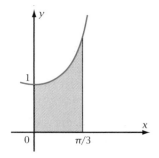

   Figure 11-4i

   a. Find the x-coordinate of the centroid of R.

   b. R is rotated about the x-axis to form a solid. Find the x-coordinate of the centroid.

   c. True or false: "The centroid of a region and the centroid of the solid formed by rotating that region about the x-axis have the same x-coordinate."

7. *Centroid of a Triangle Experiment:* Prove that the centroid of a triangle is located one-third of the way up from the base (Figure 11-4j). Then draw a triangle on cardboard, draw lines one-third of the way from each base, and cut out the triangle. If your work is done accurately, you should find that the lines intersect at one point and that you can balance the triangle on the point of a pencil placed at that point.

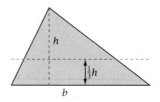

   Figure 11-4j

8. *Centroid Cut-Out Experiment:* Let R be the region under the graph of $y = x^{2/3}$ from $x = 0$ to $x = 8$, where x and y are in centimeters.

   a. Find the area of R.

   b. Find the first moment of R with respect to the x-axis.

   c. Find the first moment of R with respect to the y-axis.

   d. Find the coordinates of the centroid of R.

   e. Plot an accurate graph of R on an index card or graph paper. Mark the centroid. Then cut out the region and try to balance the region on a pencil point at the centroid.

9. *Second Moment of Area Problem:* Let R be the region under the graph of $y = \sin x$ from $x = 0$ to $x = \pi$ (Figure 11-4k), where x and y are in centimeters.

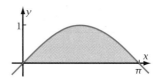

   Figure 11-4k

a. Show that the centroid of R is at $x = \pi/2$.

b. Find the second moment of area of the region with respect to the y-axis.

c. Find the radius of gyration (displacement to center of second moment) of R with respect to the y-axis.

10. *Second Moments for Plane Regions Problem:* Find the second moment of area for these shapes (Figure 11-4l).

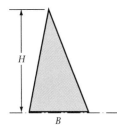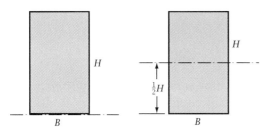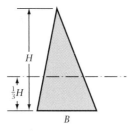

Figure 11-4l

a. A rectangle of base B and height H, with respect to an axis along the base

b. A rectangle of base B and height H, with respect to an axis through the centroid, parallel to the base

c. A triangle of base B and height H, with respect to an axis along the base

d. A triangle of base B and height H, with respect to an axis parallel to the base, one-third of the way to the opposite vertex (that is, through the centroid)

11. *Second Moments for Solid Figures:* Find the second moment of volume and radius of gyration for these solids (Figure 11-4m):

a. A cylinder of radius R and height H, with respect to its axis

b. A cone of base radius R and height H, with respect to its axis

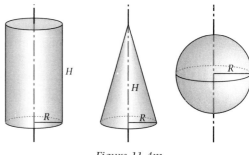

Figure 11-4m

c. A sphere of radius R, with respect to its diameter

12. *Rotation of Solids Problem:* You can measure the amount of resistance an object has to starting or stopping rotation by its second moment of mass with respect to the axis of rotation. If the density is constant, the second moment of mass equals the density times the second moment of volume. Suppose that 1000 cm^3 of clay is made into a sphere, and another 1000 cm^3 is made into a cylinder with a diameter equal to its height (Figure 11-4n). Which one has higher resistance to rotating? Justify your answer.

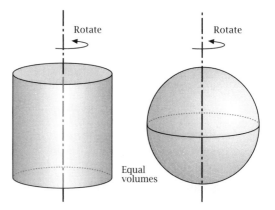

Figure 11-4n

13. *Beam Moment Problem:* The stiffness of a beam is directly proportional to the second moment of area of the beam's cross section with respect to an axis through the centroid of the cross section. In this problem you will investigate the stiffness of beams with the same cross-sectional area but with different shapes.

a. Prove that if a rectangle has base B and height H, then the second moment of area with respect to an axis through the centroid and parallel to the base is equal to $BH^3/12$.

b. Find the stiffness of a 2 in.-by-12 in. beam (Figure 11-4o) if it is

 i. Turned on edge

 ii. Lying flat

 Use k for the proportionality constant. Based on the results, explain why boards used for floor joists in houses are turned on edge rather than laid flat.

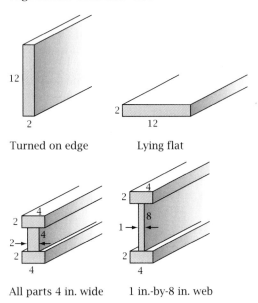

Turned on edge   Lying flat

All parts 4 in. wide   1 in.-by-8 in. web

Figure 11-4o

c. An I-beam is made with the same cross-sectional area as the 2 in.-by-12 in. beam in part b. Find its stiffness if

 i. All three parts (two flanges and one web) are 4 in. wide

 ii. The two flanges are 2 in. by 4 in., but the web is 1 in. by 8 in.

d. Does increasing an I-beam's depth without changing the cross-sectional area make much change in the beam's stiffness? What physical limitations keep people from making a beam very tall and thus very stiff?

14. *Introduction to the Theorem of Pappus:* The region R under the graph of $y = x^3$ from $x = 0$ to $x = 2$ is rotated about the y-axis to form a solid (Figure 11-4p).

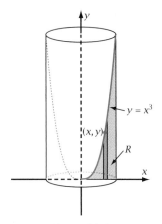

Figure 11-4p

a. Find the area of R.

b. Find the volume of the solid using vertical slices of R.

c. Find the first moment of area of R with respect to the y-axis. What do you notice about the integral?

d. Find the x-coordinate of the centroid of R.

e. A theorem of Pappus states that the volume of a solid of revolution equals the area of the region being rotated times the distance the centroid of the region travels. Show that this problem confirms the theorem.

15. *Theorem of Pappus Problem:* Pappus was a Greek mathematician who lived in Alexandria in the fourth century A.D. One of his theorems is stated here.

> **Theorem: The Theorem of Pappus for Volumes**
>
> The volume, V, of a solid of revolution is given by
>
> $$V = 2\pi \overline{R} A$$
>
> where A is the area of the region being rotated, $\overline{R}$ is the displacement from the axis of rotation to the centroid of the region, and the region is not on both sides of the axis of rotation. The quantity $2\pi \overline{R}$ is thus the distance the centroid travels as the region rotates.

In Problem 14, you saw an example of this theorem. In this problem you will use the theorem, once forward and once backward.

a. *Toroid Problem:* A toroid (Figure 11-4q) is formed by rotating a circle of radius r about an axis R units from the center of the circle, where $r \le R$. Find the volume of the toroid.

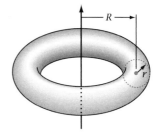

Figure 11-4q

b. *Centroid of a Semicircle:* A semicircle of radius r is rotated about its diameter to form a sphere (Figure 11-4r). You know formulas for the area of a semicircle and for the volume of a sphere. Use these facts to find the displacement from the center of a semicircle to its centroid.

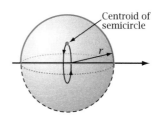

Centroid of semicircle

Figure 11-4r

16. *Theorem of Pappus Proof:* Prove the theorem of Pappus for volumes.

11-5 Force Exerted by a Variable Pressure—Center of Pressure

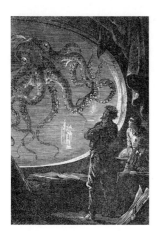

Pressure exerted by a fluid, such as air or water, is defined to be the force exerted by the fluid per unit area. For instance, water in a home's pipes is usually at a pressure of 40 to 100 psi (lb/in.^{2}). The air in a scuba diving tank is compressed to about 3000 psi. This means that each square inch of the tank's wall is pushed with a force of 3000 lb. As a result of the definition of pressure,

Force = (pressure)(area)

In many real-world situations, the pressure acting on a surface is different at various places on the surface. For instance, the pressure acting on an airplane's wings is usually higher near the middle of the wing than it is at either the

In 20,000 Leagues Under the Sea, Jules Verne foresaw the idea of a pressurized underwater vehicle.

leading or trailing edge. So the total force, which holds up the plane, must be found by integrating rather than simply by multiplying.

In this section you will learn how to calculate the force exerted by a variable pressure. You will also calculate the **moment of force** and use it to find the **center of pressure**, where the entire force could be concentrated to produce the same moment.

OBJECTIVE   Given a region and a function specifying the pressure acting on the region, calculate the total force, the moment of force, and the center of pressure.

▶ **EXAMPLE 1**   *Weir Problem:* A weir (a small dam) is to be built across a stream. When finished, the weir will have a vertical parabolic cross section the shape of the region above the graph of $y = x^2$ and below the line $y = 4$, where x and y are in feet (Figure 11-5a). The pressure at any sample point below the water's surface will be directly proportional to the distance from the surface down to that point. The proportionality constant is the density of water, 62.4 lb/ft^3.

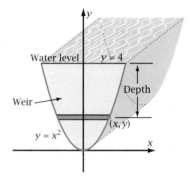

Figure 11-5a

a. Predict the total force acting on the weir when the water is all the way to the top.

b. Find the first moment of force with respect to the x-axis.

c. Find the *center of pressure*, the point on the face of the weir where the entire force could be applied to produce the same moment with respect to the x-axis.

d. The weir face is wider at the top, so you would expect the center of pressure to be more than halfway up. But the pressure is greater toward the bottom, so you would also expect the center of pressure to be less than halfway up. Based on your answer to part c, which of these two competing effects predominates?

Solution   a. Slice the region parallel to the x-axis so that the pressure will be essentially constant at any point in the slice. Pick a sample point (x, y) in the slice and on the graph. Let F stand for force, p for pressure, and A for area. By the definition of pressure,

$$dF = p \, dA$$

Pressure varies directly with depth, $(4 - y)$, therefore the pressure at the sample point will be

$$p = k(4 - y)$$

where $k = 62.4$ lb/ft^3, which will be substituted at the end to get a numerical answer. Because $dA = 2x \, dy = 2y^{1/2} \, dy$, you can write

$$dF = 2k(4 - y) \, y^{1/2} \, dy$$

To find the total force acting on the weir's face, you add all the dF's and take the limit as dy goes to zero. This process is, of course, definite integration.

$$F = \int_0^4 2k(4 - y)\, y^{1/2}\, dy$$

$$F = 2k\left(\tfrac{8}{3}y^{3/2} - \tfrac{2}{5}y^{5/2}\right)\Big|_0^4$$

$$= \frac{256k}{15} = \frac{(256)(62.4)}{15}$$

$$= 1064.96, \text{ or about } 1065 \text{ lb}$$

b. The first moment of force with respect to the x-axis is defined to be the product of the force and the displacement from the x-axis to the point where the force is acting. Fortunately, all points in the horizontal slice (Figure 11-5a) are essentially the same displacement from the x-axis as the sample point, (x, y). Using M for moment, the moment acting on the slice is

$$dM = y\, dF = y \cdot p\, dA$$

where y and p are both measured at the sample point.

$$dM = y \cdot 2k(4 - y)\, y^{1/2}\, dy \qquad \text{Substitute for } p \text{ and } dA.$$
$$= 2k(4y^{3/2} - y^{5/2})\, dy$$

$$\therefore M = 2k\int_0^4 (4y^{3/2} - y^{5/2})\, dy \qquad \text{Add the } dM\text{'s and take the limit}$$
$$\text{(that is, integrate).}$$

$$= 2k\left(\tfrac{8}{5}y^{5/2} - \tfrac{2}{7}y^{5/2}\right)\Big|_0^4$$

$$= \frac{1024k}{35} = \frac{(1024)(62.4)}{35} = 1825.645\ldots, \text{ or about } 1826 \text{ lb-ft}$$

c. By the definition of center of pressure, its vertical coordinate, $\overline{y}$, is the number for which

$$\overline{y}F = M$$

$$\therefore \overline{y} = \frac{1024k/35}{256k/15} = \frac{12}{7}, \text{ or } 1\tfrac{5}{7} \text{ ft}$$

By symmetry, the x-coordinate, $\overline{x}$, of the center of pressure is zero, so the center of pressure is at point $(0, 1\tfrac{5}{7})$.

d. The center of pressure is less than halfway up, meaning that the increasing pressure at greater depths predominates over the decreasing area. ◀

As you learned in Section 11-4, the first moment of force with respect to an axis is called torque. In part b of Example 1, 1826 lb-ft is the amount of torque exerted by the water on the dam face. It measures the amount of twisting that the force does on the dam face, tending to make it rotate about the x-axis and fall over.

Problem Set 11-5

Q1. The geometrical center of a solid is called its —?—.

Q2. The point where all the mass of a solid could be concentrated is called its —?—.

Q3. The displacement from an axis to the center of second moment with respect to that axis is called the —?—.

Q4. The process of adding parts of a physical quantity then taking the limit as the size of the parts goes to zero is called —?—.

Q5. The process of finding the antiderivative is called —?—.

Q6. Density equals —?— divided by —?—.

Q7. Simplify: $(x^{1/3})(x^{1/6})$

Q8. Integrate: $\int \sec x \, dx$

Q9. Differentiate: $y = \tan^{-1} x$

Q10. $(0.5)(e^x + e^{-x})$ is defined to be —?—.

    A. $\cosh x$ B. $\sinh x$ C. $\tanh x$

    D. Constant   E. Linear

---

1. *Trough Problem:* A trough has a vertical end in the shape of the region above the graph of $y = 2x^4$ and below the line $y = 2$, where x and y are in feet (Figure 11-5b). The trough is filled with liquid of density k, in pounds per cubic foot.

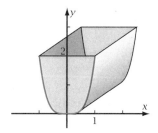

Figure 11-5b

a. Find the force acting on the end of the trough. Recall that force = (pressure)(area), and the pressure at a point varies directly with the displacement from the surface of the liquid to that point.

b. The force in part a causes a moment with respect to the x-axis. However, different points on the end of the trough are at different displacements from the x-axis. Find the moment of force with respect to the x-axis.

c. Find the center of pressure, where the entire force could be concentrated to produce the same moment with respect to the x-axis.

2. *Dam Problem:* At its narrowest point, Scorpion Gulch is 20 ft wide and 100 ft deep. A dam at this point has its vertical face in the shape of the region bounded by the graphs of $y = x^2$ and $y = 100$, where x and y are in feet.

a. Confirm that the dam is 20 ft wide at the top.

b. Find the area of the vertical dam face.

c. When the gulch is filled with water to the top of the dam, the greatest force will be exerted. It is important to know whether this force will be large enough to rip the dam from its foundations. Find this total force.

d. The dam could also fail by being pushed over. The first moment of force with respect to the x-axis is the quantity that the dam must withstand to prevent this. Find the first moment of force with respect to the x-axis.

e. How far above the bottom of Scorpion Gulch could the entire force be concentrated to produce the same moment as in part d?

3. *Ship's Bulkhead Problem:* A bulkhead on a ship is a vertical wall that separates two compartments. Bulkheads are often designed so that they will withstand the water pressure if the compartment on one side of it gets flooded. Suppose that you are hired to build a bulkhead that goes all the way across a ship (Figure 11-5c). The bulkhead is to be 40 ft wide at the top and 32 ft from bottom to top. The cross section of the ship where the bulkhead will go is in the shape of the quartic ellipse

$$\left(\frac{x}{20}\right)^4 + \left(\frac{y - 32}{32}\right)^4 = 1$$

where y is the vertical displacement, in feet, from the bottom of the ship to the sample point and x is the horizontal displacement, also in feet, from the center line of the ship to the sample point.

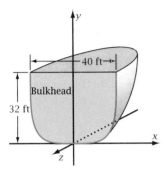

Figure 11-5c

a. Find the area of the bulkhead.

b. Find the force that you must design the bulkhead to withstand from water pressure if the compartment on one side is filled to 32 ft with seawater of density 67 lb/ft^3.

c. You also must design the bulkhead to withstand the torque caused by this force. Find the torque with respect to the x-axis, taking into account that different parts of the force act at different displacements from the axis.

d. Find the center of pressure at which the total force could be concentrated to produce the same torque.

e. Find the centroid of the bulkhead. Is the center of pressure located at the centroid?

f. When the ship is floating, the water outside is expected to come up to $y = 16$ ft. Assuming that the ship has a uniform cross section, the center of buoyancy of the ship is located at the centroid of the part of the bulkhead that lies below the water line. Find the center of buoyancy. (The center of mass of the ship must be below the center of buoyancy, or the ship will capsize. Ships often carry ballast, consisting of rock, metal scrap, lead, and so on, at the bottom for the specific purpose of lowering the center of mass.)

4. *Oil Truck Problem:* An oil truck has a tank the shape of an elliptical cylinder (Figure 11-5d). The tank is 12 ft wide and 6 ft high. Your job is to analyze the forces acting on the elliptical end of the tank.

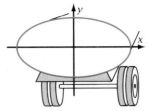

Figure 11-5d

a. Write an equation for the ellipse using axes with origin at the ellipse's center.

b. Suppose that the tank is half full. Find the force acting on the ellipse. Recall that pressure at a sample point is directly proportional to the point's displacement below the liquid's surface. The proportionality constant is the density of oil, 50 lb/ft^3. (If your answer comes out negative, see if you can find your mistake!)

5. *Airplane Wing Problem I:* An airplane wing has the shape of the region bounded by the graph of $y = 60 \cos \frac{\pi}{20} x$ and the x-axis, where x and y are in feet.

a. Find the area of the wing.

b. Assume that the pressure pushing up on the wing when the plane is in flight is constant in the y-direction, but is directly proportional to the quantity $(10 - |x|)$ in the x-direction. Find the total force acting on the wing.

c. For the plane to fly, each wing must lift 96 tons. What must the proportionality constant equal?

6. *Airplane Wing Problem II:* Suppose you have been hired by Fly-By-Night Aircraft Corporation. You are to analyze the forces that will act on the wings of a new plane. From the design department you find that, looking from the top, the wing's shape (Figure 11-5e, on the next page) is the region bounded by the graph of $y = 100 - x^2$ and the x-axis, where x and y

are in feet. The x-axis runs along the line where the wing joins the fuselage.

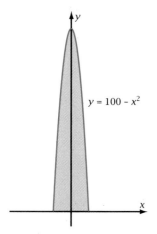

Figure 11-5e

a. The research department finds that in normal flight the pressure pushing up on the wings varies linearly with x. The pressure is 90 lb/ft² at the y-axis and 20 lb/ft² at $x = 10$. Find the total force acting on the wing.

b. Find the first moment of force with respect to the y-axis.

c. For the plane to be properly balanced, you need to know where in the x-direction the force could be concentrated to produce the same first moment of force. Find the x-coordinate of the center of pressure.

d. During a certain banking maneuver, the pressure pattern changes. It becomes directly proportional to y and does not vary in the x-direction. The pressure is 60 lb/ft² at $y = 50$. Find the total force acting on the wing.

e. Find the first moment of force in part d with respect to the x-axis.

f. The first moment of force in part e measures the amount of twisting needed to make the plane bank. How far out on the wing could the total force be concentrated and produce the same first moment with respect to the x-axis?

7. *Double-Integration Airplane Wing Problem:* The place where an airplane's wing joins the

fuselage must be designed to withstand the bending torque caused by air pressure acting on the wing. The stiffness of this joint is measured by the second moment of area of the joint with respect to a horizontal axis. Suppose that a new design of plane is to have a cross section as shown in Figure 11-5f. The equation of the curve is

$$y = 0.25(x - 4) - (x - 4)^{1/3}$$

Your mission is to calculate the second moment of area of this region with respect to the x-axis.

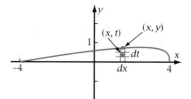

Figure 11-5f

a. Slice the region vertically into strips of width dx. Slice each strip horizontally, forming rectangles of dimensions dx by dt. Find the moment of a strip by finding the moment of a rectangle, then integrating from $t = 0$ to $t = y$. Recall that x and dx will be constants with respect to this integration.

b. Find the total second moment of the region by adding the moments of the vertical strips and taking the limit (that is, by integrating with respect to x).

8. *Double Integration Variable Pressure Problem:* A variable pressure acts on the region bounded by the curve $y = e^{-x}$ and the lines $y = 1$ and $x = \ln 5$. Find the force acting on this region if

a. The pressure is constant in the y-direction but varies directly with the square of x.

b. The pressure is constant in the x-direction but varies inversely with y.

c. The pressure varies both directly with the square of x and inversely with y. To solve this problem, you must slice one way, then slice the slice the other way. One integration gives dF. A second integration gives F. See Figure 11-5g for suggestions.

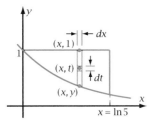

Figure 11-5g

9. Problems 7 and 8 involve *double integrals*. Why do you suppose this name is used?

10. *Floodgate Problem:* A vertical floodgate at the bottom of a dam has the shape of the region bounded by the graph of $y = 5\tan^2(\pi/8)x$ and the line $y = 5$, where x and y are in feet (Figure 11-5h). The lake behind the dam is filled with water to the level $y = 20$ ft.

a. Find the area of the floodgate.

b. Find the total force acting on the floodgate.

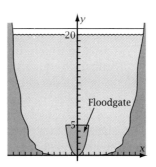

Figure 11-5h

c. The force acting on one side of the floodgate makes it difficult to open. The force with which the equipment must pull up on the gate equals the coefficient of friction between the gate and the dam multiplied by the force of the water acting on the gate. Experience shows that this force is about 10,000 lb. What does the coefficient of friction equal?

11-6 Other Variable-Factor Products

The area of a rectangular region equals its length times its width. For the region in Quadrant I bounded by the graph of $y = 4 - x^2$ (Figure 11-6a), the length and width vary. If you slice the region into horizontal strips, you are taking a small amount of length, dy, in which the width is essentially constant. If you slice into vertical strips, you are taking a small amount of width, dx, in which the length is essentially constant. In the former case, $dA = x\,dy$. In the latter case, $dA = y\,dx$. Integrating either one gives the exact area by adding the areas of the strips and taking the limit.

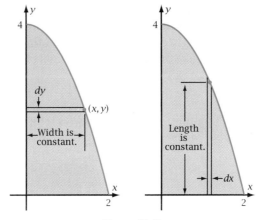

Figure 11-6a

Most applications of definite integration involve similar reasoning. A product has one or both factors that vary. You take a small amount of one quantity (slice) in which the other quantity is essentially constant, evaluate the "constant" quantity at a sample point in the strip, and find the product. The trick is in deciding which factor to slice. For example, volume = (cross-sectional area)(height). For plane slices (Figure 11-6b, left), you take a small amount of height, dy, in which the cross-sectional area, πx^2, is constant. For cylindrical shells, you take a small amount of cross-sectional area, $2\pi x\, dx$, in which the height, y, is constant (Figure 11-6b, right).

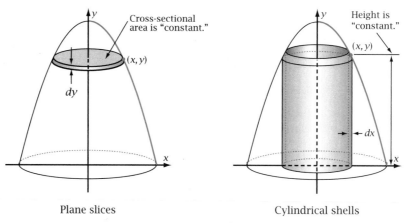

Plane slices Cylindrical shells

Figure 11-6b

Slicing may be easier one way than another. You know that work = (force)(displacement), or $W = (F)(D)$. If the whole object is moved the same amount, you should take small displacements, dD, in which the force is essentially constant, so $dW = F\, dD$. But different parts of a fluid may move different amounts. So it is preferable to take small amounts of force (that is, weight) for which the displacement is constant. In this case, $dW = D\, dF$ (Figure 11-6c).

In this section you will work problems in which one factor of a product varies. You will be expected to read and interpret the definition of each physical quantity, then translate it into the appropriate mathematics. It is the ability to see the underlying similarities in seemingly different phenomena that will enable you to make intelligent applications of mathematics.

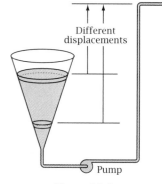

Figure 11-6c

OBJECTIVE Given a real-world situation in which a product has a factor that varies, calculate the value of the product.

Problem Set 11-6

Quick Review

Q1. $\int x^{100}\,dx = $ —?—

Q2. $\int_{-1}^{1} \tan x\,dx = $ —?—

Q3. $\int \ln x\,dx = $ —?—

Q4. Write $2\sin x\cos x$ in terms of a trigonometric function of $2x$.

Q5. Work = —?—

Q6. If $y = \tan^{-1} 3x$, then $y' = $ —?—.

Q7. $y = x^3 + 6x^2$ has a point of inflection at $x = $ —?—.

Q8. $\dfrac{d}{dx}(\sec^2 x) = $ —?—

Q9. If $y = \cos 3x$, then $d^2y/dx^2 = $ —?—.

Q10. The graph of $3x^2 - 7y^2 = 39$ is a(n) —?—.

    A. Ellipse   B. Circle   C. Parabola

    D. Hyperbola   E. Cylinder

---

1. *Heat Capacity Problem:* The number of calories (heat, as energy) required to warm a substance from temperature T_1 to temperature T_2 equals the heat capacity of the substance (calories per degree) times the change in temperature $(T_2 - T_1)$, where T is in degrees Celsius. Unfortunately, most substances have heat capacities that vary with temperature. Assume

that calculus foeride (a rare, tough substance!) has a heat capacity given by

$$C = 10 + 0.3T^{1/2}$$

where C is in calories per degree and T is in degrees. How many calories would be needed to warm a gram of calculus foeride from $100°$C to $900°$C?

2. *Phoebe's Speeding Problem:* Phoebe is caught speeding. The fine is $3.00/min for each mile per hour above the 55 mi/h speed limit. She was clocked at speeds up to 64 mi/h during a 6-min period, so the judge fines her

$$(\$3.00)(\text{time})(\text{mi/h over 55})$$
$$= (3.00)(6)(64 - 55) = \$162$$

Phoebe is good at calculus. She argues that her speed varied over the 6 min. It was 55 mi/h at $t = 0$ and $t = 6$ and was 64 mi/h only at $t = 3$. She figures that her speed, v, was

$$v = 55 + 6t - t^2$$

  a. Show that this equation gives the correct speeds at the times 0, 3, and 6 min.

  b. What should Phoebe propose to the judge as a more reasonable fine?

3. *Tunnel Problem:* The amount of money it takes to dig a tunnel equals the length of the tunnel times the cost per unit length. However, the

---

cost per unit length increases as the tunnel gets longer because of the expense of carrying workers and tools in and carrying dirt and rock out. Assume that the price per foot varies quadratically with the number of feet, x, from the beginning of the tunnel.

a. Find the particular equation for the price per foot if these prices are known.

x	Price
0	$500
100	$820
200	$1180

b. Find the cost per foot for digging at a point 700 ft from the beginning of the tunnel.

c. Find the total cost, in dollars, for digging a tunnel 1000 ft long if the workers start at one end and dig through to the other end.

d. How much money could be saved by starting the 1000-ft tunnel from both ends and meeting in the middle?

4. *Water Pipe Problem:* The flow rate of water through a pipe (cubic inches per second) equals the velocity of the water (in inches per second) times the cross-sectional area of the pipe.

a. Show that velocity times cross-sectional area gives the right units for flow rate.

b. In real pipes, the flow rate varies at different points across the pipe, with a maximum at the center and dropping to zero at the pipe walls (Figure 11-6d). Assume that the velocity, v, through any cross section of a 4-in.-diameter pipe is given by

$$v = 4 - x^2$$

where x is the number of inches from the center of the pipe and v is in inches per second. Show that the velocity really is a maximum at the center and zero at the pipe walls.

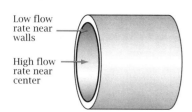

Low flow rate near walls

High flow rate near center

Figure 11-6d

c. What is the flow rate in cubic inches per second?

d. How many gallons per minute are flowing through the pipe? (There are 231 in.^{3} in a gallon.)

e. How many gallons per minute would be flowing through the pipe if all the water were moving at the maximum velocity?

f. As far as the mathematics is concerned, this problem is identical to some other kind of problem you have worked. Which kind?

5. *Wire-Pulling Problem:* Paul Hardy tries to pull down a tree. He attaches one end of a long wire to the tree and the other end to the bumper of his truck. As he slowly drives the truck away, the wire stretches tighter and tighter and finally breaks. At first the force increases linearly with x, the number of inches the wire stretches. At $x = 2$ in., the wire yields (that is, begins to break). The table shows the forces to the point where the wire breaks at $x = 5$.

x (in.)	Force (lb)	
0	0	
0.5	150	
1	300	
1.5	450	
2	600	(before yielding)
2	450	(after yielding)
2.5	470	
3	440	
3.5	420	
4	410	
4.5	390	
5	330	(breaks)

a. Plot the graph of force versus x.

b. Describe the behavior of the force function at $x = 2$.

c. Find the work done in stretching the wire from $x = 0$ to $x = 2$.

d. Find the work done in stretching the wire from $x = 2$ to $x = 5$.

e. Find the total work done in breaking the wire.

f. Is it possible for a discontinuous function to be integrable? Explain.

6. *Variable Attraction Problem:* A solid paraboloid is formed by rotating about the y-axis the

region in Quadrant I bounded by the graph of $y = 4 - x^2$ (Figure 11-6e), where x and y are in centimeters. The solid has a uniform density k, in grams per cubic centimeter.

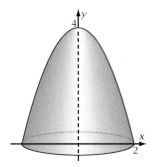

Figure 11-6e

a. Find the mass of the solid.

b. The solid is attracted by a force that is inversely proportional to the square root of the distance from the base of the solid and directly proportional to the mass. That is,

Force = (constant)(mass)($y^{-1/2}$)

But different parts of the solid are at different distances from the base. By appropriate slicing, find the total force exerted on this solid.

7. *Moment of Inertia Problem:* The second moment of mass of an object with respect to an axis is defined to be the mass times the square of the distance between the object and the axis. It is sometimes called the moment of inertia because it measures how difficult it is to start or stop rotating the solid about the axis. Find the second moment of mass of the solid in Problem 6 with respect to the y-axis.

8. *Degree-Days Problem:* Engineers who design heating and air-conditioning systems use a quantity called degree-days to measure how much above or below normal the weather has been. For example, if the temperature is 10 degrees above normal for 2 days, then the weather has been $(+10)(2) = +20$ degree-days. If it is 30 degrees below normal for half a day, the weather has been $(-30)(1/2) = -15$ degree-days. However, the temperature varies throughout the day, so degree-days should be calculated by calculus rather than by arithmetic. Suppose that one morning the temperature starts out at normal, and 6 hours later has risen 20 degrees above normal, and that any time D, in days, after the morning reading, the temperature, T, is

$$T = 20 \sin 2\pi D$$

a. Show that this equation gives the right values of T for times $D = 0$ and $D = 1/4$.

b. Find the number of degree-days between $D = 0$ and $D = 1/4$.

9. *Rocket Car Problem:* Iona Carr is building a rocket-powered car that she plans to use in racing. In this problem you are to help Iona figure out the car's speed and the distance it will travel in a given time. The car, with Iona in it and a full load of fuel, will have a mass of 2000 kg. When the engine is running it will develop 7000 N (newtons) of thrust, which means the car will be pushed with a constant force of 7000 N. You recall from physics that force = (mass)(acceleration), and 1 N is 1 kg-m/s^2. However, the car uses fuel at a rate of 5 kg/s, so its mass is decreasing.

a. Write an equation expressing mass as a function of time.

b. Write an equation expressing acceleration as a function of time.

c. The answer to part b is a differential equation because acceleration is the derivative of velocity. Solve the differential equation for velocity as a function of time if $v(0) = 0$.

d. Twenty seconds after the car starts, how fast will it be going? How far will it have gone?

10. *Field Worth Problem:* Suppose you have a tract of land the shape of the "triangular" region in Quadrant I bounded by the y-axis and by the graphs of $y = 4 - x^2$ and $y = 4x - x^2$ (Figure 11-6f, on the next page), where x and y are in kilometers. The land's value per square kilometer is directly proportional to its distance from the railroad tracks (along the y-axis), being $200 thousand per square kilometer at the point farthest from the tracks. To the nearest thousand dollars, what is the total worth of the land? How much less would it cost you than if the entire tract were worth $200 thousand per square kilometer?

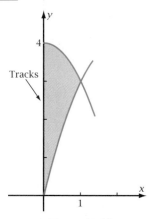

Figure 11-6f

11. *Sinusoidal Land Tract Problem:* A tract of land has the shape of the region in Quadrant I under the graph of $y = \cos x$ (Figure 11-6g). Find the total worth of the land if the worth per square unit is

a. Constant in the y-direction but directly proportional to x in the x-direction

b. Constant in the x-direction but directly proportional to y in the y-direction

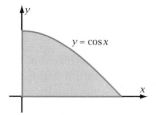

$y = \cos x$

Figure 11-6g

12. *Painted Wall Problem:* Calvin has a contract to paint the wall of a new auditorium. The wall is the shape of the region under the graph of $y = 9 - x^2$, where x and y are in meters and the x-axis runs along the ground. It is harder to paint higher up, so he charges a price per square meter that is directly proportional to the square of the distance above the ground. At a point 2 m above the ground, he charges $12 per square meter. What total amount will Calvin charge for the job?

13. *City Land Value Problem:* Suppose that you have been hired by the tax assessor's office in the town of Scorpion Gulch. You are to calculate the total worth of all land within the city limits.

The town is circular, with a radius of 3 km. You find that land is worth $10 million per square kilometer at the center of town and $1 million per square kilometer at the edge of town. For parts a–c, find the total worth of the land assuming that the price per square kilometer

a. Varies linearly with the distance from the center of town.

b. Varies exponentially with the distance from the center of town.

c. Is given by the table of values.

Distance (km)	Million $/ km²	Distance (km)	Million $/ km²
0	10	1.8	8
0.3	12	2.1	5
0.6	15	2.4	3
0.9	14	2.7	2
1.2	13	3	1
1.5	10		

d. This problem is mathematically equivalent to another type of problem you have worked. What kind of problem?

e. What real-world reason(s) can you think of to explain the pattern of the data in part c?

14. *Diving Board Problem:* Calvin sits on the end of a diving board (Figure 11-6h), exerting a pressure p, in pounds per square foot, on the diving board that varies in the x-direction according to the formula

$$p = 100[(x - 8)^{1/2} - 0.5(x - 8)]$$

The pressure is constant across the width of the board (in the z-direction). The diving board is 2 ft wide in the z-direction.

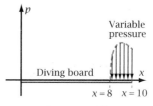

Figure 11-6h

a. To the nearest pound, how much does Calvin weigh?

b. What is the average pressure Calvin exerts in the interval between $x = 8$ and $x = 10$?

c. To the nearest pound-foot, find Calvin's first moment of force with respect to the yz-plane.

d. Calvin wishes to exert the same first moment by standing on tiptoes at some point near the end of the board. Where should he stand?

15. *Skewness Problem:* Figure 11-6i shows the graphs of

$$f(x) = 9 - x^2 \quad \text{and} \quad g(x) = -\tfrac{1}{3}x^3 - x^2 + 3x + 9$$

The region under each graph has the same area, as you will show in part b. But the region under the g graph is skewed to the right. In this problem you will calculate the **skewness**, which is used in statistics to measure how unbalanced a region is.

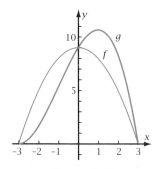

Figure 11-6i

a. Show that the only x-intercepts of both functions f and g are -3 and 3 (Figure 11-6i).

b. Show that the regions under the two graphs have equal area. How do the properties of definite integrals between symmetrical limits help you calculate the area easily?

c. At what value of x in the interval $[-3, 3]$ do the maxima of the f and g graphs occur?

d. Find the x-coordinate of the centroid of the region under the graph of function g. Recall that this is the number $\bar{x}$ such that $(\bar{x})(\text{area}) = $ first moment of area of the region with respect to the y-axis.

e. True or false: "The centroid of the region is on a vertical line through the maximum on the g graph."

f. True or false: "There is just as much area to the left of the centroid as there is to the right."

g. The **skewness** of a region is defined to be the third moment of area of the region with respect to a vertical line through the centroid. Calculate the skewness of the region under the graph of function g.

h. Show that the skewness of the region under the parabola, function f, is equal to zero. Why is the word *skewness* appropriate in this case?

i. Sketch the graph of a region with skewness that is the opposite sign from that under the graph of function g.

16. *Moment of Arc Length Problem:* You have found moments of area, mass, and volume. It is also possible to find moments of length. Figure 11-6j shows the arc of the parabola $y = x^2$ from $x = 0$ to $x = 2$. The moment, dM_y, of the arc dL with respect to the y-axis is the length, dL, times its distance, x, from the y-axis. That is,

$$dM_y = x \, dL$$

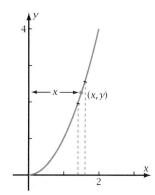

Figure 11-6j

a. Find dM_y explicitly in terms of x and dx.

b. Find the total moment of the parabolic arc with respect to the y-axis.

c. Find the length of the parabolic arc.

d. Find the x-coordinate of the centroid of the parabolic arc. This point would be the center of mass of a thin, uniform wire bent in the shape of the arc.

e. Find the surface area of the paraboloid formed by rotating the arc about the y-axis.

f. What interesting thing do you notice about the integral in part e?

17. *Another Theorem of Pappus Problem:* In Section 11-4, you learned the theorem of Pappus for volumes. The theorem states the volume of a solid of revolution equals the area of the region being rotated times the distance traveled by the centroid of that region. There is a similar theorem for surface area.

Theorem: The Theorem of Pappus for Surfaces

The area, S, of a surface of revolution is given by

$$S = 2\pi \overline{R} L$$

where L is the length of the curve being rotated, $\overline{R}$ is the displacement from the axis of rotation to the centroid of the curve, and the curve is not on both sides of the axis of rotation. The quantity $2\pi \overline{R}$ is thus the distance the centroid travels as the curve rotates.

Demonstrate that this theorem is true for the paraboloid in Problem 16, part e.

18. *Application of Pappus' Other Theorem:* A toroidal surface (like an inner tube) is formed by rotating a circle of radius r about an axis (Figure 11-6k) R units from the center of the circle. It is hard to find the surface area of the toroid directly by integration, but it is easy to do so by using the theorem of Pappus for surfaces. Find a formula for the area of a toroidal surface in terms of r and R.

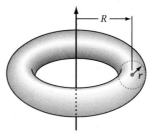

Figure 11-6k

11-7   Chapter Review and Test

In this chapter you have applied definite integration to problems involving a product of two quantities, where one of the quantities is a variable. By now you should be able to take any such situation, familiar or unfamiliar, and perform the appropriate mathematics. The ability to see similarities among seemingly dissimilar phenomena is the key to intelligent application of mathematics.

Review Problems

R0. Update your journal with what you've learned since the last entry. You should include such things as those listed here.

- The one most important thing you have learned in studying Chapter 11

- Which boxes you have been working on in the "define, understand, do, apply" table

- Physical quantities that you can calculate as products, such as work, moment, and mass

- Centroid, center of mass, center of gravity, center of pressure, and so on

- Any techniques or ideas about calculus that are still unclear

R1. *Work Problem:* Manuel Dexterity drags Bob Tail across the floor. Manuel pulls hard at first, then eases off. The force he exerts (Figure 11-7a) is given by

$$F = 30e^{-0.2x}$$

where F is in pounds and x is Bob's displacement in feet from the starting point. The work Manuel does is the force times the displacement. Find the number of foot-pounds of work he does in dragging Bob from $x = 0$ to $x = 10$.

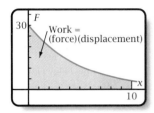

Figure 11-7a

R2. a. *Magnet Problem:* A magnet repels another magnet with a force inversely proportional to the square of their distance apart. That is, $F = k/x^2$, where F is in pounds and x is in inches. Find the work done in moving the magnets from 3 in. apart to 1 in. apart.

b. *Conical Cup Problem:* Phil puts a 10-in.-long straw into a conical cup filled to the top with root beer (Figure 11-7b). The cup has a top diameter 6 in. and a height 7 in. The root beer has a density of 0.036 lb/in.^{3}. How much work will Phil do in raising all the liquid in the cup to the top of the straw?

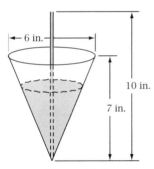

Figure 11-7b

R3. *Variable Density Problem:* The region in Quadrant I bounded by the graph of $y = 8 - x^3$ is rotated about the y-axis to form a solid. Find the mass of the solid if the density

a. Is constant in the radial direction but equal to ky in the y-direction, where k is a proportionality constant

b. Is constant in the axial direction but equals e^x in the radial direction

R4. a. *Triangle Centroid Problem:* Figure 11-7c shows a triangle of base b and height h. Write an equation for the width of the triangle in terms of y, the distance from the base to a sample point. Find the first moment of area of the region inside the triangle with respect to its base. Show that the centroid is one-third of the distance from the base to the opposite vertex.

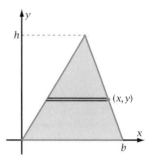

Figure 11-7c

b. *Second Moment of Volume Problem:* The region under the graph of $y = e^x$ from $x = 0$ to $x = 1$ is rotated about the y-axis to form a solid. Find the second moment of volume of the solid with respect to the y-axis.

R5. *Wind Force Problem:* A tower has the shape of a slender pyramid (Figure 11-7d). The base of the pyramid, at ground level, is a square of length 150 ft. The building is 400 ft tall. When the

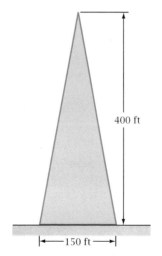

Figure 11-7d

wind blows, the pressure acting on the triangular face of the building is greater at the top than at the bottom, because the wind speed increases with altitude. Assume that the pressure due to the wind is given by

$$p = 200(1 - e^{-0.01y})$$

where p is in pounds per square foot and y is the height in feet above the ground. Assume also that the face of the building is a vertical triangle of base 150 ft and height 400 ft. Calculate the total force of the wind acting on that face.

R6. *Oil Well Problem:* Suppose you work for a company that plans to drill a well to a depth of 50,000 ft, farther than anyone has ever drilled before. Your job is to estimate the cost of drilling. From historical records you find that it costs about $30 per foot to drill at the surface and about $50 per foot at a depth of 10,000 ft.

a. Assume that the cost per foot, in dollars, varies exponentially with depth. Write the particular equation expressing cost per foot in terms of depth.

b. The total cost of the well is the cost per foot times the depth, in feet. The cost per foot varies, so you realize that this is a job for calculus! Your boss needs the estimated cost of the well. What are you going to tell him?

Concept Problems

C1. *Cubic Parabola Region Problem:* The following problems concern a region in the xy-plane of an xyz-coordinate system. The region is bounded by the graphs of $y = x^3$, $y = 8$, and $x = 0$ (Figure 11-7e).

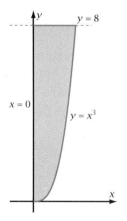

Figure 11-7e

a. Find the area of the region.

b. Find the first moment of area of the region with respect to

   i. The x-axis

   ii. The y-axis

c. Find the centroid of the region.

d. Find the volume of the solid generated by rotating the region

   i. About the x-axis

   ii. About the y-axis

   iii. About the line $x = 3$

e. Show that for each of the solids in part d, the volume is equal to the area of the region times the distance traveled by the centroid of the region as it rotates to form the solid.

f. Find the first moment of volume with respect to the xz-plane for the solid in part d, part ii.

g. Find the centroid of the solid in part d, part ii.

h. Does the y-coordinate of the centroid of the solid in part d, part ii equal the y-coordinate of the centroid of the region?

i. Find the mass of the solid in part d, part ii if the density varies directly with the square of the distance from the y-axis.

j. Find the second moment of mass of the solid in part d, part ii with respect to the y-axis.

k. Find the force acting perpendicular to the region due to a pressure equal to $(3 - x)$ in the x-direction (and constant in the y-direction).

l. Find the work done by moving the region from $z = 1$ to $z = 3$. Assume that the force in part k acts on the region when it is at $z = 1$, and that the force varies inversely with the square of z as the region moves in the z-direction and acts in the positive z-direction.

m. Suppose that the object in part d, part ii is made of a substance with heat capacity 0.3 cal/g/°C and density 5.8 g/cm³, and that its dimensions are in centimeters. Suppose also that the object is warmed from 0°C to a temperature that is constant in the radial direction but is given by $T = 10 - y$ in the axial direction. Find the amount of heat needed to cause this temperature change.

C2. *Moment vs. Volume Problem:* Show that finding the first moment of the area of a region with respect to the y-axis is mathematically equivalent to finding the volume by cylindrical shells for the solid formed by rotating that region about the y-axis, provided that the region is not on both sides of the axis of rotation.

C3. *Paraboloid Moment Conjecture Problem:* A solid paraboloid is formed by rotating about the y-axis the region in Quadrant I under the graph of $y = 9 - x^2$. Show that the first moment of volume of the solid with respect to the plane of its base equals the second moment of volume of the solid with respect to its axis. Does this property hold in general for any solid paraboloid? Justify your answer.

C4. *Infinitesimals of Higher Order:*

a. Figure 11-7f shows a lower Riemann sum for $y = mx$, where $m \neq 0$. The area of each strip is $dA \approx y\,dx$, the area of the rectangle. The length of each piece of graph is $dL \approx dx$. Both approximations become exact as Δx approaches zero. Show that on integrating from a to b, $dA \approx y\,dx$ gives the exact area of the region, but $dL \approx dx$ does not give the exact length.

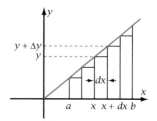

It works for area, but not for length.

Figure 11-7f

b. Figure 11-7g shows the cone formed by rotating about the x-axis the graph of $y = mx\,(m \neq 0)$ from $x = 0$ to $x = h$. Plane sections cut the cone into frustums of volume $dV \approx \pi y^2\,dx$. Each frustum has area $dS \approx 2\pi y\,dx$. Both approximations become exact as Δx approaches zero. Show that on integrating from 0 to h, $dV \approx \pi y^2\,dx$ gives the exact volume of the cone, but $dS \approx 2\pi y\,dx$ does not give the exact surface area.

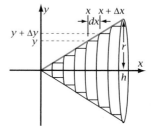

It works for volume, but not for surface area.

Figure 11-7g

c. Find the exact area of a strip in Figure 11-7f and the exact volume of a frustum in Figure 11-7g. (The volume of a frustum of height dx is $V = (\pi/3)(R^2 + Rr + r^2)(dx)$, where R is the larger radius and r is the smaller radius of the frustum.)

d. A quantity, such as $2\pi y\,dx$, that approaches zero as Δx approaches zero is called a **first-order infinitesimal**. A quantity, such as $0.5\Delta y\,dx$, that is the product of two or more first-order infinitesimals is called a **higher-order infinitesimal**. Show that the approximations $dA \approx y\,dx$ and

$dV \approx \pi y^2 \, dx$ differ from the exact values in part c only by infinitesimals of higher order.

e. You recall that the differential of arc length is $dL = \sqrt{dx^2 + dy^2}$. The approximation $dL \approx dx = \sqrt{dx^2}$ leaves out the first-order infinitesimal $\sqrt{dy^2}$. Make a conjecture about how accurate a differential of a quantity must be so that it yields the exact value when it is integrated.

f. The second-order infinitesimal $0.5\Delta y \, dx$ appears in the exact value of the area of the strip in Figure 11-7f in part d. The limit of the Riemann sum of such a higher-order infinitesimal equals zero. Give a reason for each step in this example of that statement.

$$\sum 0.5\Delta y \, dx = 0.5\Delta y \sum dx$$
$$= 0.5\Delta y(b - a)$$
$$\therefore \lim_{\Delta x \to 0} \sum 0.5\Delta y \, dx = 0.5(0)(b - a) = 0$$

Property: Infinitesimals of Higher Order

If $dQ \approx \Delta Q$ leaves out only infinitesimals of higher order, then

$$\int_a^b dQ \text{ is exactly equal to } Q$$

Chapter Test

PART 1: No calculators allowed (T1–T4)

T1. Complete the equations.

a. Work = —?— · —?—

b. —?— = density · volume

c. —?— = pressure · area

d. Moment of area with respect to the x-axis = —?— · —?—

e. —?— = volume · (displacement from y-axis)^2

f. Moment of mass with respect to yz-plane = —?— · mass

T2. Find the work done in dragging an object from $x = 1$ to $x = 4$ if the force, in pounds, exerted on the object is given by

$$F = 40x - 10x^2$$

T3. An object of mass 200 g is at a point where its moment with respect to the xz-plane is 3000 g-cm. Find $\bar{y}$, the distance from the xz-plane to the center of mass.

T4. A circle of radius 7 cm in the xy-plane has its center at the point (8, 9). Find the volume of the toroid formed by rotating this circle about the y-axis.

PART 2: Graphing calculators allowed (T5–T9)

T5. A packing case is dragged across the floor from $x = 0$ to $x = 10$ ft. As it moves, it becomes damaged, causing the force needed to move it to increase, as shown in the table. Find the exponential function that best fits these data. Use the function to find the total amount of work done.

x	Force (lb)
0	30
2	34
4	38
6	43
8	49
10	55

T6. For the region under the graph of $y = e^x$ from $x = 0$ to $x = 2$, where x and y are in inches,

a. Find the first moment of area with respect to the y-axis. Give the units in your answer.

b. Find the second moment of area with respect to the y-axis. Give the units in your answer.

c. Find the x-coordinate of the centroid of the region. Give the units in your answer.

T7. A solid is formed by rotating about the x-axis the region under the graph of $y = x^{1/2}$ from $x = 0$ to $x = 16$, where x and y are in centimeters. Its density is constant axially and equal to $3y$ radially (in grams per cubic centimeter). Find its mass.

T8. A trough 8 ft deep and 4 ft across has an end the shape of the region above the graph of $y = |x^3|$ (Figure 11-7h).

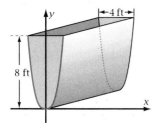

Figure 11-7h

a. Find the total force acting on the end of the trough when it is full of water of density 62.4 lb/ft³.

b. Find the center of pressure, where the force could be concentrated to produce the same moment.

T9. *Theater in the Round Problem:* A round theater is to be built with the stage at the center. The seats will be in circular rows starting at $r = 30$ ft from the center and ending at $r = b$ ft (Figure 11-7i).

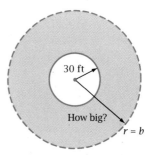

Figure 11-7i

a. The value of the seating area for any given performance is given by $v = 150/r$ (in dollars per square foot). Find the total value of the seating area in terms of the outer radius, b.

b. How big must the theater be to receive $60,000 for a performance?

The Calculus of Functions Defined by Power Series

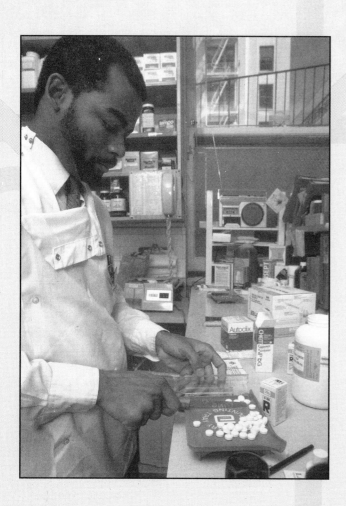

If you take regular doses of a medication, the amount in your system is the sum of what remains from the series of the doses you have taken. The limit of that series as the number of doses becomes large is important to know for determining long-term effects of the medication. You can find such limits by calculus techniques.

Mathematical Overview

How does a calculator find sines and logs when all it can do is add and multiply? In Chapter 12, you will see that these transcendental functions can be calculated as accurately as you need using infinite polynomials called power series. You will study these series in four ways.

Graphically   The icon at the top of each even-numbered page of this chapter shows that the first few terms of a power series can fit the sine function close to $x = 0$.

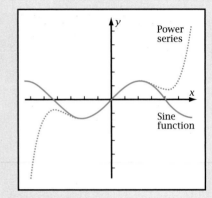

Numerically $\sin 0.6 = 0.56462473\ldots$

n	Series for sin 0.6
0	0.6
1	0.564
2	0.564648
3	0.564642445...
4	0.564642473...

Algebraically $\sin x = x - \dfrac{1}{3!}x^3 + \dfrac{1}{5!}x^5 - \ldots$, the Maclaurin series for sine

Verbally   *Perhaps the most surprising thing I have learned about power series is that sometimes they converge as more and more terms are added, and sometimes they don't. For each power series, there is an interval of x-values for which the series converges.*

12-1   Introduction to Power Series

Suppose that $f(x) = 6/(1 - x)$ (Figure 12-1a). If you divide $1 - x$ into 6, you get a polynomial that continues forever!

$$P(x) = 6 + 6x + 6x^2 + 6x^3 + 6x^4 + 6x^5 + \cdots$$

The result is called a **power series**. The word *series* indicates that an infinite number of terms are being *added*. The word *power* indicates that each term contains a power of x.

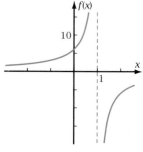

Figure 12-1a

OBJECTIVE   Find values of $P(x)$ for a given power series, and compare the results with the corresponding values of the function from which the power series originates.

The following problem set is designed to let you work toward this objective either on your own or with your study group following your test on Chapter 11.

Exploratory Problem Set 12-1

Let $f(x) = 6/(1 - x)$ and let
$P(x) = 6 + 6x + 6x^2 + 6x^3 + 6x^4 + 6x^5 + \cdots$.

1. On the same screen, plot $f(x)$ and the polynomial function $P_5(x)$ (the six terms of $P(x)$ through $6x^5$). Use a window with $x = 1$ as a grid point, an x-range of about $[-2, 2]$, and a y-range of $[-100, 100]$. Sketch the result. For what range of x-values is the graph of P_5 so close to the graph of f that you can't tell them apart? Give an x-value for which the graph of P_5 bears no resemblance to the graph of f.

2. On the same screen as Problem 1, plot the polynomial function $P_6(x)$. Does the graph of P_6 fit the graph of f for a range of x-values wider than the range of P_5?

3. Show that $P_6(0.5)$ is closer to $f(0.5)$ than $P_5(0.5)$ is, but that $P_6(2)$ is not closer to $f(2)$ than $P_5(2)$ is.

4. If the limit of the sum of a series as you add more and more terms equals the

corresponding value of $f(x)$, then the series is said to **converge** to $f(x)$. If the series does not converge to $f(x)$ or some other number, it is said to **diverge**. Make a conjecture about the interval of values of x for which the series

$$P(x) = 6 + 6x + 6x^2 + 6x^3 + 6x^4 + 6x^5 + \cdots$$

converges to $f(x) = 6/(1 - x)$.

5. Write the values of $P_0(1)$, $P_1(1), \ldots, P_4(1)$. Write the values of $P_0(-1)$, $P_1(-1), \ldots, P_4(-1)$. Explain why each series diverges. Does this result affect your conjecture in Problem 4?

6. By how much do $P_5(0.5)$ and $P_5(-0.5)$ differ from the respective values of $f(0.5)$ and $f(-0.5)$? How do these differences compare with $6x^6$, the first term of the series that is left out of the sum?

7. Each term of $P(x)$ after the first equals the preceding term multiplied by the same number, x in this case. What is this type of series called? What is the multiplier x called?

12-2   Geometric Sequences and Series as Mathematical Models

In Problem Set 12-1, you saw how the rational function $f(x) = 6/(1 - x)$ could be expanded as a series, $6 + 6x + 6x^2 + 6x^3 + \cdots$. You can generate any term after the first by multiplying the preceding term by the same number, x in this case. Thus the series fits the definition of geometric series from algebra. In this section you will reverse the process to see how to represent a geometric series, at least in some instances, as a rational function. By so doing you will be able to analyze some functions in the real world in which the function values change **discretely** (by jumps) rather than continuously.

OBJECTIVE   Given a function defined by a geometric series, explain whether you can write the series as a rational algebraic function, and if so, write an equation for that target function.

Background

A **geometric series** is defined to be a series, $t_1 + t_2 + t_3 + t_4 + \cdots$, for which there is a constant, r, called the **common ratio** such that $t_n = rt_{n-1}$ for any integer $n > 1$. The numbers $t_1, t_2, t_3, \ldots$ are called **terms** of the series, hence the letter t. The variable n is called the **term index**. If n starts at 1, the term index is the same as the term number. The **nth partial sum**, S_n, of a geometric series is the indicated sum of the first n terms. By clever algebra it is possible to derive a closed formula (no ellipsis) for S_n as a function of the first term, t_1, and the common ratio, r.

The sizes of dolls in a set of Russian Matryoshka dolls are calculated using a common ratio.

$$S_n = t_1 + t_2 + t_3 + t_4 + \cdots + t_{n-1} + t_n$$

$$S_n = t_1 + rt_1 + r^2t_1 + r^3t_1 + \cdots + r^{n-2}t_1 + r^{n-1}t_1 \qquad \text{Each term is } r \text{ times the preceding term.}$$

$$rS_n = rt_1 + r^2t_1 + r^3t_1 + r^4t_1 + \cdots + r^{n-1}t_1 + r^nt_1 \qquad \text{Multiply both sides of the equation by } r.$$

$$S_n - rS_n = t_1 - r^nt_1 \qquad \text{Subtract the third equation from the second one. The middle terms telescope.}$$

$$S_n = t_1 \cdot \frac{1 - r^n}{1 - r} \qquad \text{Solve for } S_n.$$

A Convergent Geometric Series—Drug Dosage

Suppose a person takes 500 mg of vitamin C every 8 h. Assume that at the end of any 8-h period, only 60% of the vitamin C present at the beginning of the period remains in the person's system. At the end of the first 8-h period, only 300 mg of the original 500 remains. After the second dose, the amount jumps to 800 mg. Immediately after the nth dose, the amount of vitamin C remaining in the system is the sum of the 500 mg from the last dose and the remains of each previous dose.

$$S_n = 500 + 300 + 180 + 108 + \cdots$$ 180 is 60% of 300, 108 is 60% of 180, and so forth.

$$= 500 + 500(0.6) + 500(0.6^2) + 500(0.6^3) + \cdots + 500(0.6^{n-1})$$

This series is geometric, with first term 500 and common ratio 0.6. After 10 doses,

$$S_{10} = 500 \cdot \frac{1 - 0.6^{10}}{1 - 0.6} = 1242.441\ldots$$

The person would have only about 1240 mg of vitamin C in his or her system, despite having taken a total of 5000 mg.

If the person continues taking the 500-mg doses for a long time, does the amount in the body become exceedingly high? To find out, it is instructive to look at the partial sums graphically and numerically. As shown in Figure 12-2a, the amount of vitamin C rapidly levels off toward 1250 mg. As shown in the accompanying table, the partial sums stay the same for more and more decimal places. Both the table and the graph give evidence that the sequence of partial sums converges to 1250. If the sequence of partial sums converges, then the series converges also.

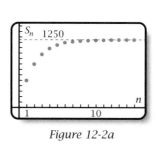

Figure 12-2a

n	S_n
25	1249.996446...
26	1249.997867...
27	1249.998720...
28	1249.999232...
29	1249.999539...
30	1249.999723...
31	1249.999834...
32	1249.999900...

From the formula for S_n, you can tell algebraically that the series converges to 1250. Take the limit of S_n as n approaches infinity.

$$\lim_{n \to \infty} S_n = \lim_{n \to \infty} \left(500 \cdot \frac{1 - 0.6^n}{1 - 0.6} \right) = 500 \cdot \frac{1}{1 - 0.6}$$ Because 0.6^n approaches zero.

$$= 1250$$

So the amount of vitamin C in the person's system never exceeds 1250 mg, no matter how long the treatment is continued.

A geometric series will converge if the common ratio, r, is a proper fraction (that is, $|r| < 1$). This is because the term r^n in the formula for S_n approaches zero as a limit as n approaches infinity. The number S to which the series converges is given by

$$S = \lim_{n \to \infty} \left(t_1 \cdot \frac{1 - r^n}{1 - r} \right) = t_1 \cdot \frac{1}{1 - r}$$

A Divergent Geometric Sequence—Compound Interest

In Chapter 7, you saw that you can use an exponential function as a mathematical model for invested money if the interest is compounded continuously. A sequence is more appropriate if the interest is compounded at discrete intervals, such as once a day, once a month, or once a quarter. Suppose you invest $500 in a savings account that pays 6% per year interest, compounded quarterly. For the first 3 months you have only the initial $500. Then $7.50 is added to the account, 1.5% of the $500 (a fourth of 6%), and you have $507.50 for the next 3 months. At the end of the second quarter the account increases by $7.6125, which is 1.5% of the $507.50.

To find a pattern in the amounts, observe how you can calculate the $507.50.

$$500 + 500(0.015) = 500(1 + 0.015) = 500(1.015) \qquad \text{After the first quarter.}$$

Repeating the computation for the second quarter without simplifying $500(1.015)$ gives

$$500(1.015) + 500(1.015)(0.015)$$
$$= 500(1.015)(1 + 0.015) = 500(1.015)^2 \qquad \text{Second quarter.}$$

The amounts are terms in a geometric sequence, with first term 500 and a common ratio 1.015.

$$500, \quad 500(1.015), \quad 500(1.015)^2, \quad 500(1.015)^3, \quad 500(1.015)^4, \quad \ldots \qquad \text{Dollars in the account.}$$

$$0 \qquad 1 \qquad 2 \qquad 3 \qquad 4 \qquad \ldots \qquad \text{Quarters.}$$

The exponent of 1.015 equals the number of quarters. After 10 years the amount would be

$$t_{40} = 500(1.015)^{40} = 907.0092\ldots \approx \$907.01 \qquad \text{Notice that } n \text{ is 40, not 10.}$$

The amount of interest earned in the 10 years would be $407.01, the difference between the $907.01 and the initial investment of $500. In this instance it is more convenient to start n at 0 instead of 1 so that the term index will equal the number of quarters elapsed.

Note that the sequence diverges because the common ratio is greater than 1. This fact is useful in the real world. For instance, it alerts bankers to the consequences of money left in dormant accounts. If the $500 had been invested by George Washington the year he died, 1799, his heirs could claim more than $106,000,000 in 2005!

Words Relating to Sequences and Series

The definitions and properties concerning sequences and series are summarized in the box on the next page.

DEFINITIONS: *Vocabulary Relating to Sequences and Series*

A **sequence** is an infinite ordered set of numbers.

> *Example:* 2, 3, 5, 7, 11, 13, 17, 19, 23, ...   The sequence of primes.

A **series** is the indicated sum of the terms of a sequence.

> *Example:* $2 + 3 + 5 + 7 + 11 + 13 + 17 + 19 + 23 + \cdots$ The series of primes.

The **terms** of a sequence or series, $t_1, t_2, t_3, \ldots, t_n, \ldots$, are the numbers that appear in the sequence or series.

The **term index** is the variable integer subscript n used to calculate the term value. If n starts at 1, the term index also equals the term number.

A **partial sum** of a series is the sum of a finite number of terms in the series.

> *Example:* $S_4 = 2 + 3 + 5 + 7 = 17$ Fourth partial sum of the series of primes.

The **sequence of partial sums** of a series is the sequence with terms that are the first partial sum, the second partial sum, the third partial sum, and so on.

> *Example:* For the series of primes, above, the sequence of partial sums is $S_1, S_2, S_3, S_4, S_5, S_6, S_7, S_8, \ldots = 2, 5, 10, 17, 28, 41, 58, 77, \ldots$.

A *sequence* **converges to** L if its nth term approaches a finite limit L as n approaches infinity. If the sequence does not converge, it is said to **diverge**.

> *Examples:* $1, 1\frac{1}{2}, 1\frac{3}{4}, 1\frac{7}{8}, 1\frac{15}{16}, \ldots$ converges to 2.
> $2, 3, 5, 7, 11, 13, \ldots$ (the sequence of primes) diverges.
> $1, 0, 1, 0, 1, 0, 1, \ldots$ diverges **by oscillation**.

A *series* **converges** if and only if its sequence of partial sums converges.

> *Example:* $1 + \frac{1}{2} + \frac{1}{4} + \frac{1}{8} + \frac{1}{16} + \cdots$ converges because its sequence of partial sums is $1, 1\frac{1}{2}, 1\frac{3}{4}, 1\frac{7}{8}, 1\frac{15}{16}, \ldots$, which converges to 2.

A **geometric series** is a series for which each term after the first term is given by $t_n = r \cdot t_{n-1}$ for some constant r, called the **common ratio**.

Properties of Geometric Series

The nth partial sum of a geometric series is given by

$$S_n = t_1 \cdot \frac{1 - r^n}{1 - r}$$

A geometric series converges if $|r| < 1$.
The number to which a convergent geometric series converges is

$$S = t_1 \cdot \frac{1}{1 - r}$$

Problem Set 12-2

Quick Review 🕐 5 min

Q1. Definition: L is the limit of $f(x)$ as x approaches infinity if and only if —?—.

Q2. "If $g(x) = \int f(x)\, dx$, then $\int_a^b f(x)\, dx = g(b) - g(a)$" is a statement of —?—.

Q3. $(d/dx)\int_a^x f(t)\, dt = f(x)$ is a statement of —?—.

Q4. "... such that $f'(c) = [f(b) - f(a)]/(b - a)$" is part of the conclusion of —?—.

Q5. Instantaneous rate of change of a function is the physical meaning of —?—.

Q6. $(d/dx)(x \cos x) = $ —?—

Q7. $\int x \cos x\, dx = $ —?—

Q8. The differential of area in polar coordinates is $dA = $ —?—.

Q9. What function is graphed in Figure 12-2b?

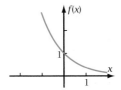

Figure 12-2b

Q10. The function —?— is graphed in Figure 12-2c.

    A. $y = \cos x$ B. $y = \sin x$

    C. $y = \sin(x - \pi/2)$ D. $y = 2 \cos x$

    E. $y = \cos x + 1$

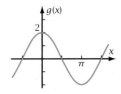

Figure 12-2c

1. Write the first few terms of the geometric series with first term 200 and common ratio -0.6. Write the corresponding partial sums. Plot the graph of the partial sums and sketch the result.

To what number does the series converge? Show this number on the graph. What is the first value of the term number, n, for which the partial sum is within 0.0001 of the limit? What can you say about the proximity of S_n to the limit for greater values of n?

2. Write the first few terms of the geometric series with first term 30 and common ratio 1.1. Write the corresponding partial sums. Give numerical and graphical evidence that the series diverges. For instance, what does S_{100} equal? What meaning does the algebraic formula $S = t_1/(1 - r)$ have for this series?

3. *Allergy Spray Dosage Problem:* Suppose a person takes a 7-μg (microgram) dose of allergy spray every 6 h. Suppose also that at the end of each 6 h, the amount of spray remaining in the body is 80% of the total amount at the beginning of that 6 h.

    a. Write a geometric series with partial sums, S_n, that represent the total amounts of spray remaining in the body just after dose number n has been taken. With which dose will the total amount in the body first exceed 20 μg? Will the total amount in the body ever reach 40 μg? Sketch a graph showing the pattern followed by the partial sums of this series.

    b. Write a sequence with terms that are the amounts of spray remaining in the body just before the nth dose. Show these terms on the graph from part a. Is there a number of doses beyond which the amount never drops below 20 μg?

    c. Sketch a dashed line on the graph from part b showing the piecewise-continuous function that represents the amount of spray remaining at times between doses.

4. *Inscribed Squares Problem:* Figure 12-2d shows an outer square of side 4 cm. The midpoints of the sides of the square are the vertices for an inscribed square. More squares are inscribed using the same pattern, *infinitely!*

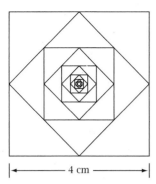

Figure 12-2d

a. Show that the perimeters of the squares form a geometric sequence.

b. The total perimeter of all the squares is a geometric series. Find the tenth partial sum of the series.

c. Does the series for the total perimeter converge, or does the total perimeter approach infinity?

d. The sum of the areas of the squares is also a geometric series. Does this series converge, or does the sum of the areas become infinite?

5. *Compound Interest Problem:* Meg A. Buck invests a million dollars in a certificate of deposit (CD) that earns 9% interest a year, compounded once a month.

a. Write a geometric sequence for the amounts the CD is worth after 0, 1, 2, and 3 mo.

b. How much will the CD be worth at the end of the first year? How much interest will have been earned?

c. How do you explain that after 12 mo the term index is 12, but there are 13 terms in the sequence?

d. The annual percentage rate (APR) an investment earns is the amount of interest for 1 yr expressed as a percentage of the investment's worth at the beginning of the year. What is the APR for Meg's CD?

e. When will Meg's CD be worth 2 million dollars?

6. *Regular Deposits Problem:* Ernest Lee invests $100 a month in an individual retirement account (IRA). The interest rate is 10.8% per year, compounded monthly. After 0 months,

Ernest has only the first $100 in the IRA. After 1 month, he has $200, plus interest on the first $100. After 2 months, he has $100 invested the last month, plus $100 with 1 month's interest for the preceding month, plus $100 with 2 months' interest.

a. Write the amount Ernest has at the end of 5 months as a partial sum of a geometric series.

b. The 5 (months) in part a is the term index. How many terms are in the partial sum? Why is the number of terms not equal to the term index?

c. How much will Ernest have after 10 years? How much of this is principal and how much is interest?

7. *Bouncing Ball Problem:* A superball is catapulted from floor level. It rises 10 ft above the floor then starts back down. On the next bounce it rises 9 ft above the floor. On each subsequent bounce it rises 90% of the maximum height of the previous bounce (Figure 12-2e).

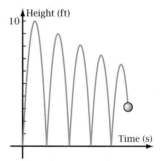

Figure 12-2e

a. The ball travels 20 ft vertically before the first bounce. Write the first few terms of the sequence of distances the ball travels between bounces.

b. Calculate the fourth partial sum of the series of distances corresponding to the sequence in part a.

c. To what number does the series in part b converge? What does this fact imply for the total distance the ball travels as it comes to rest?

d. From physics you learn that the distance an object drops from rest under the influence of gravity is $d = (1/2)gt^2$. If d is distance, in feet, and t is time, in seconds, then $g \approx 32.2$ ft/s^2. The time taken for an up-and-down cycle is twice the time to fall from a high point. How long does it take the ball to make the 20-ft first up-and-down cycle? How long does it take to make the 18-ft second cycle?

e. According to this mathematical model, the ball makes an infinite number of bounces before it comes to rest. Does this infinite number of bounces take an infinite length of time, or does the model predict that the ball eventually comes to rest? Explain.

8. *Snowflake Curve Problem:* A figure called the **snowflake curve** is generated as shown in Figure 12-2f. An equilateral triangle has the one-third points marked on each side. For the first iteration, the middle one-third of each side is erased and two line segments equal to the length of the erased segment are added to form sides of smaller equilateral triangles. In the second iteration, the process is repeated. Each old segment is replaced with four new segments, each of which is one-third as long as the segment it replaces. The snowflake curve is the figure that results from taking the limit as the number of iterations approaches infinity. This limit was first considered by Helge von Koch in 1904.

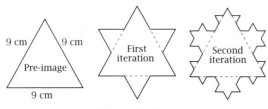

Figure 12-2f

a. Suppose the pre-image (the original triangle) has sides 9 cm long. Write the total length of the curve at the first, second, and third iterations. How can the length at one particular iteration be generated from the length at the previous iteration?

b. Does the sequence of total lengths in part a converge or diverge? What does the answer indicate about the total length of the snowflake curve? Surprising?

c. The area enclosed by each iteration is a partial sum of a geometric series. Write the first few terms of this series. Does the series converge or diverge? If it converges, find the limit to which it converges. If it diverges, explain how you know it diverges.

d. The snowflake curve is a classic example of a **fractal curve**. It is so fractured that it is more than one-dimensional but less than two-dimensional. Its dimension is a fraction equal to about 1.26. The curve is continuous everywhere but differentiable nowhere. For more about such curves, see, for example, Benoit Mandelbrot's *The Fractal Geometry of Nature*, published by W. H. Freeman and Company in 1983.

9. *Derivatives of a Power Series:* In this problem you will consider the derivatives of the power series from Problem Set 12-1 and see how they relate to the rational algebraic function from which the series was derived.

$$P(x) = 6 + 6x + 6x^2 + 6x^3 + 6x^4 + 6x^5 + \cdots$$

$$f(x) = 6/(1 - x)$$

Assume you can find the derivatives of a series by differentiating each term. Write series for $P'(x)$, $P''(x)$, and $P'''(x)$. [$P'''(x)$ is the **third derivative**, or the derivative of $P''(x)$.] Show that $P'(0)$, $P''(0)$, and $P'''(0)$ equal the corresponding values of $f'(0)$, $f''(0)$, and $f'''(0)$. How does $P^{(n)}(0)$ relate to $f^{(n)}(0)$, the ***n*th derivative**?

12-3   Power Series for an Exponential Function

In Section 12-1, you saw that you can write the function

$$f(x) = \frac{6}{1 - x}$$

as a power series,

$$P(x) = 6 + 6x + 6x^2 + 6x^3 + 6x^4 + \cdots$$

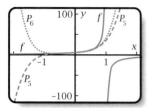

Figure 12-3a

Figure 12-3a shows that for values of x between -1 and 1, the more terms of the series used, the closer the graph of P is to the graph of f.

You can calculate values of $f(x) = 6/(1 - x)$ directly. You cannot, however, calculate values of the elementary transcendental functions directly using only a finite number of arithmetic operations, namely, $+$, $-$, $\times$, and $\div$. Fortunately, it is possible to express many of these functions as power series. You can use partial sums of these series to calculate $\sin x$, $\ln x$, e^x, and so forth to as many decimal places as you need. In this section you will derive a power series for $f(x) = 5e^{2x}$.

OBJECTIVE   Given a particular exponential function, derive a power series that fits the function for values of x close to zero.

In Problem Set 12-3, you will accomplish this objective.

Problem Set 12-3

Quick Review

Q1. Write $1/3$ as a repeating decimal.

Q2. Write $4/9$ as a repeating decimal.

Q3. Write 0.6666...(repeating) as a ratio of two integers.

Q4. Write $0.4 + 0.04 + 0.004 + 0.0004 + \cdots$ as a ratio of two integers.

Q5. Write the next term of the arithmetic series $1 + 5 + 9 + \cdots$.

Q6. Write the next term of the geometric series $1 + 5 + 25 + \cdots$.

Q7. First moment of mass = —?— times —?—.

Q8. Center of volume is called —?—.

Q9. Evaluate the integral: $\int x^{-1}\,dx$

Q10. For the function $y = \sec 2x$, $y' = $ —?—.

　　A. $\tan x$　　　　　　B. $\sec 2x \tan 2x$

　　C. $\frac{1}{2}\sec 2x \tan 2x$　　D. $2\sec 2x \tan 2x$

　　E. $2\sec^2 2x$

For Problems 1–11, let $f(x) = 5e^{2x}$.

1. Find $f'(x)$, $f''(x)$, $f'''(x)$, and $f^{(4)}(x)$, the first, second, third, and fourth derivatives of $f(x)$.

2. Function f is locally linear at $x = 0$. Let P_1 be the linear function $P_1(x) = c_0 + c_1 x$ that best fits $f(x)$ at $x = 0$. (c_0 and c_1 are used for the constants rather than b and m so that you will see a pattern later.) Find the values of c_0 and c_1 that make $P_1(0) = f(0)$ and $P_1'(0) = f'(0)$.

3. Function f is also **locally quadratic** at $x = 0$. That is, there is a quadratic function $P_2(x) = c_0 + c_1 x + c_2 x^2$ that best fits $f(x)$ at $x = 0$. Find the values of the constants c_0, c_1, and c_2 that make $P_2(0) = f(0)$, $P_2'(0) = f'(0)$, and $P_2''(0) = f''(0)$. How do the values of c_0 and c_1 compare with the corresponding values for the linear function P_1?

4. Function f is also **locally cubic** and **locally quartic** at $x = 0$. Find equations for the cubic and quartic functions,

   $$P_3(x) = c_0 + c_1 x + c_2 x^2 + c_3 x^3$$
   $$P_4(x) = c_0 + c_1 x + c_2 x^2 + c_3 x^3 + c_4 x^4$$

   that best fit $f(x)$ at $x = 0$. For these equations, $P_3(0)$ and $P_4(0)$ must equal $f(0)$, and the first three derivatives (for P_3) or the first four derivatives (for P_4) must equal the corresponding derivatives of f at $x = 0$. How do the coefficients c_0, c_1, and c_2 compare with those for the quadratic and linear functions in Problems 2 and 3?

5. Plot graphs of f, P_3, and P_4 on the same screen. Use a window with an x-range of $[-2, 2]$ and a y-range of $[-20, 100]$. Sketch the results.

6. For what range of values of x is the graph of P_4 indistinguishable from the graph of f on your grapher?

7. Show that the value of $P_4(1)$ is closer to the actual value of $f(1)$ than the value of $P_3(1)$ is.

8. In Problem 4, you should have found that for the fourth derivative, $P_4^{(4)}(0)$, to equal $f^{(4)}(0)$, the value of c_4 was

   $$24 c_4 = 80 \Rightarrow c_4 = \frac{80}{24}$$

   Write the 80 as the product of 5 and a power of 2. Write the 24 as a factorial. What pattern do you notice?

9. Show that c_3, c_2, c_1, and even c_0 follow the pattern in Problem 8.

10. Let $P(x)$ be the power series $P(x) = c_0 + c_1 x + c_2 x^2 + c_3 x^3 + c_4 x^4 + c_5 x^5 + \cdots$. Note that $P_3(x)$ and $P_4(x)$ are partial sums of this series. Make a conjecture about the values of c_5 and c_6 such that $P_5(x)$ and $P_6(x)$ best fit $f(x)$ at $x = 0$.

11. In previous courses you probably learned how to express series in Σ (sigma) notation. For instance,

   $$\sum_{n=0}^{\infty} \frac{1}{n+1} x^n = 1 + \tfrac{1}{2}x + \tfrac{1}{3}x^2 + \tfrac{1}{4}x^3 + \cdots$$

   In sigma notation you evaluate the expression $1/(n+1) \cdot x^n$ for each integer value of n starting at 0, and going to infinity, as you add the terms. Use what you have learned in this problem set to write in Σ notation the series for $P(x)$ that best fits $f(x) = 5e^{2x}$.

12-4 Power Series for Other Elementary Functions

In Sections 12-1 and 12-3, you saw that you can represent two quite different functions in similar form as power series, at least for values of x close to zero.

Rational function: $f(x) = \dfrac{6}{1 - x}$

Series: $P(x) = 6 + 6x + 6x^2 + 6x^3 + 6x^4 + 6x^5 + \cdots$

Exponential function: $f(x) = 5e^{2x}$

Series: $5 + 10x + 10x^2 + \tfrac{20}{3}x^3 + \tfrac{10}{3}x^4 + \tfrac{4}{3}x^5 + \cdots$

The two series have the same form. The only difference is the values of the coefficients, all 6s for one series and 5, 10, 10, 20/3, ... for the other.

The process of finding the right coefficients for a particular function is called **expanding the function as a power series**. If a function and the series match each other at $x = 0$, the function is said to be **expanded about $x = 0$.**

In Problem 5 of Problem Set 12-4, you will see how to expand a function as a power series about $x = 1$ rather than $x = 0$.

OBJECTIVE

Given an elementary function, find the first few terms of the power series that best fits the function, find a pattern that allows you to write more terms of the series, write the series in sigma notation, and plot the graph to see how well the series fits the function.

The general form of a power series is defined here.

DEFINITION: Power Series

You can write a power series for $f(x)$ expanded about $x = 0$ as

$$P(x) = c_0 + c_1 x + c_2 x^2 + c_3 x^3 + c_4 x^4 + c_5 x^5 + \cdots$$

where c_0, c_1, c_2, ... stand for constant coefficients.

Informally: A power series is a polynomial with an infinite number of terms.

▶ **EXAMPLE 1**

By equating derivatives, show that the following are the first three nonzero terms of the power series for $f(x) = \sin x$ expanded about $x = 0$.

$$x - \tfrac{1}{3!} x^3 + \tfrac{1}{5!} x^5$$

Solution

Let $P(x) = c_0 + c_1 x + c_2 x^2 + c_3 x^3 + c_4 x^4 + c_5 x^5 + \cdots$

For $P(x)$ to fit $f(x) = \sin x$ at $x = 0$, the function value and each derivative of $f(x)$ must equal the corresponding function value and derivative of $P(x)$ at $x = 0$. Assume that you can differentiate the series termwise.

$$P(x) = c_0 + c_1 x + c_2 x^2 + c_3 x^3 + c_4 x^4 + c_5 x^5 + c_6 x^6 + \cdots \Rightarrow P(0) \ = c_0$$
$$P'(x) = c_1 + 2c_2 x + 3c_3 x^2 + 4c_4 x^3 + 5c_5 x^4 + 6c_6 x^5 + \cdots \Rightarrow P'(0) \ = c_1$$
$$P''(x) = 2c_2 + 6c_3 x + 12c_4 x^2 + 20c_5 x^3 + 30c_6 x^4 + \cdots \ \Rightarrow P''(0) = 2c_2$$
$$P'''(x) = 6c_3 + 24c_4 x + 60c_5 x^2 + 120c_6 x^3 + \cdots \ \Rightarrow P'''(0) = 6c_3 = 3!c_3$$
$$P^{(4)}(x) = 24c_4 + 120c_5 x + 360c_6 x^2 + \cdots \ \Rightarrow P^{(4)}(0) = 24c_4 = 4!c_4$$
$$P^{(5)}(x) = 120c_5 + 720c_6 x + \cdots \ \Rightarrow P^{(5)}(0) = 120c_5 = 5!c_5$$

[Recall that 3! (three factorial) is the product of the first three counting numbers, $1 \cdot 2 \cdot 3$.]

For the function, these are the derivatives of $f(x)$.

$$f(x) = \sin x \quad \Rightarrow f(0) = 0$$
$$f'(x) = \cos x \quad \Rightarrow f'(0) = 1$$
$$f''(x) = -\sin x \Rightarrow f''(0) = 0$$
$$f'''(x) = -\cos x \Rightarrow f'''(0) = -1$$
$$f^{(4)}(x) = \sin x \quad \Rightarrow f^{(4)}(0) = 0$$
$$f^{(5)}(x) = \cos x \quad \Rightarrow f^{(5)}(0) = 1$$

Equating the function values and corresponding derivatives of f and P gives

$$c_0 = 0$$
$$c_1 = 1$$
$$2c_2 = 0 \Rightarrow c_2 = 0$$
$$3!c_3 = -1 \Rightarrow c_3 = -\frac{1}{3!}$$
$$4!c_4 = 0 \Rightarrow c_4 = 0$$
$$5!c_5 = 1 \Rightarrow c_5 = \frac{1}{5!}$$

Thus, the sum of the first three nonzero terms is

$$P(x) = x - \frac{1}{3!}x^3 + \frac{1}{5!}x^5, \text{Q.E.D.} \quad \blacktriangleleft$$

Once you have found derivatives for the series, you can remember the pattern and use it when you are called upon to expand other functions as series by equating derivatives.

PROPERTY: Derivatives of a Power Series

If $P(x) = c_0 + c_1x + c_2x^2 + c_3x^3 + c_4x^4 + c_5x^5 + \cdots + c_nx^n + \cdots$,
then $P(0) = c_0$, $P'(0) = c_1$, $P''(0) = 2!c_2, \ldots, P^{(n)}(0) = n!c_n, \ldots$.

All you have to do to expand a function f as a power series about $x = 0$ is find the values of $f(0)$, $f'(0)$, $f''(0), \ldots, f^{(n)}(0), \ldots$, set them equal to the values of $P(0)$ and derivatives, and solve for the values of c.

▶ **EXAMPLE 2**   For the series $\sin x = x - \frac{1}{3!}x^3 + \frac{1}{5!}x^5 - \cdots$

a. Demonstrate that you understand the pattern in the series by writing the next three terms.

b. Write the series using sigma notation.

Solution   a. The exponents of x are the odd integers. Each coefficient is the reciprocal of the factorial of the exponent. The signs alternate, with the first term being positive. Thus,

$$\sin x = x - \frac{1}{3!}x^3 + \frac{1}{5!}x^5 - \frac{1}{7!}x^7 + \frac{1}{9!}x^9 - \frac{1}{11!}x^{11} + \cdots$$

b. You can write the series in sigma notation this way:

$$\sum_{n=0}^{\infty}(-1)^n \frac{1}{(2n+1)!}\,x^{2n+1}$$

This symbol is read, "the sum from $n = 0$ to infinity of..." It means to let $n = 0, 1, 2, 3, \ldots$ and add the resulting terms. The secret to finding a formula for t_n (the term with index n) is to write the values of n under the terms. In this case it is helpful to start n at 0 rather than 1.

$$x - \tfrac{1}{3!}x^3 + \tfrac{1}{5!}x^5 - \tfrac{1}{7!}x^7 + \cdots$$
$$\;\;0\quad\;\;1\quad\;\;2\quad\;\;3$$

Write the values of the term index, n, under the respective terms.

By comparing the values of n with numbers in the terms, you can see that the exponent and denominator in each term are one more than twice the value of n. The factor $(-1)^n$ makes the signs alternate. The index of summation, n, could start at 0, 1, or whatever number you feel is appropriate. For instance, the answer above could be written

$$\sum_{n=1}^{\infty}(-1)^{n+1} \frac{1}{(2n-1)!}\,x^{2n-1}$$ Alternative form.   ◀

▶ **EXAMPLE 3**   Consider the power series for $\sin x$ expanded about $x = 0$.

a. Plot the sixth partial sum.

b. Find the approximate interval of x-values for which the sixth partial sum is within 0.0001 unit of the value of the sine function.

c. Find a wider interval for which the ninth partial sum is within 0.0001 unit of $\sin x$.

Solution   a. The sixth partial sum is $S_5(x)$ because the index of summation starts at $n = 0$.

$$S_5(x) = x - \tfrac{1}{3!}x^3 + \tfrac{1}{5!}x^5 - \tfrac{1}{7!}x^7 + \tfrac{1}{9!}x^9 - \tfrac{1}{11!}x^{11}$$

A time-efficient way to enter the partial sum uses the formula for the nth term that was found in Example 2 and the grapher's sequence commands. For a typical grapher you can enter the formula this way:

$$y_1 = \text{sum}\,(\text{seq}\,((-1)^\wedge n/(2n+1)! * x^\wedge(2n+1), n, 0, 5, 1))$$

The sequence command tells the grapher to generate a set of numbers using the formula inside the parentheses. The n after the comma tells the grapher that n is the index of summation. The last three numbers tell the grapher to start n at 0, end it at 5, and increase it by steps of 1 each time (thus giving six terms). The sum command tells the grapher to add the terms of the sequence it has calculated. The grapher performs this computation for each value of x in the window you specify. Figure 12-4a shows the result.

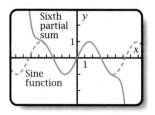

Figure 12-4a

b. A table of values of the differences between $\sin x$ and the series quickly shows the interval of x-values. Enter $(\sin x - y_1)$ as y_2. From the table you can see that if x is between -2 and 2, the absolute value of the difference is less than 0.0001. By exploring the interval between 2 and 3 with another table, stepping x by 0.1, you can find that the series is within 0.0001 unit of $\sin x$ for $-2.7 < x < 2.7$.

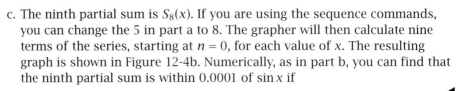

x	$\sin x - \text{sum}$
-4	$-0.0100020\ldots$
-3	$-0.0002454\ldots$
-2	$-0.00000129\ldots$
-1	$-0.000000000159\ldots$
0	0
1	$0.000000000159\ldots$
2	$0.00000129\ldots$
3	$0.0002454\ldots$
4	$0.0100020\ldots$

You can also find the answer using the solver feature of your grapher. If you have entered $\sin x$ as y_1 and the partial sum as y_2, then set $(y_1 - y_2) - 0.0001$ equal to 0. The result is $x \approx 2.7986\ldots$. By symmetry, the interval is $-2.7986\ldots < x < 2.7986\ldots$.

c. The ninth partial sum is $S_8(x)$. If you are using the sequence commands, you can change the 5 in part a to 8. The grapher will then calculate nine terms of the series, starting at $n = 0$, for each value of x. The resulting graph is shown in Figure 12-4b. Numerically, as in part b, you can find that the ninth partial sum is within 0.0001 of $\sin x$ if

$$-4.8974\ldots < x < 4.8974\ldots \qquad \blacktriangleleft$$

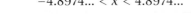

Figure 12-4b

Power series relate to functions the way decimals relate to irrational numbers. For instance,

$$\pi = 3.141592653\ldots \qquad \text{and} \qquad \sqrt{7359} = 85.7846140\ldots$$

The human mind can grasp the size of the approximation 85.78... more easily than it can grasp the exact value $\sqrt{7359}$. Similarly, a computer can calculate values of $\sin x$ from a power series more easily than it can calculate them directly from a definition of sine.

The more decimal places you use for a number such as π, the more accurate the approximation is. In many cases, the more terms you use for the partial sum of a series, the better the partial sum fits the function values. As was shown in Figure 12-4b, the ninth partial sum of the series in Example 3 seems to coincide visually with $\sin x$ for $-7 < x < 7$. The sixth partial sum in Figure 12-4a coincides only for about $-4 < x < 4$.

Problem Set 12-4

Quick Review

Q1. Sketch the graph of $y = \sin x$.

Q2. Sketch the graph of $y = \cos x$.

Q3. Sketch the graph of $y = e^x$.

Q4. Sketch the graph of $y = \ln x$.

Q5. Sketch the graph of $y = \cosh x$.

Q6. Sketch the graph of $y = \tan^{-1} x$.

Q7. In the expression $3x^5$, the number 5 is called the —?—.

Q8. In the expression $3x^5$, the number 3 is called the —?—.

Q9. The expression x^5 is called a(n) —?—.

Q10. The area of the region between the graph of $y = 9 - x^2$ and the x-axis equals —?—.

    A. 0   B. 18   C. $26\frac{2}{3}$ D. 36   E. 48

---

1. *Exponential Function Series Problem:* Consider the function $f(x) = e^x$.

  a. Show by equating derivatives that the power series expansion for e^x about $x = 0$ is

$$P(x) = 1 + x + \tfrac{1}{2!}x^2 + \tfrac{1}{3!}x^3 + \cdots$$

  b. Write the next two terms of the series.

  c. Write the series using sigma notation.

  d. Plot the fourth partial sum of the series. On the same screen, plot $y = e^x$. Use a window with an x-range of $[-3, 3]$ and a y-range of $[-2, 10]$. Sketch the result.

  e. For what interval of x-values are the two graphs indistinguishable from each other?

  f. For what interval of x-values is the fourth partial sum within 0.0001 unit of e^x?

  g. For what wider interval is the ninth partial sum of the series within 0.0001 of e^x?

2. *Cosine Function Series Problem:* Consider the function $g(x) = \cos x$.

  a. Show by equating derivatives that the power series expansion for $\cos x$ about $x = 0$ is

$$P(x) = 1 - \tfrac{1}{2!}x^2 + \tfrac{1}{4!}x^4 - \tfrac{1}{6!}x^6 + \tfrac{1}{8!}x^8 - \cdots$$

  b. Write the next three terms of the series.

  c. Write the series using sigma notation. Start the index of summation at $n = 0$.

  d. Figure 12-4c shows the graph of the fifth partial sum, $S_4(x)$. Plot this graph on your grapher. Then plot $y = \cos x$ on the same screen. Sketch both graphs.

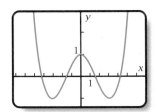

Figure 12-4c

e. Plot the graph of the eighth partial sum, $S_7(x)$. For what interval of x-values is the $S_7(x)$ graph indistinguishable from that of $y = \cos x$? Sketch the result.

f. For what interval of x is the eighth partial sum within 0.0001 unit of $\cos x$?

g. Explain why the series for $P(x)$ agrees with the properties of the cosine function.

3. *Sine Series Problem:* Let $P(x) = x - \tfrac{1}{3!}x^3 + \tfrac{1}{5!}x^5 - \tfrac{1}{7!}x^7 + \cdots$, which is the sine series.

  a. Show that $S_3(0.6)$, the fourth partial sum, is approximately equal to $\sin 0.6$.

  b. The **tail** of a power series is the series of terms left after a given partial sum. Write the value of $\sin 0.6$ to as many places as your grapher gives. Use this number to evaluate, approximately, the tail of the $P(0.6)$ series for $S_1(0.6)$, $S_2(0.6)$, and $S_3(0.6)$. Show that in each case the value of the tail of the series is *less* in magnitude than the absolute value of the *first* term of the tail.

  c. Assuming that your observation in part b about the tail of the series is correct for all values of n, determine how many terms of the series for $P(0.6)$ it would take to get a partial sum that estimates $\sin 0.6$ correct to at least 20 decimal places.

4. *Hyperbolic Sine and Cosine Series Problem:* Let

$$P(x) = \sum_{n=0}^{\infty} \frac{1}{(2n+1)!} x^{2n+1}$$

  a. Write the first four terms of the series (through $n = 3$).

  b. Show by equating derivatives that the series $P(x)$ represents $\sinh x$.

  c. Show that $S_3(0.6)$, the fourth partial sum of the series, is approximately equal to $\sinh 0.6$.

  d. For what interval of x-values is $S_3(x)$ within 0.0001 unit of $\sinh x$?

  e. Assume that the derivative of the series equals the sum of the derivatives of the terms. Differentiate each term of the series for $P(x)$ to get a series for $P'(x)$.

f. Because the derivative of $\sinh x$ is $\cosh x$, the series you found in part e should be the series for $\cosh x$. Demonstrate that this is true by showing that the value of $P'(0.6)$ is approximately equal to $\cosh 0.6$. Use the fourth partial sum of the derivative series.

g. Integrate the series for $P(x)$ term by term to get a power series for $\int P(x)\,dx$. Show that the result is the series for $\cosh x$ in part f if the integration constant is picked appropriately.

5. *Natural Log Series Problem:* Let $P(x) = (x - 1) - \frac{1}{2}(x - 1)^2 + \frac{1}{3}(x - 1)^3 - \frac{1}{4}(x - 1)^4 + \cdots$. This series is the power series for $\ln x$ *expanded about $x = 1$.*

a. By equating derivatives, show that $P(x)$ and $\ln x$ have the same function value at $x = 1$, and the same first, second, and third derivative values at $x = 1$.

b. Write the next two terms of the series.

c. Write the series using sigma notation. Start the index of summation at $n = 1$.

d. Figure 12-4d shows the graph of the fourth partial sum of the series. The graph fits $y = \ln x$ reasonably well when x is close to 1. Plot $y = \ln x$ and $S_{10}(x)$, the tenth partial sum of the series, on the same screen. Sketch.

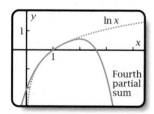

Figure 12-4d

e. By appropriate use of the TRACE or TABLE feature, compare $S_{10}(1.2)$, $S_{10}(1.95)$, and $S_{10}(3)$ with the values of $\ln 1.2$, $\ln 1.95$, and $\ln 3$. For what interval of x-values does the tenth partial sum of the series seem to fit

the ln function? Is this interval much larger than that for the fourth partial sum in Figure 12-4d?

6. *Convergence and Divergence Problem:* From Problem Set 12-1, recall that a series converges to $f(x)$ for a particular value of x if the partial sums of $P(x)$ approach the value of $f(x)$ as the number of terms in the partial sum approaches infinity. In Problem 5, the series $P(x)$ converges to $\ln x$ when $x = 1.2$ and $x = 1.95$, but the series diverges for $x = 3$. In this problem you will see why this is true.

a. Make a table of values for the first few terms of $P(3)$. What is happening to the absolute value of the terms?

b. By appropriate use of l'Hospital's rule, show that the absolute value of the nth term for $P(3)$ approaches infinity as n approaches infinity. Explain how this fact indicates that the series for $P(3)$ cannot possibly converge.

c. Make a table of values for the first few terms of $P(1.2)$. Show that these terms approach zero for a limit as n approaches infinity, and thus the series could converge.

d. In Problem 3, you observed that the value of the tail of the series for $\sin x$ that remained after the nth partial sum was smaller in absolute value than the absolute value of the first term of the tail. Is this observation true for $P(1.2)$? Justify your answer.

7. *Inverse Tangent Series Problem:* Let $f(x) = \tan^{-1} x$.

a. Let $P(x)$ be the power series

$$P(x) = \sum_{n=0}^{\infty} (-1)^n \frac{1}{2n + 1} x^{2n+1}$$

Write the first few terms of the series.

b. Plot the graph of f and the graphs of the sixth and seventh partial sums on the same screen. For what values of x do the partial sums represent the graph of f quite well? For what values of x do the partial sum graphs bear little or no resemblance to the graph of f?

12-5   Taylor and Maclaurin Series, and Operations on These Series

In Problem 5 of Problem Set 12-4, you learned that the expansion of $\ln x$ about $x = 1$ is

$$\ln x = (x - 1) - \tfrac{1}{2}(x - 1)^2 + \tfrac{1}{3}(x - 1)^3 - \tfrac{1}{4}(x - 1)^4 + \cdots$$

You can find the coefficients by equating derivatives at $x = 1$, as you will see in Example 1 of this section. In general, if function f is expanded about $x = a$, then the coefficients are

$$c_0 = f(a), \ c_1 = f'(a), \ c_2 = \frac{f''(a)}{2!}, \ c_3 = \frac{f'''(a)}{3!}, \ldots$$

The powers are $(x - a)^n$ rather than x^n because $(x - a)$ equals zero when $x = a$. The result is called the **Taylor series expansion of $f(x)$ about $x = a$** after the British mathematician Brook Taylor (1685–1731). If $a = 0$, the Taylor series is called the **Maclaurin series expansion of $f(x)$** after the Scottish mathematician Colin Maclaurin (1698–1746), although neither Taylor nor Maclaurin was the first to publish this kind of series. A partial sum of a Taylor series is called a **Taylor polynomial**. For instance, $\ln x \approx P(x) = (x - 1) - \tfrac{1}{2}(x - 1) + \tfrac{1}{3}(x - 1)^3 - \tfrac{1}{4}(x - 1)^4$ is called a fourth-degree Taylor polynomial approximating $\ln x$.

DEFINITIONS: Taylor Series and Maclaurin Series

If f is a function with differentiable derivatives, then you can write $f(x)$ as a **Taylor series** expansion about $x = a$ as follows:

$$f(x) = f(a) + f'(a)(x - a) + \frac{f''(a)}{2!}(x - a)^2 + \frac{f'''(a)}{3!}(x - a)^3 + \cdots$$
$$+ \frac{f^{(n)}(a)}{n!}(x - a)^n + \cdots$$

If $a = 0$, the series is called a **Maclaurin series**.

A partial sum of a Taylor series is called a Taylor polynomial.

OBJECTIVE   Given the Taylor series for e^x, $\sin x$, $\cos x$, $\sinh x$, $\cosh x$, $\ln x$, $1/(1 - x)$, and $\tan^{-1} x$, perform operations on these series to derive power series for related functions.

To accomplish the objective efficiently, it is a good idea to memorize the eight series on the next page, which you derived in Sections 12-1 through 12-4.

Taylor series are used in error analysis of data recorded at an acoustics lab.

Eight Basic Power Series

$$e^x = 1 + x + \tfrac{1}{2!}x^2 + \tfrac{1}{3!}x^3 + \tfrac{1}{4!}x^4 + \cdots = \sum_{n=0}^{\infty} \frac{1}{n!}x^n$$

$$\sin x = x - \tfrac{1}{3!}x^3 + \tfrac{1}{5!}x^5 - \tfrac{1}{7!}x^7 + \cdots = \sum_{n=0}^{\infty}(-1)^n \frac{1}{(2n+1)!}x^{2n+1}$$

$$\cos x = 1 - \tfrac{1}{2!}x^2 + \tfrac{1}{4!}x^4 - \tfrac{1}{6!}x^6 + \cdots = \sum_{n=0}^{\infty}(-1)^n \frac{1}{(2n)!}x^{2n}$$

$$\sinh x = x + \tfrac{1}{3!}x^3 + \tfrac{1}{5!}x^5 + \tfrac{1}{7!}x^7 + \cdots = \sum_{n=0}^{\infty} \frac{1}{(2n+1)!}x^{2n+1}$$

$$\cosh x = 1 + \tfrac{1}{2!}x^2 + \tfrac{1}{4!}x^4 + \tfrac{1}{6!}x^6 + \cdots = \sum_{n=0}^{\infty} \frac{1}{(2n)!}x^{2n}$$

$$\ln x = (x-1) - \tfrac{1}{2}(x-1)^2 + \tfrac{1}{3}(x-1)^3 - \tfrac{1}{4}(x-1)^4 + \cdots$$

$$= \sum_{n=1}^{\infty}(-1)^{n+1}\tfrac{1}{n}(x-1)^n$$

$$\frac{1}{1-x} = 1 + x + x^2 + x^3 + x^4 + \cdots = \sum_{n=0}^{\infty} x^n \quad \text{(A geometric series.)}$$

$$\tan^{-1} x = x - \tfrac{1}{3}x^3 + \tfrac{1}{5}x^5 - \tfrac{1}{7}x^7 + \cdots = \sum_{n=0}^{\infty}(-1)^n \frac{1}{2n+1}x^{2n+1}$$

▶ **EXAMPLE 1**   Show by equating derivatives that the Taylor series for $\ln x$ expanded about $x = 1$ is

$$\ln x = (x-1) - \tfrac{1}{2}(x-1)^2 + \tfrac{1}{3}(x-1)^3 - \tfrac{1}{4}(x-1)^4 + \cdots$$

Solution   So that $P(1)$ will equal c_0, $P'(1)$ will equal c_1, and so forth, the series is written in powers of $(x-1)$ instead of in powers of x as before.

$$P(x) = c_0 + c_1(x-1) + c_2(x-1)^2 + c_3(x-1)^3 + c_4(x-1)^4 + \cdots$$

[If you were to use $P(x) = c_0 + c_1 x + c_2 x^2 + \cdots$, then $P(1)$ would equal $c_0 + c_1 + c_2 + \cdots$ instead of simply c_0.] Equating derivatives gives

$$
\begin{aligned}
f(x) &= \ln x & &\Rightarrow & f(1) &= 0 & &\Rightarrow & c_0 &= 0 \\
f'(x) &= 1/x = x^{-1} & &\Rightarrow & f'(1) &= 1 & &\Rightarrow & c_1 &= 1 \\
f''(x) &= -x^{-2} & &\Rightarrow & f''(1) &= -1 & &\Rightarrow & 2!c_2 = -1 &\Rightarrow & c_2 &= -\tfrac{1}{2} \\
f'''(x) &= +2x^{-3} & &\Rightarrow & f'''(1) &= 2 & &\Rightarrow & 3!c_3 = 2 &\Rightarrow & c_3 &= \tfrac{1}{3} \\
f^{(4)}(x) &= -6x^{-4} & &\Rightarrow & f^{(4)}(1) &= -6 & &\Rightarrow & 4!c_4 = -6 &\Rightarrow & c_4 &= -\tfrac{1}{4} \\
f^{(5)}(x) &= +24x^{-5} & &\Rightarrow & f^{(5)}(1) &= 24 & &\Rightarrow & 5!c_5 = 24 &\Rightarrow & c_5 &= \tfrac{1}{5}
\end{aligned}
$$

$$\therefore\ \ln x = (x-1) - \tfrac{1}{2}(x-1)^2 + \tfrac{1}{3}(x-1)^3 - \tfrac{1}{4}(x-1)^4 + \cdots, \text{Q.E.D}$$ ◀

▶ **EXAMPLE 2**   Expand $f(x) = \sin x$ as a Taylor series about $x = \pi/3$.

Solution
$$f(x) = \sin x \quad \Rightarrow \quad f\left(\tfrac{\pi}{3}\right) = \sin \tfrac{\pi}{3} = \tfrac{\sqrt{3}}{2}$$
$$f'(x) = \cos x \quad \Rightarrow \quad f'\left(\tfrac{\pi}{3}\right) = \cos \tfrac{\pi}{3} = \tfrac{1}{2}$$
$$f''(x) = -\sin x \quad \Rightarrow \quad f''\left(\tfrac{\pi}{3}\right) = -\sin \tfrac{\pi}{3} = -\tfrac{\sqrt{3}}{2}$$
$$f'''(x) = -\cos x \quad \Rightarrow \quad f'''\left(\tfrac{\pi}{3}\right) = -\cos \tfrac{\pi}{3} = -\tfrac{1}{2}, \text{ and so forth.}$$

$$\therefore \ \sin x = \sqrt{3}/2 + (1/2)(x - \pi/3) - \left[\tfrac{\sqrt{3}/2}{2!}\right][(x - \pi/3)^2]$$
$$- \left[\tfrac{1/2}{3!}\right][(x - \pi/3)^3] + \left[\tfrac{\sqrt{3}/2}{4!}\right][(x - \pi/3)^4] + \cdots \qquad ◀$$

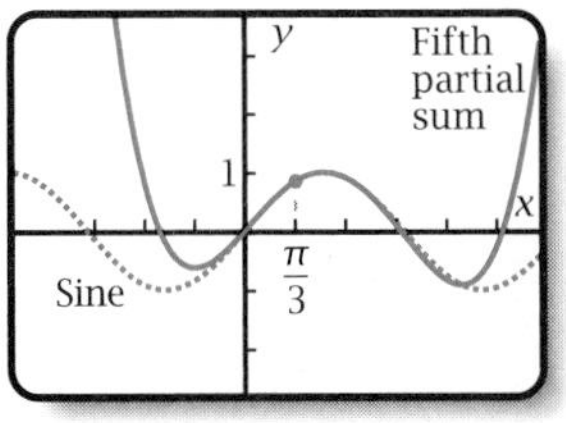

Figure 12-5a

Example 2 shows that once you have found the derivatives for the function you are expanding, you can simply substitute them directly into the formula for the Taylor series. It isn't necessary to equate derivatives.

Figure 12-5a shows that the fifth partial sum in Example 2 fits $\sin x$ well if x is in a neighborhood of $\pi/3$.

It is usually easier to derive a series by starting with one of the known series than it is to calculate derivatives of $f(x)$. Examples 3–6 show you ways to do this.

▶ **EXAMPLE 3**   Write the first few terms of the Maclaurin series for $\sin(3x)^2$.

Solution   Take the known series for $\sin x$, and replace the x with $(3x)^2$. The rest is algebra.
$$\sin(3x)^2 = (3x)^2 - \tfrac{1}{3!}[(3x)^2]^3 + \tfrac{1}{5!}[(3x)^2]^5 - \tfrac{1}{7!}[(3x)^2]^7 + \cdots$$
$$= 3^2 x^2 - \tfrac{3^6}{3!}x^6 + \tfrac{3^{10}}{5!}x^{10} - \tfrac{3^{14}}{7!}x^{14} + \cdots \qquad ◀$$

▶ **EXAMPLE 4**   Write the first few terms of the Maclaurin series for
$$g(x) = \frac{1}{1 + x^3}$$

Solution   You can derive this series by performing long division or by substituting $-x^3$ for x in the geometric series from $1/(1 - x)$.
$$g(x) = 1 + (-x^3) + (-x^3)^2 + (-x^3)^3 + (-x^3)^4 + \cdots$$
$$= 1 - x^3 + x^6 - x^9 + x^{12} - \cdots \qquad ◀$$

▶ **EXAMPLE 5**   By appropriate operations, show that the Maclaurin series for $\tan^{-1} x$ is
$$\tan^{-1} x = x - \tfrac{1}{3}x^3 + \tfrac{1}{5}x^5 - \tfrac{1}{7}x^7 + \cdots$$

Solution   If you start the process of equating derivatives, the result is
$$f(x) = \tan^{-1} x$$
$$f'(x) = \frac{1}{1 + x^2}.$$

You can expand the expression for the first derivative as a Maclaurin series by substituting $-x^2$ for x in the geometric series from $1/(1 - x)$, as in Example 4.

$$f'(x) = 1 - x^2 + x^4 - x^6 + \cdots$$

Then you can find the series for $\tan^{-1} x$ by integrating, assuming that you can integrate an infinite series termwise.

$$\tan^{-1} x = \int (1 - x^2 + x^4 - x^6 + \cdots)\, dx$$

$$= x - \tfrac{1}{3}x^3 + \tfrac{1}{5}x^5 - \tfrac{1}{7}x^7 + \cdots + C$$

Because $\tan^{-1} 0 = 0$, the constant of integration C is also zero. Thus

$$\tan^{-1} x = x - \tfrac{1}{3}x^3 + \tfrac{1}{5}x^5 - \tfrac{1}{7}x^7 + \cdots, \text{Q.E.D.}$$
◀

▶ **EXAMPLE 6**   Write a power series for $f(x) = \int_0^x t \cos t^5\, dt$. Evaluate the sixth partial sum at $x = 0.8$.

Solution   Write a series for the integrand, then integrate term by term. As in Example 5, assume that you can integrate an infinite series termwise, which is true in this case but not always. First, replace x with t^5 in the Maclaurin series for $\cos x$.

$$\cos t^5 = 1 - \tfrac{1}{2!}t^{10} + \tfrac{1}{4!}t^{20} - \tfrac{1}{6!}t^{30} + \cdots$$

Then multiply each term by t, and integrate.

$$t \cos t^5 = t - \tfrac{1}{2!}t^{11} + \tfrac{1}{4!}t^{21} - \tfrac{1}{6!}t^{31} + \cdots$$

$$f(x) = \int_0^x \left(t - \tfrac{1}{2!}t^{11} + \tfrac{1}{4!}t^{21} - \tfrac{1}{6!}t^{31} + \cdots \right) dt$$

$$= \tfrac{1}{2}t^2 - \tfrac{1}{12 \cdot 2!}t^{12} + \tfrac{1}{22 \cdot 4!}t^{22} - \tfrac{1}{32 \cdot 6!}t^{32} + \cdots \Big|_0^x$$

$$= \tfrac{1}{2}x^2 - \tfrac{1}{12 \cdot 2!}x^{12} + \tfrac{1}{22 \cdot 4!}x^{22} - \tfrac{1}{32 \cdot 6!}x^{32} + \cdots$$

To find $S_6(0.8)$, substitute and use a calculator with sufficient accuracy to calculate the total.

$$S_6(0.8) = \tfrac{1}{2}0.8^2 - \tfrac{1}{12 \cdot 2!}0.8^{12} + \tfrac{1}{22 \cdot 4!}0.8^{22} - \tfrac{1}{32 \cdot 6!}0.8^{32}$$

$$+ \tfrac{1}{42 \cdot 8!}0.8^{42} - \tfrac{1}{52 \cdot 10!}0.8^{52}$$

$$= 0.31715062893841\ldots$$
◀

Of course you could find the answer by calculating Riemann sums. But the series takes only six terms to give 13-place accuracy. As you will see in Section 12.6, it is possible to determine the accuracy of an integral if you use a series rather than a Riemann sum or other numerical methods.

▶ **EXAMPLE 7**   Let f be a function with derivatives of all orders and with values that are given approximately by the fourth-degree Taylor polynomial

$$f(x) \approx P_4(x) = 5 - 2(x - 3) + 0.6(x - 3)^2 + 0.12(x - 3)^3 - 0.08(x - 3)^4$$

a. Find the approximate value of $f(2.6)$. What assumption must you make about 2.6 for this approximation to be valid?

b. Use the pattern for equating derivatives to find $f(3), f'(3), f''(3), f'''(3)$, and $f^{(4)}(3)$.

c. Find a fourth-degree Taylor polynomial for $g(x) = f(x^2 + 3)$, and use it to find an approximation for $g(1)$, assuming the series converges if $x = 1$.

d. Use the second-derivative test to show that function g in part c has a local maximum at $x = 0$.

e. Find a fifth-degree Taylor polynomial for $h(x) = \int_0^x g(t)\,dt$.

Solution   a. $f(2.6) \approx P_4(2.6) = 5 - 2(-0.4) + 0.6(-0.4)^2 + 0.12(-0.4)^3$
$$- 0.08(-0.4)^4$$
$$= 5.886272$$

You must assume that the series converges if $x = 2.6$.

b. $f(3) = c_0 = 5$
$f'(3) = c_1 = -2$
$f''(3) = 2!c_2 = 2(0.6) = 0.12$
$f'''(3) = 3!c_3 = 6(0.12) = 0.72$
$f^{(4)}(3) = 4!c_4 = 24(-0.08) = -1.92$

c. $g(x) = 5 - 2(x^2) + 0.6(x^2)^2 + 0.12(x^2)^3$ The ellipsis indicates
$$- 0.08(x^2)^4 + \cdots$$ a *series*.
$$= 5 - 2x^2 + 0.6x^4 + 0.12x^6 - 0.08x^8 + \cdots$$
$$\therefore P_4(x) = 5 - 2x^2 + 0.6x^4$$ No ellipsis indicates
a *polynomial*.

$$g(1) \approx P_4(1) = 5 - 2 + 0.6 = 3.6$$ Assuming the series
converges if $x = 1$.

d. By equating derivatives, $g(0) = 5, g'(0) = 0$, and $g''(0) = 2!(-2) = -4$.
$\therefore g(0) = 5$ is a local maximum because the tangent is horizontal and the graph is concave down at $x = 0$.

e. $$h(x) = \int_0^x (5 - 2t^2 + 0.6t^4 + 0.12t^6 - 0.08t^8 + \cdots)\,dt$$

$$= \left(5x - \tfrac{2}{3}x^3 + \tfrac{0.6}{5}x^5 + \tfrac{0.12}{7}x^7 - \tfrac{0.08}{9}x^9 + \cdots\right) - 0$$

$$\therefore P_5(x) = 5x - \tfrac{2}{3}x^3 + \tfrac{0.6}{5}x^5$$ ◀

The techniques for finding Taylor or Maclaurin series for a given function are summarized here.

TECHNIQUES: Finding a Taylor or Maclaurin Series

You can find a Taylor or Maclaurin series for a function

- By equating derivatives
- From memory, by looking it up, or by computer algebra system
- By operating on a given or known series

    Substitute for the variable in the series.

    Multiply or divide the terms by a given expression.

    Integrate or differentiate the series termwise.

- By operating on the parent function

    Integrate or differentiate the function, then find a series.

    Perform long division, such as $6/(1 - x) = 6 + 6x + 6x^2 + 6x^3 + \cdots$.

Problem Set 12-5

Quick Review

Q1. Evaluate: 4!

Q2. Evaluate: 3!

Q3. Evaluate: 4!/4

Q4. What does n equal if $4!/4 = n!$?

Q5. If $m!/m = n!$, then $n = -?-$.

Q6. $0! = m!/m$. What does m equal?

Q7. Why does 0! equal 1?

Q8. Why is $(-1)!$ infinite?

Q9. Differentiate: $f(x) = \sqrt{x^2 - 7}$

Q10. $\int \sinh x \, dx = -?-$

    A. $\cosh x + C$

    B. $-\cosh x + C$

    C. $\dfrac{1}{\sqrt{x^2 + 1}} + C$

    D. $x \sinh^{-1} x - (x^2 + 1)^{1/2} + C$

    E. $\sinh x + C$

For Problems 1–8, write the power series from memory.

1. $f(u) = e^u$
2. $f(u) = \ln u$
3. $f(u) = \sin u$
4. $f(u) = \cos u$
5. $f(u) = \cosh u$
6. $f(u) = \sinh u$
7. $f(u) = (1 - u)^{-1}$
8. $f(u) = \tan^{-1} u$

For Problems 9–22, derive a power series for the given function. Write enough terms of the series to show the pattern.

9. $x \sin x$
10. $x \sinh x$
11. $\cosh x^3$
12. $\cos x^2$
13. $\ln x^2$
14. e^{-x^2}
15. $\int_0^x e^{-t^2} \, dt$
16. $\int_{1/3}^x \ln(3t) \, dt$
17. $\dfrac{1}{x^4 + 1}$
18. $\dfrac{9}{x^2 + 3}$
19. $\int_0^x \dfrac{1}{t^4 + 1} \, dt$
20. $\int_0^x \dfrac{9}{t^2 + 3} \, dt$
21. $\dfrac{d}{dx}(\sinh x^2)$
22. $\dfrac{d}{dx}(\cos x^{0.5})$

For Problems 23-24, write a Taylor polynomial of the given degree for the function described.

23. Fourth-degree Taylor polynomial expanded about $x = 2$, if $f(2) = -8, f'(2) = 3, f''(2) = 0.7$, $f'''(2) = 0.51$, and $f^{(4)}(2) = -0.048$

24. Fifth-degree Taylor polynomial expanded about $x = -1$, if $f(-1) = 7, f'(-1) = 2$, $f''(-1) = -0.48, f'''(-1) = 0, f^{(4)}(-1) = 0.36$, and $f^{(5)}(-1) = -0.084$

25. Let f be a function with derivatives of all orders and with values that are given approximately by the fourth-degree Taylor polynomial

$$f(x) \approx P_4(x) = 2 + 0.5(x + 1) - 0.3(x + 1)^2$$
$$- 0.18(x + 1)^3 + 0.02(x + 1)^4$$

a. Find the approximate value of $f(0.4)$. What assumption must you make about 0.4 for this approximation to be valid?

b. Use the pattern for equating derivatives to find $f(-1), f'(-1), f''(-1), f'''(-1)$, and $f^{(4)}(-1)$.

c. Find a sixth-degree Taylor polynomial for $g(x) = f(x^3 - 1)$, and use it to find an approximation for $g(1)$, assuming the series converges if $x = 1$.

d. Use the second-derivative test to determine whether $g(0)$ in part c is a local maximum or a local minimum.

e. Find a seventh-degree Taylor polynomial for $h(x) = \int_0^x g(t)\,dt$.

26. Let f be a function with derivatives of all orders and with values that are given approximately by the fourth-degree Taylor polynomial

$$f(x) \approx P_4(x) = -4 + 3(x - 2) + 0.5(x - 2)^2$$
$$- 0.09(x - 2)^3 - 0.06(x - 2)^4$$

a. Find the approximate value of $f(1)$. What assumption must you make about 1 for this approximation to be valid?

b. Use the pattern for equating derivatives to find $f(2), f'(2), f''(2), f'''(2)$, and $f^{(4)}(2)$.

c. Find a fourth-degree Taylor polynomial for $g(x) = f(x^2 + 2)$, and use it to find an approximation for $g(1)$, assuming the series converges if $x = 1$.

d. Use the second-derivative test to determine whether $g(0)$ in part c is a local maximum or a local minimum.

e. Find a fifth-degree Taylor polynomial for $h(x) = \int_0^x g(t)\,dt$.

For Problems 27-32, expand the function as a Taylor series about the given value of x. Write enough terms to reveal clearly that you have seen the pattern.

27. $f(x) = \sin x$, about $x = \pi/4$

28. $f(x) = \cos x$, about $x = \pi/4$

29. $f(x) = \ln x$, about $x = 1$

30. $f(x) = \log x$, about $x = 10$

31. $f(x) = (x - 5)^{7/3}$, about $x = 4$

32. $f(x) = (x + 6)^{4.2}$, about $x = -5$

33. Find the Maclaurin series for $\cos 3x$ by equating derivatives. Compare the answer, and the ease of finding the answer, with the series you obtain by substituting $3x$ for x in the $\cos(x)$ series.

34. Find the Maclaurin series for $\ln(1 + x)$ by equating derivatives. Compare the answer, and the ease of finding the answer, with the series you obtain by substituting $(1 + x)$ for x in the Taylor series for $\ln x$ expanded about $x = 1$.

35. *Accuracy for ln x Series Value:* Estimate $\ln 1.5$ using $S_4(1.5)$, fourth partial sum of the Taylor series. How close is your answer to the exact answer? How does the error in the series value compare with the first term of the tail of the series, t_5, which is the first term left out in the partial sum?

36. *Accuracy Interval for ln x Series:* Find the interval of values of x for which the fourth partial sum of the Taylor series for $\ln x$ gives values that are within 0.0001 unit of $\ln x$.

37. *Inverse Tangent Series and an Approximation for π:* Recall that $\tan(\pi/4) = 1$. Thus, $\tan^{-1} 1 = \pi/4$. In this problem you will use the inverse tangent series to estimate π.

a. Write the first few terms of the Maclaurin series for $\tan^{-1} 1$. Then use the appropriate features of your grapher to find the 10th partial sum of this series. Multiply by 4 to find an approximate value of π. How close does this approximation come to π?

b. Find another approximation for π using the 50th partial sum of the series in part a. Is this approximation much better than the one using the 10th partial sum?

c. By appropriate trigonometry, show that

$$\tan^{-1} 1 = \tan^{-1} \tfrac{1}{2} + \tan^{-1} \tfrac{1}{3}$$

Use the result to write $\pi/4$ as a sum of two Maclaurin series. Estimate the value of π by adding the 10th partial sums of the two series. How much better is this method for estimating π than the methods of parts a and b?

38. *Tangent Series Problem:* Recall that $\tan x = (\sin x)/(\cos x)$. Use long division to divide the Maclaurin series for $\sin x$ by the Maclaurin series for $\cos x$ to get a power series for $\tan x$. Use enough terms of both the sine and cosine series to find four terms of the tangent series. Show by calculator that the fourth partial sum for $\tan 0.2$ is close to $\tan 0.2$.

39. *Taylor Series Proof Problem:* Prove algebraically that for all positive integers n, the nth derivative of the general Taylor series equals $f^{(n)}(a)$.

40. *Historical Problem:* What were Taylor's and Maclaurin's first names? When did they live in relation to Newton and Leibniz, the inventors of calculus?

41. *Ratio of Terms Problem:* A Taylor series usually gives better and better approximations for values of a function the more terms you use. For some series this is true only for certain values of x. For instance, the series for the natural logarithm,

$$\ln x = \sum_{n=1}^{\infty} (-1)^{n+1} \frac{1}{n} (x - 1)^n$$

converges to $\ln x$ only for $0 < x \le 2$. If x is outside this **interval of convergence**, the series does not converge to a real number. It diverges

and thus cannot represent $\ln x$. In this problem you will investigate the ratio of a term in this series to the term before it. You will also try to discover a way to find, from this ratio, whether the series converges.

a. The formula for t_n, the nth term in the series for $\ln x$, is

$$t_n = (-1)^{n+1} \frac{1}{n} (x - 1)^n$$

Let r_n be the ratio $|t_{n+1}/t_n|$. Find a formula for r_n in terms of x and n.

b. Calculate r_{10} for $x = 1.2$, $x = 1.95$, and $x = 3$.

c. Let r be the limit of r_n as n approaches infinity. Find an equation for r in terms of x.

d. Evaluate r for $x = -0.1, x = 0, x = 0.9$, $x = 1.9, x = 2$, and $x = 2.1$.

e. Make a conjecture: "The series converges to $\ln x$ whenever the value of x makes r —?— and diverges whenever the value of x makes r —?—."

f. If your conjecture is correct, you can use it to show that the series converges if x is in the interval $0 < x < 2$. Check your conjecture by showing that it gives this interval.

42. *Journal Problem:* Update your journal with what you've learned since the last entry. Include such things as

- The one most important thing you've learned since the last journal entry
- The difference between a sequence and a series
- The distinction between term index and term number
- The definition of geometric series
- The meaning of power series and for what purpose it may be useful
- What it means for a series to converge and to diverge
- Anything about series that is still unclear

12-6 Interval of Convergence for a Series—The Ratio Technique

A series converges to a certain number if the limit of the nth partial sum as n approaches infinity is that number. Power series often converge if x is within 1 unit of a, the constant about which the series is expanded. For instance, the series

$$\ln x = (x - 1) - \tfrac{1}{2}(x - 1)^2 + \tfrac{1}{3}(x - 1)^3 - \tfrac{1}{4}(x - 1)^4 + \cdots$$

$$= \sum_{n=1}^{\infty}(-1)^{n+1}\tfrac{1}{n}(x - 1)^n$$

converges when $x = 1.6$. The quantity $(x - 1)$ equals 0.6, and the powers 0.6^n approach zero rapidly as n gets large. But if $x = 4$, the quantity $(x - 1)$ is 3, and the powers 3^n become infinitely large as n approaches infinity. You can see what happens from a table of values.

n	nth term, $x = 1.6$	nth term, $x = 4$
1	0.6	3
2	−0.18	−4.5
3	0.072	9
4	−0.0324	−20.25
5	0.015552	48.6
6	−0.007776	−121.5
7	0.00399908...	312.428...
⋮	⋮	⋮
20	−0.00000182...	−174339220.05

Figure 12-6a shows what happens to the partial sums of the natural logarithm series. The graph on the left shows that the partial sums for $x = 1.6$ converge rapidly to a number around 0.5 as n approaches infinity. The graph on the right shows that the partial sums for $x = 4$ diverge.

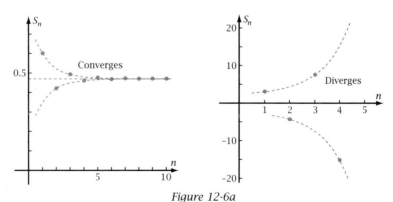

Figure 12-6a

Surprisingly, the series for $\sin x$,

$$\sin x = x - \frac{1}{3!}x^3 + \frac{1}{5!}x^5 - \frac{1}{7!}x^7 + \cdots = \sum_{n=0}^{\infty}(-1)^n\frac{1}{(2n + 1)!}x^{2n+1}$$

converges no matter how large x is! At $x = 10$, for instance, the power 10^{2n+1} is very large. But the denominator, $(2n + 1)!$, is much larger. If $n = 20$, then $(2n + 1)! = 41! = 3.3... \times 10^{49}$, which is 300 million times as large as 10^{41}. In this section you will develop a method called the **ratio technique** (sometimes called the **ratio test**) for finding algebraically the **open interval of convergence**—that is, the interval of x-values for which a power series converges.

OBJECTIVE   Given a power series in x, use the ratio technique to find the open interval of convergence.

The ratio technique is based on bounding the given series with a convergent geometric series. To see how the technique works, consider the series for $\ln x$ when $x = 1.6$.

$$\ln 1.6 = (1.6 - 1) - \tfrac{1}{2}(1.6 - 1)^2 + \tfrac{1}{3}(1.6 - 1)^3 - \tfrac{1}{4}(1.6 - 1)^4 + \cdots$$
$$= 0.6 - \tfrac{1}{2}(0.6)^2 + \tfrac{1}{3}(0.6)^3 - \tfrac{1}{4}(0.6)^4 + \cdots$$

If you take the ratios of the absolute values of adjacent terms, $|t_{n+1}/t_n|$, you get the sequence

$$\tfrac{1}{2}(0.6), \ \tfrac{2}{3}(0.6), \ \tfrac{3}{4}(0.6), \ \tfrac{4}{5}(0.6), \ \tfrac{5}{6}(0.6), \ \tfrac{6}{7}(0.6), \ \ \tfrac{7}{8}(0.6), \ \tfrac{8}{9}(0.6), \ldots$$
$$= 0.3, \quad 0.4, \quad 0.45, \quad 0.48, \quad 0.5, \quad 0.514..., \ 0.525, \ 0.5333..., \ldots$$

A given term in the $\ln 1.6$ series is formed by multiplying the preceding term by the appropriate one of these ratios. So each term is less than 0.6 times the preceding term.

Compare $|t_n|$, the absolute values of the terms in the tail of the $\ln 1.6$ series, with the terms g_n in a geometric series that is known to converge. For instance, starting at $|t_4| = 0.25(0.6)^4 = 0.0324$, and using a geometric series with common ratio between 0.6 and 1, such as $r = 0.7$, you find

n:	4	5	6	7	$\cdots$
$|t_n|$:	0.0324	0.015552	0.007776	0.0039990...	$\cdots$
g_n:	0.0324	0.02268	0.015876	0.011132	$\cdots$

Figure 12-6b graphically shows that $|t_n| < g_n$ for all $n > 4$. This is true because $|t_4| = g_4$ and $|t_n|$ decreases faster than g_n for $n > 4$.

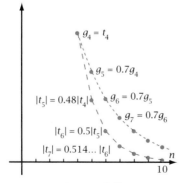

Figure 12-6b

The geometric series converges to

$$S = 0.0324 \cdot \frac{1}{1 - 0.7} = 0.108$$

So the sum of the tail of the ln 1.6 series is bounded above by 0.108. Similarly, the tail is bounded below by -0.108. Because the terms of the ln 1.6 series alternate and approach 0 as n increases, you can bound the tail of the series by numbers arbitrarily close to zero. Because the series of absolute values of terms converges, the series is said to **converge absolutely**. If a series converges absolutely, then it converges even if some of the terms are negative.

In general, a power series will converge if you can keep the ratio of the absolute values of adjacent terms less than some number R, and R is less than 1. In that case you can always find a geometric series with common ratio between R and 1 that converges and is an upper bound for the tail of the series.

One way to show that there is such a number R is to take the limit, L, of the ratios of adjacent terms. As shown in Figure 12-6c, if $L < 1$, then you can pick an epsilon small enough so that $R = L + \epsilon$ is also less than 1. Then any geometric series with common ratio r between R and 1, and with a suitable first term, will be an upper bound for the tail of the given series.

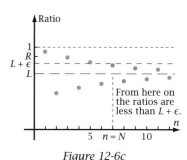

Figure 12-6c

You can use this fact as a relatively simple way to find the interval of values of x for which a series converges absolutely. Example 1 shows you how this technique is used with the series for $\ln x$.

▶ **EXAMPLE 1**   Find the interval of convergence for $\ln x = \sum\limits_{n=1}^{\infty} (-1)^{n+1} \frac{1}{n} (x - 1)^n$.

Solution

$$L = \lim_{n \to \infty} \left| \frac{t_{n+1}}{t_n} \right| = \lim_{n \to \infty} \left| \frac{(-1)^{n+2}(x-1)^{n+1}}{(n+1)} \cdot \frac{n}{(-1)^{n+1}(x-1)^n} \right|$$

$$= \lim_{n \to \infty} \left| \frac{(x-1)(n)}{n+1} \right| \qquad \text{Notice that } |-1|^{n+1} \text{ and } |-1|^{n+2} \text{ both equal 1.}$$

$$= |x - 1| \lim_{n \to \infty} \frac{n}{n+1} \qquad \text{Because } |x - 1| \text{ is independent of } n.$$

$$= |x - 1| \tfrac{1}{1} \qquad \text{By l'Hospital's rule, first embedding } t_n \text{ in a continuous function.}$$

$$= |x - 1|$$

So the series will converge if

$$|x - 1| < 1 \Rightarrow -1 < x - 1 < 1 \Rightarrow 0 < x < 2 \quad \text{Open interval of convergence.} \quad ◀$$

Note that the ratio technique also allows you to conclude that if $L > 1$, then the series diverges. You can bound the absolute values of the terms below by a divergent geometric series. If $L = 1$, as it will be at the endpoints of the interval of convergence, the ratio technique does not indicate whether the series converges or diverges. So the technique finds only the *open* interval of convergence. In the next section you will learn other techniques for proving convergence or divergence at the endpoints. Here is a statement of the ratio technique.

TECHNIQUE: The Ratio Technique for Convergence of Series

For the series $\sum\limits_{n=1}^{\infty} t_n$, if $L = \lim\limits_{n \to \infty} \left| \dfrac{t_{n+1}}{t_n} \right|$, then

i. The series converges absolutely if $L < 1$.

ii. The series diverges if $L > 1$.

iii. The series may either converge or diverge if $L = 1$.

The interval of convergence in Example 1, $0 < x < 2$, goes ± 1 unit on either side of $x = 1$, the value of x about which the series is expanded. The half-width of the interval of convergence is called the **radius of convergence**. The word *radius* is used because if x is allowed to be a complex number, the series turns out to converge for all x inside a circle of that radius (Figure 12-6d).

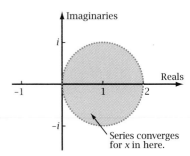

Figure 12-6d

▶ **EXAMPLE 2**   For the series $\sum\limits_{n=1}^{\infty} \dfrac{n}{3^n}(x - 5)^n$

    a. Write the first few terms.

    b. Find the interval of convergence.

    c. Find the radius of convergence.

Solution   a. $\dfrac{1}{3}(x - 5) + \dfrac{2}{9}(x - 5)^2 + \dfrac{3}{27}(x - 5)^3 + \dfrac{4}{81}(x - 5)^4 + \cdots$

    b.
$$L = \lim_{n \to \infty} \left| \frac{t_{n+1}}{t_n} \right|$$

$$= \lim_{n \to \infty} \left| \frac{(n + 1)(x - 5)^{n+1}}{3^{n+1}} \cdot \frac{3^n}{n(x - 5)^n} \right|$$

$$= \lim_{n \to \infty} \left| \frac{n + 1}{n} \cdot \frac{1}{3}(x - 5) \right|$$

$$= |x - 5| \lim_{n \to \infty} \left| \frac{n + 1}{3n} \right|$$

$$= |x - 5|\tfrac{1}{3}$$

The series will converge if $L < 1$, so make $L < 1$.

$$|x - 5|\tfrac{1}{3} < 1 \Rightarrow |x - 5| < 3 \Rightarrow -3 < x - 5 < 3 \Rightarrow 2 < x < 8$$

c. The radius of convergence is the distance from the midpoint of the interval of convergence to one of its endpoints.

Radius of convergence is 3. ◀

If x equals the number at an endpoint of the interval of convergence, the limit of the ratio of terms equals 1. The series may or may not converge in this case. In Section 12-7, you will learn tests for convergence that can be used when the ratio technique doesn't work.

Example 3 shows you that the radius of convergence of a series is zero if the limit of the ratio of terms is infinite.

▶ **EXAMPLE 3** For the series $\sum_{n=1}^{\infty} \frac{n!}{n^4}(x-3)^n$

a. Write the first few terms.

b. Show that though the first few terms decrease in value, the radius of convergence is zero.

c. For what one value of x does the series converge?

Solution a. $(x-3) + \frac{1}{8}(x-3)^2 + \frac{2}{27}(x-3)^3 + \frac{3}{32}(x-3)^4 + \cdots$

b. Note that the factorials simplify nicely when you divide adjacent terms.

$$L = \lim_{n \to \infty} \left| \frac{t_{n+1}}{t_n} \right| = \lim_{n \to \infty} \left| \frac{(n+1)!(x-3)^{n+1}}{(n+1)^4} \cdot \frac{n^4}{n!(x-3)^n} \right|$$

$$= \lim_{n \to \infty} \left| \frac{(n+1)(n!)}{n!} \cdot \left(\frac{n}{n+1} \right)^4 (x-3) \right|$$

$$= |x-3| \lim_{n \to \infty} \left| (n+1) \left(\frac{n}{n+1} \right)^4 \right|$$

Because $n/(n+1)$ goes to 1 as n approaches infinity, its fourth power also goes to 1. So the quantity inside the absolute value sign approaches the other $(n+1)$, and L is infinite for all values of x not equal to 3. The radius of convergence is thus equal to zero, Q.E.D.

c. If $x = 3$, the series becomes $0 + 0 + 0 + \cdots$, which converges to zero. So 3 is the only value of x for which the series converges. ◀

Here are the two special cases from Example 3.

SPECIAL CASES: *Zero and Infinite Radius of Convergence*

For a power series in $(x - a)^n$ with radius of convergence r

If $r = 0$, the series converges only at $x = a$.

If r is infinite, the series converges for all values of x.

Problem Set 12-6

Q1. $x - x^3/3! + x^5/5! - x^7/7! + \cdots = -?-$

Q2. $x + x^3/3! + x^5/5! + x^7/7! + \cdots = -?-$

Q3. $1 - x + x^2/2! - x^3/3! + x^4/4! - \cdots = -?-$

Q4. $1 + x + x^2/2! + x^3/3! + x^4/4! + \cdots = -?-$

Q5. $1 + x + x^2 + x^3 + x^4 + x^5 + \cdots = -?-$

Q6. Integrate: $\int \cos 2x\, dx$

Q7. Differentiate: $f(x) = \tan 3x$

Q8. Find the limit as x approaches zero of $f(x) = \cos 4x$.

Q9. Find the limit as x approaches zero of $g(x) = (1 + x)^{1/x}$.

Q10. (Force)(displacement) = $-?-$

    A. Mass   B. Work   C. Moment

    D. Total force   E. Velocity

For Problems 1–6,

    a. Write the first few terms.

    b. Find the open interval of convergence.

    c. Find the radius of convergence.

1. $\sum\limits_{n=1}^{\infty} \dfrac{n}{4^n} x^n$

2. $\sum\limits_{n=1}^{\infty} \dfrac{x^n}{n \cdot 2^n}$

3. $\sum\limits_{n=1}^{\infty} \dfrac{(2x + 3)^n}{n}$

4. $\sum\limits_{n=1}^{\infty} \dfrac{(5x - 7)^n}{2n}$

5. $\sum\limits_{n=1}^{\infty} \dfrac{n^3}{n!} (x - 8)^n$

6. $\sum\limits_{n=1}^{\infty} \dfrac{n!}{n^4} (x + 2)^n$

For Problems 7–12, show that these familiar series for the transcendental functions converge for all real values of x.

7. $\sin x = x - \frac{1}{3!}x^3 + \frac{1}{5!}x^5 - \frac{1}{7!}x^7 + \cdots$

8. $\cos x = 1 - \frac{1}{2!}x^2 + \frac{1}{4!}x^4 - \frac{1}{6!}x^6 + \cdots$

9. $\sinh x = x + \frac{1}{3!}x^3 + \frac{1}{5!}x^5 + \frac{1}{7!}x^7 + \cdots$

10. $\cosh x = 1 + \frac{1}{2!}x^2 + \frac{1}{4!}x^4 + \frac{1}{6!}x^6 + \cdots$

11. $e^x = 1 + x + \frac{1}{2!}x^2 + \frac{1}{3!}x^3 + \frac{1}{4!}x^4 + \cdots$

12. $e^{-x} = 1 - x + \frac{1}{2!}x^2 - \frac{1}{3!}x^3 + \frac{1}{4!}x^4 - \cdots$

13. Show that the series $0! + 1!x + 2!x^2 + 3!x^3 + \cdots$ converges only for the trivial case, $x = 0$.

14. Mae writes the first few terms of the series

$$\sum_{n=0}^{\infty} \frac{n!}{100^n} x^n$$

$$= 1 + 0.01x + 0.0002x^2 + 0.000006x^3 + \cdots$$

She figures that because the coefficients are getting small so fast, the series is bound to converge, at least if she picks a value of x such as 0.7, which is less than 1. Show Mae that she is wrong and that the series converges only for the trivial case, $x = 0$.

15. Amos evaluates the Maclaurin series for $\cosh 10$ and gets

$$\cosh 10 = \sum_{n=0}^{\infty} \frac{1}{(2n)!} \cdot 10^{2n}$$

$$= 1 + 50 + 416.666\ldots$$

$$+ 1388.888\ldots + \cdots$$

He figures that because the terms are increasing so fast, the series could not possibly converge. Explain Amos's mistake by showing him that the series does actually converge, even though the terms increase for a while.

16. For the Taylor series for $\ln 0.1$ expanded about $x = 1$, construct a table of values showing the term index, n; the term value, t_n; and the absolute value of the ratio of terms, $|t_{n+1}/t_n|$. Make a conjecture about what number the ratio seems to be approaching as n approaches infinity. By taking the limit of the ratio, show that your conjecture is correct, or change the conjecture.

17. *Inverse Tangent Series Problem:* The series

$$P(x) = x - \frac{x^3}{3} + \frac{x^5}{5} - \frac{x^7}{7} + \cdots$$

converges to $\tan^{-1} x$ for certain values of x.

    a. Find the open interval of convergence of the series.

    b. On the same screen, plot the graphs of $\tan^{-1} x$ and the fourth and fifth partial sums of the series. How do the graphs confirm what you found algebraically in part a?

c. Evaluate the fourth partial sum of the series for $x = 0.1$.

d. Find the value of the tail of the series after the fourth partial sum by comparing your answer to part c with the value of $\tan^{-1} 0.1$ that you obtain with your calculator.

e. Show that the remainder of the series in part d is less in magnitude than the absolute value of the first term of the tail of the series.

18. *Volume Problem:* Figure 12-6e shows the solid generated by rotating about the y-axis the region under the graph of $y = x^2 \sin 2x$ from $x = 0$ to $x = 1.5$.

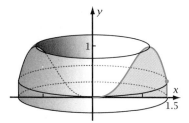

Figure 12-6e

a. Find the volume of this solid algebraically by integrating by parts. Write all the decimal places your grapher displays. Does numerical integration give precisely the same answer?

b. Find the volume again by writing the integrand as a Maclaurin series, integrating, and evaluating the fifth partial sum. Be sure to show that 1.5 is within the interval of convergence. How does the answer compare with the answer found algebraically in part a?

c. The integrated series in part b is an alternating series with terms that decrease in value and approach zero as n approaches infinity. Thus, the remainder of the series after a given partial sum is no larger in magnitude than the absolute value of the first term of the tail following that partial sum. How many terms of the series would you need to use to estimate the volume correct to ten decimal places?

19. *The Error Function:* Figure 12-6f shows the graph of

$$y = \frac{2}{\sqrt{\pi}} e^{-t^2}$$

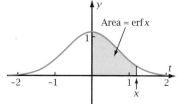

Figure 12-6f

This function is related to the **normal distribution curve**, sometimes used to "curve" grades. The area of the region under the graph from $t = 0$ to $t = x$ is called the **error function** of x, written erf x. That is,

$$\text{erf } x = \frac{2}{\sqrt{\pi}} \int_0^x e^{-t^2} dt$$

The limit of the area as x approaches infinity is equal to 1, so erf x is the fraction of the area that lies between 0 and x. The quantity erf $(x/\sqrt{2})$ is equal to the fraction of a normally-distributed population that lies within $\pm x$ standard deviations of the mean of the distribution.

You cannot use the fundamental theorem to evaluate erf x because e^{-t^2} is not the derivative of an elementary function. But you can use a power series. Let

$$f(x) = \int_0^x e^{-t^2} dt$$

a. Find the Maclaurin series for $f(x)$. Write enough terms to show the pattern clearly.

b. On the same screen, plot the sixth partial sum of the series for $f(x)$ and the value of $f(x)$ by numerical integration. Use a window with an x-range of $[-3, 3]$. For what interval of x-values does the partial-sums graph fit the numerical-integration graph reasonably well?

c. Does the series in part a converge for all values of x? Justify your answer.

d. Does erf x really seem to approach 1 as x approaches infinity? Explain.

20. *The Sine-Integral Function:* The function

$$f(x) = \int_0^x \frac{\sin t}{t}\, dt$$

is called the **sine-integral function of x,** abbreviated Si x. Because the antiderivative of $(\sin t)/t$ is not an elementary transcendental function, you cannot find values of Si x directly using the fundamental theorem. Power series give a way to do this.

a. Write a power series for the integrand by a time-efficient method. Integrate the series to find a power series for Si x.

b. Is the radius of convergence for the Si x series the same as that for the integrand series?

c. Find the third partial sum of the series for Si 0.6. How does this value compare with the value you get by numerical integration?

d. Plot the graph of Si x by numerical integration. On the same screen, plot the graph of the tenth partial sum of the series for Si x. Use a window with an x-range of $[-12, 12]$. For what interval do the partial sums seem to fit the numerical integration values reasonably well?

21. *The Root Technique:* You can show that a series of positive terms converges if *the nth root of the nth term approaches a constant less than 1 for its limit as n approaches infinity.* Figure 12-6g shows such a series. Let

$$L = \lim_{n\to\infty} \sqrt[n]{t_n}, \text{ where } L < 1$$

a. Use the definition of limit to show that for any number $\epsilon > 0$ there is a number k such that if $n > k$ then $\sqrt[n]{t_n} < L + \epsilon$.

b. Show that you can make ϵ small enough so that $L + \epsilon$ is also less than 1.

c. Show that for all integers $n > k$, $t_n < (L + \epsilon)^{n-k}$.

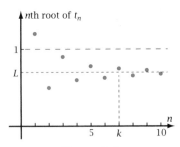

Figure 12-6g

d. Show that the tail of the series after t_n is bounded above by a convergent geometric series.

e. Explain how the reasoning in parts a–d verifies that the series converges.

The result of Problem 21 is called the **root technique,** or **root test,** and is stated here.

Technique: The Root Technique for Convergence of Series

For the series $\sum_{n=1}^{\infty} t_n$, if $L = \lim_{n\to\infty} \sqrt[n]{|t_n|}$ then

i. The series is *absolutely convergent* if $L < 1$.

ii. The series is *absolutely divergent* if $L > 1$.

iii. The series may either converge or diverge if $L = 1$.

22. *A Special Limit Problem:* To use the root technique, it helps to know the limit of the nth root of n. Let

$$L = \lim_{n\to\infty} \sqrt[n]{n}$$

Prove that $L = 1$. (Try taking $\ln L$, finding its limit, then raising e to that power to get L.)

Property: Limit of the nth Root of n

$$\lim_{n\to\infty} \sqrt[n]{n} = 1$$

23. Use the root technique to show that the open interval of convergence of the Taylor series for $\ln x$ is $0 < x < 2$.

24. Use the root technique to show that $\sum_{n=1}^{\infty} \frac{1}{n^n} x^n$ converges for all values of x.

25. Use the root technique to show that $\sum_{n=1}^{\infty} n^n x^n$ converges only for $x = 0$.

26. *"Which One Wins?" Problem:* In Problems 13 and 24, you showed that $\sum_{n=1}^{\infty} n! x^n$ converges for no values of x except $x = 0$ and that $\sum_{n=1}^{\infty} \frac{1}{n^n} x^n$ converges for all values of x. For what values of x does $\sum_{n=1}^{\infty} \frac{n!}{n^n} x^n$ converge?

12-7 Convergence of Series at the Ends of the Convergence Interval

In Section 12-6, you learned the ratio technique for finding the interval of x-values for which a power series converges. Because the limit of the ratios of term values is 1 or -1 at the endpoints of the interval, other techniques are needed to test for convergence there.

OBJECTIVE Given a series of constants for which the ratio technique is inconclusive, prove either that the series converges or that it diverges.

To accomplish this objective, it is helpful for you to consolidate your knowledge about the tail and the remainder of a series.

DEFINITIONS: Tail and Remainder of a Series

The **tail** of a series is the indicated sum of the terms remaining in the series beyond the end of a particular partial sum.

Example:

$$2 + 3 + 5 + 7 + 11 + 13 + 17 + 19 + 23 + \cdots \qquad \text{The series of primes.}$$
$$\text{4th partial sum} \mid \text{———— Tail ————} \longrightarrow$$

The **remainder** of a series, R_n, is the value of the tail after partial sum S_n, provided the tail converges.

Examples:

For $1 + \frac{1}{2} + \frac{1}{4} + \frac{1}{8} + \frac{1}{16} + \cdots$, $R_5 = \frac{1}{16}$ because $S_5 = 1\frac{15}{16}$, the series converges to 2, and $2 - 1\frac{15}{16} = \frac{1}{16}$.

For the series of primes, R_4 is infinite because the series diverges.

See the Definitions box in Section 12-2 for other vocabulary relating to series.

Convergence of Sequences

There is one major property of sequences that leads to several methods of testing for convergence. You might think at first that a sequence converges if there is an upper bound for the terms of the sequence. Not true! The sequence

$$2, 3, 2, 3, 2, 3, 2, 3, \ldots$$

is bounded above by 3 and does not converge. It diverges by oscillation.

However, a sequence such as

$$\frac{1}{2}, \frac{2}{3}, \frac{3}{4}, \frac{4}{5}, \frac{5}{6}, \ldots$$

does converge because the terms are *strictly increasing* as well as being bounded above (Figure 12-7a). The number 1 is an upper bound for the terms because the numerators are always less than the denominators. Term $t_n = n/(n+1)$ can be made arbitrarily close to 1 by picking a large enough value of n. Beyond that value of n, the terms are even closer to 1 because they are strictly increasing. So 1 is the limit of t_n as n approaches infinity.

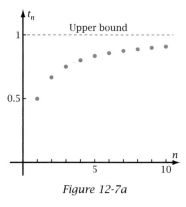

Figure 12-7a

In Problem 48 of Problem Set 12-7, you will prove that a sequence converges if its terms are bounded above and are strictly increasing.

PROPERTY: *Convergence of Sequences*

If a sequence $\{t_1, t_2, t_3, \ldots, t_n, \ldots\}$ is increasing and bounded above, then the sequence converges.

Convergence of a Series of Positive Terms—The Integral Test

If a series has all positive terms, then its partial sums are increasing, even though the terms themselves may be decreasing. The ***p*-series** with $p = 3$,

$$\sum_{n=1}^{\infty} \frac{1}{n^3} = 1 + \frac{1}{8} + \frac{1}{27} + \frac{1}{64} + \frac{1}{125} + \frac{1}{216} + \frac{1}{343} + \cdots$$

called a *p*-series because the denominators are *powers*, has partial sums

$$S_1 = 1, \ S_2 = 1.125, \ S_3 = 1.1620\ldots, \ S_4 = 1.1776\ldots, \ \ldots$$

that are increasing. The remainder, R_4, after S_4 is the value of the tail of the series,

$$R_4 = \sum_{n=5}^{\infty} \frac{1}{n^3} = \frac{1}{125} + \frac{1}{126} + \frac{1}{343} + \cdots$$

Because the partial sums are known to be increasing, all you need do to prove convergence is show that R_4 is bounded above.

The left side of Figure 12-7b shows that R_4 is bounded above by the area of the region under the graph of $f(x) = 1/x^3$, from $x = 4$ to infinity. This is true because the terms of the tail starting at $n = 5$ are **embedded** in this continuous function, the rectangles have areas equal to the term values of the tail, and function f is *decreasing,* so that the rectangles are *inscribed* in the region under the graph. So the sum of the terms from 5 to infinity is bounded above by an improper integral from 4 to infinity.

$$R_4 < \int_4^\infty \frac{1}{x^3}\, dx = \lim_{b \to \infty}\left(-\frac{1}{2}x^{-2}\Big|_4^b\right) = \lim_{b \to \infty}\left(-\frac{1}{2}b^{-2} + \frac{1}{2}\cdot 4^{-2}\right) = \frac{1}{32}$$

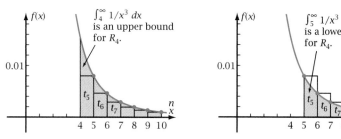

Figure 12-7b

This *p*-series converges because it is increasing and bounded above by $S_4 + 1/32$. The technique used is called the **integral test for convergence**, because you found the upper bound by comparing the tail with an improper integral.

The right side of Figure 12-7b shows that you can draw the rectangles to the other side of the *n*-values, making them *circumscribed* about the region under $f(x) = 1/x^3$, from $x = 5$ to infinity. So the sum of the terms from 5 to infinity is bounded *below* by an improper integral from 5 to infinity. Evaluating the other integral gives

$$R_4 > \int_5^\infty \frac{1}{x^3}\, dx = \frac{1}{2}\cdot 5^{-2} = \frac{1}{50} = 0.02$$

By itself, the fact that $R_4 > 0.02$ indicates nothing about convergence of the series. Combined with $R_4 < 1/32$, however, you can conclude that the remainder is between $1/50$ and $1/32$.

Not all *p*-series converge. For instance, if $p = 0.6$,

$$\sum_{n=1}^\infty \frac{1}{n^{0.6}} = \frac{1}{1^{0.6}} + \frac{1}{2^{0.6}} + \frac{1}{3^{0.6}} + \frac{1}{4^{0.6}} + \frac{1}{5^{0.6}} + \frac{1}{6^{0.6}} + \cdots$$

Figure 12-7c shows R_4 for this series, bounded below by an improper integral from 5 to infinity.

$$R_4 > \int_5^\infty \frac{1}{x^{0.6}}\, dx = \lim_{b \to \infty}\left(2.5x^{0.4}\Big|_5^b\right) = \lim_{b \to \infty}(2.5b^{0.4} - 4.759\ldots) = \infty$$

The remainder is infinite, so the series diverges.

Analysis of these two *p*-series shows that to determine convergence of a series of positive terms that decrease toward zero, you can evaluate the improper integral of the function in which you can embed the tail. If the integral converges, the series converges, and vice versa.

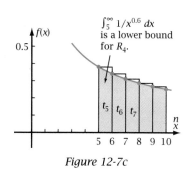

Figure 12-7c

Given $S = \sum_{n=1}^{\infty} f(n)$ and $I = \int_a^{\infty} f(x)\,dx$ where $f(x)$ decreases monotonically to 0 for all $x \geq a$, then S converges if I converges, and S diverges if I diverges.

Note that you can use the integral test without actually embedding the terms in the function being integrated. If $t_n \leq f(n)$ for all $n \geq a$ and if the integral converges, then the series converges because the integral still forms an upper bound for the tail. Similarly, if $t_n \geq f(n)$ for all $n \geq a$ and if the integral diverges, then the series diverges.

A **harmonic series** is a series in which the terms are reciprocals of the terms of an **arithmetic series**. The simplest harmonic series is the p-series with $p = 1$,

$$1 + \frac{1}{2} + \frac{1}{3} + \frac{1}{4} + \frac{1}{5} + \frac{1}{6} + \cdots = \sum_{n=1}^{\infty} \frac{1}{n}$$

You can use the integral test to show quickly that this series diverges. The terms are decreasing for all $n \geq 1$, and

$$\int_1^{\infty} \frac{1}{x}\,dx = \lim_{b \to \infty} \left(\ln x \Big|_1^b \right) = \lim_{b \to \infty} (\ln b - \ln 1) = \infty$$

So the series diverges because the integral diverges.

In Problem 4 of Problem Set 12-7, you will prove that the p-series converges if and only if $p > 1$.

The p-series $\displaystyle\sum_{n=1}^{\infty} \frac{1}{n^p} = \frac{1}{1^p} + \frac{1}{2^p} + \frac{1}{3^p} + \frac{1}{4^p} + \cdots$

converges if $p > 1$ and diverges if $p \leq 1$.

Figure 12-7d gives graphical evidence that the p-series converges for larger values of p and diverges for smaller values of p. The graph on the left shows that for $p = 3$, the terms decrease fast enough for the series to converge. The graph in the middle shows that for $p = 0.6$, the terms decrease too slowly, and the series does not converge. The graph on the right shows that for $p = 1$, the series decreases faster, but still not fast enough to converge.

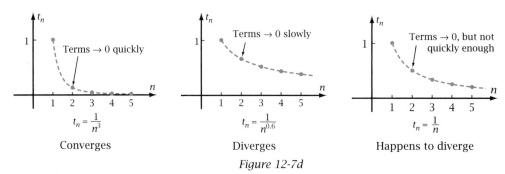

Figure 12-7d

Convergence of a Series of Positive Terms—The Comparison Test

To derive the ratio technique in Section 12-6, you found an upper bound for the given series by comparing it with a geometric series that was known to converge. This method of finding the upper bound is called the **comparison test for convergence**, which is particularly useful if you cannot integrate the embedding function. For instance, consider the **factorial reciprocal series**,

$$\sum_{n=0}^{\infty} \frac{1}{n!} = \frac{1}{0!} + \frac{1}{1!} + \frac{1}{2!} + \frac{1}{3!} + \frac{1}{4!} + \frac{1}{5!} + \frac{1}{6!} + \cdots$$

$$= 1 + 1 + \frac{1}{2} + \frac{1}{6} + \frac{1}{24} + \frac{1}{120} + \frac{1}{720} + \cdots$$

You cannot integrate the function $f(x) = 1/x!$ by techniques you know. Instead, compare this series with the convergent geometric series beginning with 1 and having common ratio $1/2$.

n:	0	1	2	3	4	5	6	
Factorial:	1	$+ 1$	$+ \frac{1}{2}$	$+ \frac{1}{6}$	$+ \frac{1}{24}$	$+ \frac{1}{120}$	$+ \frac{1}{720}$	$+ \cdots$
Geometric:	1	$+ \frac{1}{2}$	$+ \frac{1}{4}$	$+ \frac{1}{8}$	$+ \frac{1}{16}$	$+ \frac{1}{32}$	$+ \frac{1}{64}$	$+ \cdots$

$$|\text{——— Tail ———} \rightarrow$$

For $n \geq 4$, the geometric series terms are upper bounds for the factorial series terms. So R_3, the remainder starting at $n = 4$, is bounded above by

$$R_3 < \frac{1}{16} + \frac{1}{32} + \frac{1}{64} + \cdots = \frac{1}{16} \cdot \frac{1}{1 - 1/2} = \frac{1}{8}$$

Thus S, the limit of the factorial series, is bounded by

$$S = S_3 + R_3 < S_3 + \frac{1}{8} = 2\frac{19}{24} = 2.7916\ldots$$

This upper bound is reasonable because the given series is the Maclaurin series for e^x evaluated at $x = 1$ and $e^1 = 2.71828\ldots$. These steps prove that this Maclaurin series converges for $x = 1$.

Convergence of a Series of Positive Terms—The Limit Comparison Test

Two functions are said to be of the **same order** if their ratio approaches a positive finite number as x approaches infinity. For instance, suppose $f(x) = 5x^2 - 7x + 9$, $g(x) = x^2$, and $h(x) = e^x$. Taking limits of the ratios, you find by applications of l'Hospital's rule that

$$\lim_{x \to \infty} \frac{f(x)}{g(x)} = \lim_{x \to \infty} \frac{5x^2 - 7x + 9}{x^2} = \lim_{x \to \infty} \frac{10x - 7}{2x} = \lim_{x \to \infty} \frac{10}{2} = 5$$

$$\lim_{x \to \infty} \frac{g(x)}{h(x)} = \lim_{x \to \infty} \frac{x^2}{e^x} = \lim_{x \to \infty} \frac{2x}{e^x} = \lim_{x \to \infty} \frac{0}{e^x} = 0$$

$$\lim_{x \to \infty} \frac{h(x)}{g(x)} = \lim_{x \to \infty} \frac{e^x}{x^2} = \lim_{x \to \infty} \frac{e^x}{2x} = \lim_{x \to \infty} \frac{e^x}{0} = \infty$$

So functions f and g are of the same order because the ratio $f(x)/g(x)$ approaches the finite number 5. Function g is of a lower order than function h because the ratio $g(x)/h(x)$ approaches 0. Similarly, function h is of a higher order than function g because $h(x)/g(x)$ approaches infinity. This property provides a relatively simple way to test a series for convergence. You can find the limit of the ratio of the respective terms. If the functions that generate the terms are of the same order and one series converges, then so does the other one. This **limit comparison test** is summarized here. It is useful because you need consider only the terms of the series, not the partial sums.

PROPERTY: *The Limit Comparison Test for Convergence of a Series of Positive Terms*

Given that $f(n) > 0$ and $g(n) > 0$ for all $n \geq N$ (where N is a positive integer)

1. If $\displaystyle\lim_{n \to \infty} \frac{f(n)}{g(n)} = L$, and L is a real number (f and g are of the same order), then $\sum f(n)$ and $\sum g(n)$ either both converge or both diverge.

2. If $\displaystyle\lim_{n \to \infty} \frac{f(n)}{g(n)} = 0$ (f is a lower order than g) and $\sum g(n)$ converges, then $\sum f(n)$ converges also.

3. If $\displaystyle\lim_{n \to \infty} \frac{f(n)}{g(n)} = \infty$ (f is a higher order than g) and $\sum g(n)$ diverges, then $\sum f(n)$ diverges also.

Example 6 on page 631 shows how to apply the limit comparison test.

Convergence of a Series of Alternating-Sign Terms—Conditional Convergence

It is fairly easy to show graphically that a series with alternating + and − signs converges if the absolute value of the terms decreases and approaches zero for a limit. There are some surprises, however, as you will see soon.

Figure 12-7e shows partial sums of the alternating p-series

$$S = \sum_{n=1}^{\infty} \frac{(-1)^{n+1}}{n^{0.95}}$$

$$= 1 - \frac{1}{2^{0.95}} + \frac{1}{3^{0.95}} - \frac{1}{4^{0.95}} + \cdots$$

$$= 1 - 0.5176\ldots + 0.3521\ldots - 0.2679\ldots + \cdots$$

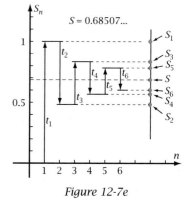

Figure 12-7e

The limit is $S = 0.68507\ldots$. Each term that is added causes the partial sums to overshoot this limit, first above then below. Because the absolute values of the terms decrease toward zero, the difference between the partial sum and the limit also approaches zero. In fact, the absolute value of this difference for a partial sum of n terms is no greater than the absolute value of the first term of the tail, $|t_{n+1}|$. For instance,

$$S_{1000} = 0.68436589..., \text{ a lower bound for } L$$

$$S_{1001} = 0.68577709..., \text{ an upper bound for } L$$

Because $S_{1001} = S_{1000} + t_{1001}$, the magnitude of the remainder, $|R_{1000}|$, is less than $|t_{1001}|$. You can find a good estimate for S by averaging the upper and lower bounds.

$$S \approx 0.5(0.68436589... + 0.68577709...) = 0.68507149...$$

So the series converges to $S \approx 0.6850...$ with an error of no more than $|t_{1001}| = 0.001411...$, or an accuracy of about three decimal places. Term t_n approaches zero as n approaches infinity, so you can make the error as small as you like by taking a sufficient number of terms.

The surprise comes if you commute, or rearrange, the terms and associate differently. For instance,

$$S = \left(1 - \frac{1}{2^{0.95}}\right) - \frac{1}{4^{0.95}} + \left(\frac{1}{3^{0.95}} - \frac{1}{6^{0.95}}\right) - \frac{1}{8^{0.95}} + \left(\frac{1}{5^{0.95}} - \frac{1}{10^{0.95}}\right) - \frac{1}{12^{0.95}} + \cdots$$

$$= 0.4823... - 0.2679... + 0.1698... - 0.1386... + 0.1045... - 0.0943... + \cdots$$

Each term of the series is used exactly once; no term is left out. The associated terms alternate in sign and decrease toward zero. As shown in Figure 12-7f, however, the partial sums approach a number S between 0.25 and 0.35, quite different from the 0.6850... shown in Figure 12-7e.

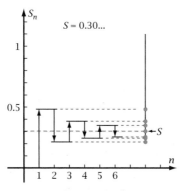

Figure 12-7f

The reason for this surprising result is that the series is really the sum of two divergent p-series,

$$\left(1 + \frac{1}{3^{0.95}} + \frac{1}{5^{0.95}} + \frac{1}{7^{0.95}} \cdots\right) + \left(-\frac{1}{2^{0.95}} - \frac{1}{4^{0.95}} - \frac{1}{6^{0.95}} + \cdots\right)$$

The first series diverges to $+\infty$ and the second one diverges to $-\infty$, so the series has the indeterminate form $\infty - \infty$. You can rearrange the series to converge to any positive or negative number or to diverge to $+\infty$ or to $-\infty$. Because of this behavior, the given series is said to be **conditionally convergent**. It converges under the *condition* that you don't rearrange the terms. This behavior would not happen if the series were *absolutely convergent*, such as the alternating *p*-series with $p = 3$,

$$\sum_{n=1}^{\infty} \frac{(-1)^{n+1}}{n^3} = 1 - \frac{1}{8} + \frac{1}{27} - \frac{1}{64} + \frac{1}{125} - \frac{1}{216} + \frac{1}{343} - \cdots$$

for which the series would still converge if you took the absolute value of each term.

Here is a summary of the definitions of conditional convergence and absolute convergence.

DEFINITIONS: Absolute and Conditional Convergence

The series $\sum_{n=1}^{\infty} t_n$ converges absolutely if $\sum_{n=1}^{\infty} |t_n|$ converges.

The series $\sum_{n=1}^{\infty} t_n$ converges conditionally if the series converges as written, but the series $\sum_{n=1}^{\infty} |t_n|$ diverges.

This box summarizes the tests for convergence in this section, presenting them in a suggested order for you to try.

PROPERTIES: Tests for Convergence of a Series of Constants

Try the tests in this order.

1. *nth term test:* If $\lim_{n \to \infty} t_n \neq 0$, then the series diverges.

2. *Alternating series test:* If (1) the signs are strictly alternating, (2) the term values are strictly decreasing in absolute value, and (3) $\lim_{n \to \infty} t_n = 0$, then the series converges. (S_n is within $|t_{n+1}|$ of the limit, S.) *Note:* To test for absolute or conditional convergence, use the definitions in the previous box.

3. *Geometric series test:* A geometric series with first term t_1 and common ratio r converges to $S = t_1 \cdot \dfrac{1}{1 - r}$ if and only if $|r| < 1$.

For series of positive terms

4. *p-series test:* If the series is a *p*-series, then it converges if $p > 1$, and it diverges if $p < 1$.

5. *Harmonic series test:* A harmonic series diverges. (A *p*-series with $p = 1$ is a special case of a harmonic series.)

6. *Integral test:* For $\sum f(n)$, find $\int_a^{\infty} f(x)\,dx$ for some positive constant a. If the integral converges, then the series converges. If the integral diverges, then the series diverges. Note that $\int_n^{\infty} f(x)\,dx < |R_n| < \int_{n+1}^{\infty} f(x)\,dx$.

(continued)

---

7. *Ratio test* (from the ratio technique): For $L = \lim_{n \to \infty} |t_{n+1}/t_n|$, if $L < 1$, the series is absolutely convergent; if $L > 1$, the series is divergent; and if $L = 1$, the test fails and the series may be either convergent or divergent.

8. *Direct comparison test:* If you can bound the tail of the series above by the terms of another series that is known to converge, then the series converges too. If you can bound the tail of the series below by another series that is known to diverge, then the series diverges too.

9. *Limit comparison test:* For two series with positive terms, find the limit $L = \lim_{n \to \infty} [f(n)/g(n)]$.

   - If L is a positive real number, then both series either converge or diverge.
   - If $L = 0$, and $\sum g(n)$ converges, then $\sum f(n)$ converges, too.
   - If $L = \infty$, and $\sum g(n)$ diverges, then $\sum f(n)$ diverges, too.

Note: You may apply these tests to the tail of the series because a series converges if and only if its tail converges.

The following examples show how to find the complete interval of convergence, including convergence at the endpoints. Examples 3-6 lead you through the necessary steps for ascertaining whether a series of constants converges, independent of whether these are at the endpoints of an interval of convergence.

▶ **EXAMPLE 1**   The open interval of convergence for the Taylor series for $\ln x$ expanded about $x = 1$ is $0 < x < 2$. Determine whether the series converges at $x = 0$ and at $x = 2$.

Solution   The series for $\ln x$ is $\ln x = (x - 1) - \frac{1}{2}(x - 1)^2 + \frac{1}{3}(x - 1)^3 - \frac{1}{4}(x - 1)^4 + \cdots$. At $x = 0$, the series becomes $-1 - \frac{1}{2} - \frac{1}{3} - \frac{1}{4} - \cdots$, which diverges because it is the opposite of the divergent harmonic series. At $x = 2$, the series becomes $1 - \frac{1}{2} + \frac{1}{3} - \frac{1}{4} + \cdots$, which converges because it meets the hypotheses of the alternating series test.

∴ the complete interval of convergence is $0 < x \leq 2$.   ◀

▶ **EXAMPLE 2**   Find the complete interval of convergence of the power series $\sum\limits_{n=0}^{\infty} \dfrac{2^n(x - 1)^n}{\ln(n + 2)}$.

Solution   By the ratio technique,

$$L = \lim_{n \to \infty} \left| \frac{2^{n+1}(x - 1)^{n+1}}{\ln(n + 3)} \cdot \frac{\ln(n + 2)}{2^n(x - 1)^n} \right|$$

$$= 2|x - 1| \lim_{n \to \infty} \left| \frac{\ln(n + 2)}{\ln(n + 3)} \right| \quad \to \quad \frac{\infty}{\infty} \qquad \text{l'Hospital's rule applies.}$$

$$= 2|x - 1| \lim_{n \to \infty} \left| \frac{n + 3}{n + 2} \right| \quad \to \quad \frac{0}{0} \qquad \text{l'Hospital's rule applies again.}$$

$$= 2|x - 1| \cdot 1 = 2|x - 1|$$

$$L < 1 \iff 2|x - 1| < 1 \iff \tfrac{1}{2} < x < \tfrac{3}{2}$$

At $x = \frac{1}{2}$, the series is

$$\frac{1}{\ln 2} - \frac{1}{\ln 3} + \frac{1}{\ln 4} - \frac{1}{\ln 5} + \cdots$$

which converges because it meets the hypotheses of the alternating series test. At $x = \frac{3}{2}$, the series is

$$\frac{1}{\ln 2} + \frac{1}{\ln 3} + \frac{1}{\ln 4} + \frac{1}{\ln 5} + \cdots$$

for which the terms starting at $1/\ln 3$ are larger than $\frac{1}{3} + \frac{1}{4} + \frac{1}{5} + \cdots$, a divergent harmonic series.

$\therefore$ the complete interval of convergence is $\frac{1}{2} \le x < \frac{3}{2}$. ◀

▶ **EXAMPLE 3**   Determine whether the series converges.

$$\sum_{n=1}^{\infty} \left(1 + \frac{1}{n^2}\right)$$

Solution   The series begins $2 + 1.25 + 1.1111\ldots + 1.0625 + \cdots$.

$\therefore$ the series diverges because t_n approaches 1, not zero, as n approaches infinity. ◀

▶ **EXAMPLE 4**   Determine whether the series converges.

$$\sum_{n=0}^{\infty} \frac{n}{n^2 + 7}$$

Solution   The series begins $0 + \frac{1}{8} + \frac{2}{11} + \frac{3}{16} + \frac{4}{23} + \frac{5}{32} + \frac{6}{43} + \cdots$, which equals

$$0 + 0.125 + 0.1818\ldots + 0.1875 + 0.17391\ldots + 0.15625 + 0.13953\ldots + \cdots$$

The series might converge because the terms are decreasing after a while and approach zero. (The n^2 in the denominator dominates the n in the numerator, as you could tell by l'Hospital's rule.) Because it's reasonably easy to integrate the given function, you can compare the series with an improper integral. The lower limit of integration can be any nonnegative number, because only the tail of the series is in question.

$$\int_0^{\infty} \frac{x}{x^2 + 7} \, dx = \lim_{b \to \infty} \int_0^b \frac{x}{x^2 + 7} \, dx$$

$$= \lim_{b \to \infty} \frac{1}{2} \ln |x^2 + 7| \Big|_0^b$$

$$= \lim_{b \to \infty} \frac{1}{2} \ln |b^2 + 7| - \frac{1}{2} \ln 7 = \infty$$

$\therefore$ the series diverges because the tail could be bounded below by a divergent integral. ◀

▶ **EXAMPLE 5**   Determine whether the series converges.

$$\sum_{n=0}^{\infty} \frac{1}{n^2 + 7}$$

Solution   The series begins $\frac{1}{7} + \frac{1}{8} + \frac{1}{11} + \frac{1}{16} + \frac{1}{23} + \cdots$.

The series might converge because the terms decrease and approach zero as a limit (t_n approaches the form $1/\infty$). Each term after the first one in the given series is smaller than the corresponding term of a convergent p-series.

$$p\text{-series:} \quad \sum_{n=1}^{\infty} \frac{1}{n^2} \quad = \quad 1 + \frac{1}{4} + \frac{1}{9} + \frac{1}{16} + \cdots$$

Value of n: $\qquad\qquad\qquad\qquad\quad 1 \quad 2 \quad 3 \quad 4$

$$\textbf{Given series:} \quad \sum_{n=0}^{\infty} \frac{1}{n^2 + 7} = \frac{1}{7} + \frac{1}{8} + \frac{1}{11} + \frac{1}{16} + \frac{1}{23} + \cdots$$

$\therefore$ the series converges because the partial sums are increasing and bounded above by the limit of a convergent p-series.   ◀

▶ **EXAMPLE 6**   Determine whether the series converges.

$$\sum_{n=1}^{\infty} \frac{1}{n^3 - 11}$$

Solution   The series begins $-\frac{1}{10} - \frac{1}{3} + \frac{1}{16} + \frac{1}{53} + \frac{1}{114} + \cdots$.

Even after the terms become positive, they are not bounded above by the p-series $\sum 1/n^3$. So use the limit comparison test.

$$L = \lim_{n\to\infty} \frac{1/(n^3 - 11)}{1/n^3} = \lim_{n\to\infty} \frac{n^3}{n^3 - 11} = 1 \qquad \text{Apply l'Hospital's rule.}$$

$\therefore$ the series converges because L is a positive real number.   ◀

Problem Set 12-7

Quick Review

Q1. $7 + 14 + 28 + 56 + \cdots$ are terms of a(n) —?— series.

Q2. The next term in the series in Problem Q1 is found from the preceding term by —?—.

Q3. The number 2 for the series in Problem Q1 is called the —?— of the series.

Q4. $(x - 1) - \frac{1}{2}(x - 1)^2 + \frac{1}{3}(x - 1)^3 - \cdots$ is the Taylor series expansion for —?—.

Q5. The first three terms in the Maclaurin series expansion of $\cos 2x$ are —?—.

Q6. The coefficient of x^6 in the Maclaurin series expansion of $f(x) = \sin x^2$ is —?—.

Q7. If the interval of convergence is $3 < x < 8$, then the radius of convergence is —?—.

Q8. The open interval of convergence for
$$\sum_{n=0}^{\infty} \frac{(x - 4)^n}{3^n} \text{ is —?—.}$$

Q9. If $\vec{r}(t) = (e^{2t})\vec{i} + (\sin 3t)\vec{j}$, then $\vec{v}(t) = $ —?—.

Q10. The volume of the solid formed by rotating the region under the graph of $y = 9 - x^2$ about the y-axis is —?—.
  A. 0   B. 36   C. 36π
  D. 40.5π E. 259.2π

1. *Vocabulary Problem 1:* Given the series

$$S = \sum_{n=1}^{\infty} (-1)^{n+1} \frac{6}{n!}$$

a. Write and add the terms of the fifth partial sum, S_5.

b. Write the first few terms of the tail of the series after the fifth partial sum.

c. The factor $(-1)^{n+1}$ makes this an alternating series, which converges by the alternating series test. What three hypotheses does the series meet that makes this test apply? Based on the tail of the series in part b, quickly find an upper bound for $|R_5|$.

d. The series is also absolutely convergent. What does this mean? What words would describe the series if it were convergent but not absolutely convergent?

e. Explain why, when you are showing absolute convergence, the partial sums are increasing even though the terms themselves are decreasing.

2. *Vocabulary Problem 2:* Write the name of the test for convergence or divergence described.

a. Find an upper bound for the partial sums by comparing the terms with a series that is known to converge.

b. Find an upper bound for the partial sums by comparing the terms with the y-values in a convergent or divergent improper integral.

c. Show divergence by showing that t_n does not approach 0 as n approaches infinity.

d. Show that the series is geometric, with common ratio $|r| < 1$.

e. Show that the limit of $|t_{n+1}/t_n| < 1$ as n approaches infinity.

f. Show that for $\sum u(n)$ and $\sum v(n)$, $\lim_{n \to \infty} [u(n)/v(n)]$ is a finite, positive number.

g. Show that the terms have the form $1/n^p$, where $p > 1$.

3. *Integral Test Problem 1:* Given the p-series

$$S = \sum_{n=1}^{\infty} \frac{1}{n^2}$$

a. Find the fifth partial sum, S_5.

b. Write the first few terms of the tail. Show graphically how you can embed the terms of the tail in a continuous function so that the area of the region under the graph of the function is an upper bound for R_5, the sum of the terms in the tail. Find this upper bound by evaluating an appropriate improper integral. Explain how your work allows you to conclude that the given p-series converges.

c. By comparison with other improper integrals, find upper and lower bounds for R_{1000}. Use the average of these upper and lower bounds to find a reasonable approximation for S. To how many decimal places is this approximation accurate?

d. How many terms of the series would it take to get a partial sum that you can guarantee is correct to six decimal places ($R_n < 0.0000005$)?

4. *Integral Test Problem 2:* Consider the p-series

$$\sum_{n=1}^{\infty} \frac{1}{n}$$

a. Write the terms of S_5, the fifth partial sum, and add them. What is the value of p for this series? What special name is given to the series?

b. Write the first few terms of the tail after S_5. Show graphically how the terms of the tail can be embedded in a continuous function so that the area of the region under the graph of the function is a lower bound for R_5, the sum of the terms in the tail. Find this lower bound by evaluating an appropriate improper integral. Explain how your work allows you to conclude that the given p-series diverges.

c. Explain why graphing the terms of the series so that the improper integral is an upper bound for the tail would indicate nothing about whether the series converges.

d. Without actually adding the terms, how many terms could you add to be sure that $S_n > 1000$? If your computer could compute and add a million terms per second, how long would it take to find this value of S_n by simple addition?

5. *Absolute vs. Conditional Convergence Problem 1:* Consider the **alternating harmonic series**

$$\sum_{n=1}^{\infty} \frac{(-1)^{n+1}}{n}$$

a. Write the first few terms of the series. Prove that the series converges by showing that it meets the three hypotheses of the alternating series test.

b. Use the integral test to prove that the series does not converge absolutely.

c. This equation is the Taylor series for $\ln x$ expanded about $x = 1$ and evaluated at $x = 2$. So the series as written converges to $\ln 2$. Find S_{1000} and S_{1001} for the series. Show that both sums are within $|t_{1001}|$ of $\ln 2$. How close to $\ln 2$ is the average of S_{1000} and S_{1001}?

d. Commute and associate the terms this way:

$$\left(1 - \tfrac{1}{2}\right) - \tfrac{1}{4} + \left(\tfrac{1}{3} - \tfrac{1}{6}\right) - \tfrac{1}{8} + \left(\tfrac{1}{5} - \tfrac{1}{10}\right) - \tfrac{1}{12} + \cdots$$

Are any terms left out? Does any term appear more than once? Evaluate the terms in parentheses first, then factor out $1/2$ from each resulting term. To what number does the series converge under this condition? What is meant by conditional convergence of a series?

6. *Absolute vs. Conditional Convergence Problem 2:* Consider the alternating p-series

$$\sum_{n=1}^{\infty} \frac{(-1)^{n+1}}{n^2}$$

a. Write the first few terms of the series. Prove that the series converges by showing that it meets the three hypotheses of the alternating series test.

b. Name an appropriate test that proves the series is absolutely convergent.

c. Explain why the ratio test (ratio technique) fails to prove that this series is absolutely convergent.

7. *Alternating Series Remainders Property Problem:* The first four terms ($n = 0, 1, 2, 3$) of the Maclaurin series for $\sin 0.6$ are $0.6 - 0.036 + 0.000648 - 0.00000555428\ldots + \cdots$

a. Show how you calculate the fourth term, t_3.

b. Calculate the second and third partial sums, S_1 and S_2, respectively.

c. Calculate R_1 and R_2, the remainders after S_1 and S_2 (the values of the tail), by finding the difference between the partial sum and the value of $\sin 0.6$ by calculator. Show that in both cases the magnitude of the remainder is less than the absolute value of the first term of the tail.

d. Use the appropriate property to prove that the series for $\sin 0.6$ converges.

8. *Sequence vs. Series Problem:* Explain why the sequence $2.3, 2.03, 2.003, \ldots, 2 + 3(0.1)^n, \ldots$ converges, but the series $2.3 + 2.03 + 2.003 + \cdots + [2 + 3(0.1^n)] + \cdots$ does not.

9. *Limit Comparison Test Problem:* Consider the series $\displaystyle \sum_{n=2}^{\infty} \frac{1}{n^2 - 1}$.

a. Write the first few terms of the series. Using the limit comparison test, prove that the series converges.

b. Why can't you prove convergence of this series by direct comparison with the p-series with $p = 2$, starting at $n = 2$?

c. Why would the series diverge if the index of summation started at 1 instead of 2?

10. *Comparison Test for the Exponential Function Series:* These are the first seven terms of the Maclaurin series for $e^{0.6}$.

$n:$	0	1	2	3	4	5	6	$\cdots$
$e^{0.6} = $	1	+ 0.6	+ 0.18	+ 0.036	+ 0.0054	+ 0.000648	+ 0.0000648	+ $\cdots$

|————5th partial sum————→| |————Tail————→|

a. Show how the seventh term, t_6, is calculated.

b. Show that the fifth partial sum, S_4, differs from the actual value of $e^{0.6}$ by more than t_5, the first term of the tail, but not by much more.

c. Show that each term of the geometric series

$$0.000648 + 0.0000648 + \cdots$$

with common ratio 0.1 is an upper bound for the corresponding term in the tail.

d. To what value does the geometric series in part c converge?

e. Based on your answer to part d, what number is an upper bound for the sum of the tail of the series? What number is an upper bound for the entire series? Show that the latter number is just above $e^{0.6}$.

11. *Limit Comparison vs. Ratio Test Problem:*

Consider the series $\sum\limits_{n=1}^{\infty} \dfrac{n}{(n-1)!}$.

a. Write the first few terms of the series. Use the ratio test to prove that the series converges.

b. Write the first few terms of this series.

$$\sum_{n=1}^{\infty} \frac{n}{n!}$$

Explain how you know from previous work that this series converges.

c. Explain why the limit comparison test is inconclusive in proving that the given series converges when comparing it with the convergent series in part b.

12. *Direct Comparison vs. Limit Comparison Test Problem:* Consider the two series

$$U = \sum_{n=1}^{\infty} \frac{2^n}{3^n + 1} \quad \text{and} \quad V = \sum_{n=1}^{\infty} \frac{2^n}{3^n - 1}$$

and the convergent geometric series

$$G = \sum_{n=1}^{\infty} \frac{2^n}{3^n}$$

a. Write the first few terms of each series. Explain why direct comparison of series U with series G proves that series U converges, but direct comparison of series V with series G fails to prove that series V converges.

b. Prove that series V converges by applying the limit comparison test, using series G in part a.

For Problems 13–32, state whether the series converges. Justify your answer.

13. $\sum\limits_{n=1}^{\infty} \dfrac{1}{n}$

14. $\sum\limits_{n=1}^{\infty} \dfrac{1}{n^2}$

15. $\sum\limits_{n=1}^{\infty} \dfrac{(-1)^n}{\sqrt{n}}$

16. $\sum\limits_{n=1}^{\infty} \dfrac{1}{\sqrt{n}}$

17. $\sum\limits_{n=1}^{\infty} \dfrac{3}{4^n}$

18. $\sum\limits_{n=1}^{\infty} \dfrac{3^n}{4^n}$

19. $\sum\limits_{n=1}^{\infty} \dfrac{1}{(2n+1)!}$

20. $\sum\limits_{n=1}^{\infty} \dfrac{1}{(-3)^n}$

21. $\sum\limits_{n=2}^{\infty} \dfrac{n^3}{n^4 - 1}$

22. $\sum\limits_{n=1}^{\infty} \dfrac{n}{n+1}$

23. $\sum\limits_{n=1}^{\infty} \dfrac{\cos n\pi}{n}$

24. $\sum\limits_{n=1}^{\infty} \sin n$

25. $\sum\limits_{n=1}^{\infty} \dfrac{1}{n}\left(\dfrac{4}{3}\right)^n$

26. $\sum\limits_{n=1}^{\infty} \left(\dfrac{7}{11}\right)^n$

27. $\sum\limits_{n=1}^{\infty} \dfrac{2^n}{n+1}$

28. $\sum\limits_{n=1}^{\infty} \dfrac{n^2 - 1}{2^n}$

29. $\sum\limits_{n=2}^{\infty} \dfrac{1}{n \ln n}$

30. $\sum\limits_{n=3}^{\infty} \dfrac{2}{n^2 + 1}$

31. $\sum\limits_{n=2}^{\infty} n!\, e^{-n}$

32. $\sum\limits_{n=2}^{\infty} \left(1 + \dfrac{1}{n}\right)^n$

For Problems 33–36, write the first few terms of the series at each end of the given open interval of convergence. From the result, find the complete interval of convergence.

33. $\sum\limits_{n=0}^{\infty} \dfrac{(x-5)^n}{4^n}$: open interval $= (1, 9)$

34. $\sum\limits_{n=1}^{\infty} \dfrac{(x-4)^n}{n \cdot 3^{2n}}$: open interval $= (1, 7)$

35. $\sum\limits_{n=1}^{\infty} \dfrac{(x+3)^n}{\sqrt{n}}$: open interval $= (-4, -2)$

36. $\sum\limits_{n=1}^{\infty} 3^n(x+2)^n$: open interval $= (-2\frac{1}{3}, -1\frac{2}{3})$

For Problems 37–46, find the interval of convergence for the given power series, including convergence or divergence at the endpoints of the interval.

37. $\displaystyle\sum_{n=1}^{\infty} n(x-3)^n$

38. $\displaystyle\sum_{n=1}^{\infty} \frac{5^n x^n}{n^2}$

39. $\displaystyle\sum_{n=1}^{\infty} \frac{x^n}{n}$

40. $\displaystyle\sum_{n=4}^{\infty} \frac{(-1)^n (x-6)^n}{n\,2^n}$

41. $\displaystyle\sum_{n=1}^{\infty} \frac{(-1)^{n+1}(x+5)^{2n}}{2n}$

42. $\displaystyle\sum_{n=1}^{\infty} \frac{(x+1)^n}{n^2}$

43. $\displaystyle\sum_{n=0}^{\infty} \frac{\ln(n+1)}{n+1} x^n$

44. $\displaystyle\sum_{n=1}^{\infty} 5(x-3)^n$

45. $\displaystyle\sum_{n=0}^{\infty} \frac{4^n}{x^n}$

46. $\displaystyle\sum_{n=1}^{\infty} \frac{1}{x^n}$

47. *Infinite Overhang Problem:* Figure 12-7g shows a pile of blocks. The top block sits so that its center of mass is exactly on the right edge of the second one down. The right edge of the third block is placed under the center of mass of the first two. The fourth is placed under the center of mass of the first three, and so on.

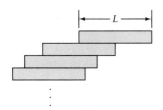

Figure 12-7g

a. Show that the overhangs of the blocks are terms in the harmonic sequence $\frac{1}{2}L, \frac{1}{4}L, \frac{1}{6}L, \frac{1}{8}L, \ldots$, where L is the length of each block. To find the centroid of a particular pile, find the sum of the moments of the blocks by summing each one's

moment with respect to the *y*-axis, then dividing by the number of blocks. You must, of course, find the centroid of one pile before you can find the centroid of the next.

b. What depth of pile is the first to have its top block projecting entirely beyond its bottom block?

c. Explain why, in theory at least, it would be possible to make a pile of blocks with *any* desired overhang, using nothing but gravity to hold the pile together.

d. If you pile up a normal 52-card deck the way the blocks are piled in this problem, by how many card-lengths would the top card be offset from the bottom card?

48. *Convergence of Sequences Proof:* Figure 12-7h shows a sequence

$$\{t_1, t_2, t_3, \ldots, t_n, \ldots\}$$

that is increasing and bounded above. Prove that the sequence converges. Use the **least upper bound postulate** (which states that any set of real numbers that is bounded above has a least upper bound) to establish the existence of a least upper bound, L; then prove that L is the limit of the sequence.

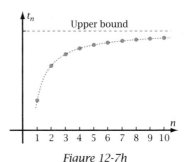

Figure 12-7h

12-8 Error Analysis for Series—The Lagrange Error Bound

From time to time in this chapter you have estimated the remainder of a series, or the value of the tail of the series after a certain number of terms. This remainder represents the error in the value of a function that you get by using a partial sum of the series. For certain alternating series, you found that the entire tail is bounded by the first term of the tail. By improper integrals, you found upper and lower bounds for the tail of some series of positive terms. In this section you will learn about the *Lagrange form* of the remainder of a Taylor series, an expression similar to the first term of the tail. Joseph Louis Lagrange (1736-1813) applied mathematics in many areas, including the motion of planets, and helped establish the French metric system.

OBJECTIVE

Given a series, determine the number of terms needed to obtain an approximation for the limit to which the series converges correct to a specified accuracy.

The Lagrange Error Bound for the Remainder

The general term of the Taylor series expansion of $f(x)$ about $x = a$ is

$$t_n = \frac{f^{(n)}(a)}{n!}(x - a)^n$$

The first term of the tail of a Taylor series is t_{n+1}. For any value of x in the interval of convergence, there is a value of c between a and x for which $R_n(x)$ is given by

$$R_n(x) = \frac{f^{(n+1)}(c)}{(n + 1)!}(x - a)^{n+1}$$

The only difference between this remainder and the first term of the tail is that $f^{(n+1)}(a)$ is replaced by $f^{(n+1)}(c)$. Usually this derivative is awkward to calculate, although you can often find an upper bound, M, for it. If this is the case, an upper bound can be found for the remainder. Finding the value of n that gives a small enough remainder allows you to determine how many terms are needed to get the desired accuracy in the partial sum representing $f(x)$. You can use this error bound to prove that the series converges to the target function, $f(x)$, not to some other value, if you can show that this error bound goes to zero as n approaches infinity.

PROPERTY: *Lagrange Form of the Remainder of a Taylor Series*

If $f(x)$ is expanded as a Taylor series about $x = a$ and x is a number in the interval of convergence, then there is a number c between a and x such that the remainder, R_n, after the partial sum, S_n, is given by the **Lagrange form**

$$R_n = \frac{f^{(n+1)}(c)}{(n + 1)!}(x - a)^{n+1}$$

If M is the maximum value of $\left| f^{(n+1)}(x) \right|$ on the interval between a and x, then the **Lagrange error bound** is

$$|R_n| \leq \frac{M}{(n + 1)!}\left| x - a \right|^{n+1}$$

The property is an extension of the mean value theorem, which concludes that there is a number c between a and x for which

$$f'(c) = \frac{f(x) - f(a)}{x - a}$$

Multiplying by $(x - a)$ gives

$$f(x) - f(a) = f'(c)(x - a)$$

As shown in Figure 12-8a, $f(x) - f(a)$ is the error in using $f(a)$ as an approximation for $f(x)$.

Solving this equation for $f(x)$ gives

$$f(x) = f(a) + f'(c)(x - a)$$

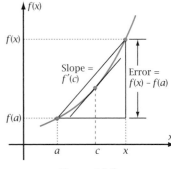

Figure 12-8a

In this form you can see that $f(a)$ is the first term ($n = 0$) of the Taylor series expansion of $f(x)$ about $x = a$, and $f'(c)(x - a)$ is the Lagrange form of the remainder after $n = 0$. You will be asked to supply the algebraic details of the derivation in Problem 21 of Problem Set 12-8.

▶ **EXAMPLE 1**   Consider the Maclaurin series for e^x.

a. Estimate e^2 using the 11th partial sum ($n = 10$).

b. Use the Lagrange form of the remainder to estimate the accuracy of using this partial sum.

c. How does the estimate of the remainder you found in part b compare with the value calculated by subtracting S_{10} from the value of e^2 on your calculator?

Solution   a. $S_{10} = \sum\limits_{n=0}^{10} \frac{1}{n!} 2^n = 7.38899470...$

b. All derivatives of e^x are equal to e^x. Because you are estimating e^2 from scratch, you should not assume that e is known to be $2.718...$. However, you know that $e < 3$. So a value of M is 3^2; so $f^{(n+1)}(x) < 9$ for all x between 0 and 2.

$$|R_{10}| < \frac{9}{11!}(2 - 0)^{11} = 0.0004617...$$

Thus, S_{10} may be off by as much as five in the fourth decimal place and, thus, should match e^2 to ± 1 in the third decimal place.

c. $e^2 = 7.38905609...$ By calculator.

$S_{10} = 7.38899470...$ From part a.

The difference is $0.00006138...$, which is significantly less than the upper bound $0.0004617...$ by the Lagrange form. Note that although the difference has zeros in the first four decimal places, the partial sum and the more precise value of e^2 still differ by one in the third decimal place.   ◀

▶ **EXAMPLE 2**

How many terms of the Maclaurin series for $\sinh x$ are needed to estimate $\sinh 4$ correct to five decimal places? Confirm your answer by subtracting the partial sum from $\sinh 4$ using your calculator.

Solution

The general term of the series is $t_n = \dfrac{1}{(2n + 1)!} x^{2n+1}$, where n is the term index.

All derivatives of $\sinh x$ are either $\cosh x$ or $\sinh x$. Both functions are increasing on the interval $[0, 4]$, with $\cosh x > \sinh x$. Thus, the derivatives are all bounded by $\cosh 4$. Because you are trying to estimate $\sinh 4$, you should not assume that you know $\cosh 4$ exactly. However, $2 < e < 3$, so $\cosh 4 < 0.5(3^4 + 2^{-4}) = 40.53125 < 41$, which means that the absolute values of the derivatives are bounded by $M = 41$.

$$|R_n| < \frac{41}{(2n + 3)!} 4^{2n+3} \qquad \text{Note that the exponent is } 2n + 3.$$

To get five-place accuracy, $|R_n|$ should have zeros in the first five decimal places and no more than 5 in the sixth place. That is, $|R_n| < 0.000005$. Using the TABLE feature,

n	$41/(2n + 3)! \cdot 4^{2n+3}$
7	0.001980319...
8	0.000092646...
9	0.00003529...

The 0.000003529... for $n = 9$ is the first value less than 0.000005. Therefore, you should use at least 10 terms (because the term index, n, starts at 0).

$$\sinh 4 = 27.289917197... \qquad \text{By calculator.}$$
$$S_9 = 27.289917108...$$

Difference $= 0.000000089...$, which is considerably less than 0.000005. ◀

▶ **EXAMPLE 3**

For the Maclaurin series for e^2 in Example 1, find the approximate value of c for which the Lagrange form equals the remainder, R_{10}.

Solution

$$R_{10} = \frac{f^{(11)}(c)}{11!}(2 - 0)^{11} = \frac{e^c}{11!} \cdot 2^{11} = 0.00006138... \qquad \text{From Example 2.}$$

$$\therefore e^c = \frac{0.00006138... \cdot 11!}{2^{11}} = 1.1965...$$

$$c = \ln 1.1965... = 0.1794..., \text{ which is between 0 and 2.} \qquad ◀$$

▶ **EXAMPLE 4**

For the Taylor series for $\ln x$ expanded about $x = 1$, how many terms would be needed in the partial sum to compute $\ln 1.4$ to five decimal places?

Solution

$$\ln 1.4 = (1.4 - 1) - \tfrac{1}{2}(1.4 - 1)^2 + \tfrac{1}{3}(1.4 - 1)^3 - \cdots$$

$$= 0.4 - 0.08 + 0.021333... - \cdots$$

Because this series meets the requirements of the alternating series test, $|R_n| < |t_{n+1}|$.

Make $\dfrac{1}{n+1}(0.4^{n+1}) < 0.000005$.

$n < 9.731... < 10$ Solve numerically for n. Because the term index starts at 1, it is equal to the term number.

Use ten terms.

As a check, $S_{10} = 0.336469445...$, and $\ln 1.4 - S_{10} = 0.00000279...$, which checks. ◀

As you saw in Section 12-7, if the terms of a series can be embedded in a decreasing continuous function, you can use the integral test to estimate the number of terms needed to get a specified accuracy for a partial sum.

▶ **EXAMPLE 5**   For the convergent p-series $S = \displaystyle\sum_{n=1}^{\infty} \frac{1}{n^{1.02}}$

     a. Find upper and lower bounds for R_{20} using appropriate improper integrals. Average the upper and lower bounds to find an estimate for R_{20} and, thus, an estimate for the limit to which the series converges, S. Find an upper bound for the error in this estimate of S.

     b. Using the technique of part a, determine the number of terms it would take to find S correctly to six decimal places ($R_n < 0.0000005$). By considering only the upper-bound integral for R_n, how many terms would you have to use to guarantee that the partial sum itself is correct to six decimal places?

Solution

     a. Figure 12-8b shows the terms of the tail after S_{20} represented as areas of rectangles with one corner embedded in the continuous decreasing function $f(x) = 1/x^{1.02}$. As shown on the left, the integral from 20 to infinity is an upper bound for R_{20}. The integral from 21 to infinity is a lower bound, as shown on the right.

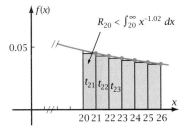 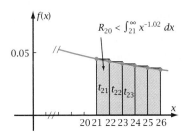

Figure 12-8b

$$R_{20} < \int_{20}^{\infty} x^{-1.02}\,dx = \lim_{b \to \infty}\left(-50x^{-0.02}\Big|_{20}^{b}\right) = 50(20^{-0.02}) = 47.092246\ldots$$

$$R_{20} > \int_{21}^{\infty} x^{-1.02}\,dx = \lim_{b \to \infty}\left(-50x^{-0.02}\Big|_{21}^{b}\right) = 50(21^{-0.02}) = 47.046315\ldots$$

$$R_{20} \approx 0.5(47.092246\ldots + 47.046315\ldots) = 47.069280\ldots$$

$$S = S_{20} + R_{20} \approx 3.509770\ldots + 47.069280\ldots$$

$$= 50.579050\ldots$$

Estimate of the limit to which the series converges.

$$\text{Error} < 0.5(47.092246\ldots - 47.046315\ldots)$$

$$= 0.022965\ldots$$

Upper bound for the error.

b. $$\text{Error} < 0.5\left(\int_{n}^{\infty} x^{-1.02}\,dx - \int_{n+1}^{\infty} x^{-1.02}\,dx\right)$$

$$= 25n^{-0.02} - 25(n+1)^{-0.02} \le 0.0000005$$

$$n = 762,698.17\ldots, \text{ which rounds up to}$$

Solve numerically for n.

762,699 terms

$$R_n < 0.0000005, \text{ if } \int_{n}^{\infty} x^{1.02}\,dx < 0.0000005$$

Use the upper bound integral for R_n.

$$\text{Set } 50n^{-0.02} = 0.0000005$$

$$n = (0.0000005/50)^{-1/0.02} = (1 \times 10^{-8})^{-50} = 1 \times 10^{400}, \text{ which is}$$

unbelievably large! ◀

Problem Set 12-8

Quick Review *5 min*

Q1. The radius of convergence of a power series can be found using the —?— technique.

Q2. A geometric series converges if and only if —?—.

Q3. The Maclaurin series for $\cos x$ converges for what values of x?

Q4. The Taylor series for $\ln x$ expanded about $x = 1$ has what radius of convergence?

Q5. Write the first four nonzero terms of the Maclaurin series for $\tan^{-1} x$.

Q6. Sketch a partial sum of the Maclaurin series for sine compared with the actual sine graph.

Q7. Evaluate the integral: $\int \sec x\,dx$

Q8. Differentiate: $y = \tan x$

Q9. Evaluate the integral: $\int \sec^2 x\,dx$

Q10. Which two people are credited with having invented calculus?

For Problems 1–4,

a. Find the indicated partial sum.

b. Use the Lagrange form of the remainder to estimate the number of decimal places to which the partial sum in part a is accurate.

c. Confirm your answer to part b by subtracting the partial sum from the calculator value.

1. $\cosh 4$ using the sixth partial sum ($n = 5$) of the Maclaurin series

2. $\sinh 5$ using the tenth partial sum ($n = 9$) of the Maclaurin series

3. e^3 using the 15th partial sum of the Maclaurin series

4. $\ln 0.7$ using eight terms of the Taylor series expanded about $x = 1$

For Problems 5–8, use the Lagrange form of the remainder to find the number of terms needed in the partial sum to estimate the function value to the specified accuracy.

5. $\sinh 2$ to six decimal places using the Maclaurin series

6. $\cosh 3$ to eight decimal places using the Maclaurin series

7. $\ln 0.6$ to seven decimal places using the Taylor series expanded about $x = 1$

8. e^{10} to five decimal places using the Maclaurin series

For Problems 9–10, calculate the value of c in the appropriate interval for which the Lagrange form of the remainder equals the remainder found by subtracting the partial sum from the function value by calculator.

9. $\cosh 2$ using five terms (S_4)

10. e^5 using 20 terms (S_{19})

For Problems 11–12, show that the hypotheses of the alternating series test apply to the function, then find the number of terms needed in the partial sum to get the specified accuracy.

11. $\cos 2.4$ to six decimal places using the Maclaurin series

12. e^{-2} to seven decimal places using the Maclaurin series

13. *p-Series Problem 1:* For the convergent p-series with $p = 3$,

$$S = \sum_{n=1}^{\infty} \frac{1}{n^3}$$

   a. Find S_{10}. Find the upper and lower bounds for R_{10}, the remainder after S_{10}, by comparison with appropriate improper integrals. Average these bounds to find an estimate for R_{10}. Use the answer to find an approximation for S, the number to which the series converges. What is the maximum error this approximation could have?

   b. Using the technique of part a, determine the number of terms you would have to add to estimate S correctly to five decimal places $(R_n < 0.000005)$. How does this number compare with the number of terms you

would have to add if you use only the upper bound for R_n to ensure that S_n itself is correct to five decimal places?

14. *p-Series Problem 2:* For the convergent p-series with $p = 1.05$,

$$S = \sum_{n=1}^{\infty} \frac{1}{n^{1.05}}$$

   a. Find S_{100}. Find the upper and lower bounds for R_{100}, the remainder after S_{100}, by comparison with appropriate improper integrals. Average these bounds to find an estimate for R_{100}. Use the answer to find an approximation for S, the number to which the series converges. What is the maximum error this approximation could have?

   b. Using the technique of part a, determine the number of terms you would have to add to estimate S correctly to five decimal places $(R_n < 0.000005)$. Why does it take more terms to get a given accuracy with $p = 1.05$, as in this problem, than it does with $p = 3$, as in Problem 13?

15. *Integral Bound Problem:* Given the series

$$S = \sum_{n=0}^{\infty} \frac{1}{x^2 + 1}$$

   a. Find the 11th partial sum, S_{10}. Show that the series converges by comparing the tail starting at t_{11} with a convergent p-series.

   b. Find the upper and lower bounds for R_{10} by embedding the tail in a continuous decreasing function and evaluating appropriate improper integrals. Average the bounds to find an approximation for S. To how many decimal places do the upper and lower bounds guarantee that the approximation for S is accurate? How many terms would it take using this method to be sure that you have four-place accuracy $(R_n < 0.00005)$?

16. *p-Series Problem 3:* The series

$$1 + \frac{1}{\sqrt{2}} + \frac{1}{\sqrt{3}} + \frac{1}{\sqrt{4}} + \cdots$$

is a p-series. Explain why the method of Problem 13 is not appropriate for estimating the remainder of this series.

17. *Geometric Series as an Upper Bound Problem:* In Example 1 of this section you saw that you could use the Lagrange form of the remainder to estimate the error in calculating e^2 using the 11th partial sum (S_{10}) of the Maclaurin series. You can also estimate the error by bounding the tail of the series with a convergent geometric series that has first term equal to t_{11} (the first term of the tail) and common ratio equal to t_{12}/t_{11}. Which method gives a better estimate of the error, the geometric series or the Lagrange remainder?

18. *Values of e^x from Values of e^{-x} Problem:* You can calculate the value of e^2 by first finding the value of e^{-2}, then taking the reciprocal. After the first few terms, the series for e^{-2} meets the hypotheses for the alternating series test. Thus, the error for any partial sum is bounded by the first term of the tail of the series after that partial sum. Estimate the error in the estimate of e^{-2} using the 11th partial sum (S_{10}). Then estimate e^2 by calculating $1/S_{10}$. Is the error in the answer any smaller than the error in using S_{10} directly for e^2, as in Example 1 of this section?

19. *Sinx for Any Argument Using a Value of x in $[0, \pi/4]$ Problem:* The Maclaurin series for $\sin x$ converges more slowly the farther x is from 0. Suppose you wanted to compute $\sin 250$.

a. The values of $\sin x$ repeat themselves with a period of 2π. Find a number b in $[0, 2\pi]$ for which $\sin b = \sin 250$.

b. Each value of $\sin x$ for x in $[0, 2\pi]$ is equal to a value of $\sin c$ for some number c in $[-\pi/2, \pi/2]$ (Figure 12-8c). Find the value of c for which $\sin c = \sin 250$.

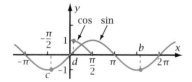

Figure 12-8c

c. Each value of $\sin x$ for x in $[-\pi/2, \pi/2]$ is equal to $\pm \sin d$ or $\pm \cos d$ for some number d in $[0, \pi/4]$ (Figure 12-8c). Find the value of d for which $\sin 250 = \pm \sin d$ or $\pm \cos d$. Demonstrate that your value of d gives the correct answer for $\sin 250$.

d. Show that you can use the technique of this problem to calculate values of $\sin x$ correctly to at least ten decimal places using just six terms of the appropriate Maclaurin series. How many terms do you need to calculate $\sin 250$ to ten places directly from the series?

e. *Project:* Write a program to calculate $\sin x$ correctly to ten decimal places by Maclaurin series using the technique of this problem. Programs similar to this are used internally by calculators to evaluate sines and cosines efficiently.

20. The National Bureau of Standards *Handbook of Mathematical Functions* lists the value of $\sin 1$ to 23 decimal places as

0.84147 09848 07896 50665 250

(The spaces are used in lieu of commas for ease of reading.) How many terms of the Maclaurin series for $\sin x$ would have to be used to get this accuracy? How many terms would it take if the technique of Problem 19 were used?

21. *Derivation of the Lagrange Form of the Remainder:* Earlier in this section you saw that the conclusion of the mean value theorem leads to a special case of the Lagrange form of the remainder. If function f has derivatives of all orders, as do exponential, trigonometric, hyperbolic, and many other functions, then the mean value theorem applies to each derivative.

a. Show that applying the mean value theorem to $f'(x)$ on the interval $[a, x]$ gives

$$f'(x) = f'(a) + f''(c)(x - a)$$

for some number c between x and a.

b. Assume that a and c are constants and x is the variable. Integrate both sides of the differential equation in part a with respect to x. Use the point $(a, f(a))$ as the initial condition. Show that you can transform the answer to

$$f(x) = f(a) + f'(a)(x - a) + \frac{1}{2} f''(c)(x - a)^2$$

c. You should recognize that the first two terms of the right side of the equation in part b are terms in the Taylor series expansion of $f(x)$ about $x = a$ and the third term is the Lagrange form of the remainder. By applying the mean value theorem to f'' on the interval $[a, x]$ and integrating twice, show that there is a number c in (a, x) for which

$$f(x) = f(a) + f'(a)(x - a) + \frac{1}{2!} f''(a)(x - a)^2$$
$$+ \frac{1}{3!} f'''(c)(x - a)^3$$

d. Without actually doing the algebra, name the mathematical technique that you could use to prove that for any integer $n > 0$, there is a number c in the interval (a, x) for which the Lagrange form of the remainder is exactly equal to the error in using the partial sum $S_n(x)$ of the Taylor series as an approximation for $f(x)$.

22. *A Pathological Function:* Figure 12-8d shows the function

$$f(x) = \begin{cases} e^{-x^{-2}}, & \text{if } x \neq 0 \\ 0, & \text{if } x = 0 \end{cases}$$

Function f has derivatives of all orders at $x = 0$, and each derivative equals zero there.

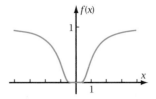

Figure 12-8d

a. By equating derivatives, show that the Maclaurin series for $f(x)$ would be $0 + 0x + 0x^2 + 0x^3 + \cdots$.

b. Show that the Maclaurin series converges for all values of x, but that it does not converge to $f(x)$ except at $x = 0$.

c. Substitute $-x^{-2}$ for x in the Maclaurin series for e^x. Write the first four terms of the power series and simplify.

d. The resulting power series is called a **Laurent series**, the name applied to a power series in which some powers can have negative exponents. By finding a partial sum of the series, make a conjecture about whether the Laurent series evaluated at $x = 2$ converges to $f(2)$.

23. *The Maclaurin Series for e^x Converges to e^x:* Problem 22 shows that a Maclaurin series can converge, but not to the target function. Use the Lagrange form of the remainder to show that the Maclaurin series for e^x does converge to e^x for all values of x by showing that the remainder of the series approaches zero as n approaches infinity.

12-9   Chapter Review and Test

In this chapter you have seen how a function can be expanded as a power series. There are two advantages to doing this. First, it allows you to calculate values of transcendental functions by doing only the four operations of arithmetic. Calculators and computers use series internally to calculate values of sines, logs, and so forth. Second, it allows you to determine how accurate a numerical integral is. If x is within the interval of convergence, you can get any desired accuracy by using enough terms of the series.

Review Problems

R0. Update your journal with what you have learned since the last entry. You should include such things as

- The one most important thing you have learned in studying Chapter 12
- Which boxes you have been working on in the "define, understand, do, apply" table
- How to write from memory some special, well-known series
- How you find a power series for a given function, either by equating derivatives or by operating on another known series
- What it means for a sequence or series to converge
- How you determine the accuracy of a function value found by series
- Any ideas about series that are still unclear

R1. Let $f(x) = \frac{9}{1-x}$ and let

$$P_n(x) = 9 + 9x + 9x^2 + 9x^3 + 9x^4$$
$$+ 9x^5 + \cdots + 9x^n$$

On the same screen, plot $f(x)$, $P_5(x)$, and $P_6(x)$. Sketch the results. For what values of x does $P_n(x)$ seem to be close to $f(x)$, and for what values of x does the graph of P bear little resemblance to the graph of f? Show that $P_6(0.4)$ is closer to $f(0.4)$ than $P_5(0.4)$ is. Show that $P_5'(0)$, $P_5''(0)$, and $P_5'''(0)$ equal $f'(0)$, $f''(0)$, and $f'''(0)$, respectively. What kind of series is $P_n(x)$ a subseries of?

R2. a. *Biceps Problem:* You start an exercise program to increase the size of your biceps. The first day you find that the circumference of each bicep increases by 3 mm. You assume that the amount of increase on each subsequent day will be only 90% of the amount of increase the day before. By how much do you predict your biceps will have increased after ten days? At the limit, what will be the total increase in each bicep?

b. *Present Value Problem:* You win $10 million in the state lottery! However, you will receive only $0.5 million now and $0.5 million a year for the next 19 years (20 payments). How much money must the state invest now so that it will have $0.5 million to pay you at the beginning of the 20th year? Assume that the state can get 10% per year in interest, compounded annually, on the investment. How much money, total, must the state invest now to make all 19 future payments? This amount is called the **present value** of your remaining $9.5 million.

R3. Let $P(x)$ be the power series $c_0 + c_1 x + c_2 x^2 + c_3 x^3 + c_4 x^4 + \cdots$. Let $f(x) = 7e^{3x}$. By equating derivatives, find the values of c_0, c_1, c_2, and c_3 that make $P(0)$, $P'(0)$, $P''(0)$, and $P'''(0)$ equal $f(0)$, $f'(0)$, $f''(0)$, and $f'''(0)$, respectively.

R4. For a–c, show that the fourth partial sum, $S_3(0.12)$, is close to $f(0.12)$.

a. $e^x = 1 + x + \frac{1}{2!}x^2 + \frac{1}{3!}x^3 + \frac{1}{4!}x^4 + \cdots$

b. $\cos x = \sum_{n=0}^{\infty} (-1)^n \frac{1}{(2n)!} x^{2n}$

c. $\sinh x = x + \frac{1}{3!}x^3 + \frac{1}{5!}x^5 + \frac{1}{7!}x^7 + \cdots$

d. Show that the 20th partial sum of

$$\sum_{n=1}^{\infty}(-1)^{n+1}\frac{1}{n}(x-1)^n$$

gives values close to $\ln x$ if $x = 1.7$, but not if $x = 2.3$.

R5. a. What is the difference between a Maclaurin series and a Taylor series?

b. Write the Maclaurin series for $\ln(x + 1)$ by performing appropriate operations on the Taylor series for $\ln x$.

c. Integrate the series in part b to find a Maclaurin series for $\int \ln(x + 1)\,dx$.

d. Show that the series in part c is equivalent to the one you would obtain by finding the antiderivative of $\ln(x + 1)$ and writing that as a Maclaurin series.

e. Write the first few terms of a power series for $\int_0^x t\cos t^2\,dt$.

f. Write $\tan^{-1}x$ as the definite integral of an appropriate function from 0 to x. Write the integrand as a Maclaurin series. Then write the first few terms of the corresponding Maclaurin series for $\tan^{-1}x$.

g. Suppose that f is a function with derivatives of all orders that are defined for all real values of x. If $f(3) = 5$, $f'(3) = 7$, $f''(3) = -6$, and $f'''(3) = 0.9$, write the first four terms of the Taylor series for $f(x)$ expanded about $x = 3$.

R6. a. Write the first few terms of $\sum_{n=1}^{\infty}(-3)^{-n}(x-5)^n$.

b. Find the open interval of convergence and radius of convergence of the series in part a.

c. Show that the Maclaurin series for $\cosh x$ converges for all values of x.

d. Write the first five terms of the Maclaurin series for $e^{1.2}$. Then calculate the error in using the fifth partial sum to approximate $e^{1.2}$ by subtracting the partial sum from $e^{1.2}$ by calculator. How does the error compare with the value of the first term in the tail of the series after the fifth partial sum?

e. On the same screen, plot the graphs of $\ln x$, the Taylor series expanded about $x = 1$ for $\ln x$ using 10 terms, and the same series

using 11 terms. Sketch the graphs. Then write a paragraph stating how the graphs relate to the interval of convergence.

R7. a. Find the tenth partial sum of the geometric series with $t_1 = 1000$ and common ratio 0.8.

b. By how much does the tenth partial sum in part a differ from the limit to which the series converges?

c. The rest of the series in part a following the tenth partial sum is called the —?— of the series.

d. The value of the rest of the series, part b, is called the —?— of the series.

e. Use appropriate improper integrals to find the upper and lower bounds for R_{10}, the remainder after the tenth partial sum, S_{10}, for the p-series

$$S = 1 + \frac{1}{8} + \frac{1}{27} + \frac{1}{64} + \cdots$$

Use the result to conclude that the series converges. Find a reasonable estimate for S. To how many decimal places can you ensure that the answer is accurate?

f. Write the first few terms of the series

$$F = \sum_{n=1}^{\infty}\frac{1}{n^3-5}$$

Use the limit comparison test, along with the results of part e, to prove that series F converges. Explain why the direct comparison test of series F to series S would not be sufficient to prove convergence of series F.

g. Show that there is a convergent geometric series that is an upper bound for the tail of the series $2/1! + 4/2! + 8/3! + 16/4! + 32/5! + \cdots$ after a suitable number of terms.

h. Show that the alternating harmonic series

$$1 - \frac{1}{2} + \frac{1}{3} - \frac{1}{4} + \frac{1}{5} - \frac{1}{6} + \cdots$$

converges conditionally to $\ln 2$, but does not converge absolutely. Show that you can rearrange the terms and regroup them so that the series converges to $0.5 \ln 2$.

i. Find an upper bound for the remainder of the series in part h after 10,000 terms.

j. Find the complete interval of convergence, including the endpoints.

i. $\sum\limits_{n=1}^{\infty} \dfrac{10^n(x-3)^n}{n^2}$

ii. $\sum\limits_{n=1}^{\infty} \dfrac{(-1)^n(x+1)^n}{n \cdot 2^n}$

k. State whether each series of constants converges. Justify your answer.

i. $\sum\limits_{n=0}^{\infty} \dfrac{10}{n!}$

ii. $\sum\limits_{n=1}^{\infty} (n^{-3} + 5^{-1})$

iii. $\sum\limits_{n=1}^{\infty} \dfrac{3 \cdot 2^n}{5^n}$

iv. $\sum\limits_{n=1}^{\infty} \dfrac{1}{n^{1/3}}$

v. $\sum\limits_{n=1}^{\infty} \dfrac{(n+3)!}{3! \cdot n! \cdot 3^n}$

R8. a. Use the Lagrange form of the remainder to estimate the error in using the fourth partial sum of the Maclaurin series to estimate cosh 2.

b. You are asked to calculate e^3 using enough terms of the Maclaurin series to get a 20-decimal-place accuracy. Use the Lagrange

form of the remainder to calculate the number of terms that you should use.

c. The Maclaurin series for cosh x converges for all values of x. Use the Lagrange form of the remainder to show that the value the series converges to when $x = 4$ really equals cosh 4.

d. Calculate, approximately, the number c in the interval (0, 0.6) for which the Lagrange form is equal to the remainder of the Maclaurin series for sinh 0.6 after the fourth partial sum $(n = 3)$.

e. Use the fact that the Taylor series for ln x is alternating if x is between 1 and 2 to find the number of terms of the series needed to compute ln 1.3 to 20 decimal places.

f. Use improper integrals to find the upper and lower bounds for R_{50}, the remainder after the 50th partial sum, for the p-series

$$S = \sum_{n=1}^{\infty} \dfrac{1}{n^4}$$

How does your work let you conclude that the series converges? Find an estimate for S using S_{50} and the upper and lower bounds for R_{50}. To how many decimal places does your work guarantee that the estimate for S is accurate?

Concept Problems

C1. *Series with Imaginary Numbers Problem:* If you substitute *ix* for *x* in the Maclaurin series for cosine and sine, you get some startling results! In this problem you will see how these results lead to another similarity between trigonometric functions and hyperbolic functions.

a. Substitute *ix* (where *i* is $\sqrt{-1}$) for *x* in the Maclaurin series for cosine. When you simplify, you should find that cos *ix* is real-valued and equals cosh x.

b. Substitute *ix* for *x* in the Maclaurin series for sine. You should find that each term has *i* as

a factor. Factor out the *i* to show that $\sin ix = i \sinh x$.

c. Substitute *ix* for *x* in the Maclaurin series for e^x. Use the result and the answers to parts a and b to show that the following formula is true.

$$e^{ix} = \cos x + i \sin x$$

d. Show that $e^{i\pi} = -1$. This one short formula combines four of the most mysterious numbers of mathematics!

C2. *Practical Calculation of Pi Problem:* In Problem Set 12-5, Problem 37, you computed π using the Maclaurin series for $\tan^{-1} x$ to compute $\pi/4$, which equals $\tan^{-1} 1$. You made use of the composite argument property from trigonometry,

$$\tan(A + B) = \frac{\tan A + \tan B}{1 - \tan A \tan B}$$

to show that

$$\tan^{-1} \tfrac{1}{2} + \tan^{-1} \tfrac{1}{3} = \tfrac{\pi}{4}$$

thus obtaining more accuracy with fewer terms. Show that the double series

$$4 \tan^{-1} \tfrac{1}{5} - \tan^{-1} \tfrac{1}{239}$$

also converges to $\pi/4$. How many terms of this series do you need in order to get π correct to the number of decimal places reported in William Shaaf's booklet *Computation of Pi* (Yale University Press, 1967), namely,

$$\pi = 3.14159\ 26535\ 89793\ 23846\ 26433$$
$$83279\ 50288\ 41971\ 69399\ 37510\ldots$$

C3. *Series Solution of a Differential Equation:* In future courses on differential equations you will learn to solve a differential equation directly in terms of a power series. This problem provides a preview of the technique used. Consider the second-order differential equation

$$y'' + 9xy = 0$$

with initial conditions $y' = 7$ and $y = 5$ when $x = 0$.

a. Assume there is a power series equal to y. That is,

$$y = \sum_{n=0}^{\infty} c_n x^n$$
$$= c_0 + c_1 x + c_2 x^2 + c_3 x^3 + c_4 x^4$$
$$+ c_5 x^5 + c_6 x^6 + \cdots$$

Assuming the series can be differentiated termwise, write equations for y' and y''.

b. Use the two initial conditions in the appropriate places to evaluate c_0 and c_1.

c. Substituting the series for y and y'' into the original differential equation and combining terms with equal powers of x gives

$$2c_2 + 6c_3 x + 12c_4 x^2 + 20c_5 x^3$$
$$+ 30c_6 x^6 + \cdots + 9x(c_0 + c_1 x + c_2 x^2$$
$$+ c_3 x^3 + c_4 x^4 + c_5 x^5 + c_6 x^6 + \cdots) = 0$$
$$2c_2 + (6c_3 + 9c_0)x + (12c_4 + 9c_1)x^2$$
$$+ (20c_5 + 9c_2)x^3 + (30c_6 + 9c_3)x^4$$
$$+ \cdots = 0$$

The right side of the equation is zero, so each coefficient on the left side must equal zero. Use this fact to calculate the values of c_2 through c_6.

d. Use the terms of the series through the sixth power to compute y when $x = 0.3$.

e. Just for fun, see if you can determine whether the series you found in part d converges when $x = 0.3$.

Chapter Test

PART 1: No calculators allowed (T1–T10)

T1. Write the first few terms and the general term of the Maclaurin series for e^{-x}.

T2. Long divide $\frac{5}{1+x}$ to get a Maclaurin power series. What special name is given to this particular kind of power series?

T3. Write the first few terms of the Taylor series for $\cos x$ expanded about $x = \pi$.

T4. Write the Lagrange form of the remainder R_5 for the series in Problem T2.

T5. Write a power series for $\sin(x^2)$. Write the answer in sigma notation.

T6. Give an example of a series that converges conditionally, but not absolutely. Explain what the "condition" refers to in the name *conditional convergence*.

T7. By equating derivatives, show that the Taylor series for $\ln x$ expanded about $x = 1$ is

$$\ln x = (x - 1) - \tfrac{1}{2}(x - 1)^2 + \tfrac{1}{3}(x - 1)^3$$
$$- \tfrac{1}{4}(x - 1)^4 + \cdots$$

T8. Show that the geometric series $1000 + 999 + \cdots$ converges, but the geometric series $0.0001 + 0.0002 + \cdots$ does not converge.

T9. Find the open interval of convergence and the radius of convergence for

$$\sum_{n=1}^{\infty} \frac{(2x - 5)^n}{3n}$$

T10. Does the series in Problem T9 converge or diverge at the endpoints of the interval of convergence? Justify your answer.

PART 2: Graphing calculators allowed (T11–T20)

T11. Let $f(x) = \int_0^x \frac{1}{1+t^3}\, dt$. Write a power series for $f(x)$.

T12. Find the complete interval of convergence of the series in Problem T11.

T13. Find an approximation for $f(0.6)$ using 20 terms of the series from Problem T11.

T14. Find an approximation for $f(0.6)$ in Problem T11 by numerical integration.

T15. Find $f(0.6)$ in Problem T11 exactly using the fundamental theorem. To how many decimal places are the answers to Problems T13 and T14 correct?

T16. Demonstrate that the error in the value of $f(0.6)$ by series in Problem T13 is less than the first term in the tail of the series.

T17. Write a power series for $\cosh x$. Express the answer in sigma notation.

T18. Use the Lagrange form of the remainder to find the number of terms of the series in Problem T17 needed to estimate $\cosh 3$ correctly to ten decimal places.

T19. For the series $S = \sum_{n=1}^{\infty} \frac{1}{n^{1.5}}$

   a. What special name is given to this series?

   b. Use the integral test to prove quickly that the series converges.

   c. Find S_{100}, the 100th partial sum.

   d. Find the upper and lower bounds for R_{100}, the remainder after S_{100}. Use the results to find a good approximation for S.

T20. What did you learn as a result of working this test that you did not know before?

12-10   Cumulative Reviews

In this section are several cumulative reviews that you may consider rehearsals for your final exam. Each review touches on most of the concepts and techniques of calculus, particularly those of the second half of the book.

Problem Set 12-10

Cumulative Review No. 1— The Dam Problem

Suppose you are hired as a mathematician by Rivera Dam Construction Company, which has been awarded a contract to build a dam across Scorpion Gulch. The following questions pertain to your part in this project. Before Mr. Rivera will allow you to work on his dams, he must be sure that you know some of the fundamental definitions, theorems, and techniques of mathematics. He asks about these in Problems 1–5.

1. There are four major concepts of calculus. Name those concepts, and state their definitions.

2. Define:

   a. Continuity of a function at a point

   b. Continuity of a function on an interval

   c. Convergence of a sequence

   d. Convergence of a series

   e. Natural logarithm

   f. The exponential a^x where $a > 0$

The Hoover Dam

3. Mr. Rivera is satisfied with your knowledge of definitions and proceeds to quiz you on your knowledge of various properties. State:

   a. The mean value theorem

   b. The intermediate value theorem

   c. The squeeze theorem

   d. The uniqueness theorem for derivatives

   e. The limit of a product property

   f. The integration by parts formula

   g. The fundamental theorem of calculus

   h. The Lagrange form of the remainder

   i. The chain rule for parametric functions

   j. The polar differential of arc length

4. To make sure you know enough algebraic techniques, Mr. Rivera asks you to find the following limits, derivatives, and integrals.

   a. $f'(x)$, if $f(x) = \int_3^x \sqrt{1 + \operatorname{sech} t}\, dt$

   b. $f'(x)$, if $f(x) = a^x$

   c. $f'(x)$, if $f(x) = x^a$

   d. $f'(x)$, if $f(x) = x^x$

   e. $\int e^{6x} \cos 3x\, dx$

   f. $\int \cosh^5 x \sinh x\, dx$

   g. $\int \sec^3 x\, dx$

   h. $\int (\sin 5x)^{-1} \cos 5x\, dx$

   i. $\displaystyle\lim_{x \to 0} \frac{\cos 7x - 1}{13x^2}$

   j. $\displaystyle\lim_{x \to 0} (1 - x)^{3/x}$

5. Mr. Rivera wants to be sure you know graphical and numerical methods.

   a. Figure 12-10a shows the slope field for

$$\frac{dy}{dx} = 0.2x - 0.3y + 0.3$$

   On a copy of the figure, sketch the particular solution containing $(1, 8)$.

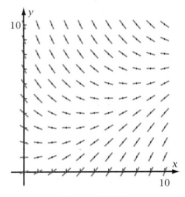

Figure 12-10a

   b. Use Euler's method with $\Delta x = 0.5$ to estimate the value of y for the solution in part a if $x = 9$. How does your answer compare with your graphical solution?

You pass your preliminary tests and start to work on the dam project. At the dam site, Scorpion Gulch has a parabolic cross section with a shape of graph $y = 0.1x^2$, where x and y are in yards. The back face of the dam (where the water will be) is vertical and lies in the xy-plane (Figure 12-10b). The front face slopes in such a way that the thickness is $z = 30$ yd at the bottom of the gulch (where $y = 0$) and $z = 10$ yd at the top of the dam. The dam is to be 40 yd high.

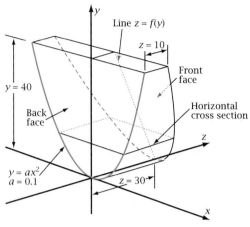

Figure 12-10b

6. Your first project is to analyze the forces that the water will exert on the vertical back face of the dam when the lake is full (that is, 40 yd deep). Assume the water density, k, is in pounds per cubic yard.

   a. Write an equation for the pressure in terms of y.

   b. Find the area of the dam's back face.

   c. Find the force exerted by the water on the back face of the dam.

   d. Find the first moment of this force with respect to the x-axis.

   e. Find the center of pressure. This is the point on the back face at which the entire force could be concentrated to produce the same first moment with respect to the x-axis.

7. Your next project is to determine some of the physical characteristics of the dam itself.

   a. Find a linear equation expressing the thickness of the dam, z, in terms of its altitude, y.

   b. At what y-value will the dam's horizontal cross-sectional area be a maximum? A minimum?

   c. A cement mixer truck holds 5 yd^3 of concrete. How many truckloads of concrete should you order when it is time to pour the dam?

d. Find the length of the joint between the dam's back face and the sides of the gulch.

8. The dam is finished. A speedboat on the lake behind the dam moves with vector equation

$$\vec{r} = (100 \cos 0.03t)\vec{i} + (50 \sin 0.03t)\vec{j}$$

where distances are in feet. How fast is the boat going when $t = 50$ s?

9. The waves from the boat displace the water's surface according to the equation $z = (\sin t)/t$. The average displacement over the time interval $[0, t]$ involves the sine integral function,

$$\mathrm{Si}\, t = \int_0^t \frac{\sin u}{u}\, du.$$

Write Si t as a power series. Use the ratio technique to determine the interval of convergence. Estimate the error in calculating Si 0.6 using just the first three nonzero terms of the series. Calculate Si 0.6 by numerical integration on your grapher.

10. The drain in the dam has a cross section in the shape of the polar curve $r = 5 + 4 \cos \theta$, where r is in feet. Find the area of the drain's cross section.

11. At time $t = 0$ h, the drain is opened. Initially water flows out at 5 million gal/h, but the rate is directly proportional to the amount of water remaining. There were 300 million gal of water behind the dam at $t = 0$. Predict the amount remaining at $t = 10$.

Cumulative Review No. 2— The Ship Problem

After graduation you apply for work at Sinkin Ship Construction Company. Mr. Sinkin gives you the following preliminary test to see how much calculus you know.

1. Define derivative.

2. Define definite integral.

3. State the mean value theorem.

4. Find $f'(x)$ if $f(x) = \int_3^x g(t)\, dt$.

5. Integrate: $\int \tanh^5 x \operatorname{sech}^2 x\, dx$

6. Integrate: $\int x \sinh 2x\, dx$

7. Integrate: $\int \dfrac{3x + 14}{(x + 3)(x - 2)}\, dx$

8. Write the Maclaurin series for $\int \dfrac{\sinh x}{x}\, dx$.

9. Find the open interval of convergence for this series.

$$\sum_{n=1}^{\infty} \frac{n(x - 5)^n}{3^n}$$

10. Evaluate the improper integral $\int_0^1 x^{-0.998}\, dx$.

11. Find the average value of $y = x^2$ on the interval $[3, 9]$.

12. Let $f(x) = x^2$. If $\delta = 0.01$, is this small enough to keep $f(x)$ within 0.08 unit of $f(4)$ when x is within δ units of 4? Justify your answer.

Mr. Sinkin is satisfied with your work on these questions and assigns you to the design team for a new ship. The hull of this ship is shown in Figure 12-10c.

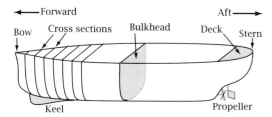

Figure 12-10c

13. The volume of the ship equals the cross-sectional area times the length. At the forward end of the ship, the cross-sectional area varies with x, the distance from the bow. Write an integral for the volume of the part of the ship from $x = 2$ to $x = 10$. Then evaluate the integral, approximately, by Simpson's rule given the cross-sectional areas shown in the table. Dimensions are in feet and square feet.

x	Area
2	153
4	217
6	285
8	319
10	343

14. The propeller will have four blades in the shape of the four-leaved rose $r = 4 \sin 2\theta$, where r is in feet. Find the area of one blade.

15. At the stern of the ship, the deck has a shape similar to the region bounded by the ellipse

$$\left(\frac{x}{5}\right)^2 + \left(\frac{y}{3}\right)^2 = 1$$

between $x = 1$ and $x = 5$. Find the area of this region.

Your next project is to analyze a vertical bulkhead (a wall) that goes across the ship. The bulkhead has the shape of the region that lies above the graph of $y = 0.0016x^4$ and below the line $y = 16$.

16. Find the area of the bulkhead.

17. The welders who will install the bulkhead need to know the length of the graph of $y = 0.0016x^4$ that forms the edge of the bulkhead. Find this length.

18. The bulkhead must be strong enough to withstand the force of the water if the compartment on one side of the bulkhead becomes flooded. Recall that force equals pressure times area and that the pressure at any point below the water's surface is proportional to that point's distance from the surface. The proportionality constant is $62.4\ \text{lb/ft}^3$, the density of water. Find to the nearest 100 lb the force exerted by the water.

A vertical keel is to extend below the bottom of the ship. When turned upside down, the keel is similar in shape to the region under the graph of $y = (\ln x)/x$ for $x \geq 1$.

19. Find the limit of y as x approaches infinity.

20. Find the x-coordinate of the maximum of the function. Justify your answer.

21. Find the x-coordinate(s) of all points of inflection of the graph.

22. Sketch the graph, consistent with your answers above.

The radar equipment needs values of natural logarithms to 20 decimal places.

23. Show that the Taylor series for $\ln x$ expanded about $x = 1$ converges for $0 < x \leq 2$.

24. How many terms of the series do you need to calculate ln 1.4 to 20 decimal places?

Your last project is analysis of the motion of the ship.

25. In linear motion the velocity of the ship is given by a differential equation with a slope field as shown in Figure 12-10d. Describe how the velocity would change if the ship started from $v = 0$ ft/s at $t = 0$ min. How would the velocity differ if somehow the ship were given an initial velocity of 50 ft/s?

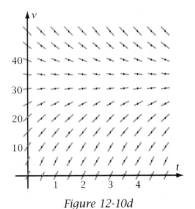

Figure 12-10d

26. In a sharp turn, the position vector of the ship is given by $\vec{r} = (\ln t)\vec{i} + (\sin 2t)\vec{j}$. Find the acceleration vector.

Cumulative Review No. 3— Routine Problems

Calculus involves four fundamental concepts. In Problems 1–4, demonstrate that you understand these concepts.

1. Demonstrate that you understand the definition of limit by sketching the graph of a function that has a limit L as x approaches c and showing an epsilon neighborhood of L and a corresponding delta neighborhood of c that is clearly smaller than is necessary.

2. Write the formal definition of derivative. Then give a graphical and a physical meaning of the derivative.

3. Write the definition of $\int f(x)\,dx$.

4. Write the definition of $\int_r^s f(t)\,dt$.

You don't use the definition of limit to find limits, so other techniques are developed.

5. There is a technique for finding interesting limits of the form $0/0$ or ∞/∞. Name this technique, and use it to evaluate

$$\lim_{x \to 0} \frac{x \cos x}{1 - e^{5x}}$$

The definition of derivative is awkward to use, so you develop shortcuts.

6. Find y' if $y = \tan(\sin 5x)$. Name the property that allows you to differentiate such composite functions.

7. Products and powers can be differentiated logarithmically. Find y' if

$$y = (5x - 3)(2x + 7)^4(x - 9)$$

8. Formulas for derivatives of inverse circular functions are found by implicit differentiation. Do this for the inverse tangent function.

In Problems 9–12, you will demonstrate your knowledge of certain basic techniques and when to use them. Integrate.

9. $\int \sin^7 x \cos x\,dx$

10. $\int \sqrt{x^2 + 9}\,dx$

11. $\int \dfrac{3x - 11}{x^2 + 2x - 3}\,dx$

12. $\int \sin^{-1} x\,dx$

Definite integrals are hard to evaluate using the definition. Fortunately, there is a theorem relating definite integrals to indefinite ones.

13. Name and state the theorem that relates definite and indefinite integrals.

14. The mean value theorem plays a key role in the proof of the theorem in Problem 13. Demonstrate that you understand the mean value theorem by sketching an appropriate graph.

You can apply the techniques of calculus to problems in the real and mathematical worlds.

15. If $f(x) = \int_3^x h(t)\,dt$, find $f'(x)$.

16. Given $f(x) = xe^{-x}$ for $x \geq 0$, find the x-coordinates of all points of inflection.

17. Find the length of the graph of $y = \sin x$ from $x = 0$ to $x = 2$.

18. The integral $\int_0^{16} x^{-3/4}\, dx$ is improper.

    a. Show that the integral converges.

    b. Use the result to find the average value of $y = x^{-3/4}$ from $x = 0$ to $x = 16$.

19. Find the area of the region inside the circle with polar equation $r = 10 \cos \theta$ from $\theta = 0.5$ to $\theta = 1$.

20. A particle travels in the xy-plane in such a way that its position vector is

    $$\vec{r} = t^2 \vec{i} + 3t^{-1} \vec{j}$$

    At time $t = 1$, what is its velocity vector? How fast is it going? Is the particle's distance from the origin increasing or decreasing? At what rate?

21. The region in Quadrant I bounded by the coordinate axes and the graph of $y = \cos x$ is rotated about the y-axis to form a solid. Find its volume.

22. A rectangle in the first quadrant has one corner at the origin and the diagonally opposite corner on the line $y = -1.5x + 6$.

    a. Show that the rectangle has maximum area when $x = 2$.

    b. If the rectangle and line are rotated about the y-axis to form a cylinder inscribed in a cone, show that a different value of x produces the maximum-volume cylinder.

23. The crew's compartment in a spaceship has an irregular shape due to all the equipment in it. Cross sections at various distances from the bottom of the compartment have areas as follows (in feet and square feet):

Distance	Area
3	51
5	37
7	41
9	63
11	59

Use Simpson's rule to estimate the volume of the crew's compartment between 3 and 11 ft.

24. The error function, used in curving grades, is defined by

$$\operatorname{erf} x = 2\pi^{-1/2} \int_0^x e^{-t^2}\, dt$$

    a. Write the first few terms of the Maclaurin series for the integral in erf x.

    b. Use the ratio technique to prove that the series for erf x converges for all values of x.

Final Examination

This section contains an examination that you can consider to be a dress rehearsal for the exam your instructor will give you. Make the rehearsal as realistic as possible by putting yourself under simulated test conditions and giving yourself a two-hour time limit. No answers are provided in the back of the book.

Calculus involves two basic concepts:

    a. Instantaneous rates of change

    b. Products in which one factor's value depends on the other factor

Both of these are based on the underlying concept of limit. They are linked by the fundamental theorem, which allows you to calculate limits of Riemann sums by using antiderivatives. On this test it is your objective to answer the questions in a way that demonstrates that you understand these concepts and their relationships.

A Guided Tour Through Calculus

1. The first problem you encountered in this book involved finding the instantaneous rate of change of a function at a given point. Find the approximate derivative of $f(x) = \sin x$ at $x = 1$ by determining how much $\sin 1$ differs from $\sin 1.1$, $\sin 1.01$, and $\sin 1.001$, and doing the appropriate division.

2. Later, you found techniques for calculating derivatives exactly. Find $f'(1)$ if $f(x) = \sin x$. Show that the three approximate values of $f'(1)$ you calculated in Problem 1 are converging toward the value of $f'(1)$ you calculated in this problem.

3. The intuitive idea of instantaneous rate of change is made precise by a formal definition of derivative. Write both forms of this definition, one for $f'(x)$ and one for $f'(c)$, where c is a particular value of x.

4. The definition of derivative in Problem 3 involves the concept of limit. The limit of $f(x)$ as x approaches c is the number you can keep $f(x)$ close to simply by keeping x close enough to c. Suppose that $f(x) = e^x$. What does the limit of $f(x)$ equal as x approaches 2? How close would you have to keep x to 2 for $f(x)$ to be within 0.1 unit of this limit?

5. The intuitive idea of closeness is made precise in the formal definition of limit. Write this definition.

6. The quantities epsilon and delta from the definition of limit appear in Problem 4. Which one is epsilon, and which is delta?

7. Your intuitive introduction to variable-factor products came from the distance = rate × time equation. Draw an appropriate graph showing that if the rate is constant, you can represent the distance as the area of a rectangle.

8. If the rate varies, you can still represent the distance as the area of the region under a graph. Suppose you have measured the rates shown in the table at the given times. Plot the graph of rate versus time. Find the distance traveled between 1.0 and 2.8 by counting squares.

Time (s)	Rate (m/s)
1.0	7
1.3	9
1.6	13
1.9	12
2.2	10
2.5	8
2.8	5

9. More recently, you have learned ways to calculate definite integrals, such as in Problem 8, without having to draw the graph and count. Find the distance using Simpson's rule. Show that it is approximately the same as the distance you found by counting squares.

10. In Problems 8 and 9, you knew no equation for rate in terms of time. If you did know such an equation, there would be other ways to calculate the distance. Suppose that $v(t) = te^{-t}$. Find the distance traveled between $t = 0$ and $t = 2$ by calculating a Riemann sum with $n = 5$ increments, taking sample points at the midpoint of each subinterval.

11. You now know the fundamental theorem of calculus, which allows you to calculate limits of Riemann sums exactly using antiderivatives. Find the exact distance traveled in Problem 10. By what percentage does the Riemann sum in Problem 10 differ from the exact value?

12. State the fundamental theorem of calculus.

13. The proof of the fundamental theorem relies on the mean value theorem. Show that the mean value theorem does apply to $f(x) = x^{2/3}$ on the inverval $[0, 1]$, in spite of f not being differentiable at $x = 0$. Calculate the point $x = c$ in the interval $[0, 1]$ at which the conclusion of the mean value theorem is true. Then plot the graph accurately, and show that the line through $(c, f(c))$ with slope $f'(c)$ really does satisfy the requirements of the mean value theorem.

14. To use the fundamental theorem, you must be reasonably good at finding indefinite integrals. Write an integral that you can evaluate by the given technique, and evaluate the integral.

    a. Partial fractions

    b. Trigonometric substitution

15. Sometimes integration by parts results in the same integral appearing on both sides of the equals sign. Show how you can handle this situation by evaluating the integral for $\int \sec^3 x \, dx$.

16. In addition to their role in evaluating definite integrals, indefinite integrals have applications in their own right. For instance, if you know how a function changes, you can find an equation for the function itself. Suppose the instantaneous rate of change of y with respect to x is directly

proportional to y. Write the appropriate differential equation and solve it to express y in terms of x.

17. Definite integrals arise from variable-factor products. The volume of a solid is equal to its cross-sectional area times its height. Sketch the solid formed by rotating about the y-axis the region in Quadrant I under the graph of $y = 4 - x^2$. Then slice the height in such a way that the cross-sectional area is (essentially) constant at any point in the slice.

18. Sketch the solid in Problem 17 again, and draw a slice using cylindrical shells. Show that this method slices the cross-sectional area so that the height is (essentially) constant at any point in the slice.

19. Recall that the moment of a quantity is the magnitude of that quantity times a distance from a reference point, line, or plane. Find the first moment of area of the region in Problem 17 with respect to the y-axis. Use the result to find the x-coordinate of the centroid of the region.

20. Once you understand the concept of variable-factor products, you can analyze any such problem, even when the quantities are unfamiliar. For instance, the number of calories needed to warm a substance from temperature T_1 to temperature T_2 equals the substance's heat capacity (in calories per degree) times the change in temperature, $T_2 - T_1$. Unfortunately, most real substances have heat capacities that vary with temperature. Assume that *calculus foeride* (a rare, tough substance!) has a heat capacity given by

$$C = 10 + 0.3T^{1/2}$$

where heat capacity, C, is in calories per degree and temperature, T, is in degrees Celsius. How many calories would you need to warm the calculus foeride from $100°$ to $900°C$?

21. You can apply definite integrals to the mathematical world. For instance, you might define a function as a definite integral. The sine-integral function is defined as

$$\text{Si } x = \int_0^x \frac{\sin u}{u}\, du$$

   a. Write an equation for Si'x.
   b. Expand the integrand as a Maclaurin series, then integrate to get a series for Si x.
   c. Evaluate Si 0.7 approximately, using the first two nonzero terms of the series.
   d. Find an upper bound for the tail of the series that is left after the first two terms. Based on this result, how many decimal places can you guarantee that your answer in part c is correct?
   e. Prove that the series for Si x converges for all values of x.

22. You can use parametric equations to apply calculus to vector functions. Suppose an object is moving in such a way that its position vector, $\vec{r}$, is given by

$$\vec{r} = (t^3)\vec{i} + (t^2)\vec{j}$$

Plot accurately the path of the object from $t = 0$ to $t = 1$. Show the location of the object when $t = 0.5$. Calculate the velocity and acceleration vectors when $t = 0.5$. Plot these two vectors with their tails at the object. Is the object speeding up or slowing down? Explain.

23. You can apply calculus to figures in polar coordinates. Figure FE-1 shows the polar graph of

$$r = \cos \theta$$

from $\theta = 0$ to $\theta = \pi/2$. Slicing the region as shown gives a wedge of angle $d\theta$. Any point on the graph that is within the angle $d\theta$ has (essentially) the same radius as at the sample point. The area of the wedge is approximately equal to the area of a sector of a circle of radius r. The sector, of course, is $d\theta/(2\pi)$ of the area of a circle of radius r. Use this information to find dA, the area of the wedge, in terms of θ. Then find the area from $\theta = 0$ to $\theta = \pi/6$ by adding the dA's and taking the limit as $d\theta$ approaches 0 (that is, definite integrating).

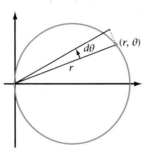

Figure FE-1

Summary of Properties of Trigonometric Functions

1. *Reciprocal*

$$\cot x = \frac{1}{\tan x} \quad \text{or} \quad \tan x \cot x = 1$$

$$\sec x = \frac{1}{\cos x} \quad \text{or} \quad \cos x \sec x = 1$$

$$\csc x = \frac{1}{\sin x} \quad \text{or} \quad \sin x \csc x = 1$$

2. *Quotient*

$$\tan x = \frac{\sin x}{\cos x} = \frac{\sec x}{\csc x}$$

$$\cot x = \frac{\cos x}{\sin x} = \frac{\csc x}{\sec x}$$

3. *Pythagorean*

$$\cos^2 x + \sin^2 x = 1$$

$$1 + \tan^2 x = \sec^2 x$$

$$\cot^2 x + 1 = \csc^2 x$$

4. *Odd-Even*

$$\sin(-x) = -\sin x \quad \text{(odd)}$$

$$\cos(-x) = \cos x \quad \text{(even)}$$

$$\tan(-x) = -\tan x \quad \text{(odd)}$$

$$\cot(-x) = -\cot x \quad \text{(odd)}$$

$$\sec(-x) = \sec x \quad \text{(even)}$$

$$\csc(-x) = -\csc x \quad \text{(odd)}$$

5. *Cofunction*

$$\cos(90° - \theta) = \sin\theta; \quad \cos\left(\frac{\pi}{2} - x\right) = \sin x$$

$$\cot(90° - \theta) = \tan\theta; \quad \cot\left(\frac{\pi}{2} - x\right) = \tan x$$

$$\csc(90° - \theta) = \sec\theta; \quad \csc\left(\frac{\pi}{2} - x\right) = \sec x$$

6. *Composite-Argument*

$$\cos(A - B) = \cos A \cos B + \sin A \sin B$$

$$\cos(A + B) = \cos A \cos B - \sin A \sin B$$

$$\sin(A - B) = \sin A \cos B - \cos A \sin B$$

$$\sin(A + B) = \sin A \cos B + \cos A \sin B$$

$$\tan(A - B) = \frac{\tan A - \tan B}{1 + \tan A \tan B}$$

$$\tan(A + B) = \frac{\tan A + \tan B}{1 - \tan A \tan B}$$

7. *Double-Argument*

$$\sin 2x = 2\sin x \cos x$$

$$\cos 2x = \cos^2 x - \sin^2 x$$

$$= 1 - 2\sin^2 x$$

$$= 2\cos^2 x - 1$$

$$\tan 2x = \frac{2\tan x}{1 - \tan^2 x}$$

$$\cos^2 x = \tfrac{1}{2}(1 + \cos 2x)$$

$$\sin^2 x = \tfrac{1}{2}(1 - \cos 2x)$$

8. *Half-Argument*

$$\sin \tfrac{1}{2}x = \pm\sqrt{\tfrac{1}{2}(1 - \cos x)}$$

$$\cos \tfrac{1}{2}x = \pm\sqrt{\tfrac{1}{2}(1 + \cos x)}$$

$$\tan \tfrac{1}{2}x = \pm\sqrt{\frac{1 - \cos x}{1 + \cos x}}$$

$$= \frac{\sin x}{1 + \cos x} = \frac{1 - \cos x}{\sin x}$$

9. *Sum and Product*

$$2\cos A \cos B = \cos(A + B) + \cos(A - B)$$

$$2\sin A \sin B = -\cos(A + B) + \cos(A - B)$$

$$2\sin A \cos B = \sin(A + B) + \sin(A - B)$$

$$2\cos A \sin B = \sin(A + B) - \sin(A - B)$$

$$\cos x + \cos y = 2\cos\tfrac{1}{2}(x + y)\cos\tfrac{1}{2}(x - y)$$

$$\cos x - \cos y = -2\sin\tfrac{1}{2}(x + y)\sin\tfrac{1}{2}(x - y)$$

$$\sin x + \sin y = 2\sin\tfrac{1}{2}(x + y)\cos\tfrac{1}{2}(x - y)$$

$$\sin x - \sin y = 2\cos\tfrac{1}{2}(x + y)\sin\tfrac{1}{2}(x - y)$$

10. *Linear Combination of Sine and Cosine*

$$A\cos x + B\sin x = C\cos(x - D),$$

where $C = \sqrt{A^2 + B^2}$ and $D = \arctan\dfrac{B}{A}$.

659

Answers to Selected Problems

CHAPTER 1

Problem Set 1-1

1. a. 95 cm
   b. From 5 to 5.1: average rate ≈ 26.34 cm/s
      From 5 to 5.01: average rate ≈ 27.12 cm/s
      From 5 to 5.001: average rate ≈ 27.20 cm/s
      The instantaneous rate of change of *d* at *t* = 5 is approximately 27.20 cm/s.
   c. Instantaneous rate would involve division by zero
   d. For *t* = 1.5 to 1.501, rate ≈ −31.42 cm/s. The pendulum is approaching the wall: The rate of change is negative, so the distance is decreasing.
   e. The instantaneous rate of change is the limit of the average rates as the time interval approaches zero. It is called the *derivative.*
   f. Before *t* = 0, the pendulum was not yet moving. For large values of *t*, the pendulum's motion will die out because of friction.

Problem Set 1-2

1. a. Increasing slowly
   b. Increasing fast
3. a. Decreasing fast
   b. Decreasing slowly
5. a. Increasing fast
   b. Increasing slowly
   c. Decreasing slowly
   d. Increasing fast
7. a. Increasing slowly
   b. Increasing slowly
   c. Increasing slowly

9. a. Increasing fast
   b. Neither increasing nor decreasing
   c. Increasing fast
   d. Increasing slowly

11. a.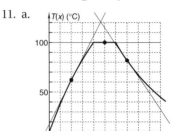

    $x = 40$: rate ≈ 1.1°/s
    $x = 100$: rate = 0°/s
    $x = 140$: rate ≈ −0.8°/s
   b. Between 0 and 80 s the water is warming up, but at a decreasing rate.
      Between 80 and 120 s the water is boiling, thus staying at a constant temperature.
      Beyond 120 s the water is cooling down, rapidly at first, then more slowly.

13. a.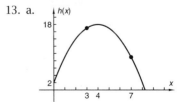

    Increasing at $x = 3$
    Decreasing at $x = 7$
   b. $h(3) = 17, h(3.1) = 17.19$
      Average rate = 1.9 ft/s

c. From 3 to 3.01: average rate = 1.99 ft/s
  From 3 to 3.001: average rate = 1.999 ft/s
  The limit appears to be 2 ft/s.

d. The derivative at $x = 7$ appears to be -6 ft/s.
  The derivative is negative because $h(x)$ is
  decreasing at $x = 7$.

15. a. Average rate = 529.902... bacteria/h

  b. $r(t) = \dfrac{200(1.2^t) - 200(1.2^2)}{t - 2}$

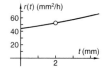

  $r(2)$ is undefined.

  c. The difference between the derivative and $r(2.01)$
  is 0.04789....
  Keep t within 0.002 unit of 2 to keep the average
  rate within 0.01 unit of the derivative.

17. a. i. -1.0 in./s   ii. 0.0 in./s   iii. 1.15 in./s

  b. 1.7 s because $y = 0$ at that time

19. a. Quadratic (or polynomial)

  b. Increasing, because the rate of change from 2.99
  to 3.01 is positive.

21. a. Exponential

  b. Increasing, because the rate of change from 1.99
  to 2.01 is positive.

23. a. Rational algebraic

  b. Decreasing, because the rate of change from 3.99
  to 4.01 is negative.

25. a. Linear (or polynomial)

  b. Decreasing, because the rate of change from 4.99
  to 5.01 is negative.

27. a. Circular (or trigonometric)

  b. Decreasing, because the rate of change from 1.99
  to 2.01 is negative.

29. Physical meaning of a derivative: instantaneous rate
  of change
  To estimate a derivative graphically: Draw a tangent
  line at the point on the graph and measure its slope.
  To estimate a derivative numerically: Take a small
  change in x, find the corresponding change in $f(x)$,
  then divide. Repeat, using a smaller change in x. See
  what number these average rates approach as the
  change in x approaches zero.
  The numerical method illustrates the fact that the
  derivative is a limit.

Problem Set 1-3

1. a. Approximately 30.8

  b. Approximately 41.8

3. a. Approximately 2.0

  b. Approximately 1.0

5. Distance ≈ 680 ft

7. Derivative = 3.42...

9. a.

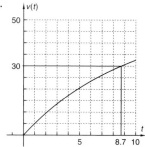

  The range is $0 \le y \le 32.5660...$.

  b. 8.6967... s

  c. Distance ≈ 150 ft; the concept used is the definite
  integral.

  d. About 3.1 (ft/s)/s; the concept used is the
  derivative. The rate of change of velocity is
  acceleration.

11. About 7.1 cm

13. See the text for the meaning of *definite integral*.

Problem Set 1-4

1. a.

  b. Integral $\approx 281{,}000$ ft
  The sum overestimates the integral because the
  trapezoids are circumscribed about the region and
  thus include more area.

  c. The units are (ft/s)(s), which equals feet, so the
  integral represents the distance the spaceship has
  traveled.

  d. Yes, it will be going fast enough because
  $v(30) \ge 27{,}000$.

3. Distance ≈ 396 ft; the plane is not in danger of
  running off the deck.

5. Answers will vary.

7. a.

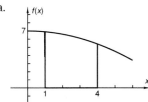

b. $T_{10} = 18.8955$
$T_{20} = 18.898875$
$T_{50} = 18.89982$
These values underestimate the integral because the trapezoids are inscribed in the region.

c. T_{10}: 0.0045 unit from the exact answer
T_{20}: 0.001125 unit from the exact answer
T_{50}: 0.00018 unit from the exact answer
T_n is first within 0.01 unit of 18.9 when $n = 7$.
$T_7 = 18.8908...$, which is 0.0091... unit from 18.9.
T_n is getting closer to 18.9 as n increases, so T_n is within 0.01 unit of 18.9 for all $n \geq 7$.

9. Area $\approx 13,808.38...$ cm^2
The estimate is too low because the trapezoids are inscribed within the ellipse. The exact area is $4400\pi = 13,823.007...$ cm^2.

11. $n = 10$: integral ≈ 21.045
$n = 100$: integral ≈ 21.00045
$n = 1000$: integral ≈ 21.0000045
Conjecture: integral $= 21$
The word is *limit*.

13. If the trapezoids are inscribed (graph concave down), the rule underestimates the integral. If the trapezoids are circumscribed (graph concave up), the rule overestimates the integral.

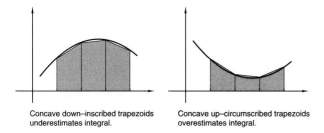

Concave down–inscribed trapezoids underestimates integral.

Concave up–circumscribed trapezoids overestimates integral.

Problem Set 1-5

1. Answers will vary.

Problem Set 1-6

R1. a. Approximately 15.4 ft

b. From 3.9 to 4: average rate ≈ -40.1 ft/s
From 4 to 4.1: average rate ≈ -29.3 ft/s

Instantaneous rate ≈ -34.7 ft/s
The distance from the water is decreasing, so he is going down.

c. Instantaneous rate ≈ 70.8

d. Going up at about 70.8 ft/s

e. Derivative

R2. a. Physical meaning: instantaneous rate of change of a function
Graphical meaning: slope of a tangent line to a function at a given point

b. $x = -4$: Decreasing fast
$x = 1$: Increasing slowly
$x = 3$: Increasing fast
$x = 5$: Neither increasing nor decreasing

c. From 2 to 2.1: average rate $= 43.6547...$
From 2 to 2.01: average rate $= 40.5614...$
From 2 to 2.001: average rate $= 40.2683...$
Differences between average rates and instantaneous rate, respectively: $3.4187...$, $0.3255...$, $0.03239...$
Yes; the derivative, the limit

d. $t = 2$: 3.25 m/s
$t = 18$: 8.75 m/s
$t = 24$: 11.5 m/s
Her velocity stays constant from 6 s to 16 s.
At $t = 24$, Mary is in her final sprint toward the finish line.

R3. Distance ≈ 23.2 ft (exact answer is $23.2422...$); the definite integral

R4. a. The graph confirms Figure 1-6c.

b. Integral ≈ 15.0 (exact answer is 15)

c. $T_6 = 14.9375$
The trapezoidal sum underestimates the integral because the trapezoids are inscribed in the region.

d. $T_{50} = 14.9991$; difference $= 0.0009$
$T_{100} = 14.999775$; difference $= 0.000225$
Yes; the limit

R5. Answers will vary.

CHAPTER 2

Problem Set 2-1

1. a. $f(2) = \dfrac{8 - 10 + 2}{2 - 2} = \dfrac{0}{0}$
No value for $f(2)$ because of division by 0

b.

x	$f(x)$
1.997	2.994
1.998	2.996
1.999	2.998
2	undefined
2.001	3.002
2.002	3.004
2.003	3.006

$f(x)$ stays close to 3 when x is kept close to 2, but not equal to 2.

c. To keep $f(x)$ within 0.0001 unit of 3, keep x within 0.00005 unit of 2. To keep $f(x)$ within 0.00001 unit of 3, keep x within 0.000005 unit of 2. To keep $f(x)$ arbitrarily close to 3, keep x within $\frac{1}{2}$ that distance of 2.

d. The discontinuity can be "removed" by defining $f(2)$ to equal 3.

3.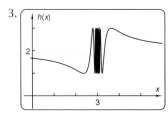

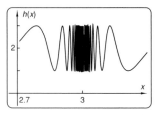

There is no limit because the graph cycles infinitely as x approaches 3.

Problem Set 2-2

1. See the text for the definition of *limit*.

3. Has a limit, 3

5. Has a limit, 3

7. Has no limit

9. Has a limit, 7

11. Has no limit

13. $\lim_{x \to 3} f(x) = 5$; for $\epsilon = 0.5$, $\delta \approx 0.2$, or 0.3

15. $\lim_{x \to 6} f(x) = 4$; for $\epsilon = 0.7$, $\delta \approx 0.5$, or 0.6

17. $\lim_{x \to 5} f(x) = 2$; for $\epsilon = 0.3$, $\delta \approx 0.5$, or 0.6

19. a. The graph should match Problem 13.
   b. $\lim_{x \to 3} f(x) = 5$
   c. Maximum $\delta = -\sin^{-1}(-0.25) = 0.25268...$
   d. Maximum $\delta = \sin^{-1}(\epsilon/2)$

21. a. The graph should match Problem 15.
   b. $\lim_{x \to 6} f(x) = 4$
   c. Maximum $\delta = 1 - (2.3/3)^3 = 0.5493...$
   d. Maximum $\delta = 1 - ((3 - \epsilon)/3)^3$

23. a. The graph should match Problem 17.
   b. $\lim_{x \to 5} f(x) = 2$
   c. Maximum $\delta = \sqrt{0.3} = 0.54772...$
   d. Maximum $\delta = \sqrt{\epsilon}$

25. a. $f(2) = \dfrac{(5)(0)}{0} = \dfrac{0}{0}$
   The graph has a removable discontinuity at $x = 2$.
   Limit $= 2^2 - 6(2) + 13 = 5$
   b. $1.951191... < x < 2.051316...$
   $\therefore$ maximum $\delta = 0.048808...$
   c.
   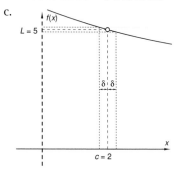

27. a. $m(t) = \dfrac{3t^2 - 48}{t - 4}$

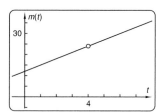

   b. The graph has a removable discontinuity at $t = 4$.
   c. Limit $= 24$ ft/s
   d. Keep t within 0.04 s of 4 s.
   e. The limit of the average velocity is the exact velocity.

1.

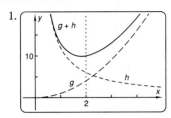

$\lim_{x\to 2} f(x) = 10, \lim_{x\to 2} g(x) = 4$, and $\lim_{x\to 2} h(x) = 6$
$\therefore \lim_{x\to 2} f(x) = \lim_{x\to 2} g(x) + \lim_{x\to 2} h(x)$, Q.E.D.

x	$f(x)$
1.97	9.9722...
1.98	9.9810...
1.99	9.9902...
2.00	10
2.01	10.0102...
2.02	10.0209...
2.03	10.0322...

All these $f(x)$ values are close to 10.

3.

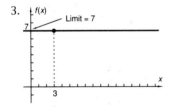

The limit is 7 because $f(x)$ is *always* close to 7, no matter what value x takes on. (It shouldn't bother you that $f(x) = 7$ for $x \neq 3$, if you think of the definition of limit for a while.)

5.

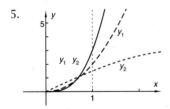

$\lim_{x\to 1} y_1 = 2, \lim_{x\to 1} y_2 = 1.5$, and $\lim_{x\to 1} y_1 \cdot y_2 = 3$
$\therefore \lim_{x\to 1} y_1 \cdot y_2 = \lim_{x\to 1} y_1 \cdot \lim_{x\to 1} y_2$, Q.E.D.

x	$y_3 = f(x)$
0.997	2.9739...
0.998	2.9825...
0.999	2.9912...
1	3
1.001	3.0087...
1.002	3.0174...
1.003	3.0262...

All these $f(x)$ values are close to $2(1.5) = 3$.

7. $\lim_{x\to 3} f(x) = \lim_{x\to 3} x^2 - 9x + 5$
$= \lim_{x\to 3} x^2 - \lim_{x\to 3} 9x + \lim_{x\to 3} 5$
Limit of a sum
$= \lim_{x\to 3} x \cdot \lim_{x\to 3} x - 9\lim_{x\to 3} x + 5$
Limit of a product, Limit of a constant
$= (3)(3) - 9(3) + 5 = -13$
Limit of x

9.

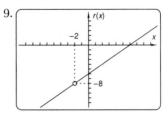

$r(-2) = \dfrac{(-2)^2 - 4(-2) - 12}{(-2) + 2} = \dfrac{0}{0}$

$r(x) = \dfrac{(x - 6)(x + 2)}{x + 2} = x - 6, x \neq -2$

$\lim_{x\to -2} r(x) = -8$

Proof:

$\lim_{x\to -2} r(x) = \lim_{x\to -2}(x - 6)$
Because $x \neq -2$
$= \lim_{x\to -2} x + \lim_{x\to -2}(-6)$
Limit of a sum
$= -2 - 6 = -8$, Q.E.D.
Limit of x

11.

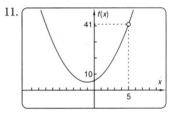

$f(-5) = \dfrac{5^3 - 3(5^2) - 4(5) - 30}{5 - 5} = \dfrac{0}{0}$

$f(x) = \dfrac{(x^2 + 2x + 6)(x - 5)}{x - 5} = x^2 + 2x + 6, x \neq 5$

$\lim_{x\to 5} f(x) = 41$

Proof:

$\lim_{x\to 5} f(x) = \lim_{x\to 5}(x^2 + 2x + 6)$
Because $x \neq 5$
$= \lim_{x\to 5} x^2 + \lim_{x\to 5}(2x) + \lim_{x\to 5} 6$
Limit of a sum
$= \lim_{x\to 5} x \cdot \lim_{x\to 5} x + 2\lim_{x\to 5} x + 6$
Limit of a product, Limit of a constant
$= 5 \cdot 5 + 2 \cdot 5 + 6 = 41$, Q.E.D.
Limit of x

13.

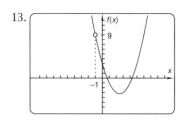

$$f(-1) = \frac{(-1)^3 - 4(-1)^2 - 2(-1) + 3}{(-1) + 1} = \frac{0}{0}$$

$$f(x) = \frac{(x^2 - 5x + 3)(x + 1)}{x + 1} = x^2 - 5x + 3, x \neq -1$$

$\lim_{x \to -1} f(x) = 9$

Proof:

$\lim_{x \to -1} f(x) = \lim_{x \to -1}(x^2 - 5x + 3)$
Because $x \neq -1$
$= \lim_{x \to -1} x^2 + \lim_{x \to -1}(-5x) + \lim_{x \to -1} 3$
Limit of a sum
$= \lim_{x \to -1} x \cdot \lim_{x \to -1} x - 5 \lim_{x \to -1} x + 3$
Limit of a product, Limit of a constant
$= (-1)(-1) + (-5)(-1) + 3 = 9$, Q.E.D.
Limit of x

15. $f(x) = \dfrac{x^3 - 3x^2 - 4x - 30}{x - 5}$

x	$f(x)$
4.990	40.8801...
4.991	40.8921...
4.992	40.9041...
4.993	40.9160...
4.994	40.9280...
4.995	40.9402...
4.996	40.9520...
4.997	40.9641...
4.998	40.9760...
4.999	40.9880...
5	undefined
5.001	41.0120...
5.002	41.0240...
5.003	41.0360...
5.004	41.0480...
5.005	41.0600...
5.006	41.0720...
5.007	41.0840...
5.008	41.0961...
5.009	41.1081...

The table shows that $f(x)$ will be within 0.1 unit of $\lim_{x \to 5} f(x) = 41$ if we keep x within 0.008 unit of 5.

17. $f(x) = \dfrac{x^2 - 5x + 6}{x^2 - 6x + 9} = \dfrac{(x - 2)(x - 3)}{(x - 3)(x - 3)} = \dfrac{x - 2}{x - 3}$

You cannot find the limit by substituting into the simplified form because the denominator still becomes zero.

19. a. $5(0)^{1/2} = 0 = v(0)$
$5(1)^{1/2} = 5 = v(1)$
$5(4)^{1/2} = 10 = v(4)$
$5(9)^{1/2} = 15 = v(9)$
$5(16)^{1/2} = 20 = v(16)$

b. $a(9) \approx 0.8333101...$
Conjecture: $a(9) = 0.8333... = 5/6$
Units of $a(t)$: (mi/h)/s

c. $a(9) = \lim_{t \to 9} \dfrac{v(t) - v(9)}{t - 9} = \lim_{t \to 9} \dfrac{5t^{1/2} - 15}{t - 9}$

$= \lim_{t \to 9} \dfrac{5(t^{1/2} - 3)}{(t^{1/2} - 3)(t^{1/2} + 3)}$

$= \lim_{t \to 9} \dfrac{5}{t^{1/2} + 3}$

$= \dfrac{5}{6}$, which agrees with the conjecture.

d. The truck went about 127 ft.

21. $-0.05994...$

23. **Proof:**

Anchor:
If $n = 1$, $\lim_{x \to c} x^1 = c = c^1$ by the limit of x.
Induction Hypothesis:
Assume that the property is true for $n = k$.
$\therefore \lim_{x \to c} x^k = c^k$
Verification for $n = k + 1$:
$\lim_{x \to c} x^{k+1} = \lim_{x \to c}(x^k \cdot x) =$
$\lim_{x \to c} x^k \cdot \lim_{x \to c} x = c^k \cdot c = c^{k+1}$
By the induction hypothesis
Conclusion:
$\therefore \lim_{x \to c} x^n = c^n$ for *all* integers $n \geq 1$, Q.E.D.

Problem Set 2-4

1. a. Has left and right limits
   b. Has no limit
   c. Discontinuous; has no limit

3. a. Has left and right limits
   b. Has a limit
   c. Continuous

5. a. Has no left and right limits
   b. Has no limit
   c. Discontinuous; no limit or $f(2)$

7. a. Has left and right limits
   b. Has a limit
   c. Discontinuous; $f(1) \neq$ limit
9. a. Has left and right limits
   b. Has a limit
   c. Discontinuous; no $f(c)$
11. Answers will vary.   13. Answers will vary.

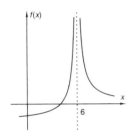

15. Answers will vary.   17. Answers will vary.

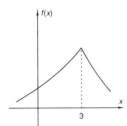
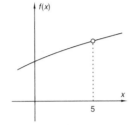

19. Answers will vary.

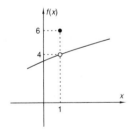

21. Discontinuous at $x = -3$
23. Discontinuous at $x = \pi/2 + \pi n$, where n is an integer
25. Discontinuous because $\lim_{x \to 2} f(x) = 2$ and $f(2) = 3$

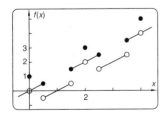

27. Discontinuous because $s(x)$ has no limit as x approaches 2 from the left (no real function values to the left of $x = 2$)

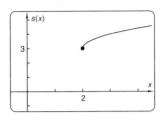

29. Discontinuous because there is no value of $h(2)$

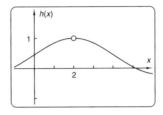

31.

c	$f(c)$	$\lim\limits_{x \to c^-} f(x)$	$\lim\limits_{x \to c^+} f(x)$	$\lim\limits_{x \to c} f(x)$	Continuous?
1	4	2	2	2	Removable
2	1	1	1	1	Continuous
4	5	5	2	None	Step
5	None	None	None	None	Infinite

33. a.

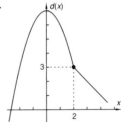

b. $\lim_{x \to 2^-} d(x) = 3$, $\lim_{x \to 2^+} d(x) = 3$; continuous

35. a.

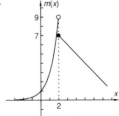

b. $\lim_{x \to 2^-} m(x) = 9$, $\lim_{x \to 2^+} m(x) = 7$; not continuous

37. $k = 2.5$

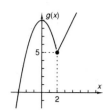

39. $k = -1/2$

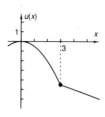

41. a. $b - 1 = a$

b. $a = -1 \Rightarrow b = 0$

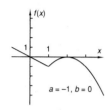

a = -1, b = 0

c. For example, $a = 1 \Rightarrow b = 2$.

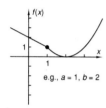

e.g., a = 1, b = 2

43. Let $T(\theta)$ = the number of seconds it takes to cross.

$$T(\theta) = \begin{cases} 24, & \text{if } \theta = 90° \\ \dfrac{40}{\sin \theta}, & \text{if } \theta \neq 90° \end{cases}$$

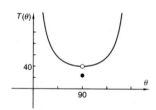

45. For any value of c, $P(c)$ is determined by addition and multiplication. Because the set of real numbers is closed under multiplication and addition, $P(c)$ will be a *unique, real* number for any real value $x = c$. $P(c)$ is the *limit* of $P(x)$ as x approaches c by the properties of the limit of a product of functions (for powers of x), the limit of a constant times a function (for multiplication by the coefficients), and the limit of a sum (for the individual terms). Therefore, P is continuous for all values of x.

Problem Set 2-5

1. $\lim_{x \to -\infty} f(x) = \infty$, $\lim_{x \to -3^-} f(x) = -4$
$\lim_{x \to -3^+} f(x) = 3$, $\lim_{x \to 1} f(x) = -\infty$
$\lim_{x \to 2} f(x) = 1$, $\lim_{x \to 3^-} f(x) = \infty$
$\lim_{x \to -3^+} f(x) = 2$, $\lim_{x \to \infty} f(x)$ does not exist.

3.

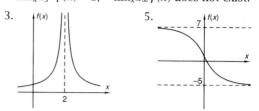

5.

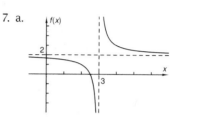

7. a.

b. $\lim_{x \to 3^+} f(x) = \infty$, $\lim_{x \to 3^-} f(x) = -\infty$,
$\lim_{x \to 3} f(x)$, none, $\lim_{x \to \infty} f(x) = 2$,
$\lim_{x \to -\infty} f(x) = 2$

c. $x = 3\frac{1}{98} = 3.0102...$

x	$f(x)$
3.01	102
3.001	1002
3.0001	10002

All these values of $f(x)$ are greater than 100. $\lim_{x \to 3^+} f(x) = \infty$ means that $f(x)$ can be kept arbitrarily far from 0 just by keeping x close enough to 3 on the positive side. There is a vertical asymptote at $x = 3$.

d. $x = 1003$

x	$f(x)$
1004	2.00099...
1005	2.00099...
1006	2.00099...

All these are within 0.001 unit of 2. $\lim_{x \to \infty} f(x) = 2$ means that you can keep $f(x)$ arbitrarily close to 2 by making the value of x arbitrarily large. $y = 2$ is a horizontal asymptote.

9. a.

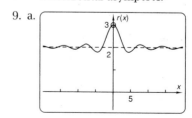

b. $\lim_{x \to \infty} r(x) = 2$

c. $r(28) = 2.00968\ldots$, which is within 0.01 unit of 2. $r(32) = 2.01723\ldots$, which is more than 0.01 unit away from 2.

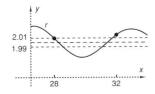

$D = 1000$

d. The line $y = 2$ is an asymptote. Even though $r(x)$ oscillates back and forth across this line, the limit of $r(x)$ is 2 as x approaches infinity, satisfying the definition of an asymptote.

e. The graph suggests that $\lim_{x \to 0} r(x) = 3$.

11. The limit is infinite. y is unbounded as x approaches infinity. If there were a number E such that $\log x < E$ for all $x > 0$, then you could let $x = 10^{2E}$ so that $\log x = \log 10^{2E} = 2E$, which is greater than E, which was assumed to be an upper bound.

13. a. The definite integral is the product of the independent and the dependent variables. Because distance = (rate)(time), the integral represents distance in this case.

b. $T_9 = 17.8060052\ldots$
$T_{45} = 17.9819616\ldots$
$T_{90} = 17.9935649\ldots$
$T_{450} = 17.9994175\ldots$

c. The exact answer is 18. It is a limit because the sums can be made as close to it as you like just by making the number of trapezoids large enough (and thus keeping their widths close to zero). The sums are smaller than the integral because each trapezoid is inscribed under the graph and thus leaves out a part of its respective strip of the region.

d. T_n is 0.01 unit from 18 when it equals 17.99. From part a, this occurs between $n = 45$ and $n = 90$. By experimentation,
$T_{66} = 17.9897900\ldots$
$T_{67} = 17.9900158\ldots$
Therefore, the approximation is within 0.01 unit of 18 for any value of $n \geq 67$.

15. Length $= 100 \sec x = 100 / \cos x$
x must be within $0.100167\ldots$ radian of $\pi / 2$.

Problem Set 2-6

1. The intermediate value theorem applies on $[1, 4]$ because f is a polynomial function, and polynomial functions are continuous for all x.

$c = 1.4349\ldots$

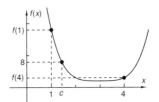

3. a. For $1 \leq y < 2$ or for $5 < y \leq 8$, the conclusion would be true. But for $2 \leq y \leq 5$, it would be false because there are no values of x in $[1, 5]$ that give these values for $f(x)$.

b. The conclusion of the theorem is true because every number y in $[4, 6]$ is a value of $g(x)$ for some value of x in $[1, 5]$.

5. Let $f(x) = x^2$.
f is a polynomial function, so it is continuous and thus the intermediate value theorem applies.
$f(1) = 1$ and $f(2) = 4$, so there is a number c between 1 and 2 such that $f(c) = 3$.
By the definition of square root, $c = \sqrt{3}$, Q.E.D.

7. The intermediate value theorem is called an existence theorem because it tells you that a number such as $\sqrt{3}$ *exists*. It does not tell you how to calculate that number.

9. Let $f(t) =$ Jesse's speed $-$ Kay's speed.
$f(1) = 20 - 15 = 5$, which is *positive*.
$f(3) = 17 - 19 = -2$, which is *negative*.
The speeds are assumed to be continuous, so f is also continuous and the intermediate value theorem applies. So there is a value of t between 1 and 3 for which $f(t) = 0$, meaning that Jesse and Kay are going at exactly the same speed at that time. The *existence* of the time tells you neither what that time is nor what the speed is.

11. You must assume that the cosine function is continuous.
$c = \cos^{-1} 0.6 = 0.9272\ldots$
Take the inverse cosine of both sides of the equation.

13. This means that a function graph has a high point and a low point on any interval in which the function is continuous.

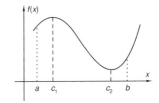

If the function is *not* continuous, there may be a point missing where the maximum or minimum would have been.

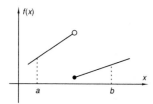

Another possibility would be a graph with a vertical asymptote somewhere between a and b.

Problem Set 2-7

R0. Answers will vary.

R1. a. $f(3) = \dfrac{0}{0}$

Indeterminate form

b.

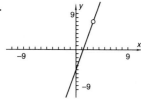

At $x = 3$ there is a removable discontinuity.

c. For 0.01, keep x within 0.0025 unit of 3.
For 0.0001, keep x within 0.000025 unit of 3.
To keep $f(x)$ within ϵ unit of 7, keep x within $\frac{1}{4}\epsilon$ unit of 3.

R2. a. $L = \lim_{x \to c} f(x)$ if and only if
for any number $\epsilon > 0$, no matter how small
there is a number $\delta > 0$ such that
if x is within δ units of c but $x \neq c$,
then $f(x)$ is within ϵ units of L.

b. $\lim_{x \to 1} f(x) = 1$
$\lim_{x \to 2} f(x)$ does not exist.
$\lim_{x \to 3} f(x) = 4$
$\lim_{x \to 4} f(x)$ does not exist.
$\lim_{x \to 5} f(x) = 3$

c. $\lim_{x \to 2} f(x) = 3$
Maximum δ: 0.6 or 0.7

d. The left side of $x = 2$ is the more restrictive.
Let $2 + \sqrt{x - 1} = 3 - 0.4 = 2.6$.
Maximum δ: 0.64

e. Let $f(x) = 3 - \epsilon$.
$2 + \sqrt{x - 1} = 3 - \epsilon$
$x = (1 - \epsilon)^2 + 1$
Let $\delta = 2 - ((1 - \epsilon)^2 + 1) = 1 - (1 - \epsilon)^2$, which is
positive for all positive $\epsilon < 1$.

R3. a. See the limit property statements in the text.

b.

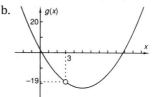

The limit of a quotient property does not apply because the limit of the denominator is 0.
$g(x) = x^2 - 10x + 2, x \neq 3$
You can cancel the $(x - 3)$ because the definition of limit says, "but not *equal* to 3."
$\lim_{x \to 3} g(x) = \lim_{x \to 3} x^2 + \lim_{x \to 3} (-10x) + \lim_{x \to 3} 2$
Limit of a sum
$= \lim_{x \to 3} x \cdot \lim_{x \to 3} x - 10 \lim_{x \to 3} x + 2$
Limit of a product, Limit of a constant times a function, and Limit of a constant
$= 3 \cdot 3 - 10(3) + 2 = -19$
Limit of x

c. $\lim_{x \to 3} f(x) = 8$, $\lim_{x \to 3} g(x) = 2$
$\lim_{x \to 3} f(x) \cdot g(x) = 8 \cdot 2 = 16$

x	$p(x)$
2.997	15.9907...
2.998	15.9938...
2.999	15.9969...
3	undefined
3.001	16.0030...
3.002	16.0061...
3.003	16.0092...

All these $p(x)$ values are close to 16.

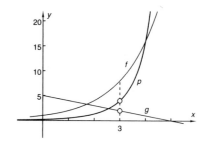

d. For 5 s to 5.1 s: average velocity $= -15.5 \, \text{m/s}$
Average velocity $= \dfrac{35t - 5t^2 - 50}{t - 5} = -5(t - 2)$, for $t \neq 5$
Instantaneous velocity $= \lim = -5(5 - 2)$
$= -15 \, \text{m/s}$
The rate is negative, so the distance above the starting point is getting smaller, which means the rock is going down.
Instantaneous velocity is a derivative.

R4. a. f is continuous at $x = c$ if and only if
    1. $f(c)$ exists
    2. $\lim_{x \to c} f(x)$ exists
    3. $\lim_{x \to c} f(x) = f(c)$
    f is continuous on $[a, b]$ if and only if f is
    continuous at every point in (a, b) and
    $\lim_{x \to a^+} f(x) = f(a)$ and $\lim_{x \to b^-} f(x) = f(b)$.

b.

c	$f(c)$	$\lim_{x \to c^-} f(x)$	$\lim_{x \to c^+} f(x)$	$\lim_{x \to c} f(x)$	Continuous?
1	None	None	None	None	Infinite
2	1	3	3	3	Removable
3	5	2	5	None	Step
4	3	3	3	3	Continuous
5	1	1	1	1	Continuous

c. i.
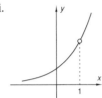
  ii.
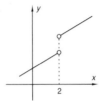

  iii.
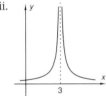
  iv.
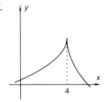

  v.

  vi.
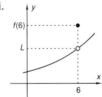

  vii.

d.
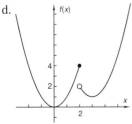

The left limit is 4 and the right limit is 2, so f is discontinuous at $x = 2$, Q.E.D.
$k = 12$

R5. a. $\lim_{x \to 4} f(x) = \infty$ means that $f(x)$ can be kept arbitrarily far from 0 on the positive side just by keeping x close enough to 4, but not equal to 4. $\lim_{x \to \infty} f(x) = 5$ means that $f(x)$ can be made to stay arbitrarily close to 5 just by keeping x large enough in the positive direction.

b. $\lim_{x \to -\infty} f(x)$ does not exist.
$\lim_{x \to -2} f(x) = 1$
$\lim_{x \to 2^-} f(x) = \infty$
$\lim_{x \to 2^+} f(x) = -\infty$
$\lim_{x \to \infty} f(x) = 2$

c. $x = 9.965$

x	$f(x)$
10	5.999023...
20	5.999999046...
30	5.99999999907...

All these $f(x)$ values are within 0.001 of 6.

d. $x = 10^{-3}$

x	$g(x)$
0.0009	$1.2345... \cdot 10^6$
0.0005	4,000,000
0.0001	$1 \cdot 10^{-8}$

All these $g(x)$ values are larger than 1,000,000.

e.

n	Trapezoidal Rule
50	467.9074...
100	467.9669...
200	467.9882...
400	467.9958...

The limit of these sums seems to be 468.
$D = 223$

R6. a. See the statement of the intermediate value theorem in the text.
The basis is the completeness axiom.
See the statement of the extreme value theorem in the text.
The word is *corollary*.

b. $f(3) = 8, f(4) = -4$
The intermediate value theorem; the continuity property
The value of x is 3.7553....

c.

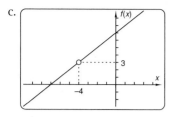

$f(x) = x + 7, x \neq 4$, so $f(-6) = 1$ and $f(-2) = 5$
Pick $y = 3$, and there is no value of x. This fact
does not contradict the intermediate value
theorem. Function f does not meet the continuity
hypothesis of the theorem.

CHAPTER 3

Problem Set 3-1

1. The graph is correct.

3.

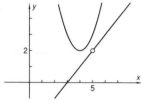

5. $r(x) = x - 3, x \neq 5$
   $f'(5) = 2$
   The derivative is the velocity of the spaceship,
   in km/min.

7. As you zoom in, the line and the graph appear to be
   the same.

Problem Set 3-2

1. See the text for the definition of *derivative*.

3. a. $f'(3) = 3.6$

   b.

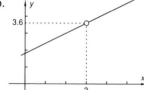

   c. and d.

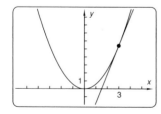

5. $f'(-2) = 1$

7. $f'(1) = -4$

9. $f'(3) = -0.7$

11. $f'(-1) = 0$

13. The derivative of a linear function equals the slope.
    The tangent line coincides with the graph of a linear
    function.

15. a. Find $f'(1) = 2$, then plot a line through point
    $(1, f(1))$ using $f'(1)$ as the slope. The line is
    $y = 2x - 1$.

    b. Near the point $(1, 1)$, the tangent line and the curve
    appear nearly the same.

    c. The curve appears to get closer and closer to the
    line.

    d. Near point $(1, 1)$ the curve looks linear.

    e. If a graph has local linearity, the graph near that
    point looks like the tangent line. Therefore, the
    derivative at that point could be said to equal the
    slope of the graph at that point.

17. a.

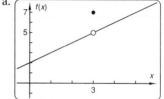

    b. Difference quotient is $m(x) = \dfrac{x - 5}{x - 3}$.

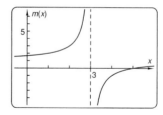

    c.

x	$f(x)$
2.997	667.66...
2.998	1001
2.999	2001
3.000	undefined
3.001	−1999
3.002	−999
3.003	−665.66...

The difference quotients are all large positive numbers on the left side of 3. On the right side they are large negative numbers. For a derivative to exist, the difference quotient must approach the *same* number as x gets closer to 3.

19. a. $f'(3) = -1$, and the tangent line on the graph has a slope of -1.

b.

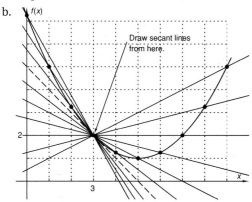

Draw secant lines from here.

As the x-distance between the point and 3 decreases, the secant lines (solid) approach the tangent line (dashed).

c. The same thing happens with secant lines from the left of $x = 3$. See the graph in part b.

d.

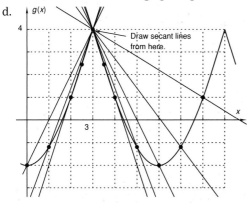

Draw secant lines from here.

e. A derivative is a limit. Because the left and right limits are unequal, there is no derivative at $x = 3$.

f. Conjecture: The numbers are π and $-\pi$.

Problem Set 3-3

1. a.

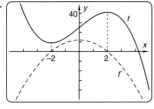

b. $f'(x)$ is positive for $-2 < x < 2$. The graph of f is increasing for these x-values.

c. $f(x)$ is decreasing for x satisfying $|x| > 2$. $f'(x) < 0$ for these values of x.

d. Where the f' graph crosses the x-axis, the f graph has a high point or a low point.

e. See the graph in part a.

f. Conjecture: f' is quadratic.

3. a.

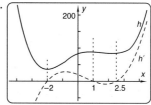

b. The h' graph looks like a cubic function graph. Conjecture: A seventh-degree function has a sixth-degree function for its derivative.

c. $h'(x) = 0$ for $x = -2, 1, 2.5$

d. If $h'(x) = 0$, the h graph has a high point or a low point. This is reasonable because if $h'(x) = 0$, the rate of change of $h(x)$ is zero, which would happen when the graph stops going up and starts going down or vice versa.

e. See the graph in part a.

5. a.

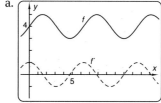

b. Amplitude = 1, period = $2\pi = 6.283...$

c. The graph of f' has amplitude 1 and period 2π.

d.

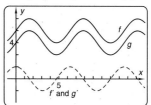

The graphs of f and g are the same shape, spaced 1 unit apart vertically. The graphs of f' and g' are identical! This is to be expected because the shapes of the f and g graphs are the same.

7.

9.

11. $\dfrac{d}{dx}(2.781...^x) = 2.781...^x$

13. a. $143.7601 \le \text{Area} \le 144.2401$
The area is within 0.2401 in.^{2} of the ideal.

   b. Keep the tile dimensions within 0.0008 in. of 12 in.

   c. The 0.02 in part b corresponds to ϵ and the 0.0008 corresponds to δ.

15. a. $f'(1) = 2$

   b. Forward: 2.31
Backward: 1.71
Symmetric: 2.01
The symmetric difference quotient is closer to the actual because it is the average of the other two, and the other two span the actual derivative.

   c. $f'(0) = -1$

   d. Forward: -0.99
Backward: -0.99
Symmetric: -0.99
All three difference quotients are equal because $f(x)$ changes just as much from -0.1 to 0 as it does from 0 to 0.1.

17. Answers will vary.

Problem Set 3-4

1. $f'(x) = 20x^3$

3. $dv/dt = -0.581t^{-84}$

5. $M'(x) = 0$

7. $dy/dx = 0.6x - 8$

9. $\dfrac{d}{dx}(13 - x) = -1$

11. $dy/dx = 2.3x^{1.3} - 10x^{-3} - 100$

13. $dv/dx = 18x - 24$

15. $f'(x) = 24x^2 + 120x + 150$

17. $P'(x) = x - 1$

19. $f'(x) = \lim_{h \to 0} \dfrac{7(x+h)^4 - 7x^4}{h}$
$= \lim_{h \to 0}(28x^3 + 42x^2h + 28xh^2 + 7h^3) = 28x^3$
By formula, $f'(x) = 7 \cdot 4x^3 = 28x^3$.

21. $v'(t) = \lim_{h \to 0} \dfrac{[10(t+h)^2 - 5(t+h) + 7] - [10t^2 - 5t + 7]}{h}$
$= \lim_{h \to 0} \dfrac{20th + 10h^2 - 5h}{h} = \lim_{h \to 0}(20t + 10h - 5)$
$= 20t - 5$
By formula, $v'(t) = 10 \cdot 2t - 5 = 20t - 5$.

23. Mae should realize that you differentiate *functions*, not values of functions. If you substitute a value for x into $f(x) = x^4$, you get $f(3) = 3^4 = 81$, which is a *new* function, $g(x) = 81$. It is the derivative of g that equals zero.
Moral: Differentiate *before* you substitute for x.

25.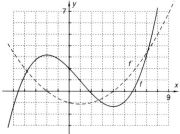

27. a.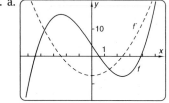

   b. The graph of f' is shown dashed in part a.

   c. There appear to be only two graphs because the exact and the numerical derivative graphs almost coincide.

   d. $f(3) = -6.2$
$f'(3) = 3.8$ (by formula)
$f'(3) \approx 3.8000004$ (depending on grapher)
The two values of $f'(3)$ are almost identical.

29. Increasing by 9/4 y-units per x-unit at $x = 4$

31. Decreasing by 1.5 y-units per x-unit at $x = 9$

33.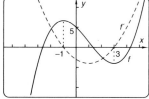

High and low points of the f graph are at the x-intercepts of the f' graph.

35. **Proof:**

$$f'(x) = \lim_{h \to 0} \frac{f(x+h) - f(x)}{h}$$

$$= \lim_{h \to 0} \frac{k \cdot g(x+h) - k \cdot g(x)}{h}$$

$$= \lim_{h \to 0} k \cdot \frac{g(x+h) - g(x)}{h}$$

$$= k \cdot \lim_{h \to 0} \frac{g(x+h) - g(x)}{h}$$

$$= k \cdot g'(x), \text{ Q.E.D.}$$

Dilating a function vertically by a factor of k results in the new function $g(x) = k f(x)$. What has been shown is that

$$\frac{d}{dx}(k f(x)) = k \frac{d}{dx} f(x)$$

That is, dilating a function vertically by a factor of k dilates the derivative function vertically by a factor of k.

37. **Proof:**

$$f'(x) = \lim_{h \to 0} \frac{(x+h)^n - x^n}{h}$$

$$= \lim_{h \to 0} \frac{x^n + n x^{n-1} h + \frac{1}{2} n(n-1) x^{n-2} h^2 + \cdots + h^n - x^n}{h}$$

$$= \lim_{h \to 0} \left[n x^{n-1} + \frac{1}{2} n(n-1) x^{n-2} h + \cdots + h^{n-1} \right]$$

$$= n x^{n-1} + 0 + 0 + \cdots + 0$$

$$= n x^{n-1}, \text{ which is from the second term in the binomial expansion of } (x+h)^n, \text{ Q.E.D.}$$

39. a. $f(x) = x^3 - 5x^2 + 5x$

b. $g(x) = f(x) + 13$ is also an answer to part a because it has the same derivative as $f(x)$. The derivative of a constant is zero.

c. The name *antiderivative* is chosen because it is an inverse operation of taking the derivative.

d. $\dfrac{d}{dx}[g(x)] = \dfrac{d}{dx}[f(x) + C] = \dfrac{d}{dx}f(x) + \dfrac{d}{dx}C = \dfrac{d}{dx}f(x)$

The word *indefinite* is used because of the unspecified constant C.

Problem Set 3-5

1. $v = 20t^3 - 7.2t^{1.4} + 7$
   $a = 60t^2 - 10.08t^{0.4}$

3. $x = -t^3 + 13t^2 - 35t + 27$

The object starts out at $x = 27$ ft when $t = 0$ s. It moves to the left to $x \approx 0.16$ ft when $t \approx 1.7$ s. It turns there and goes to the right to $x = 70$ ft when $t = 7$ s. It turns there and speeds up, going to the left for all higher values of t.

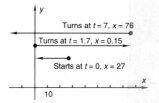

Turns at $t = 7$, $x = 76$
Turns at $t = 1.7$, $x = 0.15$
Starts at $t = 0$, $x = 27$

5. a. $v = -3t^2 + 26t - 35$
   $a = -6t + 26$

b. $t = 1$: $v(1) = -12$ and $a(1) = 20$
So x is decreasing at 12 ft/s at $t = 1$, and the object is slowing down at 20 (ft/s)/s because the velocity and acceleration are in opposite directions when $t = 1$.
$t = 6$: $v(6) = 13$ and $a(6) = -10$
So x is increasing at 13 ft/s at $t = 6$, and the object is slowing down at 10 (ft/s)/s because the velocity and acceleration are in opposite directions when $t = 6$.
$t = 8$: $v(8) = -19$ and $a(1) = -22$
So x is decreasing at 19 ft at $t = 8$, and the object is speeding up at 22 (ft/s)/s because the velocity and acceleration are in the same directions when $t = 8$.

c. At $t = 7$, x has a relative maximum. x is never negative for t in $[0, 9]$.

7. a.

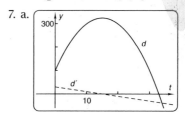

b. Velocity is positive for $0 \le t < 15$.
Calvin is going up the hill for the first 15 s.

c. At 15 s his car stopped, at a distance of 324 ft.

d. He'll be back at the bottom when $t = 33$ s.

e. The car runs out of gas 99 ft from the bottom.

9. a. $d'(t) = 18 - 9.8t$
   $d'(1) = 18 - 9.8 = 8.2$
   $d'(3) = 18 - 9.8 \cdot 3 = -11.4$
   d' is called *velocity* in physics.

b. At $t = 1$ the football is going up at 8.2 m/s.
At $t = 3$ the football is going down at 11.4 m/s.
The ball is going up when the graph slopes up and coming down when the graph slopes down.

c. $d'(4) = -21.2$, which suggests that the ball is going down at 21.2 m/s. However, $d(4) = -6.4$, which reveals that the ball has gone underground. The function gives meaningful answers in the real world only if the domain of t is restricted to values that make $d(t)$ nonnegative.

11.

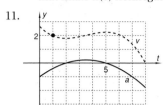

13. The average rate is defined to be the change in the dependent variable divided by the change in the independent variable (such as total distance divided by total time). Thus, the difference quotient is an average rate. The instantaneous rate is the limit of this average rate as the change in the independent variable approaches zero.

15. $\dfrac{d^2y}{dx^2} = 30x$

17. $\dfrac{d^2y}{dx^2} = 18 + 20x^3$

19. $m'(5) = 153.4979...$
$m'(10) = 247.2100...$
These numbers represent the instantaneous rate of change of the amount of money in the account. The second quantity is larger because the money is growing at a rate proportional to the amount of money in the account. Because there is more money after 10 years, the rate of increase should also be larger.
$m''(5) = 14.6299...$
$m''(10) = 23.5616...$
Both quantities are in units ($/yr)/yr.
The quantities represent the instantaneous rate of change of the instantaneous rate of change of the amount of money in the account. For example, at $t = 5$ yr, the rate of increase of the account (153.50 $/yr) is increasing at a rate of 14.63 ($/yr)/yr.

21.

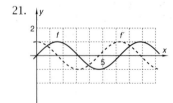

Conjecture: $y' = \cos(x)$

Problem Set 3-6

1.

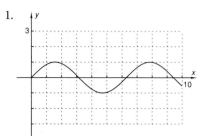

3. Conjecture: $g'(x) = 3\cos 3x$
The graph confirms the conjecture.

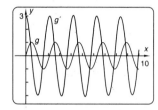

5. Conjecture: $t'(x) = 0.7x^{-0.3}\cos x^{0.7}$
The graph confirms the conjecture.

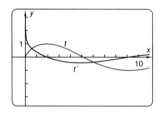

7. a. Inside: $3x$; outside: sine
   b. Inside: sine; outside: cube
   c. Inside: cube; outside: sine
   d. Inside: cosine; outside: exponential
   e. Inside: tangent; outside: reciprocal
   f. Inside: secant; outside: logarithm

Problem Set 3-7

1. a. Let $y = f(u), u = g(x)$.
   $$\frac{dy}{dx} = \frac{dy}{du} \cdot \frac{du}{dx}$$
   b. $y' = f'(g(x)) \cdot g'(x)$
   c. To differentiate a composite function, differentiate the outside function with respect to the inside function, then multiply by the derivative of the inside function with respect to x.

3. $f'(x) = -3\sin 3x$

5. $g'(x) = -3x^2 \sin(x^3)$

7. $y' = -3\cos^2 x \sin x$

9. $y' = 6\sin^5 x \cos x$

11. $y' = -18\cos 3x$

13. $\dfrac{d}{dx}(\cos^4 7x) = -28\cos^3 7x \sin 7x$

15. $f'(x) = 160\sin^{2/3} 4x \cos 4x$

17. $f'(x) = 35(5x + 3)^6$

19. $y' = -72x^2(4x^3 - 7)^{-7}$

21. $y' = -200x \cos^{99}(x^2 + 3)\sin(x^2 + 3)$

23. $\dfrac{d^2 y}{dx^2} = -100\cos 5x$

25. $f(x) = \dfrac{1}{5}\sin 5x + C$

27.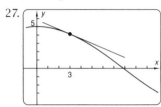

The line has the equation $y = -0.5646...x + 5.8205...$.
The line is tangent to the graph.

29. a. $\dfrac{dV}{dr} = 4\pi r^2$ (cm^3/cm), or cm^2

   b. $r = 6t + 10$

   c. $\dfrac{dr}{dt} = 6$ cm/min

   d. $\dfrac{dV}{dt} = \dfrac{dV}{dr} \cdot \dfrac{dr}{dt} = 38{,}400\pi$ cm^3/min
   dV/dr has units cm^2 and dr/dt has units cm/min, so dV/dt has units cm$^2 \cdot \dfrac{\text{cm}}{\text{min}}$, which becomes cm^3/min.

   e. $V = \dfrac{4\pi}{3}(6t + 10)^3$
   $\therefore \dfrac{dV}{dt} = 24\pi(6t + 10)^2$
   When $t = 5$, $\dfrac{dV}{dt} = 38{,}400\pi$.

Problem Set 3-8

1. a.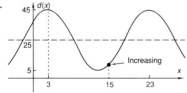

   $y(t) = 25 + 20\cos\dfrac{\pi}{10}(t - 3)$

   b. $y'(t) = -2\pi\sin\dfrac{\pi}{10}(t - 3)$

c. $y(t)$ is increasing at about 3.7 ft/s.

d. The fastest that $y(t)$ changes is 2π, or 6.28... ft/s when the seat is 25 ft above the ground.

3. a. $f(x) = 0.75 + \dfrac{2.5}{44}x$

   b. $g(x) = 0.25 + \dfrac{2.5}{44}x - 0.25\cos\dfrac{\pi}{4}x$

   c. $g'(x) = \dfrac{2.5}{44} + \dfrac{\pi}{16}\sin\dfrac{\pi}{4}x$
   $g'(9) = 0.1956...$ ft/ft
   Going up at about 0.2 vertical foot per horizontal foot
   $g'(15) = -0.0820...$ ft/ft
   Going down at about 0.08 vertical foot per horizontal foot
   A positive derivative implies $g(x)$ is getting larger and thus the child is going down. A negative derivative implies $g(x)$ is getting smaller and thus the child is going down.

   d. The steepest upward slope is 0.2531... ft/ft, and the steepest downward slope is $-0.1395...$ ft/ft.

5. Answers will vary.

7. a.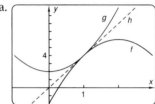

   The limits are all equal to 4.

   b. $f(x) \le g(x), \lim_{x \to 1} f(x) = \lim_{x \to 1} g(x) = 4$, and $f(x) \le h(x) \le g(x)$ for all x in a neighborhood of $x = 1$.

   c.

x	$f(x)$	$h(x)$	$g(x)$
0.95	3.795	3.8	3.805
0.96	3.8368	3.84	3.8432
0.97	3.8782	3.88	3.8818
0.98	3.9192	3.92	3.9208
0.99	3.9598	3.96	3.9602
1.00	4	4	4
1.01	4.0398	4.04	4.0402
1.02	4.0792	4.08	4.0808
1.03	4.1182	4.12	4.1218
1.04	4.1568	4.16	4.1632
1.05	4.192	4.2	4.205

   d. From the table, $\delta = 0.01$ or 0.02 will work, but 0.03 is too large. All the values of $h(x)$ are between the corresponding values of $f(x)$ and $g(x)$, and the three functions all approach 4 as a limit.

9. a. The numbers are correct.

b.

x	$(\sin x)/x$
0.05	0.99958338541...
0.04	0.99973335466...
0.03	0.99985000674...
0.02	0.99993333466...
0.01	0.99998333341...

Values are getting closer to 1.

c.-e. Answers will vary.

11. See the proof in the text.

13. a. The limit seems to be 2.

b.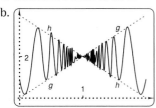

c. See the graph in part b. The lines have the equations $g(x) = x + 1$ and $h(x) = 3 - x$.

d. Prove that $\lim_{x \to 1} y = 2$.

   Proof:

   $\lim_{x \to 1} (x + 1) = 1 + 1 = 2$
   $\lim_{x \to 1} (3 - x) = 3 - 1 = 2$
   For $x < 1, g(x) \le y \le h(x)$.
   ∴ the squeeze theorem applies, and
   $\lim_{x \to 1^-} y = 2$.
   For $x > 1$, $h(x) \le y \le g(x)$.
   ∴ the squeeze theorem applies, and
   $\lim_{x \to 1^+} y = 2$.
   Both left- and right-hand limits equal 2, so
   $\lim_{x \to 1} y = 2$, Q.E.D.

e. The word *envelope* (a noun) is used because the small window formed by the two lines "envelops" (a verb) the graph of the function.

f. As $|x|$ becomes large, $(x - 1) \cdot \sin \dfrac{1}{x - 1}$

   $= \dfrac{\sin [1/(x - 1)]}{1/(x - 1)}$ takes on the form $\dfrac{\sin (\text{argument})}{(\text{argument})}$

   as the argument approaches zero. Thus the limit is 1, and y approaches $2 + 1$, which equals 3.

Problem Set 3-9

1. a. $M'(x) = 60e^{0.06x}$
      $M'(1) = 63.7101...\ \$/\text{yr}$
      $M'(10) = 109.3271...\ \$/\text{yr}$
      $M'(20) = 199.2070...\ \$/\text{yr}$

   b. $M(0) = \$1000$
      $M(1) = \$1061.84$
      $M(2) = \$1127.50$
      $M(3) = \$1197.22$

No, the amount of money in the account does not change by the same amount each year.

c. APR for 0 to 1 year: 6.184%
   APR for 1 to 2 years: 6.184%
   APR for 2 to 3 years: 6.184%
   The APR is higher than the instantaneous rate. Savings institutions may prefer to advertise the APR instead of the instantaneous rate because the APR is higher.

3. a. $A'(p) = \dfrac{-23.5}{p}$

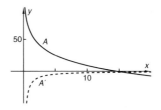

b. If the pressure is increasing, then the altitude is decreasing. The altitude is changing at -2.35 thousand feet/psi. The negative sign means that the altitude is decreasing.

c. The fact that $|A'(5)| > |A'(10)|$ means that the altitude is changing faster at 5 psi than it is at 10 psi.

d. The pressure of air at sea level is 14.5975... psi. The fact that $A(p)$ is negative for all values of p greater than 14.5975... means that if the air pressure is above 14.5975 psi, then the plane must be beneath sea level.

5. $f'(x) = 15e^{3x}$

7. $g'(x) = 4(\sin x)\, e^{\cos x}$

9. $y' = 8e^{4x} \cos (e^{4x})$

11. $f'(x) = 10/x$

13. $T' = 54/x$

15. $y' = -15 \tan 5x$

17. $u' = 3x^{-0.5} \cot x^{-0.5}$

19. $r'(x) = 1$

21. $f'(x) = (\ln 3)3^x$

23. $y' = -\ln 1.6 \sin x\, (1.6^{\cos x})$

25. $\dfrac{d^2 y}{dx^2} = -5/x^2$

27. $y'' = 0.49e^{-0.7x}$

29. $f(x) = 6e^{2x} + C$

R0. Answers will vary.

R1. a.

x	Average Rate of Change from 2 to x
1.97	$\dfrac{f(2) - f(1.97)}{0.03} = 11.82$
1.98	$\dfrac{f(2) - f(1.98)}{0.02} = 11.88$
1.99	$\dfrac{f(2) - f(1.99)}{0.01} = 11.94$
2.01	$\dfrac{f(2.01) - f(2)}{0.01} = 12.06$
2.02	$\dfrac{f(2.02) - f(2)}{0.02} = 12.12$
2.03	$\dfrac{f(2.03) - f(2)}{0.03} = 12.18$

The derivative of f at $x = 2$ is approximately 12.

b. $r(x) = \dfrac{f(x) - f(2)}{x - 2}$

$r(2)$ is of the form $\dfrac{0}{0}$.

$\lim_{x \to 2} r(x)$ appears to be 12.

c. $r(x) = x^2 + 2x + 4$ if $x \neq 2$

∴ $\lim_{x \to 2} r(x) = 12$

d. The answers in parts a, b, and c are the same.

R2. a. $f'(c) = \lim_{x \to c} \dfrac{f(x) - f(c)}{x - c}$

b. $f'(3) = \lim_{x \to 3} \dfrac{0.4x^2 - x + 5 - 5.6}{x - 3} = 1.4$

c. $m(x) = \dfrac{0.4x^2 - x + 0.6}{x - 3}$

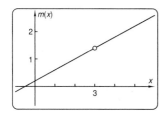

d. Line: $y = 1.4x + 1.4$

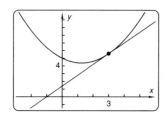

e. The line is tangent to the graph.

f. Yes, f does have local linearity at $x = 3$. Zooming in on the point $(3, 5.6)$ shows that the graph looks more and more like the line.

R3. a.

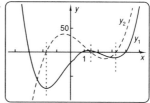

b. See the graph in part a.

c. The y_1 graph has a high point or a low point at each x-value where the y_2 graph is zero.

d.

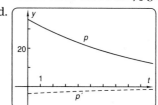

Decreasing at about 2.69 psi/h when $t = 3$
Decreasing at about 1.96 psi/h when $t = 6$
Decreasing at about 3.69 psi/h when $t = 0$
The units are psi/h.
The sign of the pressure change is negative because the pressure is decreasing.
Yes, the rate of pressure change is getting closer to zero.

R4. a. See the text for the definition of *derivative*.

b. Differentiate.

c. If $y = x^n$, then $y' = nx^{n-1}$.

d. See the solution to Problem Set 3-4, Problem 35.

e. See the proof in Section 3-4.

f. $\dfrac{dy}{dx}$ is read "dy, dx."

$\dfrac{d}{dx}(y)$ is read "$d, dx,$ of y."

Both mean the derivative of y with respect to x.

g. i. $f'(x) = \dfrac{63}{5}x^{4/5}$

ii. $g'(x) = -28x^{-5} - \dfrac{x}{3} - 1$

iii. $h'(x) = 0$

h. $f'(32) = 201.6$ exactly

The numerical derivative is equal to or very close to 201.6.

i.

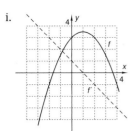

R5. a. $v = \dfrac{dx}{dt}$, or $x'(t)$

$a = \dfrac{dv}{dt}$, or $v'(t); a = \dfrac{d^2x}{dt^2}$, or $x''(t)$

b. $\dfrac{d^2y}{dx^2}$ means the second derivative of y with respect to x.

$y'' = 120x^2$

c. $f(x) = 3x^4 + C$

$f(x)$ is the antiderivative of $f(x)$, or the indefinite integral.

d. The slope is -1 at $x = 1$, 3 at $x = 5$, and 0 at $x = -1$.

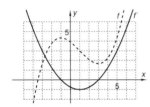

e. i. $v = -0.03t^2 + 1.8t - 25$

$a = -0.06t + 1.8$

ii. $a(15) = 0.9 \,(\text{km/s})/\text{s}$

$v(15) = -4.75 \,\text{km/s}$

The spaceship is slowing down at $t = 15$ because the velocity and the acceleration have opposite signs.

iii. The spaceship is stopped at about 21.8 s and 38.2 s.

iv. The spaceship touches the surface of Mars when $t = 50$. The velocity at that time is $-10 \,\text{km/s}$. It is a crash landing!

R6. a.

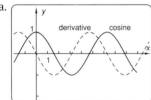

b. The graph of the derivative is the same as the sine graph, but inverted in the y-direction. Thus $(\cos x)' = -\sin x$ is confirmed.

c. $-\sin 1 = -0.841470984\ldots$

Numerical derivative ≈ -0.841470984

The two are very close!

d. Composite function

$f'(x) = -2x \sin(x^2)$

R7. a. i. $\dfrac{dy}{dx} = \dfrac{dy}{du} \cdot \dfrac{du}{dx}$

ii. $f(x) = g(h(x)) \Rightarrow f'(x) = g'(h(x)) \cdot h'(x)$

iii. The derivative of a composite function is the derivative of the outside function with respect to the inside function times the derivative of the inside function with respect to x.

b. See the derivation in the text. This derivation constitutes a proof.

Δu must be nonzero throughout the interval.

c. i. $f(x) = (x^2 - 4)^3$

$f'(x) = 3(x^2 - 4)^2 \cdot 2x = 6x(x^2 - 4)^2$

ii. $f(x) = x^6 - 12x^4 + 48x^2 - 64$

$f'(x) = 6x^5 - 48x^3 + 96x$

Expanding the answer to part i gives

$f'(x) = 6x^5 - 48x^3 + 96x$.

d. i. $f'(x) = -3x^2 \sin x^3$

ii. $g'(x) = 5 \cos 5x$

iii. $h'(x) = -6 \sin x \cos^5 x$

iv. $k'(x) = 0$

e. $f''(x) = -36 \sin 3x$

$f(x) = 4 \sin 3x + C$

$f(x)$ is the second derivative of $f(x)$.

$f(x)$ is the antiderivative, or indefinite integral, of $f(x)$.

f. When the shark is 2 ft long, it weighs 4.8 lb and gains about 2.88 lb/day.

When the shark is 10 ft long, it weighs 600 lb and gains about 72 lb/day.

The chain rule allows you to calculate dW/dt by multiplying dW/dx by dx/dt.

R8. a. $\lim_{x \to 0}[(\sin x)/x] = 1$

x	$(\sin x)/x$
-0.05	$0.99958338541\ldots$
-0.04	$0.99973335466\ldots$
-0.03	$0.99985000674\ldots$
-0.02	$0.99993333466\ldots$
-0.01	$0.99998333341\ldots$
0.00	undefined
0.01	$0.99998333341\ldots$
0.02	$0.99993333466\ldots$
0.03	$0.99985000674\ldots$
0.04	$0.99973335466\ldots$
0.05	$0.99958338541\ldots$

The values of $(\sin x)/x$ approach 1 as x approaches 0.

b. See the statement of the squeeze theorem in the text. Squeeze $(\sin x)/x$ between $\cos x$ and $\sec x$.

c. See the proof in Section 3-8.

d. $\cos x = \sin(\pi/2 - x)$
$\cos' x = \cos(\pi/2 - x)(-1) = -\sin x$, Q.E.D.

e. $d(t) = 180 + 20\cos\dfrac{\pi}{30}t$

$d'(t) = -\dfrac{2\pi}{3}\sin\dfrac{\pi}{30}t$

At 2, $t = 10$: $d'(10) \approx -1.81$ cm/s
At 3, $t = 15$: $d'(15) \approx -2.09$ cm/s
At 7, $t = 35$: $d'(35) \approx 1.05$ cm/s
At the 2 and 3, the tip is going down, so the distance from the floor is decreasing, which is implied by the negative derivatives.
At the 7, the tip is going up, as implied by the positive derivative.

R9. a. $p'(0) = -10$
$p'(10) = -3.6787...$
$p'(20) = -1.3533...$
The rates are negative because the amount of medication in your body is decreasing.
The half-life is 6.9314... h.
After two half-lives have elapsed, 25% of the medicine remains in your body.

b. i. $f'(x) = 10e^{2x}$

ii. $dy/dx = (\ln 7)7^x$

iii. $\dfrac{d}{dx}[\ln(\cos x)] = -\tan x$

iv. $\dfrac{d^2y}{dx^2} = \dfrac{-8}{x^2}$

c. $f(x) = 4e^{3x}$

d.

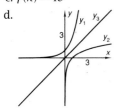

$y_1 = e^x$ is the inverse of $y_2 = \ln x$, so y_1 is a reflection of y_2 across the line $y = x$.

CHAPTER 4

Problem Set 4-1

1. $f'(x) = -3\sin x$, $g'(x) = 2\cos x$

3.

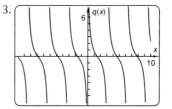

q is the cotangent function.
$q'(2) = -1.8141...$
$q(x)$ is decreasing at $x = 2$.
$f'(2)/g'(2) = 3.2775... \neq q'(2)$

5.

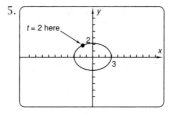

$\Delta x = -0.54466...$, $\Delta y = -0.16618...$
$dy/dx \approx \Delta y/\Delta x = 0.3051...$
At $t = 2$, $\dfrac{dy/dt}{dx/dt} = \dfrac{2\cos 2}{-3\sin 2} = 0.3051...$, which agrees with the difference quotient. Also, from the graph in Problem 4, the slope of a line tangent to the graph at the point where $t = 2$ would be positive and less than 1, which agrees with the value of 0.3051... for the difference quotient.

Problem Set 4-2

1. $f'(x) = 3x^2\cos x - x^3\sin x$

3. $g'(x) = 1.5x^{0.5}e^{2x} + 2x^{1.5}e^{2x}$

5. $dy/dx = x^6(2x + 5)^9(34x + 35)$

7. $z' = (1/x)\sin 3x + 3\ln x\cos 3x$

9. $y' = (6x + 11)^3(5x - 9)^6(330x + 169)$

11. $P' = 10x(x^2 - 1)^9(x^2 + 1)^{14}(5x^2 - 1)$

13. $a'(t) = 12\cos 3t\cos 5t - 20\sin 3t\sin 5t$

15. $y' = -3\sin(3\sin x)\cos x$

17. $d^2y/dx^2 = -36e^{6x}\sin e^{6x} - 36e^{12x}\cos e^{6x}$

19. $z' = 3x^2(5x - 2)^4\sin 6x + 20x^3(5x - 2)^3\sin 6x + 6x^3(5x - 2)^4\cos 6x$

21. **Proof:**

$y = uvw = (uv)w$
$\therefore y' = (uv)'w + (uv)w' = (u'v + uv')w + (uv)w'$
$= u'vw + uv'w + uvw'$, Q.E.D.

23. $z' = 5x^4\cos^6 x\sin 7x - 6x^5\cos^5 x\sin x\sin 7x + 7x^5\cos^6 x\cos 7x$

25. $y' = 4x^3(\ln x)^5\sin x\cos 2x + 5x^3(\ln x)^4\sin x\cos 2x + x^4(\ln x)^5\cos x\cos 2x - 2x^4(\ln x)^5\sin x\sin 2x$

27. a. $v(t) = e^{-0.1t}[-0.3\cos(\pi t) - 3\pi \sin(\pi t)]$

   b. $v(2) = 0.2456...$, so there is not a high point at $t = 2$. $v(t) = 0$ when $t = 1.9898...$, so there is a high point when $t = 1.9898$.

29. **Proof:**

   For any function, the chain rule gives
   $\frac{d}{dx}f(-x) = f'(-x) \cdot (-1) = -f'(-x)$.
   For an odd function, $\frac{d}{dx}f(-x) = \frac{d}{dx}(-f(x)) = -f'(x)$.
   $\therefore -f'(-x) = -f'(x)$ or $f'(-x) = f'(x)$,
   and the derivative is an even function.
   For an even function, $\frac{d}{dx}f(-x) = \frac{d}{dx}f(x) = f'(x)$.
   $\therefore -f'(-x) = f'(x)$ or $f'(-x) = -f'(x)$,
   and the derivative is an odd function, Q.E.D.

31. **Proof (by induction on n):**

   If $n = 1$, then $f_1(x) = x^1$, which implies that
   $f_1'(x) = 1 = 1x^0$, which anchors the induction. Assume
   that for some integer $n = k > 1, f_k'(x) = kx^{k-1}$.
   For $n = k + 1, f_{k+1}(x) = x^{k+1} = (x^k)(x)$. By the
   derivative of a product property,
   $f_{k+1}'(x) = (x^k)'(x) + (x^k)(x)' = (x^k)'(x) + x^k$.
   Substituting for $(x^k)'$ from the induction hypothesis,
   $f_{k+1}'(x) = (kx^{k-1})(x) + x^k = kx^k + x^k = (k+1)x^k = (k+1)x^{(k+1)-1}$, completing the induction.
   $\therefore f_n'(x) = nx^{n-1}$ for all integers ≥ 1, Q.E.D.

33. a.
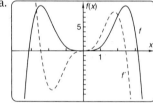

   b. $f'(x) = 3x^2 \sin x + x^3 \cos x$
      The graph in part a is correct.

   c. The numerical derivative graph duplicates the algebraic derivative graph, as in part a, thus showing that the algebraic derivative is correct.

35. a. $\frac{dA}{dt} = \frac{dL}{dt} \cdot W + L \cdot \frac{dW}{dt}$

   b. At $t = 4$, $dA/dt = 7.132...$, so A is increasing.
      At $t = 5$, $dA/dt = -4.949...$, so A is decreasing.

Problem Set 4-3

1. $f'(x) = \dfrac{3x^2 \sin x - x^3 \cos x}{\sin^2 x}$

3. $g'(x) = \dfrac{3\cos^2 x \,(-\sin x) \cdot \ln x - \cos^3 x \cdot (1/x)}{x^{10}}$

5. $y' = \dfrac{10 \cos 10x \cos 20x + 20 \sin 20x \sin 10x}{\cos^2 20x}$

7. $y' = \dfrac{-27}{(6x + 5)^2}$

9. $\dfrac{dz}{dx} = -\dfrac{(8x + 1)^5(120x + 141)}{(5x - 2)^{10}}$

11. $Q' = \dfrac{3x^2 e^{x^2} \sin x - e^{x^2} \cos x}{\sin^2 x}$

13. $\dfrac{d}{dx}(60x^{-4/3}) = -80x^{-7/3}$

15. $r'(x) = \dfrac{-36}{x^4} = -36x^{-4}$

17. $v'(x) = \dfrac{7 \sin 0.5x}{\cos^2 0.5x}$

19. $r'(x) = \dfrac{-1}{x^2} = -x^{-2}$

21. $W'(x) = 150x^2\,(x^3 - 1)^4$

23. $T'(x) = \sec^2 x$ ("T" is for "tangent function.")

25. $C'(x) = -\csc x \cot x$ ("C" is for "cosecant function.")

27. a. $v(1) = 500$ mi/h
      $v(2) = 1000$ mi/h
      $v(3) = 1000/0$. No value for $v(3)$.

   b. $a(t) = \dfrac{1000}{(3 - t)^2}$
      $a(1) = 250$ (mi/h)/h
      $a(2) = 1000$ (mi/h)/h
      $a(3) = 1000/0$. No value for $a(3)$.

   c. Units are (mi/h)/h, or mi/h^2.

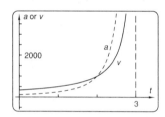

   d. Range is $0 \leq t < 1.585...$.

29. $f'(x) = \dfrac{1}{(2x + 5)^2}$
   $f'(4) = 0.005917159...$

   For 4.1, the difference quotient gives $0.005827505...$.
   For 4.01, the difference quotient gives $0.005908070...$.
   For 4.001, the difference quotient gives $0.005916249...$.
   Difference quotients are approaching $f'(4)$.

31. **Proof:**

   Let $n = -p$, where p is a positive integer.
   $\therefore y = x^{-p} = \dfrac{1}{x^p}$
   $\therefore y' = \dfrac{0 \cdot x^p - 1 \cdot px^{p-1}}{x^{2p}}$ because p is a positive integer.
   $\qquad = -\dfrac{px^{p-1}}{x^2 p} = -px^{p-1-2p} = -px^{-p-1}$
   Replacing $-p$ with n gives $y' = nx^{n-1}$, Q.E.D.

33. Answers will vary.

1. $f'(x) = 5 \sec^2 5x$

3. $y' = 7x^6 \sec x^7 \tan x^7$

5. $g'(x) = -11e^{11x} \csc^2(e^{11x})$

7. $r'(x) = -\cot x$

9. $(d/dx)(y) = 20 \tan^4 4x \sec^2 4x$

11. $(d/dx)(\sec x \tan x) = \sec x \tan^2 x + \sec^3 x$

13. $y' = \sec^2 x - \csc^2 x$

15. $y' = \sec x \tan x$

17. $y' = \dfrac{5}{x \cot 14x} + \dfrac{70 \ln 7x}{\cos^2 14x}$

19. $w' = 3 \sec^2(\sin 3x) \cdot \cos 3x$

21. $S'(x) = 0$

23. $A'(x) = 2x \cos x^2$

25. $F'(x) = 2 \sin x \cos x$

27. $d^2y/dx^2 = 2 \sec^2 x \tan x$

29. $y = \cot x = \dfrac{\cos x}{\sin x} \Rightarrow$

$y' = \dfrac{-\sin x \cdot \sin x - \cos x \cdot \cos x}{\sin^2 x}$

$= \dfrac{-1}{\sin^2 x} = -\csc^2 x$

or:

$y = \dfrac{1}{\tan x} = (\tan x)^{-1} \Rightarrow$

$y' = -1 \cdot (\tan x)^{-2} \cdot \sec^2 x = -\csc^2 x$

31. a. See the graph in part b.

b. $f(x) = \tan x \Rightarrow f'(x) = \sec^2 x$

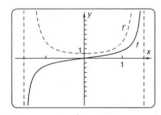

The predicted graph should be close to the actual one.

c. $\dfrac{\tan 1.01 - \tan 0.99}{2(0.01)} = 3.42646416...$

$\tan' 1 = \sec^2 1 = (1/\cos 1)^2 = 3.42551882...$
The difference quotient is within 0.001 of the actual value.

33. a. $y/10 = \tan x \Rightarrow y = 10 \tan x$, Q.E.D.

b. At $x = 1$, y is increasing at about 34.3 ft/radian, which is 0.5978... ft/degree.

c. At $y = 535$, y is increasing at about 28,632.5 ft/radian.

35. a. $y = \sin x + C$

b. $y = -\dfrac{1}{2} \cos 2x + C$

c. $y = \dfrac{1}{3} \tan 3x + C$

d. $y = -\dfrac{1}{4} \cot 4x + C$

e. $y = 5 \sec x + C$

Problem Set 4-5

1. See Figure 4-5d in the text.

3. See Figure 4-5d in the text.

5. The principal branch of the inverse cotangent function goes from 0 to π so that the function will be continuous.

7. $\sin(\sin^{-1} 0.3) = 0.3$

9. $y = \sin^{-1} x \Rightarrow \sin y = x \Rightarrow \cos y \cdot y' = 1 \Rightarrow$

$y' = \dfrac{1}{\cos y} = \dfrac{1}{\sqrt{1 - x^2}}$, Q.E.D.

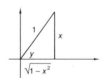

[Because $\sin y = $ (opposite leg)/(hypotenuse), put x on the opposite leg and 1 on the hypotenuse. Adjacent leg $= \sqrt{1 - x^2}$, and $\cos y = $ (adjacent)/(hypotenuse).]

11. $y = \csc^{-1} x \Rightarrow \csc y = x \Rightarrow -\csc y \cot y \cdot y' \Rightarrow$

$y' = -\dfrac{1}{\csc y \cot y} = -\dfrac{1}{x\sqrt{x^2 - 1}}$ if $x > 0$

If $x < 0$, then y is in Quadrant IV. So both $\csc y$ and $\cot y$ are negative, and thus their product is positive.

$\therefore y' = -\dfrac{1}{|x|\sqrt{x^2 - 1}}$, Q.E.D.

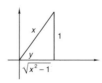

[Because $\csc y = $ (hypotenuse)/(opposite leg), put x on the hypotenuse and 1 on the opposite leg. Adjacent leg $= \sqrt{x^2 - 1}$, and $\csc y = x$ and $\cot y = $ (adjacent)/(opposite).]

13. $y' = \dfrac{4}{\sqrt{1-16x^2}}$

15. $y' = -\dfrac{0.5e^{0.5x}}{1+e^x}$

17. $y' = \dfrac{1}{|x|\sqrt{x^2-9}}$

19. $y' = -\dfrac{10x}{\sqrt{1-25x^4}}$

21. $g'(x) = 2\sin^{-1}x \cdot \dfrac{1}{\sqrt{1-x^2}}$

23. $v' = \sin^{-1}x$. The surprise is that you now have seen a formula for the antiderivative of the inverse sine function.

25. a. $\tan\theta = x/100$, so $\theta = \tan^{-1}(x/100)$, Q.E.D.

b. $\dfrac{d\theta}{dx} = \dfrac{100}{10000+x^2}$, $\dfrac{d\theta}{dt} = \dfrac{100}{10000+x^2}\cdot\dfrac{dx}{dt}$

c. The truck is going 104 ft/s ≈ 71 mi/h.

27.

x	Numerical Derivative*	Algebraic Derivative
-0.8	$-1.666671...$	$-1.666666...$
-0.6	$-1.250000...$	-1.25
-0.4	$-1.091089...$	$-1.091089...$
-0.2	$-1.020620...$	$-1.020620...$
0	$-1.000000...$	-1
0.2	$-1.020620...$	$-1.020620...$
0.4	$-1.091089...$	$-1.091089...$
0.6	$-1.250000...$	-1.25
0.8	$-1.666671...$	$-1.666666...$

*The precise value for the numerical derivative will depend on the tolerance to which the grapher is set.

29. a. $y = \sin^{-1}x \Rightarrow \sin y = x \Rightarrow \cos y \cdot y' = 1 \Rightarrow$
$y' = \dfrac{1}{\cos y}$, Q.E.D.

b. $y' = \dfrac{1}{\cos(\sin^{-1}x)} = \dfrac{1}{\cos(\sin^{-1}0.6)} = 1.25$
$y' = \dfrac{1}{\sqrt{1-x^2}} = \dfrac{1}{\sqrt{1-0.6^2}} = 1.25$, Q.E.D.

c. $y = f^{-1}(x) \Rightarrow f(y) = x \Rightarrow f'(y)\cdot\dfrac{d}{dx}(y) = 1 \Rightarrow$
$\dfrac{d}{dx}(y) = \dfrac{1}{f'(y)} \Rightarrow \dfrac{d}{dx}(f^{-1}(x)) = \dfrac{1}{f'(f^{-1}(x))}$, Q.E.D.

d. If $f(x) = 10$, then $x = 2$. So $h(10) = 2$.
Because $h(x) = f^{-1}(x)$ and $f'(x) = 3x^2+1$,
$h'(10) = \dfrac{1}{f'(h(10))} = \dfrac{1}{f'(2)} = \dfrac{1}{3\cdot2^2+1} = \dfrac{1}{13}$.

1. Continuous

3. Neither

5. Neither

7. Both

9. Neither

11. Continuous

13. a.

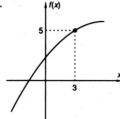

b. Equations will vary.

15. a.

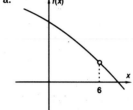

b. Equations will vary.

17. a.

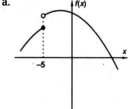

b. Equations will vary.

19. a.

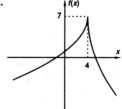

b. Equations will vary.

21. Continuous

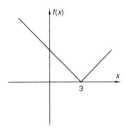

23. Both

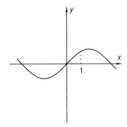

25. $a = -1.5, b = 2.5$

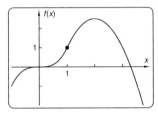

27. $a = -0.5, b = 16$

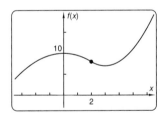

29. $a = 0.5671..., b = 1.7632...$

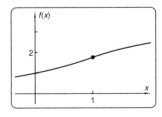

31. a. $a = -3, b = -1.25, c = 0, d = 0$
    b. $k = -1.1875$

33. $f'(x) = \begin{cases} 2x, & \text{if } x < 2 \\ 2x, & \text{if } x > 2 \\ \text{undefined, if } x = 2 \end{cases}$

Taking the left and right limits gives
$\lim_{x \to 2^-} f'(x) = 2 \cdot 2 = 4$
$\lim_{x \to 2^+} f'(x) = 2 \cdot 2 = 4$
Using the definition of derivative, taking the limit
from the left, $f'(x) = \lim_{x \to 2^-} \frac{x^2 + 1 - 4}{x - 2} \Rightarrow \frac{1}{0}$, which is
infinite. The same thing happens from the right. As
this graph shows, the secant lines become vertical as
x approaches 2 from either side.

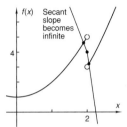

Thus f is not differentiable at $x = 2$, even though the
right and left limits of $f'(x)$ are equal to each other.
The function must be continuous if it is to have a
chance of being differentiable.

35. a. $y = mx + b \Rightarrow y' = m$, which is independent of x.
      $\therefore$ linear functions are differentiable for all x.
      $\therefore$ linear functions are continuous for all x.
    b. $y = ax^2 + bx + c \Rightarrow y' = 2ax + b$, which exists for
      all x by the closure axioms.
      $\therefore$ quadratic functions are differentiable for all x.
      $\therefore$ quadratic functions are continuous for all x.
    c. $y = 1/x = x^{-1} \Rightarrow y' = -x^{-2}$, which exists for all
      $x \neq 0$ by closure and multiplicative inverse axioms.
      $\therefore$ the reciprocal function is differentiable for all
      $x \neq 0$.
      $\therefore$ the reciprocal function is continuous for all
      $x \neq 0$.
    d. $y = x \Rightarrow y' = 1$, which is independent of x.
      $\therefore$ the identity function is differentiable for all x.
      $\therefore$ the identity function is continuous for all x.
    e. $y = k \Rightarrow y' = 0$, which is independent of x.
      $\therefore$ constant functions are differentiable for all x.
      $\therefore$ constant functions are continuous for all x.

Problem Set 4-7

1. $\dfrac{dy}{dx} = \dfrac{3 \cos 3t}{4t^3}, \dfrac{d^2y}{dx^2} = \dfrac{-9t \sin 3t - 9 \cos 3t}{16t^7}$

3. a.

t	x	y
-3	-1	-6
-2	0	-1
-1	1	2
0	2	3
1	3	2
2	4	-1
3	5	-6

b.

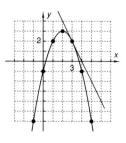

c. If $t = 1$, $dy/dx = -2$ and $(x, y) = (3, 2)$.
   The line through the point $(3, 2)$ with slope -2 is tangent to the graph. See part b.

d. $y = 3 - (x - 2)^2$
   This is the Cartesian equation of a parabola because only one of the variables is squared.

e. By direct differentiation, $dy/dx = -2(x - 2)$.
   At $(x, y) = (3, 2)$, $dy/dx = -2$, which agrees with part c.

5. a. The grapher confirms the figure in the text.

   b. $\dfrac{dy}{dx} = \dfrac{5 \cos t}{-3 \sin t}$

   c. If $t = \pi/4$, $(x, y) = (3\sqrt{2}/2, 5\sqrt{2}/2)$ and $dy/dx = -5/3$.

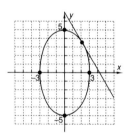

   The line is tangent to the graph.

   d. False. The line from $(0, 0)$ to $(2.1..., 3.5...)$ does not make an angle of $45°$ with the x-axis. (This shows that the t in parametric functions is not the same as the θ in polar coordinates.)

   e. The tangent line is horizontal at $(0, 5)$ and $(0, -5)$. The tangent line is vertical at $(3, 0)$ and $(-3, 0)$. See the graph in part c.

   f. $(x/3)^2 + (y/5)^2 = 1$, which is a standard form of the equation of an ellipse centered at the origin with x-radius 3 and y-radius 5.

7. a.

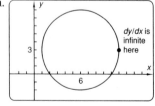

b. $dy/dx = -\cot t$

c. $dy/dx = 0$ if $t = 0.5\pi, 1.5\pi, 2.5\pi, \ldots$.
   dy/dx is infinite if $t = 0, \pi, 2\pi, \ldots$.
   At a point where dy/dx is infinite, dx/dt must be 0. Because this happens where $t = \pi/2 + n\pi$, $dy/dx = 5 \cos t = 0$ at those points.

d. $\left(\dfrac{x - 6}{5}\right)^2 + \left(\dfrac{y - 3}{5}\right)^2 = 1$

e. This is an equation of a circle centered at $(6, 3)$ with radius 5.
   The 6 and 3 added in the original equations are the x- and y-coordinates of the center, respectively. The coefficients, 5, for cosine and sine in the original equations are the x- and y-radii, respectively. Because the x- and y-radii are equal, the graph is a circle.

9. a. The grapher confirms the figure in the text.

   b. $\dfrac{dy}{dx} = \dfrac{\cos t - \cos 2t}{-\sin t - \sin 2t}$

   c. Cusps occur where both dx/dt and $dy/dt = 0$. A graphical solution shows that this occurs at $t = 0, 2\pi/3, 4\pi/3, 2\pi, \ldots$. (A cusp could also occur if dx/dt and $dy/dt = 0$, but, for this figure, there is no such place.)

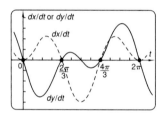

   At $t = 0, 2\pi, \ldots$, the tangent appears to be horizontal. At $t = 2\pi/3, 4\pi/3, 8\pi/3, 10\pi/3, \ldots$, there appears to be a tangent line but not a horizontal one.
   dy/dx approaches $-1.732...$ as t approaches $2\pi/3$.

11. a. The grapher confirms the figure in the text.

    b. $dx/dy = \tan t$

    c. At $t = \pi$, $dy/dt = \tan \pi = 0$.
       The string will be pointing straight up from the x-axis. The diagram shows that the tangent to the graph is horizontal at this point.

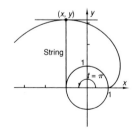

13. Answers will vary.

15. a. The grapher confirms the figure in the text.

b.
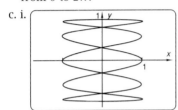

If n is an even number, the graph comes to endpoints and retraces its path, making two complete cycles as t goes from 0 to 2π.
If n is an odd number, the graph does not come to endpoints. It makes one complete cycle as t goes from 0 to 2π.

c. i.
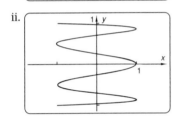

ii.
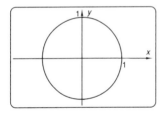

d. If $n = 1$, the graph is a circle.

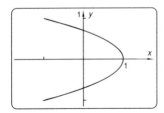

If $n = 2$, the graph is a parabola.

e. Jules Lissajous (1822–1880) lived in France. Nathaniel Bowditch (1773–1838) lived in Massachusetts.

Problem Set 4-8

1. $y' = -\dfrac{3x^2}{28y^3}$

3. $y' = -\dfrac{y \ln y}{x}$

5. $y' = \dfrac{2 \cos 2x - 1 - y}{x + 1}$

7. $y' = y^{0.5}/x^{0.5}$

9. $y' = \dfrac{-ye^{xy}}{xe^{xy} - \sec^2 y}$

11. $y' = \dfrac{1 - 15x^{14}y^{20}}{1 + 20x^{15}y^{19}}$

13. $y' = \dfrac{\cos x \sin x}{\cos y \sin y}$

15. $y' = -\dfrac{y}{x}$

17. $y' = \sec y$

19. $y' = -\sin y \tan y$

21. $y' = -\dfrac{1}{\sin y} = -\dfrac{1}{\sqrt{1 - x^2}}$

23. $y = x^{11/5} \Rightarrow y^5 = x^{11} \Rightarrow 5y^4 \cdot y' = 11x^{10} \Rightarrow$

$y' = \dfrac{11x^{10}}{5y^4} = \dfrac{11x^{10}}{5(x^{11/5})^4} = \dfrac{11x^{10}}{5x^{44/5}} = \dfrac{11}{5}x^{6/5}$,

which is the answer obtained using the derivative of a power formula, Q.E.D.

25. a. At $(-6, 8), (-6)^2 + 8^2 = 100$, which shows that $(-6, 8)$ is on the graph, Q.E.D.

b. $dy/dx = -x/y$. At $(-6, 8), dy/dx = 0.75$.
A line at $(-6, 8)$ with slope 0.75 is tangent to the graph, showing that the answer is reasonable.

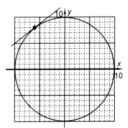

c. $\dfrac{dy}{dx} = -\dfrac{\cos t}{\sin t}$

At $x = -6, t = \cos^{-1}(-0.6)$

$\sin[\cos^{-1}(-0.6)] = 0.8$

$\therefore \dfrac{dy}{dx} = -\dfrac{-0.6}{0.8} = 0.75$,

which agrees with part b, Q.E.D.

27. a. $dy/dx = -x^2/y^2$

At $x = 0, dy/dx = 0$, so the tangent line is horizontal.

At $x = 2, dy/dx = -0.2732...$, so the tangent line has a small negative slope.

At $x = 4, dy/dx$ is infinite, so the tangent line is vertical.

These values are all consistent with the graph.

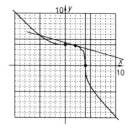

b. $dy/dx = -1$

c. As x approaches infinity, dy/dx approaches -1.

d. The name comes from analogy with the equation of a circle, such as $x^2 + y^2 = 64$.

Problem Set 4-9

1. $\dfrac{dr}{dt} = \dfrac{6}{\pi r}$

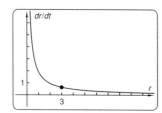

$\dfrac{dr}{dt} = \dfrac{2}{\pi} = 0.6366...$ mm/h when $r = 3$ mm.

$\dfrac{dr}{dt}$ varies inversely with the radius.

3. The length of the major axis is decreasing at $12/\pi$ cm/s.

5. Let $y =$ Milt's distance from home plate and $x =$ Milt's displacement from third base.

$\dfrac{dy}{dt} = \dfrac{-20x}{\sqrt{x^2 + 90^2}}$

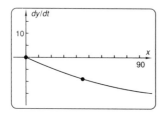

At $x = 45, \dfrac{dy}{dt} \approx -8.9$ ft/s (exact: $-4\sqrt{5}$).

At $x = 0, \dfrac{dy}{dt} = 0$ ft/s, which is reasonable because Milt is moving perpendicular to his line from home plate.

7. a. Let $L =$ length, $W =$ width, and $H =$ depth (in meters).

$\dfrac{dH}{dt} = -\dfrac{20}{LW^2}(0.1) - \dfrac{20}{L^2W}(-0.3)$

b. The depth is increasing at 0.02 m/s.

9. a. Let $x =$ distance from the bottom of the ladder to the wall, $y =$ distance from the top of the ladder to the floor, and $v =$ velocity of the weight.

$v = \dfrac{x}{\sqrt{400 - x^2}}\dfrac{dx}{dt}$

b. $v = -0.6123...$ ft/s (exact: $-\sqrt{6}/4$)

c. v is infinite.

11. a. $16.2\pi = 50.8938... \approx 50.9$ m^3/h

b. i. $= \dfrac{-50}{144\pi} = -0.1105... \approx -0.11$ m/h

ii. $-\infty$

c. i. $\dfrac{dV}{dt} = -0.25\sqrt{h}$

ii. -0.2 m^3/h

iii. $-0.4317... \approx -0.43$ m/h

13. a. Let $\omega =$ angular velocity in radians per day.

$\omega_E = \dfrac{2\pi}{365}, \ \omega_M = \dfrac{2\pi}{687}$

$\dfrac{d\theta}{dt} = \dfrac{644\pi}{250755} \approx 0.00807$ radian/day

b. The period is $778.7422... \approx 778.7$ days. The planets are at their closest position 779 days later, on October 14, 2005 (or October 15 if the planets were aligned later than about 6:11 a.m. back on August 27, 2003).

c. $D = \sqrt{28530 - 26226\cos\theta}$ million miles

d. Answers will vary depending on today's date.

e. No. The maximum occurs at $\theta \approx 0.8505...$, or $48.7...°$.

f.

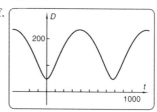

It is not a sinusoid.

R0. Answers will vary.

R1. a. $x = g(t) = t^3 \Rightarrow g'(t) = 3t^2$
$y = h(t) = \cos t \Rightarrow h'(t) = -\sin t$
If $f(t) = g(t) \cdot h(t) = t^3 \cos t$, then, for example,
$f'(1) = 0.7794...$ by numerical differentiation.
$g'(1) \cdot h'(1) = 3(1^2) \cdot (-\sin 1) = -2.5244...$.
$\therefore f'(t) \neq g'(t) \cdot h'(t)$, Q.E.D.

b. If $f(t) = g(t)/h(t) = t^3/\cos t$, then, for example,
$f'(1) = 8.4349...$ by numerical differentiation.
$g'(1)/h'(1) = 3(1^2)/(-\sin 1) = 3.5651...$.
$\therefore f'(t) \neq g'(t)/h'(t)$, Q.E.D.

c. $y = \cos t$
$x = t^3 \Rightarrow t = x^{1/3} \Rightarrow y = \cos(x^{1/3})$
$\dfrac{dy}{dx} = -\sin(x^{1/3}) \cdot \dfrac{1}{3}x^{-2/3}$
At $x = 1$, $\dfrac{dy}{dx} = -\sin 1 \cdot \dfrac{1}{3} = -0.280490...$.
If $x = 1$, then $t = 1^{1/3} = 1$.
$\therefore \dfrac{dy/dt}{dx/dt} = \dfrac{-\sin t}{3t^2} = \dfrac{-\sin 1}{3} = -0.280490...$,
which equals dy/dx, Q.E.D.

R2. a. If $y = uv$, then $y' = u'v + uv'$.

b. See the proof of the product formula in the text.

c. i. $f'(x) = 7x^6 \ln 3x - x^6$

ii. $g'(x) = \cos x \cos 2x - 2 \sin x \sin 2x$

iii. $h'(x) = 15(3x - 7)^4 (5x + 2)^2(8x - 5)$

iv. $s'(x) = -e^{-x}(x^8) + 8e^{-x}x^7$

d. $f(x) = (3x + 8)(4x + 7)$

i. $f'(x) = 3(4x + 7) + (3x + 8)(4) = 24x + 53$

ii. $f(x) = 12x^2 + 53x + 56$
$f'(x) = 24x + 53$

R3. a. If $y = u/v$, then $y' = \dfrac{u'v - uv'}{v^2}$.

b. See the proof of the quotient formula in the text.

c. i. $f'(x) = \dfrac{10x \cos 10x - 5 \sin 10x}{x^6}$

ii. $g'(x) = \dfrac{18(2x + 3)^8(5x - 11)}{(9x - 5)^5}$

iii. $h'(x) = -1500x^2(100x^3 - 1)^{-6}$

d. $y = 1/x^{10}$
As a quotient:
$y' = \dfrac{0 \cdot x^{10} - 1 \cdot 10x^9}{x^{20}} = \dfrac{-10}{x^{11}} = -10x^{-11}$
As a power:
$y = x^{-10} \Rightarrow y' = -10x^{-11}$

e. $t'(x) = \sec^2 x$
$t'(1) = 3.4255...$

f. $m(x) = \dfrac{t(x) - t(1)}{x - 1} = \dfrac{\tan x - \tan 1}{x - 1}$

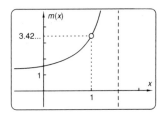

x	$m(x)$
0.997	3.40959...
0.998	3.41488...
0.999	3.42019...
1	undefined
1.001	3.43086...
1.002	3.43622...
1.003	3.44160...

The values get closer to 3.4255... as x approaches 1 from either side, Q.E.D.

R4. a. i. $y' = 7 \sec^2 7x$

ii. $y' = -4x^3 \csc^2(x^4)$

iii. $y' = e^x \sec e^x \tan e^x$

iv. $y' = -\csc x \cot x$

b. See the derivation in the text for $\tan' x = \sec^2 x$.

c.

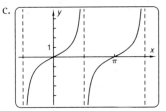

The graph is always sloping upward, which is connected to the fact that $\tan' x$ equals the square of a function and is thus always positive.

d. $f'(t) = 7 \sec t \tan t$
$f'(1) = 20.17...$
$f'(1.5) = 1395.44...$
$f'(1.57) = 11038634.0...$
There is an asymptote in the secant graph at $t = \pi/2 = 1.57079...$. As t gets closer to this value, secant changes very rapidly!

R5. a. i. $y' = \dfrac{3}{1 + 9x^2}$

ii. $\dfrac{d}{dx}(\sec^{-1} x) = \dfrac{1}{|x|\sqrt{x^2 - 1}}$

iii. $c'(x) = \dfrac{-2 \cos^{-1} x}{\sqrt{1 - x^2}}$

b.

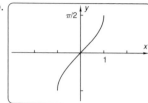

$$y'(0) = \frac{1}{\sqrt{1-0^2}} = 1, \text{ which agrees with the graph.}$$

$$y'(1) = \frac{1}{\sqrt{1-1^2}} = \frac{1}{0}, \text{ which is infinite.}$$

The graph becomes vertical as x approaches 1 from the negative side.

$y'(2)$ is undefined because $f(2)$ is not a real number.

R6. a. *Differentiability implies continuity.*

b. i.

ii.

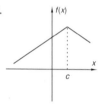

iii. No such function

iv.

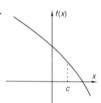

c. i.

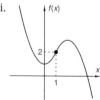

ii. f is continuous at $x = 1$ because right and left limits are both equal to 2, which equals $f(1)$.

iii. f is differentiable. Left and right limits of $f'(x)$ are both equal to 2, and f is continuous at $x = 2$.

d. $a = 1, b = 0$

R7. a. $\dfrac{dy}{dx} = \dfrac{3t^2}{2e^{2t}}, \dfrac{d^2y}{dx^2} = \dfrac{3t - 3t^2}{2e^{4t}}$

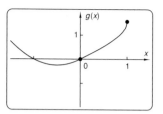

b. $\dfrac{dy}{dx} = \dfrac{\sin t + t\cos t}{\cos t - t\sin t}$

Where the graph crosses the positive x-axis, $t = 0, 2\pi, 4\pi, 6\pi, \dots$.

If $t = 6\pi$, then $x = 6$ and $y = 0$.

$\therefore (6, 0)$ *is* on the graph.

If $t = 6\pi$, then $dy/dx = 6\pi$.

So the graph is *not* vertical where it crosses the x-axis. It has a slope of $6\pi = 18.84\dots$.

c. $x = 20\sin\dfrac{\pi}{10}(t - 3)$

$y = 25 + 20\cos\dfrac{\pi}{10}(t - 3)$

When $t = 0, dy/dt = 5.0832\dots$, so the Ferris wheel is going up at about 5.1 ft/s.

When $t = 0, dx/dt = 3.6931\dots$, so the Ferris wheel is going right at about 3.7 ft/s.

dy/dx is first infinite at $t = 8$ s.

R8. a. $y = x^{8/5} \Rightarrow y^5 = x^8 \Rightarrow$

$$5y^4y' = 8x^7 \Rightarrow y' = \frac{8x^7}{5y^4} = \frac{8x^7}{5(x^{8/5})^4} = \frac{8}{5}x^{3/5}$$

Using the power rule directly:

$$y = x^{8/5} \Rightarrow y' = \frac{8}{5}x^{3/5}$$

b. $\dfrac{dy}{dx} = \dfrac{4.5x^{3.5} - y^4\cos(xy)}{3y^2\sin(xy) + xy^3\cos(xy)}$

c. i. At $(2, 2), dy/dx = 2$.

At $(2, -2), dy/dx = -2$.

Lines at these points with these slopes are tangent to the graph.

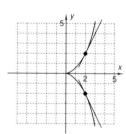

ii. At $(0, 0), dy/dx$ has the indeterminate form $0/0$, which is consistent with the cusp.

iii. $x = 4$

R9. The glass was moving 1.3 cm/s slower than Rover.

CHAPTER 5

Problem Set 5-1

1. $f(1000) = 24$ \$/ft
$f(4000) = 84$ \$/ft
The price increases because it is harder and slower to drill at increasing depths.

3. $R_6 = 143750$, so the cost is about \$143,750. R_6 is close to T_6.

5. $g(x) = 20x + \frac{1}{3}(0.000004)x^3 + C$
$g(4000) - g(1000) = 144000$, which is approximately the conjectured value of the definite integral!
Another name for *antiderivative* is *indefinite integral*.

Problem Set 5-2

1. $y = 21.6x - 48.6$
$x = 3.1$: Error $= 0.11042$
$x = 3.001$: Error $= 0.0000108...$
$x = 2.999$: Error $= 0.0000107...$

3. a. $y = 2x - 1$
The graph shows zoom by factor of 10.

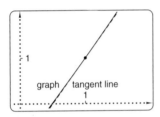

Local linearity describes the property of the function because if you keep x close to 1 (in the "locality" of 1), the curved graph of the function looks like the straight graph of the tangent line.

b.

x	$f(x)$	y	Error, $f(x) - y$
0.97	0.9409	0.94	0.0009
0.98	0.9604	0.96	0.0004
0.99	0.9801	0.98	0.0001
1	1	1	0
1.01	1.0201	1.02	0.0001
1.02	1.0404	1.04	0.0004
1.03	1.0609	1.06	0.0009

The table shows that for x-values close to 1 (the point of tangency), the tangent line is a close approximation to the function values.

5. a. Let A be the number of radians in θ degrees.
By trigonometry, $\tan A = \frac{x}{100} \Rightarrow A = \tan^{-1}\frac{x}{100}$.
Because 1 radian is $180/\pi$ degrees,
$\theta = \frac{180}{\pi}\tan^{-1}\frac{x}{100}$, Q.E.D.

b. $d\theta = \dfrac{1.8/\pi}{1 + (x/100)^2}dx$
$x = 0$: $d\theta = 0.5729... \, dx$
$x = 10$: $d\theta = 0.5672... \, dx$
$x = 20$: $d\theta = 0.5509... \, dx$

c. The error is $0.1492...°$, which is about 1.3%.

d. $0.5729...$ is approximately 0.5, so multiplying by it is approximately equivalent to dividing by 2.
For a 20% grade: 10°; compared with the actual angle of $11.309...°$, an error of about 11.6%
For a 100% grade: 50°; compared with the actual angle of 45°, an error of about 11.1%

7. a. $0.8219...$, or about 82 cents

b. $dm = 6000(0.05/365) \, e^{(0.05/365)t} \, dt$
For $t = 0$ and $dt = 1$: $dm = 0.8219...$, the same as part a
For $t = 0$ and $dt = 30$: $dm = 24.6575... \approx \$24.66$
For $t = 0$ and $dt = 60$: $dm = 49.3150... \approx \$49.32$

c. $t = 1$: $\Delta m = 0.8219...$, almost exactly equal to dm
$t = 30$: $\Delta m = 24.7082...$, about 5 cents higher than dm
$t = 60$: $\Delta m = 49.5182...$, about 20 cents higher than dm
As t increases, $t \, dm$ is a less accurate approximation for Δm.

9. $dy = 21x^2 \, dx$

11. $dy = 28x^3(x^4 + 1)^6 \, dx$

13. $dy = (6x + 5) \, dx$

15. $dy = -1.7e^{-1.7x} \, dx$

17. $dy = 3\cos 3x \, dx$

19. $dy = 3\tan^2 x \sec^2 x \, dx$

21. $dy = (4\cos x - 4x\sin x) \, dx$

23. $dy = (x - 1/4) \, dx$

25. $dy = \dfrac{-\sin(\ln x)}{x}dx$

27. $y = 5x^4 + C$

29. $y = -(1/4)\cos 4x + C$

31. $y = (2/7)(0.5x - 1)^7 + C$

33. $y = \tan x + C$

35. $y = 5x + C$

37. $y = 2x^3 + 5x^2 - 4x + C$

39. $y = (1/6)\sin^6 x + C$

41. a. $dy = 6(3x + 4)(2x - 5)^2(5x - 1) \, dx$
b. $dy = -60.48$
c. $\Delta y = -60.0218...$
d. -60.48 is close to $-60.0218...$.

Problem Set 5-3

1. $\frac{1}{11}x^{11} + C$

3. $-\frac{4}{5}x^{-5} + C$

5. $\sin x + C$

7. $\frac{4}{7}\sin 7x + C$

9. $\frac{5}{0.3}e^{0.3x} + C$

11. $\frac{4^m}{\ln 4} + C$

13. $\frac{1}{12}(4v + 9)^3 + C$

15. $-\frac{1}{20}(8 - 5x)^4 + C$

17. $\frac{1}{7}\sin^7 x + C$

19. $-\frac{1}{5}\cos^5 \theta + C$

21. $\frac{1}{3}x^3 + \frac{3}{2}x^2 - 5x + C$

23. $\frac{1}{7}x^7 + 3x^5 + 25x^3 + 125x + C$

25. $e^{\sec x} + C$

27. $\tan x + C$

29. $\frac{1}{8}\tan^8 x + C$

31. $-\frac{1}{9}\csc^9 x + C$

33. $D(t) = 40t + \frac{10}{3}t^{3/2}$
$D(10) = 505.4092... \approx 505$ feet

35. **Proof:**

Let $h(x) = \int f(x)\,dx + \int g(x)\,dx$.
By the derivative of a sum property,
$h'(x) = \frac{d}{dx}\int f(x)\,dx + \frac{d}{dx}\int g(x)\,dx$.
By the definition of indefinite integral applied twice
to the right side of the equation,
$h'(x) = f(x) + g(x)$.
By the definition of indefinite integral applied in the
other direction,
$h(x) = \int [f(x) + g(x)]\,dx$.
By the transitive property, then,
$\int [f(x) + g(x)]\,dx = \int f(x)\,dx + \int g(x)\,dx$, Q.E.D.

37. a. Integral ≈ 50.75

   b. Integral ≈ 50.9375

   c. As shown in Figures 5-3c and 5-3d, the Riemann
      sum with six increments has smaller regions
      included above the graph and smaller regions

excluded below the graph, so the Riemann sum
should be closer to the integral.

   d. Conjecture: The exact value is 51.
      By the trapezoidal rule with $n = 100$, integral $\approx$
      51.00045, which agrees with the conjecture.

   e. The object traveled 51 ft.
      Average velocity = 17 ft/min

Problem Set 5-4

1. $R_6 = 20.9375$

3. $R_8 = 23.97054...$

5. $R_5 = 0.958045...$

7. $U_4 = 1.16866...$, $M_4 = 0.92270...$, $T_4 = 0.95373...$,
   $L_4 = 0.73879...$
   $\therefore M_4$ and T_4 are between L_4 and U_4, Q.E.D.

9. $\int_1^5 \ln x\,dx$ is underestimated by the trapezoidal rule
   and overestimated by the midpoint rule.

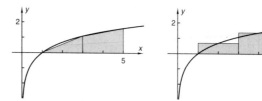

11. a. $x = 1, \pi/2, 2, 3, 4,$ and 6

   b. $x = 0, 1, 3, 4, 3\pi/2,$ and 5

   c. $U_6 = 21.71134...$, $L_6 = 14.53372...$

13. a. The program should give the values listed in
      the text.

   b. $L_{100} = 20.77545$, $L_{500} = 20.955018$
      L_n seems to be approaching 21.

   c. $U_{100} = 21.22545$, $U_{500} = 21.045018$
      U_n also seems to be approaching 21.
      f is integrable on $[1, 4]$ if L_n and U_n have the same
      limit as n approaches infinity.

   d. The trapezoids are circumscribed around the
      region under the graph and thus contain more
      area (see figure on the left). For rectangles, the
      "triangular" part of the region that is left out has
      more area than the "triangular" part that is
      included because the "triangles" have equal bases
      but unequal altitudes (see figure on the right).

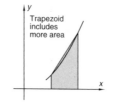

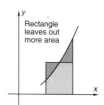

15. To evaluate $\int_0^2 x^3\,dx$, find an upper sum using the sample points $1 \cdot \frac{2}{n}, 2 \cdot \frac{2}{n}, 3 \cdot \frac{2}{n}, \ldots, n \cdot \frac{2}{n}$.

$$U_n = \left(\frac{2}{n}\right)\left(1 \cdot \frac{2}{n}\right)^3 + \left(\frac{2}{n}\right)\left(2 \cdot \frac{2}{n}\right)^3$$
$$+ \left(\frac{2}{n}\right)\left(3 \cdot \frac{2}{n}\right)^3 + \cdots + \left(\frac{2}{n}\right)\left(n \cdot \frac{2}{n}\right)^3$$
$$= \frac{4}{n^2} \cdot (n+1)^2 = 4\left(1 + \frac{1}{n}\right)^2$$
$$\lim_{n\to\infty} U_n = 4(1+0)^2 = 4$$

Problem Set 5-5

1. See the text statement of the mean value theorem.

3.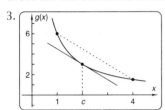

$c = 2$
The tangent at $x = 2$ parallels the secant line.

5.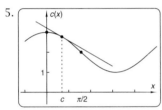

$c = 0.69010\ldots$
The tangent at $x = 0.69010\ldots$ parallels the secant line.

7.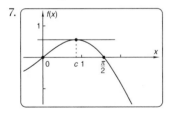

$c = 0.86033\ldots$
The horizontal line at $x = 0.86033\ldots$ is tangent.

9.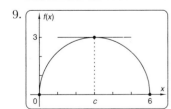

$[0, 6]$
$c = 3$
The horizontal line at $x = 3$ is tangent.

11. a. \$74,357.52. Surprising!

   b. Average rate $\approx 1,467.15$ \$/yr

   c. $d'(0) \approx 86.18$ \$/yr
   $d'(50) \approx 6,407.96$ \$/yr
   The average of these is 3,247.07 \$/yr, which does not equal the average in part b.

   d. $t \approx 32.893\ldots$ yr
   This time is not halfway between 0 and 50.

13. See Figure 5-5d.

15.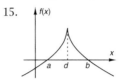

17.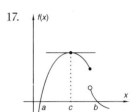

19. $f(1) = -3 \neq 0$
The conclusion is not true.
$f'(2) = 0$, but 2 is not in the interval $(0, 1)$.

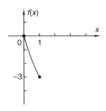

21. $f(2) = -4 \neq 0$
The conclusion is not true.
$f'(2) = 0$, but 2 is not in the *open* interval $(0, 2)$.

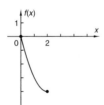

23. $f(3) = -3 \neq 0$
The conclusion is true.
$f'(2) = 0$ and 2 is in the interval $(0, 3)$.

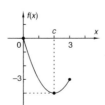

25. $f(0)$ does not exist.
The conclusion is not true.
$f'(x)$ never equals 0.

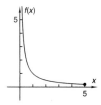

27. f is not differentiable at $x = 3$.
The conclusion is not true.
$f'(x)$ never equals 0.

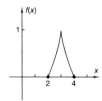

29. g is discontinuous at $x = 2$. Thus, the hypotheses of the mean value theorem are not met. The conclusion is not true for [1, 3] because the tangent line would have to contain $(2, g(2))$, as shown in the left figure. The conclusion is true for (1, 5), because the slope of the secant line is 1, and $g'(x) = 1$ at $x = 3$, which is in the interval (1, 5). See the figure on the right.

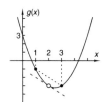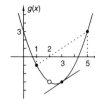

31. a. $f(x) = \begin{cases} 3x - 3, & \text{if } x \geq 3 \\ x + 3, & \text{if } x < 3 \end{cases}$

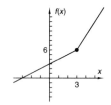

b. f is continuous at $x = 3$ because the right and left limits both equal 6.
f is not differentiable at $x = 3$ because the left limit of $f'(x)$ is 1 and the right limit of $f'(x)$ is 3.

c. f is not differentiable at $x = 3$, which is in (1, 6). The secant line has slope 11/5. The tangent line has slope either 1 or 3, and thus is never 11/5.

d. f is integrable on [1, 6]. The integral equals 41.5, the sum of the areas of the two trapezoids shown in the figure below.

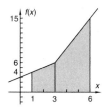

33. a. The grapher graph agrees with Figure 5-5l.

b. $f'(x) = -2x^2 + 10x - 8\pi \sin[2\pi(x - 5)]f'(5) = 0$
Because the derivative at $x = 5$ is 0, the tangent line at $x = 5$ is horizontal. This is consistent with $x = 5$ being a high point on the graph.

c.

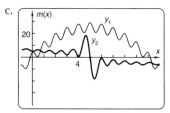

x	$m(x)$	x	$m(x)$
3.0	2	5.5	-16.5
3.5	6.8333...	6.0	-1
4.0	1	6.5	$-6.833...$
4.5	16.5	7.0	-2
5.0	undefined		

The difference quotient is positive when x is less than 5 and negative when x is greater than 5.

d. In the proof of Rolle's theorem, the left limit of the difference quotient was shown to be positive or zero and the right limit was shown to be negative or zero. The unmentioned hypothesis is *differentiability* on the interval (a, b). The function f is differentiable. There is a value of $f'(5)$, so both the left and right limits of the difference quotient must be equal. This number can only be zero, which establishes the conclusion of the theorem. The conclusion of Rolle's theorem *can* be true even if the hypotheses aren't met. For instance, $f(x) = 2 + \cos x$ has zero derivatives every π units of x, although $f(x)$ is never equal to zero.

35. The hypotheses of the mean value theorem state that f should be differentiable in the *open* interval (a, b)

694

and continuous at $x = a$ and $x = b$. If f is differentiable in the *closed* interval $[a, b]$, it is automatically continuous at $x = a$ and $x = b$, because differentiability implies continuity.

37. By the definition of antiderivative (indefinite integral), $g(x) = \int 0\, dx$ if and only if $g'(x) = 0$. Any other function f for which $f'(x) = 0$ differs from $g(x)$ by a constant. Thus, the antiderivative of zero is a constant function, Q.E.D.

39. The hypotheses of Rolle's theorem say that f is *differentiable* on the open interval (a, b). Differentiability implies continuity, so f is also *continuous* on (a, b). Combining this fact with the hypothesis of continuity at a and b allows you to conclude that the function is continuous on the *closed* interval $[a, b]$.

41. Answers will vary.

Problem Set 5-6

1. a. $I = 10/3 = 3.33333...$
   The $+C$ and $-C$ add up to zero.

   b.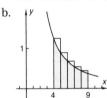

   c. $U_5 = 3.80673199...$, $L_5 = 2.92710236...$
   Average $= 3.36691717...$
   The average overestimates the integral, $3.33333...$. This is because the graph is concave up, and thus the area above each lower rectangle is less than half the difference between each upper rectangle and lower rectangle.

   d. $M_{10} = 3.32911229...$
   $M_{100} = 3.33329093...$
   $M_{1000} = 3.33333290...$
   The sums are converging to $10/3$, the exact value of I.

3. See the text statement of the fundamental theorem.

5. See the text proof of the fundamental theorem.

7. Distance $= \int_0^8 (100 - 20(t + 1)^{1/2})\, dt = 453\frac{1}{3}$ feet

9. a.   b.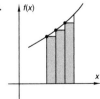

c. 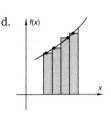   d.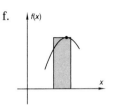

e. 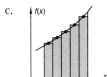   f.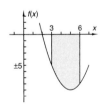

Problem Set 5-7

1. 21

3. 125

5. 1116

7. 30

9. $10\frac{2}{3}$

11. $4\frac{1}{3}$

13. 8

15. $(7/6)\sqrt{3} - 1/2 = 1.52072...$

17. 7.5

19. $\frac{1}{4}(\sin^4 2 - \sin^4 1) = 0.045566...$

21. $\frac{1}{3}(\sin 0.6 - \sin 0.3) = 0.0897074...$

23. 20

25. No value

27. Integral $= -(\text{area})$

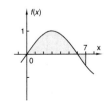

29. Integral $\neq$ area

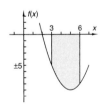

31. −7

33. 25

35. Cannot be evaluated

37.

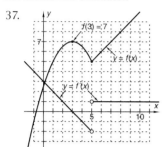

39. Converse: "If $\int_a^b f(x)\,dx < \int_a^b g(x)\,dx$,
then $f(x) < g(x)$ for all x in $[a, b]$."
The converse can be shown to be false by any
counterexample in which the area of the region
under the g graph is greater than the area under
the f graph, but in which the g graph touches or
crosses the f graph somewhere in $[a, b]$. One
counterexample is $f(x) = 1.5$ and $g(x) = 2 + \cos x$
on $[0, 2\pi]$.

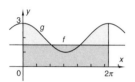

Problem Set 5-8

1. a.

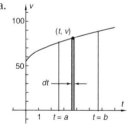

$dy = (55 + 12t^{0.6})\,dt$

b. $\int_0^1 (55 + 12t^{0.6})\,dt = 62.5$ mi
$\int_1^2 (55 + 12t^{0.6})\,dt \approx 70.2$ mi

c. $\int_0^2 (55 + 12t^{0.6})\,dt = 132.735...$, which equals the
sum of the two integrals above.

d. Approximately 4.134 h

e. At the end of the trip you are going about
83 mi/h.

3. a.

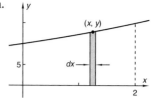

b. $dA = y\,dx = 10e^{0.2x}\,dx$

c. $50e^{0.4} - 50$

d. 24.59123...
The region is approximately a trapezoid with
height 2 and bases 10 and $y(2) \cdot y(2) = 12.2809...$,
so the area of the trapezoid is $2/2(10 + 14.9182...) = 24.9182...$.

5. a. $dA = [x + 2 - (x^2 - 2x - 2)]\,dx$
The top and bottom of the strip are not horizontal,
so the area of the strip is slightly different from
dA. As dx approaches zero, the difference
between dA and the area of the strip gets smaller.

b. $125/6 = 20.8333...$

c. $R_{100} = 20.834375$ (checks)

7. a.

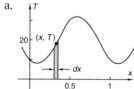

b. $dW = 0.6x\,dx$
24.3 inch-pounds

c. The region under F from $x = 0$ to $x = 9$ is a
triangle with base 9 and height $F(9) = 5.4$. So the
area is $1/2 \cdot 9 \cdot 5.4 = 24.3$.

d. Inch-pounds

9. a.

b. $dD = [20 - 12 \cos 2\pi(x - 0.1)]\,dx$

c. $7.75482... \approx 7.75$ degree-days

d. From noon to midnight: $12.24517... \approx 12.25$ degree-days
From one midnight to the next: 20 degree-days

696

11. a.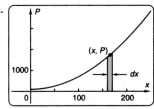

b. $dC = (100 + 0.06x^2)\, dx$
$C = 100b + 0.02b^3$

c. 0 m to 100 m: \$30,000
100 m to 200 m: \$150,000
0 m to 200 m: \$180,000
$\int_0^{200} P\, dx = \int_0^{100} P\, dx + \int_{100}^{200} P\, dx,$
which shows that the sum of integrals with the same integrand applies.

13.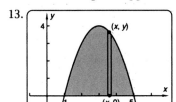

$A = \int_1^5 (-x^2 + 6x - 5)\, dx = 10\frac{2}{3}$

15.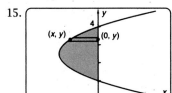

$A = \int_1^4 (-y^2 + 5y - 4)\, dy = 4\frac{1}{2}$

17.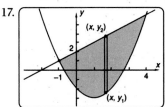

$A = \int_{-1}^4 (-x^2 + 3x + 4)\, dx = 20\frac{5}{6}$

19.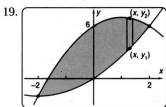

$A = \int_{-2}^2 (-1.5x^2 + 6)\, dx = 16$

21.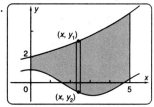

$A = \int_0^5 (2e^{0.2x} - \cos x)\, dx = 10e - \sin 5 - 10 = 18.1417...$

23.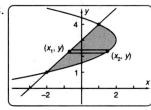

$A = \int_1^4 (-y^2 + 5y - 4)\, dy = 4\frac{1}{2}$

25.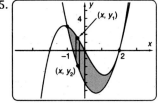

$A = \int_{-1}^2 (x^3 - 3x^2 + 4)\, dx = 6\frac{3}{4}$

27. You can always tell the right way because the height of the strip should be positive. This will happen if you take (larger value) − (smaller value). In this case, if you slice vertically it's line minus curve (see graph). For curve minus line, you'd get the opposite of the right answer.

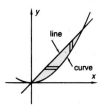

Note that if you slice horizontally, it would be curve minus line.

29.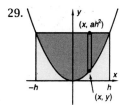

$A = \int_{-h}^{h}(ah^2 - ax^2)\,dx = \frac{4}{3}ah^3$

Area of rectangle $= 2h(ah^2) = 2ah^3$

$\therefore \dfrac{\text{area of region}}{\text{area of rectangle}} = \dfrac{(4/3)ah^3}{2ah^3} = \dfrac{2}{3}$, Q.E.D.

Area of region $= \dfrac{2}{3}(20)(60) = 800$

31. $A = 20\frac{5}{6} = 20.8333...$
    $R_{10} = 20.9375$
    $R_{100} = 20.834375$
    $R_{1000} = 20.83334375$
    The Riemann sums seem to be approaching the exact answer.

Problem Set 5-9

1. a. $dV = \pi x^2\,dy = \pi(9 - y)\,dy$
   b. $V = 40.5\pi = 127.2345...$
   c. $R_{100} = 127.2345...$
   d. The volume of the circumscribed cylinder is $9(\pi \cdot 3^2) = 254.4690...$. Half of this is $127.2345...$, which is equal to the volume of the paraboloid.

3. a. $dV = \pi y^2\,dx = 9\pi e^{-0.4x}\,dx$
   b. $V = -22.5\pi e^{-2} + 22.5\pi = 61.1195...$
      The midpoint Riemann sum R_{100} gives $61.1185...$, which is close to the answer found using integration.
   c. Slice perpendicular to the axis of rotation, so slice vertically if rotating about the x-axis and horizontally if rotating about the y-axis.

5. $V = 1640\pi = 5152.2119...$

7. $V = \dfrac{6141}{11}\pi = 1753.8654...$

9. $V = \pi(22.25 - 1.25e^{2.4}) = 26.6125...$ ft^3
   The midpoint Riemann sum R_{100} gives $26.6127...$, which is close to the answer found using integration.

11. $V = \dfrac{1}{6}\pi = 0.5325...$

13. a. $V = 5.76\pi = 18.09557...$
    b. $R_{10} = 5.7312\pi$
       $R_{100} = 5.75971...\pi$
       $R_{1000} = 5.7599971...\pi$
       Values are getting closer to $V = 5.76\pi$.

15. $V = 70.4\pi = 221.168...$

17.

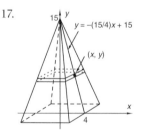

$y = -(15/4)x + 15$

(x, y)

$dV = (2x)^2\,dy = 64\left(1 - \dfrac{1}{15}y\right)^2 dy$

$V = 320$ cm^2

The circumscribed rectangular box has volume 960, so the pyramid is one-third the volume of the circumscribed rectangular box, Q.E.D.

The volume of a pyramid is one-third the volume of the circumscribed rectangular box, just as the volume of a cone is one-third the volume of the circumscribed cylinder.

19. a. $dV = \dfrac{1}{2}y^2\,dx = \dfrac{1}{2}x^{1.2}\,dx$
    b. $V = \dfrac{1}{4.4} \cdot 4^{2.2} = 4.7982...$
       The midpoint Riemann sum R_{100} gives $4.7981...$, which is close to the answer found using integration.
    c. The volume would double, to $9.5964...$.

21. $V = \dfrac{8\sqrt{3}}{15} = 0.9237...$

23. a. $y = \dfrac{1}{2}x$
    b. $z = \sqrt{36 - x^2}$
    c. $dV = y \cdot 2z \cdot dx = (36 - x^2)^{1/2}(x\,dx)$
       $V = 72$ in.^{3}

25. A cone of radius r and altitude h can be generated by rotating about the x-axis the line $y = \frac{r}{h}x$ from $x = 0$ to h.

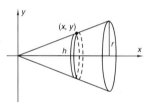

$dV = \pi y^2\,dx = \pi\dfrac{r^2}{h^2}x^2\,dx$

$V = \dfrac{\pi r^2}{h^2}\int_0^h x^2\,dx = \dfrac{\pi r^2}{h^2} \cdot \dfrac{1}{3}x^3\Big|_0^h = \dfrac{1}{3}\pi r^2 h$, Q.E.D.

27. A sphere can be generated by rotating the circle $x^2 + y^2 = r^2$ about the y-axis.
    Slicing perpendicular to the y-axis gives
    $dV = \pi x^2\,dy = \pi(r^2 - y^2)\,dy$.
    $V = \int_{-r}^{r} \pi(r^2 - y^2)\,dy = \pi(r^2 y - \frac{1}{3}y^3)\Big|_{-r}^{r}$
    $= \pi(r^3 - \frac{1}{3}r^3) - \pi(-r^3 + \frac{1}{3}r^3) = \frac{4}{3}\pi r^3$, Q.E.D.

29. $V = \int_0^{600}\left(50y + y^2 \cot\dfrac{52\pi}{180}\right)dx$
    $V \approx S_{20} = 1{,}647{,}388.8... \approx 1{,}647{,}389$ yd^3
    Cost $\approx \$19{,}793{,}324$

Problem Set 5-10

1. $\int_{0.3}^{1.4} \cos x\,dx \approx 0.6899295233...$

3. $\int_0^3 2^x\,dx \approx 10.09886529...$

5. $\int_{0.3}^{1.4} \cos x \, dx = \sin x \Big|_{0.3}^{1.4} = \sin 1.4 - \sin 0.3 =$
0.6899295233...

For the ten digits of the answer shown by calculator, there is no difference between this solution and the solution to Problem 1.

7.

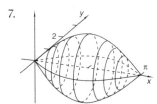

$\int_0^\pi \pi \sin^2 x \, dx = 4.9348...$
We cannot compute this integral algebraically because we do not know an antiderivative for $\sin^2 x$.

9. a.

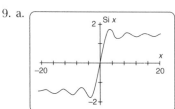

b. 1

c. Answers will vary depending on the grapher used. The TI-83 gives Si 0.6 = 0.58812881 using TRACE or 0.588128809608 using TABLE. Both of these values are correct to as many decimal places as the NBS values.

d. Si x seems to approach a limit of about 1.57.

e.

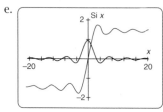

The graph of $f(x)$ is positive, and greatest when x is between $-\pi$ and π, which agrees with the large positive slope of the Si x graph in this region. Each place where the Si x graph has a high or low point, the graph of $f(x)$ has a zero, corresponding to the zero slope of the Si x graph.

11. a. Distance = 3.444... ≈ 3.4 nautical miles

b. $T_6 = 3.7333... ≈ 3.7$ nautical miles

c. The answer by Simpson's rule should be closer, because the graph is represented by curved segments instead of straight ones.

13. a.

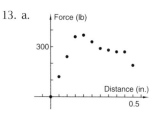

b. Work = 132.8333... ≈ 132.8 inch-pounds

15. a. Simpson's rule will give a more accurate answer because the function $y = \sin x$ is approximated better by quadratic functions than by straight lines.

b. $S_4 = 2.0045...$
$T_4 = 1.8961...$
$I = 2$
S_4 is closer to 2 than T_4.

17. Using a Simpson's rule program, the mass of the spleen is 171.6 cm³.

19. a.

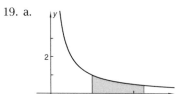

As x varies, the area beneath the curve $y = 1/t$ from $t = 1$ to $t = x$ varies also.

b. Using the power formula on $\int t^{-1} dt$ gives $\frac{t^0}{0}$. Division by 0 is undefined, so this approach does not work.

c.

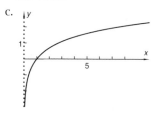

The graph resembles $y = \ln x$. The value of $f(x)$ is negative for $x < 1$ because for these values the lower limit of integration is larger than the upper limit, resulting in negative values for dx.

d. $f(2) = 0.6931...$
$f(3) = 1.0986...$
$f(6) = 1.7917...$
$f(2) + f(3) = f(2 \cdot 3)$
This is a property of logarithmic functions.

Problem Set 5-11

R0. Answers will vary.

R1. a. $T_3 = 2[22 + 2(26.9705...) + 2(30.7846...) + 34] = 343.0206...$

T_3 underestimates the integral because $v(t)$ is concave down, so trapezoids are inscribed under the curve.

b. $R_3 = 344.4821\ldots$

This Riemann sum is close to the trapezoidal rule sum.

c. $T_{50} = 343.9964\ldots$, $T_{100} = 343.9991\ldots$

Conjecture: The exact value of the integral is 344.

d. $g(16) - g(4) = 344$

This is the value the trapezoidal rule sums are approaching.

R2. a. $l(x) = -\pi x + \pi$

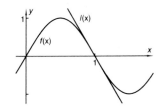 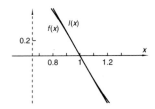

As you zoom in, you see that $f(x)$ is very close to the line $l(x)$ for values near $x = 1$.

$x = 1.1$: Error $= 0.0051\ldots$

$x = 1.001$: Error $= 5.1677\ldots \times 10^{-9}$

b. i. $dy = -10\csc^5 2x \cot 2x\, dx$

ii. $dy = (x^4 + x^{-4})\, dx$

iii. $dy = -12(7 - 3x)^3\, dx$

iv. $dy = -1.5e^{-0.3x}\, dx$

v. $dy = 4/x\, dx$

c. i. $y = \sec x + C$

ii. $y = \dfrac{1}{18}(3x + 7)^6 + C$

iii. $y = 5x + C$

iv. $y = -e^{-0.2x} + C$

v. $y = \dfrac{6^x}{\ln 6} + C$

d. i. $dy = (2x + 5)^{-1/2}\, dx$

ii. $dy = 25^{-1/2} \cdot 0.3 = 0.06$

iii. $\Delta y = 0.059644\ldots$

iv. 0.06 is close to $0.059644\ldots$.

R3. a. See the text definition of indefinite integral.

b. i. $7.2x^{5/3} + C$

ii. $\dfrac{1}{7}\sin^7 x + C$

iii. $\dfrac{1}{3}x^3 - 4x^2 + 3x + C$

iv. $4e^{3x} + C$

v. $\dfrac{7^x}{\ln 7} + C$

R4. a. See the text definition of integrability.

b. See the text definition of definite integral.

c. i. $U_6 = 2.845333\ldots$

ii. $L_6 = 1.872703\ldots$

iii. $M_6 = 2.209073\ldots$

iv. $T_6 = 2.359018\ldots$

d.

e.

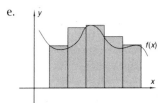

R5. a. The hypothesis is the "if" part of a theorem, and the conclusion is the "then" part.

b. $1.12132\ldots \approx 1.12$ s

c. $[0, 4]$, $c = 1$

At $x = 0$, $g'(0)$ takes the form $1/0$, which is infinite. Thus, g is not differentiable at $x = 0$.

However, the function need not be differentiable at the endpoints of the interval, just on the open interval.

d. For a function to be continuous on a *closed* interval, the limit needs to equal the function value only as x approaches an endpoint from *within* the interval. This is true for function f at both endpoints, but not true for function g at $x = 2$. The graphs show that the conclusion of the mean value theorem is true for f, but not for g.

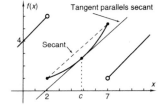

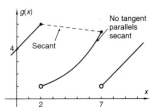

e. See the text derivation of Rolle's theorem.

f.

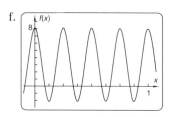

$$x = \frac{1}{8}, \frac{1}{4}, \frac{3}{8}, \frac{1}{2}, \frac{5}{8}, \frac{3}{4}, \text{ and } \frac{7}{8}$$

g. If $r'(x) = s'(x)$ for all x in an interval, then
$r(x) = s(x) + C$, where C is any constant.

R6. a. $R_3 = 12.4$, which is the exact value of the integral.

b. Integral $= 92/3 = 30.6666...$

c. $T_{100} = 30.6656$, which is close to $92/3$.

d. $M_{10} = 30.72$
$M_{100} = 30.6672$
$M_{1000} = 30.666672$
These Riemann sums are approaching $92/3$.

R7. a. i. $4/5$

ii. $(1/12)(19)^6 - (1/12)(12)^6 = 3671658.0833...$

iii. $2 - 5\pi = -13.7079...$

iv. 48

v. $\dfrac{78}{\ln 3} = 70.9986...$

b.

The integral is negative because each y-value in the Riemann sum is negative.

c. Integral $= 80$

d.

The total area is equal to the sum of the two areas.

R8. a.

$dy = 150t^{0.5}dt$
$y = \int_0^9 150t^{0.5}dt = 2700$ ft
For $[0, 4]$, $y = 800$.

For $[4, 9]$, $y = 1900$.
$2700 = 800 + 1900$, Q.E.D.

b. $dA = x\,dy = e^y\,dy$
$\int_0^{\ln 4} e^y\,dy = 3$

R9. a. $V = 2.5\pi(e^{1.6} - 1) = 31.0470...$

b. $V = \dfrac{2}{9}\pi = 0.6981...$

c. $V = 8\pi \approx 25.1327...$
The right circular cone of altitude 6 and radius 2
also has volume $\frac{1}{3}\pi \cdot 2^2 \cdot 6 = 8\pi$.

R10. a. Integral $= 6.0913...$

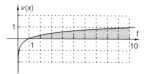

The integral is reasonable because counting
squares gives approximately 6.

b. $dW = (1000 + 50x)(4 - 0.2x^2)\,dx$
$\int_0^3 (1000 + 50x)(4 - 0.2x^2)\,dx = 10897.5$

c. $\int_3^5 v(t)\,dt$
Integral $= 1/3(0.2)(29 + 4(41) + 2(50) + 4(51) + 2(44) + 4(33) + 2(28) + 4(20) + 2(11) + 4(25) + 39) = 67.6$
Values of velocity are more likely to be connected by
smooth curves than straight lines, so the quadratic
curves given by Simpson's rule will be a better fit
than the straight lines given by the trapezoidal rule.

CHAPTER 6

Problem Set 6-1

1. The integral would be $(1/0)P^0$, which involves
division by zero.

N	Integral
1000	0
1500	0.4054...
2000	0.6931...
2500	0.9162...
500	−0.6931...
100	−2.3205...

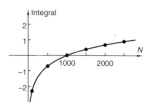

The graph resembles a logarithmic function.

3. $P(20) = 2718.2818..., P(0) = 1000$

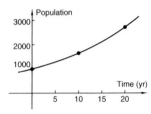

The graph resembles an exponential function.

Problem Set 6-2

1. $y' = 1/x$

3. $f'(x) = 5/x$

5. $h'(x) = -12/x$

7. $r'(t) = 3/t$

9. $y' = (1/x)(\ln 4x + \ln 6x)$ or $\dfrac{\ln 24x^2}{x}$

11. $y' = \dfrac{\ln 3x - \ln 11x}{x(\ln 3x)^2}$ or $\dfrac{\ln(3/11)}{x(\ln 3x)^2}$

13. $p' = (\cos x)(\ln x) + (\sin x)(1/x)$

15. $y' = -\sin(\ln x) \cdot (1/x)$

17. $y' = -\tan x$ (Surprise!)

19. $T'(x) = \sec^2(\ln x) \cdot (1/x)$

21. $y' = -3(3x + 5)^{-2}$

23. $y' = 4x^3 \ln 3x + x^3$

25. $y' = -1/x$

27. $7 \ln|x| + C$

29. $\dfrac{1}{3} \ln|x| + C$

31. $\dfrac{1}{3} \ln|x^3 + 5| + C$

33. $-\dfrac{1}{6} \ln|9 - x^6| + C$

35. $\ln|1 + \sec x| + C$

37. $\ln|\sin x| + C$

39. $\ln 8 = 2.0794...$

41. $\ln 30 = 3.4011...$

43. $\dfrac{2}{3}(\ln 28 - \ln 9) = 0.7566...$

45. $\dfrac{1}{6}(\ln x)^6 + C$

47. $f'(x) = \cos 3x$

49. $\tan^3 x$

51. $f'(x) = 2x \cdot 3^{x^2}$

53. $h'(x) = 3\sqrt{1 + (3x - 5)^2}$

55. Fundamental theorem:
   $5 \ln 3 - 5 \ln 1 = 5 \ln 3 = 5.493061...$
   Midpoint Riemann sum: $M_{100} = 5.492987...$
   Trapezoidal rule: $T_{100} = 5.493209...$
   Numerical integration: $5.493061...$

57. a.

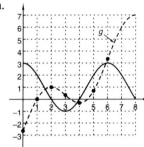

   b. $h'(2) = -4$

59. $\int_{1000}^{N}(1/P)\,dP = \ln N - \ln 1000$
   $N \approx 1649$ people

61. a. $a = -35.934084..., b = 9.050741...$
   $d(f) = -35.9340... + 9.0507... \ln f$
   b.

f	$d(f)$ cm	d' (part c)
53	0	0.1707...
60	1.1227...	0.1508...
70	2.5197...	0.1292...
80	3.7265...	0.1131...
100	5.7461...	0.0905...
120	7.3962...	0.0754...
140	8.7914...	0.0646...
160	10.0	0.0565...

   Measured distances will vary. The measured distances should be close to the calculated distances.

   c. $d'(f) = 9.0507.../f$. See table in part b.
   d. $d'(f)$ is in cm/10 kHz.
   e. $d'(f)$ decreases as f gets larger; this is consistent with the spaces between the numbers getting smaller as f increases.

63. Answers will vary.

Problem Set 6-3

1. $\ln 6 + \ln 4 = 3.17805...$
   $\ln 24 = 3.17805...$

3. $\ln 2001 - \ln 667 = 1.09861...$
   $\ln(2001/667) = 1.09861...$

5. $3 \ln 1776 = 22.44635...$
   $\ln(1776^3) = 22.44635...$

7. See the text proof of the uniqueness theorem.

9. **Proof:**

Let $f(x) = \ln(x/b)$, $g(x) = \ln x - \ln b$ for $x, b > 0$.
Then $f'(x) = (b/x)(1/b) = 1/x$, and
$g'(x) = (1/x) - 0 = 1/x$.
$\therefore f'(x) = g'(x)$ for all $x > 0$.
$f(b) = \ln(b/b) = \ln 1 = 0$
$g(b) = \ln b - \ln b = 0$
$\therefore f(b) = g(b)$.
$\therefore f(x) = g(x)$ for all $x > 0$ by the uniqueness theorem.
$\therefore \ln(x/b) = \ln x - \ln b$ for all $x > 0$.
$\therefore \ln(a/b) = \ln a - \ln b$ for all $a > 0$ and $b > 0$, Q.E.D.

11. **Proof:**

$\ln(a/b) = \ln(a \cdot b^{-1}) = \ln a + \ln b^{-1}$
$= \ln a + (-1)\ln b = \ln a - \ln b$
$\therefore \ln(a/b) = \ln a - \ln b$, Q.E.D.

13. $\ln x = \displaystyle\int_1^x \frac{1}{t}\, dt$

15. $f'(x) = 1/(x \ln 3)$
$f'(5) = 0.182047\ldots$
The graph shows that the tangent line at $x = 5$ has a small positive slope.

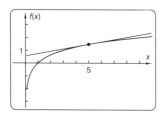

17. $g'(x) = 40/x$

19. $T'(x) = (\cot x)/(\ln 5)$

21. $p'(x) = (2\ln x)/(x \ln 5)$

23. $f'(x) = 3/x - \cot x$

25. $\dfrac{d}{dx}(\ln x^{3x}) = 3\ln x + 3$

27. a. $dy/dx = 7(-0.9^x)(\ln 0.9)$
$x = 0$: $dy/dx = 0.737\ldots$ mi/h
$x = 1$: $dy/dx = 0.663\ldots$ mi/h
$x = 5$: $dy/dx = 0.435\ldots$ mi/h
$x = 10$: $dy/dx = 0.257\ldots$ mi/h
The lava is slowing down.

b. $x = (1/\ln 0.9)[\ln(2 - y/7)]$

c. $\dfrac{dx}{dy} = \dfrac{9.491221\ldots}{14 - y}$
$y = 10$: $dx/dy = 2.372\ldots$ h/mi

d. $x = 10$: $dx/dy = 3.888651\ldots$

e. $3.888\ldots$ is the reciprocal of $0.257\ldots$, the value of dy/dx when $x = 10$, not when $y = 10$.

29. The intersection point is at $x = 2.7182818\ldots$, which is approximately e.

1. a. $a = 60000$, $k = 1.844\ldots$
$R(t) = 60000e^{1.844\ldots t}$

b. Approximately 607 million rabbits

c. About 5.6 yr earlier, or in about 1859

3. a. $m'(t) = 1000(1.06)^t (\ln 1.06)$
$m'(0) = 58.27$ \$/yr
$m'(5) = 77.98$ \$/yr
$m'(10) = 104.35$ \$/yr

b. $m(0) = \$1000.00$
$m(5) = \$1338.23$
$m(10) = \$1790.85$
The rates are increasing. \$338.23 is earned between 0 and 5 yr; \$452.62 is earned between 5 and 10 yr. This agrees with the increasing derivatives shown in part a.

c. $m'(t)/m(t) = \ln 1.06$, a constant

d. $m(1) = 1060.00$, so you earn \$60.00. The rate starts out at only \$58.27/yr, but has increased enough by year's end to make the total for the year equal to \$60.00.

5. $e = \lim_{n\to\infty}\left(1 + \frac{1}{n}\right)^n$ and $e = \lim_{n\to 0}(1 + n)^{1/n}$
When you substitute ∞ for n in the first equation, you get the indeterminate form 1^∞. When you substitute 0 for n in the second equation, you also get the indeterminate form 1^∞.

n	$(1 + 1/n)^n$
100	$2.70481\ldots$
1000	$2.71692\ldots$
10000	$2.71814\ldots$

n	$(1 + n)^{1/n}$
0.01	$2.70481\ldots$
0.001	$2.71692\ldots$
0.00001	$2.71826\ldots$

7. $y' = -2001e^{-3x}$

9. $g'(x) = x^{-7}e^x(-6 + x)$

11. $s'(t) = e^t \tan t + e^t \sec^2 t$

13. $v' = 0$

15. $y' = \dfrac{(1/x)e^x - \ln x \cdot e^x}{e^{2x}}$

17. $y' = -7e^{\cos x}\sin x$

19. $y' = 20$

21. $y' = -20x$

23. $h'(x) = 42$

25. $y' = e^x + e^{-x}$

27. $y' = 40x^4\, e^{x^5}$

29. $f'(x) = 10^{-0.2x}(-0.2 \ln 10)$

31. $h'(x) = 1000(1.03^x) \ln 1.03$

33. $m'(x) = 5^x \cdot x^7(\ln 5 + 7/x)$

35. $y' = 2 \ln x \cdot x^{\ln x - 1}$

37. $y' = (\cos 2x)^{3x} [3 \ln(\cos 2x) - 6x \tan 2x]$

39. $y' = \dfrac{8x + 33}{(4x - 7)(x + 10)}$

41. $y' = \dfrac{(270 - 105x)(10 + 3x)^9}{(4 - 5x)^4}$

43. $\ln x$

45. $2x(\ln \cos x^2)$

47. $49e^{7x}$

49. $\dfrac{1}{7}e^{7x} + C$

51. $\dfrac{1.05^x}{\ln 1.05} + C$

53. $5e^{0.2x} + C$

55. $e^{\tan x} + C$

57. $150x^2 + C$

59. $-\dfrac{1}{404}(1 - e^{4x})^{101} + C$

61. Fundamental theorem:
$e^2 - e^{-2} - e^{-1} + e^1 = 9.604123...$
Numerically: integral $\approx 9.604123...$ (checks)

63. Answers will vary.

Problem Set 6-5

1. Limit $= 10/3$

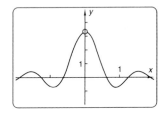

3. Limit $= 1$

5. Limit $= 1/2$

7. Limit $= \infty$

9. Limit $= 0$

11. Limit $= e/5$

13. Limit $= -26.4329...$

15. Limit $= \infty$

17. Limit $= 3/4$

19. Limit $= 1/4$

21. Limit $= 1$

23. Limit $= 1$

25. Limit $= 1$

27. Limit $= e^3 = 20.0855...$

29. Limit $= 1/2$

31.

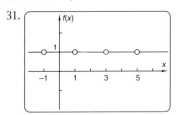

Where secant and tangent are defined, the Pythagorean properties tell us that $f(x) = 1$.

33. Limit $= e^k$

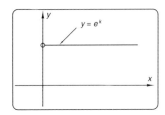

The graph is a horizontal line $y = e^k$ defined for $x > 0$.
By the definition of a power,
$f(x) = x^{k/(\ln x)} = (x^k)^{1/\ln x} = (e^{k \ln x})^{1/\ln x} = e^k$

35. a. For yearly compounding, $m(t) = 1000(1 + 0.06)^t$.
For semiannual compounding,
$m(t) = 1000(1 + 0.06/2)^{2t}$ because there are two compounding periods per year, each of which gets half the interest rate.

b. $m(t) = 1000(1 + 0.06/n)^{nt}$
Limit $= 1000e^{0.06t}$
When interest is compounded continuously,
$m(t) = 1000e^{0.06t}$.

c. 5 yr: \$11.63
20 yr: \$112.98
50 yr: \$1665.38

d. $m(t) = 1000e^{0.07t}$

37. Answers will vary.

Problem Set 6-6

1. $y' = 3/(3x + 4)$

3. $y' = 3$

5. $y' = -5 \tan x$

7. $y' = -\tan(\tan x) \sec^2 x$

9. $y' = -(1/x) \sin(\ln x)$

11. $y' = 7e^{7x}$

13. $y' = 5x^4$

15. $y' = -e^x \sin e^x$

17. $y' = 5x^4 e^{x^5}$

19. $y' = \dfrac{e^x}{\cos y} = \dfrac{e^x}{\sqrt{1 - e^{2x}}}$

21. $y' = 1/x$

23. $y' = 1/x$

25. $y' = 2^x \ln 2$

27. $y' = 2x$

29. $y' = x^x (\ln x + 1)$

31. $y' = xe^x$

33. $y' = \frac{1}{2}(e^x + e^{-x})$

35. $y' = 5^x \ln 5$

37. $y' = \dfrac{x^{-8}}{\ln 2}(-7 \ln x + 1)$

39. $y' = e^{-2x}(-2 \ln 5x + 1/x)$

41. $y' = 1/x$

43. $y' = 1/x$

45. $y' = \dfrac{7}{x \ln 5}$

47. $y' = e^{\sin x} \cos x$

49. $y' = 0$

51. $y' = \cos x$

53. $y' = -\csc x \cot x$

55. $y' = \sec^2 x$

57. $\frac{1}{4} e^{4x} + C$

59. $\frac{1}{4} e^{x^4} + C$

61. $\frac{1}{6} (\ln x)^6 + C$

63. $\dfrac{5^x}{\ln 5} + C$

65. $\ln x$

67. $\dfrac{2^x}{\ln 2} + C$

69. $3 \ln |x| + C$

71. $\frac{1}{10} (\ln x)^{10} + C$

73. $\frac{1}{2} x^2 + C$

75. C

77. $\frac{1}{2} \ln |\sec 2x + \tan 2x| + C$

79. $\frac{1}{4} \ln |\sin 4x| + C$

81. Limit $= 0$

83. Limit $= \pi/2$

85. Limit $= \dfrac{125}{6} = 20.8333...$

87. Limit $= 1$

89. Limit $= e^{-3/2} = 0.2231...$

Problem Set 6-7

R0. Answers will vary.

R1. a. $dM/dt = 0.06M \Rightarrow (1/M)\, dM = 0.06\, dt$
When $t = 0$, $M = 100$, and when $t = 5$, M is unknown.
$\therefore \int_{100}^{x} \frac{1}{M} dM = \int_{0}^{5} 0.06\, dt$, Q.E.D.

b. $x \approx 134.9858...$

c. \$34.99

R2. a. Integrating x^{-1} by the power rule results in division by zero: $(1/0)x^0 + C$.

b. If $g(x) = \int_a^x f(t)\, dt$ and $f(x)$ is continuous in a neighborhood of a, then $g'(x) = f(x)$.
$\ln x = \int_1^x \frac{1}{t}\, dt$
$\dfrac{d}{dx}(\ln x) = \dfrac{d}{dx}\left(\int_1^x \frac{1}{t}\, dt \right) = \dfrac{1}{x}$

c. i. $y' = (3/x)(\ln 5x)^2$

ii. $f'(x) = 9/x$

iii. $y' = -\csc(\ln x) \cot(\ln x) \cdot (1/x)$

iv. $g'(x) = 2x \csc x^2$

d. i. $\ln |\sec x| + C$

ii. $10(\ln 3 - \ln 2) = 4.0546...$

iii. $\frac{1}{3} \ln |x^3 - 4| + C$

e.

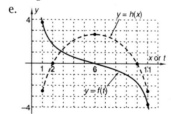

f. i. $y(100) \approx 70$ names; 70% remembered
$y(1) = 1$ name; 100% remembered

ii. $y'(100) = 0.505$ name/person
$y'(1) = 1$ name/person

iii. Assume that Paula has not forgotten any names as long as $x - y < 0.5$. After meeting 11 people she remembers about $10.53... \approx 11$ names, but after meeting 12 people she remembers about $11.44... \approx 11$ names.

R3. a. i. See the text definition of logarithm.

ii. See the text definition of $\ln x$ as a definite integral.

iii. See the text statement of the uniqueness theorem.

iv. See the text proof.

v. See the solution to Problem 10 in Lesson 6-3.

b. i. $e = \lim_{n \to 0}(1 + n)^{1/n}$ or $e = \lim_{n \to \infty}(1 + 1/n)^n$

ii. $\log_b x = \dfrac{\ln x}{\ln b}$

c. i. $y' = \dfrac{1}{x \ln 4}$

ii. $f'(x) = -\dfrac{\tan x}{\ln 2}$

iii. $y' = \log_5 9$

R4. a. i.

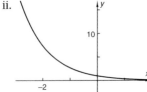

ii.

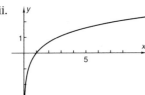

iii.

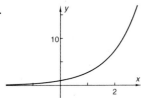

b. i. $f'(x) = 1.4x^{0.4} e^{5x} + 5x^{1.4} e^{5x}$

ii. $g'(x) = -2e^{-2x} \cos(e^{-2x})$

iii. $\dfrac{d}{dx}(e^{\ln x}) = 1$

iv. $y' = (\ln 100)\, 100^x$

v. $f'(x) = 0.74 \ln 10 \cdot 10^{0.2x}$

vi. $r'(t) = t^{\tan t}\left(\sec^2 t \ln t + \dfrac{\tan t}{t}\right)$

c. $y' = (120x - 90)(5x - 7)^2 (3x + 1)^4$

d. i. $-5e^{-2x} + C$

ii. $-e^{\cos x} + C$

iii. $-10e^{-0.2} + 10e^{0.2} = 4.0267...$

iv. $\dfrac{10^{0.2x}}{0.2 \ln 10} + C$

e. i. The exposure is the product of $C(t)$ and t, where $C(t)$ varies. Thus, a definite integral must be used.

ii. $E(x) = 937.5(-e^{-0.16x} + 1)$
$E(5) = 516.2540...$ ppm $\cdot$ days
$E(10) = 748.2220...$ ppm $\cdot$ days

As x grows very large, $E(x)$ seems to approach 937.5.

iii. $E'(x) = 150e^{-0.16x}$
$E'(5) = 67.3993...$ ppm (or ppm $\cdot$ days per day)
$E'(10) = 30.2844...$ ppm

f. i. The maximum concentration is about 150 ppm at about 2 h.

ii. $C'(1) \approx 58.7$ ppm/h
$C'(5) \approx 24.2$ ppm/h
The concentration is increasing if $C'(t)$ is positive and decreasing if it is negative.

iii. For about 6 h, from $t \approx 0.2899...$ to $t \approx 6.3245...$

iv. The concentration peaks sooner at a lower concentration and stays above 50 ppm for a much shorter time.

R5. a. Limit = $-2/5$

b. Limit = 3

c. Limit = 0

d. Limit = $e^{-2/\pi} = 0.529077...$

e. Limit = 48

f. Limit = 1

g. Examples of indeterminate forms:
$0/0, \infty/\infty, 0 \cdot \infty, 0^0, 1^\infty, \infty^0, \infty - \infty$

R6. a. i. $y' = 4(1/\sin 7x) \cdot \cos 7x \cdot 7 = 28 \cot 7x$

ii. $y' = x^{-4}e^{2x}(2x - 3)$

iii. $y' = -\sin(2^x) \cdot 2^x \ln 2$

iv. $y' = \dfrac{4}{x \ln 3}$

b. i. $(-1/1.7)e^{-1.7x} + C$

ii. $(1/\ln 2)\, 2^{\sec x} + C$

iii. $\ln(5 + \sin x) + C$ (No absolute value is needed.)

iv. $\ln 5$

c. i. Limit = ∞

ii. Limit = $e^{-3} = 0.0497...$

Problem Set 6-8

1. $f'(3) = 5.5496...$

2. $\int_{10}^{50} g(x)\, dx \approx 200$

3. $L = \lim_{x \to c} f(x)$ if and only if for any $\epsilon > 0$ there is a $\delta > 0$ such that if x is within δ units of c but not equal to c, $f(x)$ is within ϵ units of L.

4.

5. $f'(x) = \lim_{h\to 0} \dfrac{f(x+h) - f(x)}{h}$

    or $f'(c) = \lim_{x\to c} \dfrac{f(x) - f(c)}{x - c}$

6. $f(x) = x^3$

    $f'(x) = \lim_{h\to 0} \dfrac{(x+h)^3 - x^3}{h}$

    $= \lim_{h\to 0} \dfrac{x^3 + 3x^2h + 3xh^2 + h^3 - x^3}{h}$

    $= \lim_{h\to 0}(3x^2 + 3xh + h^2) = 3x^2$, Q.E.D.

7. $f'(5) = 75$

    $\Delta x = 0.01: f'(5) \approx 75.0001$

    $\Delta x = 0.001: f'(5) \approx 75.000001$

    The symmetric differences are getting closer to 75 as Δx gets closer to zero.

8. $f'(7) = 3/8$

9. A line with slope 3/8 is tangent to the graph at $x = 7$.

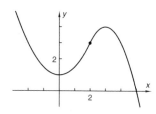

10. a. $y' = 2e^{2x}\cos 3x - 3e^{2x}\sin 3x$

    b. $q'(x) = \dfrac{1}{x\tan x} - \dfrac{\ln x}{\cos^2 x}$

    c. $\dfrac{d^2}{dx^2}(5^x) = (\ln 5)^2 5^x$

11. $a = 1/2$ and $b = -5$

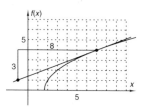

12. $U_6 = 24.875$

13. $M_{10} = 20.9775$, $M_{100} = 20.999775$

    The sums seem to be approaching 21.

14. a. $-\dfrac{1}{6}\cos^6 x + C$

    b. $\ln |x| + C$

    c. $-\ln |\cos x| + C$

    d. $\ln |\sec x + \tan x| + C$

    e. $\frac{2}{9}(3x - 5)^{3/2} + C$

15. Integral = 21, as conjectured in Problem 13.

16. If f is differentiable on (a, b) and continuous at $x = a$ and $x = b$, then there is a number $x = c$ in (a, b) such that $f'(x) = \frac{f(b) - f(a)}{b - a}$.

17. $y^7 = x^9$

    $7y^6 y' = 9x^8$

    $y' = \dfrac{9x^8}{7y^6} = \dfrac{9x^8}{7(x^{9/7})^6} = \dfrac{9}{7}x^{8 - 54/7} = \dfrac{9}{7}x^{2/7}$

    This answer is the same as the answer found using the derivative of a power formula.

18. If x^{-1} were the derivative of a power, then the power would have to be x^0. But $x^0 = 1$, so its derivative equals 0, not x^{-1}. Thus, x^{-1} is not the derivative of a power, Q.E.D.

19. $f'(x) = \cos(3\tan x) \cdot \sec^2 x$

20. $f(x) = \int_1^x dt \Rightarrow f'(x) = 1/x$, Q.E.D.

21. **Proof:**

    Let $f(x) = \ln x^a$ and $g(x) = a\ln x$.

    Then $f'(x) = \frac{1}{x^a}\cdot ax^{a-1} = a\cdot\frac{1}{x} = \frac{a}{x}$ and

    $g'(x) = a\cdot\frac{1}{x} = \frac{a}{x}$.

    $\therefore f'(x) = g'(x)$ for all $x > 0$.

    $f(1) = \ln(1^a) = \ln 1 = 0$ and

    $g(1) = a\ln 1 = 0$

    $\therefore f(1) = g(1)$

    $\therefore f(x) = g(x)$ for all $x > 0$, and thus $\ln x^a = a\ln x$ for all $x \geq 0$, Q.E.D.

22. $\dfrac{dy}{dx} = \dfrac{3\cos t}{-5\sin t} = \dfrac{-3}{5}\cot t$

23. At $t = 2$, $\frac{dy}{dx} = \frac{3\cos 2}{-5\sin 2} = \frac{-3}{5}\cot 2 = 0.2745\ldots$

    The graph shows that a line with slope $0.27\ldots$ at point $(-2.08\ldots, 2.72\ldots)$ (the point at which $t = 2$) is tangent to the curve.

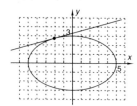

24. $v = \dfrac{1}{1 + t^2}$

    $a = -\dfrac{2t}{(1 + t^2)^2}$

25. Limit = 3/5

26. $L = \lim_{n \to 0} (1 + n)^{1/n} \to 1^{\infty}$

$\ln L = \lim_{n \to 0} \left[\frac{1}{n} \ln(1 + n) \right] \to \infty \cdot 0$

$= \lim_{n \to 0} \frac{\ln(1 + n)}{n} \to \frac{0}{0}$

$= \lim_{n \to 0} \frac{1/(1 + n)}{1} = 1$

$\therefore L = e^1 = e$, Q.E.D.

27. $\frac{dz}{dt} = \frac{-20}{\sqrt{5}} = -8.9442\ldots$ ft/s

The distance z is decreasing.

28. Integral $= 556\frac{2}{3}$

29. $dv = \pi y^2 = \frac{\pi r^2}{h^2} x^2 \, dx$

$\therefore V = \int_0^h \frac{\pi r^2}{h^2} x^2 \, dx = \frac{\pi r^2}{h^2} \cdot \frac{1}{3} x^3 \big|_0^h = \frac{1}{3} \pi \frac{r^2}{h^2} (h^3 - 0^3) = \frac{1}{3} \pi r^2 h$, Q.E.D.

30. Answers will vary.

CHAPTER 7

Problem Set 7-1

1. $D(0) = 500$
$D(10) = 895.42$
$D(20) = 1603.57$

3. $R(0) = 0.0582689081\ldots$
$R(10) = 0.0582689081\ldots$
$R(20) = 0.0582689081\ldots$

5. $f(x) = a \cdot b^x \Rightarrow f'(x) = a \cdot (\ln b) \cdot b^x = (\ln b)(a \cdot b^x) = (\ln b) \cdot f(x)$
So, $f'(x)$ is directly proportional to $f(x)$.

Problem Set 7-2

1. a. B = number of millions of bacteria; t = number of hours
$dB/dt = kB \Rightarrow \int dB/B = \int k \, dt$
$B = C_1 e^{kt}$

   b. $B = 5e^{(1/3)\ln(7/5)t} = 5e^{0.112157\ldots t}$ or $5\left(\frac{7}{5}\right)^{t/3}$

   c.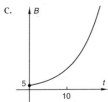

   d. About 74 million

   e. $t = 47.2400\ldots$
   About 47 h after start

3. a. F = number of mg; t = number of minutes
$dF/dt = kF$
$F = 50e^{-0.025541\ldots t} = 50(0.6)^{t/20}$

   b.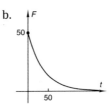

   c. 10.8 mg

   d. $t = 347.4323\ldots$
   About 5 h 47 min

5. a. $dC/dt = kC$

   b. $C = 0.00372e^{-0.0662277\ldots t}$

   c. Either $C = 0.015 \Rightarrow t = -21.05\ldots$, which is before the poison was inhaled, or $t = -20 \Rightarrow C = 0.0139\ldots$, which is less than 0.015.
   $\therefore$ the concentration never was that high.

   d.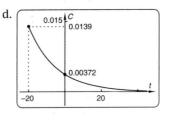

   e. $t = 10.4661\ldots$
   About 10.5 h

7. $dM/dt = kM \Rightarrow M = Ce^{kt}$, where C is the initial investment. $\therefore M$ varies exponentially with t.
Let i = interest rate as a decimal.
$dM/dt = Ck \cdot e^{kt}$
At $t = 0$, $dM/dt = Ci$
$\therefore Ci = Ck \cdot e^0 \Rightarrow k = i \Rightarrow M = Ce^{it}$
Examples:
$1000 at 7% for 5 yr: $1419.07
$1000 at 7% for 10 yr: $2013.75
$1000 at 14% for 5 yr: $2013.75
$1000 at 14% for 10 yr: $4055.20
Leaving the money twice as long has the *same* effect as doubling the interest rate. Doubling the amount invested obviously doubles the money at any particular time. But doubling either the time or the interest rate will always eventually yield more than doubling the investment, once t is high enough. For example, at an interest rate of 7%, doubling the time or interest rate will yield more than doubling the investment after 9 years 11 months.

9. $dy/dx = 0.3y$
$\int \frac{dy}{y} = 0.3 \int dx$
$\ln|y| = 0.3x + C$
$|y| = e^{0.3x+C} = e^{0.3x} \cdot e^C$
$y = \pm e^C \cdot e^{0.3x} = C_1 e^{0.3x} \quad -4 = C_1 e^0 \Rightarrow C_1 = -4$,
showing that C_1 can be negative.
$\therefore y = -4e^{0.3x}$

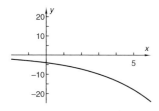

11. $dy/dx = ky \Rightarrow \int dy/y = \int k\,dx \Rightarrow \ln|y| = kx + C_1$
$|y| = e^{kx+C1} \Rightarrow y = Ce^{kx} \Rightarrow y(0) = Ce^{k \cdot 0} \Rightarrow C = y(0)$
$\Rightarrow y = y_0 e^{kx}$, Q.E.D.

Problem Set 7-3

1. a. $dM/dt = 100 - S$

   b. $S = kM \Rightarrow dM/dt = 100 - kM$

   c. $M = \dfrac{100}{k}(1 - e^{-kt})$

   d. $M = 5000(1 - e^{-0.02t})$

   e.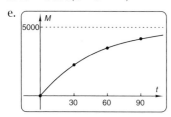

   f. $t = 30$: \$2255.94 (\$3000 in, \$744.06 spent)
   $t = 60$: \$3494.03 (\$6000 in, \$2505.97 spent)
   $t = 90$: \$4173.51 (\$9000 in, \$4826.49 spent)

   g. $t = 365$: $M = 4996.622... \approx \$4996.62$ in the account
   $dM/dt = 0.06755...$
   M is increasing at about \$0.07 per day.

   h. $\lim_{t \to \infty} M = 5000$

3. a. $E = RI + L(dI/dt)$

   b. $I = \dfrac{E}{R}[1 - e^{-(R/L)t}]$

   c. $I = 11(1 - e^{-0.5t})$

   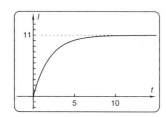

   d. i. $t = 1$: $I = 4.3281... \approx 4.33$ amps
   ii. $t = 10$: $I = 10.9258... \approx 10.93$ amps
   iii. $t = \infty$: $I = 11$ amps

   e. $t = -2\ln 0.05 \approx 6$ s

5. a. $\dfrac{dV}{dt} = kV^{1/2}$

   b. $V = \left(\dfrac{kt + C}{2}\right)^2$
   V varies quadratically with t.

   c. $V = (t - 14)^2$

   d. False. Because $dV/dt = 2t - 28$, the water flows out at 28 ft^3/min only when $t = 0$. For instance, at $t = 5$, $dV/dt = -18$, which means water flows out at only 18 ft^3/min. So, it takes longer than 7 min to empty the tub.

   e. The tub is empty at $t = 14$ min.

   f.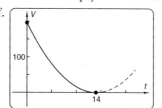

   g. See Problem Set 7-7, C4.

7. a. $n = 1, k = 1, C = -3$: $y = \pm 0.04978...e^x$

   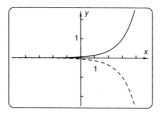

   b. $n = 0.5, k = 1, C = -3$: $y = \frac{1}{4}(x - 3)^2$

   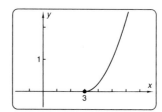

   (Note: $x \geq 3$ because $y^{0.5}$ is a positive number.)

   c. $n = -1, k = 1, C = -3$: $y = \pm\sqrt{2x - 6}$

   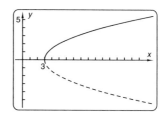

$n = -2, k = 1, C = -3 \Rightarrow y = \sqrt[3]{3x - 9}$

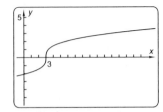

d. For $n > 1$, $\frac{dy}{dx} = ky^n \Rightarrow \int y^{-n} dy = k \int dx$

$\Rightarrow -\frac{y^{-(n-1)}}{n-1} = kx + C$ because $n > 1$

so $y = \frac{-1}{\sqrt[n-1]{(n-1) \cdot (kx + C)}}$,

which has a vertical asymptote at $x = -C/k$ because the denominator $= 0$ for this point. Note that the radical will involve a $\pm$ sign when the root index is even (for example, when n is odd). For $n = 2, k = 1, C = -3$: $y = -(x - 3)^{-1}$

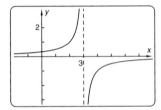

For $n = 3, k = 1, C = -3$: $y = \frac{-1}{\pm\sqrt{2x - 6}}$

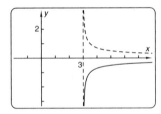

e. $n = 0, k = 1, C = -3$: $y = x - 3$, which is a linear function.

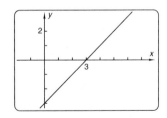

9. $T \approx 70 + 102.26...(1 - e^{-0.02933...t})$
Answers will vary.

1. a. At $(3, 5)$, $dy/dx = 3/10 = 0.3$
   At $(-5, 1)$, $dy/dx = -5/2 = -2.5$
   On the graph, the line at $(3, 5)$ slopes upward with a slope less than 1. At $(-5, 1)$, the line slopes downward with a slope much steeper than -1.

   b. The figure looks like one branch of a hyperbola opening in the y-direction. (The lower branch shown on the graph is also part of the solution, but you are not expected to find this graphically.)

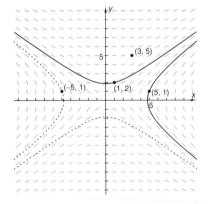

   c. See the graph from part b. The figure looks like the right branch of a hyperbola opening in the x-direction. (The left branch is also part of the solution, but you are not expected to find this graphically.)

   d. $x^2 - 2y^2 = 23$
   This is the particular equation of a hyperbola opening in the x-direction, which confirms the observations in part c.

3. a. At $(3, 2)$, $\frac{dy}{dx} = -\frac{3}{(2)(2)} = -0.75$
   At $(1, 0)$, $\frac{dy}{dx} = -\frac{1}{(2)(0)}$, which is infinite.

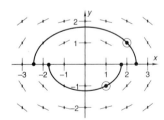

   b. See graph from part a. The figures resemble half-ellipses.

   c. $y = -\sqrt{1.5 - 0.5x^2}$ (Use the negative square root because of the initial condition.)
   The graph agrees with part b.
   The equation can be transformed to $0.5x^2 + y^2 = 1.5$, which is the equation of an ellipse because x^2 and y^2 have the same sign but unequal coefficients.

5.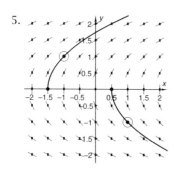

7.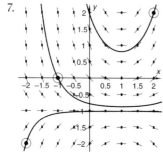

9. a.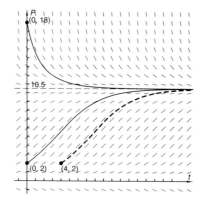

b. $\dfrac{dy}{dx} = -0.2xy$

Evidence: At (1, 1) the slope was given to be -0.2, which is true for this differential equation. As x or y increases from this point, the slope gets steeper in the negative direction, which is also true for this differential equation. In Quadrants I and III the slopes are all negative, and in Quadrants II and IV they are all positive.

11. a. Initial condition (0, 2)

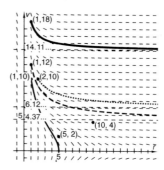

b. See graph in part a with initial condition (4, 2). The graph is the same as that in part a, but shifted over 4 months. This behavior is to be expected because dP/dt depends only on P, not on t, and both initial conditions have the same value of P.

c. See graph in part a with initial condition (0, 18). The population is *decreasing* to the same asymptote, $P = 10.5$, as in parts a and b.

d. The asymptote at $P = 10.5$ indicates that the island can sustain only 1050 rabbits. If the population is lower than that, it increases. If the population is higher than that, it decreases. The number 10.5 is a value of P that makes dP/dt equal zero. Note that there is another asymptote at $P = 0$, which also makes dP/dt equal zero.

13. a.

$ma = \dfrac{mg}{r^2}$ By hypothesis.

$\dfrac{dv}{dt} = \dfrac{g}{r^2}$ Divide by m; $a = \dfrac{dv}{dt}$.

$\dfrac{dv}{dr} \cdot \dfrac{dr}{dt} = \dfrac{g}{r^2}$ Chain rule.

$\dfrac{dv}{dr} \cdot v = \dfrac{g}{r^2}$ $v = \dfrac{dr}{dt}$ (r = distance).

$\dfrac{dv}{dr} = \dfrac{g}{r^2 v}$ Divide by v.

b. $\dfrac{dv}{dr}(5, 2) = -1.2488$

$\dfrac{dv}{dr}(1, 10) = -6.244$

$\dfrac{dv}{dr}(10, 4) = -0.1561$

These slopes agree with those shown.

c. Initial condition (1, 10)
The spaceship is about 4 earth-radii, or about 25,000 km, above the surface.

d. See graph in part c with initial condition (1, 12). The graph levels off between 4 and 5 km/s.

e. See graph in part c with initial condition (1, 18). The graph levels off at $v \approx 14$ km/s. Here the spaceship loses about 4 km/s of velocity, whereas it loses 7 or 8 km/s when starting at 12 km/s.

Both cases lose the same amount of kinetic energy, which is proportional to v^2 (the change in v^2 is the same in both cases).
The precise value of v can be found.

f. See graph in part c with initial condition $(2, 10)$. The graph levels off at about 6 km/s, so the spaceship does escape. Alternatively, note that the solution through $(2, 10)$ lies above the solution through $(1, 12)$.

Problem Set 7-5

1. a. For $(1, 3)$: $dy = -(1/3)(0.5) = -0.1666\ldots$, so $y \approx 3 - 0.1666\ldots = 2.8333\ldots$ at $x = 1.5$.
   For $(1.5, 2.8333\ldots)$: $dy = -(1.5/2.8333\ldots)(0.5) = -0.2647\ldots$, so $y \approx 2.8333\ldots - 0.2647\ldots = 2.5686\ldots$ at $x = 2$.

x	y
0	3.2456...
0.5	3.1666...
1	3
1.5	2.8333...
2	2.5686...
2.5	2.1793...
3	1.6057...

The Euler's y-values overestimate the actual values because the tangent lines are on the convex side of the graph, and the convex side is upward. As x gets farther from 1, the size of the error increases.

   b. $y = \sqrt{10 - x^2}$ (Use the positive square root.)
   The particular solution stops at the x-axis because points on the circle below the x-axis would lead to two values of y for the same value of x, making the solution not a function.
   At $x = 3$, Euler's solution overestimates the actual value by $0.6057\ldots$.

3.

x	dy/dx	dy	y
2	3	0.6	1
2.2	5	1.0	1.6
2.4	4	0.8	2.6
2.6	1	0.2	3.4
2.8	-3	-0.6	3.6
3	-6	-1.2	3.0
3.2	-5	-1.0	1.8
3.4	-3	-0.6	0.8
3.6	-1	-0.2	0.2
3.8	1	0.2	0.0
4	2	0.4	0.2

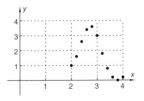

You cannot tell whether the last value of y is an overestimate or an underestimate because the convex side of the graph is downward in some places and upward in other places.

5. Answers will vary.

7. a-b.

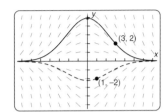

9. a.

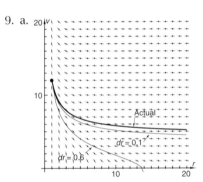

The graph with $dr = 0.6$ shows that velocity changes from positive to negative between $r = 13$ and $r = 14$.

   b. See graph in part a with $dr = 0.1$.

   c. $v = \sqrt{124.88r^{-1} + 19.12}$
   When $r = 20$, $v \approx 5.0362\ldots$
   Because the graph is concave up (convex side down), the Euler's solution underestimates the actual velocity. The first increment, where the graph is steep, makes a large error that accumulates as the iterations continue, putting the graph into a region of the slope field from which the spacecraft would not escape Earth's gravity.

   d. Let v_1 be the initial velocity at $r = 1$. Solving for C gives
   $$0.5v_1^2 = 62.44 + C$$
   $$C = 0.5v_1^2 - 62.44$$
   If $v_1 < \sqrt{2(62.44)}$, then C is negative, making $v = \sqrt{128.44r^{-1} + C}$ an imaginary number when r is large enough. If $v_1 > \sqrt{2(62.44)}$, then C is positive, making v a positive real number for all positive values of r. (The asymptote is at $v = \sqrt{C}$.)

11. a. For $x \leq 5$, the radicand $25 - x^2$ is nonnegative, giving a real-number answer for y. For $x > 5$, the radicand is negative, giving no real solution.

b. The slope field shows that the graph will be concave up (convex side down), making the Euler's tangent lines lie below the graph, leading to an underestimate.
At $x = 4.9, y = -0.6\sqrt{25 - 4.9^2} = -0.5969...$
The Euler's solution at $x = 4.9$ is $-0.8390...$, which is an underestimate because $-0.8390...$
$< -0.5969...$, but is reasonably close to the actual value.

c. The Euler's solutions for the given points are

x	y
5.1	$-0.3425...$
5.2	$0.1935...$
5.3	$-0.7736...$
6.6	$26.9706...$

From 5.1 to 5.2, $dy = 0.5360...$, indicating that the graph is still taking upward steps.
From 5.2 to 5.3, $dy = -0.9672...$, indicating that the graph takes a relatively large downward step. The sign change in dy happens whenever the prior Euler's y-value changes sign. The graph starts over on another ellipse representing a different particular solution. At $x = 6.6$, Euler's method predicts a very large upward step.

d. Euler's method can predict values that are outside the domain, which are inaccurate.

Problem Set 7-6

1. a. dB/dt is proportional to B, which means that the larger the population is, the faster it grows. But dB/dt is also proportional to $(30 - B)/30$, which means that the closer B is to 30, the slower it grows. When $0 < B < 30, dB/dt$ is positive because B is positive and $(30 - B)$ is positive. When $B > 30$, dB/dt is negative because B is positive and $(30 - B)$ is negative.

b.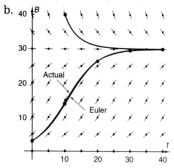

For the initial condition $(0, 3)$, the population grows, leveling off at $B = 30$. For the initial condition $(10, 40)$, the population drops because it is starting out above the maximum sustainable value (carrying capacity).

c.

t	B
0	3
10	13.8721...
20	26.4049...
30	29.5565...
40	29.9510...

See the graph in part b. Euler's points and the graphical solution are close to each other.

d. $B = \frac{30}{1 + 9e^{-0.21t}}$
At $t = 20, B = \frac{30}{1 + 9e^{-4.2}} = 26.4326...$.
The Euler's value, $B \approx 26.4049...$, is very close to this precise value.

e. $\frac{d}{dB}\left(\frac{dB}{dt}\right) = -0.014B + 30(0.007)$
Derivative $= 0$ if $-0.014B + 30(0.007) = 0$, which is true if and only if $B = 15$.
This value is halfway between $B = 0$ and $B = 30$. The "point of inflection" is $(10.4629..., 15)$.

3. a. $k = \frac{89}{1680} = 0.0529..., M = 178$
Ajax expects to sell 178,000 CDs based on this mathematical model.

b.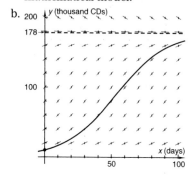

The slope field has horizontal slope lines at about $y = 178$, thus confirming $M = 178$.

c. $y = \frac{178}{1 + 16.8e^{-0.05297...x}}$
See the graph in part b. The graph follows the slope lines.

d. At $x = 50, y = 81.3396...$
At $x = 51, y = 83.6844...$
They expect to sell about 2354 CDs on the 51st day.

e. Ajax will have sold 89,000 CDs at the point of inflection, which occurs at $x \approx 53.2574...$, or on the 54th day.

5. a. At $t = 5.5$, $F \approx 1.7869... \approx 2$ fish left.
At $t = 5.6$, $F \approx -11.0738...$, meaning no fish are left.

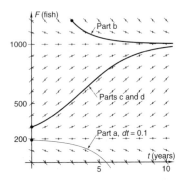

The fish are predicted to become extinct in just over 5.5 yr.

b. See the graph in part a, with initial condition $(3, 1200)$. The graph shows that the fish population will decrease because the initial condition is above the 1000 maximum sustainable, then level off at 1000.

c. See the graph in part a, with initial condition $(0, 300)$. The graph shows that the population rises slowly at first, then faster, eventually slowing down as the population approaches the 1000 maximum sustainable.

d. $F = \frac{800}{1 + 7e^{-(13/25)t}} + 200$
See the graph in part a. The graph shows that the sketch from part c reasonably approximates this precise algebraic solution.

7. a–b.

Year	P	$\Delta P / \Delta t$	$(\Delta P / \Delta t)/P$
1940	131.7		
1950	151.4	2.38	0.01571...
1960	179.3	2.59	0.01444...
1970	203.2	2.36	0.01161...
1980	226.5	2.275	0.01004...
1990	248.7		

You can't find $\Delta P / \Delta t$ for 1940 and 1990 because you don't know values of P both before and after these values.

c. $(\Delta P / \Delta t)/P \approx 0.02802596... - 0.0000792747...P$
The correlation coefficient, r, is greater for a linear function than for a logarithmic, exponential, or power function.

d. $\frac{dP}{dt} = P(0.02802596... - 0.0000792747...P)$

e.

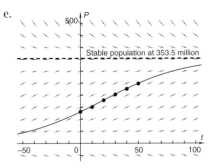

f.

Year	t	Euler	Actual*
1890	−50	44.6...	62.9
1900	−40	56.9...	76.0
1910	−30	71.7...	92.0
1920	−20	89.2...	105.7
1930	−10	109.3...	122.8
1940	0	131.7	131.7
1950	10	155.4...	151.4
1960	20	180.1...	179.3
1970	30	204.7...	203.2
1980	40	228.2...	226.5
1990	50	249.9...	248.7
2000	60	269.3...	281.4
2010	70	286.1...	
2020	80	300.2...	
2030	90	311.8...	
2040	100	321.1...	

*Data from *The World Book Encyclopedia*

g. Predicted ultimate population ≈ 353.5 million
Differential equation: $P = 353.5...$ makes $dP/dt = 0$.
Slope field: $P = 353.5...$ is a horizontal asymptote.

h. See graph in part e. Data do follow the solution.

i. Answers will vary.

j. Actual data are given in the table in part f. Answers will vary.

k. The predicted population for 2010 from part f is 286.1... million. Using 486.1 million as an initial condition in 2010 gives the following predictions:

Year	t	Euler
2010	70	446.1...
2020	80	444.5...
2030	90	417.7...
2040	100	399.7...
2050	110	387.1...

The logistic model predicts that the population will *drop*, approaching the ultimate value of 353.5 million from *above*. This behavior shows up in the slope field of part e because the slopes are *negative* for populations above 353.5.

9. $\frac{dR}{dt} = k_1 R \Rightarrow \frac{dR}{R} = k_1 dt \Rightarrow \ln|R| = k_1 t + C$
$\Rightarrow |R| = e^C e^{k_1 t} \Rightarrow R = C_1 e^{k_1 t}$

R is increasing because $k_1 > 0$.

11. $\dfrac{dR}{dt} = k_1 R - k_3 RF$

$\dfrac{dF}{dt} = -k_2 F + k_4 RF$

13. $\dfrac{dF}{dR} = \dfrac{11.25}{28} = 0.4017...$

15. The populations vary periodically, and the graph is cyclical. The fox population reaches its maximum 1/4 cycle after the rabbit population reaches its maximum.

17. $\dfrac{dF}{dR} = \dfrac{11.25}{-21} = -0.5357...$

19. The populations now spiral to a fixed point. The rabbit population stabilizes at the same value as in Problem 16, $R = 40$ (4000 rabbits), which is surprising. The stable fox population decreases from 25 to 15.

21. Initial condition (70, 15)

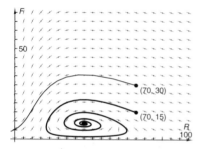

23. See graph in Problem 21 with initial condition (70, 30). With this many foxes and hunters chasing rabbits, the rabbits become extinct. At this point the foxes have been reduced to just 5. After the rabbits become extinct, the foxes decrease exponentially with time, eventually becoming extinct themselves.

Problem Set 7-7

R0. Answers will vary.

R1.

t	$P(t)$	$P'(t)$	$P'(t)/P(t)$
0	35	$-0.7070...$	$-0.2020...$
10	28.5975...	$-0.5777...$	$-0.2020...$
20	23.3662...	$-0.4720...$	$-0.2020...$

$\frac{P'(t)}{P(t)} = 35(0.98^t) = \ln 0.98 = -0.2020...$, which is a constant, Q.E.D.

R2. a. V = speed in mi/h, t = time in s
$\frac{dV}{dt} = kV$

b. $\int \frac{dV}{V} = k \int dt$
$\ln|V| = kt + C \Rightarrow |V| = e^{kt+C} = e^C \cdot e^{kt} \Rightarrow V = C_1 e^{kt}$
C_1 can be positive or negative, so the absolute value sign is not needed for V. In the real world,

V is positive, which also makes the absolute value sign unnecessary.

c. $V = 400 e^{0.005578...t}$

d. $t = 112.68... \approx 113$ s

R3. a. $y = (3x + C)^2$

b. $y = (3x - 4)^2$

c.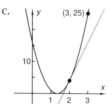

d. $dy/dx = 12$
See graph in part c.
A line through (2, 4) with slope 12 is tangent to the graph.

e. i. $dN/dt = 100 - kN$
$N = 2210.6...(1 - e^{-0.045236t})$

ii. About 1642 names

iii. Her brain saturates at about 2211 names.

iv. $t \approx 27$ days

R4. a. At (2, 5), $dy/dx = -1.75$
At (10, 16), $dy/dx = 0.675$
The slopes at points (2, 5) and (10, 16) agree with these numbers.

b.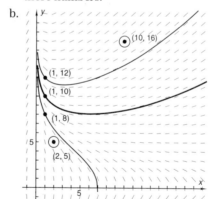

The solution containing point (1, 8) crosses the x-axis near $x = 7$, converges asymptotically to the y-axis as x approaches zero, and is symmetric across the x-axis. The solution containing (1, 12) goes to infinity as x goes to infinity.

c. See the graph in part b with initial condition (1, 10). The solution containing (1, 10) behaves more like the one containing (1, 12), although a slight discrepancy in plotting may make it seem to go the other way.

R5. a.

x	$y(\Delta x = 1)$	$y(\Delta x = 0.1)$
1	9	9
2	7.227...	7.707...
3	6.205...	6.949...
4	5.441...	6.413...
5	4.794...	5.999...
6	4.200...	5.662...
7	3.616...	5.377...
8	3.007...	5.130...
9	2.326...	4.910...
10	1.488...	4.712...
11	0.2185...	4.529...
12	−8.091...	4.359...
13		4.199...
14		4.045...
15		3.896...
16		3.750...
17		3.604...
18		3.457...
19		3.306...
20		3.150...
⋮		⋮
28.9		0.1344...
29		−0.3810...

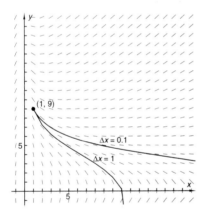

b. See the table in part a, $\Delta x = 0.1$. See the graph in part a. Each shows a different pattern.

c. The accuracy far away from the initial condition is very sensitive to the size of the increment. For instance, in part a the first step takes the graph so far down that it crosses the x-axis before running off the edge of the grid. The greater accuracy with $\Delta x = 0.1$ shows that the graph actually crosses the x-axis after $x = 20$.

d. Continuing the computations in part c, the graph crosses the x-axis close to $x = 28.9$. See the table in part a.

R6. a.

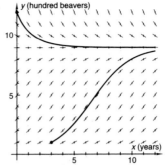

The population is decreasing because it is above the maximum sustainable, 900 beavers ($y = 9$). By Euler's method, $y \approx 9.3598...$, or about 936 beavers, at $x = 3$ years.

b. See the graph in part a. The graph shows that the population is expected to increase slowly, then more rapidly, then more slowly again, leveling off asymptotically toward 900. This happens because the initial population of 100 is below the maximum sustainable.

c. $y = \frac{9}{1 + 48.3971...e^{-0.6x}}$
The population increases fastest at $x = 6.4657... \approx 6.5$ years.

d. $\frac{dy}{dx} = \frac{-0.5(x-6)}{(y-7)}$
$dy = 0$ when $x = 6$ and $dx = 0$ when $y = 7$.
So the stable point is $(6, 7)$, corresponding to the present population of 600 Xaltos natives and 7000 yaks.

e. Initial condition $(9, 7)$

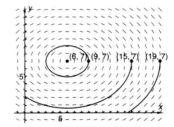

Suddenly there are too many predators for the number of prey, so the yak population declines. Because y is decreasing from $(9, 7)$, the graph follows a clockwise path.

f. See the graph in part e with initial condition $(19, 7)$. The graph crosses the x-axis at $x \approx 14.4$, indicating that the yaks are hunted to extinction. (The Xaltos would then starve or become vegetarian!)

g. See the graph in part e with initial condition $(15, 7)$. The graph never crosses the x-axis, but crosses the y-axis at $y \approx 2.3$, indicating that the yak population becomes so sparse that the predators become extinct. (The yak population would then explode!)

Problem Set 7-8

1. $v(t)\,dt$ represents the distance traveled in time dt.

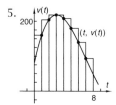

2. Definite integral

3. 1280 mi

4. $M_{100} = 1280.0384, M_{1000} = 1280.000384$
   The Riemann sums seem to be approaching 1280 as n increases. Thus, the 1280 that was found by purely algebraic methods seems to give the correct value of the limit of the Riemann sum.

5.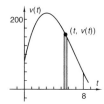

6. Any Riemann sum is bounded by the corresponding lower and upper sums. That is, $L_n \le R_n \le U_n$.
   By the definition of integrability, the limits of L_n and U_n are equal to each other and to the definite integral. By the squeeze theorem, then, the limit of R_n is also equal to the definite integral.

7. Definition: $\int_a^b f(x)\,dx = \lim_{\Delta x \to 0} L_n = \lim_{\Delta x \to 0} U_n$
   provided that the two limits are equal.
   Fundamental theorem: If f is integrable on $[a, b]$ and $g(x) = \int f(x)\,dx$, then $\int_a^b f(x)\,dx = g(b) - g(a)$.
   Or: If $F(x) = \int_a^x f(t)\,dt$, then $F'(x) = f(x)$.

8. Numerically, the integral equals 1280.
   By counting, there are approximately 52 squares. Thus, the integral $\approx 52(25)(1) = 1300$.

9. $\Delta t = 0.1$: $v'(4) \approx -19.9$ (mi/min)/min
   $\Delta t = 0.10$: $v'(4) \approx -19.9999$ (mi/min)/min

10. $f'(c) = \lim_{x \to c} \dfrac{f(x) - f(c)}{x - c}$ or
    $f'(x) = \lim_{\Delta x \to 0} \dfrac{f(x + \Delta x) - f(x)}{\Delta x}$

11. $v'(4) = -20$

12. Slowing down
    $v'(4) < 0 \Rightarrow$ velocity is decreasing.

13. Draw a line with slope -20 through point $(4, 208)$. The line is tangent to the graph.

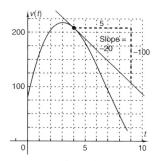

14. Acceleration

15. $v'(t) = 0 \iff t = 3.041\ldots$ or $10.958\ldots$
    So the maximum is *not* at exactly $t = 3$.

16. $v''(t) = 6t - 42$

17. At $x = 2$, $\frac{dy}{dt} = -0.9e^{-1} = -0.3310\ldots$
    y is decreasing at about 0.33 units per second.

18. At $x = 2$, $\frac{dz}{dt} = -0.04391\ldots$
    z is decreasing at about 0.044 unit per second.

19. $\dfrac{dm}{dt} = km$

20. $\int \dfrac{dm}{m} = k\int dt \Rightarrow \ln|m| = kt + C \Rightarrow$
    $|m| = e^{kt+C} \Rightarrow m = C_1 e^{kt}$

21. Exponentially

22. General

23. $m = 10{,}000(1.09)^t$

24. False. The rate of increase changes as the amount in the account increases. At $t = 10$,
    $m = 10{,}000(1.09)^{10} \approx 23{,}673.64$.
    The amount of money would grow by \$13,673.64, not just \$9,000.

25. Integral ≈ 1022

26. By symmetric difference quotient, $y' \approx 1.75$

27. See the text statement of Rolle's theorem.

28.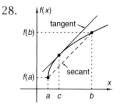

29.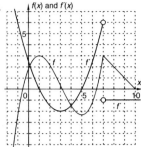

30.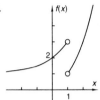

Step discontinuity at $x = 1$

31. $g'(x) = \frac{1}{3}x^{-2/3}(4x - 1)$
$g'(0)$ is undefined because $0^{-2/3}$ takes on the form $1/0^{2/3}$ or $1/0$.

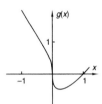

32. $x = 3.0952...$ or $x = 7.5606...$
$A = 0.8787...$

33. $V = 8.0554...$

34. Initial conditions $(0, 3)$ and $(10, 4)$

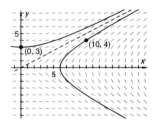

35. See the graph in Problem 34. Any initial condition for which $y = 0.5x$, such as $(2, 1)$, gives the asymptote.

36. $x^2 - 4y^2 = 36$ or $y = \pm 0.5\sqrt{x^2 - 36}$

37. $y = 4.30842...$

38. At $(10, 4)$, $\frac{dy}{dx} = 0.25 \cdot \frac{10}{4} = 0.625$.
Using $\Delta x = 0.5$, $y(10.5) \approx 4.3125$, which is close to the exact value of $4.30842...$.

39. $\frac{d}{dx}(\sin^{-1} x^3) = \frac{3x^2}{\sqrt{1 - x^6}}$

40. $dy/dx = -\sec t = -y$

41. $-\frac{1}{3}\ln|4 - 3x| + C$

42. $h'(x) = 5^x \ln 5$

43. Limit $= -4.5$

44. There is a removable discontinuity at $(0, -4.5)$.

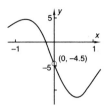

45. Answers will vary.

46. Answers will vary.

CHAPTER 8

Problem Set 8-1

1. $f'(x) = 3x^2 - 12x + 9$

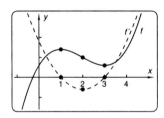

$g'(x) = 3x^2 - 12x + 15$

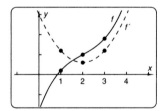

$h'(x) = 3x^2 - 12x + 12$

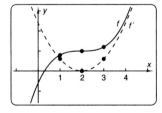

Positive derivative $\Rightarrow$ increasing function
Negative derivative $\Rightarrow$ decreasing function

Zero derivative ⇒ function could be at a high point or a low point, but not always.

3. $g''(x) = (d/dx)(3x^2 - 12x + 15) = 6x - 12$
   $h''(x) = (d/dx)(3x^2 - 12x + 12) = 6x - 12$
   All the second derivatives are the same!

5. Points of inflection occur where the first derivative graph reaches a minimum.
   Points of inflection occur where the second derivative graph crosses the x-axis.

Problem Set 8-2

1.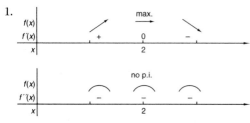

3.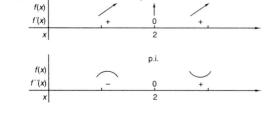

5.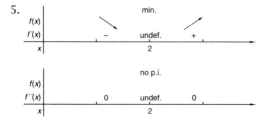

7.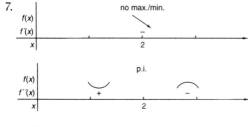

9.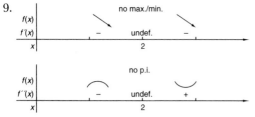

11.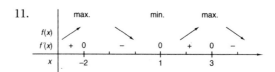

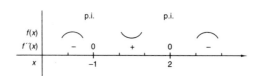

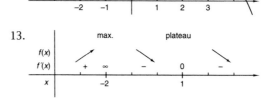

13.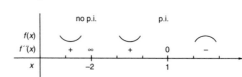

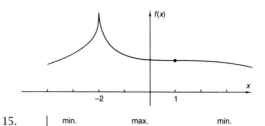

15.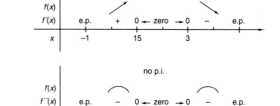

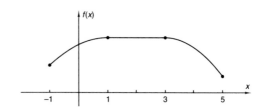

17.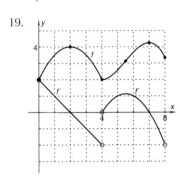

19.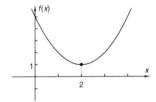

21. $f'(2) = 0 \Rightarrow$ critical point at $x = 2$
$f''(2) = -7.3890... < 0$
$\therefore$ local maximum at $x = 2$

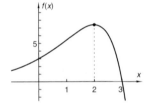

23. $f'(2) = 0 \Rightarrow$ critical point at $x = 2$
$f''(2) = 2 > 0$
$\therefore$ local minimum at $x = 2$

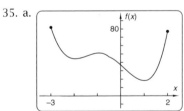

25. $f'(2) = 0 \Rightarrow$ critical point at $x = 2$
$f''(2) = 0$, so the test fails.
$f'(x)$ goes from positive to positive as x increases through 2, so there is a plateau at $x = 2$.

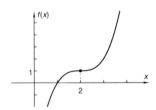

27. a. Critical points for $f(x)$: $x = -1, \, 0, \, 1$
Critical points for $f'(x)$: $x = 0, \, \pm\sqrt{1/2}$

b. The graph begins after the f-critical point at $x = -1$; the f'-critical point at $x = -\sqrt{1/2}$ is shown, but is hard to see.

c. $f'(x)$ is negative for both $x < 0$ and $x > 0$.

29. a. Critical points for $f(x)$: $x = 1$
Critical points for $f'(x)$: $x = 2$

b. Because $f(x)$ approaches its horizontal asymptote ($y = 0$) from above, the graph must be concave up for large x; but the graph is concave down near $x = 1$, and the graph is smooth; somewhere the concavity must change from down to up.

c. No. $e^{-x} \neq 0$ for all x, so $xe^{-x} = 0 \iff x = 0$.

31. a. Critical points for $f(x)$: $x = -2, 0$
Critical points for $f'(x)$: $x = 1$ ($f'(0)$ is undefined, so f' has no critical point at $x = 0$.)

b. The y-axis ($x = 0$) is a tangent line because the slope approaches $-\infty$ from both sides.

c. There is no inflection point at $x = 0$ because concavity is down for both sides, but there is an inflection point at $x = 1$.

33. a.

Max. $(2.5, 7.6)$; min. $(0.8, 4.9)$; p.i. $(1.7, 6.3)$
No global maximum or minimum

b. $f'(x) = -3x^2 + 10x - 6$
$f'(x) = 0 \iff x = \frac{1}{3}(5 \pm \sqrt{7}) = 2.5485...$ or $0.7847...$
$f''(x) = -6x + 10$; $f''(x) = 0 \iff x = \frac{5}{3} = 1.6666...$

c. $f''(0.7847...) = 5.2915... > 0$, confirming local minimum

d. Critical and inflection points occur only where f, f', or f'' is undefined (no such points exist) or is zero (all such points are found above).

35. a.

Max. $(-3, 82), (-1, 50), (2, 77)$; min. $(-2, 45)$, $(1, 18)$; p.i. $(-1.5, 45.7)$, $(0.2, 32.0)$
Global maximum is $(-3, 82)$, global minimum is $(1, 18)$.

b. $f'(x) = 12(x + 2)(x - 1)(x + 1)$
$f'(x) = 0 \iff x = -2, -1, 1$
$f'(x)$ is undefined $\iff x = -3, 2$
$f''(x) = 12(3x^2 + 4x - 1)$
$f''(x) = 0 \iff x = -\frac{1}{3}(2 \pm \sqrt{7}) = 0.2152...$ or $-1.5485...$
$f''(x)$ is undefined $\iff x = -3, 2$

c. $f''(-2) = 36 > 0$, confirming local minimum

d. Critical and inflection points occur only where f, f', or f'' is undefined (at endpoints) or is zero (all such points are found above).

37. $f''(x) = 6ax + 2b \Rightarrow f''(x) = 0$ at $x = -b/(3a)$
Because the equation for $f''(x)$ is a line with nonzero slope, $f''(x)$ changes sign at $x = -b/(3a)$, so there is a point of inflection at $x = -b/(3a)$.

39. $f(x) = -\frac{1}{2}x^3 + \frac{9}{2}x^2 - \frac{15}{2}x - \frac{5}{2}$

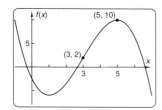

41. a. $f'(-0.8) = 1.92$
$f'(-0.5) = 0.75$
$f'(0.5) = 0.75$
$f'(0.8) = 1.92$

b. The slope seems to be decreasing from -0.8 to -0.5; $f''(x) = 6x < 0$ on $-0.8 \leq x \leq -0.5$, which confirms that the slope decreases.
The slope seems to be increasing from 0.5 to 0.8; $f''(x) = 6x > 0$ on $0.5 \leq x \leq 0.8$, which confirms that the slope increases.

c. The curve lies above the tangent line.

43. a.

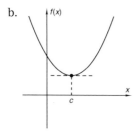

b.

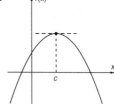

c.

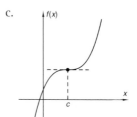

d.

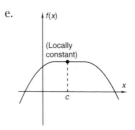

e.

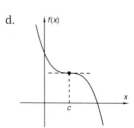

45. $f(0) = 1$ and $g(0) = 1$
$f'(0) = 0.06$ and $g'(0) = 0.06$
$f''(0) = 0.0036$ and $g''(0) = 0.0036$
$f'''(0) = 0.000216$ and $g'''(0) = 0.000216$
But $f(10) = e^{0.6} = 1.822... \neq g(10) = 1.816$;
$f'(10) = 0.109... \neq g'(10) = 0.1068$

47. Answers will vary.

Problem Set 8-3

1. Width: 150 ft; length: 100 ft

3. a. $20 \leq x \leq 93.3333...$
b. $A(x) = 22500 - 450x + 4.25x^2$

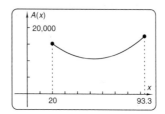

c. The graph shows a maximum at endpoint $x = 93.3333...$, so the greatest area $\approx 17,522$ ft^2.

5. a. 6.32 cm square by 3.16 cm deep

   b. Conjecture: An open box with square base of side length x and fixed surface area A will have maximum volume when the base length is twice the height, which occurs when $x = \sqrt{A/3}$.

7. The minimum cost is \$523.47.
   Justifications will vary.

9. The closest point to the origin is $(x, y) \approx (-0.4263, 0.6529)$.
   Justifications will vary.

11. The shortest ladder has length $5\sqrt{5} \approx 11.18$ ft.
   Justifications will vary.

13. The maximum volume occurs with rectangle 400 mm wide (radius) and 200 mm high.

15. a. $V = 141.2185\pi = 443.6510\ldots$ cm^3

   b. $A = 2\pi(141.2185r^{-1} + r^2)$

   c. Minimum at $r = \sqrt[3]{70.60925} = 4.1332\ldots$
   $h = 2\sqrt[3]{70.60925} = 8.2664\ldots$
   So radius ≈ 4.1 cm, and height ≈ 8.3 cm.
   Because altitude = 2× radius, height = diameter. So the minimal can is neither tall and skinny nor short and fat.

   d. The normally proportioned can is taller and thinner than the minimal can.
   $1.465\ldots \approx 1.5\%$ of metal would be saved

   e. The savings is about \$6.4 million!

17. a. $r = \sqrt[3]{43.75} = 3.5236\ldots, h = \sqrt[3]{43.75} = r$
   The minimal cup has $r \approx 3.52$ cm, $h \approx 3.52$ cm.

   b. $d:h = 2r:h = 2:1$

   c. The savings is about \$754,000/yr. Proposals will vary.

   d. $\pi r^2 h = V \Rightarrow h = (V/\pi)r^{-2}$
   Minimize $A(r) = \pi r^2 + 2\pi rh = \pi r^2 + 2Vr^{-1}$.
   $A'(r) = 2\pi r - 2Vr^{-2} = 0$ at $r = \sqrt[3]{V/\pi}$
   $A''(r) = 2\pi + 4Vr^{-3} > 0$ for all $r > 0$, so this is a minimum.
   Minimal cup has $r = \sqrt[3]{V/\pi}$,
   $h = (V/\pi)(V/\pi)^{-2/3} = \sqrt[3]{V/\pi} = r$

19. Maximum area $= 1.1221\ldots$ at $x \approx 0.8603\ldots$

21. a. Limit $= 1/2$

   b. Maximum area does not exist, but the area approaches a limit of $1/2$ as x approaches 0.

23. a. The maximum rectangle has width $= 2\sqrt{3}$ and length $= 6$.
   Justifications will vary.

   b. The maximum rectangle has width $= 2$ and length $= 8$.
   Justifications will vary.

   c. No. The maximum-area rectangle is $2\sqrt{3}$ by 6. The maximum-perimeter rectangle is 2 by 8.

25. a. $V(x) = 2\pi x^2 \sqrt{100 - x^2}$

   b. The maximal cylinder has radius $= \frac{10\sqrt{6}}{3} = 8.1649\ldots$, height $= \frac{20\sqrt{3}}{3} = 11.5470\ldots$, and volume $= \frac{4000\pi\sqrt{3}}{9} = 2418.39\ldots$.
   Justifications will vary.

   c. Height = radius $\cdot \sqrt{2}, V_c = V_s/\sqrt{3}$

27. a. The maximum lateral area occurs at radius $x = 2.5$ cm.

   b. The maximum total area is with the degenerate cylinder consisting only of the top and bottom, radius 5 and altitude 0.
   Justifications will vary.

29. Maximum volume $= 32\pi\sqrt{3} \approx 174.1$ m^3;
   radius $= \sqrt{32/3} \approx 3.27$ m; height $= \sqrt{27} \approx 5.20$ m
   Justifications will vary.

31. a. If $f(c)$ is a local maximum, then $f(x) - f(c) \le 0$ for x in a neighborhood of c.
   For x to the left of $c, x - c < 0$.
   Thus $\frac{f(x) - f(c)}{x - c} \ge 0$ and
   $f'(c) = \lim_{x \to 0^-} \frac{f(x) - f(c)}{x - c} \ge 0$.
   For x to the right of $c, x - c > 0$.
   Thus $\frac{f(x) - f(c)}{x - c} \le 0$ and
   $f'(c) = \lim_{x \to 0^+} \frac{f(x) - f(c)}{x - c} \le 0$.
   Therefore, $0 \le f'(c) \le 0$.
   Because $f'(c)$ exists, $f'(c) = 0$ by the squeeze theorem, Q.E.D.

   b. If f is not differentiable at $x = c$, then $f'(c)$ does not exist and thus cannot equal 0.

   c. The converse would say that if $f'(c) = 0$, then $f(c)$ is a local maximum. This statement is false because $f(c)$ could be a local minimum or a plateau point.

33. Answers will vary.

Problem Set 8-4

1. a. $dV = 2\pi(4x - x^3)\, dx$

   b. $V = 8\pi = 25.1327\ldots$

   c. $V = 8\pi = 25.1327\ldots$, which is the same answer.

3. $V = 85.5\pi \approx 268.6061\ldots$
   Circumscribed cylinder has volume 329.8..., which is a reasonable upper bound for the calculated volume.

5. $V = 64\pi \approx 201.0619\ldots$
   Circumscribed cylinder has volume 301.5..., which is a reasonable upper bound for the calculated volume.

7. $V = 11.6\pi \approx 36.4424\ldots$
   Circumscribed cylinder has volume 65.9..., which is a reasonable upper bound for the calculated volume.

9. $V = 69.336\pi \approx 217.8254\ldots$
   Circumscribed cylinder has volume 226.1..., which is a reasonable upper bound for the calculated volume.

11. $V = 145\frac{5}{6}\pi \approx 458.1489...$
Circumscribed cylinder has volume 769.6..., which is a reasonable upper bound for the calculated volume.

13. $V = 51\frac{3}{7}\pi \approx 161.5676...$
Circumscribed cylinder has volume 376.9..., which is a reasonable upper bound for the calculated volume.

15. $V = 124.2\pi \approx 390.1858...$
Circumscribed cylinder has volume 1055.5..., which is a reasonable upper bound for the calculated volume.

17. $V \approx 163.8592...$
Circumscribed cylinder has volume 316.1..., which is a reasonable upper bound for the calculated volume.

19. $V = 11.6\pi \approx 36.4424...$, which agrees with the answer to Problem 7.

21. $V = 19.2\pi = 60.3185789...$
$R_8 = 19.3662109...\pi = 60.8407460...$
$R_{100} = 19.2010666...\pi = 60.3219299...$
$R_{1000} = 19.2000106...\pi = 60.3186124...$
R_n is approaching 19.2π as n increases.

23. a. $dV = 180\pi \cos^2 t \sin t \, dt$
$V = 60\pi = 188.4955...$

b. $V = 60\pi \approx 188.4955...$, which agrees with the volume found in part a.

c. $V = 210\pi^2 \approx 2072.6169...$

Problem Set 8-5

1. a.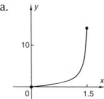

b. $L \approx 6.7848...$

c. $L \approx 6.7886...$

3. a.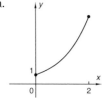

b. $L \approx 14.4394...$

c. $L \approx 14.4488...$

5. a.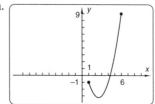

b. $L \approx 15.8617...$

c. The low point is $(2.5, -3.25)$. Chords from $(1, -1)$ to $(2.5, -3.25)$ and from $(2.5, -3.25)$ to $(6, 9)$ have combined length 15.4..., which is a reasonable lower bound for L.

7. a.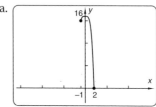

b. $L \approx 18.2470...$

c. Chords from $(-1, 15)$ to $(0, 16)$ and from $(0, 16)$ to $(2, 0)$ have combined length 17.5..., which is a reasonable lower bound for L.

9. a.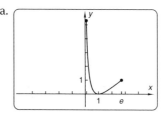

b. $L \approx 7.6043...$

c. Chords from $x = 0.1$ to $x = 1$ and from $x = 1$ to $x = e$ have combined length 7.3658..., which is a reasonable lower bound for L.

11. a.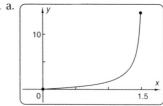

b. $L \approx 14.4488...$

c. The distance between the endpoints is 14.1809..., which is a reasonable lower bound for L.

13. a.
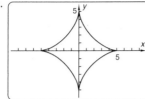

b. $L = 30$

c. Circle of radius 5 has circumference 31.4152..., which is close to the calculated value of L.

15. a.
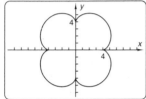

b. $L = 40$

c. The maximum/minimum values of x and y are $\pm 3\sqrt{3}$. A circle of radius $3\sqrt{3}$ has circumference 32.6483..., which is close to the calculated value of L.

17. a.
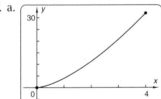

b. $L = \frac{1}{54}(145^{3/2} - 1) \approx 32.3153...$

c. The chord connecting the endpoints has length 32.2490..., which is a reasonable lower bound for L.

19. a.
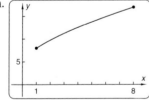

b. $L = 8\sqrt{8} - 5\sqrt{5} \approx 11.4470...$

c. Distance between endpoints is $\sqrt{130} = 11.4017...$, which is a reasonable lower bound for L.

21. $y = \dfrac{53}{441000}x^2 + 220$

$L \approx 4372.0861... \approx 4372$ feet

23. Outer ellipse: $L \approx 692.5791... \approx 692.6$ m

Inner ellipse: $L \approx 484.4224... \approx 484.4$ m

25. $L = 4\dfrac{2}{3} = 4.6666...$

27. The spiral is generated as t goes from 0 to 7π.

$L \approx 77.6508...$

29.

A	L
0	6.283185... $(= 2\pi)$
1	7.640395...
2	10.540734...
3	13.974417...

Doubling A doubles the amplitude of the sinusoid. However, it less than doubles the length of the sinusoid, for much the same reason that doubling one leg of a right triangle does not double the hypotenuse. In the limit as A approaches infinity, doubling A approaches doubling the length.

31. The function $y = (x - 2)^{-1}$ has a vertical asymptote at $x = 2$, which is in the interval $[1, 3]$. So the length is infinite. Mae's partition of the interval skips over the discontinuity, as shown in the graph.

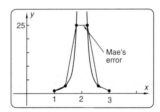

33. Answers will vary.

Problem Set 8-6

1. a. $S = \int_0^3 2\pi x \sqrt{1 + x^2}\, dx \approx 64.1361...$

b. The inscribed cone has lateral surface area 50.9722..., which is a reasonable lower bound for S.

c. $S = \frac{2}{3}\pi(10\sqrt{10} - 1) = 64.1361...$, which agrees with the answer to part a.

3. $S = 9.0242...$

5. $S = 15.5181...$

7. $S = 77.3245...$

9. $S = \frac{4\pi}{3}(1.25^{3/2} - 0.125) = 5.3304...$

11. $S = 4\frac{155}{256}\pi = 14.4685...$

13. $S = 49.5\pi = 155.5088...$

15. $S = 101\frac{5}{18}\pi = 318.1735...$

17. a. $dS = 10\pi\, dx$

b. i. $S_{0,1} = 10\pi$

ii. $S_{1,2} = 10\pi$

iii. $S_{2,3} = 10\pi$

iv. $S_{3,4} = 10\pi$

v. $S_{4,5} = 10\pi$

c. The two features exactly balance each other.

19. Pick a sample point in the spherical shell at radius r from the center. The surface area at the sample point is $4\pi r^2$. The volume of the shell is approximately (surface area)(thickness).
$dV = 4\pi r^2 \cdot dr$
$V = \int_0^R 4\pi r^2 \, dr = \frac{4}{3}\pi r^3 \big|_0^R = \frac{4}{3}\pi R^3$, Q.E.D.

21. $S = \dfrac{\pi}{6a^2}[(1 + 4a^2 r^2)^{3/2} - 1]$

23. $S = \int_0^\pi 6\pi \sin t \sqrt{(-5\sin t)^2 + (3\cos t)^2} \, dt$
$\approx 165.7930...$
Using the Cartesian equation,
$dL = \sqrt{1 + 0.36x^2(25 - x^2)^{-1}} \, dx$
At $x = \pm 5$, dL involves divison by zero, which makes it difficult to use the Cartesian equation for finding the arc length of an ellipse.

25. A circle of radius L has area πL^2 and circumference $2\pi L$. The circumference of the cone's base is $2\pi R$. The arc length of the sector of the circle of radius L must be equal to this, so the sector is $(2\pi R)/(2\pi L) = R/L$ of the circle and has surface area $S = \pi L^2(R/L) = \pi R L$, Q.E.D.

Problem Set 8-7

1. a. $A = 50\pi \approx 157.0796...$
   b. The calculated area is twice the area of the circle because the circle is traced out twice as θ increases from 0 to 2π. Although r is negative for $\pi < \theta < 2\pi$, dA is positive because r is squared.

3. a. The calculator graph confirms that the text figure is traced out once as θ increases from 0 to 2π.
   b. $A = 20.5\pi \approx 64.4026...$
   c. $L \approx 28.8141...$

5. a. The calculator graph confirms that the text figure is traced out once as θ increases from 0 to 2π.
   b. $A = 53.5\pi \approx 168.0752...$
   c. $L \approx 51.4511...$

7. a. The calculator graph confirms that the text figure is traced out once as θ increases from 0 to 2π.
   b. $A = 37.5\pi \approx 117.8097...$
   c. $L = 40$

9. a. The graph makes one complete cycle as θ increases from 0 to π.

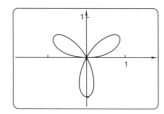

b. $A = 0.25\pi \approx 0.7853...$
c. $L \approx 6.6824...$

11. Right loop: $-\pi/4 \le \theta \le \pi/4$
    Area of both loops: 49

13. Intersections: $\theta = \cos^{-1}(2/3) = \pm 0.8410... + 2\pi n$
    $A = 26\cos^{-1}(2/3) - (4/3)\sqrt{5} \approx 18.8863...$

15. a. $L \approx 89.8589...$
    b. $A = \dfrac{13}{16}\pi^3 = 25.1925...$

17. a.

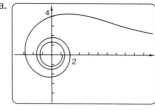

$L \approx 31.0872...$

b.

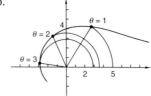

$A(1) = A(2) = A(3) = 12.5$
In general, $A(\theta) = 12.5$, which is independent of the value of θ.

19. Answers will vary.

21. a. Slope ≈ -1.5
    b. $\dfrac{dy}{dx} = \dfrac{\sin\theta + \theta\cos\theta}{\cos\theta - \theta\sin\theta}$
    At $\theta = 7$, $dy/dx = -1.54338...$, which confirms the answer found graphically.

Problem Set 8-8

R0. Answers will vary.
R1. a.

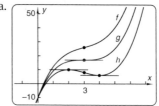

b. $f'(x) = 3x^2 - 18x + 30$, $f''(x) = 6x - 18$
   $g'(x) = 3x^2 - 18x + 27$, $g''(x) = 6x - 18$
   $h'(x) = 3x^2 - 18x + 24$, $h''(x) = 6x - 18$
c. $h(x)$ has $h'(x) = 0$ at $x = 2$ and $x = 4$.
   At $x = 2$ there is a local maximum.
   At $x = 4$ there is a local minimum.

d. $g(x)$ has a horizontal tangent at $x = 3$, but no maximum or minimum.

e. Each function has a point of inflection at $x = 3$, where the second derivative, $6x - 18$, equals zero.

R2. a.

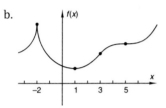

b.

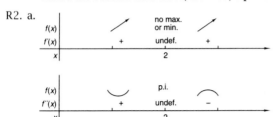

c. i. $f'(x) = \frac{2}{3}x^{-1/3} - 1$, $f''(x) = -\frac{2}{9}x^{-4/3}$

ii. Zooming in shows that there is a local minimum cusp at $(0, 0)$ and a local maximum with zero derivative at $x \approx 0.3$.

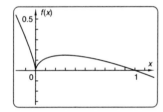

$f'(x) = 0$ at $x = (2/3)^3 = 8/27$, and $f'(x)$ is undefined at $x = 0$, thus locating precisely the minimum and maximum found by graph. Because there are no other critical values of x, there are no other maximum or minimum points.

iii. $f''(x)$ is undefined at $x = 0$, and $f''(x) < 0$ everywhere else; f'' never changes sign, so there are no inflection points.

iv. $f(0) = 0$, $f(8/27) = 4/27$, $f(5) = -2.0759\ldots$
Global maximum at $(8/27, 4/27)$
Global minimum at $(5, -2.0759\ldots)$

d. Local minimum at $(0, 0)$, local maximum at $(2, 0.5413\ldots)$, and points of inflection at $(0.5857\ldots, 0.1910\ldots)$ and $(3.4142\ldots, 0.3835\ldots)$

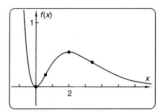

R3. a. The optimal battery has cells $\sqrt{70/12} = 2.4152\ldots$ cm wide and $10\sqrt{12/70} = 4.1403\ldots$ cm long, giving a battery of overall dimensions about 14.5 in. by 4.1 in., which is longer and narrower than the typical battery. Minimal wall length does *not* seem to be a major consideration in battery design.

b. The maximum rectangle has $x = \sqrt[3]{16/5} = 1.4736\ldots$ and $y = 4.8$.

R4. a. $V = 0.4\pi \approx 1.2566\ldots$

b. $V = 0.4\pi \approx 1.2566\ldots$, which is the same.

c. i. $V = 8\pi \approx 25.1327\ldots$

ii. $V = 51.2\pi \approx 160.8495\ldots$

iii. $V = 55\frac{7}{15}\pi \approx 174.2536\ldots$

iv. $V = 64\pi \approx 201.0619\ldots$

R5. a. $L = \int_{-1}^{2} \sqrt{1 + (2x)^2}\, dx \approx 6.1257\ldots$

b. $L = \frac{2}{6.75}(21.25^{3/2} - 1) = 28.7281\ldots$
Distance between the endpoints is $\sqrt{10^2 + 26^2} = 27.8567\ldots$, so the answer is reasonable.

c. $L \approx 25.7255\ldots$

R6. a. $S = \frac{\pi}{27}(145^{3/2} - 1) = 203.0436\ldots$
The disk of radius 8 has area $64\pi = 201.0619\ldots$, so the answer is reasonable.

b. $S = \int_{0}^{1} 2\pi(\tan x + 1)\sqrt{1 + \sec^4 x}\, dx \approx 20.4199\ldots$

c. $S \approx 272.0945\ldots$

R7. a. $L \approx 32.4706\ldots$

b. $A = \frac{7.5}{6}\pi^3 = 38.7578\ldots$

CHAPTER 9

Problem Set 9-1

1. $V = 3.5864...$

3. $\int f'(x)\,dx = \int x \cos x\,dx + \int \sin x\,dx$

5. $V = \pi^2 - 2\pi$

7. The method involves working separately with the different "parts" of the integrand. The function $f(x) = x \sin x$ was chosen because one of the terms in its derivative is $x \cos x$, which is the original integrand. See Section 9-2.

Problem Set 9-2

1. $-x \cos x + \sin x + C$

3. $\dfrac{1}{4} x e^{4x} - \dfrac{1}{16} e^{4x} + C$

5. $-\dfrac{21}{25} e^{-5x} - \dfrac{1}{5} x e^{-5x} + C$

7. $\dfrac{1}{4} x^4 \ln x - \dfrac{1}{16} x^4 + C$

9. $x^2 e^x - 2x e^x + 2e^x + C$

11. $x \ln x - x + C$

Problem Set 9-3

1. $\dfrac{1}{2} x^3 e^{2x} - \dfrac{3}{4} x^2 e^{2x} + \dfrac{3}{4} x e^{2x} - \dfrac{3}{8} e^{2x} + C$

3. $-x^4 \cos x + 4x^3 \sin x + 12x^2 \cos x - 24x \sin x$
$- 24 \cos x + C$

5. $\dfrac{1}{2} x^5 \sin 2x + \dfrac{5}{4} x^4 \cos 2x - \dfrac{5}{2} x^3 \sin 2x - \dfrac{15}{4} x^2 \cos 2x$
$+ \dfrac{15}{4} x \sin 2x + \dfrac{15}{8} \cos 2x + C$

7. $-\dfrac{1}{2} e^x \cos x + \dfrac{1}{2} e^x \sin x + C$

9. $\dfrac{5}{34} e^{3x} \sin 5x + \dfrac{3}{34} e^{3x} \cos 5x + C$

11. $\dfrac{1}{8} x^8 \ln 3x - \dfrac{1}{64} x^8 + C$

13. $\dfrac{\ln 7}{5} x^5 + C$

15. $\dfrac{1}{6} \sin^6 x + C$

17. $\dfrac{2}{3} x^3 (x + 5)^{3/2} - \dfrac{4}{5} x^2 (x + 5)^{5/2} + \dfrac{16}{35} x(x + 5)^{7/2}$
$- \dfrac{32}{315} (x + 5)^{9/2} + C$

19. $5x \ln x - 5x + C$

21. $\dfrac{1}{2} x^4 e^{x^2} - x^2 e^{x^2} + e^{x^2} + C$

23. $\dfrac{1}{2} x^2 (\ln x)^3 - \dfrac{3}{4} x^2 (\ln x)^2 + \dfrac{3}{4} x^2 \ln x - \dfrac{3}{8} x^2 + C$

25. $\dfrac{1}{10} x^2 (x^2 + 1)^5 - \dfrac{1}{60} (x^2 + 1)^6 + C$

27. $\dfrac{1}{2} \cos x \sin x + \dfrac{1}{2} x + C$

29. $\dfrac{1}{2} \sec x \tan x + \dfrac{1}{2} \ln |\sec x + \tan x| + C$

31. $x \log_3 x - \dfrac{1}{\ln 3^x} + C$

33. $-\cos x + C$

35. $-\ln |\csc x + \cot x| + C$

37. $-\ln |\cos x| + C$

39. For the first integral, Wanda integrated $\cos x$ and differentiated x^2, but in the second integral she plans to differentiate $\int \cos x\,dx$ and integrate $2x$, effectively canceling out what she did in the first part. She will get $\int x^2 \cos x\,dx = x^2 \sin x - x^2 \sin x + \int x^2 \cos x\,dx$, which is true, but not very useful!

41. After two integrations by parts, $\int e^x \sin x\,dx = -e^x \cos x + e^x \sin x - \int e^x \sin x\,dx$, but after two more integrations, $\int e^x \sin x\,dx = -e^x \cos x + e^x \sin x + e^x \cos x - e^x \sin x + \int e^x \sin x\,dx$. Two integrations produced the original integral with the opposite sign (which is useful), and two more integrations reversed the sign again to give the original integral with the same sign (which is not useful).

43.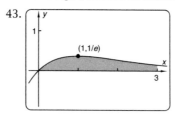

Maximum at $x = 1$
$A = -4e^{-3} + 1 = 0.8008...$

45. $V = 5\pi (\ln 5)^2 - 10\pi \ln 5 + 8\pi = 15.2589...$

47. For integration by parts, $\int u\,dv = uv - \int v\,du$. Applying limits of integration gives

$$\int_c^d u\,dv = uv \Big|_{u=a}^{u=b} - \int_a^b v\,du$$
$$= (bd - ac) - \int_a^b v\,du$$

The quantity $(bd - ac)$ is the area of the "L-shaped" region, which is the area of the larger rectangle minus the area of the smaller one. Thus, the integral of $u\,dv$ equals the area of the L-shaped region minus the area represented by the integral of $v\,du$.

49. $\int \sin^7 x \, dx = -\sin^6 x \cos x + 6\int \sin^5 x \cos^2 x \, dx$

$\quad = -\sin^6 x \cos x + 6\int \sin^5 x \, dx - 6\int \sin^7 x \, dx$

$\quad \int \sin^7 x \, dx = -\dfrac{1}{7} \sin^6 x \cos x + \dfrac{6}{7}\int \sin^5 x \, dx$

$\quad \int \sin^7 x \, dx = -\dfrac{1}{7} \sin^6 x \cos x - \dfrac{6}{35} \sin^4 x \cos x$

$\quad - \dfrac{8}{35} \sin^2 x \cos x - \dfrac{16}{35} \cos x + C$

Problem Set 9-4

1. $\int \sin^9 x \, dx = -\dfrac{1}{9} \sin^8 x \cos x + \dfrac{8}{9}\int \sin^7 x \, dx$

3. $\int \cot^{12} x \, dx = -\dfrac{1}{11} \cot^{11} x - \int \cot^{10} x \, dx$

5. $\int \sec^{13} x \, dx = \dfrac{1}{12} \sec^{11} x \tan x + \dfrac{11}{12}\int \sec^{11} x \, dx$

7. $\int \cos^n x \, dx = \cos^{n-1} x \sin x + (n-1)\int \cos^{n-2} x \sin^2 x \, dx$

$\quad = \cos^{n-1} x \sin x + (n-1)\int \cos^{n-2} x (1 - \cos^2 x) \, dx$

$\quad = \cos^{n-1} x \sin x + (n-1)\int \cos^{n-2} x \, dx$

$\quad \ - (n-1)\int \cos^n x \, dx$

$\quad n \int \cos^n x \, dx = \cos^{n-1} x \sin x + (n-1)\int \cos^{n-2} x \, dx$

$\quad \int \cos^n x \, dx = \dfrac{1}{n} \cos^{n-1} x \sin x + \dfrac{n-1}{n}\int \cos^{n-2} x \, dx$

9. $\int \tan^n x \, dx = \int \tan^{n-2} x \tan^2 x \, dx$

$\quad = \int \tan^{n-2} x (\sec^2 x - 1) \, dx$

$\quad = \int \tan^{n-2} x \sec^2 x \, dx - \int \tan^{n-2} x \, dx$

$\quad = \dfrac{1}{n-1} \tan^{n-1} x - \int \tan^{n-2} x \, dx$

11. $\int \csc^n x \, dx = -\csc^{n-2} x \cot x - (n-2)\int \csc^{n-2} x \cot^2 x \, dx$

$\quad = -\csc^{n-2} x \cot x - (n-2)\int \csc^{n-2} x (\csc^2 x - 1) \, dx$

$\quad = -\csc^{n-2} x \cot x - (n-2)\int \csc^n x \, dx$

$\quad \ + (n-2)\int \csc^{n-2} x \, dx (n-1)\int \csc^n x \, dx$

$\quad = -\csc^{n-2} x \cot x + (n-2)\int \csc^{n-2} x \, dx$

$\quad \int \csc^n x \, dx = -\dfrac{1}{n-1} \csc^{n-2} x \cot x + \dfrac{n-2}{n-1}\int \csc^{n-2} x \, dx$

13. $-\dfrac{1}{5} \sin^4 x \cos x - \dfrac{4}{15} \sin^2 x \cos x - \dfrac{8}{15} \cos x + C$

15. $-\dfrac{1}{5} \cot^5 x + \dfrac{1}{3} \cot^3 x - \cot x - x + C$

17. $\dfrac{1}{3} \sec^2 x \tan x + \dfrac{2}{3} \tan x + C$

19. a. $y = \cos x$ is on top; $y = \cos^3 x$ is in the middle; $y = \cos^5 x$ is on the bottom.

b. For $y = \cos x$, area $\approx 2.0000...$
For $y = \cos^3 x$, area $\approx 1.3333...$
For $y = \cos^5 x$, area $\approx 1.06666...$

c. $A_1 = 2, A_3 = 4/3, A_5 = 16/15$

d. Based on the graphs, the area under $\cos x$ should be greater than that under $\cos^3 x$, which in turn is greater than the area under $\cos^5 x$. This is exactly what happens with the calculated answers:
$A_1 > A_3 > A_5$.

e.

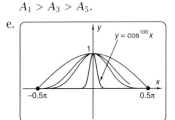

f. Yes; $\lim_{n \to \infty} \int_{-\pi/2}^{\pi/2} \cos^n x \, dx = 0$

21. $\int \sec^3 x \, dx = \dfrac{1}{2} \sec x \tan x + \dfrac{1}{2} \ln | \sec x + \tan x| + C$
The answer is half the derivative of secant plus half the integral of secant.

23. $\dfrac{d}{dx}\left[-\dfrac{1}{3a}(\cos ax)(\sin^2 ax + 2) \right]$

$\quad = \dfrac{1}{3}\left[\sin^3 ax + 2 \sin ax \, (\sin^2 ax) \right]$

$\quad = \sin^3 ax$

$\quad \therefore \int \sin^3 ax \, dx = -\dfrac{1}{3a}(\cos ax)(\sin^2 ax + 2) + C,$ Q.E.D.

Problem Set 9-5

1. $-\cos x + \dfrac{2}{3} \cos^3 x - \dfrac{1}{5} \cos^5 x + C$

3. $\dfrac{1}{9} \sin 9x - \dfrac{1}{9} \sin^3 9x + \dfrac{1}{15} \sin^5 9x - \dfrac{1}{63} \sin^7 9x + C$

5. $\dfrac{1}{15} \sin^5 3x + C$

7. $-\dfrac{1}{56} \cos^7 8x + \dfrac{1}{72} \cos^9 8x + C$

9. $-\dfrac{1}{3} \cos^3 x + \dfrac{2}{5} \cos^5 x - \dfrac{1}{7} \cos^7 x + C$

11. $\dfrac{1}{2} x + \dfrac{1}{4} \sin 2x + C$

13. $\dfrac{1}{2} x - \dfrac{1}{20} \sin 10x + C$

15. $\dfrac{1}{3} \tan^3 x + \tan x + C$

17. $-\dfrac{1}{42} \cot^7 6x - \dfrac{1}{10} \cot^5 6x - \dfrac{1}{6} \cot^3 6x - \dfrac{1}{6} \cot 6x + C$

19. $\dfrac{1}{11} \tan^{11} x + C$

21. $\dfrac{1}{10} \sec^{10} x + C$

23. $(\sec^{10} 20) x + C$

25. $\dfrac{1}{2} \sin 2x + C$

27. $-\cot x + C$

29. $\dfrac{1}{2} \sec x \tan x + \dfrac{1}{2} \ln |\sec x + \tan x| + C$

31. a. $\dfrac{5}{16} \sin 5x \sin 3x + \dfrac{3}{16} \cos 5x \cos 3x + C$

   b. $\displaystyle\int_0^{2\pi} \cos 5x \sin 3x \, dx = 0$
   Because the integral finds the area above minus the area below, this calculation shows the two areas are equal.

33. $V = \pi^2 / 2$

35. $A = \pi + 2 + 10\sqrt{2} + \frac{25}{8}\pi = 29.1012...$, which agrees with the numerical answer.

37. Answers will vary.

Problem Set 9-6

1. $\dfrac{49}{2} \sin^{-1} \dfrac{x}{7} + \dfrac{1}{2} x \sqrt{49 - x^2} + C$

3. $\dfrac{1}{2} x \sqrt{x^2 + 16} + 8 \ln \left| \sqrt{x^2 + 16} + x \right| + C$

5. $\dfrac{1}{2} x \sqrt{9x^2 - 1} - \dfrac{1}{6} \ln \left| 3x + \sqrt{9x^2 - 1} \right| + C$

7. $\sin^{-1} \dfrac{x}{\sqrt{17}} + C$

9. $\ln \left| \sqrt{x^2 + 1} + x \right| + C$

11. $\dfrac{1}{4} x^3 \sqrt{x^2 - 9} - \dfrac{9}{8} x \sqrt{x^2 - 9} - \dfrac{81}{8} \ln \left| x + \sqrt{x^2 - 9} \right| + C$

13. $\dfrac{1}{4} x (1 - x^2)^{3/2} + \dfrac{3}{8} \sin^{-1} x + \dfrac{3}{8} x \sqrt{1 - x^2} + C$

15. $\dfrac{1}{9} \tan^{-1} \dfrac{x}{9} + C$

17. a. $\sqrt{x^2 + 25} + C$

   b. $\displaystyle\int \dfrac{x \, dx}{\sqrt{x^2 + 25}} = \dfrac{1}{2} \int (x^2 + 25)^{-1/2}(2x \, dx) =$
   $\sqrt{x^2 + 25} + C$, which agrees with part a. Moral: Always check for an *easy* way to integrate before trying a more sophisticated technique!

19. $\sin^{-1} \dfrac{x - 5}{3} + C$

21. $\ln \left| x + 4 + \sqrt{x^2 + 8x - 20} \right| + C$

23. $50 \sin^{-1} 0.8 + 25 \sin(2 \sin^{-1} 0.8) - 50 \sin^{-1}(-0.3)$
   $- 25 \sin[2 \sin^{-1}(-0.3)] = 99.9084...$
   Numerical integration: $99.9084...$

25. $L = \dfrac{5}{2} \sqrt{901} + \dfrac{1}{12} \ln \left| \sqrt{901} + 30 \right| = 75.3828...$
   Numerical integration: $L = 75.3828...$

27. $x^2 + y^2 = r^2 \Rightarrow y = \pm\sqrt{r^2 - x^2}, x = 0$ at $y = \pm r$
   $dA = 2y \, dx = 2\sqrt{r^2 - x^2} \, dx$
   $A = 2 \displaystyle\int_{-r}^{r} \sqrt{r^2 - x^2} \, dx$

Let $\dfrac{x}{r} = \sin \theta \Rightarrow \sqrt{r^2 - x^2} = r \cos \theta, \theta = \sin^{-1} \dfrac{x}{r}$
$\Rightarrow A = 2 \displaystyle\int_{-\pi/2}^{\pi/2} r \cos \theta \cdot r \cos \theta \, d\theta = \pi r^2$, Q.E.D.

29. Rotating about the y-axis: $V = \frac{4}{3}\pi a^2 b$
   Rotating about the x-axis: $V = \frac{4}{3}\pi a b^2$

31. $V = \dfrac{256}{3}\pi = 268.0825...$

33. $A = \pi ab$, as in Problem 28.
   With this method, you get $\int \sin^2 t \, dt$ directly. With trigonometric substitution in Problem 28, you get $\int \cos^2 t \, dt$, *indirectly*.

35. For the sine and tangent substitution, the ranges of the inverse sine and inverse tangent make the corresponding radical positive. For the secant substitution, the situation is more complicated, but still gives an answer of the same algebraic form as if x had been only positive.

Problem Set 9-7

1. $4 \ln |x - 1| + 7 \ln |x - 2| + C$

3. $\dfrac{7}{2} \ln |x + 2| + \dfrac{3}{2} \ln |x - 4| + C$

5. $-7 \ln |x + 5| + 7 \ln |x + 2| + C$

7. $2 \ln |x + 1| + 3 \ln |x - 7| + 4 \ln |x + 2| + C$

9. $-\ln |x + 3| + 2 \ln |x + 1| + 3 \ln |x - 2| + C$

11. $x^3 + \dfrac{5}{2} x^2 - 7x + 2 \ln |x - 1| + C$

13. $\dfrac{1}{2} \ln |x^2 + 1| + 2 \tan^{-1} x + 3 \ln |x + 4| + C$

15. $\ln |x + 5| + 3 \ln |x + 1| + 2(x + 1)^{-1} + C$

17. $-\dfrac{1}{2}(x - 2)^{-2} + C$

19. a. $y = \dfrac{1000}{1 + 99 e^{-2t}}$

   b. $y(1) = 69.4531... \approx 69$ students
   $y(4) = 967.8567... \approx 968$ students
   $y(8) = 999.9888... \approx 1000$ students
   Everyone knows by the end of the day!

   c. The rumor was spreading fastest when $t = \frac{1}{2} \ln 99 = 2.2975...$ h, when 500 students had heard the rumor.

   d.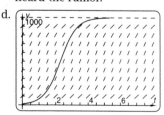

The curve follows the slope-field pattern.

21. $A = 5 \ln \dfrac{b-1}{b+4} + 5 \ln 6$

$A(7) = 5 \ln \dfrac{6}{11} + 5 \ln 6 = 5.9281\dots$

$\lim_{b \to \infty} A(b) = \lim_{b \to \infty} 5 \ln \dfrac{1}{1} + 5 \ln 6$ (l'Hospital's rule)

$= 5 \ln 6 = 8.9587\dots$,

so the area does approach a finite limit.

23. a. $\dfrac{1}{2} \ln |x - 2| + \dfrac{1}{2} \ln |x - 4| + C$

b. $\ln \sqrt{x^2 - 6x + 8} + C$

c. $\dfrac{1}{2} \ln |x^2 - 6x + 8| + C$

d. Each can be transformed to $\ln \sqrt{x^2 - 6x + 8} + C$.

Problem Set 9-8

1. $x \tan^{-1} x - \dfrac{1}{2} \ln |x^2 + 1| + C$

3. $x \cos^{-1} x - \sqrt{1 - x^2} + C$

5. $x \sec^{-1} x - \operatorname{sgn} x \ln \left| x + \sqrt{x^2 - 1} \right| + C$

7. $4 \tan^{-1} 4 - \dfrac{\pi}{4} - \dfrac{1}{2} \ln \dfrac{17}{2} = 3.4478\dots$

Numerical integration gives 3.4478…. The answers are the same to at least 4 decimal places.

9. Both methods give $A = 1$.

Problem Set 9-9

1.

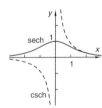

3. $f'(x) = 3 \tanh^2 x \operatorname{sech}^2 x$

5. $\dfrac{1}{6} \cosh^6 x + C$

7. $g'(x) = -\operatorname{csch} x \coth x \sin x + \operatorname{csch} x \cos x$

9. $\dfrac{1}{4} \tanh 4x + C$

11. $h'(x) = 3x^2 \coth x - x^3 \operatorname{csch}^2 x$

13. $\ln (\cosh 3) - \ln (\cosh 1) = 1.875547\dots$

15. $q'(x) = \dfrac{5 \cosh 5x \ln 3x - x^{-1} \sinh 5x}{(\ln 3x)^2}$

17. $\cosh 1 - \sinh 1 = e^{-1} = 0.36787\dots$

19. $y' = \dfrac{12}{\sqrt{16x^2 + 1}}$

21. $\dfrac{1}{5} x \tanh^{-1} 5x + \dfrac{1}{10} \ln \left| 1 - (5x)^2 \right| + C$

23. $0.5x\sqrt{x^2 + 9} + 4.5 \sinh^{-1} \dfrac{x}{3} + C$

25. a. The horizontal force is given by the vector $(h, 0)$ and the vertical force is the vector $(0, v)$, so their sum, the tension vector, is the vector (h, v), which has slope $\dfrac{v}{h}$. Because the tension vector points along the graph, the graph's slope, y', also equals $\dfrac{v}{h}$.

b. $v =$ weight of chain below $(x, y) = s \cdot w$

$\Rightarrow y' = \dfrac{v}{h} = \dfrac{s \cdot w}{h} = \dfrac{w}{h} \cdot s$

c. $d(y') = \dfrac{w}{h} \, ds = \dfrac{w}{h} \sqrt{1 + (y')^2} \, dx$

d. $\sinh^{-1} y' = \dfrac{w}{h} x + C$

e. $C = 0$

f. $\sinh^{-1} y' = \dfrac{w}{h} x + 0 \Rightarrow y' = \sinh \dfrac{w}{h} x$

g. $\dfrac{dy}{dx} = \sinh \dfrac{w}{h} x \Rightarrow \int dy = \int \sinh \dfrac{w}{h} x \, dx \Rightarrow$

$y = \dfrac{h}{w} \cosh \dfrac{w}{h} x + C$

27. a. $y = 500 \cosh \dfrac{1}{500} x + 110 - 500 \cosh 0.3$

$y(0) = 87.3307\dots \approx 87.3$ ft

b. $L = 1000 \sinh 0.3 = 304.5202\dots \approx 304.5$ ft

Weight $= 243.6162\dots \approx 243.6$ lb

c. The maximum tension is at the ends, at 150 ft.

$T(150) = 400 \cosh 0.3 = 418.1354\dots \approx 418.1$ lb

d. $h = 901.3301\dots \approx 901.3$ lb

29. a. $S = 5.07327\dots \approx 5.07 \text{ft}^2$

b. Cost = \$578.35

c. $V = 1.25317\dots \approx 1.253 \text{ ft}^3$

31. a. $H'(1) = -\operatorname{csch} 1 \coth 1 = -1.1172855\dots$

b. $H'(1) \approx -1.11738505\dots$

The answers differ by 0.0000995…, which is about 0.0089% of the actual answer.

33. By parts: $\dfrac{2}{3} e^x \cosh 2x - \dfrac{1}{3} e^x \sinh 2x + C$

By transforming to exponential form: $\dfrac{1}{6} e^{3x} + \dfrac{1}{2} e^{-x} + C$

Transforming to exponential form is easier!

35. a. $\cosh^2 x - \sinh^2 x = \left(\dfrac{e^x + e^{-x}}{2}\right)^2 - \left(\dfrac{e^x - e^{-x}}{2}\right)^2 = 1$

   b. $\dfrac{1}{\cosh^2 x}(\cosh^2 x - \sinh^2 x) = \dfrac{1}{\cosh^2 x}$

     $\Rightarrow 1 - \tanh^2 x = \operatorname{sech}^2 x$

   c. $\dfrac{1}{\sinh^2 x}(\cosh^2 x - \sinh^2 x) = \dfrac{1}{\sinh^2 x}$

     $\Rightarrow \coth^2 x - 1 = \operatorname{csch}^2 x$

37. a. On the circle, $L = \displaystyle\int_{\cos 2}^{1} \dfrac{du}{\sqrt{1 - u^2}} = -\cos^{-1} 1 + 2 = 2$

     On the hyperbola, $L = \cosh 2 - 1 = 2.762\dots$. So the length of the curve is greater than 2, Q.E.D.

   b. The area of the triangle that circumscribes the sector is $0.5(2 \sinh 2 \cosh 2) = \sinh 2 \cosh 2$. The area of the region between the upper and lower branches of the hyperbola from $u = 1$ to $u = \cosh 2$ is $A = 2\displaystyle\int_0^2 \sinh^2 t\, dt \approx 11.644958\dots$.
Thus the area of the sector is $\cosh 2 \sinh 2 - 11.644958\dots = 2$, Q.E.D.

   c. By definition of the circular functions, x is the length of the arc from $(1, 0)$ to $(\cos x\ \sin x)$. So the total arc has length $2x$. The circumference of a unit circle is 2π and its area is π. Thus $A_{\text{sector}} = \dfrac{2x}{2\pi}\pi = x$, Q.E.D.

   d. Area of circumscribing triangle $= \cosh x \sinh x$
Area between branches $= \sinh x \cosh x - x$
Area of the sector $= \cosh x \sinh x - (\sinh x \cosh x - x) = x$, Q.E.D.

Problem Set 9-10

1. a. It might converge because the integrand approaches zero as x approaches infinity.

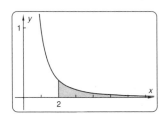

   b. The integral converges to $\frac{1}{2}$.

3. a. It might converge because the integrand approaches zero as x approaches infinity.

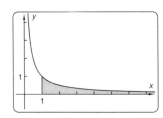

   b. The integral diverges.

5. a. It might converge because the integrand approaches zero as x approaches infinity.

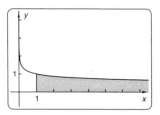

   b. The integral diverges.

7. a. It might converge because the integrand becomes infinite only as x approaches zero.

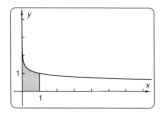

   b. The integral converges to 1.25.

9. a. It might converge because the integrand approaches zero as x approaches infinity.

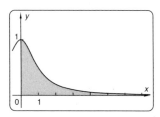

   b. The integral converges to $\pi/2$.

11. a. It might converge because the integrand becomes infinite only as x approaches 0 or 1.

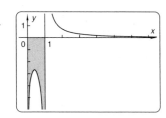

   b. The integral diverges.

13. a. It might converge because the integrand approaches zero as x approaches infinity.

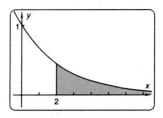

b. The integral converges to $2.5e^{-0.8} = 1.1233...$.

15. a. It does not converge because the integrand is undefined for $x < 0$.

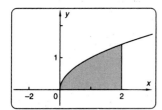

b. Not applicable

17. a. It might converge because the integrand seems to approach zero as x approaches infinity.

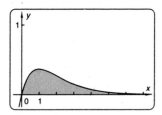

b. The integral converges to 1.

19. a. It diverges because the integrand does not approach zero as x approaches infinity.

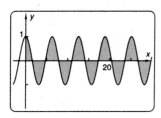

b. Not applicable

21. As $b \to \infty$, $\int_0^b \cos x \, dx$ oscillates between -1 and 1 and never approaches a limit. Similarly, $\int_0^b \sin x \, dx$ oscillates between 0 and 2.

23. a. $A \to \infty$. The area does *not* approach a finite limit because the improper integral diverges.

b. The volume converges to π.

c. The volume diverges, because $\lim\limits_{b \to \infty} \int_1^b 2\pi \, dx = \infty$.

d. False. The volume could approach a constant as in part b or become infinite as in part c.

25. a. $f(1) = 1, f(2) = 2, f(3) = 6$

b. Conjecture:
$f(4) = 4f(3) = 24 = 4!$
$f(5) = 5f(4) = 120 = 5!$
$f(6) = 6f(5) = 720 = 6!$

c. $f(x) = xf(x - 1)$, Q.E.D.

d. Part a shows that $f(1) = 1 = 1!$.
Part c shows that $f(n) = nf(n - 1) =$
$n(n - 1)f(n - 2) = n(n - 1)(n - 2)...(2)(1) = n!$, Q.E.D.

e. $\int_0^b t^3 e^{-t} \, dt < 0.000001$ for $b \geq 24$.

f. $0.5! \int_0^\infty t^{0.5} e^{-t} dt \approx \int_0^{24} t^{0.5} e^{-t} \, dt \approx 0.886227311...$
From the graphs, $t^{0.5} e^{-t} < t^3 e^{-t}$ for $x \geq 24$.
The error in 0.5! from stopping at $b = 24$ is the area under the "tail" of the graph from $b = 24$.
Error $= \int_{24}^\infty t^{0.5} e^{-t} \, dt < \int_{24}^\infty t^3 e^{-t} \, dt < 0.000001$
The difference between the tabulated value of 0.5! and the value calculated here is $0.8862269255 - 0.866227311... = -0.000000386$, which is less in absolute value than 0.000001. Note, however, that the difference is *negative* because the calculated value is larger than the tabulated value. This observation means either that the tabulated value is incorrect or that there is more inaccuracy in the numerical integration algorithm than there is in the error caused by dropping the tail of the integral.

g. $1.5! = 1.5(0.5!) = 1.3293...$
$2.5! = 2.5(1.5!) = 3.3233...$
$3.5! = 3.5(2.5!) = 11.6317...$

h. $0! = \int_0^\infty t^0 e^{-t} \, dt = 1$, Q.E.D.

i. $(-1)! = 0!/0$, which is infinite. So $(-2)!$ and $(-3)!$, which equal $(-1)!/(-1)$ and $(-2)!/(-2)$, are also infinite. However,
$(-0.5)! = 0.5!/(0.5) = 1.77245...$
$(-1.5)! = (-0.5)!/(-0.5) = -3.54490...$
$(-2.5)! = (-1.5)!/(-1.5) = 2.36327...$,
all of which are finite.

j. $0.5! = \dfrac{\sqrt{\pi}}{2} = 0.886226925...$

27. a. $\int_1^3 \left(2^x - \dfrac{|x - 2|}{x - 2} \right) dx$
$= \int_1^2 \left(2^x - \dfrac{|x - 2|}{x - 2} \right) dx + \int_2^3 \left(2^x - \dfrac{|x - 2|}{x - 2} \right) dx$

b. $\lim\limits_{b \to 2^-} \int_1^b \left(2^x - \dfrac{|x - 2|}{x - 2} \right) dx$
$+ \lim\limits_{a \to 2^+} \int_a^3 \left(2^x - \dfrac{|x - 2|}{x - 2} \right) dx$

c. The integral converges to $6/\ln 2 = 8.6561...$.

d. The integral is defined by dividing the interval into Riemann partitions and adding the subintervals. But the Riemann partitions may be chosen so that the discontinuities are at endpoints of subintervals. Then the subintervals corresponding to each continuous piece may be added separately.

e. False. Some discontinuous functions (notably, piecewise continuous functions) are integrable.

Problem Set 9-11

1. $y' = 3 \sec 3x \tan^2 3x + 3 \sec^3 3x$

3. $\frac{1}{4}x \sinh 4x - \frac{1}{16} \cosh 4x + C$

5. $f'(x) = -3(3x + 5)^{-2}$

7. $\frac{1}{3} \ln |3x + 5| + C$

9. $t'(x) = 20 \tan^4 4x \sec^2 4x$

11. $\frac{1}{2}x - \frac{1}{2} \sin x \cos x + C$

13. $y' = \dfrac{23}{(x + 2)^2}$

15. $6x - 23 \ln |x + 2| + C$

17. $f'(t) = \dfrac{t}{\sqrt{1 + t^2}}$

19. $\frac{1}{2}t\sqrt{1 + t^2} + \frac{1}{2} \ln \left| \sqrt{1 + t^2} + t \right| + C$

21. $y' = 3x^2 e^x + x^3 e^x = x^2 e^x (3 + x)$

23. $x^3 e^x - 3x^2 e^x + 6x e^x - 6e^x + C$

25. $f'(x) = (1 - x^2)^{-1/2}$

27. $x \sin^{-1} x + \sqrt{1 - x^2} + C$

29. $-\frac{1}{6} \ln |x + 5| + \frac{1}{6} \ln |x - 1| + C$

31. $\ln \left| x + 2 + \sqrt{x^2 + 4x - 5} \right| + C$

33. $f'(x) = \operatorname{sech}^2 x$

35. $\ln |\cosh x| + C$

37. $y' = e^{2x}(2 \cos 3x - 3 \sin 3x)$

39. $\frac{3}{13} e^{2x} \sin 3x + \frac{2}{13} e^{2x} \cos 3x + C$

41. $g'(x) = x^2(3 \ln 5x + 1)$

43. $\frac{1}{4}x^4 \ln 5x - \frac{1}{16}x^4 + C$

45. $y' = \dfrac{x}{(x + 2)(x + 3)(x + 4)}[x^{-1} - (x + 2)^{-1} - (x + 3)^{-1} - (x + 4)^{-1}]$

47. $-\ln |x + 2| + 3 \ln |x + 3| - 2 \ln |x + 4| + C$

49. $y' = -3 \cos^2 x \sin^2 x + \cos^4 x$

51. $-\frac{1}{4} \cos^4 x + C$

53. $\sin x - \frac{1}{3} \sin^3 x + C$ or $\frac{1}{3} \cos^2 x \sin x + \frac{2}{3} \sin x + C$

55. $\frac{1}{4} \cos^3 x \sin x + \frac{3}{8} \cos x \sin x + \frac{3}{8}x + C$

57. $g'(x) = 12x^3(x^4 + 3)^2$

59. $\frac{1}{13}x^{13} + x^9 + \frac{27}{5}x^5 + 27x + C$

61. $\frac{1}{16}(x^4 + 3)^4 + C$

63. $\frac{1}{5}x^5 + 3x + C$

65. $f'(x) = (x^4 + 3)^3$

67. $e^2 = 7.3890...$

69. $r'(x) = x e^x + e^x$

71. $q'(x) = \dfrac{-1 - \ln x}{x^2}$

73. $\frac{1}{2}(\ln x + 2)^2 + C$

75. $f'(x) = 2x e^{x^2}$

77. $\frac{1}{2} e^{x^2} + C$

79. $\frac{1}{2}x^2 e^{x^2} - \frac{1}{2} e^{x^2} + C$

81. $\dfrac{b}{a^2 + b^2} e^{ax} \sin bx + \dfrac{a}{a^2 + b^2} e^{ax} \cos bx + C$ (for a, b not both 0); $x + C$ (for $a = b = 0$)

83. $\frac{1}{2}x - \frac{1}{4c} \sin 2cx + C$ (for $c \neq 0$); C (for $c = 0$)

85. $f'(x) = \dfrac{ad - bc}{(cx + d)^2}$ (for c, d not both 0); (undefined for $c = d = 0$)

87. $\dfrac{ax}{c} + \dfrac{bc - ad}{c^2} \ln |cx + d| + C$ (for $c \neq 0$); $\dfrac{a}{2d}x^2 + \dfrac{b}{d}x + C$ (for $c = 0, d \neq 0$); (undefined for $c = d = 0$)

89. $\sqrt{x^2 + a^2} + C$

91. $\ln \left| \sqrt{x^2 + a^2} + x \right| + C$

93. $f'(x) = 2x \sin ax + ax^2 \cos ax$

95. $-\frac{1}{a}x^2 \cos ax + \frac{2}{a^2}x \sin ax + \frac{2}{a^3} \cos ax + C$ (for $a \neq 0$); C (for $a = 0$)

97. $\frac{1}{a} \cosh ax + C$ (for $a \neq 0$); C (for $a = 0$)

99. $x \cos^{-1} ax - \frac{1}{a}\sqrt{1 - (ax)^2} + C$ (for $a \neq 0$); $\frac{\pi}{2}x + C$ (for $a = 0$)

101. $2(1 + \sqrt{x}) - 2 \ln \left| 1 + \sqrt{x} \right| + C$ or $2\sqrt{x} - 2 \ln \left| 1 + \sqrt{x} \right| + C_1$

103. $\frac{4}{3}\left(1 + \sqrt[4]{x}\right)^3 - 6\left(1 + \sqrt[4]{x}\right)^2 + 12\left(1 + \sqrt[4]{x}\right)$
$- 4\ln\left(1 + \sqrt[4]{x}\right) + C$

or $\frac{4}{3}\left(\sqrt[4]{x}\right)^3 - 2\left(\sqrt[4]{x}\right)^2 + 4\sqrt[4]{x} - 4\ln\left|1 + \sqrt[4]{x}\right| + C_1$

105. $\ln\left(\sqrt{e^x + 1} - 1\right) - \ln\left(\sqrt{e^x + 1} + 1\right) + C$

107. a. Substitute $t = x/2$.

    b. Multiply the right side of each equation by
$\dfrac{1 + \tan^2(x/2)}{1 + \tan^2(x/2)}$, and simplify.

    c. Substitute $x = 2\tan^{-1} u$ into each equation.

    d. $\displaystyle\int \frac{1}{1 + \cos x}\,dx$

$\displaystyle= \int \frac{1}{1 + \frac{1-u^2}{1+u^2}} \cdot \frac{2\,du}{1 + u^2}$

$\displaystyle= \int \frac{2\,du}{(1 + u^2) + (1 - u^2)} = \int du, \text{ Q.E.D.}$

    e. $\tan(x/2) + C$

109. $-\cot(x/2) + C$

111. $-\cot(x/2) - x + C$

Problem Set 9-12

1. Answers will vary.

Problem Set 9-13

R0. Answers will vary.

R1. $f'(x) = \cos x - x \sin x$

$\displaystyle\int x \sin x\,dx = \sin x - x \cos x + C$

$\displaystyle\int_1^4 x \sin x\,dx = \sin x - x \cos x\,\Big|_1^4$
$= \sin 4 - 4\cos 4 - \sin 1 + \cos 1 = 1.5566...$

Numerically, $\displaystyle\int_1^4 x \sin x\,dx \approx 1.5566...$

R2. $-\dfrac{5}{2}x \cos 2x + \dfrac{5}{4}\sin 2x + C$

R3. a. $\dfrac{1}{2}x^3 \sin 2x + \dfrac{3}{4}x^2 \cos 2x - \dfrac{3}{4}x \sin 2x - \dfrac{3}{8}\cos 2x + C$

    b. $-\dfrac{3}{25}e^{4x} \cos 3x + \dfrac{4}{25}e^{4x} \sin 3x + C$

    c. $\dfrac{1}{2}x^2(\ln x)^2 - \dfrac{1}{2}x^2 \ln x + \dfrac{1}{4}x^2 + C$

    d. $V = \dfrac{16}{3}\pi \ln 2 - \dfrac{14}{9}\pi = 6.7268...$

R4. a. $\displaystyle\int \cos^{30} dx = \frac{1}{30}\cos^{29} x \sin x + \frac{29}{30}\int \cos^{28} x\,dx$

    b. $\dfrac{1}{5}\sec^4 x \tan x + \dfrac{4}{15}\sec^2 x \tan x + \dfrac{8}{15}\tan x + C$

    c. $\displaystyle\int \tan^n x\,dx = \frac{1}{n-1}\tan^{n-1} x - \int \tan^{n-2} x\,dx$

R5. a. $\sin x - \dfrac{2}{3}\sin^3 x + \dfrac{1}{5}\sin^5 x + C$

    b. $\dfrac{1}{5}\tan^5 x + \dfrac{2}{3}\tan^3 x + \tan x + C$

    c. $\dfrac{1}{2}x - \dfrac{1}{28}\sin 14x + C$

    d. $\dfrac{1}{2}\sec x \tan x + \dfrac{1}{2}\ln|\sec x + \tan x| + C$

    e. $(\tan^9 32)x + C$

    f. $A = \dfrac{113}{8}\pi + 64 - 36\sqrt{2} = 57.4633...$

R6. a. $\dfrac{1}{2}x\sqrt{x^2 - 49} - \dfrac{49}{2}\ln\left|x + \sqrt{x^2 - 49}\right| + C$

    b. $\dfrac{1}{2}(x - 5)\sqrt{x^2 - 10x + 34}$
$+ \dfrac{9}{2}\ln\left|\sqrt{x^2 - 10x + 34} + x - 5\right| + C$

    c. $\sin^{-1}\dfrac{x}{2} + \dfrac{1}{2}x\sqrt{1 - 0.25x^2} + C$

    d. $A = 25(\sin^{-1} 0.8 - \sin^{-1} 0.6) = 7.0948...$

R7. a. $\ln|x + 1| + 5\ln|x - 4| + C$

    b. $3\ln|x - 1| + 4\ln|x + 2| - 2\ln|x - 3| + C$

    c. $5\ln|x| + \tan^{-1}\dfrac{x}{3} + C$

    d. $\ln\left|x^2(x + 4)^3\right| + \dfrac{1}{x + 4} + C$

    e. $y = 3 + \dfrac{5}{1 + 0.25e^{0.5x}}$

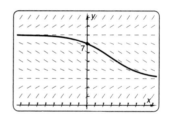

R8. a.

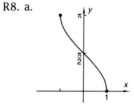

    b. $f'(x) = \dfrac{1}{|x|\sqrt{9x^2 - 1}}$

    c. $x\tan^{-1} 5x - \dfrac{1}{10}\ln\left|1 + 25x^2\right| + C$

    d. $A = 1$

R9. a.

b.

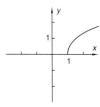

c. $h'(x) = -x^2 \operatorname{sech} x \tanh x + 2x \operatorname{sech} x$

d. $f'(x) = \dfrac{5}{\sqrt{25x^2 + 1}}$

e. $\dfrac{1}{3} \ln |\cosh 3x| + C$

f. $x \cosh^{-1} 7x - \dfrac{1}{7}\sqrt{49x^2 - 1} + C$

g. $\cosh^2 x - \sinh^2 x$
$$= \dfrac{1}{2}(e^x + e^{-x})^2 - \dfrac{1}{2}(e^x - e^{-x})^2$$
$$= \dfrac{1}{4}(e^{2x} + 2 + e^{-2x}) - \dfrac{1}{4}(e^{2x} - 2 + e^{-2x})$$
$$= 1, \text{ Q.E.D.}$$

h. $y = 2.5269\ldots \cosh \dfrac{t}{2.5269\ldots} + 5 - 2.5269\ldots$
$y(10) = 68.5961\ldots$
$y = 20 \Rightarrow x = \pm 6.6324\ldots$

R10. a. The integral converges to 5.

b. The integral diverges.

c. The integral converges to 6.

d. The integral converges to $\dfrac{10}{3} = 3.333\ldots$.

e. $\int_1^\infty x^{-p}\, dx$ converges if $p > 1$; otherwise, it diverges.

R11. a. $f'(x) = \sin^{-1} x + \dfrac{x}{\sqrt{1 - x^2}}$

b. $\dfrac{1}{2}x^2 \sin^{-1} x - \dfrac{1}{4}\sin^{-1} x + \dfrac{1}{4}x\sqrt{1 - x^2} + C$

c. $e^x \operatorname{sech}^2 e^x$

d. $-\ln |x| + \dfrac{1}{2}\ln |x - 1| + \dfrac{1}{2}\ln |x + 1| + C$

e. $f'(x) = -x(1 - x^2)^{-1/2}$

f. $\dfrac{1}{2}\sin^{-1} x + \dfrac{1}{2}x\sqrt{1 - x^2} + C$

g. $g'(x) = 2\ln x \cdot \dfrac{1}{x}$

h. $\dfrac{1}{2}x^2 \ln x - \dfrac{1}{4}x^2 + C$

R12. For $\int(9 - x^2)^{-1/2} x\, dx$, the $x\, dx$ can be transformed to the differential of the inside function by multiplying by a *constant*, $-\dfrac{1}{2}\int(9 - x^2)^{-1/2}(-2x\, dx) = -(9 - x^2)^{1/2} + C$, and thus has no inverse sine.

For $\int(9 - x^2)^{-1/2}\, dx$, transforming the dx to the differential of the inside function, $-2x\, dx$, requires multiplying by a *variable*. Because the integral of a product does not equal the product of the two integrals, you can't divide on the outside of the integral by $-2x$. So a more sophisticated technique must be used, in this case, trigonometric substitution. As a result, an inverse sine appears in the answer:
$\int(9 - x^2)^{-1/2}\, dx = \sin^{-1}\dfrac{x}{3} + C$.

CHAPTER 10

Problem Set 10-1

1. $5.3955\ldots \approx 5.40$ min

3. $100.0231\ldots \approx 100.0$ ft upstream

5. $\int_0^{10}\left|100\, e^{t\ln 0.8} - 30\right| dt = 203.6452\ldots \approx 203.6$ ft

Problem Set 10-2

1. a. Positive on $[0, 2)$; negative on $(2, 6]$

b. $[0, 2): 14\frac{2}{3}$ ft; $(2, 6]: 26\frac{1}{3}$ ft

c. Displacement $= -12$ ft; distance $= 41\frac{1}{3}$ ft

d. Displacement $= 14\frac{2}{3} + (-26\frac{2}{3}) = -12$ ft;
distance $= 14\frac{2}{3} + 26\frac{2}{3} = 41\frac{1}{3}$ ft

e. -4 (ft/s)/s

3. a. Positive on $(8, 11]$; negative on $[1, 8)$

b. $[1, 8): -4.9420\ldots \approx 4.94$ km;
$(8, 11]: 4.7569\ldots \approx 4.76$ km

c. Displacement $= -0.1850\ldots \approx -0.19$ km;
distance $= 9.6990\ldots \approx 9.70$ km

d. Displacement $= -4.9420\ldots + 4.7569\ldots = -0.1850\ldots$
≈ -0.19 km; distance $= -(-4.9420\ldots) + 4.7569\ldots$
$= 9.6990\ldots \approx 9.70$ km

e. $0.1851\ldots \approx 0.19$ (km/h)/h

5. $v(t) = \dfrac{2}{3}t^{3/2} - 18$; displacement $= -14\frac{14}{15}$ ft;
distance $= 179\frac{7}{15}$ ft

7. $v(t) = -6\cos t - 3$; displacement
$= -9.4247\ldots \approx -9.42$ km; distance
$= 13.5338\ldots \approx 13.53$ km

9. a. 4 s

b. Displacement $= 1\frac{1}{3}$ ft

c. Distance $= 4$ ft

11. a. Displacement = 300 ft

b. Distance = 500 ft

13. a.

t_{end} s	a_{av} (mi/h)/s	v_{end} mi/h	v_{av} mi/h	s_{end} mi
0	—	0	—	0
5	2.95	14.75	7.375	0.0102...
10	3.8	33.75	24.25	0.0439...
15	1.75	42.5	38.125	0.0968...
20	0.3	44	43.25	0.1569...
25	0	44	44	0.2180...
30	0	44	44	0.2791...
35	0	44	44	0.3402...
40	−0.2	43	43.5	0.4006...
45	−0.9	38.5	40.75	0.4572...
50	−2.6	25.5	32	0.5017...
55	−3.5	8	16.75	0.525
60	−1.6	0	4	0.5305...

b. At $t = 60$, $v_{end} = 0$, ∴ the train is at rest.

c. The train is just starting at $t = 0$; its acceleration must be greater than zero to get it moving, even though it is stopped at $t = 0$. Acceleration and velocity are different quantities; the velocity can be zero but changing, which means the acceleration is nonzero.

d. Zero acceleration means the velocity is constant, but not necessarily zero.

e. 0.5305... ≈ 0.53 mi between stations

15. a. $a = \dfrac{dv}{dt} \Rightarrow v = \int a\, dt = at + C; v = v_0$ when
$t = 0 \Rightarrow C = v_0 \Rightarrow v = v_0 + at$

b. $v = \dfrac{ds}{dt} \Rightarrow s = \int v\, dt = \int (v_0 + at)\, dt = v_0 t + \dfrac{1}{2}at^2 + C;$
$s = s_0$ when $t = 0 \Rightarrow C = s_0 \Rightarrow s = v_0 t + \dfrac{1}{2}at^2 + s_0$

Problem Set 10-3

1. a. $y_{av} = 41$

b. The rectangle has the same area as the shaded region.

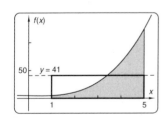

c. $c = 3.4028...$

3. a. $y_{av} = 2.0252...$

b. The rectangle has the same area as the shaded region.

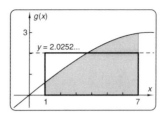

c. $c = 3.7053...$

5. a. $y_{av} = 2\frac{1}{6}$

b. The rectangle has the same area as the shaded region.

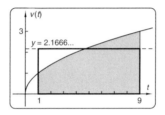

c. $c = 4\frac{25}{36}$

7. $y_{av} = \dfrac{1}{3}ak^2$

9. $y_{av} = \dfrac{1}{k}a(e^k - 1)$

11. $v(25) = 120$ ft/s; displacement = 2500 ft;
$v_{av} = 100$ ft/s

13. Consider an object with constant acceleration a, for a time interval $[t_0, t_1]$.
$v(t) = \int a\, dt = at + C$
At $t = t_0$, $v(t) = v_0 \Rightarrow v_0 = at_0 + C \Rightarrow C = v_0 - at_0$.
∴$v(t) = at + v_0 - at_0 = v_0 + a(t - t_0)$ $v_{av} =$
$$\dfrac{\int_{t_0}^{t_1} [v_0 + a(t - t_0)]\, dt}{t_1 - t_0}$$
$$= \dfrac{1}{t_1 - t_0}\left[v_0 t_1 + \dfrac{1}{2}a(t_1 - t_0)^2 - v_0 t_0 - \dfrac{1}{2}a(t_0 - t_0)^2 \right]$$
$$= v_0 + \dfrac{1}{2}a(t_1 - t_0)$$
The average of v_0 and v_1 is
$$\dfrac{1}{2}(v_0 + v_1) = \dfrac{1}{2}(v_0 + v_0 + a(t_1 - t_0))$$
$$= v_0 + \dfrac{1}{2}a(t_1 - t_0).$$
∴v_{av} = the average of v_0 and v_1, Q.E.D.

15. a. $y_{av} = 52$, or $52,000$; cost of inventory = $260.00

b. At $x = 12$ they may have had a large sale, dropping the inventory from $70,000 to $40,000. There is no day on which the inventory is worth $52,000.

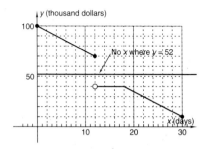

17. $y_{av} = 12.875 \approx 12.9°C$ (by the trapezoidal rule) The average of the high and low temperatures is $14°C$, which is higher than the actual average. Averaging high and low temperatures is easier than finding the average by calculus, but the latter is more realistic for such applications as determining heating and air-conditioning needs.

19. Average $= \dfrac{2A}{\pi}$; $A = 55\pi = 172.78\ldots$ V; The average value of one arc of $y = \sin x$ is $\dfrac{1}{\pi - 0} \int_0^\pi \sin x\, dx = \dfrac{2}{\pi}$, and $y = \sin x$ has a maximum value of 1. A horizontal stretch does not affect the average value. Write a proportion to find the maximum of a sinusoidal curve with an average value of 110: $\dfrac{2/\pi}{1} = \dfrac{110}{m}$, so $m = 55\pi$.

Problem Set 10-4

1. Ann should swim toward a point about 21.8 m downstream.

3. The pipeline should be laid 600 m along the road from the storage tanks, then straight across the field to meet the well.

5. a. For the minimal path, $x = 100/\sqrt{21}$.
   $\therefore \ \sin\theta = \dfrac{x}{\sqrt{50^2 + x^2}} = 0.4 = 2/5$, Q.E.D.
   b. For the minimal path, $x = 400$.
   $\sin\theta = \dfrac{x}{\sqrt{300^2 + x^2}} = 0.8 = 40/50$, Q.E.D.

7. $\sin\theta = \dfrac{12}{13}$ $x = 30\tan\left(\sin^{-1}\dfrac{12}{13}\right) = 72$
   She should swim $100 - 72 = 28$ m, then dive. The algebraic solution is easier than before because no algebraic calculus needs to be done. Mathematicians seek general solutions to find patterns that allow easier solution of similar problems.

9. A graph or table of times for paths close to the optimum shows that a near miss will have virtually no effect on the minimum cost.

11. $47.8809\ldots \approx 47.9$ yd from the perpendicular line from the ship to the shore

13. $T = \dfrac{1}{v_1}\sqrt{a^2 + x^2} + \dfrac{1}{v_2}\sqrt{b^2 + (k - x)^2}$
   $T' = \dfrac{x}{v_1\sqrt{a^2 + x^2}} - \dfrac{k - x}{v_2\sqrt{b^2 + (k - x)^2}}$
   $\sin\theta_1 = \dfrac{x}{\sqrt{a^2 + x^2}}$, $\sin\theta_2 = \dfrac{k - x}{\sqrt{b^2 + (k - x)^2}}$
   $\therefore \ T' = \dfrac{1}{v_1}\sin\theta_1 - \dfrac{1}{v_2}\sin\theta_2$
   For the minimal path, $T' = 0$. Thus,
   $\dfrac{1}{v_1}\sin\theta_1 = \dfrac{1}{v_2}\sin\theta_2$
   $\dfrac{\sin\theta_1}{\sin\theta_2} = \dfrac{v_1}{v_2}$, Q.E.D.

15. Answers will vary.

Problem Set 10-5

1. Minimum is $D(1) = 2$, or 2000 mi; maximum is $D(3) = 3\frac{1}{3}$, or about 3333 mi

3. $x = 0.5$

5. a.
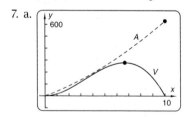

b. Fran should study for 3 h.

c. i. $G(4)$ is about 1 point less.

   ii. $G(2)$ is about 1 or 2 points less.

7. a.

b. Maximum V: $r = 6\frac{2}{3}$ in., $h = 2$ in.; maximum A: $r = 10$ in., $h = 0$ in. The maximum volume and maximum area do not occur at the same radius.

9. $-0.1286\ldots \approx 0.129$ m/min

11. a. $w = 1000 + 15t$ (lb); $p = 0.90 - 0.01t$ ($/lb)
$A = 900 + 3.5t - 0.15t^2$ ($)
b. $A'(11\frac{2}{3}) = 0$
A at $t = 11\frac{2}{3}$ is a maximum, not a minimum, because $\frac{dA}{dt}$ goes from positive to negative there.
c. $920.42

Problem Set 10-6

1.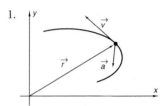

3. a. $\vec{v}(1) = -0.8186...\vec{i} + 3.7560...\vec{j}$
x is decreasing at approximately 3.84 cm/s.
b. 2.43 cm; 2.78 cm
c. $\vec{a} = (-2e^t \sin t)\vec{i} + (2e^t \cos t)\vec{j}$

5. a. $\vec{v}(t) = (6 \cos 0.6t)\vec{i} + (-4.8 \sin 1.2t)\vec{j}$
$\vec{a}(t) = (-3.6 \sin 0.6t)\vec{i} + (-5.76 \cos 1.2t)\vec{j}$
b. $\vec{r}(0.5) = 2.9552...\vec{i} + 3.3013...\vec{j}$
$\vec{v}(0.5) = 5.7320...\vec{i} - 2.7102...\vec{j}$
$\vec{a}(0.5) = -1.0638...\vec{i} - 4.7539...\vec{j}$

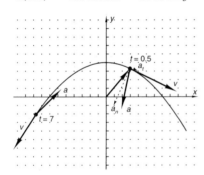

These vectors make sense because the head of $\vec{r}$ is on the graph, $\vec{v}$ is tangent to the graph, and $\vec{a}$ points to the concave side of the graph.
c. The object is speeding up, because the angle between $\vec{a}$ and $\vec{v}$ is acute.
d. $\vec{a}(0.5) \cdot \vec{v}(0.5) = 6.7863...$, so the angle is acute.
$\vec{a}_t(0.5) = 0.9676...\vec{i} - 0.4575...\vec{j}$
$\vec{a}_n(0.5) = -2.0314...\vec{i} - 4.2964...\vec{j}$
See the graph in part b.
e. The object is speeding up at $1.0703... \approx 1.07$ (ft/s)/s.
f. $\vec{r}(7) = -8.7157...\vec{i} - 2.0771...\vec{j}$
$\vec{v}(7) = -2.9415...\vec{i} - 4.1020...\vec{j}$
$\vec{a}(7) = 3.1376...\vec{i} + 2.9911...\vec{j}$
See the graph in part b.

The object is slowing down, because the angle between $\vec{a}$ and $\vec{v}$ is obtuse.
g. $\vec{a}(0) \cdot \vec{v}(0) = 0$, so $\vec{a}(0)$ and $\vec{v}(0)$ are perpendicular. This means the object is neither slowing down nor speeding up at $t = 0$.

7. a.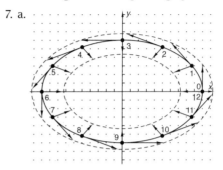

b. See the graph in part a.
c. For $\vec{r} + \vec{v}$, show that $\left(\frac{x}{10}\right)^2 + \left(\frac{y}{6}\right)^2 = 1 + \left(\frac{\pi}{6}\right)^2$, which is the equation of an ellipse.
d. See the graph in part a.
e. See the graph in part a. The direction of each acceleration vector is the opposite of the corresponding position vector and thus directed toward the origin.

9. a. $\vec{r}(x) = x\vec{i} + x^2\vec{j}$; $\vec{v}(x) = \frac{dx}{dt}\vec{i} + 2x\frac{dx}{dt}\vec{j}$
b. $\vec{v}(2) = -3\vec{i} - 12\vec{j}$
Speed $= |\vec{v}(2)| = \sqrt{153} \approx 12.4$ cm/s
c.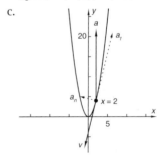

This is reasonable because $\vec{v}(2)$ points along the curve to the left, indicating that x is decreasing.
d. $\vec{a}(x) = 18\vec{j}$; $\vec{a}(2) = 18\vec{j}$
See the graph in part c.
e. $\vec{a}_t(2) = 4.2352...\vec{i} + 16.9411...\vec{j}$
$\vec{a}_n(2) = -4.2352...\vec{i} + 1.0588...\vec{j}$
See the graph in part a.
$\vec{a}_t(2)$ is parallel to the curve. $\vec{a}_n(2)$ is normal to the curve and points inward to the concave side.
f. The object is slowing down, when $x = 2$ because the angle between $\vec{a}(2)$ and $\vec{v}(2)$ is obtuse, as shown by the graph and by the fact that the dot

product is negative. Also, $\vec{a}_t(2)$ points in the opposite direction of $\vec{v}(2)$.

g. $dL = \sqrt{1 + 4x^2}\,dx$

$\dfrac{dx}{dt} = \dfrac{5}{\sqrt{17}} \approx 1.21$ cm/s

11. 12.0858... ft

13. a. $\vec{v}(t) = -130\,\vec{i} - 32t\,\vec{j}$

b. $\vec{r}(t) = (-130t + 60.5)\,\vec{i} + (-16t^2 + 8)\,\vec{j}$

c. The ball reaches the plate at $t = 60.5/130 \approx 0.46$ s. The ball passes 4.5346... ft above the plate, which is slightly above the strike zone.

d. At $t = 0$, $dx/dt = 200 \cos 15°$, $dy/dt = 200 \sin 15°$;
$\vec{r}(t) = (200t \cos 15°)\,\vec{i} + (-16t^2 + 200t \sin 15° + 3)\,\vec{j}$

e. Phyllis will make the home run because the ball is about 41.6 ft above the wall when $x = 400$.

15. a. $d = 90 + 150 \cos t$

b. $\vec{v}(1) = -212.1270...\,\vec{i} - 13.7948...\,\vec{j}$;
speed = 212.5750... $\approx$ 212.6 cm/s

c. $\vec{a}(1) = 76.2168...\,\vec{i} - 348.5216...\,\vec{j}$
$\vec{a}_t(1) = 53.3266...\,\vec{i} + 3.4678...\,\vec{j}$
$\vec{a}_n(1) = 22.8902...\,\vec{i} - 351.9894...\,\vec{j}$
Annie is slowing down at 53.4392... $\approx$ 53.4 cm/s.

17. a. $\vec{v}(t) = (5 - 12 \cos t)\,\vec{i} + (-12 \sin t)\,\vec{j}$;
$\vec{a}(t) = (12 \sin t)\,\vec{i} + (-12 \cos t)\,\vec{j}$

b. $\vec{v}(2.5) = 14.6137...\,\vec{i} - 7.1816...\,\vec{j}$;
$\vec{a}(2.5) = 7.1816...\,\vec{i} + 9.6137...\,\vec{j}$

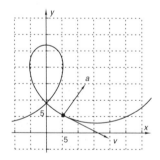

c. $\vec{a}_t(2.5) = 1.9791...\,\vec{i} - 0.9726...\,\vec{j}$;
$\vec{a}_n(2.5) = 5.2024...\,\vec{i} + 10.5863...\,\vec{j}$

d. $\vec{v}(2.5)$ is reasonable because its graph points along the path in the direction of motion. $\vec{a}(2.5)$ is reasonable because it points toward the concave side of the path. The roller coaster is traveling at $|\vec{v}(2.5)| = 16.2830...$ ft/s. Its speed is increasing at 2.2052... ft/s² because the scalar projection of $\vec{a}(2.5)$ onto $\vec{v}(2.5) = 2.2052...$.

e. $\vec{a}(0 + 2\pi n) = 0\,\vec{i} - 12\,\vec{j}$, which points straight down.
$\vec{a}(\pi + 2\pi n) = 0\,\vec{i} + 12\,\vec{j}$, which points straight up.

f. 78.7078... $\approx$ 78.7 ft

19. $\vec{v}(1) = (8 \cos 0.8)\,\vec{i} + (-6 \sin 0.6)\,\vec{j} + 3\,\vec{k}$;
$\vec{a}(1) = (-6.4 \sin 0.8)\,\vec{i} + (-3.6 \cos 0.6)\,\vec{j} - 1.5\,\vec{k}$
$\vec{a}(1) \cdot \vec{v}(1) = -20.0230...$, so the object is slowing down because the dot product is negative.

Problem Set 10-7

R0. Answers will vary.

R1. Popeye's velocity becomes positive at $t = 9$ s. They have moved 9 ft closer to the sawmill.
From $t = 0$ to $t = 25$: displacement = $8\frac{1}{3}$ ft; distance = $26\frac{1}{3}$ ft

R2. a. i.
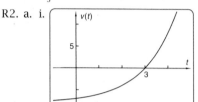

ii. Displacement = $-3.8022... \approx -3.8$ cm

iii. Distance = $10.8853... \approx 10.9$ cm

b. $a_{av} = (a_{end} + a_{begin})/2$; $v_{end} = v_{begin} + a_{av} \cdot 5$ s

t	a	a_{av}	v_{end}	
0	2	—	30	speeding up
5	8	5	55	speeding up
10	1	4.5	77.5	speeding up
15	0	0.5	80	neither
20	−10	−5	55	slowing down
25	−20	−15	−20	slowing down

R3. a. i. $v_{av} = 2/\pi = 0.6366...$

ii. $v_{av} = 0$

iii. $v_{av} = 0$

b. i. The average on [0, 6] is 18.

ii. The rectangle has the same area as the shaded region.

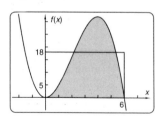

iii. The average of the two values of $f(x)$ at the endpoints is zero, not 18.

R4. a. Turning at a point about 467 ft from the intersection of the two sidewalks takes the minimum time, although it takes only a second longer to head straight for the English building.

b. The minimum cost is \$122,000, obtained by going 7.5 km along the beach, then cutting across to the island. This path saves about \$29,600 over the path straight to the island.

R5. a. i. Maximum acceleration = 9 at $t = 3$; minimum acceleration = -40 at $t = 10$

ii. Maximum velocity = 36 at $t = 6$; minimum velocity = $-33\frac{1}{3}$ at $t = 10$

iii. Maximum displacement = $182\frac{1}{4}$ at $t = 9$; minimum displacement = 0 at $t = 0$

b. i. Let t = number of days Dagmar has been saving. Let $V(t)$ = real value (in constant day zero pillars) of money in account after t days. $V(t) = 50t(0.5^{0.005t})$

ii. Dagmar's greatest purchasing power will be after about 289 days because $V'(t)$ goes from positive to negative at $t = 288.5390...$.

R6. a. i. and ii.

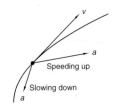

Speeding up

Slowing down

b. i. $\vec{r}(1) = 7.7154...\vec{i} + 3.5256...\vec{j}$
$\vec{v}(1) = 5.8760...\vec{i} + 4.6292...\vec{j}$
$\vec{a}(1) = 7.7154...\vec{i} + 3.5256...\vec{j}$

ii.

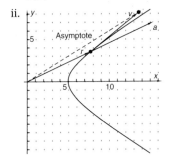

iii. Speed = $7.4804... \approx 7.48$ units/min
The object is speeding up at $8.2423... \approx 8.24$ units/min^2.

iv. $4.5841... \approx 4.58$ units

v. $\vec{r}(t) + \vec{v}(t) =$
$(5\cosh t + 5\sinh t)\vec{i} + (3\sinh t + 3\cosh t)\vec{j}$
The y-coordinate is 0.6 times the x-coordinate, so the head lies on $y = 0.6x$, one asymptote of the hyperbola.

CHAPTER 11

Problem Set 11-1

1.

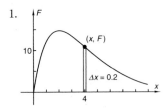

$F = 10.8268... \approx 10.83$ lb in the strip
$W = 2.1653... \approx 2.17$ ft-lb

3. Integral = 69.1289...

5. $W = 80$ ft-lb

Problem Set 11-2

1. 800 ft-lb

3. $50k$

5. $1,396,752.0937... \approx 1.4$ million ft-lb

7. a. $117,621,229... \approx 117.6$ million ft-lb
b. $250,925,288.4... \approx 250.9$ million ft-lb

9. a. $1504.7320... \approx 1504.7$ in.-lb
b. $-566.9574... \approx 567$ in.-lb
c. $937.7746... \approx 937.8$ in.-lb
d. Carnot (kar-NŌ); Nicolas Léonard Sadi Carnot, 1796–1832, was a French physicist and a pioneer in the field of thermodynamics.

Problem Set 11-3

1. a. $m = 8.1419...k$
b. $m = 108.1103...$

3. a. $m = 40.5\pi k$
b. $m = 546.75\pi k$
c. $m = 105.3\pi k$
d. The solid in part b has the largest mass.

5. a. Prediction: The cone on the left, with higher density at its base, has greater mass because higher density is in the larger part of the cone.
b. For the cone on the left, $m = 1305\pi$ oz.
For the cone on the right, $m = 1035\pi$ oz.
∴ the cone on the left has the higher mass, as predicted in part a.

7. $m = \frac{16}{21}\pi k = 2.3935...k$

9. $m = 4.82\pi \approx 15.14$ g

11. a. $m = \dfrac{1}{2}\pi r^4 k$

  b. $m = \dfrac{1}{4}\pi^2 r^4 k$

  c. $m = \pi k r^4$

13. $m = 8.6261\ldots$

Problem Set 11-4

1. a. $V = 40.5\pi$

  b. $M_{xz} = 121.5\pi$

  c. The centroid is at $(0, 3, 0)$.

3. a. $m = 170.1375\ldots k$

  b. $M_{xz} = 612.4952\ldots k$

  c. The center of mass is at $(0, 3.6, 0)$.

  d. False. The centroid is at $(0, 3, 0)$, but the center of mass is at $(0, 3.6, 0)$.

5. a. $\bar{x} = 1.3130\ldots$

  b. $\bar{x} = 1.5373\ldots$

  c. False. For the solid, $\bar{x}$ is farther from the yz-plane.

7. Construct axes with the origin at a vertex and the x-axis along the base, b.

  Slice the triangle parallel to the x-axis.

  The width of a strip is $b - \dfrac{b}{h}y$.

  $dA = \left(b - \dfrac{b}{h}y\right)dy$

  $dM_x = y\,dA = \left(by - \dfrac{b}{h}y^2\right)dy$

  $M_x = \displaystyle\int_0^h \left(by - \dfrac{b}{h}y^2\right)dy = \dfrac{1}{2}by^2 - \dfrac{b}{3h}y^3 \Big|_0^h = \dfrac{1}{6}bh^2$

  $\bar{y} \cdot A = M_x \Rightarrow \bar{y} = \dfrac{\frac{1}{6}bh^2}{\frac{1}{2}bh} = \dfrac{1}{3}h$, Q.E.D.

9. a. Slice the region parallel to the y-axis so that each point in a strip is about x units from the y-axis, where x is at the sample point (x, y).

  $dA = y\,dx = \sin x\,dx$

  $A = \displaystyle\int_0^\pi \sin x\,dx = 2$

  $dM_y = x\,dA = x \sin x\,dx$

  $M_y = \displaystyle\int_0^\pi x \sin x\,dx = 3.1415\ldots = \pi$

  $\bar{x} \cdot A = M_y \Rightarrow \bar{x} = \dfrac{\pi}{2}$, Q.E.D.

  b. $M_{2y} = 5.8696\ldots$

  c. $\bar{x} = 1.7131\ldots$

11. a. $M = \dfrac{1}{2}\pi H R^4$, $\bar{r} = \dfrac{1}{\sqrt{2}}R$

  b. $M = \dfrac{1}{10}\pi H R^4$, $\bar{r} = \sqrt{0.3}\,R$

  c. $M = \dfrac{8}{15}\pi R^5$, $\bar{r} = \sqrt{0.4}\,R$

13. a. Set up axes with the x-axis through the centroid.
  $dM_2 = y\,dA = y^2 \cdot B\,dy$

  $M_2 = B \displaystyle\int_{-0.5H}^{0.5H} y^2\,dy = \dfrac{1}{3}y^3 \Big|_{-0.5H}^{0.5H} = \dfrac{1}{12}BH^3$, Q.E.D.

  b. i. Stiffness $= 288k$

  ii. Stiffness $= 8k$
    A board on its edge is 36 times stiffer.

  c. i. Stiffness $= 160k$

  ii. Stiffness $= 448k$ (2.8 times stiffer!)

  d. Increasing the depth does seem to increase stiffness greatly, but making the beam *very* tall would also make the web *very* thin, perhaps too thin to withstand much force.

15. a. $V = 2\pi^2 r^2 R$

  b. $\bar{r} = \dfrac{4}{3\pi}r$

Problem Set 11-5

1. a. $F = 2.8444\ldots k$

  b. $M_x = 2.1880\ldots k$

  c. The center of pressure is at $\left(0, \frac{10}{13}\right)$.

3. a. $A = 1186.6077\ldots \approx 1186.6 \text{ ft}^2$

  b. $F = 1{,}199{,}294.1645\ldots \approx 1.199$ million lb

  c. $M_x = 13{,}992{,}028.2564\ldots \approx 13.992$ million lb-ft

  d. The center of pressure is at about $(0, 11.67)$ ft.

  e. The centroid is at about $(0, 16.92)$ ft.
    The centroid is different from the center of pressure.

  f. The center of buoyancy is at about $(0, 8.66)$ ft.

5. a. $A = 763.9437\ldots \approx 763.9 \text{ ft}^2$

  b. $F = 4863.4168\ldots k$

  c. Make $k \geq 0.0197\ldots$ tons/ft^2.

7. a. $dM_{2x} = \dfrac{1}{3}(0.25(x-4) - (x-4)^{1/3})^3\,dx$

  b. $M_{2x} = 0.5333\ldots$

9. The integrals in Problems 7 and 8 can be written in the form

  $\displaystyle\int_{x=a}^{x=b} \int_{t=c}^{t=d} f(x, t)\,dt\,dx$

  The result is called a double integral because two integrals appear.

Problem Set 11-6

1. 13,200 calories

3. a. $P(x) = 0.002x^2 + 3x + 500$

  b. $P(700) = \$3580/\text{ft}$

  c. The cost is about \$2,666,667.

  d. The savings is about \$1,250,000.

5. a.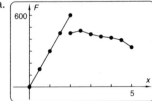

b. F has a step discontinuity at $x = 2$.

c. $W = 600$ in.-lb

d. $W \approx 1266\frac{2}{3}$ in.-lb

e. Total work $\approx 1866\frac{2}{3}$ in.-lb

f. Yes, a piecewise-continuous function such as this one can be integrable. See Problem 27 in Problem Set 9-10 (Improper Integrals).

7. $M_{2y} = 33.5103...k$ g-cm^2

9. a. $m = 2000 - 5t$

b. $a = 1400(400 - t)^{-1}$

c. $v(t) = 1400 \ln \dfrac{400}{|400 - t|}$

d. $v(20) = 71.8106... \approx 71.81$ m/s;
$s = 711.9673... \approx 712.0$ m

11. a. $W = 0.5707...k$

b. $W = 0.3926...k$

13. a. $W = 113.0973... \approx \$113.1$ million

b. $W = 71.4328... \approx \$71.4$ million

c. $W = 163.9911... \approx \$164.0$ million

d. This problem is equivalent to volume by cylindrical shells, in which the value of the land per square unit takes the place of the altitude of the cylinder. It is also equivalent to the water flow in Problem 4 of this problem set.

e. Answers will vary.

15. a. $f(x) = 9 - x^2 = (3 - x)(3 + x) = 0$ only at $x = \pm 3$
$g(x) = -\frac{1}{3}x^3 - x^2 + 3x + 9 = -\frac{1}{3}(x - 3)(x + 3)^2 = 0$
only at $x = \pm 3$.

b. $A_f = \displaystyle\int_{-3}^{3} (9 - x^2)\, dx = 36$

$A_g = \displaystyle\int_{-3}^{3} \left(-\frac{1}{3}x^3 - x^2 + 3x + 9\right) dx = 36$

To simplify algebraic integration, you could use

$A_f = 2\displaystyle\int_{0}^{3} (9 - x^2)\, dx$

$A_g = 2\displaystyle\int_{0}^{3} (9 - x^2)\, dx$, where the odd terms integrate to zero between symmetrical limits. Thus, the two integrals are identical.

c. The maximum of f is at $x = 0$; the maximum of g is at $x = 1$.

d. $\bar{x} = 0.6$

e. False. For the symmetrical region under the graph of f, the centroid is on the line through the high point. But for the asymmetrical region under the graph of g, the high point is at $x = 1$ and the centroid is at $x = 0.6$.

f. False: area to left $= 17.1072$ and area to right $= 18.8928$

g. $S = -17.7737...$

h. By symmetry, the centroid of the area under f is on the y-axis, so $\bar{x} = 0$. Then
$dS = x^3 dA = x^3(9 - x^2)\, dx$
$S = \displaystyle\int_{-3}^{3} x^3(9 - x^2)\, dx = 0$ (odd function integrated between symmetrical limits).
The "skewness" being zero reflects the symmetry of this region. It is not skewed at all.

i.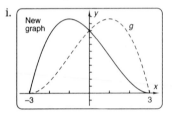

17. In Problem 16, $R = \bar{x} = 1.2390...$ and $L = 4.6467...$.
$2\pi RL = 2\pi(1.2390...)(4.6467...) = 36.1769...$, which equals S, Q.E.D.

Problem Set 11-7

R0. Answers will vary.

R1. $W = 129.6997... \approx 129.7$ ft-lb

R2. a. $W = -\frac{2}{3}k$ ft-lb

b. $W = 11.2814... \approx 11.28$ in.-lb

R3. a. $m = 57.6\pi k$

b. $m = 64\pi$

R4. a. $W = b - \dfrac{b}{h}y$; $M_x = \dfrac{1}{6}bh^2$; $A = \dfrac{1}{2}bh \Rightarrow \bar{y} = \dfrac{1}{3}h$, Q.E.D.

b. $M_{2y} = 3.5401...$

R5. $F = 3,736,263.2708... \approx 3.736$ million lb

R6. a. $r(x) = 30\left(\dfrac{5}{3}\right)^{x/10000}$
(or $r(x) = 30e^{-\ln 0.6 \cdot x/10000} = 30e^{0.00005108256...x}$)

b. $C = 6,965,243.17... \approx \$6.965$ million

CHAPTER 12

Problem Set 12-1

1.

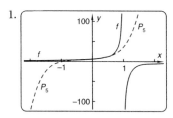

The graph of P_5 fits the graph of f reasonably well for about $-0.8 < x < 0.6$.
The graph of P_5 bears no resemblance to the graph of f at $x = 2$ and at $x = -2$, for example.

3. $P_5(0.5) = 11.8125, P_6(0.5) = 11.90625,$ and $f(0.5) = 12$
∴ $P_6(0.5)$ is closer to $f(0.5)$ than $P_5(0.5)$ is.
$P_5(2) = 378, P_6(2) = 762,$ and $f(2) = -6$
∴ $P_6(2)$ is not closer to $f(2)$ than $P_5(2)$ is.

5. $P_0(1) = 6 \qquad P_0(-1) = 6$
$P_1(1) = 12 \qquad P_1(-1) = 0$
$P_2(1) = 18 \qquad P_2(-1) = 6$
$P_3(1) = 24 \qquad P_3(-1) = 0$
$P_4(1) = 30 \qquad P_4(-1) = 6$

For $x = 1$, the sums just keep getting larger and larger as more terms are added. For $x = -1$, the sums oscillate between 0 and 6. In neither case does the series converge. If the answer to Problem 4 includes $x = 1$ or $x = -1$, the conjecture would have to be modified.

7. A geometric series; the common ratio

Problem Set 12-2

1. Series: $200 - 120 + 72 - 43.2 + 25.92 - 15.552 + \cdots$
Sums: $200, 80, 152, 108.8, 134.72, 119.168, \ldots$

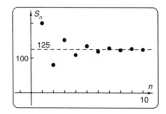

The series converges to 125.
S_n will be within 0.0001 unit of 125 for all values of $n \geq 28$.

3. a. Series: $\sum_{n=1}^{\infty} 7(0.8^{n-1}) = 7 + 5.6 + 4.48 + \cdots$
Sums: $7, 12.6, 17.08, 20.664, 23.5312, \ldots$
The amount first exceeds 20 μg at the 4th dose.
The total amount never exceeds 40 μg.

See closed dots in the graph below.

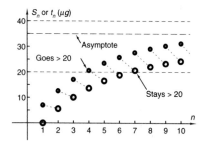

b. $0, 5.6, 10.08, 13.664, 16.5312, \ldots$
See the graph in part a. The open circles show the partial sums just before a dose.
After the 7th dose, the amount remains above 20 μg.

c. See the graph in part a.

5. a.

Months	Dollars
0	1,000,000.00
1	1,007,500.00
2	1,015,056.25
3	1,022,669.17

b. Worth is \$1,093,806.90; interest is \$93,806.90

c. The first deposit is made at time $t = 0$, the second at time $t = 1$, and so forth, so at time $t = 12$, the term index is 13.

d. 9.3806...% APR

e. After 93 months

7. a. Sequence: $20, 18, 16.2, 14.58, 13.122, \ldots$

b. $S_4 = 20 + 18 + 16.2 + 14.58 = 68.78$ ft

c. $S = 200$. The ball travels 200 ft before it comes to rest.

d. 20-ft first cycle: $t = 1.5762\ldots$ s
18-ft second cycle: $t = 2.1147\ldots$ s

e. The series of times converges to 30.7155, so the model predicts that the ball comes to rest after about 43.4 s.

9. $P'(0) = 6$ and $f'(0) = 6; P''(0) = 12$ and $f''(0) = 12; P'''(0) = 36$ and $f'''(0) = 36$
Conjecture: $P^{(n)}(0) = f^{(n)}(0)$ for all values of n.

Problem Set 12-3

1. $f'(x) = 10e^{2x}; f''(x) = 20e^{2x}; f'''(x) = 40e^{2x};$
$f^{(4)}(x) = 80e^{2x}$

3. $c_0 = 5; c_1 = 10; c_2 = 10$
c_0 and c_1 are the same as for $P_1(x)$.

5.

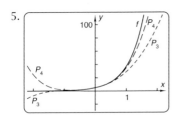

7. $P_3(1) = 31.6666666...$
   $P_4(1) = 35.0000000...$
   $f(1) = 5e^2 = 36.9452804...$
   $\therefore P_4(1)$ is closer to $f(1)$ than $P_3(1)$, Q.E.D.

9. $c_3 = \dfrac{20}{6} = \dfrac{5 \cdot 2^3}{3!}; c_2 = \dfrac{20}{2} = \dfrac{5 \cdot 2^2}{2!};$

   $c_1 = \dfrac{10}{1} = \dfrac{5 \cdot 2^1}{1!}; c_0 = 5 = \dfrac{5 \cdot 2^0}{0!}(0! = 1)$

11. $P(x) = \sum\limits_{n=0}^{\infty} \dfrac{5 \cdot 2^n}{n!} x^n$

Problem Set 12-4

1. a.

n	$f^{(n)}(x)$	$f^{(n)}(0)$	$P^{(n)}(0)$	c_n
0	e^x	1	c_0	1
1	e^x	1	c_1	1
2	e^x	1	$2!c_2$	$\dfrac{1}{2!}$
3	e^x	1	$3!c_3$	$\dfrac{1}{3!}$

   $\therefore P(x) = 1 + x + \dfrac{1}{2!}x^2 + \dfrac{1}{3!}x^3 + \cdots$, Q.E.D.

   b. $\cdots + \dfrac{1}{4!}x^4 + \dfrac{1}{5!}x^5 + \cdots$

   c. $\sum\limits_{n=0}^{\infty} \dfrac{1}{n!}x^n$

   d.

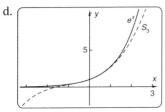

   e. The interval is about $-1 < x < 1$.

   f. The interval is $-0.2237... < x < 0.2188...$.

   g. The interval is $-1.5142... < x < 1.4648...$.

3. a. $S_3(0.6) = 0.564642445...$
      $\sin 0.6 = 0.564642473...$
      $\therefore S_3(0.6) \approx \sin 0.6$, Q.E.D.

b. $\sin 0.6 - S_1(0.6) = 0.0006424733...$
   $t_2 = 0.000648$
   $\sin 0.6 - S_2(0.6) = -0.00000552660...$
   $t_3 = -0.00000555428...$
   $\sin 0.6 - S_3(0.6) = 0.0000000276807...$
   $t_4 = 0.0000000277714...$
   In each case the tail is less in magnitude than the absolute value of the first term of the tail, Q.E.D.

c. Use at least 9 terms.

5. a. $P(1) = 0 = f(1)$
      $P'(1) = 1 = f'(1)$
      $P''(1) = -1 = f''(1)$
      $P'''(1) = 2 = f'''(2)$, Q.E.D.

   b. $\cdots + \dfrac{1}{5}(x-1)^5 - \dfrac{1}{6}(x-1)^6 + \cdots$

   c. $P(x) = \sum\limits_{n=1}^{\infty} (-1)^{n+1} \cdot \dfrac{1}{n}(x-1)^n$

   d.

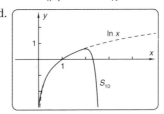

   e. $S_{10}(1.2) = 0.182321555...$
      $\ln 1.2 = 0.182321556...$
      $S_{10}(1.95) = 0.640144911...$
      $\ln 1.95 = 0.667829372...$
      $S_{10}(3) = -64.8253968...$
      $\ln 3 = 1.0986122...$
      $S_{10}(x)$ fits $\ln x$ in about $0 < x < 2$. This is a wider interval of agreement than that for the fourth partial sum, which looks like about $0.3 < x < 1.7$. $S_{10}(1.2)$ and $\ln 1.2$ agree through the 8th decimal place. The values of $S_{10}(1.95)$ and $\ln 1.95$ agree only to 1 decimal place. The values of $S_{10}(3)$ and $\ln 3$ bear no resemblance to each other.

7. a. $P(x) = x - \dfrac{1}{3}x^3 + \dfrac{1}{5}x^5 - \dfrac{1}{7}x^7 + \cdots$

   b.

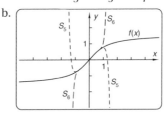

Both partial sums fit the graph of f very well for about $-0.9 < x < 0.9$. For $x > 1$ and $x < -1$, the partial sums bear no resemblance to the graph of f.

1. $1 + u + \dfrac{1}{2!}u^2 + \dfrac{1}{3!}u^3 + \dfrac{1}{4!}u^4 + \dfrac{1}{5!}u^5 + \cdots$

3. $u - \dfrac{1}{3!}u^3 + \dfrac{1}{5!}u^5 - \dfrac{1}{7!}u^7 + \dfrac{1}{9!}u^9 - \dfrac{1}{11!}u^{11} + \cdots$

5. $1 + \dfrac{1}{2!}u^2 + \dfrac{1}{4!}u^4 + \dfrac{1}{6!}u^6 + \dfrac{1}{8!}u^8 + \dfrac{1}{10!}u^{10} + \cdots$

7. $1 + u + u^2 + u^3 + u^4 + u^5 + \cdots$

9. $x^2 - \dfrac{1}{3!}x^4 + \dfrac{1}{5!}x^6 - \dfrac{1}{7!}x^8 + \dfrac{1}{9!}x^{10} - \cdots$

11. $1 + \dfrac{1}{2!}x^6 + \dfrac{1}{4!}x^{12} + \dfrac{1}{6!}x^{18} + \dfrac{1}{8!}x^{24} + \cdots$

13. $(x^2 - 1) - \dfrac{1}{2}(x^2 - 1)^2 + \dfrac{1}{3}(x^2 - 1)^3 - \cdots$

$\left(\text{or } 2\ln x = 2(x-1) - (x-1)^2 + \dfrac{2}{3}(x-1)^3 - \cdots\right)$

15. $x - \dfrac{1}{3}x^3 + \dfrac{1}{5} \cdot \dfrac{1}{2!}x^5 - \dfrac{1}{7} \cdot \dfrac{1}{3!}x^7 + \dfrac{1}{9} \cdot \dfrac{1}{4!}x^9 - \cdots$

17. $1 - x^4 + x^8 - x^{12} + x^{16} - \cdots$

19. $x - \dfrac{1}{5}x^5 + \dfrac{1}{9}x^9 - \dfrac{1}{13}x^{13} + \dfrac{1}{17}x^{17} - \cdots$

21. $2x + x^5 + \dfrac{1}{12}x^9 + \dfrac{1}{360}x^{13} + \cdots$

23. $P_4(x) = -8 + 3(x - 2) + 0.35(x - 2)^2 + 0.085(x - 2)^3$
$- 0.002(x - 2)^4$

25. a. $f(0.4) = 1.694912$
We must assume that the series converges for $x = 0.4$.

b. $f(-1) = c_0 = 2$
$f'(-1) = c_1 = 0.5$
$f''(-1) = 2!c_2 = 2(-0.3) = -0.6$
$f'''(-1) = 3!c_3 = 6(-0.18) = -1.08$
$f^{(4)}(-1) = 4!c_4 = 24(0.02) = 0.48$

c. $g(x) \approx 2 + 0.5x^3 - 0.3x^6; g(1) = 2.2$

d. $g'(0) = 0$ and $g''(0) = 0$, so the second derivative test does not give enough information. So test $g'(x)$ on either side of $x = 0$. $g(x)$ is increasing on both sides of $x = 0$, so $g(0)$ is neither a maximum nor a minimum.

e. $h(x) = 2x + 0.125x^4 - \dfrac{3}{70}x^7$

27. $\dfrac{\sqrt{2}}{2} + \dfrac{\sqrt{2}}{2}\left(x - \dfrac{\pi}{4}\right) - \dfrac{\sqrt{2}}{2 \cdot 2!}\left(x - \dfrac{\pi}{4}\right)^2$
$- \dfrac{\sqrt{2}}{2 \cdot 3!}\left(x - \dfrac{\pi}{4}\right)^3 + \dfrac{\sqrt{2}}{2 \cdot 4!}\left(x - \dfrac{\pi}{4}\right)^4$
$+ \dfrac{\sqrt{2}}{2 \cdot 5!}\left(x - \dfrac{\pi}{4}\right)^5 - \cdots$

29. $(x - 1) - \dfrac{1}{2}(x - 1)^2 + \dfrac{1}{3}(x - 1)^3 - \dfrac{1}{4}(x - 1)^4 + \cdots$

31. $-1 + \dfrac{7}{3}(x - 4) - \dfrac{7 \cdot 4}{3^2 2!}(x - 4)^2 + \dfrac{7 \cdot 4 \cdot 1}{3^3 3!}(x - 4)^3$
$- \dfrac{7 \cdot 4 \cdot 1 \cdot (-2)}{3^4 4!}(x - 4)^4$
$+ \dfrac{7 \cdot 4 \cdot 1 \cdot (-2)(-5)}{3^5 5!}(x - 4)^5 - \cdots$

33. Both give $\cos 3x = 1 - \dfrac{9}{2!}x^2 + \dfrac{81}{4!}x^4 - \dfrac{729}{6!}x^6 + \cdots$.
Substitution gives the answer more easily in this case.

35. $S_4(1.5) = 0.40104166\ldots; \ln 1.5 = 0.40546510\ldots$
Error $= 0.00442344\ldots$
Fifth term $= \dfrac{1}{5}(1.5 - 1)^5 = 0.00625$
The error is smaller in absolute value than the first term of the tail.

37. a. $4 S_9(1) = 3.04183961\ldots$
$\pi = 3.14159265\ldots$
The error is about 3%.

b. $4 S_{49}(1) = 3.12159465\ldots$
$\pi = 3.14159265\ldots$
The error is about 0.6%.

c. $p = 4 S_9 = 4\displaystyle\sum_{n=0}^{9} \dfrac{(-1)^n}{2n + 1}\left[\left(\dfrac{1}{2}\right)^{2n+1} + \left(\dfrac{1}{3}\right)^{2n+1}\right]$
$= 3.14159257\ldots$
$\pi = 3.14159265\ldots$
The answer differs from π by only 1 in the 7th decimal place. The improvement in accuracy is accounted for by the fact that the inverse tangent series converges much more rapidly for $x = 1/2$ and $x = 1/3$ than it does for $x = 1$.

39. Define $a_i(x) = \dfrac{f^{(i)}(a)}{i!}(x - a)^i$, the ith term of the general Taylor series. So, $f(x) = \sum_{i=0}^{\infty} a_i(x)$.
We must assume $\dfrac{d^n}{dx^n}\sum_{i=0}^{\infty} a_i(x) = \sum_{i=0}^{\infty}\dfrac{d^n}{dx^n}a_i(x)$, that is, the nth derivative of an infinite series is the infinite sum of the nth derivatives of the individual terms.
For $i < n$, $\dfrac{d^n}{dx^n}a_i(x) = \dfrac{f^{(i)}(a)}{i!} \cdot \dfrac{d^n}{dx^n}(x - a)^i$
$= \dfrac{f^{(i)}(a)}{i!} \cdot 0 = 0$.
For $i = n$, $\dfrac{d^n}{dx^n}a_i(x) = \dfrac{d^n}{dx^n}a_n(x) = \dfrac{f^{(n)}(a)}{n!} \cdot \dfrac{d^n}{dx^n}(x - a)^n$
$= \dfrac{f^{(n)}(a)}{n!} \cdot n!(x - a)^0$.
For $i > n$, $\dfrac{d^n}{dx^n}a_i(x) = \dfrac{f^{(i)}(a)}{i!} \cdot \dfrac{d^n}{dx^n}(x - a)^i$
$= \dfrac{f^{(i)}(x)}{i!}i \cdot (i - 1)(i - 2)\ldots(i - n + 1)(x - a)^{i-n} = 0$
for $x = a$.
So, $\dfrac{d^n}{dx^n}a_i(a) = 0$ for $i < n$ and $i > n$, and
$\dfrac{d^n}{dx^n}a_n(a) = f^{(n)}(a)$.

Thus, $\frac{d^n}{dx^n} \sum_{i=0}^{\infty} a_i(x)$ evaluated at $x = a$ is
$\frac{d^n}{dx^n} a_n(a) = f^{(n)}(a)$.

41. a. $r_n = \dfrac{n}{n + 1}|x - 1|$

b. $r_{10} = \dfrac{2}{11}$ for $x = 1.2$

$r_{10} = \dfrac{9.5}{11}$ for $x = 1.95$

$r_{10} = \dfrac{20}{11}$ for $x = 3$

c. $r = |x - 1|$

d. $r = 1.1$ for $x = -0.1$
$r = 1$ for $x = 0$
$r = 0.9$ for $x = 0.1$
$r = 0.9$ for $x = 1.9$
$r = 1$ for $x = 2$

e. The series converges to $\ln x$ whenever the value of x makes $r < 1$ and diverges whenever the value of x makes $r > 1$.

f. $r = |x - 1| < 1 \Rightarrow -1 < (x - 1) < 1 \Rightarrow 0 < x < 2$

Problem Set 12-6

1. a. $\dfrac{1}{4}x + \dfrac{2}{16}x^2 + \dfrac{3}{64}x^3 + \dfrac{4}{256}x^4 + \dfrac{5}{1024}x^5 + \cdots$

b. The open interval of convergence is $(-4, 4)$.

c. The radius of convergence $= 4$.

3. a. $(2x + 3) + \dfrac{(2x + 3)^2}{2} + \dfrac{(2x + 3)^3}{3} + \dfrac{(2x + 3)^4}{4} + \cdots$

b. The open interval of convergence is $(-2, -1)$.

c. The radius of convergence $= \frac{1}{2}$.

5. a. $(x - 8) + \dfrac{8}{2}(x - 8)^2 + \dfrac{27}{6}(x - 8)^3 + \dfrac{64}{24}(x - 8)^4 + \cdots$

b. The series converges for all values of x.

c. The radius of convergence is infinite.

7. $L = x^2 \cdot 0 < 1$ for all x and the series converges for all x.

9. $L = x^2 \cdot 0 < 1$ for all x and the series converges for all x.

11. $L = |x| \cdot 0 < 1$ for all x and the series converges for all x.

13. $L = |x| \cdot \infty$
$L = \infty$ for all $x \neq 0$; $L = 0$ at $x = 0$.
$\therefore$ the series converges only for $x = 0$.

15. $\cosh 10 = \sum_{n=0}^{\infty} \dfrac{1}{(2n)!} 10^{2n}$

$L = 10^2 \cdot 0 = 0 < 1 \Rightarrow$ series converges

17. a. The open interval of convergence is $(-1, 1)$.

b.

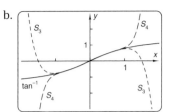

The graphs fit very well for $-1 < x < 1$. The partial sums diverge from $\tan^{-1} x$ for x outside this interval.

c. $S_3(0.1) = 0.09966865238095\ldots$

d. Tail $= 0.00000000011021\ldots$

e. The first term of the tail is $0.00000000011111\ldots$, which is larger than the tail.

19. a. $f(x) = x - \dfrac{1}{3}x^3 + \dfrac{1}{5 \cdot 2!}x^5 - \dfrac{1}{7 \cdot 3!}x^7 + \dfrac{1}{9 \cdot 4!}x^9 - \dfrac{1}{11 \cdot 5!}x^{11} + \cdots$

b.

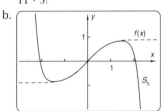

The partial sum is reasonably close for approximately $-1.5 < x < 1.5$.

c. The series converges for all x because $L = x^2 \cdot 1 \cdot 0 < 1$ for all x.

d. Erf x does seem to be approaching 1 as x increases, as shown by the following table generated by numerical integration.

x	erf x
1	$0.8427007929\ldots$
2	$0.9953222650\ldots$
3	$0.9999779095\ldots$
4	$0.9999999845\ldots$
5	$0.9999999999\ldots$

21. a. Given $L = \lim_{n \to \infty} \sqrt[n]{t_n}$ where $L < 1$. By the definition of limit as $n \to \infty$, there is a number $k > 0$ for any $\epsilon > 0$ such that if $n > k$, then $\sqrt[n]{t_n}$ is within ϵ units of L. Thus, $\sqrt[n]{t_n} < L + \epsilon$, Q.E.D.

b. $L < 1 \Rightarrow 1 - L > 0$
So, take any $\epsilon < 1 - L$
$\Rightarrow L + \epsilon < L + 1 - L$
$\Rightarrow L + \epsilon < 1$

c. For all integers $n > k$,
$0 \leq \sqrt[n]{t_n} < L + \epsilon \Rightarrow 0 \leq t_n < (L + \epsilon)^n$

and $(L + \epsilon)^n < (L + \epsilon)^{n-k}$ for all $n > k$ because
$L + \epsilon < 1$, so $0 \le t_n < (L + \epsilon)^{n-k}$, Q.E.D.

d. Because $0 \le t_n < (L + \epsilon)^{n-k}$ for all $n > k$, it follows
that the tail after t_n satisfies
$0 \le t_{n+1} + t_{n+2} + t_{n+3} + \cdots$
$< (L + \epsilon)^{n+1-k} + (L + \epsilon)^{n+2-k} + (L + \epsilon)^{n+3-k} + \cdots$
$= (L + \epsilon)^{n+1-k}[1 + (L + \epsilon) + (L + \epsilon)^2 + \cdots]$
which converges because $L + \epsilon < 1$.

e. The tail of the series is increasing and is bounded
above by
$(L + \epsilon)^{n+1-k}[1 + (L + \epsilon) + (L + \epsilon)^2 + \cdots]$
$= \dfrac{(L + \epsilon)^{n+1-k}}{1 - (L + \epsilon)}$
So the series converges, Q.E.D.

23. The open interval of convergence is $(0, 2)$.

25. $L = 0$ if $x = 0$ and is infinite if $x \ne 0$.
∴ the series converges only if $x = 0$.

Problem Set 12-7

1. a. $S_5 = 6 - 3 + 1 - \dfrac{1}{4} + \dfrac{1}{20} = 3.8$

b. Tail $= -\dfrac{6}{6!} + \dfrac{6}{7!} - \dfrac{6}{8!} + \cdots$

c. Hypotheses: (1) signs are strictly alternating,
(2) $|t_n|$ are strictly decreasing, and
(3) $\lim_{n \to \infty} t_n = 0$.
Upper bound is $1/120$.

d. Absolute convergence means that $\sum_{n=1}^{\infty} |t_n|$
converges.
If a convergent series were not absolutely
convergent, it would be called *conditionally
convergent*.

e. When you show absolute convergence, you find the
partial sums of $|t_n|$. The parital sums must be
increasing because $|t_n|$ is positive. $|t_n|$ is
decreasing because the series is convergent.

3. a. $S_5 = 1.463611\ldots$

b. Tail $= \dfrac{1}{36} + \dfrac{1}{49} + \dfrac{1}{64} + \cdots$
The graph shows the tail bounded above by
$\int_5^{\infty} (1/x^2)\,dx = 0.2$.

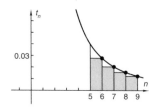

The series converges because the sequence of
partial sums is increasing and bounded above
by 0.2.

c. $1/1001 < R_{1000} < 1/1000$
$S = 1.644934\ldots$
The answer is correct to at least three decimal
places.

d. About 20 million terms

5. a. $1 - \dfrac{1}{2} + \dfrac{1}{3} - \dfrac{1}{4} + \cdots$
The series converges because it meets these
hypotheses:
(1) strictly alternating signs, (2) strictly decreasing
$|t_n|$, and (3) $t_n \to 0$.

b. $\int_1^{\infty} (1/x)\,dx = \lim_{b \to \infty} (\ln b - \ln 1) = \infty$
The series $\sum_{n=1}^{\infty} |t_n|$ diverges, so the given series
does not converge absolutely.

c. $S_{1000} = 0.692647\ldots$; $S_{1001} = 0.693646\ldots$;
$\ln 2 = 0.693147\ldots$
$|S_{1000} - \ln 2| = 0.0004997\ldots$, $|S_{1001} - \ln 2| =$
$0.0004992\ldots$, $|t_{1001}| = 1/1001 = 0.00009900\ldots$
∴ both partial sums are within $|t_{1001}|$ of $\ln 2$.

d. No term is left out.
No term appears more than once.
The series converges to $\frac{1}{2} \ln 2$.
Conditional convergence means that whether the
series converges, and, if so, what value it
converges to, depends on the condition that you
do not rearrange the terms.

7. a. $t_3 = -\dfrac{1}{7!} 0.6^7 = -0.00000555428571\ldots$

b. $S_1(0.6) = 0.6 - \dfrac{1}{3!} 0.6^3 = 0.564$

$S_2(0.6) = 0.6 - \dfrac{1}{3!} 0.6^3 + \dfrac{1}{5!} 0.6^5 = 0.564648$

c. $R_1 = \sin 0.6 - S_1(0.6) = 0.0006424\ldots$
$R_2 = \sin 0.6 - S_2(0.6) = -0.0000055266\ldots$
$|t_2| = 0.000648$
∴ $|R_1| < |t_2|$
$|t_3| = 0.0000055542\ldots$
∴ $|R_2| < |t_3|$

d. The terms are strictly alternating in sign.
The terms are strictly decreasing in absolute value.
The terms approach zero for a limit as $n \to \infty$.
∴ the series converges by the alternating
series test.

9. a. $\dfrac{1}{3} + \dfrac{1}{8} + \dfrac{1}{15} + \dfrac{1}{24} + \cdots$
Compare with the p-series with
$p = 2 \lim_{n \to \infty} \dfrac{1/(n^2 - 1)}{1/n^2} = \lim_{n \to \infty} \dfrac{n^2}{n^2 - 1} = 1$, a positive
real number.
∴ the series converges by the limit comparison
test.

b. The *p*-series with $p = 2$ begins $\frac{1}{4} + \frac{1}{9} + \frac{1}{16} + \frac{1}{25}$.
These terms form a lower bound, not an upper bound, so the direct comparison test fails.

c. If *n* started at 1, the first term would be 1/0, which is infinite.

11. a. $\frac{1}{1} + \frac{2}{1} + \frac{3}{2} + \frac{4}{6} + \frac{5}{24} + \cdots$

$L = \lim_{n \to \infty} \left| \frac{n+1}{n!} \cdot \frac{(n-1)!}{n} \right| = 0$

∴ the series converges because $L < 1$.

b. $\frac{1}{1} + \frac{2}{2} + \frac{3}{6} + \frac{4}{24} + \cdots = 1 + 1 + \frac{1}{2!} + \frac{1}{3!} + \cdots$

This is the Maclaurin series that converges to e^1.

c. $L = \infty$

∴ the test fails because the limit of the ratio is infinite.

13. Divergent harmonic series

15. It converges because it is an alternating series meeting the three hypotheses.

17. It converges because it is a geometric series with common ratio 1/4, which is less than 1 in absolute value.

19. Converges by comparison with:
Geometric series with $t_0 = 1$ and $r = 1/6$

21. Diverges; use an integral test or compare with a harmonic series

23. It converges because it is an alternating series meeting the three hypotheses.

25. It diverges by the ratio test.

27. It diverges because t_n does not approach zero.

29. It diverges by the integral test.

31. It diverges because t_n does not approach zero.

33. $x = 1: -1 + 1 - 1 + 1 - 1 + \cdots$
Diverges by the *n*th term test.
$x = 9: 1 + 1 + 1 + 1 + 1 + \cdots$
Diverges by the *n*th term test.
Complete interval is $(1, 9)$.

35. $x = -4: -\frac{1}{\sqrt{1}} + \frac{1}{\sqrt{2}} - \frac{1}{\sqrt{3}} + \frac{1}{\sqrt{4}} - \cdots$
Converges by the alternating series test.
$x = -2: -\frac{1}{\sqrt{1}} + \frac{1}{\sqrt{2}} + \frac{1}{\sqrt{3}} + \frac{1}{\sqrt{4}} + \cdots$
Diverges. *p*-series with $p = 0.5$, which is less than 1.
Complete interval is $[-4, -2)$.

37. The interval of convergence is $(2, 4)$.

39. The interval of convergence is $[-1, 1)$.

41. The interval of convergence is $[-6, -4]$.

43. The interval of convergence is $[-1, 1]$.

45. The intervals of convergence are $(-\infty, -4)$ and $(4, \infty)$.

47. a. Assume all the blocks have equal mass = *m* with the center of mass at the center of the block and equal length = *L*.
Write H_n = the distance the *n*th block overhangs the $(n + 1)$th block. ($n = 1$ for the top block.)
Note that according to the rule, H_n = the distance between the rightmost edge of the *n*th block and the center of mass of the pile of the top *n* blocks.
Now the center of mass of the *n*th block is $\frac{1}{2}L$ units from its rightmost edge, and the center of mass of pile of the top $n - 1$ blocks is 0 units from (i.e., right on top of) the edge of the *n*th block according to the rule.
Therefore the center of mass of the pile of the top *n* blocks is $\frac{1}{nm}\left[\frac{1}{2}L \cdot m + 0 \cdot (n-1)m\right]$ units from the edge of the *n*th block, that is, $H_n = \frac{1}{2n}L$, Q.E.D.

b. 5 blocks

c. To make a pile with overhang *H*, find an *n* such that $1 + \frac{1}{2} + \frac{1}{3} + \cdots + \frac{1}{n-1} > \frac{2H}{L}$ (this is possible since the harmonic series diverges to infinity). Then a stack of *n* blocks will have total overhang.
$H_1 + \cdots + H_{n-1}$

$= \frac{1}{2}L + \frac{1}{4}L + \frac{1}{6}L + \cdots + \frac{1}{2(n-1)}L$

$= \frac{1}{2}L\left(1 + \frac{1}{2} + \frac{1}{3} + \cdots + \frac{1}{n+1}\right) < \frac{1}{2}L \cdot \frac{2H}{L} = H$

(The achieved overhang is greater than *H*, so one may pull blocks slightly back—moving blocks back can only make the pile more stable—until the overhang equals *H* exactly.)

d. Slightly more than two-and-a-quarter card lengths

Problem Set 12-8

1. a. $S_5(4) = 27.2699118\ldots$

b. $S_5(4)$ is within 2 of cosh 4 in the units digit.

c. $\cosh 4 - S_5(4) = 0.0383\ldots$, which is well within the upper bound found by Lagrange's form.

3. a. $S_{14}(3) = 20.0855234\ldots$

b. $S_{14}(3)$ is within 3 units of e^3 in the 4th decimal place.

c. $e^3 = 20.085536923\ldots$
$S_{14}(3) = 20.085523458\ldots$
$e^3 - S_{14}(3) = 0.00001346\ldots$, which is within the upper bound found by Lagrange's form.

5. At least 7 terms ($n = 6$)

7. At least 32 terms

9. $c = \cosh^{-1} 1.0309\ldots = 0.2482\ldots$, which is between 0 and 2.

11. $\cos 2.4 = 1 - 2.88 + 1.3824 - 0.2654208 + 0.0273004... - \cdots$

The terms are strictly alternating. They are decreasing in absolute value after t_1, and they approach zero for a limit as $n \to \infty$.
Therefore the hypotheses of the alternating series test apply.
8 terms ($n = 7$)

13. a. $S_{10} = 1.19753198...$
$0.00413223... < R_{10} < 0.005$
$S \approx 1.20209810...$
Maximum error: 0.00043388... (about 3 decimal places)

b. By using the method of part a, you find you should use 46 terms.
By calculating exactly, you would use approximately 317 terms, considerably more than the 46 terms with the other method.

15. a. $S_{10} = 1.9817928...$
The series converges because the terms of the tail starting at t_1 are bounded above by the convergent p-series with $p = 2$.

b. $0.0906598... < R_{10} < 0.0996686...$
$S \approx 2.0769570...$
2 decimal places
100 terms

17. By Lagrange form, $|R_{10}| < 0.0004617...$;
by geometric series, $|R_{10}| < 0.00006156...$
The geometric series gives a better estimate of the remainder than does the Lagrange form.

19. a. $b = 4.9557730...$ radians

b. $c = -1.32741228...$ radians

c. $d = 0.243384039...$ radians

d. $|R_5(x)| < |t_6(\pi/4)| = 3.8980... \times 10^{-13}$,
which is small enough to guarantee that $\sin x$ will be correct to 10 decimal places.
Direct calculation would take about 349 terms.

e. Answers will vary.

21. a. There is a number $x = c$ in (x, a) such that
$$f''(c) = \frac{f'(x) - f'(a)}{x - a}$$
$\Rightarrow f'(x) = f'(a) + f''(c)(x - a)$, Q.E.D.

b. $f(x) = f'(a)x + f''(c) \cdot \frac{1}{2}(x - a)^2 + f(a) - f'(a)a$
$f(x) = f(a) + f'(a)(x - a) + \frac{1}{2}f''(c)(x - a)^2$, Q.E.D.

c. There is a number $x = c$ in (a, x) such that
$$f'''(c) = \frac{f''(x) - f''(a)}{x - a}$$
$\Rightarrow f''(x) = f''(a) + f'''(c)(x - a)$
Integrate once to get $f'(a)$.
$f'(x) = f'(a) + f''(a)(x - a) + \frac{1}{2}f'''(c)(x - a)^2$

Integrate again to get $f(x)$.
$f(x) = f'(a)x + \frac{1}{2!}f''(a)(x - a)^2 + \frac{1}{3!}f'''(c)(x - a)^3$
$+ f(a) - f'(a)a$
$f(x) = f(a) + f'(a)(x - a) + \frac{1}{2!}f''(a)(x - a)^2$
$+ \frac{1}{3!}f'''(c)(x - a)^3$, Q.E.D.

d. Mathematical induction

23. Using the Lagrange form of the remainder, the value of e^x is given *exactly* by
$e^x = \sum_{n=0}^{k} \frac{1}{n!}x^n + R_k(x)$, where $R_k(x) = \frac{f^{(k+1)}(c)}{(k+1)!}x^{k+1}$
and c is between 0 and x.
$$|R_k(x)| \le \frac{M}{(k+1)!}|x|^{k+1}$$
Because all derivatives of e^x equal e^x, the value of M for any particular value of x is also e^x, which is less than 3^x if $x \ge 0$ (or 1 if $x < 0$).
$\lim_{k \to \infty} |R_k(x)| < \lim_{k \to \infty} \frac{3^x}{(k+1)!}|x|^{k+1}$, which
approaches 0 as $k \to \infty$ by the ratio technique.
Because the remainder approaches zero as n approaches infinity, e^x is given exactly by
$e^x = \sum_{n=0}^{\infty} \frac{1}{n!}x^n$, Q.E.D.

Problem Set 12-9

R0. Answers will vary.

R1.

$P_5(x)$ and $P_6(x)$ are close to $f(x)$ for x between about -0.7 and 0.6 and bear little resemblance to $f(x)$ beyond ± 1.
$P_5(0.4) = 14.93856$
$P_6(0.4) = 14.975424$
$f(0.4) = 15$
$\therefore P_6(0.4)$ is closer to $f(0.4)$ than $P_5(0.4)$ is, Q.E.D.
$P_5(0) = 9 = f(0)$
$P_5'(0) = 9 = f'(0)$
$P_5''(0) = 18 = f''(0)$
$P_5'''(0) = 54 = f'''(0)$
$P_n(x)$ is a subseries of a geometric series.

R2. a. About 19.5 mm increase in 10 days
About 30 mm increase eventually

b. The state must invest \$81,754.00 now in order to make the last payment.
It must invest \$4,182,460.05 now to make all 19 payments.

R3. $c_0 = 7, c_1 = 21, c_2 = 31.5, c_3 = 31.5$

R4. a. $e^{0.12} = 1.127496851...$
$S_3(0.12) = 1.127488$, which is close to $e^{0.12}$.

b. $\cos 0.12 = 0.9928086358538...$
$S_3(0.12) = 0.9928086358528$, which is close.

c. $\sinh(0.12) = 0.1202882074311...$
$S_3(0.12) = 0.1202882074310...$, which is close.

d. $\ln 1.7 = 0.530628251...$
$S_{20}(1.7) = 0.530612301...$, which is close.
$\ln 2.3 = 0.83290912...$
$S_{20}(2.3) = -4.42067878...$, which is not close.

R5. a. A Maclaurin series is a Taylor series expanded about $x = 0$.

b. $\ln(x + 1) = x - \dfrac{1}{2}x^2 + \dfrac{1}{3}x^3 - \dfrac{1}{4}x^4 + \cdots$

c. $\displaystyle\int \ln(x+1)\,dx = \dfrac{1}{2}x^2 - \dfrac{1}{3 \cdot 2}x^3 + \dfrac{1}{4 \cdot 3}x^4 - \cdots + C$

d. $\displaystyle\int \ln(x+1)\,dx = (x+1)\ln(x+1) - (x+1) + C_1$
$= x\ln(x+1) + \ln(x+1) - x + C\,(C = C_1 - 1)$
$= \dfrac{1}{2}x^2 - \dfrac{1}{3 \cdot 2}x^3 + \dfrac{1}{4 \cdot 3}x^4 - \cdots + C,$
which is the same as the series in part c.

e. $\displaystyle\int_0^x t\cos t^2\,dt$
$= \dfrac{1}{2}x^2 - \dfrac{1}{6 \cdot 2!}x^6 + \dfrac{1}{10 \cdot 4!}x^{10} - \dfrac{1}{14 \cdot 6!}x^{14} + \cdots$

f. $\tan^{-1} x = \displaystyle\int_0^x \dfrac{1}{1 + t^2}\,dt$
$= \displaystyle\int_0^x [1 - t^2 + (t^2)^2 - (t^2)^3 + (t^2)^4 - \cdots]\,dt\,(|t| \le 1)$
$= x - \dfrac{1}{3}x^3 + \dfrac{1}{5}x^5 - \dfrac{1}{7}x^7 + \dfrac{1}{9}x^9 - \cdots$

g. $f(x) = 5 + 7(x-3) - 3(x-3)^2 + 0.15(x-3)^3 + \cdots$

R6. a. $\displaystyle\sum_{n=1}^{\infty} = (-3)^{-n}(x-5)^n$
$= -\dfrac{1}{3}(x-5) + \dfrac{1}{9}(x-5)^2 - \dfrac{1}{27}(x-5)^3 + \cdots$

b. The open interval of convergence is (2, 8).
The radius of convergence is 3.

c. $L = x^2 \cdot 0 < 1$ for all x.
The series converges for all x, Q.E.D.

d. $e^{1.2} = 1 + 1.2 + \dfrac{1}{2!}(1.2)^2 + \dfrac{1}{3!}(1.2)^3$
$+ \dfrac{1}{4!}(1.2)^4 + \cdots$
Error $= e^{1.2} - S_4(1.2) = 0.02571692...$
First term of tail is $t_5 = 0.020736$
The error is greater than t_5, but not much greater.

e.

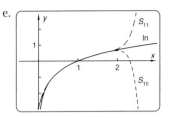

The open interval of convergence is (0, 2). Both partial sums fit ln well within this interval. Above $x = 2$, the partial sums diverge rapidly to $\pm\infty$. Below $x = 0$, the partial sums give answers, but there are no real values for $\ln x$.

R7. a. $S_{10} = 4463.129088$

b. $S - S_{10} = 536.870912$, which differs from the limit by about 10.7%.

c. "Tail"

d. "Remainder"

e. $0.004132... < R_{10} < 0.005$
The series converges because the tail is bounded above by 0.005.
$S = 1.202098...$, accurate to about three decimal places.

f. $-\dfrac{1}{4} + \dfrac{1}{3} + \dfrac{1}{22} + \dfrac{1}{59} + \cdots$
$L = \displaystyle\lim_{n\to\infty} \dfrac{1/(n^3 - 5)}{1/n^3} = \lim_{n\to\infty} \dfrac{n^3}{n^3 - 5} = 1$
$\therefore$ the series converges because L is a positive real number.
The terms of the F series begin
$\dfrac{1}{1} + \dfrac{1}{8} + \dfrac{1}{27} + \dfrac{1}{64} + \cdots$
Although the F series converges, its terms (after t_1) are less, not greater, than the corresponding terms of the S series, so the comparison test is inconclusive.

g. $2/1! + 4/2! + 8/3! + 16/4! + 32/5! + = 2 + 2 +$
$1.3333... + 0.6666... + 0.2666... + \cdots = \displaystyle\sum_{n=1}^{\infty} 2^n/n!$
The terms are decreasing starting at t_2, which can be seen numerically above or algebraically by the fact that the next term is formed by multiplying the numerator by 2 and the denominator by more than 2.
R_1 is bounded by the geometric series with first term 2 and common ratio $1.3333.../2 = 2/3$. Because |common ratio| is less than 1, the geometric series converges (to $2/(1 - 2/3) = 6$). Thus, the tail after the first partial sum is bounded above by a convergent geometric series, Q.E.D.

h. The given series converges because, as written, it meets the three hypotheses of the alternating series test. It does not converge absolutely because replacing all $-$ signs with $+$ signs gives the divergent harmonic series.

The given series is the Taylor series for $\ln x$ expanded about $x = 1$ and evaluated at $x = 2$. The remainders approach zero, so the series converges to $\ln 2$.

The series can be rearranged like this:

$$= \frac{1}{2}\left(\frac{1}{1} - \frac{1}{2} + \frac{1}{3} - \frac{1}{4} + \frac{1}{5} - \frac{1}{6} + \cdots\right)$$

The series in parentheses is the original series that converges to $\ln 2$. So the series as rearranged converges to $0.5 \ln 2$, Q.E.D.

  i. 1/10001

  j. i. [2.9, 3.1]

    ii. $(-3, 1]$

  k. i. The tail after S_0 is bounded above by the convergent geometric series with first term 10 and common ratio 0.5. Thus, the series converges. (Other justifications are possible.)

    ii. It diverges because $t_n \rightarrow 0.2$, not 0, as $n \rightarrow \infty$.

    iii. Converges. The general term can be rewritten $3(2/5)^n$, so the series is a convergent geometric series with common ratio $r = 2/5$.

    iv. Diverges. p-series with $p = 1/3$, which is not greater than 1.

    v. Converges by the ratio technique.

$$\lim_{n \to \infty} \left| \frac{(n+4)!}{3! \cdot (n+1)! \cdot 3^{n+1}} \cdot \frac{3! \cdot n! \cdot 3^n}{(n+3)!} \right|$$

$$= \frac{1}{3} \lim_{n \to \infty} \frac{n+4}{n+1} = \frac{1}{3} \cdot 1 < 1$$

R8. a. The error is less than 0.03.

  b. At least 34 terms ($n = 33$)

  c. Using the Lagrange form of the remainder, the value of $\cosh 4$ is given *exactly* by

$$\cosh 4 = \sum_{n=0}^{k} \frac{1}{(2n)!} \cdot 4^{2n} + R_k(4), \text{ where}$$

$$R_k(4) = \frac{f^{(2k+2)}(c)}{(2k+2)!} \cdot 4^{2k+2} \text{ and } c \text{ is between 0 and 4.}$$

$$|R_k(4)| \le \frac{M}{(2k+2)!} \cdot 4^{2k+2}$$

$$\lim_{k \to \infty} |R_k(4)| = 0$$

Because the remainder approaches zero as k approaches infinity, $\cosh 4$ is given exactly by

$$\cosh 4 = \sum_{n=0}^{\infty} \frac{1}{(2n)!} \cdot 4^{2n}, \text{ Q.E.D.}$$

  d. $c = \cosh^{-1} 1.00328... = 0.0809...$, which is in the interval (0, 0.6).

  e. At least 35 terms

f. $0.000002512... < R_{50} < 0.000002666...$

The series converges because the sequence of partial sums is increasing, and the tail after S_{50} is bounded above by 0.000002512....
$S = 1.082323235...$
About 7 decimal places

Problem Set 12-10

Cumulative Review Number 1

1. Limit, derivative, indefinite integral, definite integral: See the definitions in the text.

2. a. Continuity of a function at a point: See Section 2-4.

  b. Continuity of a function on an interval: See Section 2-4.

  c. Convergence of a sequence: A sequence converges if and only if $\lim_{n \to \infty} t_n$ exists.

  d. Convergence of a series: A series converges if and only if the sequence of partial sums converges.

  e. Natural logarithm: See Section 3-9.

  f. Exponential: $a^x = e^{x \ln a}$

3. a. Mean value theorem: See Section 5-5.

  b. Intermediate value theorem: See Section 2-6.

  c. Squeeze theorem: See Section 3-8.

  d. Uniqueness theorem for derivatives: See Section 6-3.

  e. Limit of a product property: See Section 2-3.

  f. Integration by parts formula: See Section 9-2.

  g. Fundamental theorem of calculus: See Section 5-6.

  h. Lagrange form of the remainder: See Section 12-8.

  i. Chain rule for parametric functions: See Section 4-7.

  j. Polar differential of arc length: See Section 8-7.

4. a. $f'(x) = \sqrt{1 + \operatorname{sech} x}$

  b. $f'(x) = a^x \ln a$

  c. $f'(x) = ax^{a-1}$

  d. $f'(x) = x^x \ln x + x^x$

  e. $\frac{1}{15} e^{6x} \sin 3x + \frac{2}{15} e^{6x} \cos 3x + C$

  f. $\frac{1}{6} \cosh^6 x + C$

  g. $\frac{1}{2} \sec x \tan x + \frac{1}{2} \ln |\sec x + \tan x| + C$

  h. $\frac{1}{5} \ln |\sin 5x| + C$

  i. Limit $= -\frac{49}{26}$

  j. Limit $= e^{-3} = 0.04978...$

5. a.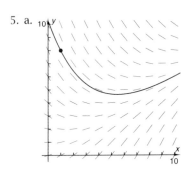

b. If $x = 9$, $y \approx 5.413...$, which agrees with the graph.

6. a. $p = k(40 - y)$

b. $A = 1066.6...$ yd^2

c. $F = 17,066.6...k$ lb

d. $M = 292,571.4...k$ lb-yd

e. The center of pressure is at $(0, 17\frac{1}{7})$.

7. a. $z = 30 - 0.5y$

b. Maximum at $y = 20$; minimum at $y = 0$

c. 3840 truckloads

d. $L = 92.9356... \approx 92.9$ yd

8. Speed $= 2.9943... \approx 2.99$ ft/s

9. Si $t = t - \dfrac{1}{3.3!}t^3 + \dfrac{1}{5.5!}t^5 - \dfrac{1}{7.7!}t^7 + \cdots +$

$\dfrac{(-1)^n}{(2n + 1)(2n + 1)!}t^{2n+1} + \cdots$

$L < 1$ for all values of t, so the series converges for all values of t.

The third partial sum is $S_2(0.6) = 0.5881296$ is correct within ± 1 in the sixth decimal place.

Si $0.6 \approx 0.588128809...$

10. $A = 103.6725... \approx 103.7$ ft^2 (exactly 33π)

11. At $t = 10$, $V = 253.9445... \approx 253.9$ million gal.

Cumulative Review Number 2

1. Derivative: See Sections 3-2 and 3-4.

2. Definite integral: See Section 5-4.

3. Mean value theorem: See Section 5-5.

4. $f'(x) = g(x)$

5. $\dfrac{1}{6}\tanh^6 x + C$

6. $\dfrac{1}{2}x \cosh 2x - \dfrac{1}{4}\sinh 2x + C$

7. $-\ln|x + 3| + 4\ln|x - 2| + C$

8. $x + \dfrac{1}{3.3!}x^3 + \dfrac{1}{5.5!}x^5 + \dfrac{1}{7.7!}x^7 + \cdots + C$

9. The open interval of convergence is $2 < x < 8$.

10. 500

11. $\bar{y} = 39$

12. $f(4) = 16$
$f(3.99) = 15.9201$, which is within 0.08 unit of 16.
$f(4.01) = 16.0801$, which is not within 0.08 unit of 16.
Thus, $\delta = 0.01$ is not small enough to keep $f(x)$ within 0.08 unit of 4.

13. $V = \displaystyle\int_2^{10} ; A\, dx \approx 2140$ ft^3

14. $A = 6.2831 \approx 6.28$ ft^2 (exactly 2π)

15. $A = 17.6021... \approx 17.6$ square units

16. $A = 256$ ft^2

17. $L = 42.5483... \approx 42.55$ ft

18. $F = 113,595.73... \approx 113,600$ lb

19. Limit $= 0$

20. There is a maximum at $x = e$ because y' goes from positive to negative there.

21. There is a point of inflection at $x \approx 4.48$ ft because y'' changes sign there.

22.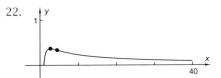

23. $\ln x = (x - 1) - \dfrac{1}{2}(x - 1)^2 + \dfrac{1}{3}(x - 1)^3 - \cdots +$

$\dfrac{(-1)^{n+1}}{n}(x - 1)^n + \cdots$

$L = |x - 1|$

$L < 1 \iff 0 < x < 2$

At $x = 0$, the series is $-1 - \frac{1}{2} - \frac{1}{3} - \frac{1}{4} - \cdots$, which is a divergent harmonic series.

At $x = 2$, the series is $1 - \frac{1}{2} + \frac{1}{3} - \frac{1}{4} + \cdots$, which converges because it meets the three hypotheses of the alternating series test.

$\therefore$ interval of convergence is $0 < x \leq 2$, Q.E.D.

24. 46 terms

25. If the velocity is 0 ft/s at time $t = 0$, the ship speeds up, approaching approximately 34 ft/s asymptotically as t increases.
If the velocity is 50 ft/s at time $t = 0$, the ship slows down, again approaching 34 ft/s asymptotically as t increases.

26. $\vec{a} = (-1/t^2)\vec{i} + (-4\sin 2t)\vec{j}$

Cumulative Review Number 3

1.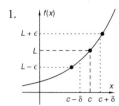

2. See Sections 3-2 and 3-4 for definitions of *derivative*.
   Graphical meaning: slope of tangent line
   Physical meaning: instantaneous rate of change

3. $g(x) = \int f(x)\,dx$ if and only if $g'(x) = f(x)$.

4. $\int_r^s f(t)\,dt = \lim_{\Delta t \to 0} L_n = \lim_{\Delta t \to 0} U_n$ where L_n and U_n are lower and upper Riemann sums, respectively, provided the two limits are equal.

5. l'Hospital's rule
   limit $= -0.2$

6. $y' = \sec^2(\sin 5x) \cdot 5\cos 5x$; chain rule

7. $y' = y\left(\dfrac{5}{5x-3} + \dfrac{8}{2x+7} + \dfrac{1}{x-9}\right)$

8. $y' = \dfrac{1}{1+x^2}$

9. $\dfrac{1}{8}\sin^8 x + C$

10. $\dfrac{1}{2}x\sqrt{x^2+9} + \dfrac{9}{2}\ln\left|\sqrt{x^2+9}+x\right| + C$

11. $5\ln|x+3| - 2\ln|x-1| + C$

12. $x\sin^{-1}x + \sqrt{1-x^2} + C$

13. Fundamental theorem of calculus
    See Section 5-6 for statement.

14. See Figure 5-5b.

15. $f'(x) = h(x)$

16. The only point of inflection is at $x = 2$.

17. $L \approx 2.3516...$

18. a. $\int_0^{16} x^{-3/4}\,dx = \lim_{a \to 0^+} 4x^{1/4}\Big|_a^{16} = \lim_{a \to 0^+}(8 - 4a^{1/4}) = 8$, so the interval converges to 8.

    b. Average value $= 0.5$

19. $A \approx 13.3478...$

20. $\vec{v}(1) = 2\vec{i} - 3\vec{j}$
    Speed $= \sqrt{13} = 3.6055...$
    The distance from the origin is decreasing at $2.2135...$.

21. $V \approx 3.5864...$

22. a. $A' = 0 \iff x = 2$
    $A(0) = 0, A(4) = 0, A(2) > 0$
    Thus, the maximum area is at $x = 2$, Q.E.D.

    b. The maximum-volume cylinder is at $x = 2\frac{2}{3}$.

23. $V \approx 394\frac{2}{3}$ ft^3

24. a. $f(x) = \int_0^x e^{-t^2}\,dt$
    $= x - \dfrac{1}{3}x^3 + \dfrac{1}{5 \cdot 2!}x^5 - \dfrac{1}{7 \cdot 3!}x^7 + \dfrac{1}{9 \cdot 4!}x^9 - \cdots$

    b. $L = x^2 \cdot 0 < 1$ for all values of x, and thus the series converges for all values of x, Q.E.D.

Glossary

The following are descriptions of the major terms used in calculus, along with references to page numbers where formal definitions and statements can be found. For references to the many other significant terms used in this text, please see the index.

Absolute maximum (pp. 95, 372, 379): The highest value that a function attains. Also called the *global maximum*.

Absolute minimum (p. 379): The lowest value that a function attains. Also called the *global minimum*.

Absolutely convergent (p. 628): Describes a series that converges even if all terms are made positive.

Acceleration (pp. 93, 313): The instantaneous rate of change of velocity.

Alternating harmonic series (p. 635): A harmonic series in which the terms have alternating signs.

Antiderivative (pp. 92, 96, 198, 313): $g(x)$ is an antiderivative of $f(x)$ if and only if $g'(x) = f(x)$. Also called an *indefinite integral*.

Average value of a function (p. 512): The integral of $f(x)$ from $x = a$ to $x = b$ divided by the quantity $(b - a)$.

Calculus: Literally means "calculation." Evolved from the same root word as "calcium" because centuries ago calculations were done using pebbles (calcium carbonate). An appendix to Isaac Newton's *Principia* is entitled *The Calculus of Infinitesimals*, which means calculating with quantities that approach zero as x approaches a particular value.

Carrying capacity (pp. 348, 352): The maximum population that can be sustained by a particular environment.

Catenary (p. 469): The shape formed by a chain that hangs under its own weight.

Center of mass (pp. 558): The point at which the mass of a solid can be concentrated to produce the same first moment of mass.

Center of pressure (p. 568): The point at which the entire force on a surface can be concentrated to produce the same force as the first moment of force.

Centroid (p. 558): An object's geometric center, found by dividing the first moment of area or volume with respect to an axis by the area or volume.

Chain rule (pp. 103, 312): The method for finding the derivative of a composite function, namely, the derivative of the outside function with respect to the inside function multiplied by the derivative of the inside function with respect to x.

Composite function (pp. 101, 102, 312): A function of the form $f(g(x))$, where function g is performed on x, then function f is performed on $g(x)$.

Concave up (or down) (pp. 371, 379): The graph of a function (or a portion of the graph between two asymptotes or two points of inflection) is called *concave up* when its "hollowed out" side faces up; it is called *concave down* when its "hollowed out" side faces down.

Conditionally convergent (p. 627): Describes a series that is not absolutely convergent, but may converge to different numbers depending on the order in which the terms are arranged.

Constant of integration (p. 96): Two antiderivatives of the same function differ by a constant at most. The constant term of an antiderivative equation is called the *constant of integration.*

Continuity (p. 47): A function is continuous at $x = c$ if and only if $f(c)$ is the limit of $f(x)$ as x approaches c.

Converge (pp. 483, 589, 593): An improper integral or series converges if it approaches a finite number as a limit.

Corner (p. 379): A point on a graph at which the function is continuous, but there is a step change in the first derivative.

Critical point (pp. 372, 379): A point on a graph at which the first derivative is either zero or undefined.

Cusp (pp. 47, 379): A point on a graph at which the function is continuous, but the derivative is discontinuous.

Cylindrical shells (p. 395): The slicing of a solid of revolution into thin shells so that each point in the shell is virtually the same distance from the axis of rotation as the sample point is. Cylindrical shells are used for setting up integrals for calculating the volume, mass, moment, and so on of a solid object.

Definite integral (pp. 15, 207, 313): *Physical meaning:* The product (dependent variable)(change in independent variable) for a function in which the dependent variable may take on different values as the independent variable changes throughout an interval. *Graphical meaning:* The area of the region under the graph of $f(x)$ from the x-value at the beginning of an interval to the x-value at the end of that interval.

Dependent variable (p. 4): A variable whose values depend on the value of another variable.

Derivative (pp. 4, 10): *Physical meaning:* The derivative of a function f at $x = c$ is the instantaneous rate of change of $f(x)$ with respect to x at $x = c$. *Graphical meaning:* The slope of the line tangent to the graph at $x = c$.

Difference quotient (p. 74): The ratio (change in $f(x)$)/(change in x). The limit of a difference quotient as the change in x approaches zero is the derivative.

Differentiability (p. 154): The property possessed by a function at $x = c$ if $f'(c)$ exists. Function f is differentiable on an interval if and only if $f'(x)$ exists for all values of x in that interval.

Differential (p. 190): If $y = f(x)$, then the differential dx is the same quantity as Δx, a change in x; and the differential dy is equal to $f'(x)\,dx$. Thus, the quotient $dy \div dx$ is equal to the derivative, $f'(x)$. The differential dy is also the change in y along a tangent to the graph, rather than along the graph itself.

Differential calculus: A term for calculus of derivatives only.

Differential equation (pp. 226, 269, 318): An equation that contains the derivative of a function. A **solution** of a differential equation is a function whose derivative is the differential equation.

Differentiation (p. 87): The process of finding the derivative of a function.

Discrete points (p. 60): Points that are disconnected.

Displacement (pp. 93, 501): The directed distance of an object from a given reference point at a given time.

Diverge (pp. 483, 589, 593): An improper integral or series diverges if it does not approach a finite number as a limit.

Dot product (p. 524): The dot product of two vectors is the product of their magnitudes and the cosine of the angle formed when the vectors are placed tail-to-tail.

e (p. 289): A naturally occurring constant equal to 2.71828... used as the base for the natural logarithm and natural exponential function to make the calculus of these functions simpler.

Euler's method (p. 341): A numerical method for solving a given differential equation by assuming the graph follows tangent segments for short distances from point to point.

Even function (pp. 136, 230): A function f that has the property $f(-x) = f(x)$ for all x in its domain.

Existence theorem (p. 62): A type of theorem that asserts the existence of a number or other mathematical object having a certain property or satisfying a required condition.

Explicit relation (p. 169): A function for which $f(x)$ is given in terms of x and constants only. For instance, $f(x) = 5x^2$ gives $f(x)$ explicitly in terms of x.

Extreme value theorem (pp. 60, 63): The theorem that if f is continuous on the closed interval $[a, b]$, then f has a maximum and a minimum on $[a, b]$.

Function (p. 4): A relationship between two variable quantities for which there is exactly one value of the dependent variable for each value of the independent variable in the domain.

Fundamental theorem of calculus (pp. 271, 313): The theorem that tells how to calculate exact values of definite integrals by using indefinite integrals. In its alternate form, the theorem tells how to find the derivative of a definite integral between a fixed lower limit of integration and a variable upper limit of integration (sometimes called the fundamental theorem of *integral* calculus).

General solution (p. 318): A family of functions that contains all possible solutions to a given differential equation.

Global maximum (pp. 372, 379): See **absolute maximum**.

Global minimum (pp. 372, 379): See **absolute minimum**.

Grapher: A graphing calculator or computer used to generate graphs of given functions.

Harmonic series (p. 624): A series in which successive terms are reciprocals of the terms in an arithmetic sequence.

Heaviside method (p. 462): A method for transforming a rational expression into partial fractions.

Hyperbolic functions (p. 469): Functions with properties similar to the trigonometric (circular) functions, but defined by points on a unit hyperbola rather than by points on a unit circle.

Hypothesis (p. 212): The "if" part(s) of an "if/then" statement.

Image theorem (pp. 60, 63): The theorem that if f is continuous on the interval $[a, b]$, then the image of f on $[a, b]$ is all real numbers between the minimum of $f(x)$ and the maximum of $f(x)$ on $[a, b]$, inclusive.

Implicit differentiation (pp. 118, 150): The process of differentiating without first getting the dependent variable explicitly in terms of the independent variable.

Implicit relation (p. 169): A relationship between two variables in which operations may be performed on the dependent variable as well as the independent variable. For example, $x^2 + y^2 = 25$.

Improper integral (pp. 482, 483): A definite integral in which either one or both limits of integration are infinite or the integrand is undefined for some value of x between the limits of integration, inclusive.

Indefinite integral (pp. 92, 96, 313): See **antiderivative**.

Independent variable (p. 4): A variable whose values are not dependent on the value of another variable.

Indeterminate form (pp. 33, 289): An expression that has no direct meaning as a number, for example, $0/0$, 0^0, ∞/∞, and so on.

Infinite discontinuity (pp. 46, 56): The feature of a graph or equation that occurs where $f(x)$ increases without bound as x approaches some value c. Also called *vertical asymptote*.

Infinite form (p. 56): An expression of the form (nonzero)/(zero), which indicates a value that is larger than any real number.

Infinitesimal (p. 583): A quantity that approaches zero as Δx approaches zero, such as dy, dA, dV, and so on.

Initial condition (p. 319): A given value of x and $f(x)$ used to find the constant of integration in the solution to a differential equation.

Integrability (p. 207): The property possessed by a function if the definite integral exists on a given interval.

Integrable (p. 207): A function is integrable on interval $[a, b]$ if its integral exists on that interval.

Integral calculus: A term for calculus of integrals only.

Integration: The process of finding either the definite integral or the indefinite integral of a function.

Integration by parts (p. 437): An algebraic method for finding the antiderivative of a product of two functions.

Intermediate value theorem (p. 60): The theorem that, for continuous functions, given any number y between $f(a)$ and $f(b)$, there is a number $x = c$ between a and b for which $f(c) = y$.

Interval of convergence (p. 614): The interval of values of x for which a given power series converges.

Invertible (pp. 147, 148): A function is invertible if its inverse relation is also a function.

Lagrange form of the remainder of a Taylor series (p. 636): A way to find an upper bound on the error introduced by using only a finite number of terms of a Taylor series to approximate the value of a function. The remainder is bounded by a multiple of the first term of the tail of the series after a given partial sum.

l'Hospital's rule (pp. 296, 314): A property for finding limits in the form $0/0$ or ∞/∞ by taking the derivative of the numerator and the denominator (sometimes spelled *l'Hôpital's* rule).

Limit (pp. 4, 10, 33): A number that a function value $f(x)$ approaches, becoming arbitrarily close to it, as x approaches either a specific value or infinity.

Limits of integration (p. 207): The values a and b in the expression $\int_a^b f(x)\,dx$.

Linearization of a function (p. 191): The process of finding the linear function that best fits function f for values of x close to $x = c$. This function is $y = f(c) + f'(c)(x - c)$, or, equivalently, $y = f(c) + f'(c)\,dx$.

Local linearity (pp. 76, 190): A function is locally linear at $x = c$ if values of the function are approximated well by values of the tangent line for values of x near c.

Local maximum: See **relative maximum**.

Local minimum: See **relative minimum**.

Logarithmic differentiation (p. 291): An implicit differentiation process in which the natural logarithm of a function is taken before differentiating, usually so that variables can be removed from exponents.

Logistic differential equation (pp. 349, 352): An equation that expresses the rate of change of a population that is constrained by a carrying capacity.

Logistic function (logistic equation) (pp. 339, 352): A solution of the logistic differential equation, used to model population growth where there is a maximum sustainable population.

Maclaurin series (p. 605): A Taylor series expanded about $x = 0$. Sine, cosine, exponential, and hyperbolic functions can be calculated using only the operations of arithmetic by first expanding the function as a Maclaurin series.

Magnitude (p. 523): The length of a vector or the size of a quantity. The magnitude of a vector is also called its *norm*.

Maximum: See **relative maximum**.

Mean value theorem (pp. 211, 313): The theorem that expresses sufficient conditions for a function graph to have a tangent line parallel to a given secant line at a value of $x = c$ between the endpoints of the secant line.

Minimum: See **relative minimum**.

Moment (p. 558): The product of a quantity such as force or mass and the power of a distance from a point, line, or plane at which that quantity is located.

Natural exponential function (p. 288): An exponential function with base e.

Natural logarithm function (p. 313): A logarithmic function with base e. The natural logarithm function is the classic example of a function defined as a definite integral between a fixed lower limit and a variable upper limit: $\int_1^x (1/t)\,dt$.

Neighborhood (p. 108): An open interval that contains a given domain value.

Newton's method (p. 182): A method of finding the zero of a function.

Norm (pp. 265, 523): See **magnitude**.

Normally distributed (p. 256): A distribution of data that is symmetric, with most data points close to the mean and fewer farther away from the mean.

Odd function (p. 136): A function f that has the property $f(-x) = -f(x)$ for all x in its domain.

One-sided limit (pp. 47, 48): The value that a function $f(x)$ approaches as x approaches a value c from only one direction.

Order of magnitude (p. 301): A range of magnitude, from some value to ten times that value.

Parameter (p. 161): The independent variable in a parametric function.

Parametric chain rule (pp. 161, 313): The method for finding the derivative of a function defined parametrically, namely, the derivative of y with respect to t divided by the derivative of x with respect to t.

Parametric function (pp. 131, 160): A function in which two variables each depend on a third variable. For example, the x- and y-coordinates of a moving object might both depend on time.

Partial fractions (p. 351): Simple rational expressions that can be summed to form a given rational algebraic expression.

Partial sum (p. 593): The nth partial sum of a series is the sum of the first n terms of the series.

Piecewise function (pp. 48, 313): A function defined by different rules for different intervals of its domain.

Plane slices (finding volumes by) (p. 245): A technique for finding the volume of a solid by slicing it into slabs of approximate volume $dV = (\text{area})\,(dx)$ or $(\text{area})\,(dy)$ then integrating to find the total volume.

Plateau point (pp. 372, 379): A point on a graph at which the derivative is zero, but the point is not a maximum or minimum.

Point of inflection (pp. 372, 379): A point at which a graph changes from concave up to concave down or vice versa. Points of inflection occur where the second derivative of a function is either zero or undefined.

Power series (p. 589): A series (with an infinite number of terms) in which each term contains a power of the independent variable.

Principal branch (p. 148): The branch of an inverse trigonometric relation with range restricted to make the corresponding inverse trigonometric function.

Radius of convergence (p. 616): The distance from the midpoint of the interval of convergence to one of its ends.

Ratio technique (ratio test) (p. 616): A technique for determining the interval of convergence for a power series by finding the values of x for which the absolute value of the ratio of adjacent terms can be kept less than 1.

Reduction formula (p. 445): A formula whereby a complicated antiderivative can be expressed in terms of a simpler antiderivative of the same form.

Related rates (p. 314): A type of problem in which an unknown rate is calculated, given one or more known rates.

Relative maximum (pp. 95, 372, 379): A function has a local maximum at $(c, f(c))$ if $f(c)$ is greater than all other values in a neighborhood of $x = c$. Also called *maximum* or *local maximum*.

Relative minimum (pp. 372, 379): A function has a local minimum at $(c, f(c))$ if $f(c)$ is less than all other values in a neighborhood of $x = c$. Also called *minimum* or *local minimum*.

Removable discontinuity (pp. 33, 35, 46): If a function is discontinuous at $x = c$, but may be made continuous there by a suitable definition of $f(c)$, then the discontinuity is removable. For instance, $f(x) = (x^2 - 25)/(x - 5)$ is discontinuous at $x = 5$ because of division by zero, but the discontinuity can be removed by defining $f(5) = 10$.

Riemann sum (pp. 189, 206, 313): A sum of the form $\sum f(x)\, dx$ in which each term of the sum represents the area of a rectangle of altitude $f(x)$ and base dx. A Riemann sum gives an approximate value for a definite integral. The limit of a Riemann sum as dx approaches zero is the basis for the formal definition of a definite integral.

Rolle's theorem (p. 211): The property that expresses sufficient conditions for a function graph to have a horizontal tangent for some value of $x = c$ between two zeros of the function.

Sample point (p. 205): A point x in a subinterval for which a term of a Riemann sum, $f(x)\, dx$, is found; or the corresponding point $(x, f(x))$ on the graph of f itself.

Scalar quantity (p. 523): A quantity, such as time, speed, or volume, that has magnitude but no direction.

Separating the variables (pp. 269, 318): The most elementary technique for transforming a differential equation so that it can be solved.

Simpson's rule (pp. 254, 314): A numerical way of approximating a definite integral by replacing the graph of the integrand with segments of parabolas, then summing the areas of the regions under the parabolic segments. The technique is similar to the trapezoidal rule, except that the graph is replaced by segments of quadratic functions rather than by segments of linear functions.

Slab (p. 245): A slice of a solid, taken perpendicular to an axis.

Slope field (p. 334): A graphical representation of the slope specified by a differential equation at each grid point in a coordinate system. A slope field, which can be generated by a grapher, allows graphical solutions of differential equations.

Solid of revolution (p. 245): A solid formed by rotating a planar region about an axis.

Speed (p. 93): The absolute value of velocity.

Squeeze theorem (p. 108): The theorem that if $f(x)$ is always between the values of two other functions and the two other functions approach a common limit as x approaches c, then $f(x)$ also approaches that limit.

Step discontinuity (pp. 35, 46): A discontinuity that occurs at $x = c$ if $f(x)$ approaches different numbers from the right and from the left as x approaches c.

Tail (p. 624): The terms of a series remaining after a given partial sum; or the sum of these terms.

Tangent line (pp. 75, 190): A tangent line to a curve at a given point is the line that passes through $(x, f(x))$ and has the same slope as the curve at that point.

Taylor series (p. 605): A power series representing a function as non-negative integer powers of $(x - a)$. The coefficients of the terms are such that each order derivative of the series equals the corresponding order derivative of the function at the point at which $x = a$.

Term index (pp. 590, 593): The value n in the term t_n.

Terminal velocity (p. 340): The maximum velocity that a falling object can attain.

Theorems of Pappus (p. 567): For volume, Volume = (area of rotated region)(distance traveled by centroid). For surfaces, Area = (length of rotated arc)(distance traveled by centroid).

Transcendental (p. 289): Describes a number that is irrational and cannot be expressed using a finite number of algebraic operations.

Trapezoidal rule (p. 20): A numerical way of approximating a definite integral by slicing the region under a graph into trapezoids and summing the areas of the trapezoids. The technique is similar to Simpson's rule, but the graph is replaced by segments of linear functions rather than by segments of quadratic functions.

Trigonometric substitution (p. 455): An algebraic method for finding an antiderivative in which the integrand involves quadratics or square roots of quadratics.

Uniqueness theorem for derivatives (p. 313): The theorem that if two functions have identical derivatives everywhere in an interval and have at least one point in common, then they are the same function.

Vector (p. 523): A quantity that has both magnitude and direction. Position, velocity, and acceleration vectors are used to analyze motion in two or three dimensions.

Velocity (pp. 93, 313): The instantaneous rate of change of displacement.

Vertical asymptote (pp. 46, 54): A vertical line $x = c$ that the graph of a function does not cross because the limit of $f(x)$ as x approaches c is infinite.

Washers (p. 244): The slicing of a solid of revolution into thin slices so that each point in the washer is virtually the same distance from a plane perpendicular to the axis of rotation as is the sample point. Washers are used to set up integrals for calculating the volume, mass, moment, and so on of a solid object.

Index of Problem Titles

General Index

derivatives (*continued*)
  definition as a function, 79, 86
  definition at a point, 74
  of a derivative. *See* second derivative
  difference quotients and, 74
  differentials compared to, 192-194
  of exponential functions, 115-117,
    120, 201-202, 290-291
  of geometric series, 596
  grapher calculation of, 78-81
  graphing, 7-10, 73, 78-81
  of hyperbolic functions, 471-472
  of implicit relations, 169-171
  of inverse hyperbolic functions,
    473-474
  of inverse trigonometric functions,
    149-151
  of logarithmic functions, 118-119,
    272-273, 274, 284-285
  mean value. *See* mean value theorem
    for derivatives
  meaning of, 10
  notation for, 85-86, 89
  numerical computation of, 8-10
  of parametric functions, 160-163
  physical meaning of, 3-4
  of power functions, 85-89
  of product of two functions,
    132-134
  of quotient of two functions,
    137-139
  second. *See* second derivative
  summary of techniques, 489
  symbols for, 74, 79, 85-86, 89
  tangent line slope and, 75
  techniques for. *See* differentiation
  third, 385, 596
  of trigonometric functions, 100-101,
    103, 108-109, 142, 143
  uniqueness theorem for, 281-283,
    760
  of vector functions, 525-531
difference quotient, 74-75, 756
  defined, 74
  formulas for, 84
  graphing, 75, 78-79
  Rolle's theorem and, 214
differentiability, 756
  continuity and, 154-157
  contrapositive of, 155
  defined, 154
  definition at a point, 154
  definition on an interval, 154
  local linearity and, 190-191
  mean value theorem and, 212-213
  of piecewise functions, 156-157
  Rolle's theorem and, 213
differential equations, 226, 756
  Euler's method and, 341-343
  logistic, 348-352, 357, 758
  population and, 269, 318-319, 460
  separating the variables to solve,
    318-320, 324-325
  series solution of, 647

slope fields and, 333-336
solving, 318-320, 324-328
differentials
  of arc length, 403
  defined, 193, 756
  derivative quantities compared to,
    192-194
  integration by parts and, 435
  of volume for cylindrical shells, 396
differentiation
  chain rule. *See* chain rule
  of composite functions, 101, 102-105
  of constant function, 88
  of constant times a function, 88
  defined, 87, 756
  of exponential functions, 115-117
  implicit. *See* implicit differentiation
  inverse trigonometric functions,
    149-151
  logarithmic, 291, 758
  of logarithmic functions, 117-119
  of power function, 87-88
  of sum of two functions, 88
  of trigonometric functions, 108-109,
    142-143
  See also derivatives
dilating a function, 92
dilation, 110, 115
Diocles, 182
direct proportion property of
    exponential functions, 317
  converse of, 324
direction fields. *See* slope fields
discontinuity
  continuity compared to, 45
  infinite, 46, 53, 56, 757
  removable, 33, 35, 46, 56, 759
  step, 35, 46, 48-49, 759
discrete points, 60, 756
disks, 243
displacement, 93, 756
  as antiderivative, 99
  average velocity and, 508-509
  distance distinguished from, 501
  integrals for, 502-503
  moment and. *See* moment
  time intervals and, 502-505
distance
  displacement distinguished from,
    501
  integrals for, 502-503
  time intervals and, 502-503
  vector equations and, 531
ditto marks, 445
divergence of improper integrals, 482,
    483, 756
divergence of series, 589, 593, 756
  geometric, 592
  by oscillation, 593
domain endpoints, 373, 380
dot products, 524-525, 756
double-argument properties
  for hyperbolic functions, 480
  for sine and cosine, 450-451

doubly curved surface, 408
"drowning swimmer" problems, 516
drug dosage, 591-592

E

e, 116-117, 756
Earth, 557
elementary functions, 433
elementary transcendental functions,
    466, 597
ellipse(s)
  area of, 251
  graphing, 161
  length of, 404
elliptic integral, 404
embedded terms, 622
endpoints, 373, 380
energy, 550
equal vectors, 523
equal-derivative theorem, 220
  converse of, 220
equations, general vs. particular, 6
equiangular spiral, 422-423
error function of *x*, 256-257, 619
escape velocity, 347
Euler, Leonhard, 341
Euler's method, 341-343, 756
evaluation of an integral, 197
even functions, 136, 229-230, 756
existence theorem, 62, 756
expanding the function as a power
    series, 599
explicit relations, 169, 756
exponential functions
  continued exponentiation, 309
  definition with *e* expressing,
    289-290
  derivative of, 115-117, 120, 201-202,
    290-291
  direct proportion property of, 317,
    324
  general equation for, 6
  indefinite integrals of, 201-202, 290
  indeterminate forms, 298, 300
  logarithmic differentiation of, 291
  natural. *See* natural exponential
    functions
  power functions compared to, 115
  power series for, 597, 598-599
  properties of, summarized, 304
exponents
  constant, 85
  fractional, 91, 171
  inverse function notation and,
    147-148
  negative, 91
extended mean value theorem, 104
extreme value theorem, 63, 756

F

factorial function, defined, 487
factorial reciprocal series, 626

family of functions, 318
first-order infinitesimal, 583
force
  moment of. *See* moment
  variable. *See* work
  variable pressure, 567–569
forward difference quotient, 78–79,
  84
four-leaved rose, 419
fractal curves, 596
fractional exponents, 91, 171
fractions, partial, 351, 460–462,
  463–464, 758
frustums of cones, 408–409, 414
Fuller, Buckminster, 557
functions, 4, 756
  absolute value. *See* absolute value
  average value of, 509–510, 755
  composite, 101, 102–105, 199–201
  continuity of. *See* continuity
  derivative. *See* derivative
  dilation of, 92, 110
  even, 136, 229–230, 756
  exponential. *See* exponential
    functions
  factorial, 487
  gamma, 487
  hyperbolic. *See* hyperbolic functions
  inside, 101, 102, 104, 193, 199–200
  inverse. *See* inverse functions
  invertible, 147, 148, 757
  linear. *See* linear functions
  linearization of, 190–191, 757–758
  locally quadratic, 598
  logarithmic. *See* logarithmic functions
  logistic, 348–352, 357, 758
  notation, 6
  odd, 136, 229–230, 758
  outside, 101, 102, 104, 296
  piecewise. *See* piecewise functions
  polynomial. *See* polynomial functions
  power. *See* power functions
  power series for. *See* power series
  quadratic. *See* quadratic functions
  rational algebraic. *See* rational
    algebraic functions
  of same order, 625–626
  signum, 52
  sine-integral, 252, 619–620
  strictly increasing or decreasing, 209
  summary of types of, 6
  transformation of, 109–110, 115
  trigonometric. *See* trigonometric
    functions
  vector, 523, 525–531
fundamental theorem of calculus, 757
  constant Riemann sums and,
    222–224
  defined, 224
  derivative of an integral form,
    270–271
  proof of, 224–225
  properties of definite integrals and,
    228–231

G
gamma function, defined, 487
Gaussian distributions, 256–257
general solution, 318–319, 757
geometric series
  convergent, 591–592, 593, 625
  defined, 590, 593
  derivatives of, 596
  divergent, 592
  properties of, 593
  See also power series
global maximum (absolute maximum),
  95, 372, 379, 755
global minimum (absolute minimum),
  372, 379, 755
Gompertz function, 362
graphers, 757
  continuity approximate on, 60
  definite integrals by, 252
  derivative calculation on, 78–81
  difference quotients and, 78–79
  discrete points on, 60
  power series on, 601, 602
graphs and graphing
  acceleration, 504–505
  asymptotes and, 55
  continuity, 46
  convergent series, 622–623, 624
  of cubic functions, 371–374
  cusps, 47, 379
  of definite integrals, 15–16, 231,
    234–236, 253–254
  of derivatives, 7–10, 73, 78–81
  difference quotient, 75, 78–79
  differential equation solutions,
    333–336, 341–343
  discontinuities, 46
  implicit relations, 170
  infinite limits, 53
  of inverse trigonometric functions,
    149
  limits, 34–37
  local linearity, 76, 191
  number-line, 373
  parabolas. *See* parabolas
  parametric equations, 161–163
  polar coordinates, 414
  of power series, 601–602
  sigmoid, 362
  sinusoids. *See* sinusoids
  sketching derivatives, 89
  "under" the graph, defined, 15
  velocity, 503

H
h, 86
harmonic series, 624, 757
heat capacity, 330
heat transfer coefficient, 330
Heaviside method, 461–462, 757
Heaviside, Oliver, 462
Hooke's law, 238
horizontal asymptotes, 54

horizontal dilation, 110
horizontal translation, 110
hyperbolic cosine, 469–470, 475–476,
  478, 606
hyperbolic functions, 757
  cosine, 469–470, 475–476, 478, 606
  definitions of, 470–471
  derivatives of, 471–472
  double-argument properties of,
    480
  equation of hanging chain or cable,
    478
  indefinite integrals of, 472–475
  inverse, 473–475
  power series for, 606, 638
hyperboloids, 413, 459
hypotenuse, negative, 150
hypotheses, 212–213, 757

I
identity function, limit of, 42
image theorem, 63, 757
implicit differentiation, 118, 150, 757
  related rates and, 174–176
  technique of, 171
implicit relations, 169–171, 757
improper integrals, 481–484, 623, 630,
  757
increments, 205
indefinite integrals
  composite functions, 199–201
  constant times a function, 198–199,
    230, 231
  defined, 96, 197
  of exponential functions, 201–202,
    290
  of hyperbolic functions, 472–473
  of inverse hyperbolic functions,
    474–475
  of inverse trigonometric functions,
    466–468
  notation for, 197–198, 199
  of power function, 200–201
  power series for, 608–609, 619–620
  of rational functions, 460–462,
    463–464
  reciprocal functions, 259, 269, 270,
    274
  reduction formulas and, 444–447
  of sum of two functions, 199,
    230, 231
  summary of techniques, 489
  symbol for, 197, 198
  See also integration techniques
independent variable, 4, 103, 757
indeterminate form, 33, 56, 757
  approaching numbers other than 1,
    108
  divergent series and, 627
  e as, 289
  infinity in, 59, 300, 301
  l'Hospital's rule and, 295–298
  limit theorems and, 40–43

indeterminate form (*continued*)
  limits of, finding, 42-43, 295-298
  removable discontinuity of, 33, 35, 46, 56, 759
  sine and, 107-108
  types of, summarized, 301
  zero as exponent as, 298, 300
  zero times infinity as, 59
inductance, 329
inequalities
  limits and, 107-108
  Riemann sums and, 207
inertia, moment of, 563, 577
infinite, 56
infinite discontinuity, 46, 53, 56, 757
infinite form, 56, 757
infinite limits, 52-56
  definitions of, 54
  graphs of, 53
  l'Hospital's rule and, 298
  notation for, 53
  orders of magnitude and, 301
  piecewise functions, 56
  vertical asymptotes, 41, 46, 53, 54, 760
infinite radius of convergence, 617
infinitesimal, 583, 757
infinitesimals of higher order, 397, 583-584
inflection point, 371, 372-380, 758
initial condition, 319, 757
inside functions
  defined, 102
  derivatives of composite functions and, 101, 102, 104
  differential of, 193
  indefinite integrals and, 199-200
instantaneous rate, 3-4, 7
  See also derivative
integrability, 207, 757
integrable, 207, 757
integral sign, 197
integral test for convergence, 622-624, 630
integrals. *See* definite integrals; indefinite integrals; integration techniques
integrand, 197
integration, 197, 757
integration by parts, 757
  choosing the parts for, 437
  constant of integration and, 435, 436, 438
  of inverse trigonometric functions, 467-468
  of natural logarithm function, 441
  rapid repeated, 438-441
  reappearance of original, 439-440
  reassociate factors between steps, 440-441
  reduction formulas and, 444-446
  technique, 434-436
  trigonometric properties and, 439-440

integration techniques
  by partial fractions, 460-462, 463-464
  by parts. *See* integration by parts
  rational functions of sin *x* and cos *x* by $u = \tan(x/2)$, 492
  rationalizing algebraic substitutions, 491-492
  by trigonometric substitution, 454-457
intercept property of natural logarithms, 282
intermediate value theorem, 60-61, 757
  converse of, 61-62
  corollary of (image theorem), 63
interval of convergence, 612, 613-614, 757
  See also convergence of series
inverse functions
  defined, 119, 146-148
  derivative of, general formula, 153
  hyperbolic, 473-475
  inverse relations vs., 147
  linear functions, 146-148
  natural exponential/logarithmic, 117-118, 119, 289
  notation for, 147-148
  properties of, 119
  trigonometric. *See* inverse trigonometric functions
inverse hyperbolic sine of *x*, 473
inverse relations, 147-148
inverse trigonometric functions, 148-151
  definitions of, 149
  derivatives of, 149-151
  graphs of, 149
  integration of, 466-468
  inverse relations vs., 147, 148
  power series for, 606, 607-608
  principal branch and, 148
  range restriction and, 148
invertible functions, 147, 148, 757
involute of a circle, 167

J

joules, 550
journal, 24-25, 493-494

K

Kepler's second law, 421
Kepler's third law, 421
knots, 257
Koch, Helge von, 596

L

Lagrange form of the remainder of a Taylor series, 635-640, 757
Lagrange, Joseph Louis, 636

Laurent series, 643
least upper bound postulate, 635
Leibniz, Gottfried, 89
lemma, 63
length, moments of, 579
length of plane curves. *See* plane curve length
l'Hospital, G. F. A. de, 295
l'Hospital's rule, 295-298, 314, 757
limaçon
  arc length of, 417
  area of, 415-416
  defined, 414
limit comparison test, 625-626, 631
limit theorems
  as distributive properties, 64
  limit of a constant function, 42
  limit of a constant times a function, 42
  limit of a product of two functions, 40-41, 42
  limit of a sum of two functions, 40-41, 42
  limit of the identity function (limit of *x*), 42
limits, 4, 757
  convergent series and, 591, 593, 615-616, 620, 625-626
  definition, 34
  definition, algebraic (absolute value), 67
  definition, verbal, 10
  derivatives, 10
  e as indeterminate, 289
  of exponential functions, 116
  graphical approaches to, 34-37
  inequalities and, 107-108
  infinite. *See* infinite limits
  of integral exact value, 20
  of *n*th root of *n*, 620
  one-sided, 47-49, 56, 758
  positive values in, 36-37
  of Riemann sum, 207-208
  secant line and, 77-78
  squeeze theorem, 108, 759
  and zero numerator/denominator. *See* indeterminate form
  See also limit theorems; limits of integration
limits of integration, 207-208, 228, 758
  function of *x* as upper, 275-276
  reversal of, 228-229, 231
  variable upper, 270-276
  variable upper and lower, 310
limniscate of Bernoulli, 419
linear combination of power functions, 85
linear functions
  general equation for, 6
  inverses of, 146-148
  linearization of a function and, 190-191
  tangent line as, 190

integration of even powers, 451
integration of odd powers, 449-450
integration of squares, 450-451
inverse. *See* inverse trigonometric
    functions
inverse relations of, 147, 148
limits and, 107-108
parametric functions, 160-163
power series for, 599-602, 606, 607
reduction formulas for, 444-445, 446,
    447
summary of properties of, 659-660
transformation of, 109-111
trigonometric substitution, 454-457,
    460, 760

U

$u = \tan(x/2)$ substitution, 492
undefined, 56
    See also indeterminate form
"under" the graph, 15
uniqueness theorem for derivatives,
    281-282, 760
    product property of ln and, 282-283
    proof that ln is a logarithm and, 283
unit circle, 470
unit hyperbola, 470
unit vectors, 524
units
    of acceleration, 93
    of torque, 550
    of velocity, 93
    of work, 547, 550
upper limits of integration, 207-208, 228

V

variables
    dependent, 4, 6, 756

independent, 4, 103, 757
    separating the, 269, 318-320,
      324-325, 759
vector functions, 523, 525-531
vector projection, 525
vector quantity, 523
vectors, 523-530, 760
    acceleration, 523, 525-526, 528-530
    addition/subtraction with, 524
    components of, 524, 525-526, 530
    dot products of, 524-525
    magnitude of, 523
    symbols for, 523-524
    three-dimensional, 537
    velocity, 523, 525-526
velocity, 92-96
    as antiderivative, 99
    average, 508-509, 526, 527-528
    critical points of, 372
    defined, 93, 760
    as derivative, 93, 96
    derivative of. *See* acceleration
    displacement and, 505
    escape, 347
    improper integrals and, 481-482
    as integral of acceleration, 504
    negative vs. positive signs of, 93-94,
      502-503
    speed compared to, 18, 93
    terminal, 98, 340, 347, 759
    units of, 93
    vector of, 523, 525-526
vertical asymptotes, 41, 46, 53, 54,
    760
vertical dilation, 109-110, 110, 115
volume
    centroid, 559-560
    of a cone, 175
    finding by cylindrical shells, 395-398,
      574

finding by plane slices, 242-246, 395,
    574
finding by spherical shells, 412
first moment of, 559
general strategy for definite integrals,
    574
maximizing, 386-387
second moment of, 563
of a sphere, 13
theorem of Pappus for, 567

W

washers, 244, 760
Witch of Agnesi, 167
"with respect to," 103, 199
work, 548
    definition, 548
    move part of object whole
      displacement, 549-550
    move whole object part of the
      displacement, 548-549
    units of, 547

Y

$y = \operatorname{argsinh} x$, 473
Yates, Robert C., 182

Z

zero
    denominator. *See* discontinuity
    infinitesimals of higher order, 397,
      583-584
    numerator and denominator. *See*
      indeterminate form
    radius of convergence, 617
zero vector, 523
zeros of a function, Newton's method
    for finding, 182-183

Photograph Credits